Integrated Soil and Water Management: Selected Papers from 2016 International SWAT Conference

Special Issue Editors

Karim Abbaspour
Raghavan Srinivasan
Saeid Ashraf Vaghefi
Monireh Faramarzi
Lei Chen

MDPI • Basel • Beijing • Wuhan • Barcelona • Belgrade

MDPI

Special Issue Editors

Karim Abbaspour
Eawag, Swiss Federal Institute for Aquatic Science and Technology
Switzerland

Raghavan Srinivasan
Texas A&M University
USA

Saeid Ashraf Vaghefi
Eawag, Swiss Federal Institute for Aquatic Science and Technology
Switzerland

Monireh Faramarzi   Lei Chen
University of Alberta   Beijing Normal University
Canada   China

Editorial Office
MDPI AG
St. Alban-Anlage 66
Basel, Switzerland

This edition is a reprint of the Special Issue published online in the open access journal *Water* (ISSN 2073-4441) from 2016–2018 (available at: http://www.mdpi.com/journal/water/special_issues/SWAT2016).

For citation purposes, cite each article independently as indicated on the article page online and as indicated below:

Lastname, F.M.; Lastname, F.M. Article title. *Journal Name* **Year**, *Article number*, page range.

First Edition 2018

ISBN 978-3-03842-815-2 (Pbk)
ISBN 978-3-03842-816-9 (PDF)

Table of Contents

About the Special Issue Editors

Karim Abbaspour is in the field of hydrology and soil physics with a background in civil engineering and mathematics. He has done extensive work in modelling of flow and transport in soils, groundwater, and catchments. On the mathematical side, he has developed an internationally used SWAT-CUP program for calibration of SWAT with an uncertainty analysis. Dr. Abbaspour has modelled water resources of Africa, Iran, Black Sea Basin, and the European continent. He has studied the impact of climate change on water resources of Iran, Africa, the Black Sea, and California. Dr. Abbaspour has also produced research works on global groundwater quality projects. He and his colleagues have studied the contamination of groundwater with arsenic and fluoride.

Raghavan Srinivasan, a professor in the departments of ecosystem science and management, and biological and agricultural engineering, at Texas A&M University has continued to transfer his sphere of knowledge and experience to his students over the past 17 years. By undertaking research over 25 years, he has had the opportunity to share current spatial science technology, as applied to water resources management, with society by providing leadership, technical knowledge, and service through participation in local, national, and international organizations. Dr. Srinivasan led a continental scale project called Hydrologic Unit Model for the United States (HUMUS) that provided information for the 1997 Resources Conservation Act (RCA) funded by the USDA. He has developed the interfaces and associated national and international databases to make use of watershed models SWAT on various GIS platforms such as GRASS, ARCVIEW, ArcGIS, and QGIS. He was responsible for integrating SWAT as one of the water quality models into EPA's BASINS framework. Dr. Srinivasan is currently leading several national and international watershed modeling projects including the EPA HAWQS project, which is a nationwide water quality modeling system. Dr. Srinivasan has published more than 200 peer-reviewed publications, 11 book chapters, and several hundred national and international conferences and professional meeting presentations.

Saeid Ashraf Vaghefi is researcher in data science, climatology, and hydrology with a background in civil engineering and mathematics. His research focus is on the big data process and the implementation of data science on water and environmental issues. He is in the developer team of the Climate Change Toolkit (CCT) program and the 2w2e website (www.2w2e.com).

Monireh Faramarzi received her PhD from the Swiss Federal Institute of Technology ETHZ in 2010 and she was the recipient of the best ETH Ph.D. Dissertation Medal in 2011. She has been Assistant Professor at the University of Alberta and Campus Alberta Innovates Chair in Watershed Science since 2016. She seeks to address water supply and demand issues and to study water availability, water reliability, and scarcity, and the water demand of different sectors (e.g. agriculture, oil and gas, municipal, industry, and environment). She will apply models to analyze alternative management options to realize opportunities and risks during periods of water surplus and scarcity in the past and in the future, linking climate change effects of water supply-demand to economic prospects.

Lei Chen, Ph.D., Associate Professor at Beijing Normal University, China. Lei Chen focuses on the mechanism of the water quality response and control technique of nonpoint source (NPS) pollution, and has made progress in the simulation tool-control technique and the reduction strategy of NPS pollution. Lei Chen has taken charge of two National Science Foundation projects and also become the principal investigator at the Beijing Normal University. As the first/corresponding and second author, the applicant has published 34 SCI papers (including 16 TOP SCI papers) in high-impact scientific journals like *J Hydro* (7), *Water Resour Res*, *HESS* (2) and *Water Res*. The applicant has published one book as a first author. The applicant was invited to be the associate editor for an SCI journal (*IJABE*) and the guest editor of the Special Issue for two SCI magazines (*Eco Eng & Water*).

Preface to "Integrated Soil and Water Management: Selected Papers from 2016 International SWAT Conference"

We are facing a time of uncertain environmental challenges. The ever-increasing trend in atmospheric CO_2 is changing the climatic dynamics, which will influence all aspects of our life. Although events such as US decision on climate change cast a shadow over the reality of climate change, some trivial facts and their grave consequences cannot be ignored or disputed. These facts and their chain reactions are: Rising CO_2 will increase earth's temperature; this will cause metling of water stored in ice caps and permafrosts, increase capacity of atmosphere to store water, and increase evapotranspiration; these will speed up the climatic dynamics putting more water in the atmosphere causing heavier rainfalls depleting water resource of the atmosphere in a shorter time, which will cause droughts in other places and floods in others. The net effect will be chaotic periods of colds and warms, droughts and floods. We rely on models to quantify these changes and to predict them, therefore, our models need to be sound accounting correctly for the basic physical laws. As models are simplifications of reality, practicing "correct neglect" is essential in developing useful models. Furtheremore, the models need to be calibrated, regionalized correctly, and their uncertainties quantified before they can be reliably used. The current special issue of Water is a step forward in creating, testing, and implementing models to predict the future changes. In this publication, issues related to climate change, model calibration/uncertainty, and effects of human activity in watersheds are modeled and discussed. Much further work needs to be done and we hope that younger scientists are inspired by the current work.

Karim Abbaspour, Raghavan Srinivasan, Saeid Ashraf Vaghefi, Monireh Faramarzi and Lei Chen
Special Issue Editors

water

MDPI

Editorial

A Guideline for Successful Calibration and Uncertainty Analysis for Soil and Water Assessment: A Review of Papers from the 2016 International SWAT Conference

Karim C. Abbaspour [1,*], Saeid Ashraf Vaghefi [1] and Raghvan Srinivasan [2]

[1]   Eawag, Swiss Federal Institute of Aquatic Science and Technology, 8600 Duebendorf, Switzerland;
       seyedsaeid.ashrafvaghefi@eawag.ch
[2]   Spatial Science Laboratory, Texas A & M University, College Station, TX 77845, USA; r-srinivasan@tamu.edu
*   Correspondence: karim.abbaspour@eawag.ch

Received: 26 November 2017; Accepted: 20 December 2017; Published: 22 December 2017

Abstract: Application of integrated hydrological models to manage a watershed's water resources are increasingly finding their way into the decision-making processes. The Soil and Water Assessment Tool (SWAT) is a multi-process model integrating hydrology, ecology, agriculture, and water quality. SWAT is a continuation of nearly 40 years of modeling efforts conducted by the United States Department of Agriculture (USDA) Agricultural Research Service (ARS). A large number of SWAT-related papers have appeared in ISI journals, building a world-wide consensus around the model's stability and usefulness. The current issue is a collection of the latest research using SWAT as the modeling tool. Most models must undergo calibration/validation and uncertainty analysis. Unfortunately, these sciences are not formal subjects of teaching in most universities and the students are often left to their own resources to calibrate their model. In this paper, we focus on calibration and uncertainty analysis highlighting some serious issues in the calibration of distributed models. A protocol for calibration is also highlighted to guide the users to obtain better modeling results. Finally, a summary of the papers published in this special issue is provided in the Appendix.

Keywords: calibration; validation; uncertainty analysis; sensitivity analysis; pre-calibration analysis; SWAT-CUP

1. Introduction

This special issue on "Integrated Soil and Water Management" deals with the application of the Soil and Water Assessment Tools (SWAT) [1] to a range of issues in watershed management. A total of 27 papers attest to the importance of the subject and the high level of research being conducted all over the globe. A common factor in almost all the published papers is the calibration/validation and uncertainty analysis of the models. Of the 27 papers published in this issue, 20 are calibrated with SWAT-CUP [2–4]. As the credibility of a model's performance is in the calibration/validation and uncertainty results, we devote this overview paper to the outstanding issues in model calibration and uncertainty analysis.

Steps for building a hydrologic model include: (i) creating the model with a hydrologic program, such as in our case, ArcSWAT; (ii) performing sensitivity analysis; (iii) performing calibration and uncertainty analysis; (iv) validating the model, and in some case; (v) performing risk analysis. Here we discuss these steps and highlight some outstanding issues in the calibration of large-scale watershed models. A protocol for calibrating a SWAT model with SWAT-CUP is also proposed. Finally, we briefly review all papers published in this special issue in the Appendix.

To avoid any confusion and for the sake of standardizing the SWAT model calibration terminology, we summarized the definition of some common terms in Table 1.

Table 1. Definition of some terminologies.

Terminology	Definition
SWAT	An agro-hydrological program for watershed management.
Model	A hydrologic program like SWAT becomes a model only when it reflects specifications and processes of a region.
Watershed	A hydrologically isolated region.
Sub-basin	A unit of land within a watershed delineated by an outlet.
Hydrologic response unit (HRU)	The smallest unit of calculation in SWAT made up of overlying elevation, soil, land-use, and slope.
Parameter	A model input representing a process in the watershed.
Variable	A model output.
Deterministic model	A model that takes a single-valued input and produces a single-valued output.
Stochastic model	A model that takes parameters in the form of a distribution and produces output variables in the form of a distribution also. SWAT and most other hydrologic models are deterministic models.

Next to the terms in Table 1, the term *sensitivity analysis* refers to the identification of the most important influence factor in the model. Sensitivity analysis is important from two points of view: First, parameters represent processes, and sensitivity analysis provides information on the most important processes in the study region. Second, sensitivity analysis helps to decrease the number of parameters in the calibration procedure by eliminating the parameters identified as not sensitive. Two general types of sensitivity analysis are usually performed. These are *one-at-a-time* (OAT) or *local sensitivity* analysis, and *all-at-a-time* (AAT) or *global sensitivity* analysis. In OAT, all parameters are held constant while changing one to identify its effect on some model output or objective function. In this case, only a few (3–5) model runs are usually sufficient (Figure 1). In the AAT, however, all parameters are changing; hence, a larger number of runs (500–1000 or more, depending on the number of parameters and procedure) are needed in order to see the impact of each parameter on the objective function. Both procedures have limitations and advantages. Limitation of OAT is that sensitivity of one parameter is often dependent on the values of other parameters, which are all fixed to values whose accuracy is not known. The advantage of OAT is that it is simple and quick. The limitation of AAT is that parameter ranges and the number of runs affect the relative sensitivity of the parameters. The advantage is that AAT produces more reliable results. In SWAT-CUP, OAT is used to directly compare the impact of three to five parameter values on the output signal (Figure 1), whereas AAT uses a multiple regression approach to quantify sensitivity of each parameter:

$$g = \alpha + \sum_{i=1}^{n} \beta_i b_i \tag{1}$$

where g is the objective function value, α is the regression constant, and β is the coefficient of parameters. A t-test is then used to identify the relative significance of each parameter b. The sensitivities given above are estimates of the average changes in the objective function resulting from changes in each parameter, while all other parameters are changing. This gives relative sensitivities based on linear approximations and, hence, only provides partial information about the sensitivity of the objective function to model parameters. In this analysis, the larger in absolute value the value of t-stat, and the smaller the p-value, the more sensitive the parameter.

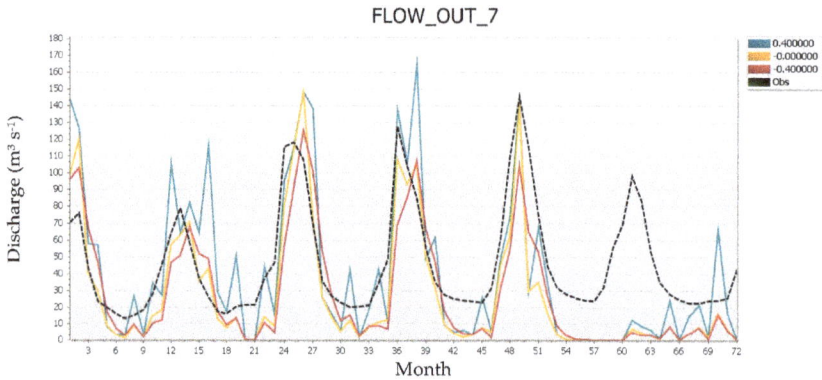

Figure 1. Sensitivity of discharge to three different values of CN2 in *one-at-a-time* (OAT) analysis.

The term *calibration* refers to a procedure where the difference between model simulation and observation are minimized. Through this procedure, it is hoped that the regional model correctly simulates true processes in the physical system (Figure 2).

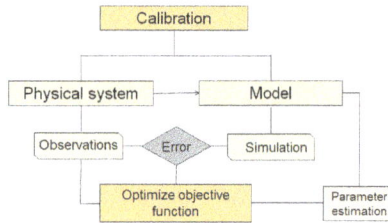

Figure 2. Conceptualization of model calibration.

Mathematically, calibration boils down to optimization of an objective function, i.e.,

$$\text{Min} : g(\boldsymbol{\theta}) = \sum_{j=1}^{v} [w_j \sum_{i=1}^{n_j} (x_o - x_s)_i^2], \tag{2}$$

or,

$$\text{Max} : g(\boldsymbol{\theta}) = \sum_{j}^{v} [w_j (1 - \frac{\sum_{i=1}^{n_j} (x_o - x_s)_i^2}{\sum_{i=1}^{n_j} (x_o - \overline{x}_o)_i^2})], \tag{3}$$

where g is the objective function, $\boldsymbol{\theta}$ is a vector of model parameters, x_o is an observed variable, x_s is the corresponding simulated variable, v is the number of measured variables to be used to calibrate the model, w_j is weight of the jth variable, and n_j is the number of measured observations in the jth variable. The case for $v > 1$ if often referred to as *multi-objective* calibration containing, in our case, variables such as discharge, nitrate, sediment, etc. A large number of different objective function formulations exist in the literature, 11 of which are used in SWAT-CUP.

Calibration is inherently subjective and, therefore, intimately linked to model output uncertainty. Parameter estimation through calibration is concerned with the problem of making inferences about physical systems from measured output variables of the model (e.g., river discharge, sediment concentration, nitrate load, etc.). This is attractive because the direct measurement of parameters describing the physical system is time consuming, costly, tedious, and often has limited applicability. Uncertainty stems from the fact that nearly all measurements are subject to some error, models are simplifications of reality, and the inferences are usually statistical in nature. Furthermore, because one can only measure a limited number of (noisy) data and because physical systems are usually modeled by continuum equations, no calibration can lead to a single parameter set or a single output. In other words, if there is a single model that fits the measurements there will be many of them. This is an old concept known as the *non-uniqueness problem* in the optimization literature. Our goal in calibration is, then, to characterize the set of models, mainly through assigning distributions (uncertainties) to the parameters that fit the data to satisfy our assumptions as well as other prior information [3].

The term *uncertainty analysis* refers to the propagation of all model input uncertainties (mapped in the parameter distribution) to model outputs. Input uncertainties can stem from the lack of knowledge of physical model inputs such as climate, soil, and land-use, to model parameters and model structure. Identification of all acceptable model solutions in the face of all input uncertainties can, therefore, provide us with model uncertainty expressed in SWAT-CUP as 95% prediction uncertainty (95PPU) (Figure 3).

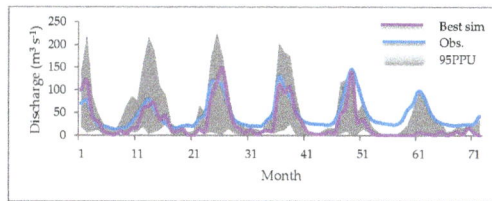

Figure 3. Illustration of model output uncertainty expressed as 95% prediction uncertainty (95PPU) as well as measured and best simulated discharge variable.

To compare the 95PPU band with, for example, a discharge signal, we devised two statistics referred to as *p-factor* and *r-factor* [2,3]. *p-factor* is the percentage of measured data bracketed by the 95PPU band. These measurements are within the simulation uncertainty of our model; hence, they are simulated well and accounted for by the model. Subsequently, (1-*p-factor*) represent the measured data not simulated well by the model, in other words, (1-*p-factor*) is the *model error*. *r-factor* is a measure of the thickness of the 95PPU band and is calculated as the average 95PPU thickness divided by the standard deviation of the corresponding observed variable:

$$r - factor_j = \frac{\frac{1}{n_j} \sum_{t_i=1}^{n_j} \left(x_s^{t_i,97.5\%} - x_s^{t_i,2.5\%} \right)}{\sigma_{oj}} \tag{4}$$

where $x_s^{t_i,97.5\%}$ and $x_s^{t_i,2.5\%}$ are the upper and lower boundary of the 95PPU at time-step t and simulation i, n_j is the number of data points, and σ_{oj} is the standard deviation of the jth observed variable.

Validation is used to build confidence in the calibrated parameters. For this purpose, the calibrated parameter ranges are applied to an independent measured dataset, without further changes. The analyst is required to do one iteration with the same number of simulations as in the last calibration iteration. Similar to calibration, validation results are also quantified by the *p-factor*, *r-factor*, and the objective function value. It is important that the data in validation period meets more or less the same physical criteria as the calibration period. For example, climate and land-use of the validation period should pertain to the same kind of climate and land-uses as the calibration period. Also, if for example, river discharge is used to calibrate the model, then the average and variance of discharges in the two periods should more or less be the same.

Risk analysis is a step usually ignored in most hydrological modeling. We often build a model, calibrate it and report the model uncertainty, but do not take the next step to analyze the problem. For example, we simulate nitrate concentration in the rivers or in the groundwater, or quantify soil erosion and soil losses, but do not go further to quantify their consequences in the environment or on human health. One impediment is usually the existence of uncertainty in the outputs. A large body of literature exists on decision-making under uncertainty, but there is still no standard and easy way of communicating the uncertainty to decision-makers. Environmental scientist researchers and engineers should pay more attention to this problem. One way forward would be to transform the uncertainty to risk. A monitory risk value is more tangible to a decision-maker than uncertainty. The risk can be calculated as the probability of failure (or loss) multiplied by the cost of failure (or loss):

$$Risk = \Pr(F).Cost(F) \tag{5}$$

To demonstrate, assume that we are interest in calculating the risk of soil loss due to erosion. To calculate the probability of soil loss, we propagate the parameter ranges that were obtained during calibration by performing an iteration of, for example, 1000 simulations. Using the "No_Observation" option for extraction, we extract the sediment loss from a sub-basin of interest (Table 2, column 1). Next, we can calculate the cost of soil loss in ways that could include loss of fertilizer, loss of crop yield, loss of organic matter, etc. [5–7]. Here, we assumed a cost of 10 $ tn^{-1} to replenish the loss of fertilizer (Table 2, column 2).

Table 2. Statistics of cumulative distribution for soil loss resulting from model uncertainty.

	Before Terracing				After Terracing			
Soil Loss (tn ha^{-1})	Cost of Soil Loss ($ ha^{-1})	Prob. of Soil Loss	Risk of Soil Loss ($ ha^{-1})	Soil Loss (tn ha^{-1})	Cost of Soil Loss ($ ha^{-1})	Prob. of Soil Loss	Risk of Soil Loss ($ ha^{-1})	Gain ($ ha^{-1})
513	5130	0.29	1501	209	2090	0.41	460	1041
534	5340	0.14	747	219	2190	0.59	241	506
601	6010	0.14	841	258	2580	0.72	464	376
668	6680	0.09	601	296	2960	0.78	414	187
735	7350	0.06	441	335	3350	0.86	335	106
802	8020	0.05	481	373	3730	0.91	261	220
869	8690	0.05	434	411	4110	0.94	206	229
936	9360	0.06	562	450	4500	0.95	180	382
1003	10,030	0.05	502	488	4880	0.98	244	258
1070	10,700	0.06	642	527	5270	1.00	211	431
Expectation			6751				3016	3735

In the "echo_95ppu_No_Obs.txt" file of SWAT-CUP, one can find the probability distribution of soil loss (Table 2, column 3). In this example, we have an uncertainty on soil loss in the range of (513 tn ha^{-1} to 1070 tn ha^{-1}). It is important to realize that this range is the model solution. Most researchers here search for one number to carry their research forward. But the model, because of uncertainty, never just has one number as the solution. The risk can then be calculated by Equation (5) as the product of cost of soil loss by the probability of soil loss (Table 2, column 3).

To carry the example forward, assume that with the help of terracing we can cut down on soil loss. Implementing this management option in SWAT and running an iteration as before, we obtain the new loss soil and its probability distribution (Table 2). We can again calculate the risk of soil loss after terracing and calculate the *Gain* or profit of terracing as:

$$Gain = Risk_b - Risk_a \tag{6}$$

where *b* and *a* stand for before and after terracing. In the last row of Table 2, the expected values are reported, where the expected value of gain as a result of terracing is calculated to be 3735 $ ha^{-1}. If the cost of terracing is less than this amount, then there is a profit in terracing. The same type of analysis can be done with different best-management options (BMP) in SWAT and the most profitable one selected.

2. Outstanding Calibration and Uncertainty Analysis Issues

Calibration of watershed models suffers from a number of conceptual and technical issues, which we believe require a more careful consideration by the scientific community. These include: (1) inadequate definition of the base model; (2) parameterization; (3) objective function definition;

(4) use of different optimization algorithms; (5) non-uniqueness; and (6) model conditionality on the face of the above issues. Two other issues having an adverse effect on calibration include: (7) time constraints; and (8) modeler's inexperience and lack of sufficient understanding of model parameters. In the following, a short discussion of these issues is presented.

2.1. Inadequate Definition of the Base Model

An important setback in model calibration is to start the process with an inadequate model. Failure to correctly setup a hydrologic model may not allow proper calibration and uncertainty analyses, leading to inaccurate parameter estimates and wrong model prediction. To build a model with an accurate accounting of hydrological processes, a data-discrimination procedure is needed during model building. This includes: (i) identifying the best data set (e.g., climate, land-use, soil) from among, at times, many available data sources; (ii) accounting for important processes in accordance with the "correct neglect" principle where only ineffective processes are ignored in the model. Often important processes, which are usually ignored, include: springs, potholes, glacier/snow melts, wetlands, reservoirs, dams, water transfers, and irrigation. Accounting for these measures, if they exist, leads to a better physical accounting of hydrological processes, which significantly improves the overall model performance. This avoids unnecessary and arbitrary adjustment of parameters to compensate for the missing processes in the model structure.

In this special issue, Kamali et al. [8] address the issue of the existence of many datasets and their effects on the assessment of water resources. They combined 4 different climate data with 2 different land-use maps to build 8 different models that they calibrated and validated. These models led to different calibrated parameter sets, which consequently led to different quantification of water resources in the region of study (Figure 4).

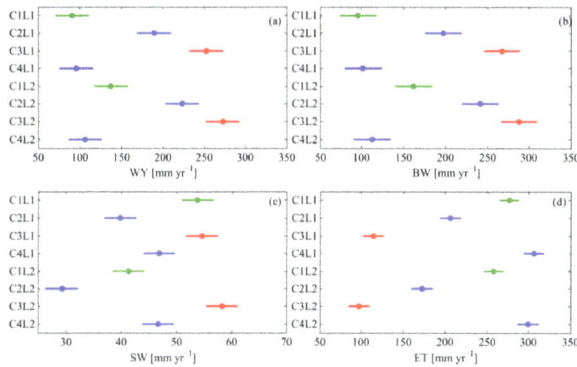

Figure 4. Range of four water resources components. (**a**) WY = water yield; (**b**) BW = blue water; (**c**) SW = soil water; (**d**) ET = evapotranspiration obtained from eight calibrated models. C represents a climate data set, and L represents a land-use dataset. (Source: Kamali et al. [8]).

2.2. Parameterization

There are two issues in parameterization: (1) which parameters to use; and (2) how to regionalize the parameters. Not all SWAT parameters are relevant to all sub-basins, and not all should be used simultaneously to calibrate the model. For example, rainfall is a driving variable and should not be fitted with other parameters at the same time. Similarly, snow-melt parameters (SFTMP, SMTMP, SMFMX, SMFMN, TIMP) and canopy storage (CANMX), which introduce water into the system, should not be calibrated simultaneously with other parameters as they will cause identifiability problems. These parameters should be fitted first and fixed to their best values and then removed from further calibration.

The other issue deals with regionalization of the parameters. That is, how to distinguish between hydraulic conductivity of the same soil unit when it is under forest, as opposed to being under pasture or agriculture. For this purpose, a scheme is introduced in SWAT-CUP where a parameter can be regionalized to the HRU level using the following assignment:

$$x\_\_<parname>.<ext>\_\_<hydrogrp>\_\_<soltext>\_\_<landuse>\_\_<subbsn>\_\_<slope>$$

where *x* is an identifier to indicate the type of change to be applied to the parameter (*v* is value change; *a* adds an increment to existing value; and *r* is for relative change of spatial parameters and it multiplies the existing parameter value by (1 + an increment)). *<parname>* is SWAT parameter name, *<ext>* is SWAT file extension code; *<hydrogrp>* is hydrologic group; *<soltext>* is soil texture; *<landuse>* is land-use type; *<subbsn>* is sub-basin number(s); and *<slope>* is the slope as it appears in the header line of SWAT input files. Any combination of the above factors can be used to calibrate a parameter. The analyst, however, must decide on the detail of regionalization as on the one hand, a large number of parameters could result, and on the other hand, by too much lumping, the spatial heterogeneity of the region may be lost. This balance is not easy to determine, and the choice of parameterization will affect the calibration results (see [9,10] for a discussion). Detailed information on spatial parameters is indispensable for building a correct watershed model. A combination of measured data and spatial analyses techniques using pedotransfer functions, geostatistical analysis, and remote sensing data would be the way forward.

2.3. Use of Different Objective Functions

There are a large number of objective functions with different properties that could be used in model calibration [11,12]. The problem with the choice of objective functions is that they can produce statistically similar and good calibration and validation results, but with quite different parameter ranges. This adds a level of uncertainty to the calibration process, which could make the calibration exercise meaningless. In this special issue, Hooshmand et al. [13] used SUFI-2 with 7 different objective functions to calibrate discharge in Salman Dam and Karkhe Basins in Iran. They found that after calibration, each objective function found an acceptable solution, but at a different location in the parameter space (Figure 5).

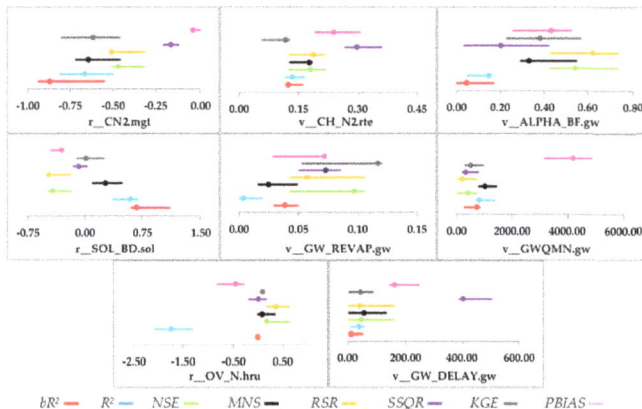

Figure 5. Uncertainty ranges of calibrated parameters using different objective functions for a project in Karkheh River Basin, Iran. The points in each line show the best value of parameters, r_ refers to a relative change where the current values are multiplied by (one plus a factor from the given parameter range), and v_ refers to the substitution by a value from the given parameter range. (Source: Hooshmand et al. [13]).

2.4. Use of Different Optimization Algorithms

Yang et al. [14] showed that different calibration algorithms converge to different calibrated parameter ranges. They used SWAT-CUP to compare Generalized Likelihood Uncertainty Estimation (GLUE) [15], Parameter Solution (ParaSol) [16], Sequential Uncertainty Fitting (SUFI-2) [2–4], and Markov chain Monte Carlo (MCMC) [17–19] methods in an application to a watershed in China. They found that these different optimization algorithms each found a different solution at different locations in the parameter spaces with roughly the same discharge results. In this special issue, Hooshmand et al. [13] also showed that use of SUFI-2, GLUE, and ParaSol resulted in the identification of different parameters ranges, with similar calibration/validation results, which led to simulation of significantly different water-resources estimates (Figure 6).

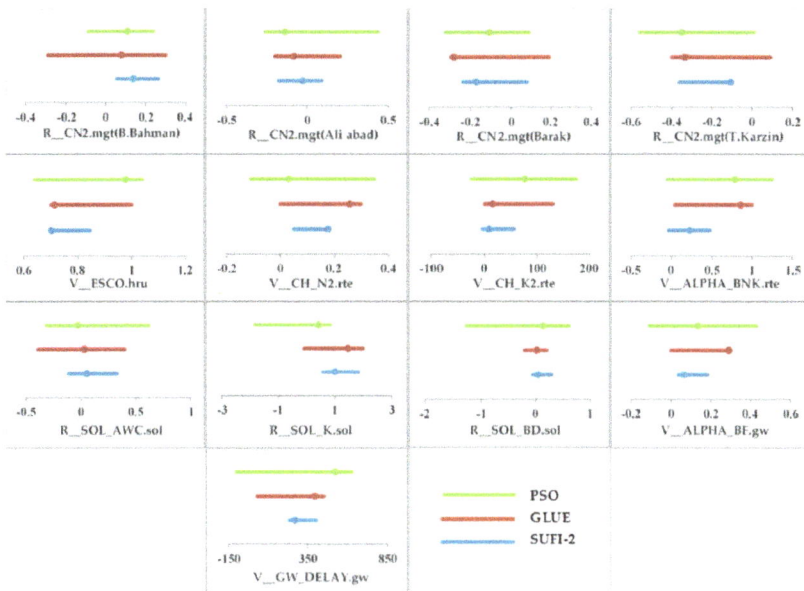

Figure 6. Uncertainty ranges of the parameters based on all three methods applied in Salman Dam Basin, Iran. The points in each line show the best value of the parameters, r_ refers to a relative change where the current values are multiplied by one plus a factor from the given parameter range, and v_ refers to the substitution by a value from the given parameter range. (Source: Hooshmand et al. [13]).

2.5. Calibration Uncertainty or Model Non-Uniqueness

As mentioned before, a single parameter set results in a single model signal in a deterministic model application. In an inverse application (i.e., calibration), the measured variable could be reproduced with thousands of different parameter sets. This non-uniqueness is an inherent property of model calibration in distributed hydrological applications. An example is shown in Figure 7 where two very different parameter sets produce signals similar to the observed discharge.

We can visualize non-uniqueness by plotting the response surface of the objective function versus two calibrating parameters. As an example, Figure 8 shows the inverse of an objective function, based on the mean square error, plotted against CN2 and GW-REVAP in an example with 2400 simulations. In this example, CN2, GW-REVAP, ESCO, and GWQMN were changing simultaneously. Size and distribution of all the acceptable solutions (1/goal > 0.8) are shown in darker shade. This multi-modal attribute of the response surface is the reason why each algorithm or each objective function finds a different good solution.

GW_Delay	CH_N2	CN2	REVAPMN	SOL_SW	NS
3.46	0.0098	50	0.8	0.11	0.90
0.34	0.131	20	2.4	0.23	0.94

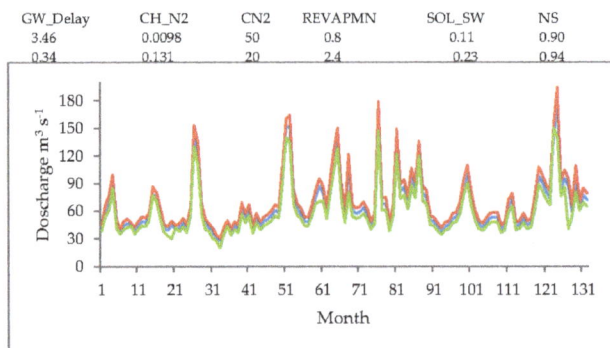

Figure 7. Example of parameter non-uniqueness showing two similar discharge signals based on quite different parameter values.

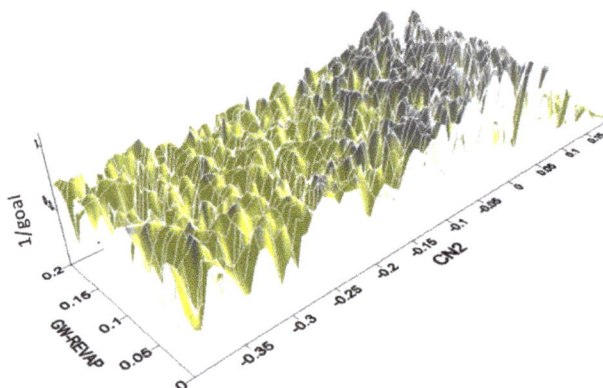

Figure 8. The "multimodal" behavior of the objective function response surface. All red-colored peaks have statistically the same value of objective function, which occur at the different regions in the parameter space.

To limit the non-uniqueness problem, we should: (i) include more variables in the objective function (e.g., discharge, ET or crop yield, nutrient loads, etc.); (ii) use multiple outlets for calibration; and (iii) constrain the objective function with soft data (i.e., knowledge of local experts on nutrient and sediment loads from different land-uses, etc.). The downside of this is that a lot of data must be measured for calibration. The use of remote-sensing data, when it becomes practically available, could be extremely useful. In fact, the next big jump in watershed modeling will be made as a result of advances in remote-sensing data availability.

2.6. Calibrated Model Conditionality

A model calibrated for a discharge station at the outlet of a watershed should not be expected to provide good discharge results for outlets inside the watershed. The outlets inside should be "regionally" calibrated for the contributing sub-basins. Also, a model calibrated for discharge, should not be expected to simulate water quality. Calibrated parameters are always expressed as distributions to reflect the model uncertainty. In other words, they are always "conditioned" on the model assumptions and inputs, as well as the methods and data used for model calibration. Hence, a model calibrated, for example, for discharge, may not be adequate for prediction of sediment; or for application to another region; or for application to another time period.

A calibrated model is, therefore, always: (i) non-unique; (ii) subjective; (iii) conditional; and subsequently (iv) limited in the scope of its use. Hence, important questions arise as to: "When is a watershed model truly calibrated?" and "For what purpose can we use a calibrated watershed model?" For example: What are the requirements of a calibrated watershed model if we want to do land-use change analysis? Or, climate change analysis? Or, analysis of upstream/downstream relations in water allocation and distribution? Or, water quality analysis? Can any single calibrated watershed model address all these issues, or should there be a series of calibrated models each fitted to a certain purpose? We hope that these issues can be addressed more fully by research in this field.

Conditionality is, therefore, an important issue with calibrated models. Calibrated parameters (θ) are conditioned on the base model parameterization (p), variables used to calibrate the model (v), choice of objective function (g) and calibration algorithm (a), as well as weights used in a multi-objective calibration (w), the type and number of data points used for calibration (d), and calibration-validation dates (t), among other factors. Mathematically, we could express a calibrated model M as:

$$M = M(\theta|p, v, g, a, w, d, t, \ldots) \tag{7}$$

To obtain an *unconditional* calibrated model, the parameter set θ must be integrated over all factors. This may make model uncertainty too large for any practical application. Hence, a model must always be calibrated with respect to a certain objective, which makes a calibrated model only applicable to that objective.

2.7. Time Constraint

Time is often a major impediment in the calibration of large-scale and detailed hydrologic models. To overcome this, most projects are run with fewer simulations, resulting in less-than-optimum solutions. To deal with this problem, a parallel processing framework was created in the Windows platform [20] and linked to SUFI-2 in the SWAT-CUP software. In this methodology, calibration of SWAT is parallelized, where the total number of simulations is divided among the available processors. This offers a powerful alternative to the use of grid or cloud computing. The performance of parallel processing was judged by calculating speed-up, efficiency, and CPU usage (Figure 9).

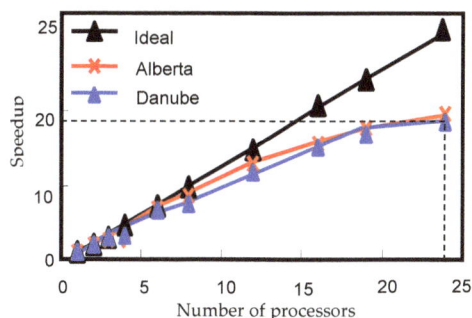

Figure 9. The speed-up achieved for different Soil and Water Assessment Tools (SWAT) projects. The number of processors on the horizontal axis indicates the number of parallel jobs submitted. The figure shows that most projects could be run 10 times faster with about 6–8 processors. (Source: Rouholahnejad, et al. [20]).

2.8. Experience of the Modeler

The success of a calibration process depends on the accuracy of the mathematical model and the procedures chosen for the calibration as already noted. However, the experience of the modeler plays an important role, and in this sense, calibration can be described as an art as well as a science [21–24].

3. A Protocol for Calibration of Soil and Water Assessment Tools (SWAT) Models

Calibration of watershed models is a long and often tedious process of refining the model for processes and calibrating parameters. We should usually expect to spend as much time calibrating a model as we take to build the model. To calibrate the model we suggest using the following general approach (also see Abbaspour et al. [2]).

3.1. Pre-Calibration Input Data and Model Structure Improvement

Build the SWAT model in ArcSWAT or QSWAT using the best parameter estimates based on the available data, literature, and the analyst and local expertise. There is always more than one dataset (e.g., soil, land-use, climate, etc.) available for a region. Test each one and choose the best dataset to proceed. It should be noted that for calibration, the performance of the initial model should not be too drastically different from the measured data. If the initial simulation is too different, often calibration might be of little help. Therefore, one should include as much of the important processes in the model as possible. There may be processes not included in SWAT (e.g., wetland, glacier melt, micronutrients, impact of salinity on crop yield), or included in SWAT but with unavailable data (e.g., reservoir operation, water withdrawal, and water transfers); or with available data, but unknown to the modeler. This requires a good knowledge of the watershed, which may be gained from literature or local experts, or by using the "Maps" option in SWAT-CUP, which can recreate the sub-basins and rivers on the Microsoft's Bing Maps (Figure 10).

At this stage, also check the contribution of rainfall, snow parameters, rainfall intercept, inputs from water treatment plants, and water transfers. A pre-calibration run of these parameters is necessary to identify their best values, and then to fix them in the model without further change.

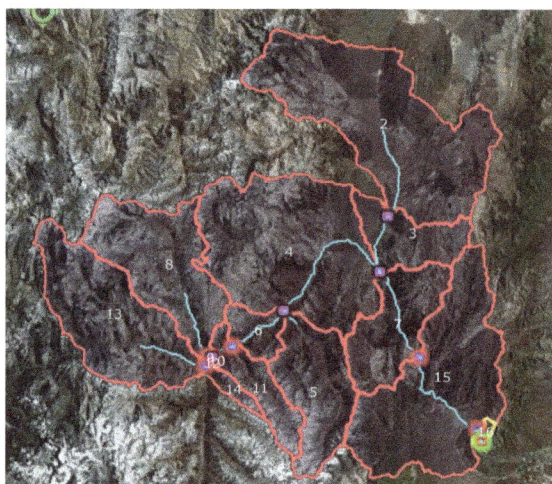

Figure 10. The Maps option of SWAT-CUP can be used to see details of the watershed under investigation, such as dams, wrongly placed outlets, glaciers, high agricultural areas, etc.

3.2. Identify the Parameters to Optimize

Based on the performance of the initial model at each outlet station, relevant parameters in the upstream sub-basins are parameterized using the guidelines in Abbaspour et al. [2]. This procedure results in regionalization of the parameters.

3.3. Identify Other Sensitive Parameters

Next to parameters identified in step 2, use the one-at-a-time sensitivity analysis to check the sensitivity of other relevant parameters at each outlet. To set the initial ranges of parameters to be optimized, some experience and hydrological knowledge are required on the part of the analyst. In addition to the initial ranges, user-defined "absolute parameter ranges" should also be set if necessary. These are the upper and lower limits of what is physically a meaningful range for a parameter at a site.

3.4. Running the Model

After the model is parameterized and the ranges are assigned, the model is run some 300–1000 times, depending on the number of parameters, model's execution time, and system's capabilities. SUFI-2 is an iterative procedure and does not require too many runs in each iteration. Usually, 3–4 iterations should be enough to attain a reasonable result. Parallel processing can be used to greatly improve the runtime. PSO and GLUE need a larger number of iterations and simulations (100–5000).

3.5. Perform Post-Processing

After simulations in each iteration are completed; the post-processing option in SWAT-CUP calculates the objective function and the 95PPU for all observed variables in the objective function. New parameter ranges are suggested by the program for another iteration [2,3].

3.6. Modifying the Suggested New Parameters

The new parameters may contain values outside the desired or physically meaningful ranges. The suggested values should be modified by the user guiding the parameters in a certain desired direction, or to make sure that they are within the absolute parameter ranges. Use the new parameters to run another iteration until desired values of *p-factor*, *r-factor*, and the objective function is reached.

Author Contributions: All authors formulated the concept. Karim C. Abbaspour wrote the initial version, Saeid Ashraf Vaghefi and Raghvan Srinivasan contributed by editing and suggesting revisions.

Appendix A

Summary of the papers in the "Integrated Soil and Water Management" special issue (Table A1).

1,2—Chambers et al. [25,26] evaluated two different approaches for modeling historical and future streamflow and stream temperature in Rhode Island, USA. They found that between 1980–2009, the number of stressful events (i.e., high or low flows occurring simultaneously with stream temperatures exceeding 21 °C) increased by 55% and average streamflow increased by 60%. For future scenarios, however, the chance of stressful events increases on average by 6.5% under the low-emission scenario and by 14.2% under the high-emission scenario relative to historical periods.

3—Chen et al. [27] studied the effect of different fertilization and irrigation schemes on water and nitrate leaching from the root zone. They concluded that N-application based on soil testing helps to improve groundwater quality.

4—Cuceloglu et al. [28] simulated the water resources of Istanbul and quantified the spatial distribution of the region. Their results show that the annual blue-water potential of Istanbul is 3.5 billion m^3, whereas the green-water flow and storage are 2.9 billion m^3 and 0.7 billion m^3, respectively.

5—Ding et al. [29] showed that total nitrogen load increased with increasing slope in a Yangtze River-cultivated Regosol soil.

6—Fabre et al. [30] studied the effect of permafrost degradation as a result of increasing temperature in the largest Arctic river system: the Yenisei River in central Siberia, Russia. Once the climate data and soil conditions were adapted to a permafrost watershed, the calibration results showed SWAT was able to estimate water fluxes at a daily time step, especially during unfrozen periods.

7—Fant et al. [31] evaluated the effect of climate change on water quality of the continental US. They report that under the business-as-usual emissions scenario, climate change is likely to cause economic impacts ranging from 1.2 to 2.3 (2005 billion) USD/year in 2050 and 2.7 to 4.8 (2005 billion) USD/year in 2090 across all climate and water-quality models.

8—Gharib et al. [32] investigated the combined effect of threshold selection and Generalized Pareto Distribution parameter estimation on the accuracy of flood quantiles. With their method, one third of the stations showed significant improvement.

9—Grusson et al. [33] studied the influence of spatial resolution of a gridded weather (16- and 32-km SAFRAN grids) dataset on SWAT output. They reported better performance of these data relative to measured station data.

10—Hooshmand et al. [13] investigated the impact of the choice of objective function and optimization algorithm on the calibrated parameters. They reported that different objective functions and algorithms produce acceptable calibration results, however, with significantly different parameters, which produce significantly different water-resources estimates. This adds another level of uncertainty on model prediction.

11—Kamali et al. [8] studied the impact of different databases on the water-resources estimates and concluded that while different databases may produce similar calibration results, the calibrated parameters are significantly different for different databases. They highlighted that "As the use of any one database among several produces questionable outputs, it is prudent for modelers to pay more attention to the selection of input data".

12—Kamali et al. [34] analyzed characteristics and relationships among meteorological, agricultural, and hydrological droughts using the Drought Hazard Index derived from a SWAT application. They quantified characteristics such as severity, frequency, and duration of droughts using the Standardized Precipitation Index (SPI), Standardized Runoff Index (SRI), and Standardized Soil Water Index (SSWI) for historical (1980–2012) and near future (2020–2052) periods. They concluded that the duration and frequency of droughts will likely decrease in SPI. However, due to the impact of rising temperature, the duration and frequency of SRI and SSWI will intensify in the future.

13—Lee et al. [35] studied the impacts of the upstream Soyanggang and Chungju multi-purpose dams on the frequency of downstream floods in the Han River basin, Korea. They concluded that the two upstream dams reduce downstream floods by approximately 31%.

14—Li et al. [36] studied the effect of urban non-point source pollution on Baiyangdian Lake in China. They found that the pollutant loads for Pb, Zn, TN and TP accounted for about 30% of the total amount of pollutant load.

15—Ligaray et al. [37] studied the fate and transport of Malathion. They used a modified three-phase partitioning model in SWAT to classify the pesticide into dissolved, particle-bound, and dissolved organic carbon (DOC)-associated pesticide. They found that the modified model gave a slightly better performance than the original two-phase model.

16—Lutz at al. [38] evaluated the impact of a buffer zone on soil erosion. Their results indicated that between 0.2 to 1% less sediment could reach the Itumbiara reservoir with buffer strip provision, which would have an important effect on the life of the dam.

17—Marcinkowski et al. [39] studied the effect of climate change on the hydrology and water quality in Poland. They predicted an increase in TN losses.

18—Paul et al. [40] determined the response of SWAT to the addition of an onsite wastewater-treatment systems on nitrogen loading into the Hunt River in Rhode Island. They concluded that using the treatment systems data in the SWAT produced a better calibration and validation fit for total N.

19—Qi et al. [41] compared SWAT to a simpler Generalized Watershed Loading Function (GWLF) model. The performances of both models were assessed via comparison between simulated and measured monthly streamflow, sediment yield, and total nitrogen. The results showed that both models were generally able to simulate monthly streamflow, sediment, and total nitrogen loadings during the simulation period. However, SWAT produced more detailed information, while GWLF could produce better average values.

20—Rouholahnejad et al. [42] investigated the impact of climate and land-use change on the water resources of the Black Sea Basin. They concluded in general that the ensemble of the climate scenarios show a substantial part of the catchment will likely experience a decrease in freshwater resources by 30% to 50%.

21—Senent-Aparicio et al. [43] investigated the effect of climate change on water resources of the Segura River Basin and concluded that water resources were expected to experience a decrease of 2–54%.

22,23—Seo et al. [44,45] used SWAT to simulate hydrologic behavior of Low Impact Developments (LID) such as the installation of bioretention cells or permeable pavements. They report that application of LID practices decreases surface runoff and pollutant loadings for all land-uses. In addition, post-LID scenarios generally showed lower values of surface runoff, lower nitrate in high-density urban land-use, and lower total phosphorus in conventional medium-density urban areas.

24—Tan et al. [46] investigated the accuracy of three long-term gridded data records: APHRODITE, PERSIANN-CDR, and NCEP-CFSR. They concluded that the APHRODITE and PERSIANN-CDR data often underestimated extreme precipitation and streamflow, while the NCEP-CFSR data produced dramatic overestimations.

25—Vaghefi et al. [47] coupled SWAT to MODSIM, which is a program for optimization of water distribution. They concluded in their study that the coupled SWAT-MODSIM approach improved the accuracy of SWAT outputs by considering the water allocation derived from MODSIM.

26—Wangpimool et al. [48] studied the effect of Para Rubber Expansion of the water balance of Loei Province in Thailand. They found that displacement of original local field crops and disturbed forest land by Para rubber production resulted in an overall increase in evapotranspiration of roughly 3%.

27—White et al. [49] describe the development of a national (US) database of preprocessed climate data derived from monitoring stations applicable to USGS 12-digit watersheds. The authors conclude that the data described in this work are suitable for the intended SWAT and APEX application and also suitable for other modeling efforts, and are freely provided via the web.

Table A1. Summary of the papers published in the special issue.

	Important Processes	Location	Country	Calibration	Water Quality	Crop	Author
1	Thermal stress	Rhode Island	USA	SWAT-CUP	Temp.		Chambers et al.
2	Fish Habitat	Rhode Island	USA	SWAT-CUP	Temp.		Chambers et al.
3	Leaching	North China Plain	China		NO_3	Maize	Chen et al.
4	Water resources	Istanbul	Turkey	SWAT-CUP			Cuceloglu et al.
5	TN loss	Yangtze River	China		TN		Ding et al.
6	Permafrost	Central Siberia	Russia	SWAT-CUP			Fabre et al.
7	Climate change	Continental US	USA		WT, DO, TN, TP		Fant et al.
8	Flooding	Alberta	Canada	SWAT-CUP			Gharib et al.
9	Gridded rainfall	Garonne River watershed	France	SWAT-CUP			Grusson et al.
10	Uncertainty issues	Karkheh River Basin	Iran	SWAT-CUP			Houshmand et al.
11	Drought, Climate change	Karkheh River Basin	Iran	SWAT-CUP			Kamali et al.
12	Uncertainty	Karkheh River Basin	Iran	SWAT-CUP			Kamali et al.
13	Flooding	Paldang Dam	Korea	manual			Lee et al.
14	Non-point source pollution	Baoding City	China	manual	TN, TP		Li et al.
15	Pesticide	Pagsanjan-Lumban Basin,	Philippines	SWAT-CUP			Ligaray et al.
16	Buffer strip	Itumbiara city	Brazil	manual	sediment		Lutz et al.
17	Climate change	Upper Narew, Barycz	Poland	SWAT-CUP	Sediment, TN		Marcinkowski et al.
18	Non-point source pollution	Hunt River Rhode Island	USA	SWAT-CUP	TN		Paul et al.
19	SWAT, GWLF comparison	Tunxi and the Hanjiaying basins	China	SWAT-CUP	Sediment, TN		Qi et al.
20	Climate-Land-use Change	Black Sea Basin	Europe	SWAT-CUP			Rouholahnejad et al.
21	Climate change	Segura River Basin	Spain	SWAT-CUP			Senent-Aparicio et al.
22	Optimal design	Clear Creek watershed (TX)	USA	SWAT-CUP			Seo et al.
23	Water quality	Clear Creek watershed (TX)	USA	SWAT-CUP			Seo et al.
24	Gridded rainfall	Kelantan River Basin	Malaysia	SWAT-CUP			Tan et al.
25	Coupling SWAT-MODSIM	Karkheh River Basin	Iran	SWAT-CUP		Wheat, maize	Vaghefi et al.
26	Water balance	Loei Province	Thailand	SWAT-CUP		Para-rubber	Wangpimool et al.
27	Climate data	USA	USA	SWAT-CUP			White et al.

References

1. Arnold, J.G.; Srinivasan, R.; Muttiah, R.S.; Williams, J.R. Large area hydrologic modeling and assessment. Part I: Model development. *J. Am. Water Resour. Assoc.* **1998**, *34*, 73–89. [CrossRef]

2. Abbaspour, K.C.; Rouholahnejad, E.; Vaghefi, S.; Srinivasan, R.; Yang, H.; Klove, B. A continental-scale hydrology and water quality model for Europe: Calibration and uncertainty of a high-resolution large-scale SWAT model. *J. Hydrol.* **2015**, *524*, 733–752. [CrossRef]

3. Abbaspour, K.C.; Yang, J.; Maximov, I.; Siber, R.; Bogner, K.; Mieleitner, J.; Zobrist, J.; Srinivasan, R. Modelling hydrology and water quality in the pre-alpine/alpine Thur watershed using SWAT. *J. Hydrol.* **2007**, *333*, 413–430. [CrossRef]

4. Abbaspour, K.C.; Johnson, A.; van Genuchten, M.T. Estimating uncertain flow and transport parameters using a sequential uncertainty fitting procedure. *Vadose Zone J.* **2004**, *3*, 1340–1352. [CrossRef]

5. Dechen, F.; Carmela, S.; Telles, T.S.; Guimaraes, M.D.; de Fatima, M.; De Maria, I.C. Losses and costs associated with water erosion according to soil cover rate. *Bragantia* **2015**, *74*, 224–233. [CrossRef]

6. Gulati, A.; Rai, S.C. Cost estimation of soil erosion and nutrient loss from a watershed of the Chotanagpur Plateau, India. *Curr. Sci.* **2014**, *107*, 670–674.

7. Mcqueen, A.D.; Shulstad, R.N.; Osborn, C.T. Controlling agricultural soil loss in Arkansas north lake Chicot watershed—A cost-analysis. *J. Soil Water Conserv.* **1982**, *37*, 182–185.

8. Kamali, B.; Houshmand Kouchi, D.; Yang, H.; Abbaspour, K.C. Multilevel Drought Hazard Assessment under Climate Change Scenarios in Semi-Arid Regions—A Case Study of the Karkheh River Basin in Iran. *Water* **2017**, *9*, 241. [CrossRef]

9. Pagliero, L.; Bouraoui, F.; Willems, P.; Diels, J. Large-Scale Hydrological Simulations Using the Soil Water Assessment Tool, Protocol Development, and Application in the Danube Basin. *J. Environ. Qual.* **2014**, *4*, 145–154. [CrossRef]

10. Whittaker, G.; Confesor, R.; Di Luzio, M.; Arnold, J.G. Detection of overparameterization and overfitting in an automatic calibration of swat. *Trans. ASABE* **2010**, *53*, 1487–1499. [CrossRef]

11. Moriasi, D.N.; Arnold, J.G.; Van Liew, M.W.; Bingner, R.L.; Harmel, R.D.; Veith, T.L. Model evaluation guidelines for systematic quantification of accuracy in watershed simulations. *Trans. ASABE* **2007**, *50*, 885–900. [CrossRef]

12. Krause, P.; Boyle, D.P.; Bäse, F. Comparison of different efficiency criteria for hydrological model assessment. *Adv. Geosci.* **2005**, *5*, 89–97. [CrossRef]

13. Houshmand Kouchi, D.; Esmaili, K.; Faridhosseini, A.; Sanaeinejad, S.H.; Khalili, D.; Abbaspour, K.C. Sensitivity of Calibrated Parameters and Water Resource Estimates on Different Objective Functions and Optimization Algorithms. *Water* **2017**, *9*, 384. [CrossRef]

14. Yang, J.; Abbaspour, K.C.; Reichert, P.; Yang, H. Comparing uncertainty analysis techniques for a SWAT application to Chaohe Basin in China. *J. Hydrol.* **2008**, *358*, 1–23. [CrossRef]

15. Beven, K.; Binley, A. The future of distributed models: Model calibration and uncertainty prediction. *Hydrol. Process* **1992**, *6*, 279–298. [CrossRef]

16. Van Griensven, A.; Meixner, T. Methods to quantify and identify the sources of uncertainty for river basin water quality models. *Water Sci. Technol.* **2006**, *53*, 51–59. [CrossRef] [PubMed]

17. Kuczera, G.; Parent, E. Monte Carlo assessment of parameter uncertainty in conceptual catchment models: The Metropolis algorithm. *J. Hydrol.* **1998**, *211*, 69–85. [CrossRef]

18. Marshall, L.; Nott, D.; Sharma, A. A comparative study of Markov chain Monte Carlo methods for conceptual rainfall–runoff modeling. *Water Resour. Res.* **2004**, *40*. [CrossRef]

19. Yang, J.; Reichert, P.; Abbaspour, K.C.; Yang, H. Hydrological Modelling of the Chaohe Basin in China: Statistical Model Formulation and Bayesian Inference. *J. Hydrol.* **2007**, *340*, 167–182. [CrossRef]

20. Rouholahnejad, E.; Abbaspour, K.C.; Vejdani, M.; Srinivasan, R; Schulin, R.; Lehmann, A. Parallelization framework for calibration of hydrological models. *Environ. Model. Softw.* **2012**, *31*, 28–36. [CrossRef]

21. Ostfeld, A.; Salomons, E.; Ormsbee, L.; Uber, J.; Bros, C.M.; Kalungi, P.; Burd, R.; Zazula-Coetzee, B.; Belrain, T.; Kang, D.; et al. Battle of the Water Calibration Networks. *J. Water Resour. Plan. Manag.* **2012**, *138*, 523–532. [CrossRef]

22. Hollaender, H.M.; Bormann, H.; Blume, T.; Buytaert, W.; Chirico, G.B.; Exbrayat, J.F.; Gustafsson, D.; Hoelzel, H.; Krausse, T.; Kraft, P.; et al. Impact of modellers' decisions on hydrological a priori predictions. *Hydrol. Earth Syst. Sci.* **2014**, *18*, 2065–2085. [CrossRef]

23. Freni, G.; Mannina, G.; Viviani, G. Uncertainty in urban stormwater quality modelling: The effect of acceptability threshold in the GLUE methodology. *Water Res.* **2008**, *42*. [CrossRef] [PubMed]

24. Whittemore, R. Is the time right for consensus on model calibration guidance? *J. Environ. Eng.* **2001**, *127*, 95–96. [CrossRef]

25. Chambers, B.M.; Pradhanang, S.M.; Gold, A.J. Assessing Thermally Stressful Events in a Rhode Island Coldwater Fish Habitat Using the SWAT Model. *Water* **2017**, *9*, 667. [CrossRef]

26. Chambers, B.M.; Pradhanang, S.M.; Gold, A.J. Simulating Climate Change Induced Thermal Stress in Coldwater Fish Habitat Using SWAT Model. *Water* **2017**, *9*, 732. [CrossRef]

27. Chen, S.; Sun, C.; Wu, W.; Sun, C. Water Leakage and Nitrate Leaching Characteristics in the Winter Wheat–Summer Maize Rotation System in the North China Plain under Different Irrigation and Fertilization Management Practices. *Water* **2017**, *9*, 141. [CrossRef]

28. Cuceloglu, G.; Abbaspour, K.C.; Ozturk, I. Assessing the Water-Resources Potential of Istanbul by Using a Soil and Water Assessment Tool (SWAT) Hydrological Model. *Water* **2017**, *9*, 814. [CrossRef]

29. Ding, X.; Xue, Y.; Lin, M.; Jiang, G. Influence Mechanisms of Rainfall and Terrain Characteristics on Total Nitrogen Losses from Regosol. *Water* **2017**, *9*, 167. [CrossRef]

30. Fabre, C.; Sauvage, S.; Tananaev, N.; Srinivasan, R.; Teisserenc, R.; Miguel Sánchez Pérez, J. Using Modeling Tools to Better Understand Permafrost Hydrology. *Water* **2017**, *9*, 418. [CrossRef]

31. Fant, C.; Srinivasan, R.; Boehlert, B.; Rennels, L.; Chapra, S.C.; Strzepek, K.M.; Corona, J.; Allen, A.; Martinich, J. Climate Change Impacts on US Water Quality Using Two Models: HAWQS and US Basins. *Water* **2017**, *9*, 118. [CrossRef]

32. Gharib, A.; Davies, E.G.R.; Goss, G.G.; Faramarzi, M. Assessment of the Combined Effects of Threshold Selection and Parameter Estimation of Generalized Pareto Distribution with Applications to Flood Frequency Analysis. *Water* **2017**, *9*, 692. [CrossRef]

33. Grusson, Y.; Anctil, F.; Sauvage, S.; Miguel Sánchez Pérez, J. Testing the SWAT Model with GriddedWeather Data of Different Spatial Resolutions. *Water* **2017**, *9*, 54. [CrossRef]

34. Kamali, B.; Abbaspour, K.C.; Yang, H. Assessing the Uncertainty of Multiple Input Datasets in the Prediction of Water Resource Components. *Water* **2017**, *9*, 709. [CrossRef]

35. Lee, J.E.; Heo, J.H.; Lee, J.; Kim, N.W. Assessment of Flood Frequency Alteration by Dam Construction via SWAT Simulation. *Water* **2017**, *9*, 264. [CrossRef]

36. Li, C.; Zheng, X.; Zhao, F.; Wang, X.; Cai, Y.; Zhang, N. Effects of Urban Non-Point Source Pollution from Baoding City on Baiyangdian Lake, China. *Water* **2017**, *9*, 249. [CrossRef]

37. Ligaray, M.; Kim, M.; Baek, S.; Ra, J.S.; Chun, J.A.; Park, Y.; Boithias, L.; Ribolzi, O.; Chon, K.; Cho, K.H. Modeling the Fate and Transport of Malathion in the Pagsanjan-Lumban Basin, Philippines. *Water* **2017**, *9*, 451. [CrossRef]

38. Luz, M.P.; Beevers, L.C.; Cuthbertson, A.J.S.; Medero, G.M.; Dias, V.S.; Nascimento, D.T.F. The Mitigation Potential of Buffer Strips for Reservoir Sediment Yields: The Itumbiara Hydroelectric Power Plant in Brazil. *Water* **2016**, *8*, 489. [CrossRef]

39. Marcinkowski, P.; Piniewski, M.; Kardel, I.; Szcześniak, M.; Benestad, R.; Srinivasan, R.; Ignar, S.; Okruszko, T. Effect of Climate Change on Hydrology, Sediment and Nutrient Losses in Two Lowland Catchments in Poland. *Water* **2017**, *9*, 156. [CrossRef]

40. Paul, S.; Cashman, M.A.; Szura, K.; Pradhanang, S.M. Assessment of Nitrogen Inputs into Hunt River by OnsiteWastewater Treatment Systems via SWAT Simulation. *Water* **2017**, *9*, 610. [CrossRef]

41. Qi, Z.; Kang, G.; Chu, C.; Qiu, Y.; Xu, Z.; Wang, Y. Comparison of SWAT and GWLF Model Simulation Performance in Humid South and Semi-Arid North of China. *Water* **2017**, *9*, 567. [CrossRef]

42. Rouholahnejad, E.; Abbaspour, K.C.; Lehmann, A. Water Resources of the Black Sea Catchment under Future Climate and Landuse Change Projections. *Water* **2017**, *9*, 598. [CrossRef]

43. Senent-Aparicio, J.; Pérez-Sánchez, J.; Carrillo-García, J.; Soto, J. Using SWAT and Fuzzy TOPSIS to Assess the Impact of Climate Change in the Headwaters of the Segura River Basin (SE Spain). *Water* **2017**, *9*, 149. [CrossRef]

44. Seo, M.; Jaber, F.; Srinivasan, R.; Jeong, J. Evaluating the Impact of Low Impact Development (LID) Practices on Water Quantity and Quality under Different Development Designs Using SWAT. *Water* **2017**, *9*, 193. [CrossRef]

45. Seo, M.; Jaber, F.; Srinivasan, R. Evaluating Various Low-Impact Development Scenarios for Optimal Design Criteria Development. *Water* **2017**, *9*, 270. [CrossRef]

46. Tan, M.L.; Gassman, P.W.; Cracknell, A.P. Assessment of Three Long-Term Gridded Climate Products for Hydro-Climatic Simulations in Tropical River Basins. *Water* **2017**, *9*, 229. [CrossRef]

47. Vaghefi, S.A.; Abbaspour, K.C.; Faramarzi, M.; Srinivasan, R.; Arnold, J.G. Modeling Crop Water Productivity Using a Coupled SWAT–MODSIM Model. *Water* **2017**, *9*, 157. [CrossRef]

48. Wangpimoo, W.; Pongput, K.; Tangtham, N.; Prachansri, S.; Gassman, P.W. The Impact of Para Rubber Expansion on Streamflow and OtherWater Balance Components of the Nam Loei River Basin, Thailand. *Water* **2017**, *9*, 1. [CrossRef]

49. White, M.J.; Gambone, M.; Haney, E.; Arnold, J.; Gao, J. Development of a Station Based Climate Database for SWAT and APEX Assessments in the US. *Water* **2017**, *9*, 437. [CrossRef]

water

MDPI

Article

Assessing Thermally Stressful Events in a Rhode Island Coldwater Fish Habitat Using the SWAT Model

Britta Chambers [1], Soni M. Pradhanang [1,*] and Arthur J. Gold [2]

[1]   Department of Geosciences, University of Rhode Island, Kingston, RI 02881, USA; britta_anderson@uri.edu
[2]   Department of Natural Resource Sciences, University of Rhode Island, Kingston, RI 02881, USA; agold@uri.edu
*   Correspondence: spradhanang@uri.edu; Tel.: +1-401-874-5980

Received: 30 June 2017; Accepted: 29 August 2017; Published: 4 September 2017

Abstract: It has become increasingly important to recognize historical water quality trends so that the future impacts of climate change may be better understood. Climate studies have suggested that inland stream temperatures and average streamflow will increase over the next century in New England, thereby putting aquatic species sustained by coldwater habitats at risk. In this study we evaluated two different approaches for modeling historical streamflow and stream temperature in a Rhode Island, USA, watershed with the Soil and Water Assessment Tool (SWAT), using (i) original SWAT and (ii) SWAT plus a hydroclimatological model component that considers both hydrological inputs and air temperature. Based on daily calibration results with six years of measured streamflow and four years of stream temperature data, we examined occurrences of stressful conditions for brook trout (*Salvelinus fontinalis*) using the hydroclimatological model. SWAT with the hydroclimatological component improved modestly during calibration (NSE of 0.93, R^2 of 0.95) compared to the original SWAT (NSE of 0.83, R^2 of 0.93). Between 1980–2009, the number of stressful events, a moment in time where high or low flows occur simultaneously with stream temperatures exceeding 21 °C, increased by 55% and average streamflow increased by 60%. This study supports using the hydroclimatological SWAT component and provides an example method for assessing stressful conditions in southern New England's coldwater habitats.

Keywords: SWAT model; coldwater fish; stream temperature; hydroclimatological model; water quality; hydrology

---

1. Introduction

Stream temperatures in the New England region of the United States have been increasing steadily over the past 100 years [1]. Over the next century, freshwater ecosystems in New England are expected to experience continued increase in mean daily stream temperatures and an increase in the frequency and magnitude of extreme flow events due to warmer, wetter winters, earlier spring snowmelt, and drier summers [1–9]. As the spatial and temporal variability of stream temperatures play a primary role in distributions, interactions, behavior, and persistence of coldwater fish species [7,10–16], it has become increasingly important to understand historical patterns of change so that a comparison can be made when projecting the future effects of climate changes on local ecosystems.

This study used the Soil and Water Assessment Tool (SWAT) [17] developed by United States Department of Agriculture to generate historical streamflow and stream temperature data, followed by an assessment of the frequency of "stressful events" affecting the Rhode Island native brook trout (*Salvelinus fontinalis*). Brook trout, a coldwater salmonid, is a species indicative of high water quality and is also of interest due to recent habitat and population restoration efforts by local environmental groups and government agencies [18,19]. This fish typically spawns in the fall and lays eggs in redds (nests) deposited in gravel substrate. The eggs develop over the winter months and hatch from late winter to early spring. However, the life-cycle of brook trout is heavily influenced by the degree and timing of temperature changes [11,20]. High stream temperatures cause physical stress including slowed metabolism and decreased growth rate, adverse effects on critical life-cycle stages such as spawning or migration triggers, and in extreme cases, mortality [7,21–24]. Distribution is also affected as coldwater fish actively avoid water temperatures that exceed their preferred temperature by 2–5 °C [25,26]. Studies have shown that optimal brook trout water temperatures remain below 20 °C. Symptoms of physiological stress develop at approximately 21 °C [21], and temperatures above 24 °C have been known to cause mortality in this species [11].

Flow regime is another central factor in maintaining the continuity of aquatic habitat throughout a stream network [22,27–32]. While temperature is often cited as the limiting factor for brook trout, the flow regime has considerable importance [33]. Alteration of the flow regime can result in changes in the geomorphology of the stream, the distribution of food producing areas as riffles and pools shift, reduced macroinvertebrate abundance and more limited access to spawning sites or thermal refugia [20,34,35]. Reductions in flow have a negative effect on the physical condition of both adult brook trout and young-of-year. Nuhfer et al. (2017) studied summer water diversions in a groundwater fed stream and found a significant decline in spring-to-fall growth of adult and young-of-year brook trout when 75% flow reductions occurred. The consequences of lower body mass are not always immediately apparent. Adults may suffer higher mortality during the winter months following the further depletion of body mass due to the rigors of spawning. Poor fitness of spawning adults may result in lower quality or reduced abundance of eggs [20]. Velocity of water through the stream reach may affect sediment and scouring of the stream bed and banks, reducing the availability of nest sites.

To address the importance of both stream temperature and flow regime, stressful events are defined herein as any day where either high or low flow occurs simultaneously with stream temperatures above 21 °C. High and low flows will be considered as those values in the 25-percent and 75-percent flow exceedance percentiles (Q25, Q75) of the 30-year historical flow on record at the study site, i.e., Cork Brook in north-central Rhode Island (Figure 1). These temperature and flow parameters were also chosen in part due to their regional applicability since many efforts are being made to conserve coldwater fish habitats in Rhode Island [18].

Analytical tools can be employed to generate models showing the effects of atmospheric temperatures on stream temperatures [8,36–41]. This study uses SWAT to simulate historical streamflow and stream temperature data. Then, a hydroclimatological stream temperature SWAT component created by Ficklin et al. [36] is incorporated to demonstrate its applicability in southern New England watersheds. This component reflects the combined influence of meteorological conditions and hydrological inputs, such as groundwater and snowmelt, on water temperature within a stream reach. Previous studies have shown that the hydroclimatological component can be used in small watersheds [36] and in New England [42]. Lastly, the generated stream temperature and streamflow data are analyzed to understand the frequency of stressful conditions for coldwater habitats in Cork Brook.

The results provide a site-specific approach to identifying critical areas in watersheds for best management practices with the goal of maintaining or improving water quality for both human consumption and aquatic habitat. In this study, the hydroclimatological component more accurately predicted stream temperatures at the study site. Between 1980 and 2009, the percent chance of stressful conditions occurring on a given day due to low streamflow levels and higher stream temperatures have increased at Cork Brook. A total of 98% of all stressful events simulated between 1980 and 2009 occurred during the low flow period rather than the high flow period. Knowing how water resources have historically responded to climate change and providing managers the most efficient analytical tools available will help identify habitats that have historically been less susceptible to unfavorable conditions. If climate trends continue as expected, decisions to protect a habitat based on its known resilience may have a large impact on how resources and preservation efforts will be allocated.

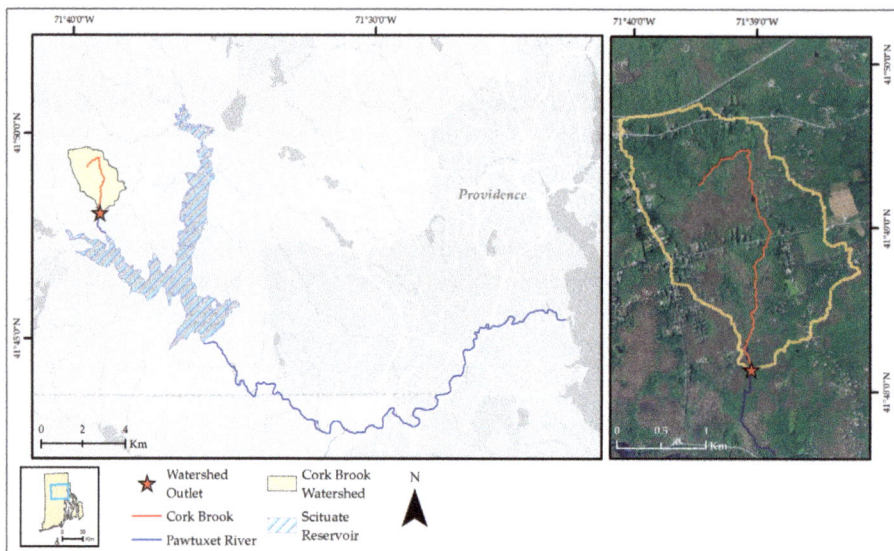

Figure 1. The Cork Brook watershed empties into the Scituate Reservoir, the main drinking water supply for the City of Providence, Rhode Island, USA.

2. Materials and Methods

The selected study site was Cork Brook in Scituate, Rhode Island. This small forested watershed is a tributary to the Scituate Reservoir, which is part of the larger Pawtuxet River basin beginning in north-central Rhode Island and eventually flowing into Narragansett Bay. The Scituate Reservoir is the largest open body of water in the state and is the main drinking water source to the City of Providence. Human disturbance within the Cork Brook watershed is minimal, and most of the land cover is undeveloped forest and brushland; however, a portion (14%) of the land use is classified as medium density residential. USGS station number 01115280 is located approximately four km downstream from the headwaters and been continuously recording streamflow at the site since 2008 and stream temperature since 2001 [43]. The mean daily discharges at the gauge are historically lowest in September (0.025 m^3/s) and highest in March (0.27 m^3/s), with an annual average of approximately 0.11 m^3/s. Average daily stream temperature is estimated at 7.8 °C since 2001.

This study uses the hydrologic and water quality model SWAT for simulating streamflow and stream temperature. SWAT is a well-established, physically-based, semi-distributed hydrologic model created by the United States Department of Agriculture (USDA) in 1998 [17]. The model is capable of simulating on a continuous daily, monthly and long-term time-step and incorporates the effects of climate, plant and crop growth, surface runoff, evapotranspiration, groundwater flow, nutrient loading, land use and in-stream water routing to predict hydrologic response and simulate discharge, sediment and nutrient yields from mixed land use watersheds [17,44–46]. As a distributed parameter model, SWAT divides a watershed into hydrologic response units (HRUs) exhibiting homogenous land, soil and slope characteristics. Surface water runoff and infiltration volumes are estimated using the modified soil conservation service (SCS) 1984 curve number method, and potential evapotranspiration is estimated using the Penman-Monteith method [47,48].

The Rhode Island Geographic Information System (RIGIS) database is the main source for the spatial data used as model inputs [49]. RIGIS is a public database managed by both the Rhode Island government and private organizations. Typical SWAT model inputs in ArcSWAT [50] include topography, soil characteristics, land cover or land use and meteorological data. Information collected for this study includes the following: 2011 Land use/land cover data derived from statewide 10-m resolution National Land Cover Data imagery [51]; soil characteristics collected from a geo-referenced digital soil map from the Natural Resource Conservation Service (NRCS) Soil Survey Geographic database (SSURGO) [52]; and topography information extracted from USGS 7.5-min digital elevation models (DEMs) with a 10-m horizontal, 7-m vertical resolution. Based on the spatial data provided, the seven-km^2 Cork Brook watershed was delineated into four subbasins and 27 HRU units using land use, soil and slope thresholds of 20%, 10% and 5%. Regional meteorological data from 1979 to 2014 including long term precipitation and temperature records were recorded by a National Climate Data Center weather station near the study site; the data were downloaded from Texas A&M University's global weather data site [53,54].

The SWAT Calibration and Uncertainty Program (SWAT-CUP), Sequential Uncertainty Fitting Version 2 (SUFI-2) [55,56], was used to conduct sensitivity analysis, calibration and model validation on stream discharge from the output hydrograph. Performance was measured using the coefficient of determination and Nash-Sutcliffe Efficiency (NSE) and percent bias (PBIAS). The coefficient of determination (R^2) identifies the degree of collinearity between simulated and measured data, and NSE was used as an indicator of acceptable model performance. R^2 values range from 0 to 1 with a larger R^2 value indicating less error variance. NSE is a normalized statistic that determines the relative magnitude of the residual variance compared to the measured data variance [57]. NSE ranges from $-\infty$ to 1; a value at or above 0.50 generally indicates satisfactory model performance [44,58–60]. This evaluation statistic is a commonly used objective function for reflecting the overall fit of a hydrograph. Percent bias is the relative percentage difference between the averaged modeled and measured data time series over (n) time steps with the objective being to minimize the value [61]. The model was validated by using calibrated parameters and performance checked using NSE, R^2 and percent bias.

The most recent version of SWAT (2012) estimates stream temperature from a relationship developed by Stefan and Preud'homme [17,62] that calculates the average daily water temperature based on the average daily ambient air temperature. Ficklin et al. developed another approach using a hydroclimatological component, which calculates stream temperature based on the combined influence of air temperature and hydrological inputs, such as streamflow, throughflow, groundwater inflow and snowmelt. Once the Cork Brook model was calibrated for streamflow, the hydroclimatological component was incorporated. A separate analysis of groundwater contributions to stream discharge was conducted for Cork Brook using an automated method for estimating baseflow [63]. An estimated 60% of stream discharge at Cork Brook is contributed to baseflow as opposed to overland flow. Therefore, incorporating the hydroclimatological component into the model may provide a more accurate prediction of stream temperature.

The main Equations (1) and (2) for water temperature (T_w) created by Ficklin et al. are listed below and described in the sequential paragraph:

$$T_w = T_{W_{initial}} + (T_{air} - T_{initial})K(TT), \text{ if } T_{air} > 0, \tag{1}$$

$$T_w = T_{W_{initial}} + \left[(T_{air} + \varepsilon) - T_{W_{initial}}\right]K(TT), \text{ if } T_{air} < 0, \tag{2}$$

where T_{air} is the average daily temperature, $K(1/h)$ is a bulk coefficient of heat transfer ranging from 0 to 1, TT is the travel time of water through the subbasin (h) and ε is an air temperature coefficient. The ε coefficient is an important component because it allows the water temperature to rise above 0 °C when the air temperature is below 0 °C. If air temperature is less than 0 °C, the model will set the stream temperature to 0.1 °C. These details are further discussed in the results section of the paper. The source code for the Ficklin model was downloaded from Darren Ficklin's research webpage at Indiana State University [64]. No additional spatial data were required for the added component and no additional streamflow calibration was necessary because discharge outputs were unchanged. Stream temperature parameters associated with the hydroclimatological model component were calibrated manually with the stream temperature data recorded at USGS Gauge 01115280. The same performance metrics (NSE and R^2) were used to determine model reliability for temperature simulation.

Upon model calibration and validation, output data simulated by SWAT with the hydroclimatological component were processed to determine the occurrence of stressful conditions in Cork Brook from 1980 to 2009. As previously discussed, a stressful event for this study is defined as any day where both temperature and flow extremes occur. This study used the Q25 and Q75 flow exceedance percentiles as indicators because of their general use in the field of hydrology [65–67] and their ecohydrological importance to coldwater fish including brook trout [11,28,30,33,68]. The most critical period for the species is typically the lowest flows of late summer to winter, and a base flow of <25% is considered poor for maintaining quality trout habitat [11,33]. A Q75 represents the lowest 25% of all daily flow rates, and a Q25 exceedance characterizes the highest 25% of all daily flow rates. Flow-exceedance probability, or flow-duration percentile, is a well-established method and generally computed using the following equation:

$$P = 100 \times \left(\frac{M}{n+1}\right)$$ (3)

where P is the probability that a given magnitude will be equaled or exceeded (percent of time), M is the ranked position (dimensionless) and n is the number of events for period of record [67]. For the stressful event analysis, the exceedance probability and average daily stream temperature for each date were identified. If the day fell into the Q25 or Q75 percentile, and if the stream temperature was greater than 21 °C, then the day was tagged as being a thermally stressful event.

3. Results and Discussion

3.1. Model Calibration & Validation

3.1.1. Stream Discharge

The initial model was run for the entire period of precipitation and rainfall data availability (1979–2014) and then calibrated in SWAT-CUP using a portion of the existing observed streamflow data from the USGS gauge. The model was calibrated for daily streamflow over a two-year time-span from 2009 to 2010 (Figures 2 and 3) due to a limited availability in observed data (2008–present). The model was validated for the year 2012 because the 2011 data showed evidence of discharge misreading and 2013 weather data were incomplete. The hydrological parameters producing the best overall fit of the modeled hydrograph to the observed hydrograph are summarized in Table 1, and the statistical results of daily streamflow calibration and validation are shown in Table 2.

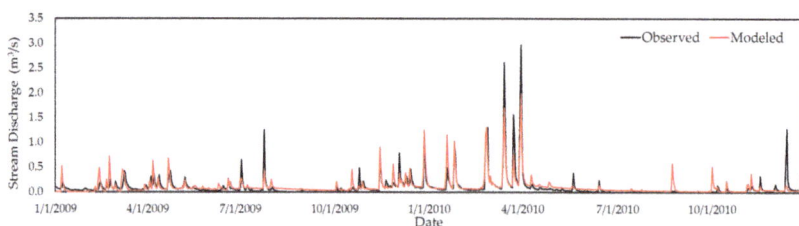

Figure 2. A simulated 2009–2010 hydrograph produced by the calibrated Cork Brook SWAT model compared to observed data from USGS Gauge 01115280.

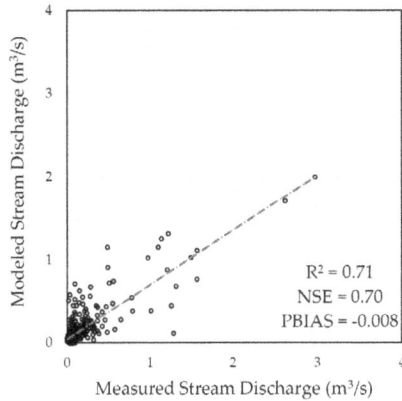

Figure 3. Streamflow scatterplot of modeled and observed average daily streamflow from USGS gauge 0111528 during 2009–2010.

The most sensitive parameters in model calibration were primarily related to groundwater and soil characteristics. The alpha-BF (baseflow) recession value was one of the most effective parameters and had a small value of 0.049. The alpha baseflow factor is a recession coefficient derived from the properties of the aquifer contributing to baseflow; large alpha factors signify steep recession indicative of rapid drainage and minimal storage whereas low alpha values suggest a slow response to drainage [63,69]. The threshold depth of water in the shallow aquifer (GWQMN) was sensitive in model calibration and the depth of water is relatively shallow (0.6 m). This is the threshold water level in the shallow aquifer for groundwater contribution to the main channel to occur. Optimal groundwater delay was short, i.e., 1.2 days. Since groundwater accounts for the majority of stream discharge within Cork Brook, the sensitivity of soil and groundwater parameters was expected. Other factors were incorporated based on the small size of the watershed, such as surface lag time, slope length, steepness and lateral subsurface flow length, and the presence of snow at the site in the winter, such as snowmelt and snowpack temperature factors.

Table 1. Range of values for the most sensitive parameters in SWAT streamflow calibration using SWAT-CUP. The parameter is listed by name and SWAT input file type, definition and the range of values that were selected for the model.

Parameter	Definition	Value Range	Units
CN2.mgt	SCS runoff curve number	−0.40–0.75	-
ALPHA_BF.gw	Baseflow alpha factor	0.0–0.10	1/Days
GW_DELAY.gw	Groundwater delay	0.0–7.0	Days
GWQMN.gw	Depth of water in shallow aquifer for return flow	200–1000	mm
v__SMTMP.bsn	Snowmelt base temperature	−0.5–2.0	°C
ESCO.hru	Soil evaporation compensation factor	0.15–0.65	-
EPCO.hru	Plant uptake compensation factor	0.15–65	-
SLSOIL.hru	Slope length for lateral subsurface flow	0.0–150.0	m

Table 2. Statistical results produced by SWAT-CUP for daily stream discharge using the parameters listed in Table 1.

Streamflow	R^2	NSE	PBIAS
Calibration	0.70	0.71	−0.01
Validation	0.54	0.50	0.03

3.1.2. Stream Temperature

Once the initial SWAT model was satisfactorily calibrated and validated for discharge, the hydroclimatological component was added to the SWAT files and the model was run using both the basic SWAT approach and the revised stream temperature program. The hydroclimatological temperature model had no effect on stream discharge; therefore, the discharge was not re-calibrated. The hydroclimatological model was manually calibrated for stream temperature by changing several variables in the basin file associated with the hydroclimatological component: K, lag time and seasonal time periods in Julian days (Table 3). The K variable represents the relationship between air and stream temperature and ranges from 0 to 1. As K approaches 1, the stream temperature is approximately the same as air temperature, and as K decreases, the stream water is less influenced by air temperature [36]. The temperature outputs are also sensitive to the lag time, a calibration parameter corresponding to the effects of delayed surface runoff and soil water into the stream. Stream temperature was calibrated using observed data recorded by the USGS gauge from 2010 to 2011 and validated from 2012 to 2013.

Table 3. Hydroclimatological SWAT calibration parameters for daily stream temperature. Time period is in Julian days and the lag unit is days.

Time Period	Alpha	Beta	Phi	K	Lag Time
1–180	1.0	1.0	1.0	1.0	4
181–270	1.0	1.0	0.8	0.8	2
271–330	1.0	1.0	0.8	0.8	2
331–366	1.0	1.0	1.0	0.7	4

The above parameters produced satisfactory stream temperature calibration statistics for the hydroclimatological model, as summarized in Table 4. During the winter and spring, the stream temperature is roughly the same as the air. In the summer and fall, the K value is decreased and the stream temperature is less affected by air temperature. This may be due to extensive tree shading [36], which is in agreement for Cork Brook as it is a relatively small watershed that is predominantly forested [70]. The lag time is relatively short throughout the year and is similar to the surface and groundwater delay parameters set during stream discharge calibration. The Ficklin et al. approach generated comparable R^2 value but a higher NSE than the basic SWAT approach. More importantly, the hydroclimatological model better predicted the occurrence of stressful stream temperatures compared to the original SWAT model during the calibration and validation periods (Figure 4). Therefore, since stream temperature is the main driving component in which a situation is considered stressful for brook trout, the hydroclimatological model appears less likely to over-predict stressful conditions than the original SWAT model.

Table 4. Statistical results of the daily stream temperature calibration. The average recorded stream temperature at the USGS gauge is 7.8 °C.

Model Type	R^2	NSE	Mean Stream Temperature
Basic SWAT Calibration	0.93	0.83	12.5 °C
Basic SWAT Validation	0.94	0.83	12.9 °C
Ficklin Calibration	0.95	0.93	9.9 °C
Ficklin Validation	0.96	0.94	10.0 °C

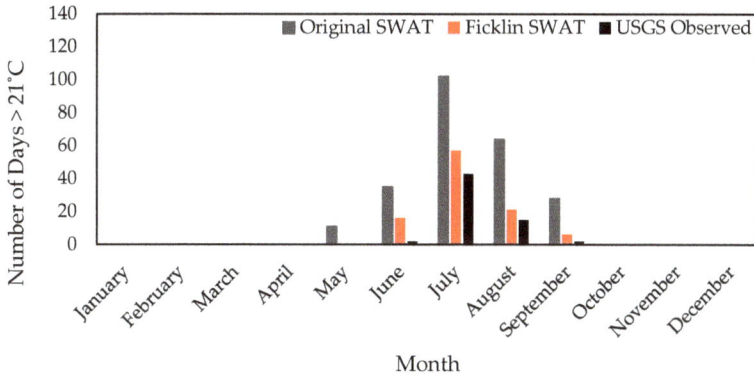

Figure 4. The number of days per month that stream temperatures exceeded the stress threshold of 21 °C during the stream temperature calibration and validation periods (2010–2013).

3.2. Stream Conditions and Stressful Event Analysis

The SWAT model incorporating the added hydroclimatological component was used for stressful event analysis, as it proved to be more accurate than the basic SWAT model. The model predicted an increase in the magnitude of stream discharge increases by each decade between 1980 and 2009, as shown in Figure 5, although the shape of the flow duration curve stayed relatively consistent. The simulated stream discharge rates increased as well, averaging 0.06 m^3/s in 1980–1989, 0.08 m^3/s in 1990–1999 and 0.10 m^3/s between 2000 and 2009. The maximum streamflow fluctuated, 1.74 m^3/s in 1980–1989, 2.75 m^3/s in 1990–1999 and 1.93 m^3/s between 2000 and 2009. Several existing studies have examined how the climate has changed over the last 30-years in New England. Since 1970, Rhode Island's annual precipitation has increased 6–11%. Fewer days with snow cover and earlier ice-out dates are occurring [71,72]. A large-scale regional study [1] collected climate and streamflow data from 27 USGS stream gauges recorded for a historical average of 71 years throughout the New England region. The study indicated that there were increases over time in annual maximum streamflows. The stream discharge results produced by the Cork Brook model align well with what has been observed statewide and across New England and support claims that certain effects of climate change are already beginning to take place.

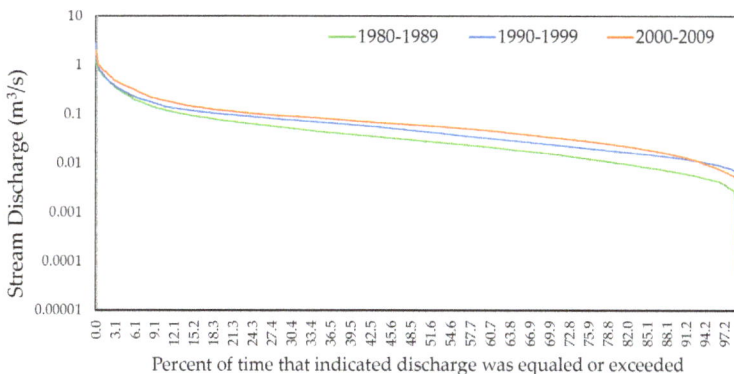

Figure 5. Simulated flow duration curves by decade generated by the SWAT model with the hydroclimatological component. Stream discharge is equal to zero at the 100th percentile.

As water temperatures increase due to global warming, brook trout may benefit from sustained flows which will prevent stream temperatures from rising further and help ensure that downstream habitat remains connected to headwaters. On the other hand, a sustained increase in flow magnitude can change the geomorphology and may not be beneficial for aquatic species during the spawning season when flows are normally lower [30]. An increase in stream discharges during the low flow season may put redds at risk of destruction from sedimentation or sheer velocity. Changes in streamflow magnitude may also increase turbidity or redistribute riffle and pool habitat throughout the stream reach. This may decrease the availability of suitable habitat as brook trout prefer stream reaches with an approximate 1:1 pool-riffle ratio [11]. Pool and riffle redistribution can also affect the type and quantity of local macroinvertebrate populations. Since warming temperatures will have an impact on body condition as fish enter the winter months, the available food supply can become an even more critical factor as the climate changes.

To identify the number of stressful events simulated by the model, output data were analyzed by decade (1980–1989, 1990–1999 and 2000–2009) and over the entire 30-year period. The percent chance that a stressful event would occur on any given day throughout the time period was also calculated. These results are shown in Table 5 below.

Table 5. Stressful event analysis of SWAT with the hydroclimatological component. Shown is the percent chance that of the 3653 days per each decade and 10,958 days between 1980 and 2009, a day with any type of stress will occur, a day with flow stress will occur, a day with temperature stress will occur and the percent chance of an event.

Date	Indicator	Any Type of Stress	Stream Temp. >21 °C	Q25 or Q75 Flow	Stressful Event
1980–1989	Days	2066	252	1814	84
	% Chance	*56.6*	*6.9*	*49.7*	*2.3*
1990–1999	Days	2049	228	1821	122
	% Chance	*56.1*	*6.2*	*49.8*	*3.3*
2000–2009	Days	2007	196	1811	131
	% Chance	*54.9*	*5.4*	*49.6*	*3.6*
1980–2009	Days	6142	676	5466	338
	% Chance	*56.0*	*6.2*	*49.9*	*3.1*

The model predicted an increase in the number of stressful events between 1980 and 2009 with the greatest change taking place between the first decade (1980–1989) and the second decade (1990–2009). It is interesting to note that although the model predicted an increase in number of stressful events between 1980 and 2009, the number of temperature stress days and the number of flow stress days generally decreased between decades (Table 5). Figure 6 have been created to gain a better understanding of how the co-occurrence of temperature stress and the flow stress has changed in Cork Brook.

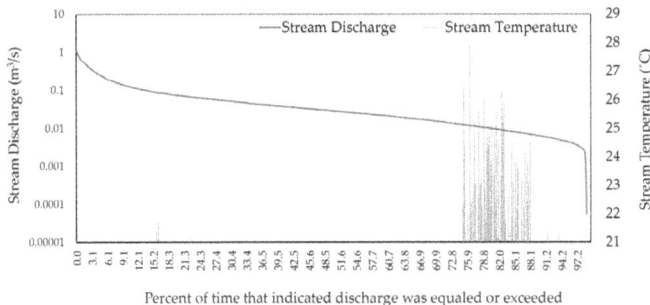

(a)

Figure 6. *Cont.*

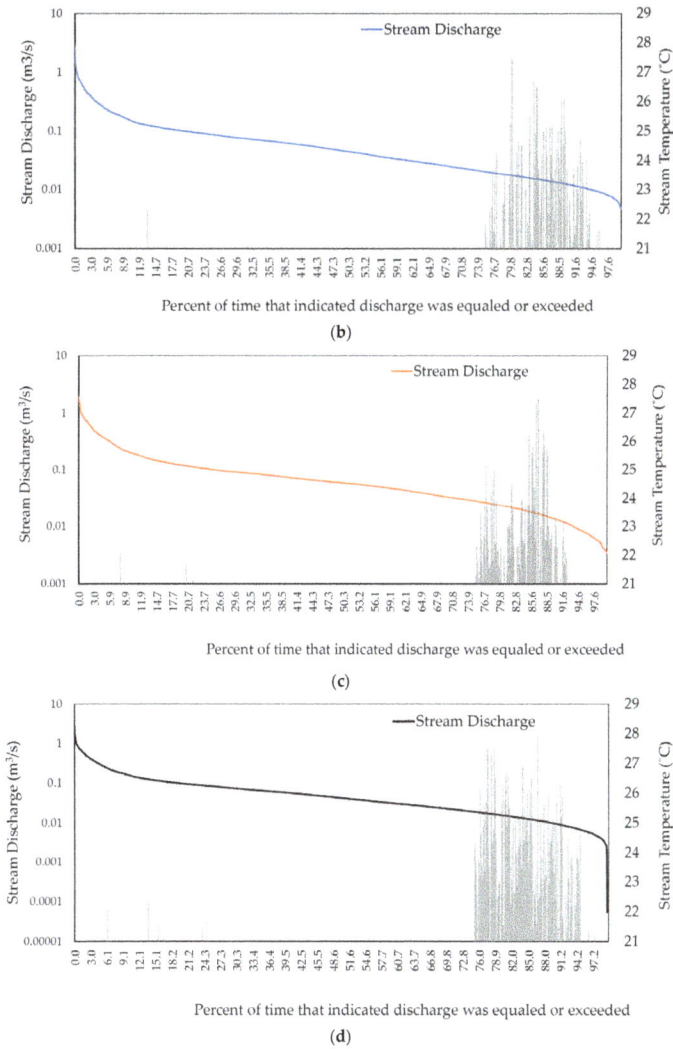

Figure 6. Cork Brook simulated flow duration curve and stream temperatures for SWAT with the hydroclimatological component over three decades. (**a**) 1980–1989, (**b**) 1990–1999 (**c**) 2000–2009 and (**d**) 1980–2009. The secondary *y*-axis begins at 21 °C and any temperatures that are not above the stressful threshold are not shown in the figures. The stream temperatures in the Q25–Q75 range are omitted from each figure.

The graphs show that of all 338 stressful events simulated between 1980 and 2009, only seven events occurred within the Q25 flow percentiles. The remaining events simulated by the model occurred when flows were within the Q75–Q97 flow percentile because lower, slower flows are exposed to air longer, causing them to increase or decrease in temperature more easily. The fact that there were no stressful events above the Q97 flow percentiles is most likely attributed to groundwater inputs. During the dry or low flow periods in summer and fall, baseflow will be the primary input to groundwater fed streams. Because the hydroclimatological model component takes the groundwater temperature into consideration, the lowest discharge amounts the model simulates will likely be baseflow driven and therefore cooler than water that is continuously exposed to ambient air temperatures. This is good news for coldwater fish species which spawn in the fall or those that begin their migration into headwaters during the low flow season, as the chance of exposure to high temperatures is lessened from groundwater contributions.

The greatest change in the number of stressful events occurred between the first and second decades where the count of stressful events increased from 84 in 1980–1989 to 122 in 1990–1999. Comparing Figure 6, the stressful events stretch from Q75 to Q87 in 1980–1989, whereas in 1990–1999 the events extend into the Q96 percentile. This shows that a combination of flow and temperature should be taken into consideration when making management decisions or evaluating the quality of aquatic habitat. For instance, managers can be reassured that withdrawing water during Q25 flows will not be as harmful to fish as withdrawing during Q75 flows. During drought years, it may become tempting to withdraw additional groundwater resources. However, the knowledge that groundwater can help reduce the occurrence of stressful events to fish during low flows may influence a manager's choice. Because Cork Brook is upstream from the Scituate Reservoir, water resource management decisions are especially applicable to this watershed.

Further analysis of 2-, 7- and 10-day moving averages at the lowest 25th percentile flow suggested that the majority of high stream temperatures are occurring during the 2-day low flow conditions as opposed to the 7-day and 10-day low flows (Figure 7). Such details can have important implications for aquatic species. Brook trout have been observed to tolerate higher stream temperatures provided their physical habitat remains stable [34]. If the co-occurrence of temperature and flow stresses last longer, then physiological stresses to individual trout may become more apparent. The data simulated from 1980 to 2009 provide a helpful baseline for comparing future projections and will help determine if the resilience of local brook trout populations may become strained under future climate conditions.

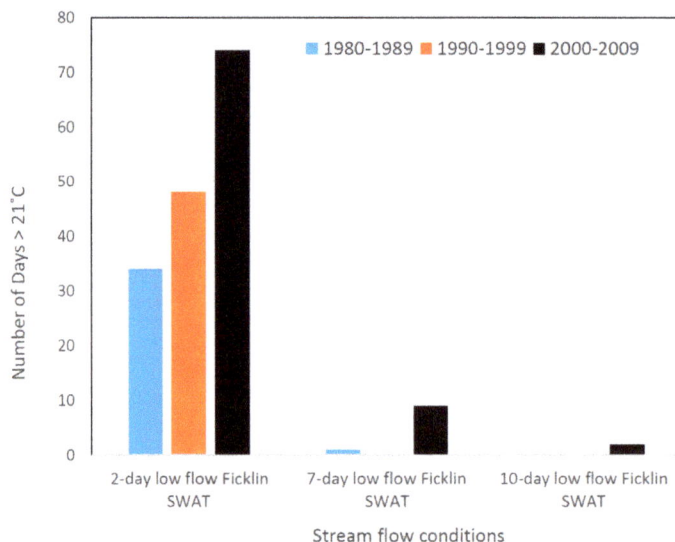

Figure 7. The number of days per decade that stream temperatures exceeded the stress threshold of 21 °C during the 2-, 7- and 10-day moving averages at the lowest 25th flow percentile.

4. Conclusions and Future Work

Since the hydroclimatological model was shown to be more accurate, future research projects should consider using the new component in similar watersheds throughout the region for both historical and climate change assessments. This study found that the long-term historical stream temperature data recorded by the USGS gauge at Cork Brook were necessary for model calibration. Therefore, scientists should have a reliable set of observed stream temperature data to calibrate and validate the stream temperature output, especially if studying ecosystems that are particularly sensitive to temperature related parameters. Other related future work may include applying the methodology to other types of temperature-sensitive aquatic organisms such as certain macroinvertebrate species. Macroinvertebrates form part of the base of the food chain, and fluctuations in their population or distributions throughout a stream reach can impact higher trophic level species that prey on these organisms.

Another consideration for future work is to limit the stressful event analysis to the spring and summer months when brook trout are more sensitive to warmer stream temperatures. Also, a study could be conducted to see if stressful events occur sequentially. This study took a wider approach by examining how stream temperatures and streamflow vary throughout the entire year. This timeframe was chosen for several reasons. First, since this is the only study of its kind within these watersheds, we did not have enough information to say with certainty that no changes to stream temperature or streamflow would occur during the fall and winter. In fact, some scientists predict that by the end of the century Rhode Island will have a climate similar to that of Georgia [71], in which case stream temperatures would almost certainly increase during the winter months. Second, while stream temperatures and streamflow during the winter months are not as critical for brook trout compared to the summer, winter conditions do effect embryo development. For instance, the length of embryo incubation during the winter ranges from 28 to 45 days depending on the temperature of the stream water [11]. Lastly, while this study focused on brook trout, our hope is that the methodology can be applied to other types of aquatic species that may be sensitive to stream conditions during other seasons.

The purpose of this study was to gain a better understanding of the historical conditions in coldwater habitat using SWAT. We successfully showed that SWAT with the hydroclimatological component is more accurate than the original SWAT model at this forested, baseflow driven watershed in Rhode Island. Moreover, thermally stressful event identification is a functional approach to analyzing model output. The data simulated from 1980 to 2009 provide a helpful baseline for comparing future projections by combining two important indicators for the survival of coldwater species.

Acknowledgments: We would like to acknowledge Thomas Boving of the University of Rhode Island and Jameson Chace of Salve Regina University for their insight and review of this Master's thesis project. This research project is supported by S-1063 Multistate Hatch Project.

Author Contributions: B.C. and S.M.P. conceived and designed the experiments; B.C. performed the experiments; B.C., S.M.P. and A.J.G. analyzed the data; B.C. wrote the paper.

Conflicts of Interest: The authors declare no conflict of interest.

References

1. Hodgkins, G.A.; Dudley, R.W.; Huntington, T.G. Changes in the timing of high river flows in New England over the 20th century. *J. Hydrol.* **2003**, *278*, 244–252. [CrossRef]
2. Hayhoe, K.; Wake, C.P.; Huntington, T.G.; Luo, L.; Schwartz, M.D.; Sheffield, J.; Wood, E.; Anderson, B.; Bradbury, J.; DeGaetano, A.; et al. Past and future changes in climate and hydrological indicators in the US Northeast. *Clim. Dyn.* **2007**, *28*, 381–407. [CrossRef]
3. Eaton, J.G.; Scheller, R.M. Effects of climate warming on fish thermal habitat in streams of the United States. *Limnol. Oceanogr.* **1996**, *41*, 1109–1115. [CrossRef]
4. Mohseni, O.; Stefan, H.G.; Eaton, J.G. Global warming and potential changes in fish habitat in U.S. Streams. *Clim. Chang.* **2003**, *59*, 389–409. [CrossRef]
5. Woodward, G.; Perkins, D.M.; Brown, L.E. Climate change and freshwater ecosystems: Impacts across multiple levels of organization. *Philos. Trans. R. Soc. B Biol. Sci.* **2010**, *365*, 2093–2106. [CrossRef] [PubMed]

6. Jiménez Cisneros, B.E.; Oki, T.; Arnell, N.W.; Benito, G.; Cogley, J.G.; Doll, P.; Jiang, T.; Mwakalila, S.S. 2014: Freshwater Resources. In *Climate Change 2014: Impacts, Adaptation and Vulnerability. Part A: Global and Sectoral Aspects*; Contribution of Working Group II to the Fifth Assessment Report of the Intergovernmental Panel on Climate Change; Cambridge University Press: Cambridge, UK; New York, NY, USA, 2014; p. 40.

7. Whitney, J.E.; Al-Chokhachy, R.; Bunnell, D.B.; Caldwell, C.A.; Cooke, S.J.; Eliason, E.J.; Rogers, M.; Lynch, A.J.; Paukert, C.P. Physiological basis of climate change impacts on North American inland fishes. *Fisheries* **2016**, *41*, 332–345. [CrossRef]

8. Mohseni, O.; Erickson, T.R.; Stefan, H.G. Sensitivity of stream temperatures in the United States to air temperatures projected under a global warming scenario. *Water Resour. Res.* **1999**, *35*, 3723–3733. [CrossRef]

9. Van Vliet, M.T.H.; Franssen, W.H.P.; Yearsley, J.R.; Ludwig, F.; Haddeland, I.; Lettenmaier, D.P.; Kabat, P. Global river discharge and water temperature under climate change. *Glob. Environ. Chang.* **2013**, *23*, 450–464. [CrossRef]

10. Brett, J.R. Some principles in the thermal requirements of fishes. *Q. Rev. Biol.* **1956**, *31*, 75–87. [CrossRef]

11. Raleigh, R.F. *Habitat Suitability Index Models: Brook Trout*; 82/10.24; U.S. Fish and Wildlife Service: Washington, DC, USA, 1982.

12. Fry, F.E.J. The effect of environmental factors on the physiology of fish. In *Fish Physiology*; Hoar, W.S., Randall, D.J., Eds.; Academic Press: Cambridge, MA, USA, 1971; Volume 6, pp. 1–98.

13. Hokanson, K.E.F.; McCormick, J.H.; Jones, B.R.; Tucker, J.H. Thermal requirements for maturation, spawning, and embryo survival of the brook trout, *Salvelinus fontinalis*. *J. Fish. Res. Board Can.* **1973**, *30*, 975–984. [CrossRef]

14. Milner, N.J.; Elliott, J.M.; Armstrong, J.D.; Gardiner, R.; Welton, J.S.; Ladle, M. The natural control of salmon and trout populations in streams. *Fish. Res.* **2003**, *62*, 111–125. [CrossRef]

15. Goniea, T.M.; Keefer, M.L.; Bjornn, T.C.; Peery, C.A.; Bennett, D.H.; Stuehrenberg, L.C. Behavioral thermoregulation and slowed migration by adult fall chinook salmon in response to high Columbia River water temperatures. *Trans. Am. Fish. Soc.* **2006**, *135*, 408–419. [CrossRef]

16. Peterson, J.T.; Kwak, T.J. Modeling the effects of land use and climate change on riverine smallmouth bass. *Ecol. Appl.* **1999**, *9*, 1391–1404. [CrossRef]

17. Arnold, J.G.; Srinivasan, R.; Muttiah, R.S.; Williams, J.R. Large area hydrologic modeling and assessment part I: Model development. *JAWRA J. Am. Water Resour. Assoc.* **1998**, *34*, 73–89. [CrossRef]

18. Erkan, D.E. *Strategic Plan for the Restoration of Anadromous Fishes to Rhode Island Coastal Streams*; Rhode Island Department of Environmental Management, Division of Fish and Wildlife: Providence, RI, USA, 2002.

19. WPWA. Maximum Daily Stream Temperature in the Queen River Watershed and Mastuxet Brook Summer 2006. Wood-Pawcatuck Watershed Association. Available online: http://www.wpwa.org/reports/2006TemperatureStudy.pdf (accessed on 1 October 2016).

20. Hakala, J.P.; Hartman, K.J. Drought effect on stream morphology and brook trout (*Salvelinus fontinalis*) populations in forested headwater streams. *Hydrobiologia* **2004**, *515*, 203–213. [CrossRef]

21. Chadwick, J.J.G.; Nislow, K.H.; McCormick, S.D. Thermal onset of cellular and endocrine stress responses correspond to ecological limits in brook trout, an iconic cold-water fish. *Conserv. Physiol.* **2015**, *3*, cov017. [CrossRef] [PubMed]

22. Letcher, B.H.; Nislow, K.H.; Coombs, J.A.; O'Donnell, M.J.; Dubreuil, T.L. Population response to habitat fragmentation in a stream-dwelling brook trout population. *PLoS ONE* **2007**, *2*, e1139. [CrossRef] [PubMed]

23. Lee, R.M.; Rinne, J.N. Critical thermal maxima of five trout species in the southwestern United States. *Trans. Am. Fish. Soc.* **1980**, *109*, 632–635. [CrossRef]

24. Bjornn, T.; Reiser, D. Habitat requirements of salmonids in streams. *Am. Fish. Soc. Spec. Publ.* **1991**, *19*, 138.

25. Kling, G.W.; Hayhoe, K.; Johnson, L.B.; Magnuson, J.J.; Polasky, S.; Robinson, S.K.; Shuter, B.J.; Wander, M.M.; Wuebbles, D.J.; Zak, D.R. *Confronting Climate Change in the Great Lakes Region: Impacts on Our Communities and Ecosystems*; Union of Concerned Scientists: Cambridge, MA, USA; Ecological Society of America: Washington, DC, USA, 2003; p. 92.

26. Magnuson, J.J.; Crowder, L.B.; Medvick, P.A. Temperature as an ecological resource. *Am. Zool.* **1979**, *19*, 331–343. [CrossRef]

27. Vannote, R.L.; Minshall, G.W.; Cummins, K.W.; Sedell, J.R.; Cushing, C.E. The river continuum concept. *Can. J. Fish. Aquat. Sci.* **1980**, *37*, 130–137. [CrossRef]

28. Bunn, S.E.; Arthington, A.H. Basic principles and ecological consequences of altered flow regimes for aquatic biodiversity. *Environ. Manag.* **2002**, *30*, 492–507. [CrossRef]

29. Freeman, M.C.; Pringle, C.M.; Jackson, C.R. Hydrologic connectivity and the contribution of stream headwaters to ecological integrity at regional scales. *JAWRA J. Am. Water Resour. Assoc.* **2007**, *43*, 5–14. [CrossRef]

30. Poff, N.L.; Allan, J.D. Functional organization of stream fish assemblages in relation to hydrological variability. *Ecology* **1995**, *76*, 606–627. [CrossRef]

31. Poff, N.L.; Allan, J.D.; Bain, M.B.; Karr, J.R.; Prestegaard, K.L.; Richter, B.D.; Sparks, R.E.; Stromberg, J.C. The natural flow regime. *BioScience* **1997**, *47*, 769–784. [CrossRef]

32. Bassar, R.D.; Letcher, B.H.; Nislow, K.H.; Whiteley, A.R. Changes in seasonal climate outpace compensatory density-dependence in eastern brook trout. *Glob. Chang. Biol.* **2016**, *22*, 577–593. [CrossRef] [PubMed]

33. DePhilip, M.; Moberg, T. *Ecosystem Flow Recommendations for the Susquehanna River Basin*; The Nature Conservancy: Harrisburg, PA, USA, 2010.

34. Nuhfer, A.J.; Zorn, T.G.; Wills, T.C. Effects of reduced summer flows on the brook trout population and temperatures of a groundwater-influenced stream. *Ecol. Freshw. Fish* **2017**, *26*, 108–119. [CrossRef]

35. Walters, A.W.; Post, D.M. An experimental disturbance alters fish size structure but not food chain length in streams. *Ecology* **2008**, *89*, 3261–3267. [CrossRef] [PubMed]

36. Ficklin, D.L.; Luo, Y.; Stewart, I.T.; Maurer, E.P. Development and application of a hydroclimatological stream temperature model within the soil and water assessment tool. *Water Resour. Res.* **2012**, *48*. [CrossRef]

37. Hayhoe, K.; Wake, C.; Anderson, B.; Liang, X.-Z.; Maurer, E.; Zhu, J.; Bradbury, J.; DeGaetano, A.; Stoner, A.M.; Wuebbles, D. Regional climate change projections for the northeast USA. *Mitig. Adapt. Strateg. Glob. Chang.* **2008**, *13*, 425–436. [CrossRef]

38. Isaak, D.J.; Wollrab, S.; Horan, D.; Chandler, G. Climate change effects on stream and river temperatures across the northwest U.S. from 1980–2009 and implications for salmonid fishes. *Clim. Chang.* **2012**, *113*, 499–524. [CrossRef]

39. Mohseni, O.; Stefan, H.G. Stream temperature/air temperature relationship: A physical interpretation. *J. Hydrol.* **1999**, *218*, 128–141. [CrossRef]

40. Null, S.; Viers, J.; Deas, M.; Tanaka, S.; Mount, J. Stream temperature sensitivity to climate warming in California's Sierra Nevada. In Proceedings of the AGU Fall Meeting Abstracts, San Francisco, CA, USA, 13–17 December 2010.

41. Preud'homme, E.B.; Stefan, H.G. *Relationship between Water Temperatures and Air Temperatures for Central US Streams*; EPA/600/R-92/243; U.S. Environmental Protection Agency: Washington, DC, USA, 2002.

42. Brennan, L. Stream Temperature Modeling: A Modeling Comparison for Resource Managers and Climate Change Analysis. Master's Thesis, University of Massachusetts, Amherst, MA, USA, 2015.

43. US Geological Survey (USGS). *U.G.S. National Water Information System Web Interface*, 2015 ed.; US Geological Survey: Reston, VA, USA, 2017.

44. Douglas-Mankin, K.R.; Srinivasan, R.; Arnold, J.G. Soil and water assessment tool (SWAT) model: Current developments and applications. *Trans. ASABE* **2010**, *53*, 1423–1431. [CrossRef]

45. Gassman, P.W.; Reyes, M.R.; Green, C.H.; Arnold, J.G. The soil and water assessment tool: Historical development, applications, and future research directions. *Trans. ASABE* **2007**, *50*, 1211–1250. [CrossRef]

46. Neitsch, S.L.; Arnold, J.G.; Kiniry, J.R.; Williams, J.R. *Soil and Water Assessment Tool Theoretical Documentation Version 2009*; Texas Water Resources Institute: College Station, TX, USA, 2011.

47. Penman, H.L. Estimating evaporation. *Eos Trans. Am. Geophys. Union* **1956**, *37*, 43–50. [CrossRef]

48. Monteith, J.L. Evaporation and environment. *Symp. Soc. Exp. Biol.* **1965**, *19*, 205–234. [PubMed]

49. University of Rhode Island. *Rhode Island Geographic Information System (RIGIS)*; University of Rhode Island Rhode Island: Kingston, RI, USA, 2016. Available online: http://www.rigis.org (accessed on 3 September 2017).

50. Texas A&M University. *Arcswat Software*; ArcSWAT 2012.10.19; Texas A&M University: College Station, TX, USA, 2012. Available online: http://swat.tamu.edu/ (accessed on 3 September 2017).

51. Homer, C.G.; Dewitz, J.A.; Yang, L.; Jin, S.; Danielson, P.; Xian, G.; Coulston, J.; Herold, N.D.; Wickham, J.; Megown, K. Completion of the 2011 national land cover database for the conterminous United States—Representing a decade of land cover change information. *Photogramm. Eng. Remote Sens.* **2015**, *81*, 345–354.

52. Rhode Island Geographic Information System (RIGIS). Data Distribution System SOIL_soils. Available online: http://www.rigis.org/geodata/soil/Soils16.zip (accessed on 5 September 2016).
53. Saha, S.; Moorthi, S.; Wu, X.; Wang, J.; Nadiga, S.; Tripp, P.; Behringer, D.; Hou, Y.-T.; Chuang, H.-Y.; Iredell, M.; et al. The ncep climate forecast system version 2. *J. Clim.* **2014**, *27*, 2185–2208. [CrossRef]
54. Texas A&M University. NCEP Global Weather Data for Swat. Available online: http://swat.tamu.edu/ (accessed on 3 January 2016).
55. Abbaspour, K.C. *SWAT Calibration and Uncertainty Program—A User Manual*; Swat-Cup 2012; Swiss Federal Institute of Aquatic Science and Technology, Eawag: Duebendorf, Switzerland, 2013.
56. Abbaspour, K. *User Manual for Swat-Cup, Swat Calibration and Uncertainty Analysis Programs*; Swiss Federal Institute of Aquatic Science and Technology, Eawag: Duebendorf, Switzerland, 2007.
57. Nash, J.E.; Sutcliffe, J.V. River flow forecasting through conceptual models part I—A discussion of principles. *J. Hydrol.* **1970**, *10*, 282–290. [CrossRef]
58. Moriasi, D.N.; Arnold, J.G.; Liew, M.W.V.; Bingner, R.L.; Harmel, R.D.; Veith, T.L. Model evaluation guidelines for systematic quantification of accuracy in watershed simulations. *Trans. ASABE* **2007**, *50*, 885–900. [CrossRef]
59. Singh, J.; Knapp, H.V.; Arnold, J.G.; Demissie, M. Hydrological modeling of the iroquois river watershed using HSPF and SWAT. *JAWRA J. Am. Water Resour. Assoc.* **2005**, *41*, 343–360. [CrossRef]
60. Liew, M.W.V.; Veith, T.L.; Bosch, D.D.; Arnold, J.G. Suitability of swat for the conservation effects assessment project: Comparison on usda agricultural research service watersheds. *J. Hydrol. Eng.* **2007**, *12*, 173–189. [CrossRef]
61. Pradhanang, S.M.; Mukundan, R.; Schneiderman, E.M.; Zion, M.S.; Anandhi, A.; Pierson, D.C.; Frei, A.; Easton, Z.M.; Fuka, D.; Steenhuis, T.S. Streamflow responses to climate change: Analysis of hydrologic indicators in a New York city water supply watershed. *JAWRA J. Am. Water Resour. Assoc.* **2013**, *49*, 1308–1326. [CrossRef]
62. Stefan, H.G.; Preud'homme, E.B. Stream temperature estimation from air temperature. *JAWRA J. Am. Water Resour. Assoc.* **1993**, *29*, 27–45. [CrossRef]
63. Arnold, J.G.; Allen, P.M.; Muttiah, R.; Bernhardt, G. Automated base flow separation and recession analysis techniques. *Ground Water* **1995**, *33*, 1010–1018. [CrossRef]
64. Ficklin, D.L. *Swat Stream Temperature Executable Code*; Indiana State University: Terre Haute, IN, USA, 2012.
65. Pyrce, R. *Hydrological Low Flow Indices and Their Uses*; Watershed Science Centre (WSC) Report; Trent University: Peterborough, ON, Canada, 2004.
66. Smakhtin, V.U. Low flow hydrology: A review. *J. Hydrol.* **2001**, *240*, 147–186. [CrossRef]
67. Ahearn, E.A. *Flow Durations, Low-Flow Frequencies, and Monthly Median Flows for Selected Streams in Connecticut through 2005*; US Department of the Interior: Washington, DC, USA; US Geological Survey: Reston, VA, USA, 2008.
68. Armstrong, D.S.; Richards, T.A.; Parker, G.W. *Assessment of Habitat, Fish Communities, and Streamflow Requirements for Habitat Protection, Ipswich River, Massachusetts, 1998–99*; Department of the Interior, US Geological Survey: Reston, VA, USA, 2001.
69. Arnold, J.G.; Allen, P.M. Automated methods for estimating baseflow and ground water recharge from streamflow records. *JAWRA J. Am. Water Resour. Assoc.* **1999**, *35*, 411–424. [CrossRef]
70. Johnson, S.L. Factors influencing stream temperatures in small streams: Substrate effects and a shading experiment. *Can. J. Fish. Aquat. Sci.* **2004**, *61*, 913–923. [CrossRef]
71. Wake, C.P.; Keeley, C.; Burakowski, E.; Wilkinson, P.; Hayhoe, K.; Stoner, A.; LaBrance, J. *Climate Change in Northern New Hampshire: Past, Present and Future*; Climate Solutions New England: Durham, NH, USA, 2014.
72. Wake, C. *Rhode Island's Climate: Past and Future Changes*; Climate Solutions New England: Durham, NH, USA, 2014; p. 2.

water

MDPI

Article

Simulating Climate Change Induced Thermal Stress in Coldwater Fish Habitat Using SWAT Model

Britta M. Chambers, Soni M. Pradhanang * and Arthur J. Gold

Department of Geosciences, University of Rhode Island, Kingston, RI 02881, USA;
britta_anderson@uri.edu (B.M.C.); agold@uri.edu (A.J.G.)
* Correspondence: spradhanang@uri.edu; Tel.: +1-401-874-5980

Received: 30 June 2017; Accepted: 15 September 2017; Published: 25 September 2017

Abstract: Climate studies have suggested that inland stream temperatures and average streamflows will increase over the next century in New England, thereby putting aquatic species sustained by coldwater habitats at risk. This study uses the Soil and Water Assessment Tool (SWAT) to simulate historical streamflow and stream temperatures within three forested, baseflow-driven watersheds in Rhode Island, USA followed by simulations of future climate scenarios for comparison. Low greenhouse gas emission scenarios are based on the 2007 International Panel on Climate Change Special Report on Emissions Scenarios (SRES) B1 scenario and the high emissions are based on the SRES A1fi scenario. The output data are analyzed to identify daily occurrences where brook trout (*Salvelinus fontinalis*) are exposed to stressful events, defined herein as any day where Q25 or Q75 flows occur simultaneously with stream temperatures exceeding 21 °C. Results indicate that under both high- and low-emission greenhouse gas scenarios, coldwater fish species such as brook trout will be increasingly exposed to stressful events. The percent chance of stressful event occurrence increased by an average of 6.5% under low-emission scenarios and by 14.2% under high-emission scenarios relative to the historical simulations.

Keywords: SWAT model; coldwater habitat; stream temperature; water quality; hydrology; climate change

---

1. Introduction

Concerns have arisen regarding the impact of warming stream temperatures on brook trout (*Salvelinus fontinalis*) habitat due to climate change. Over the next century, freshwater ecosystems in the New England region of the United States are expected to experience a continued increase in mean daily stream temperatures and an increase in the frequency and magnitude of extreme high flow events due to warmer, wetter winters, earlier spring snowmelt, and drier summers [1–9]. As the spatial and temporal variability of stream temperatures play a primary role in distributions, interactions, behavior, and persistence of coldwater fish species [7,10–16], it has become increasingly important to understand what challenges freshwater fisheries managers will face because of climate change. Analytical models such as the Soil and Water Assessment Tool (SWAT) [17] can be used to estimate the effects of climate change on stream temperatures and aquatic species [5,18–24]. Several studies have used global climatic model output or temperature and precipitation variations to drive hydrologic and stream temperature models for the United States [25] and worldwide [8]. This study uses both SWAT and global climate data downscaled for New England [3,4,26–28] to simulate the effects of increasing air temperatures and changes to regional rainfall patterns on coldwater fish habitats in southern New England watersheds.

The SWAT model was used to generate historical and future stream temperature and streamflow data, followed by an assessment of the frequency of "stressful events" affecting the Rhode Island native brook trout. Brook trout, a coldwater salmonid, is a species indicative of high water quality and is also of interest due to recent habitat and population restoration efforts by local environmental groups and government agencies [29,30]. This fish typically spawns in the fall and lays eggs in redds (nests) deposited in gravel substrate. Eggs develop over the winter months and hatch from late winter to early spring [11,12,31]. The life cycle of brook trout, however, is heavily influenced by the degree and timing of temperature changes. High stream temperatures cause physical stress including slowed metabolism and decreased growth rate, adverse effects on critical life-cycle stages such as spawning or migration triggers, and in extreme cases, mortality [7,10,32–35]. Distribution is also affected as coldwater fish actively avoid water temperatures that exceed their preferred temperature by 2–5 °C [36,37]. Studies have shown that optimal brook trout water temperatures are below 20 °C, symptoms of physiological stress develop at approximately 21 °C [33] and temperatures above 24 °C have been known to cause mortality in this species [12].

Flow regime is another critical factor in maintaining the continuity of aquatic habitats throughout a stream network [35,38–43]. While temperature is often cited as the limiting factor for brook trout, the flow regime has equal importance [44]. Alteration of the flow regime can result in changes in the geomorphology of the stream and the distribution of food-producing areas as riffles and pools shift. Changes in the distribution of riffles and pools can cause a decrease in food-producing areas, reduced macroinvertebrate abundance and more limited access to spawning sites or thermal refugia [12,31,45,46]. Reductions in flow have a negative effect on the physical condition of both adult brook trout and young-of-year. Nuhfer et al. found a significant decline in spring-to-fall growth of brook trout when 75% flow reductions occurred [45]. The consequences of lower body mass are not always immediately apparent. Adults may suffer higher mortality during the winter months following the further depletion of body mass due to the rigors of spawning. Poor fitness of spawning adults may result in lower quality or abundance of eggs and a decline in hatching during the late winter to early spring [31]. Velocity of water through the stream reach can affect sediment and scouring of the stream bed and banks, reducing the availability of nest sites or, in the event of low flows, cause water temperatures to rise.

To address the importance of both stream temperature and flow regime, "stressful events" are defined herein as days where either high or low flow occurs simultaneously with stream temperatures exceeding 21 °C. For the purpose of this study, high and low flows will be considered as the values exceeding the 25-percent and 75-percent percentiles (Q25, Q75) for both historical and future simulated SWAT model output. Two Wood-Pawcatuck River headwater subbasins, the Queen River and the Beaver River, were selected as study sites due to their pristine, undisturbed aquatic habitat. A third pristine watershed, Cork Brook, was chosen as a study site because of its association with the Scituate Reservoir, which supplies drinking water to the city of Providence, Rhode Island. This study incorporated two climate change scenarios for future stream conditions at the three project sites. Low greenhouse gas emission scenarios are based on the 2007 International Panel on Climate Change Special Report on Emissions Scenarios (SRES) B1 scenario and the high-emission scenarios are based on the SRES A1fi scenario. Model output was analyzed over four time periods: historical (1980–2009), short term (2010–2039), medium term (2040–2067) and long term (2070–2099) to understand how coldwater habitat in these watersheds will react to climate change over the next century. Results provide a site-specific approach for watershed managers trying to determine the types and distribution of future habitat risk to coldwater species. As the demands for water quality and quantity increase for wildlife and human consumption over the next century, new evaluation techniques will help anticipate and solve unprecedented challenges. In the Wood-Pawcatuck and Cork Brook watersheds, the anticipated challenges may include an increase in stressful conditions.

2. Materials and Methods

2.1. Study Sites

Three watersheds were selected to achieve the objective: Queen River, Beaver River and Cork Brook. The Queen and Beaver watersheds lie adjacent to each other within the Wood-Pawcatuck watershed in southern Rhode Island (Figure 1). In its entirety, the 800 km^2 watershed is comprised of seven drainage basins and two major rivers. The upper reaches of the Wood-Pawcatuck watershed trend towards undisturbed rural environments. The watershed becomes increasingly urban and impaired towards the downstream reaches before emptying into Little Narragansett Bay. This watershed supports native brook trout populations, high-quality wildlife habitat and a species diversity that is unique for a watershed of this scale in southern New England [30,47–56]. The effect of climate change on stream water quality is a serious concern in Wood-Pawcatuck watershed and many non-profit organizations, recreational fishing groups and government agencies have taken interest in the long-term survival of local brook trout.

The Beaver River and the Queen River watersheds cover areas of approximately 23 and 52 km^2, respectively. Many similarities exist between the two subbasins. Land use is primarily forest although wetlands and agriculture make up a small portion of each watershed. Both are HUC 12 river headwaters to the larger Pawcatuck river and each watershed hosts nature preserves owned and managed by The Nature Conservancy [54,57]. Continuous and permanent United States Geological Survey (USGS) gauges have been recording flow data for several decades within each river [58]. The Beaver River USGS gauge (number 01117468) is located near Usquepaug, RI where it intersects State Highway 138, or approximately 5.8 km upstream from its confluence with the Pawcatuck River. The gauge has been in continual operation since 1974. Mean daily discharges at the Beaver River gauge are typically lowest in September (0.02 m^3/s) and highest in April (1.04 m^3/s), with annual mean daily discharge of 0.59 m^3/s. USGS gauge station (number 01117370) is located on the intersection between the Queen River and Liberty Road, and data has been recording since 1998. Discharges at the Queen River gauge are higher, historically lowest in August (0.039 m^3/s) and highest in March (2.08 m^3/s) with mean daily discharges of approximately 1.06 m^3/s. A separate analysis of groundwater contributions to stream discharge was conducted using an automated method for estimating baseflow [59]. A noteworthy difference between the two watersheds is the baseflow contributions to each river, 93% within the Beaver River and 78% for the Queen River.

The third study site is Cork Brook in Scituate, Rhode Island. This small forested watershed is a tributary to the Scituate Reservoir, which is part of the larger Pawtuxet River basin beginning in north-central Rhode Island and eventually flowing into Narragansett Bay. The Scituate Reservoir is the largest open body of water in the state and is the main drinking water source to the city of Providence. Cork Brook is approximately four km long and covers an area of approximately seven km^2. Human disturbance within the watershed is minimal and most of the land use within the watershed is undeveloped forest and brushland, although a portion (14%) of land use within the watershed is classified as medium density residential. USGS station number 01115280 is located on Rockland Road near Clayville, RI, which has been continuously recording streamflow at the site since 2008 [58]. A primary difference between the Cork Brook and Wood-Pawcatuck watersheds is size and the stream discharge amounts. The mean daily discharges at the gauge are historically lowest in September (0.025 m^3/s), highest in March (0.27 m^3/s) and annually average approximately 0.11 m^3/s. Average daily stream temperature is estimated at 7.8 °C since 2001. An important similarity to the Beaver and Queen watersheds is groundwater contribution; baseflow contributes the majority (60%) of stream discharges.

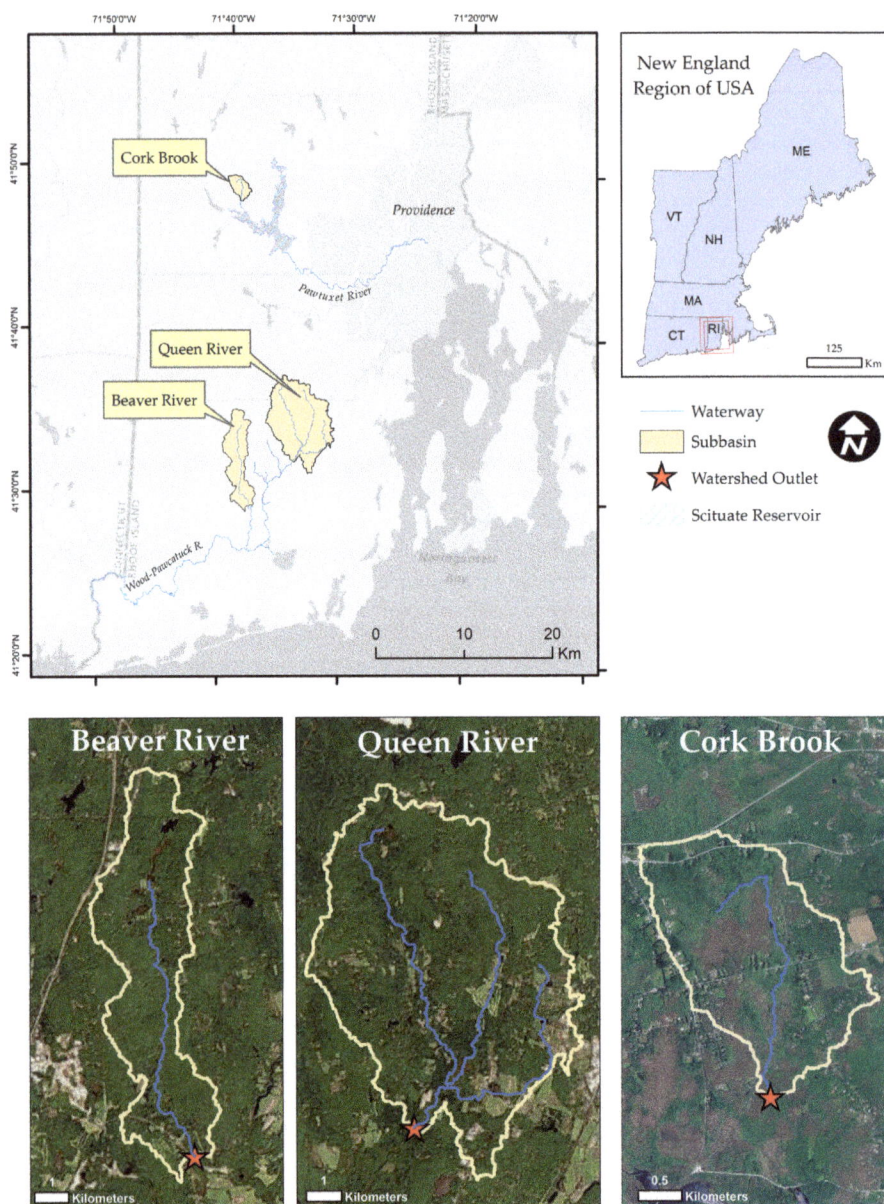

Figure 1. The study sites include the Beaver River, Queen River and Cork Brook watersheds in Rhode Island, USA.

2.2. Model Setup

This study uses the hydrologic and water quality model SWAT for simulating streamflow and stream temperature. SWAT is a well-established, physically-based, semi-distributed hydrologic model created by the United States Department of Agriculture (USDA) in 1998 [17,60–62]. Surface water runoff and infiltration volumes are estimated using the modified soil conservation service (SCS) 1984 curve number method, and potential evapotranspiration is estimated using the Penman–Monteith method [63,64]. Stream temperature is estimated from air temperature based on a linear regression method developed by Stefan and Prued'homme [17,65]:

$$T_w(t) = 5.0 + 0.75 T_{air}(t - \delta) \tag{1}$$

where (T_W) represents average daily water temperature (°C), (T_{air}) represents and average daily air temperatures (°C). Time (t) and lag (δ) are in days. Water temperatures follow air temperatures closely, the time lag for a shallow stream is expected to be on the order of a few hours due to the thermal inertia of the water [65]. The average relationship indicates that when the daily air temperature is close to 0 °C then the water will be approximately 5 °C warmer. When the daily air temperature is below 20 °C the water temperature is likely to be greater than the air temperature [65]. The Rhode Island Geographic Information System (RIGIS) database is the main source for the spatial data used as model inputs [66]. RIGIS is a public database managed by both the RI government and private organizations. Typical SWAT model inputs in ArcSWAT [67] include topography, soil characteristics, land cover or land use and meteorological data. Information collected for this study includes the following: 2011 Land use/land cover data derived from statewide 10-m resolution National Land Cover Data imagery [68]; soil characteristics collected from a geo-referenced digital soil map from Natural Resource Conservation Service (NRCS) Soil Survey Geographic database (SSURGO) [69]; and topography information extracted from USGS 7.5-min digital elevation models (DEMs) with a 10-m resolution. Regional meteorological data from 1979 to 2014 including long term precipitation and temperature statistics were downloaded from Texas A&M University's global weather data site [70,71].

Based on the spatial data provided, SWAT delineated the watersheds into HRU units, which are represented as a percentage of the subbasin area. The user sets a soil, land and slope threshold based on the level of heterogeneity within a watershed and when a parcel of land meets or exceed all thresholds a HRU is created. The Beaver River basin is generally homogenous and was delineated into five subbasins and 12 HRUs using land, soil and slope thresholds of 20%. The Queen River has more variability in the type and distribution of land use throughout the watershed. This watershed was delineated into eight subbasins and 17 HRUs using land, soil and slope thresholds of 25%, 20% and 20%. Cork Brook was delineated in SWAT to create four subbasins and 27 HRUs using land, soil and slope thresholds of 20%, 10% and 5%. The soil types and elevation changes are variable throughout the Cork Brook watershed and as such these thresholds were reduced to capture basin heterogeneity.

2.3. Model Calibration & Validation

The SWAT Calibration and Uncertainty Program (SWAT-CUP), Sequential Uncertainty Fitting Version 2 (SUFI-2) [72,73], was used to conduct sensitivity analysis, calibration and model validation on daily stream discharge from the output hydrograph. Performance was measured using coefficient of determination and Nash-Sutcliffe Efficiency (NSE) and percent bias (PBIAS). Coefficient of determination (R^2) identifies the degree of collinearity between simulated and measured data and NSE was used as an indicator of acceptable model performance. R^2 values range from 0 to 1 with a larger R^2 value indicating less error variance. NSE is a normalized statistic that determines the relative magnitude of the residual variance compared to the measured data variance [74]. NSE ranges from $-\infty$ to 1; a value at or above 0.50 generally indicates satisfactory model performance [60,75–77]. This evaluation statistic is a commonly used objective function for reflecting the overall fit of a hydrograph. Percent bias is the relative percentage difference between the averaged modeled and measured data time series over (n) time steps with the objective being to minimize the value [78]. The model was validated by using calibrated parameters and performance checked using NSE, R^2 and percent bias.

Each model was run for the entire period of precipitation and rainfall data availability (1979–2014) and then calibrated for daily streamflow in SWAT-CUP via SUFI-2 using a portion of the existing observed data at each associated USGS gauge. The Beaver River and Queen River watersheds were calibrated over the same five-year time span from 2000 to 2005 due to data availability and avoidance of natural streamflow anomalies in 2010 and 2006. Validation occurred from 2007 to 2008 in both the Beaver and Queen River watersheds. Meanwhile, the Cork Brook model was calibrated for streamflow over a shorter two-year time-span from 2009 to 2010 due to a limited availability in observed discharge data (2008–present). The Cork Brook model was validated for the year 2012 because the 2011 data showed evidence of discharge anomalies and 2013 weather data were incomplete. The modeled hydrographs versus the observed hydrographs are shown in Figures 2–4 and the statistical results of calibration and validation are shown in Tables 1 and 2.

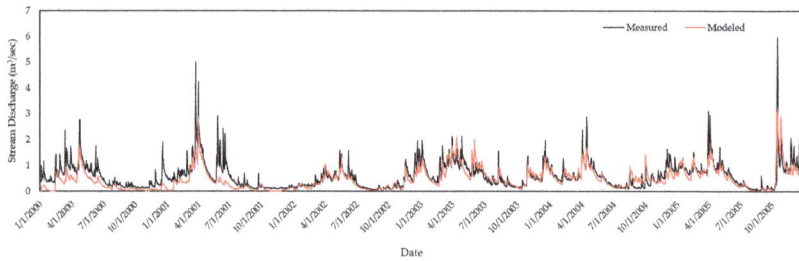

(a)

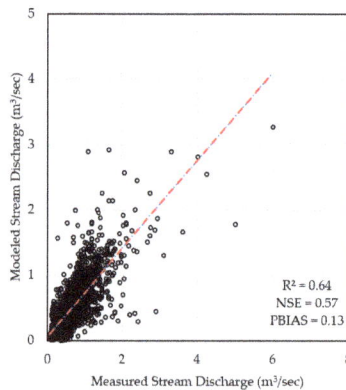

(b)

Figure 2. (a) Hydrograph and (b) scatterplot of observed versus SWAT modeled streamflow at Beaver River USGS gauge 01117468 during calibration years 2000–2005.

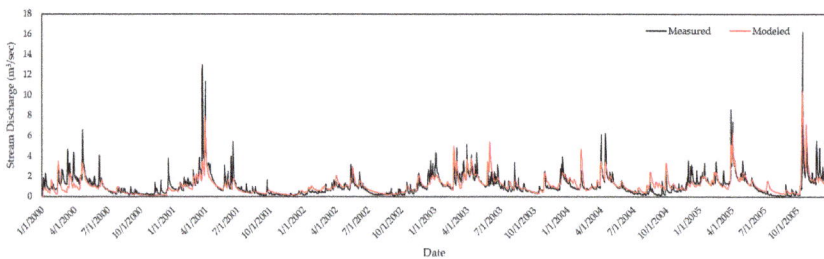

(a)

Figure 3. *Cont.*

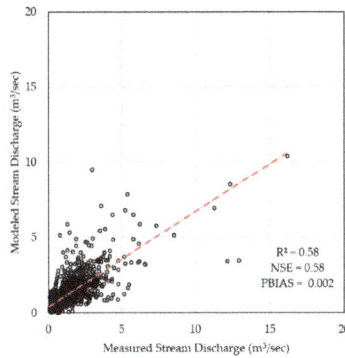

(b)

Figure 3. (**a**) Hydrograph and (**b**) scatterplot of observed versus SWAT modeled streamflow at Queen River USGS gauge 01117370 during calibration years 2000–2005.

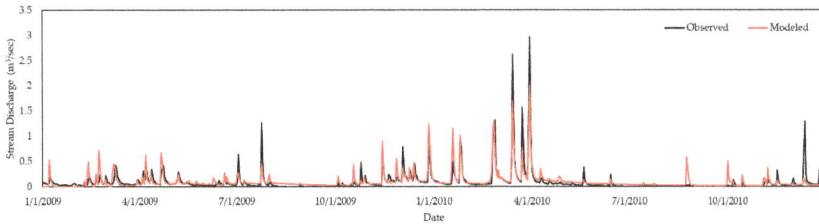

(a)

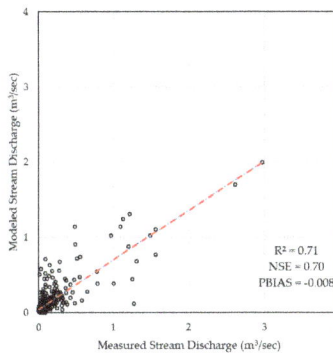

(b)

Figure 4. (**a**) Hydrograph and (**b**) scatterplot of observed versus SWAT modeled streamflow at Cork Brook USGS gauge 01115280 during calibration years 2009–2010.

Table 1. Statistical results of daily streamflow calibration produced by SWAT-CUP.

Watershed	R^2	NSE	PBIAS
Beaver River	0.64	0.57	0.13
Queen River	0.58	0.58	0.002
Cork Brook	0.70	0.71	−0.01

Table 2. Statistical results of daily streamflow validation produced by SWAT-CUP.

Streamflow	R^2	NSE	PBIAS
Beaver River	0.66	0.60	0.13
Queen River	0.60	0.59	0.003
Cork Brook	0.54	0.50	0.03

The most sensitive parameters in daily streamflow calibration are summarized in Table 3 and were primarily related to groundwater and soil characteristics. The alpha-BF (baseflow) recession value was one of the most effective parameters for all three models and the values were all very small. The alpha baseflow factor is a recession coefficient derived from the properties of the aquifer contributing to baseflow; large alpha factors signify steep recession indicative of rapid drainage and minimal storage whereas low alpha values suggest a slow response to drainage [59,79]. Alpha-bnk (bankflow) was another sensitive parameter, which is simulated with a recession curve like that used for groundwater. For this parameter, a high value at all three sites indicates a flat recession curve, which is similar to the alpha-bf value that specifies a slow response to drainage. The threshold depth of groundwater in the shallow aquifer (GWQMN) is small and very similar between all three models, less than a meter within each. This is the threshold water level in the shallow aquifer for groundwater contribution to the main channel to occur. Since groundwater accounts for the majority of stream discharge at all sites the sensitivity of soil and groundwater parameters was expected.

Table 3. Range of values for ten most sensitive parameters during daily streamflow calibration using SWAT-CUP for (a) Beaver River, (b) Queen River and (c) Cork Brook. Parameters are listed by name and SWAT input file type, definition and the range of values that were selected for the model.

Parameter	Definition	Value Range	Units
\multicolumn{4}{c}{(a) Beaver River parameters for daily streamflow calibration.}			
CN2.mgt	SCS runoff curve number	−60–75	-
ALPHA_BF.gw	Baseflow alpha factor	0.0–0.10	1/Days
GW_DELAY.gw	Groundwater delay	0.0–10	Days
TIMP.bsn	Snowpack temperature lag factor	−1.5–2.0	-
ALPHA_BNK.rte	Baseflow alpha factor for bank storage	0.50–1.0	Days
OV_N.hru	Manning's (n) value for overland flow	1.0–30	-
SLSUBBSN.hru	Average slope length	10–50	m
\multicolumn{4}{c}{(b) Queen River parameters for daily streamflow calibration.}			
CN2.mgt	SCS runoff curve number	60–75	-
ALPHA_BF.gw	Baseflow alpha factor	0.0–0.10	1/Days
GW_REVAP.gw	Groundwater revap coefficient	0.02–0.15	Days
GW_DELAY.gw	Groundwater delay	0.0–10.0	Days
GWQMN.gw	Depth of water in shallow aquifer for return flow	150–1000	mm
TIMP.bsn	Snowpack temperature lag factor	0.0–1.0	-
ALPHA_BNK	Baseflow alpha factor for bank storage	0.5–1.0	Days
\multicolumn{4}{c}{(c) Cork Brook parameters for daily streamflow calibration.}			
CN2.mgt	SCS runoff curve number	−60–75	-
ALPHA_BF.gw	Baseflow alpha factor	0.0–0.10	1/Days
GW_DELAY.gw	Groundwater delay	0.0–7.0	Days
GWQMN.gw	Depth of water in shallow aquifer for return flow	200–1000	mm
SMTMP.bsn	Snowmelt base temperature	−0.5–2.0	°C
ESCO.hru	Soil evaporation compensation factor	0.15–0.65	-
EPCO.hru	Plant uptake compensation factor	0.15–0.65	-
SLSOIL.hru	Slope length for lateral subsurface flow	0.0–150.0	m

2.4. Climate Change Variables

The anticipated change in average air temperature and precipitation over short term (2010–2039), medium term (2040–2069) and long term (2070–2099) time-spans for low and high greenhouse gas (GHG) scenarios were incorporated and compared to the historical period (1980–2009). Low greenhouse gas emission scenarios are based on the 2007 International Panel on Climate Change SRES B1 scenario and the high emissions are based on the SRES A1fi scenario. The B1 scenario is a situation where economic growth incorporates clean, ecologically friendly technology and GHG emissions levels return to pre-industrial concentrations, estimated at CO_2 levels of 300 parts per million (ppm). The high-emission scenario (A1fi) is a scenario based on fossil fuel intensive technologies for worldwide economic growth resulting in CO_2 levels reaching 940 ppm.

Climate variables in the calibrated SWAT subbasin input files (.sub) were edited to simulate the future scenarios. The default carbon dioxide (CO_2) concentration, relative rainfall adjustment and temperature increases (°C) are 330 parts per million (ppm), zero and zero respectively. The default values within all .sub files for each model were replaced with climate change variables. The variables used in this study are based on values published by Wake et al. at the University of New Hampshire [26–28], which were generated from four global climatic models downscaled to the New England region. Two of the published climate grids for Rhode Island were adopted and modified for this study and four different CO_2 levels were used. SWAT output for all low-emission scenarios is based on 330 ppm (the lower limit in the SWAT program code) and the RI climate grid change factors in Table 4a,c. For the high-emission alternatives, the short, medium and long-term SWAT climate change simulations were run with CO_2 levels at 540, 740 and 940 ppm, respectively, in addition to the RI climate grid change factors in Table 4b,d.

Table 4. Climate change variables adopted and modified from Wake et al., 2014 [27] for (**a,b**) Kingston, RI (Beaver River and Queen River) and (**c,d**) North Foster, RI (Cork Brook). Low emissions (**a,c**) based on SRES A1fi scenario and high emissions (**b,c**) based on SRES B1 scenario. Temperatures (Temp.) listed as degree (°C) increase, averaged from the published minimum and maximum temperatures. Precipitation (Precip.) values listed as a relative change computed based on the published values.

Indicator	January	February	March	April	May	June	July	August	September	October	November	December
				(**a**) Low Emissions–Kingston, RI							
Short-term Temp.	0.97	0.97	1.42	1.42	1.42	0.83	0.83	0.83	0.36	0.36	0.36	0.97
Med-term Temp.	1.50	1.50	2.47	2.47	2.47	1.58	1.58	1.58	0.56	0.56	0.56	1.50
Long-term Temp.	2.17	2.17	3.25	3.25	3.25	1.97	1.97	1.97	0.83	0.83	0.83	2.17
Short-term Precip.	8.76	8.76	9.80	9.80	9.80	17.9	17.9	17.9	5.59	5.59	5.59	8.76
Med-term Precip.	14.3	14.3	10.3	10.3	10.3	17.9	17.9	17.9	6.90	6.90	6.90	14.3
Long-term Precip.	14.9	14.9	16.3	16.3	16.3	18.6	18.6	18.6	10.6	10.6	10.6	14.9
				(**b**) High Emissions–Kingston, RI							
Short-term Temp.	0.97	0.97	0.83	0.83	0.83	1.11	1.11	1.11	1.00	1.00	1.00	0.97
Med-term Temp.	2.22	2.22	2.36	2.36	2.36	3.06	3.06	3.06	3.00	3.00	3.00	2.22
Long-term Temp.	3.83	3.83	4.28	4.28	4.28	5.22	5.22	5.22	4.92	4.92	4.92	3.83
Short-term Precip.	8.09	8.09	14.2	14.2	14.2	12.5	12.5	12.5	4.93	4.93	4.93	8.09
Med-term Precip.	10.0	10.0	15.8	15.8	15.8	12.5	12.5	12.5	6.2	6.2	6.2	10.0
Long-term Precip.	22.3	22.3	22.0	22.0	22.0	10.2	10.2	10.2	8.16	8.16	8.16	22.3
				(**c**) Low Emissions–North Foster, RI							
Short-term Temp.	1.00	1.00	1.42	1.42	1.42	0.97	0.97	0.97	0.39	0.39	0.39	1.00
Med-term Temp.	1.58	1.58	2.53	2.53	2.53	1.81	1.81	1.81	0.58	0.58	0.58	2.22
Long-term Temp.	2.22	2.22	3.33	3.33	3.33	2.25	2.25	2.25	0.81	0.81	0.81	2.22
Short-term Precip.	10.6	10.6	11.3	11.3	11.3	16.9	16.9	16.9	6.62	6.62	6.62	10.6
Med-term Precip.	12.9	12.9	11.9	11.9	11.9	17.4	17.4	17.4	10.1	10.1	10.1	12.9
Long-term Precip.	16.2	16.2	15.6	15.6	15.6	17.4	17.4	17.4	11.8	11.8	11.8	16.2
				(**d**) High Emissions–North Foster, RI							
Short-term Temp.	0.97	0.97	0.89	0.89	0.89	1.22	1.22	1.22	0.89	0.89	0.89	0.97
Med-term Temp.	2.22	2.22	2.50	2.50	2.50	3.28	3.28	3.28	2.78	2.78	2.78	2.22
Long-term Temp.	3.86	3.86	4.47	4.47	4.47	5.50	5.50	5.50	4.64	4.64	4.64	3.86
Short-term Precip.	6.29	6.29	10.8	10.8	10.8	15.7	15.7	15.7	2.08	2.08	2.08	6.29
Med-term Precip.	8.84	8.84	11.3	11.3	11.3	18.0	18.0	18.0	2.76	2.76	2.76	8.84
Long-term Precip.	17.7	17.7	20.0	20.0	20.0	17.4	17.4	17.4	5.37	5.37	5.37	17.7

2.5. Stressful Event Identification

Upon model calibration, validation and incorporation of climate change variables, output data for both model versions were processed to predict the occurrence of stressful conditions in all three watersheds from 1980 to 2099. As previously discussed, a stressful event for this study is defined as any day where both temperature and flow extremes occur. This study used the Q25 and Q75 flow exceedance percentiles as indicators because of their common use [80–82] and ecohydrological importance to brook trout. The most critical period for the species is typically the lowest flows of late summer to winter and a base flow of <25% is considered poor for maintaining quality trout habitat [12,44]. A Q25 exceedance characterizes the highest 25% of all daily flow rates and Q75 represents the lowest 25% of all daily flow rates [82]. For the stressful event analysis, the exceedance probability and average daily stream temperature for each date were identified. If the day fell into the Q25 or Q75 percentile, and if the stream temperature was greater than 21 °C, then the day was tagged as being a thermally stressful event.

3. Results and Discussion

3.1. Historical Conditions

The modeled average daily stream temperature was nearly the same at all three sites. The average daily discharge, however, was different at all three sites and corresponded to watershed area, with the highest discharge within the Queen River (largest watershed) and the lowest discharge within Cork Brook (smallest watershed) (Table 5). This is in agreement with the observed data, the Queen River had the highest discharge followed by the Beaver River and Cork Brook. The calibrated model for each watershed was first run over the entire 30-year period (1980–2009) (Table 5) to understand the percent chance that a stressful event will occur on a given day. Of the three study sites, the Queen River had the highest percent chance that a stressful event would occur on any given day and the Beaver River had the lowest percent chance (Table 6).

Table 5. The average stream temperature simulated by SWAT 1980–2009.

Watershed	Average Daily Stream Temp. (°C)	Average Daily Discharge (m³/s)
Beaver River	13.0	0.38
Queen River	13.0	1.0
Cork Brook	12.5	0.081

Table 6. Stressful event analysis of SWAT simulation for the three study sites.

Date	Watershed	Indicator	Any Type of Stress	Stream Temp. > 21 °C	Q25 or Q75 Flow	Stressful Event
	Beaver River	Days	6416	959	5457	511
		% Chance	58.6%	8.8%	49.8%	4.7%
1980–2009	Queen River	Days	6506	959	5547	700
		% Chance	59.4%	8.8%	50.6%	5.5%
	Cork Brook	Days	6875	1409	5466	551
		% Chance	62.7%	12.9%	49.9%	4.4%

The frequency of stress events in the three watersheds are similar (Table 6). The chances of any type of stress occurring within the watersheds vary by just 1.1%. One difference between Cork Brook and the Pawcatuck watersheds is the number of days with stream temperatures greater than 21 °C. The Beaver River and the Queen River have the same number of days with temperature stress because the air temperature for each model was collected from the same weather station. The number of days with stream temperature greater than 21 °C at Cork Brook is 46% higher than the Pawcatuck watersheds. This may be attributed to the low discharge levels at Cork Brook (0.081 m³/s) because lower, slower flows are exposed to air longer causing them to increase or decrease in temperature more readily. This interpretation is illustrated in Figures 5–7, which show the distribution of high stream temperatures within the Q25 and Q75 percentiles for each watershed. For all watersheds, a greater number of stressful events occurred during periods of low flow rather than periods of high flow.

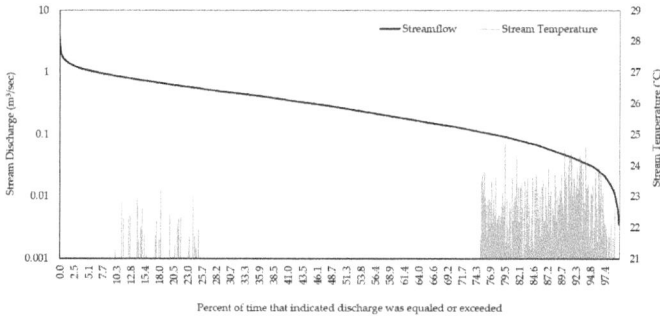

Figure 5. Beaver River simulated historical flow duration curve and stream temperatures. The secondary y-axis begins at 21 °C and any temperatures that are not above the stressful threshold are not shown in the figure. The stream temperatures in the Q25–Q75 range are omitted from the figure.

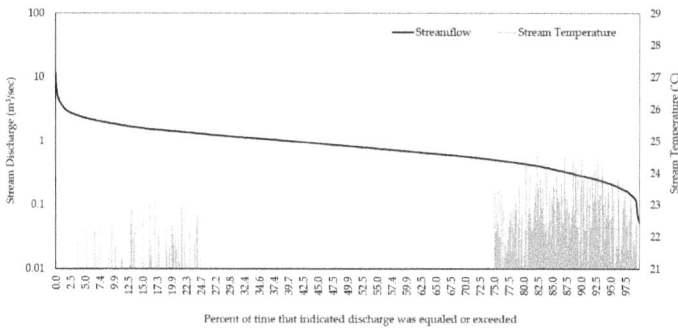

Figure 6. Queen River simulated historical flow duration curve and stream temperatures. The secondary y-axis begins at 21 °C and any temperatures that are not above the stressful threshold are not shown in the figure. The stream temperatures in the Q25–Q75 range are omitted from the figure.

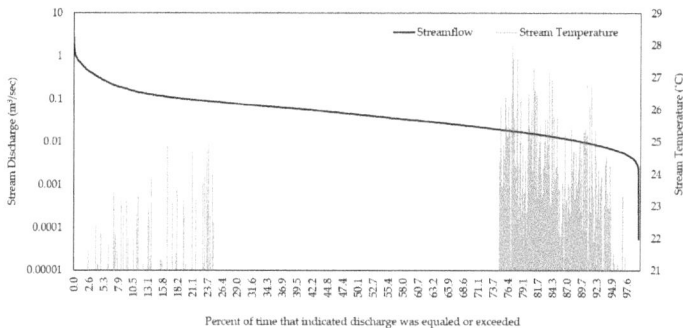

Figure 7. Cork Brook simulated historical flow duration curve and stream temperatures. The secondary y-axis begins at 21 °C and any temperatures that are not above the stressful threshold are not shown in the figure. The stream temperatures in the Q25–Q75 range are omitted from the figure.

Lastly, it is interesting to note the occurrences of stressful events within each watershed. Even though the Queen River has the same number of temperature stress days as the Beaver River, a difference of only 90 flow stress days increased the percent chance of stressful event occurrences from 4.7% in the Beaver River to 5.5% chance in the Queen River. This shows that a combination of flow and temperature should be taken into consideration when making management decisions or evaluating the quality of aquatic habitat. Such details can have important implications for aquatic species. Brook trout have been observed to tolerate higher stream temperatures provided their physical habitat remains stable [45]. If the co-occurrence of temperature and flow stresses increases, then physiological stresses to individual trout may become more apparent. The data simulated from 1980 to 2009 provide a baseline for comparing future projections, and will help determine if the resilience of local brook trout populations may become strained under future climate change conditions by combining two important indicators for survival.

3.2. Future Projections

New England is predicted to experience a warmer and wetter climate due to global warming [3]. Since 1970, Rhode Island specifically has had the average maximum and minimum air temperatures increase by 1.2 °C annually and by 2020–2099 it is expected that there will be hotter summers with 12–44 more days above 50 °C [26]. Also since 1970, the frequency and magnitude of extreme precipitation events has increased and annual precipitation has increased 6–11%. By 2020–2099, annual precipitation averages will increase by 18–20% and a two-fold increase in extreme precipitation events is expected to occur. A decrease in snow cover is anticipated and Rhode Island may have 20–32 fewer snow covered days [26].

3.2.1. Stream Discharge and Stream Temperature

Within the Beaver and Queen Rivers the simulated stream discharge change was much greater for high CO_2 emission scenarios 2010–2099 than for low CO_2 emission scenarios, a change of 3.4 °C as opposed to 1.6 °C, respectively. Discharges between the two Wood-Pawcatuck subbasins were different and a greater change was observed in the Beaver River subbasin. In the Beaver River, under the low-emission scenario 2010–2099 the discharges increased by 23% related to historical discharges and under the high-emission scenario increased by 71%. In the Queen River, under the low-emission scenario 2010–2099 the discharges increased by 19% of historical discharge levels and under the high-emission scenario increased by 49%. This is interesting because groundwater inputs are greater in the Beaver River (93%) than in the Queen River (78%). In the New England region, baseflow contributions have shown an upward trend likely linked to increasing precipitation [83] and climate change may be impacting storage by increasing the volume of water held in groundwater or as soil moisture within the basin. When storage is exceeded, the upper streamflow quantiles may be affected [84]. Brook trout can benefit from increased baseflow. Groundwater inflow can cool stream water [85], especially when flows are lower in the summer months [86]. Brook trout rely on groundwater seeps as refugia from increased stream temperatures and to keep developing embryos submerged in cool water [12].

An increase in stream temperature and streamflow was also seen in Cork Brook. Stream temperature increased by 1.6 °C between 2010 and 2099 under the low-emission scenario and 3.5 °C under the high-emission scenario, very similar to the degree changes in the Pawcatuck watersheds. Between 2010 and 2099, discharges increased by 20% under the low-emission scenario and 60% under the high-emission scenario. While not exact, the changes in discharge at Cork Brook for the low-emission scenario are more similar to the changes within the Queen River based on percent increase although under the high-emission scenario Cork Brook is the median between the Beaver River and Queen River. Overall, the SWAT streamflow projections in the three watersheds align well with climate change predictions for New England under the low-emission simulations and exceed predictions under the high-emission simulations [26]. The modeled average daily stream temperature and average daily stream discharge increased at all sites for both low and high CO_2 emission scenarios (Table 7).

Table 7. Average stream temperature and streamflow simulated with climate change variables for (a) Beaver River, (b) Queen River and (c) Cork Brook. High and low CO_2 emission scenarios projected for short (2010–2039), medium (2040–2069) and long-term (2070–2099). Unchanged historical results included for reference.

Scenario	Date	Average Daily Stream Temp. (°C)	Average Daily Discharge (m³/s)
		(a)	
Beaver River Historical	1980–2009	13.0	0.38
Beaver River Low Emissions	2010–2039	13.6	0.44
	2040–2069	14.2	0.45
	2070–2099	14.6	0.47
Beaver River High Emissions	2010–2039	13.7	0.49
	2040–2069	15.0	0.53
	2070–2099	16.4	0.65
		(b)	
Queen River Historical	1980–2009	13.0	1.0
Queen River Low Emissions	2010–2039	13.6	1.14
	2040–2069	14.2	1.16
	2070–2099	14.6	1.19
Queen River High Emissions	2010–2039	13.7	1.20
	2040–2069	15.0	1.27
	2070–2099	16.4	1.49
		(c)	
Cork Brook Historical	1980–2009	12.5	0.081
Cork Brook Low Emissions	2010–2039	13.2	0.09
	2040–2069	13.25	0.10
	2070–2099	14.11	0.10
Cork Brook High Emissions	2010–2039	13.25	0.10
	2040–2069	14.52	0.10
	2070–2099	15.97	0.13

3.2.2. Flow Regime

The flow duration curves for each watershed were compared to historical streamflow (1980–2009) and future long term (2070–0299) scenarios to assess the flow conditions at the end of the century (Figures 8–10). The curve for each watershed under the low emission scenarios changed very little in shape even though the stream discharges were increased. Under the high-emissions scenario the magnitude of discharges also increases but in the Beaver River and Cork Brook, the shape of the rating curve became flatter in the Q50–Q75 percentiles. A flat curve generally indicates that flows are sustained throughout the year and can be caused by factors such as groundwater contributions to the stream reach.

As water temperatures increase due to global warming, brook trout may benefit from sustained flows that will prevent stream temperatures from rising further and help ensure that downstream habitat remains connected to headwaters. From this perspective, the Beaver River and Cork Brook may provide better future trout habitat in comparison to the Queen River, which saw little change to the shape of the rating curve. On the other hand, a sustained increase in flow magnitude can change the geomorphology and may not be beneficial for aquatic species during the spawning season when flows are historically lower [41]. An increase in stream discharges during the low flow season may put redds (nests) at risk of destruction from sedimentation or sheer velocity. Changes in streamflow magnitude may also increase turbidity or redistribute riffle and pool habitat throughout the stream reach. This may decrease the availability of suitable habitat as brook trout prefer stream reaches with an approximate 1:1 pool-riffle [12]). Pool and riffle redistribution can also affect the type and quantity of local macroinvertebrate populations. Since warming temperatures will have an impact on trout body condition as fish enter the winter months, the available food supply can become an even more critical factor as the climate changes.

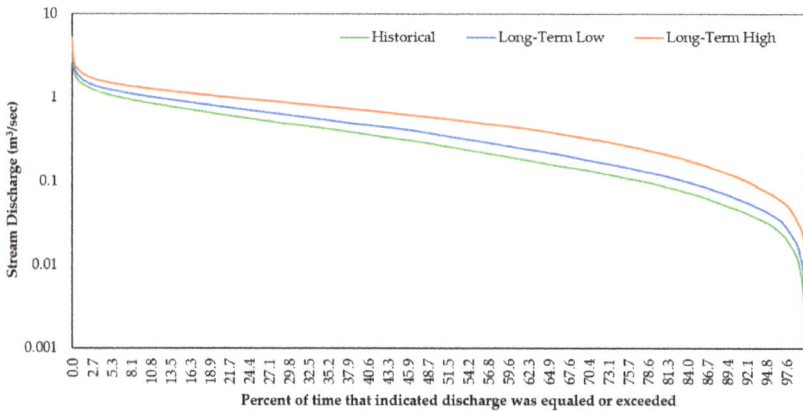

Figure 8. Beaver River flow duration curves simulated for high and low CO_2 emission scenarios by the end of the long-term (2070–2099). Unchanged historical results included for reference.

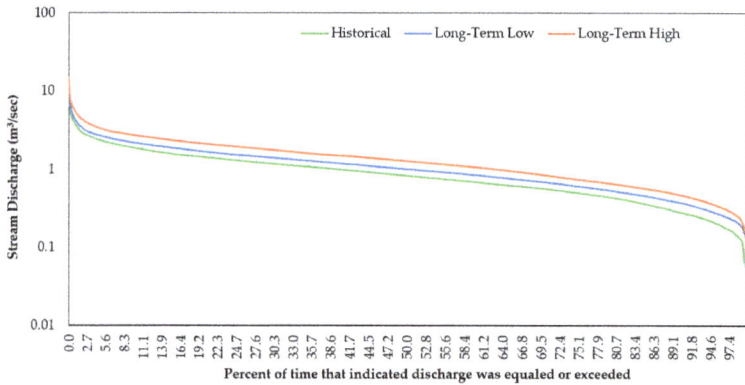

Figure 9. Queen River flow duration curves simulated for high and low CO_2 emission scenarios by the end of the long-term (2070–2099). Unchanged historical results included for reference.

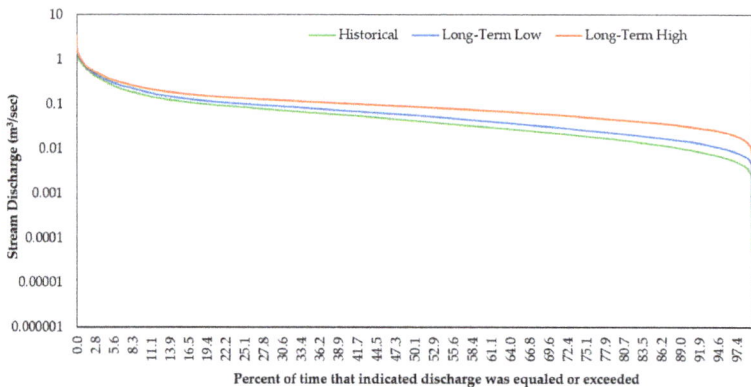

Figure 10. Cork Brook flow duration curves simulated for high and low CO_2 emission scenarios by the end of the long-term (2070–2099). Unchanged historical results included for reference.

3.2.3. Timing of Stream Temperatures

The model predicted that between 1980 and 2099 stream temperatures in all watersheds will increase by 1.6 °C under the low-emission scenario or 3.4 °C under the high-emission scenarios (Table 7). Further analysis was conducted to assess if the temporal distribution of stream temperatures has changed throughout the year. In the Beaver and Queen River watersheds no change to the timing of high stream temperatures was observed and high temperatures continued to occur primarily in July–September (Figure 11a). In the Cork Brook watershed, however, the model predicted that the occurrence of high stream temperatures will increase and will occur as early as April by the end of the century under both high- and low-emission scenarios (Figure 11b). In all watersheds, the number of days with stressful temperatures during the low-emission scenario increased only slightly compared to historical observations. The number of occurrences per month increased under the high-emission scenario for all watersheds compared to historical simulations.

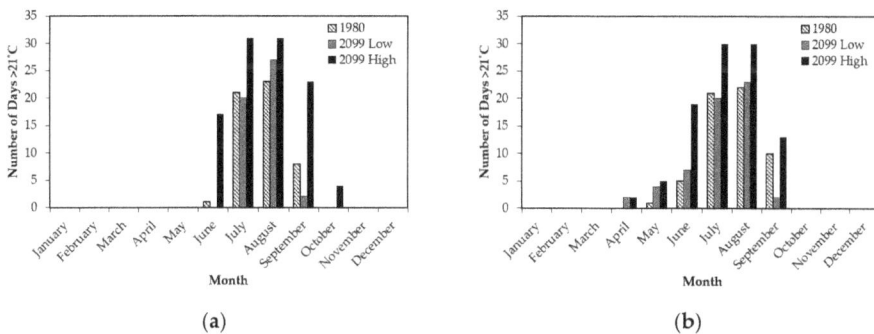

Figure 11. The number of days per month that stream temperatures exceeded the stress threshold in 1980, 2099 under low CO_2 emissions and 2099 under high CO_2 emissions in (**a**) the Beaver and Queen Rivers which had the same weather station and (**b**) Cork Brook.

Stream temperatures reaching the stressful threshold sooner in the year will have implications for those coldwater species in Cork Brook. A shift in the timing of high stream temperatures can influence the development of both young-of-year and adult individuals. Embryos develop over winter and the length of incubation is temperature dependent; 45 days for development at 10 °C, 165 days at 2.8 °C and 28 days at 14.8 °C [12]. Higher temperatures earlier in the spring will mean that fish experience physiological stress sooner and may not be able to survive until the spawning period in late fall when stress will be relieved by cooler temperatures. Additionally, because brook trout avoid warmer water and are rarely found in streams with 60 days mean temperatures above 20 °C [7,33], changes to the temporal distribution of stream temperatures will likely have an effect on the spatial distribution of trout [7,10–16].

3.2.4. Stressful Event Analysis

The results of the stressful event analysis are summarized in Table 8 over 30-year increments. There are few notable differences between the three watersheds when the data were assessed over these 30-year increments. An analysis in 10-year increments, however, yielded greatly different results (Appendix A). Of the three sites between 1980 and 2099, the Queen River watershed had the greatest (i.e. maximum) number of stressful days and percent chance of an event occurring under both low CO_2 emissions (7 of 12 decades) and high CO_2 emissions (8 of 12 decades). Under low-emission scenarios, the Beaver River had the maximum count just once and under the high-emission scenario the Cork Brook watershed had the maximum count once. Under the low-emission scenario, the difference in percent chance of a stressful event occurring from 1980–1989 compared to 2090–2099 was calculated as

4.6% in the Beaver River, 6.7% in the Queen River and 8.4% in Cork Brook. Under the high-emission scenario, the difference in percent chance of a stressful event occurring from 1980–1989 compared to 2090–2099 is 13.4% in the Beaver River, 14.8% in the Queen River and 14.3% in Cork Brook

Table 8. Percent chance of a stressful event occurring under future climate scenarios. Results for each watershed by 30-year increments. High and low CO_2 emission scenarios projected for short (2010–2039), medium (2040–2069) and long-term (2070–2099). Unchanged historical results included for reference.

Date	Emission Scenario	Unit	Beaver	Queen	Cork
1980–2009	Historical	% Chance	4.7	5.5	4.4
	Historical		4.7	5.5	4.4
2010–2039	Low	% Chance	6.2	6.9	6.5
	High		7.2	7.9	7.2
2040–2069	Low	% Chance	7.9	8.5	7.1
	High		12.4	13.1	11.3
2079–2099	Low	% Chance	9.0	9.8	8.6
	High		16.1	16.8	15.2

The Beaver River has a lower change in stressful event chance than the other watersheds for both low-emission and high-emission climate change scenarios. This may be because it has the greatest percent of groundwater contributions and streams that are groundwater fed receive inputs that are less exposed to ambient air temperatures. The benefits of groundwater inputs are greater under the low-emission scenario and less effective under the high-emission scenarios. For instance, the watershed with the least amount of baseflow (Cork Brook) has a change in percent chance that is more than double that of the watershed with the highest baseflow (Beaver River). Under the high-emission scenario, however, the change in percent chance is less distributed and the Beaver River and Cork Brook differ by just 1%. Groundwater temperatures are expected to follow projected increases in mean annual air temperature from climate warming [86]. Under the high-emission scenario, this effect may be more prominent allowing for less dampening of in-stream temperatures by baseflow.

The number of stressful events under the high-emission scenario is greater than the number of events under the low-emission scenario for every decade since 2010, in every watershed (Figures 12–14). The graphs also show that for future simulations the number of events in any given decade is higher than the previous decade except for 2060–2069 in the Queen River and 2070–2079 in the Beaver River and Cork Brook. Additionally, it should be noted that there is a minor disconnect between the historical trend and the short-term future simulations; In the Queen River and in Cork Brook Cork there is a higher occurrence between 2000–2009 than there is 2010–2019. The timing of the decrease is likely a result of shifting the model from the regular SWAT code to SWAT with added climate variables, rather than the simulation itself.

Of the three watersheds, the Beaver River and Cork Brook are most likely to provide resilient habitat for brook trout as the local water conditions change due to global warming. Under low-emission scenarios, the Beaver River more frequently displayed the lower percent chance of a stressful event occurring and under the high-emission scenario Cork Brook more frequently had the lowest percent chance by the end of the century. Under both the high- and low-emission scenarios, the chance of stressful events occurring was consistently predicted to be greater in the Queen River. Possible causes of this difference are the larger size of the Queen River watershed and the two tributaries located upstream of the watershed outlet. Fisherville Brook and Queen's Fort Brook are two waterways that discharge into the Queen River (Figure 1). The Queen's Fort Brook flows along the eastern side of the watershed through the agricultural area and Fisherville Brook is located along the western side of the watershed where the slope is steeper. Additionally, the main stem of the Queen River itself flows through a large golf course in the middle of the watershed. The tributaries and the main stem come into closer contact with the heterogeneous areas of the basin and may be able to capture additional effects of climate change not seen in the other watersheds. This is not to say that coldwater habitat restoration is not worthwhile in the Queen River, rather that more effort will be needed to restore or maintain brook trout populations in this watershed.

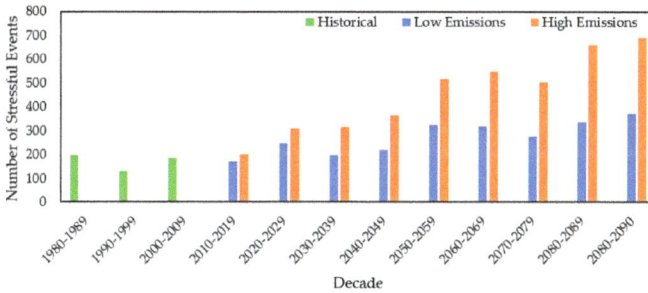

Figure 12. Number of stressful events predicted in the Beaver River watershed between 1980 and 2099 under historical conditions, low CO_2 emissions and high CO_2 emission scenarios.

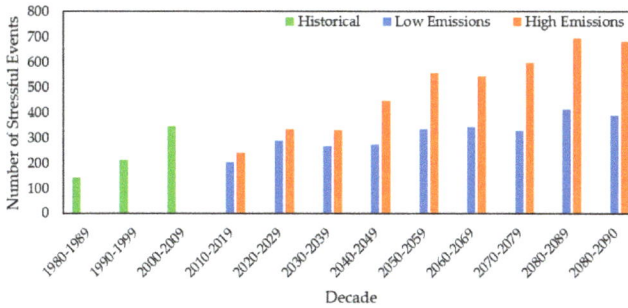

Figure 13. Number of stressful events predicted in the Queen River watershed between 1980 and 2099 under historical conditions, low CO_2 emissions and high CO_2 emission scenarios.

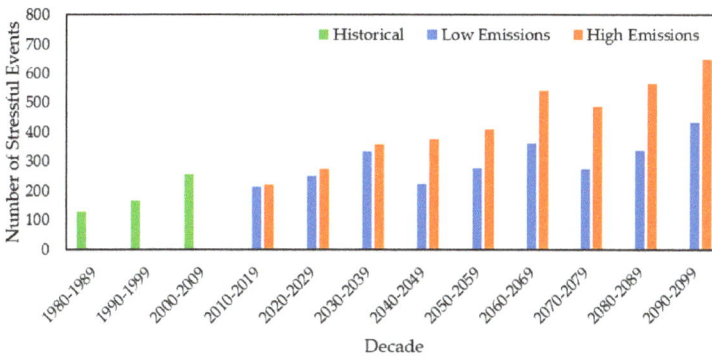

Figure 14. Number of stressful events predicted in the Cork Brook watershed between 1980 and 2099 under historical conditions, low CO_2 emissions and high CO_2 emission scenarios.

Stream temperatures in all three watersheds were simulated to increase under both low CO_2 and high CO_2 emission scenarios. It is challenging to discern from this study if stream temperatures in the Beaver River or the Queen River differ significantly because the UGSG gauges at the basin outlet do not record stream temperature and the weather station data used in SWAT simulations was the same for both watersheds. Simulated results do show, however, that stream temperatures will increase through the end of the century by either 1.6 °C under low emissions or 3.4 °C under high emissions in these two watersheds. One-way resource managers can buffer this effect is by preserving existing canopy cover along the riparian corridor. Forest harvesting can increase solar radiation in the riparian zone as well as wind speed and exposure to air advected from clearings, typically causing increases in stream water temperature regimes [87,88]. Additionally, managers may also advocate for preserving groundwater resources that discharge to the streams because baseflow will help regulate stream temperatures, especially if the global low CO_2 emission scenario is achieved.

4. Conclusions and Future Work

To help managers identify which areas within a watershed are in the greatest need of protection, a subbasin analysis could be conducted. For instance, both Wood-Pawcatuck basins are home to small preserves managed by The Nature Conservancy. Setting up the model so that a subbasin outlet (as opposed to the watershed outlet) is located within each preserve will allow for assessing site specific conditions when it is not practical to create a model on a small scale. If model output shows that historically these preserves have changed very little, and that future simulations predict minimal change, then managers can put efforts and financial resources towards other preserves that are in greater need.

Another consideration for future work is to limit the stressful event analysis to the spring and summer months when brook trout are more sensitive to warmer stream temperatures. Also, a study could be conducted to see if stressful events occur sequentially. This study took a wider approach by examining how stream temperatures and streamflow vary throughout the entire year. This timeframe was chosen for several reasons. First, since this is the only study of its kind within these watersheds we did not have enough information to say with certainty that no changes to stream temperature or streamflow would occur during the fall and winter. In fact, some scientists predict that by the end of the century Rhode Island will have a climate similar to that of South Carolina and Georgia [26], in which case stream temperatures would almost certainly increase during the winter months. Second, while stream temperatures and streamflow during the winter months are not as critical for brook trout compared to the summer, winter conditions do effect embryo development. For instance, the length of embryo incubation during the winter ranges from 28 to 45 days depending on the temperature of the stream water [12]. Lastly, while this study focused on brook trout, our hope is that the methodology can be applied to other types of aquatic species that may be sensitive to stream conditions during other seasons.

Finally, since all three of these watersheds are baseflow driven, using a model approach that considers the influence of groundwater discharges on stream temperatures would be valuable. A study conducted by Ficklin et al. developed a hydroclimatological SWAT component that incorporates the effects of both air temperatures and hydrological inputs, such as groundwater, on stream temperatures. Previous studies have shown that the hydroclimatological component can be used in small watersheds [89] and in New England [90]. Since the hydroclimatological model component takes the groundwater temperature into consideration, the stream reach will receive inputs that are less exposed to ambient air and therefore cooler during the summer and slightly warmer than the air during the winter. Using a SWAT model with this component may produce more accurate stream temperature results in streams that are baseflow driven.

The purpose of this study was to gain a better understanding of the effects of climate change on coldwater habitat using SWAT. We successfully showed that SWAT can be used to simulate both historical and future climate scenarios in forested, baseflow-driven watersheds in Rhode Island. Moreover, thermally stressful event identification can be a functional approach to analyzing model output. The results indicate that climate change will have a negative effect on coldwater fish species in these types of ecosystems, and that the resiliency of local populations will be tested as stream conditions will likely become increasingly stressful.

Acknowledgments: A thank you to Thomas Boving of the University of Rhode Island and Jameson Chace of Salve Regina University for their insight and review of this Master's thesis project. We would also like to thank Khurshid Jahan for her help organizing the stream temperature data and creating a template for data processing. We would also like to thank S-1063 Multistate Hatch Grant for supporting this research.

Author Contributions: B.M.C. and S.M.P. conceived and designed the experiments; B.M.C. performed the experiments; B.M.C., S.M.P. and A.J.G. analyzed the data; B.M.C. wrote the paper.

Conflicts of Interest: The authors declare no conflict of interest.

Appendix A

Table A1. Stressful event results for each watershed by decade. High and low CO_2 emission scenarios projected for short (2010–2039), medium (2040–2069) and long-term (2070–2099). Unchanged historical results included for reference.

Date	Emission Scenario	Unit	Beaver	Queen	Cork
1980–1989	Low	Days	200	141	127
		% Chance	*5.5%*	*3.9%*	*3.5%*
	High	Days	200	141	127
		% Chance	*5.5%*	*3.9%*	*3.5%*
1990–1999	Low	Days	130	213	168
		% Chance	*3.6%*	*5.8%*	*4.6%*
	High	Days	130	213	168
		% Chance	*3.6%*	*5.8%*	*4.6%*
2000–2009	Low	Days	185	346	256
		% Chance	*5.1%*	*9.5%*	*7.0%*
	High	Days	185	346	256
		% Chance	*5.1%*	*9.5%*	*7.0%*
2010–2019	Low	Days	172	141	216
		% Chance	*4.7%*	*3.9%*	*5.9%*
	High	Days	203	238	221
		% Chance	*5.6%*	*6.5%*	*6.0%*
2020–2029	Low	Days	249	213	252
		% Chance	*6.8%*	*5.8%*	*6.9%*
	High	Days	308	334	276
		% Chance	*8.4%*	*9.1%*	*7.6%*
2030–2039	Low	Days	200	346	335
		% Chance	*5.5%*	*9.5%*	*9.2%*
	High	Days	317	330	358
		% Chance	*8.7%*	*9.0%*	*9.8%*
2040–2049	Low	Days	221	273	223
		% Chance	*6.0%*	*7.5%*	*6.1%*
	High	Days	364	445	375
		% Chance	*10.0%*	*12.2%*	*10.0%*
2050–2059	Low	Days	325	334	278
		% Chance	*8.9%*	*9.1%*	*7.6%*
	High	Days	516	555	410
		% Chance	*14.1%*	*15.2%*	*11.0%*

Table A1. *Cont.*

Date	Emission Scenario	Unit	Beaver	Queen	Cork
2060–2069	Low	Days	320	343	363
		% Chance	*8.8%*	*9.4%*	*9.9%*
	High	Days	547	543	540
		% Chance	*15.0%*	*14.9%*	*14.8%*
2070–2079	Low	Days	276	326	276
		% Chance	*7.6%*	*8.9%*	*7.6%*
	High	Days	502	597	487
		% Chance	*13.7%*	*16.3%*	*13.3%*
2080–2089	Low	Days	337	412	338
		% Chance	*9.2%*	*11.3%*	*9.3%*
	High	Days	662	694	566
		% Chance	*18.1%*	*19.0%*	*15.5%*
2090–2099	Low	Days	370	389	433
		% Chance	*10.1%*	*10.6%*	*11.9%*
	High	Days	692	682	649
		% Chance	*18.9%*	*18.7%*	*17.8%*

References

1. Hodgkins, G.A.; Dudley, R.W.; Huntington, T.G. Changes in the timing of high river flows in New England over the 20th Century. *J. Hydrol.* **2003**, *278*, 244–252. [CrossRef]
2. Eaton, J.G.; Scheller, R.M. Effects of climate warming on fish thermal habitat in streams of the United States. *Limnol. Oceanogr.* **1996**, *41*, 1109–1115. [CrossRef]
3. Hayhoe, K.; Wake, C.; Anderson, B.; Liang, X.-Z.; Maurer, E.; Zhu, J.; Bradbury, J.; DeGaetano, A.; Stoner, A.M.; Wuebbles, D. Regional climate change projections for the Northeast USA. *Mitig. Adapt. Strateg. Glob. Chang.* **2008**, *13*, 425–436. [CrossRef]
4. Hayhoe, K.; Wake, C.P.; Huntington, T.G.; Luo, L.; Schwartz, M.D.; Sheffield, J.; Wood, E.; Anderson, B.; Bradbury, J.; DeGaetano, A.; et al. Past and future changes in climate and hydrological indicators in the US Northeast. *Clim. Dyn.* **2007**, *28*, 381–407. [CrossRef]
5. Mohseni, O.; Erickson, T.R.; Stefan, H.G. Sensitivity of stream temperatures in the United States to air temperatures projected under a global warming scenario. *Water Resour. Res.* **1999**, *35*, 3723–3733. [CrossRef]
6. Woodward, G.; Perkins, D.M.; Brown, L.E. Climate change and freshwater ecosystems: Impacts across multiple levels of organization. *Philos. Trans. R. Soc. B Biol. Sci.* **2010**, *365*, 2093–2106. [CrossRef] [PubMed]
7. Whitney, J.E.; Al-Chokhachy, R.; Bunnell, D.B.; Caldwell, C.A.; Cooke, S.J.; Eliason, E.J.; Rogers, M.; Lynch, A.J.; Paukert, C.P. Physiological basis of climate change impacts on North American Inland Fishes. *Fisheries* **2016**, *41*, 332–345. [CrossRef]
8. Van Vliet, M.T.H.; Franssen, W.H.P.; Yearsley, J.R.; Ludwig, F.; Haddeland, I.; Lettenmaier, D.P.; Kabat, P. Global river discharge and water temperature under climate change. *Glob. Environ. Chang.* **2013**, *23*, 450–464. [CrossRef]
9. Jiménez Cisneros, B.E.; Oki, T.; Arnell, N.W.; Benito, G.; Cogley, J.G.; Doll, P.; Jiang, T.; Mwakalila, S.S. 2014: Freshwater resources. In *Climate Change 2014: Impacts, Adaptation and Vulnerability. Part A: Global and Sectoral Aspects. Contribution of Working Group II to the Fifth Assessment Report of the Intergovernmental Panel on Climate Change*; Kundzewicz, Z., Ed.; Cambridge University Press: Cambridge, UK; New York, NY, USA, 2014; Volume 40.
10. Brett, J.R. Some principles in the thermal requirements of fishes. *Q. Rev. Biol.* **1956**, *31*, 75–87. [CrossRef]
11. Fry, F.E.J. The effect of environmental factors on the physiology of fish. In *Fish Physiology*; Hoar, W.S., Randall, D.J., Eds.; Academic Press: Cambridge, MA, USA, 1971; pp. 1–98.
12. Raleigh, R.F. Habitat Suitability Index Models: Brook trout. In *FWS/OBS*; United States Fish and Wildlife Service: Washington, DC, USA, 1982.

13. Hokanson, K.E.F.; McCormick, J.H.; Jones, B.R.; Tucker, J.H. Thermal requirements for maturation, spawning, and embryo survival of the brook trout, salvelinus fontinalis. *J. Fish. Res. Board Can.* **1973**, *30*, 975–984. [CrossRef]

14. Milner, N.J.; Elliott, J.M.; Armstrong, J.D.; Gardiner, R.; Welton, J.S.; Ladle, M. The natural control of salmon and trout populations in streams. *Fish. Res.* **2003**, *62*, 111–125. [CrossRef]

15. Peterson, J.T.; Kwak, T.J. Modeling the effects of land use and climate change on riverine smallmouth bass. *Ecol. Appl.* **1999**, *9*, 1391–1404. [CrossRef]

16. Goniea, T.M.; Keefer, M.L.; Bjornn, T.C.; Peery, C.A.; Bennett, D.H.; Stuehrenberg, L.C. Behavioral thermoregulation and slowed migration by adult fall chinook salmon in response to high columbia river water temperatures. *Trans. Am. Fish. Soc.* **2006**, *135*, 408–419. [CrossRef]

17. Arnold, J.G.; Srinivasan, R.; Muttiah, R.S.; Williams, J.R. Large area hydrologic modeling and assessment part I: Model development 1. *JAWRA J. Am. Water Resour. Assoc.* **1998**, *34*, 73–89. [CrossRef]

18. Isaak, D.J.; Wollrab, S.; Horan, D.; Chandler, G. Climate change effects on stream and river temperatures across the northwest U.S. from 1980–2009 and implications for salmonid fishes. *Clim. Chang.* **2012**, *113*, 499–524. [CrossRef]

19. Luo, Y.; Ficklin, D.L.; Liu, X.; Zhang, M. Assessment of climate change impacts on hydrology and water quality with a watershed modeling approach. *Sci. Total Environ.* **2013**, *450*, 72–82. [CrossRef] [PubMed]

20. Mohseni, O.; Stefan, H.G. Stream temperature/air temperature relationship: A physical interpretation. *J. Hydrol.* **1999**, *218*, 128–141. [CrossRef]

21. Null, S.; Viers, J.; Deas, M.; Tanaka, S.; Mount, J. Stream temperature sensitivity to climate warming in California's Sierra Nevada. In *AGU Fall Meeting Abstracts*; Utah State University: Logan, UT, USA, 2010.

22. Preud'homme, E.B.; Stefan, H.G. *Relationship between Water Temperatures and Air Temperatures for Central US Streams*; St. Anthony Falls Hydraulic Lab, Minnesota University: Minneapolis, MN, USA, 1992; 146p.

23. Anandhi, A.; Frei, A.; Pierson, D.C.; Schneiderman, E.M.; Zion, M.S.; Lounsbury, D.; Matonse, A.H. Examination of change factor methodologies for climate change impact assessment. *Water Resour. Res.* **2011**, *47*, 1–10. [CrossRef]

24. Saila, S.; Cheeseman, M.; Poyer, D. *Maximum Stream Temperature Estimation from Air Temperature Data and Its Relationship to Brook Trout (Salvelinus Fontinalis) Habitat Requirements in Rhode Island*; Wood-Pawcatuck Watershed Association: Hope Valley, RI, USA, 2004.

25. Mohseni, O.; Stefan, H.G.; Eaton, J.G. Global warming and potential changes in fish habitat in U.S. Streams. *Clim. Chang.* **2003**, *59*, 389–409. [CrossRef]

26. Climate Solutions New England. *Rhode Island's Climate Past and Future Changes*; Sustainability Institute at the University of New Hampshire: Durham, NH, USA, 2016. Available online: http://clim-map.csrcdev.com/files/Rhode%20Island%20(Total)%20Fact%20Sheet.pdf (accessed on 1 March 2017).

27. Wake, C.; Large, S. *Climate Grids for Rhode Island*; Sustainability Institute at the University of New Hampshire: Durham, NH, USA, 2014. Available online: clim-map.csrcdev.com/files/Rhode%20Island%20(Total)%20Climate%20Grids.pdf (accessed on 1 March 2017).

28. Wake, C.P.; Keeley, C.; Burakowski, E.; Wilkinson, P.; Hayhoe, K.; Stoner, A.; LaBrance, J. *Climate Change in Northern New Hampshire: Past, Present and Future*; Sustainability Institute at the University of New Hampshire: Durham, NH, USA, 2014. Available online: http://scholars.unh.edu/sustainability (accessed on 1 March 2017).

29. Erkan, D.E. *Strategic Plan for the Restoration of Anadromous Fishes to Rhode Island Coastal Streams*; Rhode Island Department of Environmental Management, Division of Fish and Wildlife: Wakefield, RI, USA, 2002.

30. WPWA; Saila, S.; Cheeseman, M.; Poyer, D. *Assessing Habitat Requirements for Brook Trout (Salvelinus fontinalis) in Low Order Streams*; Wood-Pawcatuck Watershed Association: Hope Valley, RI, USA, 2004.

31. Hakala, J.P.; Hartman, K.J. Drought effect on stream morphology and brook trout (*Salvelinus fontinalis*) populations in forested headwater streams. *Hydrobiologia* **2004**, *515*, 203–213. [CrossRef]

32. Bjornn, T.; Reiser, D. Habitat requirements of salmonids in streams. *Am. Fish. Soc. Spec. Publ.* **1991**, *19*, 138.

33. Chadwick, J.J.G.; Nislow, K.H.; McCormick, S.D. Thermal onset of cellular and endocrine stress responses correspond to ecological limits in brook trout, an iconic cold-water fish. *Conserv. Physiol.* **2015**, *3*, cov017. [CrossRef] [PubMed]

34. Lee, R.M.; Rinne, J.N. Critical thermal maxima of five trout species in the southwestern United States. *Trans. Am. Fish. Soc.* **1980**, *109*, 632–635. [CrossRef]

35. Letcher, B.H.; Nislow, K.H.; Coombs, J.A.; O'Donnell, M.J.; Dubreuil, T.L. Population response to habitat fragmentation in a stream-dwelling brook trout population. *PLoS ONE* **2007**, *2*, e1139. [CrossRef] [PubMed]

36. Kling, G.W.; Hayhoe, K.; Johnson, L.B.; Magnuson, J.J.; Polasky, S.; Robinson, S.K.; Shuter, B.J.; Wander, M.M.; Wuebbles, D.J.; Zak, D.R. *Confronting Climate Change in the Great Lakes Region: Impacts on Our Communities and Ecosystems*; Union of Concerned Scientists: Cambridge, MA, USA; Ecological Society of America: Washington, DC, USA, 2003; p. 92.

37. Magnuson, J.J.; Crowder, L.B.; Medvick, P.A. Temperature as an ecological resource. *Am. Zool.* **1979**, *19*, 331–343. [CrossRef]

38. Vannote, R.L.; Minshall, G.W.; Cummins, K.W.; Sedell, J.R.; Cushing, C.E. The river continuum concept. *Can. J. Fish. Aquat. Sci.* **1980**, *37*, 130–137. [CrossRef]

39. Bunn, S.E.; Arthington, A.H. Basic principles and ecological consequences of altered flow regimes for aquatic biodiversity. *Environ. Manag.* **2002**, *30*, 492–507. [CrossRef]

40. Freeman, M.C.; Pringle, C.M.; Jackson, C.R. Hydrologic connectivity and the contribution of stream headwaters to ecological integrity at regional scales1. *JAWRA J. Am. Water Resour. Assoc.* **2007**, *43*, 5–14. [CrossRef]

41. Poff, N.L.; Allan, J.D. Functional organization of stream fish assemblages in relation to hydrological variability. *Ecology* **1995**, *76*, 606–627. [CrossRef]

42. Poff, N.L.; Allan, J.D.; Bain, M.B.; Karr, J.R.; Prestegaard, K.L.; Richter, B.D.; Sparks, R.E.; Stromberg, J.C. The natural flow regime. *Bioscience* **1997**, *47*, 769–784. [CrossRef]

43. Bassar, R.D.; Letcher, B.H.; Nislow, K.H.; Whiteley, A.R. Changes in seasonal climate outpace compensatory density-dependence in eastern brook trout. *Glob. Chang. Biol.* **2016**, *22*, 577–593. [CrossRef] [PubMed]

44. DePhilip, M.; Moberg, T. *Ecosystem Flow Recommendations for the Susquehanna River Basin*; The Nature Conservancy: Harrisburg, PA, USA, 2010.

45. Nuhfer, A.J.; Zorn, T.G.; Wills, T.C. Effects of reduced summer flows on the brook trout population and temperatures of a groundwater-influenced stream. *Ecol. Freshw. Fish* **2017**, *26*, 108–119. [CrossRef]

46. Walters, A.W.; Post, D.M. An experimental disturbance alters fish size structure but not food chain length in streams. *Ecology* **2008**, *89*, 3261–3267. [CrossRef] [PubMed]

47. Saila, S.; Burgess, D.; Cheeseman, M.; Fisher, K.; Clark, B. *Interspecific Association, Diversity, and Population Analysis of Fish Species in the Wood-Pawcatuck Watershed*; Wood-Pawcatuck Watershed Association: Hope Valley, RI, USA, 2003.

48. Fulweiler, R.W.; Nixon, S.W. Export of nitrogen, phosphorus, and suspended solids from a southern New England watershed to little Narragansett bay. *Biogeochemistry* **2005**, *76*, 567–593. [CrossRef]

49. Poyer, D.; Hetu, M. *Study of Maximum Daily Stream Temperature of Select Streams in the Pawcatuck Watershed Summer 2005*; Wood-Pawcatuck Watershed Association: Hope Valley, RI, USA, 2005.

50. Dickerman, D.C.; Ozbilgin, M.M. Hydrogeology, water quality, and ground-water development alternatives in the Beaver-Pasquiset ground-water reservoir, Rhode Island. In *Water-Resources Investigations Report*; US Geological Survey: Reston, VA, USA, 1985.

51. Kliever, J.D. Hydrologic data for the Usquepaug-Queen river basin, Rhode Island. In *Open-File Report*; US Geological Survey: Reston, VA, USA, 1995.

52. Liu, T.; Merrill, N.H.; Gold, A.J.; Kellogg, D.Q.; Uchida, E. Modeling the production of multiple ecosystem services from agricultural and forest landscapes in Rhode Island. *Agric. Resour. Econ. Rev.* **2013**, *42*, 251–274. [CrossRef]

53. Poyer, D.; Hetu, M. *Maximum Daily Stream Temperature in the Queen River Watershed and Mastuxet Brook Summer 2006*; Wood-Pawcatuck Watershed Association: Hope Valley, RI, USA, 2006.

54. The Nature Conservancy (TNC). Beaver River Preserve. Places We Protect 2017; The Nature Conservancy Is a Nonprofit, Charitable Organization under Section 501(c)(3). Available online: https://www.nature.org/ourinitiatives/regions/northamerica/unitedstates/rhodeisland/placesweprotect/beaver-river-preserve.xml (accessed on 1 May 2017).

55. Armstrong, D.S.; Parker, G.W. *Assessment of Habitat and Streamflow Requirements for Habitat Protection, Usquepaug-Queen River, Rhode Island, 1999–2000*; USGS, Ed.; DTIC Document; DTIC: Denver, CO, USA, 2003.

56. Tefft, E. Factors affecting the distribution of brook trout (salvelinus fontinalis) in the wood-pawcatuck watershed of Rhode Island. In *Department of Natural Resources*; University of Rhode Island: Kingston, RI, USA, 2013.

57. The Nature Conservancy (TNC). Queen's River Preserve. Places We Protect 2017; The Nature Conservancy Is a Nonprofit, Charitable Organization under Section 501(c)(3). Available online: https://www.nature.org/ourinitiatives/regions/northamerica/unitedstates/rhodeisland/placesweprotect/queens-river-preserve.xml (accessed on 1 May 2017).

58. US Geological Survey (USGS). *National Water Information System Web Interface*; US Geological Survey: Reston, VA, USA, 2017.

59. Arnold, J.G.; Allen, P.M. Automated methods for estimating baseflow and ground water recharge from streamflow records. *JAWRA J. Am. Water Resour. Assoc.* **1999**, *35*, 411–424. [CrossRef]

60. Douglas-Mankin, K.R.; Srinivasan, R.; Arnold, J.G. Soil and Water Assessment Tool (SWAT) Model: Current developments and applications. *Trans. ASABE* **2010**, *53*, 1423–1431. [CrossRef]

61. Gassman, P.W.; Reyes, M.R.; Green, C.H.; Arnold, J.G. The soil and water assessment tool: Historical development, applications, and future research directions. *Trans. ASABE* **2007**, *50*, 1211–1250. [CrossRef]

62. Neitsch, S.L.; Arnold, J.G.; Kiniry, J.R.; Williams, J.R. *Soil and Water Assessment Tool Theoretical Documentation Version 2009*; Texas Water Resources Institute: College Station, TX, USA, 2011.

63. Penman, H.L. Estimating evaporation. *EOS Trans. Am. Geophys. Union* **1956**, *37*, 43–50. [CrossRef]

64. Monteith, J.L. Evaporation and environment. *Symp. Soc. Exp. Biol.* **1965**, *19*, 4.

65. Stefan, H.G.; Preud'homme, E.B. Stream Temperature estimation from air temperature. *JAWRA J. Am. Water Resour. Assoc.* **1993**, *29*, 27–45. [CrossRef]

66. RIGIS. *Rhode Island Geographic Information System*; University of Rhode Island: Kingston, RI, USA, 2016.

67. Texas A&M University. *Arcswat Software*; Texas A&M University: College Station, TX, USA, 2012. Available online: http://swat.tamu.edu/ (accessed on 19 Octorber 2012).

68. Homer, C.G.; Dewitz, J.A.; Yang, L.; Jin, S.; Danielson, P.; Xian, G.; Coulston, J.; Herold, N.D.; Wickham, J.; Megown, K. Completion of the 2011 National Land Cover Database for the conterminous United States-Representing a decade of land cover change information. *Photogramm. Eng. Remote Sens.* **2015**, *81*, 345–354.

69. Rhode Island Geographic Information System (RIGIS) Data Distribution System SOIL_Soils. Available online: http://www.rigis.org/geodata/soil/Soils16.zip (accessed on 5 September 2016).

70. Saha, S.; Moorthi, S.; Wu, X.; Wang, J.; Nadiga, S.; Tripp, P.; Behringer, D.; Hou, Y.-T.; Chuang, H.-Y.; Iredell, M.; et al. The NCEP climate forecast system version 2. *J. Clim.* **2014**, *27*, 2185–2208. [CrossRef]

71. Texas A&M, U. NCEP Global Weather Data for Swat. Available online: https://globalweather.tamu.edu (accessed on 3 January 2016).

72. Abbaspour, K.C. *SWAT-CUP 2012*; SWAT Calibration and Uncertainty Program—A User Manual; Swiss Federal Institute of Aquatic Science and Technology Eawag: Duebendorf, Switzerland, 2013.

73. Abbaspour, K. *User Manual for SWAT-CUP, SWAT Calibration and Uncertainty Analysis Programs*; Swiss Federal Institute of Aquatic Science and Technology Eawag: Duebendorf, Switzerland, 2007.

74. Nash, J.E.; Sutcliffe, J.V. River flow forecasting through conceptual models part I—A discussion of principles. *J. Hydrol.* **1970**, *10*, 282–290. [CrossRef]

75. Moriasi, D.N.; Arnold, J.G.; Liew, M.W.V.; Bingner, R.L.; Harmel, R.D.; Veith, T.L. Model evaluation guidelines for systematic quantification of accuracy in watershed simulations. *Trans. ASABE* **2007**, *50*, 885–900. [CrossRef]

76. Pradhanang, S.M.; Mukundan, R.; Schneiderman, E.M.; Zion, M.S.; Anandhi, A.; Pierson, D.C.; Frei, A.; Easton, Z.M.; Fuka, D.; Steenhuis, T.S. Streamflow responses to climate change: Analysis of hydrologic indicators in a New York City water supply watershed. *J. Am. Water Resour. Assoc.* **2013**, *49*, 1308–1326. [CrossRef]

77. Liew, M.W.V.; Veith, T.L.; Bosch, D.D.; Arnold, J.G. Suitability of SWAT for the conservation effects assessment project: Comparison on USDA agricultural research service watersheds. *J. Hydrol. Eng.* **2007**, *12*, 173–189. [CrossRef]

78. Arnold, J.G.; Allen, P.M.; Muttiah, R.; Bernhardt, G. Automated base flow separation and recession analysis techniques. *Ground Water* **1995**, *33*, 1010–1018. [CrossRef]

79. Singh, J.; Knapp, H.V.; Arnold, J.G.; Demissie, M. Hydrological modeling of the Iroquois river watershed using HSPF and SWAT. *JAWRA J. Am. Water Resour. Assoc.* **2005**, *41*, 343–360. [CrossRef]

80. Pyrce, R. *Hydrological Low Flow Indices and Their Uses*; Watershed Science Centre (WSC) Report; Watershed Science Center, Trent University: Peterborough, ON, Canada, 2004.

81. Smakhtin, V.U. Low flow hydrology: A review. *J. Hydrol.* **2001**, *240*, 147–186. [CrossRef]
82. Ahearn, E.A. *Flow Durations, Low-Flow Frequencies, and Monthly Median Flows for Selected Streams in Connecticut through 2005*; US Department of the Interior, US Geological Survey: Reston, VA, USA, 2008.
83. Demaria, E.M.C.; Palmer, R.N.; Roundy, J.K. Regional climate change projections of streamflow characteristics in the Northeast and Midwest U.S. *J. Hydrol. Reg. Stud.* **2016**, *5*, 309–323. [CrossRef]
84. Douglas, E.M.; Vogel, R.M.; Kroll, C.N. Trends in floods and low flows in the United States: Impact of spatial correlation. *J. Hydrol.* **2000**, *240*, 90–105. [CrossRef]
85. Ficklin, D.L. *SWAT Stream Temperature Executable Code*; Indiana State University: Terre Haute, IN, USA, 2012.
86. Meisner, J.D.; Rosenfeld, J.S.; Regier, H.A. The role of groundwater in the impact of climate warming on stream salmonines. *Fisheries* **1988**, *13*, 2–8. [CrossRef]
87. Moore, R.D.; Spittlehouse, D.L.; Story, A. Riparian microclimate and stream temperature response to forest harvesting: A Review. *J. Am. Water Resour. Assoc.* **2005**, *41*, 813–834. [CrossRef]
88. Rishel, G.B.; Lynch, J.A.; Corbett, E.S. Seasonal stream temperature changes following forest harvesting. *J. Environ. Qual.* **1982**, *11*, 112–116. [CrossRef]
89. Ficklin, D.L.; Luo, Y.; Stewart, I.T.; Maurer, E.P. Development and application of a hydroclimatological stream temperature model within the Soil and Water Assessment Tool. *Water Resour. Res.* **2012**, *48*. [CrossRef]
90. Brennan, L. Stream Temperature Modeling: A modeling comparison for resource managers and climate change analysis. In *Environmental and Water Resources Engineering*; University of Massachusetts: Amherst, MA, USA, 2015.

water

MDPI

Article

Water Leakage and Nitrate Leaching Characteristics in the Winter Wheat–Summer Maize Rotation System in the North China Plain under Different Irrigation and Fertilization Management Practices

Shufeng Chen [1,2], Chengchun Sun [1], Wenliang Wu [2,*] and Changhong Sun [1]

[1] Beijing Municipal Research Institute of Environmental Protection, Beijing 100037, China; shufengchen@163.com (S.C.); shandong.baozi@163.com (Che.S.); wuwl@cau.edu.cn (Cha.S.)
[2] College of Resources and Environmental Sciences, China Agricultural University, Beijing 100193, China
*   Correspondence: wuwenl@cau.edu.cn; Tel.: +86-10-62732387

Academic Editor: Karim Abbaspour
Received: 16 October 2016; Accepted: 14 February 2017; Published: 22 February 2017

Abstract: Field experiments were carried out in Huantai County from 2006 to 2008 to evaluate the effects of different nitrogen (N) fertilization and irrigation management practices on water leakage and nitrate leaching in the dominant wheat–maize rotation system in the North China Plain (NCP). Two N fertilization (NF_1, the traditional one; NF_2, fertilization based on soil testing) and two irrigation (IR_1, the traditional one; IR_2, irrigation based on real-time soil water content monitoring) management practices were designed in the experiments. Water and nitrate amounts leaving the soil layer at a depth of 2.0 m below the soil surface were calculated and compared. Results showed that the IR_2 effectively reduced water leakage and nitrate leaching amounts in the two-year period, especially in the winter wheat season. Less than 10 percent irrigation water could be saved in a dry winter wheat season, but about 60 percent could be saved in a wet winter wheat season. Besides, 58.8 percent nitrate under single NF_2IR_1 and 85.2 percent under NF_2IR_2 could be prevented from leaching. The IR_2 should be considered as the best management practice to save groundwater resources and prevent nitrate from leaching. The amounts of N input play a great role in affecting nitrate concentrations in the soil solutions in the winter wheat–summer maize rotation system. The NF_2 significantly reduced N inputs and should be encouraged in ordinary agricultural production. Thus, nitrate leaching and groundwater contamination could be alleviated, but timely N supplement might be needed under high precipitation condition.

Keywords: water leakage; nitrate leaching; maize; winter wheat; optimized nitrogen fertilization; optimized irrigation

1. Introduction

The North China Plain (NCP) is a major grain-producing region in China with a long-term average annual precipitation of 550 mm, most of which occurs from June to September, and an average yearly crop evapotranspiration of 850 mm. To obtain higher grain yields, a large amount of fertilizer and water were supplied by farmers in this region. It was reported that the average annual amount of nitrate input was above 500 kg $N \cdot ha^{-1}$ in the NCP [1]. In some high-yielding farmlands, the input even reached as high as 600 kg $N \cdot ha^{-1}$ [2]. Farmers' traditional N fertilization practice usually causes high nitrate losses because of excessive N input. Research on field-scale N balances found that about 50% of N applied under traditional fertilization management practice was not accounted for by crop removal [3]. With excess surface water application, soluble nitrate can leach below the root zone to underlying groundwater, causing possible contamination of drinking water [4]. Excessive N input and inefficient application of irrigation water have caused heavy groundwater contamination in the NCP. Between the years 2002 and 2007, the nitrate concentration in shallow groundwater doubled [5]. Besides, a lack of surface water in this area led to the excessive extraction of groundwater for agricultural production, bringing about an annual 1.5 m drop in the groundwater level [6,7]. Water shortage and groundwater pollution have become important environmental concerns in the NCP. Improving water and N management in the farmland to reduce both groundwater exploitation and farmland nitrate leaching is of great importance to protect the groundwater resources.

Limited irrigation and reduced N application may be effective measures to reduce farmland nitrate leaching and to improve water and N use efficiencies. Reduction of the applied N fertilizer rate to an optimized rate can reduce soil nitrate leaching [8–10]. Nowadays, the optimized irrigation technique based on real-time monitoring of soil water content has been widely used in agricultural production and provides us an irrigation control measure [11]. It could attain the dual objectives of water-saving and high yield. Furthermore, the optimized N fertilization based on soil testing is a fertilization management practice that quantitatively supplies N fertilizer to ensure crop growth. It increases N use efficiencies of crops by a comprehensive consideration of available N in soil and crop N demand in different stages [11]. However, there have been few studies focused on the environmental effects of both the irrigation based on real-time monitoring of soil water content and the optimized N fertilization based on soil testing in the typical rotation system in the NCP so far.

The amounts of water leakage and nitrate leaching in the soil were the keys to analyze the effects of irrigation and fertilization management practices on nitrate leaching characteristics. Both water leakage and nitrate leaching were related to irrigation/precipitation, the rate and type of N fertilizer applied, the variety of crops, and other environmental factors. Major methods for direct measurement on water leaching amounts and nitrate leaching amounts include large lysimeter, leakage plate, tensiometers and so on [12–17]. Certainly, each method has its limitations and advantages. Currently, the method of using a tensiometer combined with Darcy's law [18] to determine the nitrate content in the soil solution has been extensively applied.

Huantai County, Shandong Province, China, is the first county reaching 1000 kg/ha (wheat and maize) per year in the north Yangtze River. The winter wheat–summer maize rotation system is the basic cropping system in this region. To acquire a high crop yield, farmers depend heavily on the use of groundwater for irrigation in addition to a high N fertilizer application. Inadequate fresh irrigation supply in this region demands careful use and less contamination of all the available water resources. In this paper, both the water leakage and nitrate leaching amounts under different irrigation and fertilization management practices were obtained and analyzed. Results of this study would be useful to improve water and N fertilization management in the NCP to achieve a sustainable development of agriculture.

2. Materials and Methods

2.1. Site Description

Our experimental fields are located in Maojia Village of Xingcheng town, Huantai County (Figure 1). Experiment fields with relatively uniform soil types and hydrogeological conditions (groundwater level 15–20 m below soil surface) were chosen to ensure similar background. The experiments were carried out from to June 2006 to July 2007. Before this, a 1-year pre-trial from June 2005 to May 2006, the same as the formal experiment, had been conducted. During the experimental period, rainfall from June 2005 to May 2008 in the county was also collected (Figure 2).

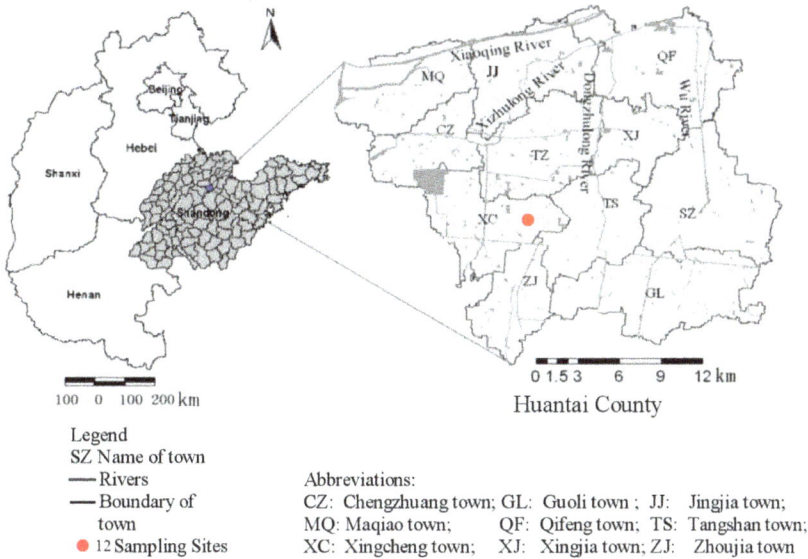

Figure 1. Location of the study area and the sampling sites.

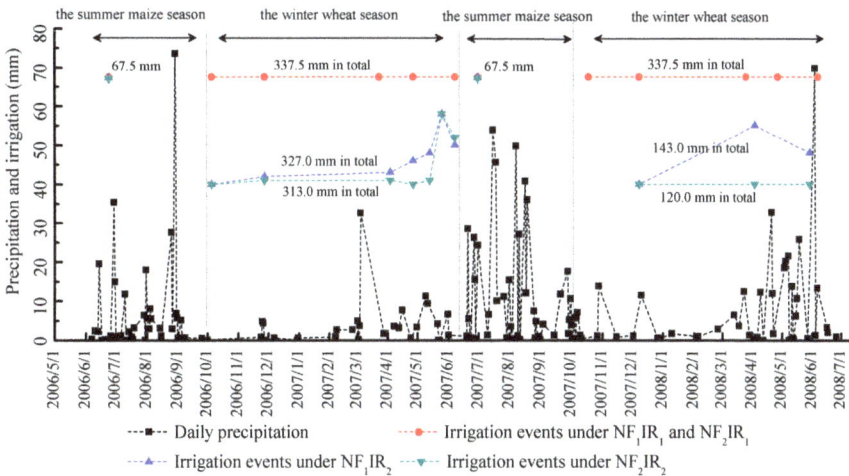

Figure 2. Daily rainfall and irrigation events from June 2005 to May 2008 in Huantai County.

2.2. Experiment Design

There were two irrigation management practices to meet the water requirements of the crops: the local traditional irrigation (IR_1) and the optimized irrigation (IR_2). Two fertilization management practices were also designed to meet the N requirements: the local traditional N fertilization (NF_1) and the optimized N fertilization (NF_2). A combination of the two factors (NF_1IR_1, NF_2IR_1, NF_1IR_2, NF_2IR_2) were applied in 12 plots with three replicates.

The traditional irrigation management practice is a local famer's practice in the high-yielding winter wheat and summer maize system of the NCP. In this situation, irrigation events are usually arranged at the planting stage, before the over-wintering stage, at the regreening stage, at the shooting stage, and before the harvesting stage for winter wheat. Under the optimized irrigation management practice, sprinkler irrigation was used to keep plant-available soil water content (the difference in water content between measured field capacity and permanent wilting point in laboratory) between 40% and 85% according to the growth stages of the crop within defined soil depths [19]. Notably, because precipitation was relatively high and temporally even during the summer maize season, no extra irrigation is needed in our experiments. However, an irrigation event with 67.5 mm at the heading stage of summer maize under the two irrigation management practices was applied. The detailed irrigation dates and amounts during the winter wheat season are shown in Figure 2.

The traditional N fertilization management practice also represents a local farmer's practices in cropping. The traditional N fertilization management received 150 kg N·ha^{-1} (urea) at planting (incorporation after broadcasting) and another 150 kg N·ha^{-1} (urea) at the regreening stage (broadcast followed by irrigation) for winter wheat. As for summer maize, it received 100 kg N·ha^{-1} (urea) as first topdressing fertilizer (broadcast followed by irrigation or before rain) at the three-leaf stage, shooting stage, and heading stage, respectively. The rate and time of N fertilization in the optimized N fertilization management practice was based on an improved N_{min} method, which considered the synchronization of crop nitrate demand and soil nitrate supply [11]. It is required to ensure the growth of crops in different growth stages. The rates of phosphorus and potassium fertilizer were the same in all the management practices, applied as basal fertilizer once before winter wheat is sown. The rate of phosphorus was 375 kg P_2O_5·ha^{-1}, and potassium was 225 kg K_2O·ha^{-1}. Detailed fertilization dates and rates are shown in Table 1.

Table 1. N fertilization dates and rates under different management practices.

Items		Crop Stage	Fertilizer Application Rate/kg N·ha^{-1}			
Season	Date		NF_1IR_1	NF_2IR_1	NF_1IR_2	NF_2IR_2
2006, Summer maize	2-June	three-leaf stage	100		100	38
	28-June	shooting stage	100	35	100	
	23-July	heading stage	100		100	38
	Subtotal		300	35	300	38
2006–2007, Winter wheat	5-October	planting stage	150		150	
	20-March	regreening stage	150	40	150	35
	16-April			22		18
	Subtotal		300	62	300	53
2007, Summer maize	26-May	three-leaf stage	100		100	
	9-June			63		57
	26-June	shooting stage	100		100	
	28-July	heading stage	100		100	
	Subtotal		300	63	300	57
2007–2008, Winter wheat	16-October	planting stage	150		150	
	20-March	regreening stage	150	44	150	36
	16-April			30		31
	Subtotal		300	74	300	67
Total			1200	234	1200	215

2.3. Observation Items and Methods

Water leakage and nitrate leaching amounts are main data needed to realize the analysis. Water and nitrate monitoring experiments combined with Darcy's Law [20] were combined to estimate the leaching amounts. Though requiring many parameters and less accuracy, this method is cost-saving and includes both upward and downward movement of water. It has obvious advantages in multi-treatment field experiments.

The tensiometer was used to obtain the soil water potentials (mm, sum of soil matrix and gravimetric potentials) in these experiments. Considering the root system of the crops, mainly distributed within a 2.0 m depth soil layer, the tensiometer installation depths in the soil profile were accordingly set at 1.8 m and 2.0 m depth in each field. The soil volumetric water contents of soil layers (0–2.0 m) were measured by a neutron probe (calibrated for about ten times at the beginning of the experiments) at 20 cm intervals. Both the soil volumetric water contents and tensiometer readings were recorded once a week. The soil water storage (0 to 2.0 m depth) of the experimental sites was calculated based on soil volumetric water content and layers of soil thickness; the thickness of each layer of soil in our study was 0.20 m. Field water capacity was calculated from the average profile water content for two 3-day periods just after irrigation or significant rainfall in summer. In this paper, it is 54.3 mm.

Suction cups (i.e., soil solution samplers) were installed at 2.0 m depth below the soil surface to obtain soil solutions. The soil solutions were collected biweekly. A portable vacuum pump was applied up to 80 kPa suction for 24 h to collect all solutions in the suction cups. When the soil was too dry to collect samples, soil samples were collected instead and extracted with 0.01 M $CaCl_2$ solution. Nitrate concentrations of all the soil solutions were determined by a continuous flow analytical system (TRAACS 2000 system, Bran and Luebbe, Norderstedt, Germany). Water from both precipitation and irrigation events was also sampled to determine nitrate concentrations, and then multiplied by water amounts to obtain the nitrate input to the soil.

Meteorological data such as precipitation were obtained from an automatic weather station of the local water conservancy bureau. In winter, observations of both water leakage and nitrate leaching were suspended because of freezing.

Water leakage amount at the 2.0 m depth of the soil profile was calculated by Darcy's Law combined with the Van Genuchten model [20–22]. The Van Genuchten model was used to obtain the parameters in the soil water characteristic curve. The obtained parameters were needed to determine $K(h)$. Then, $K(h)$ was used in Darcy's Law, which was used to calculate the water transport. The formulas are listed as follows:

$$q_{200}(t) = K(h) \times (H_{180}(t) - H_{200}(t))/\Delta H, \tag{1}$$

$$\begin{cases} \theta(h) = \theta_r + \dfrac{\theta_s - \theta_r}{[1+|ah|^n]^m} & , \text{ when } h < 0 \\ \theta(h) = \theta_s & , \text{ when } h = 0 \end{cases} \tag{2}$$

$$K(h) = K_s \left(\dfrac{\theta - \theta_r}{\theta_s - \theta}\right)^l \left[1 - \left(1 - \left(\dfrac{\theta - \theta_r}{\theta_s - \theta}\right)^{1/m}\right)^m\right]^2 \tag{3}$$

where $q_{200}(t)$ is the water leakage amount at 2.0 m depth of the soil profile during t (mm·day^{-1}); $K(h)$ is the hydraulic conductivity (mm·day^{-1}); h is the matrix potential (mm) measured once a week; $H_{180}(t)$ and $H_{200}(t)$ are soil water potentials (mm) at 1.8 m and 2.0 m depth of the soil profile during t (measured once a week in this paper), respectively; t is the measurement period (7 days here); ΔH is the distance between the 1.8 m and 2.0 m depths (mm); θ_s, θ_r, and $\theta(h)$ are saturated, residual, and measured soil volumetric water content, respectively (%); θ_r, θ_s, α, m, n, l are parameters of the soil water characteristic curve (analyzed by the Van Genuchten model). The soil layer at 1.6 m to 2.0 m depth was chosen, and finally, these parameters were found to be 0.11, 0.39, 0.059, 0.324, 1.48, and 0.5, respectively; K_S is the saturated hydraulic conductivity (24.6 mm·day^{-1}, measured in laboratory by the constant head method [23].

$$Q(T) = \int_0^T q_{200}(t)dt \tag{4}$$

where $Q(T)$ is the water leakage amount at 2.0 m depth of the soil during a period T (mm); T is the measurement cycle (days, growth season of winter wheat or summer maize in this paper).

Nitrate leaching amounts were calculated as the product of the nitrate concentrations and the corresponding $Q(T)$ through the 2.0 m depth soil profile as follows:

$$N(T) = \int_0^T C(t)\, q_{200}(t)/100\, dt \tag{5}$$

Duncan's multiple range test was used to determine significant differences in means of water leakages and nitrate leaching amounts among the four management practices [24]. It is a method of multiple comparisons in which the group means are ranked from smallest to largest (a, b, c, etc.).

3. Results

3.1. Water Leakage at 2.0 m Depth

In the summer maize growth season of 2006, no severe water leakage occurred at 2.0 m depth of the soil under the traditional and the optimized fertilization management practice (Figure 3a). The precipitation during the whole summer maize growth season of 2006 was temporally evenly distributed and about 200 mm less than the average annual precipitation (1962–2007) (Figure 4). Under all the management practices, the soil water storage in 2.0 m soil depth soil was, most of the time, lower than the field capacity (Figure 3b). The precipitation could only meet the requirement of summer maize growth. Only minor water leakage occurred in the period from August to September in 2006 (Figure 3a).

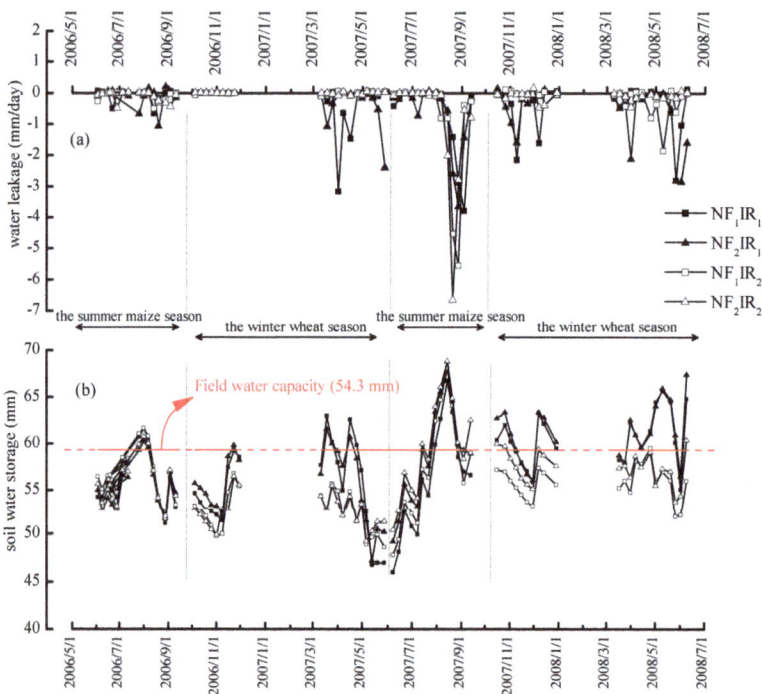

Figure 3. Water leakage and soil water storage (**a**) water leakage at a depth of 2.0 m under different fertilizer and irrigation management practices; (**b**) soil water storage in the 0–2.0 m soil profile.

During the winter wheat growth season of 2006–2007, due to the drought in the past summer maize season, no water leakage occurred in winter (Figure 3a). In the spring of 2007, no water leakage occurred under the IR_2 management practices, but water leakages were observed under the IR_1 management practices (Figure 3a). Furthermore, the amounts were 39.5 mm under NF_1 and

52.1 mm under NF_2 management practice, respectively (Table 2). The corresponding leakage rates (leakage/(irrigation + precipitation), the same below) were 8.8% and 11.6%, respectively, which showed significant difference. The data indicated that even under the drought condition in the previous summer maize growth season, the traditional irrigation could still lead to farmland water leakage in the next winter wheat growth season. The traditional irrigation events in the spring of 2007 and the relatively higher precipitation in March (Figure 4) resulted in rapid increased soil water storage (five irrigation events in Figure 2), which exceeded the field capacity, eventually leading to water leakage. However, under the IR_2 management practice, the optimized irrigation amount prevented soil water storage from exceeding the field capacity, which could greatly reduce the risk of water leakage (Figure 3b). Although there was little difference in total irrigation amount between IR_1 and IR_2 (Table 2), small quantities in the high frequency (seven irrigation events in Figure 2) of IR_2 can maintain the soil water content in a reasonable range, which might have reduced the possibility of water leakage into deeper soil layers. In addition, the more irrigation occurences, the more water evaporation loss might have happened during the irrigation process.

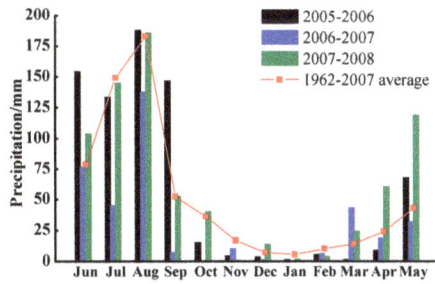

Figure 4. Monthly precipitation from June 2005 to May 2008 in Huantai County.

Table 2. Soil water balance under different irrigation and fertilizer management practices.

Management Practices		NF_1IR_1	NF_2IR_1	NF_1IR_2	NF_2IR_2
2006, Summer maize	Precipitation (mm)	266.1	266.1	266.1	266.1
	Irrigation (mm)	67.5	67.5	67.5	67.5
	Water leakage amount (mm)	8.1d	10.1bc	9.2c	11.8a
	Water leakage rate */%	2.4	3.0	2.8	3.6
2006–2007, Winter wheat	Precipitation (mm)	111.3	111.3	111.3	111.3
	Irrigation (mm)	337.5	337.5	327.0	313.0
	Water leakage amount (mm)	39.5b	52.1a	0	0
	Water leakage rate/%	8.8	11.6	0	0
2007, Summer maize	Precipitation (mm)	486.5	486.5	486.5	486.5
	Irrigation (mm)	67.5	67.5	67.5	67.5
	Water leakage (mm)	52.0c	88.0a	57.9c	72.0b
	Water leakage rate/%	9.4	15.9	10.5	13.0
2007–2008, Winter wheat	Precipitation (mm)	322.6	322.6	322.6	322.6
	Irrigation (mm)	337.5	337.5	143.0	120.0
	Water leakage (mm)	106.0b	134.1a	37.1d	61.1c
	Water leakage rate/%	16.1	20.3	8.0	13.8
Total water leakage amount (mm)		205.6b	284.0a	104.2d	145.2c
Total water leakage rate/%		10.3	14.2	5.8	8.3

Notes: data in the same line with a same letter means no significant difference in means according to Duncan's multiple range test ($p < 0.05$); * presents the percent of water leakage that accounts for the sum of precipitation and irrigation.

In the summer maize growth season of 2007, precipitation of 434.1 mm from June to August had led to the continuously increased soil water storage (Figure 3b). Also, heavy water leakages were observed. Total water leakages were 52.0 mm, 88.0 mm, 57.9 mm and 72.0 mm under NF_1IR_1, NF_2IR_1, NF_1IR_2 and NF_2IR_2, respectively, and the corresponding water leakage rates were 9.4%, 15.9%, 10.5% and 13.0% (Table 2). Under the same irrigation management practice, water leakages under NF_2 management practice were obviously higher than those under NF_1 management practice. Nutrient leaching due to frequent rainfalls and untimely topdressing under NF_2 management practice might have a negative effect on crop growth, thus more water leakage occurred. During the winter wheat growth season of 2007–2008, excessive precipitation (Figures 2 and 4) caused severe water leakage (Figure 3a). The water leakages were 106.0, 134.1, 37.1 and 61.1 mm under NF_1IR_1, NF_2IR_1, NF_1IR_2 and NF_2IR_2, respectively. Furthermore, the corresponding leakage rates were 9.4%, 15.9%, 10.5% and 13.0%, respectively (Table 2).

Under the same fertilization management practice, the water leakages under IR_2 were significantly less than those under IR_1 (Table 2), illustrating that the optimized irrigation could sharply reduce water leakage. Under the same irrigation management practice, water leakage under NF_1 was less than that under NF_2 (Table 2). During this season, the precipitation was 107.0 mm higher than the average annual value (Figure 4). Thus, water was adequate for wheat growth, and nutrient supply had become one of the key factors for the wheat growth. Furthermore, the NF_1 provided more adequate N fertilizer for wheat growth, and then increased the water consumption and reduced the risk of water leakage.

Besides, under IR_1, water leakage mainly occurred before the wintering, the regreening, and the harvesting stages, especially at the harvesting stage (Figure 3a). Under the rich precipitation conditions, the soil water storage often exceeded the field water capacity (Figure 3b), directly resulting in the water leakages. Under IR_2, the irrigation amounts were 143.0 and 120.0 mm under NF_1IR_2 and NF_2IR_2, respectively. The soil water storage rarely exceeded the field water capacity, which greatly reduced the risk of water leakage occurring (Figure 3b).

During four crop growth seasons over two years, the total water leakages were 205.6, 284.0, 104.2, and 145.2 mm and the total water leakage rates were 10.3%, 14.2%, 5.8% and 8.3% under NF_1IR_1, NF_2IR_1, NF_1IR_2, NF_2IR_2 management practices, respectively (Table 2). There was a significant difference between any two management practices.

3.2. Nitrate Concentrations of the Soil Solutions at a Depth of 2.0 m

Table 1 showed that NF_2 significantly reduced N inputs. Due to the 1-year pre-trial and less N inputs of the NF_2 management practice, the nitrate concentrations of the soil solutions at a depth of 2.0 m in the four crop growth seasons were obviously lower than those under NF_1 (Figure 5). It should be noticed that the nitrate concentrations under both NF_2IR_1 and NF_2IR_2 management practices had shown a decreasing trend from 2006 to 2008, while the two under NF_1IR_1 and NF_1IR_2 tended to increase. Under the NF_2, the groundwater may face less nitrate contamination risk. Under the same NF_1 or NF_2 management practices, the nitrate concentrations of IR_2 were higher than those under IR_1 management, suggesting that higher water input under IR_1 might have diluted nitrate concentrations.

During the whole winter wheat and summer maize season from 2006 to 2007, less precipitation (Figure 2) and only minor water leakage (Figure 3a) indicated that both N fertilization and the irrigation management practices had no obvious effect on nitrate concentration changes. Nitrate concentrations contained a stable level in the four management practices, respectively.

In July and August of 2007 in the summer maize season, heavy water leakage occurred (Figure 3a) and nitrate concentrations of NF_1IR_1 and NF_1IR_2 increased sharply at a maximum of 114 and 121 mg $N\cdot L^{-1}$, respectively (Figure 5). The 434.1 mm precipitation from June to August in 2007 (Figure 2) led to soil water storage that was higher than the field water capacity (Figure 3b). The nitrate in the soil moved from the upper soil layer down to the 2.0 m profile along with the water movement. However, under the NF_2, nitrate concentrations of both NF_2IR_1 and NF_2IR_2 were at a lower level, ranging from 27–37 mg $N\cdot L^{-1}$ to 31–42 mg $N\cdot L^{-1}$, respectively (Figure 5).

During the winter wheat season from 2007 to 2008, nitrate concentrations of NF_2IR_2 hovered between 18 and 28 mg·L^{-1} and those of NF_2IR_1 between 21 and 33 mg·L^{-1} (Figure 5). Nitrate concentrations under NF_2 were only about one-third of those under NF_1. This showed that nitrate contents at the 2.0 m depth of the soil profile under the optimized nitrate fertilization were effectively controlled, though the values still exceeded the drinking water standard in China (20 mg N·L^{-1}).

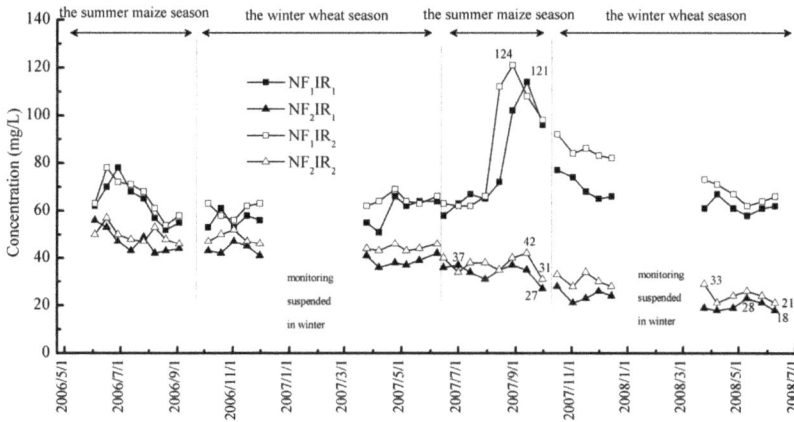

Figure 5. Nitrate concentrations of soil solutions at a depth of 2.0 m under different N and irrigation management practices.

3.3. Nitrate Leaching Amounts from the Soil Profile at a Depth of 2.0 m

From 2006 to 2007, during the maize season, minor nitrate leaching occurred in all the four management practices (Figure 6 and Table 3). During the following winter wheat growth season, minor nitrate leaching occurred under NF_1IR_1 and NF_2IR_1, and the nitrate leaching amounts were 23.0 and 24.0 kg N·ha^{-1}, respectively (Table 3). Almost no nitrate leaching was observed under NF_1IR_2 and NF_2IR_2. Because water movement is the driving force of nitrate leaching, the lower amount of water leakage (Figure 3a) under IR_2 effectively minimized nitrate leaching in this season.

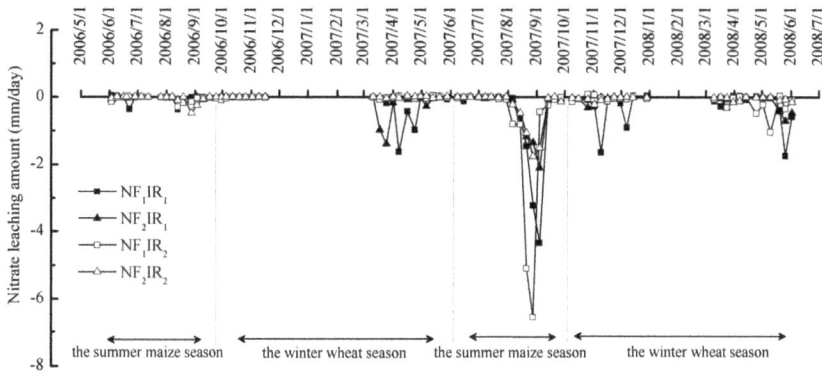

Figure 6. Nitrate leaching amounts under different N and irrigation management practices during the experiment. Negative values are represented by moving downward only.

Table 3. Summary of nitrate inputs from rainfall, irrigation water and N fertilizer applied, and nitrate leaching in both wheat–maize cropping years.

	Items		Management Practices	NF_1IR_1	NF_2IR_1	NF_1IR_2	NF_2IR_2
2006, Summer maize season	Nitrate input/kg N·ha^{-1}		from fertilization	300.0	35.0	300.0	38.0
			from precipitation	5.2	5.2	5.2	5.2
			from irrigation	4.8	4.8	4.8	4.8
			Total	310.0	45.0	310.0	48.0
		Leaching amount/kg N·ha^{-1}		4.6a	5.3a	5.4a	5.6a
		Nitrate leaching rate/%		1.5	11.8	1.7	11.6
2006–2007, winter wheat season	Nitrate input/kg N·ha^{-1}		from fertilization	300.0	62.0	300.0	53.0
			from precipitation	2.5	2.5	2.5	2.5
			from irrigation	21.1	21.1	20.1	14.5
			Total	323.6	85.6	322.6	70.0
		Leaching amount/kg N·ha^{-1}		23.3a	24.0a	-	-
		Nitrate leaching rate/%		7.1	28.0	-	-
2007, summer maize season	Nitrate input/kg N·ha^{-1}		from fertilization	300	63	300	57
			from precipitation	8.6	8.6	8.6	8.6
			from irrigation	4.2	4.2	4.2	4.2
			Total	312.8	75.8	312.8	69.8
		Leaching amount/kg N·ha^{-1}		71.0b	41.5c	98.4a	38.6c
		Nitrate leaching rate/%		22.7	54.8	31.5	55.2
2007–2008, winter wheat season	Nitrate input/kg N·ha^{-1}		from fertilization	300.0	74.0	300.0	67.0
			from precipitation	4.4	4.4	4.4	4.4
			from irrigation	21.2	21.2	8.6	7.5
			Total	325.6	99.6	313.0	78.9
		Leaching amount/kg N·ha^{-1}		74.2a	30.6b	27.5b	11.0c
		Nitrate leaching rate/%		22.8	30.7	8.8	13.9
Total leaching amount/kg N·ha^{-1}				172.8a	101.4c	131.2b	55.1d
Total Nitrate leaching rate/%				13.6	33.1	10.4	20.7

Note: data in the same line with the same letter means no significant difference in means according to Duncan's multiple range test ($p < 0.05$).

During the summer maize season in 2007, serious nitrate leaching occurred under all the four management practices (Figure 6). The total leaching amounts of NF_1IR_1, NF_2IR_1, NF_1IR_2, and NF_2IR_2 were 71.0, 41.5, 98.4 and 38.6 kg N·ha^{-1}, respectively (Table 3). The leaching amounts were far greater than those in 2006. Serious water leakage for excessive rainfall that occurred in this season may account for this case (Figures 2 and 3a). Under the same NF_1 management practice, the nitrate leaching amount under IR_2 was significantly higher than that under IR_1. Comparably, under the same NF_2 management, the nitrate leaching amount under IR_2 and IR_1 was quite similar. We concluded that under NF_1IR_2, higher soil residual nitrate that existed in the former winter season led to higher nitrate concentrations in the soil solutions (Figure 5). Therefore, the nitrate leaching amount under this management practice was the maximum in condition of high rainfalls (Figure 2). Interestingly, nitrate leaching amounts under NF_2IR_1 and NF_2IR_2 were approximately half of those under NF_1IR_1 and NF_1IR_2. Though more water leakages were observed under NF_2IR_1 and NF_2IR_2 (Table 2), lower nitrate concentrations in the soil solutions (Figure 5) prevented higher nitrate from leaching. Besides, from the perspective of nitrate leaching rates (leaching/(fertilization + irrigation + precipitation), the same below), about 55% of nitrate input leached from the soil under the two NF_2 management practices (Table 3). Consequently, timely N topdressing should be given more attention to increase crop growth under the NF_2 management practice.

During the winter wheat season from 2007 to 2008, a slightly higher N input than the last summer maize season (Table 1) was adopted under NF_2 for the heavy water leakage (Figure 3a). Figure 6 showed that relatively sufficient precipitation in winter led to nitrate leaching under the four combinations. The leaching amounts of NF_1IR_1 and NF_2IR_1 were 74.2 and 30.6 kg N·ha^{-1}, respectively (Table 3). By contrast, those of NF_1IR_2 and NF_2IR_2 management practices were 27.5 and 11.0 kg N·ha^{-1}, respectively (Table 3). The IR_2 helped to reduce both water leakage and nitrate leaching. Besides, under the same irrigation management practice, the leaching amount under NF_2IR_1 was about half of that under NF_1IR_1, and the leaching amount under NF_2IR_2 was about one-third of that under NF_1IR_2.

The optimized N fertilization based on the soil testing measure could also prevent farmland nitrate from leaching effectively. Under the two effects of both N management and irrigation management, as much as 85.2% (= 1 − 10.97/74.21) of the nitrate leaching amount was prevented from contaminating the groundwater.

4. Discussion

After the two-year experiment, water leakages under the four management practices were compared. The obtained water leakage rate under the traditional irrigation was in line with the reported results by Zhang et al. [25]. Judging from effects of preventing water leakage, the combination of NF_1IR_2 was the best management practice. The magnitude effects followed the order of $NF_1IR_2 > NF_2IR_2 > NF_1IR_1 > NF_2IR_1$ (Table 2). It is the IR_2 management practice that prevented water leakage from leaching effectively. Inversely, the NF_2 management practice tended to aggravate water leakage, especially under high precipitation condition. Nutrient leaching due to frequent rainfalls and untimely topdressing under NF_2 management practice (about 55% of nitrate input had leached on the basis of much less N input) might have resulted in poorer crop growth, and more water leakage. Guidance for timely nitrogen topdressing should be considered.

From the effects of nitrogen and irrigation management practices on nitrate concentrations at 2.0 m depth of the soil profile, it is easy to find that concentration gaps caused by nitrogen input are higher than those caused by irrigation (Figure 5). The amounts of nitrogen input play a great role in decreasing nitrate concentrations in the soil solutions. Reduction in the traditional nitrogen fertilization should be considered in this area. Increasing evidence has reported that nitrate leaching occurred in low-rainfall regions, episodically during extraordinarily wet periods [26,27]. In our study, the total nitrate leaching amounts under NF_1IR_1 and NF_1IR_2 were 172.8 and 131.2 kg N·ha^{-1}, respectively (Table 3). These results were supported by other studies in this region [25,28]. The continuous two-year observations in this experiment also showed us the elevated nitrate concentrations in the relatively wet summer maize and winter wheat season, especially under the traditional nitrogen fertilization management practice. Thus, soil nitrate movement to a deeper soil layer was an important pathway that caused nitrate losses in this winter wheat–summer maize rotation system.

Though heavier water leakage might be caused by nitrogen input based on soil testing, NF_2 management practice is still an alternative for farmers here to effectively decrease nitrate leaching amounts and prevent groundwater contamination. Lower nitrate concentrations in the soil solution at 2.0 m depth prevented heavier nitrate from leaching (Table 3), especially when it is combined with the IR_2 management practice. Under NF_2IR_2, the least 55.0 kg N·ha^{-1} over the two years was observed. However, the effects of NF_2 on crop growth should be given more attention. The IR_2 had significant positive effects on both water and nitrate loss in this area.

It is reported that excessive irrigation would lead to decreased crop water use efficiency [29]. Management practices that adjust water application to crop needs also reduced nitrate leaching by a mean of 80% without a reduction in crop yield [30]. In this study, less than 10 percent irrigation water could be saved in a dry winter wheat season (from 2006 to 2007), but about 60 percent could be saved in a wet winter wheat season (from 2007 to 2008) under the IR_2 management practice.

Single optimized fertilization might not work in preventing nitrate leaching in the dry winter wheat season (Table 3), but in the wet winter wheat season, a reduction of 58.8 percent nitrate leaching was acquired in the wet winter wheat season. A combination of NF_2 and IR_2 management practice could reduce 85.2 percent of nitrate leaching, compared with NF_1IR_1 (Table 3). Thus, optimized management practices may reduce nitrate leaching risk and could enhance environmental sustainability.

The best management practices to minimize nitrate contamination of groundwater in this area include the use of an efficient irrigation method and an optimized fertilization method. The use of an optimized irrigation method combined with soil-testing nitrogen fertilization reduced nitrate input and prevented nitrate leaching over the 2-year period. However, the economic and environmental costs due to applying too much groundwater might not induce farmers to embrace the new technological changes without a detailed cost-effective analysis of equipment and endeavor.

5. Conclusions

Water shortage and groundwater pollution due to excessive water use and nitrogen inputs are important environmental concerns in the NCP. The development of better irrigation and nitrogen fertilization management practices that minimize groundwater pollution is crucial. Field results showed that irrigation water, nitrogen fertilizer use, and nitrate leaching could be decreased substantially by applying the optimized water management in comparison with traditional methods. The optimized nitrogen fertilization management could effectively reduce nitrate concentrations of the soil solutions at 2.0 m depth, and then prevent nitrate leaching. However, it should be carefully managed to achieve stable grain yields. Further study is needed to find a way to maintain grain output while adopting the optimized nitrogen fertilization management practice when precipitation is too high. The optimized irrigation management practice should be considered to save groundwater resources.

Acknowledgments: This research was supported by The Key Projects in the National Science & Technology Pillar Program during the Eleventh Five-Year Plan Period (Project No. 2006BAD17B05), Sino-US Science and Technology Cooperation Project (Project No. 2009DFA91790), and List Compilation of Priority Pollutants in Beijing Surface Water (Project No. 1541STC60334).

Author Contributions: Shufeng Chen and Wenliang Wu conceived and designed the experiments; Shufeng Chen performed the experiments; Chengchun Sun analyzed the data and wrote the paper; Wenliang Wu and Changhong Sun contributed reagents and materials.

Conflicts of Interest: The authors declare no conflict of interest.

References

1. Zhang, W.L.; Tian, Z.X.; Zhang, N.; Li, X.Q. Nitrate pollution of groundwater in northern China. *Agric. Ecosyst. Environ.* **1996**, *59*, 223–231. [CrossRef]
2. Liu, G.D.; Wu, W.L.; Zhang, J. Regional differentiation of non-point source pollution of agriculture-derived nitrate nitrogen in groundwater in northern China. *Agric. Ecosyst. Environ.* **2005**, *107*, 211–220. [CrossRef]
3. Karlen, D.L.; Kramer, L.A.; Logsdon, S.D. Field-scale nitrogen balances associated with long-term continuous corn production. *Agron. J.* **1998**, *90*, 644–650. [CrossRef]
4. Mahvi, A.H.; Nouri, J.; Babaei, A.A.; Nabizadeh, R. Agricultural activities impact on groundwater nitrate pollution. *Int. J. Environ. Sci. Technol.* **2013**, *2*, 41–47. [CrossRef]
5. Chen, S.; Wu, W.; Hu, K.; Li, W. The effects of land use change and irrigation water resource on nitrate contamination in shallow groundwater at county scale. *Ecol. Complex.* **2010**, *7*, 131–138. [CrossRef]
6. Hu, Y.; Moiwo, J.P.; Yang, Y.; Han, S.; Yang, Y. Agricultural water-saving and sustainable groundwater management in Shijiazhuang Irrigation District, North China Plain. *J. Hydrol.* **2010**, *393*, 219–232. [CrossRef]
7. Sun, H.; Shen, Y.; Yu, Q.; Flerchinger, G.N.; Zhang, Y.; Liu, C.; Zhang, X. Effect of precipitation change on water balance and WUE of the winter wheat summer maize rotation in the North China Plain. *Agric. Water Manag.* **2010**, *97*, 1139–1145. [CrossRef]
8. Power, J.F.; Wiese, R.; Flowerday, D. Managing nitrogen for water quality—Lessons from management systems evaluation area. *J. Environ. Qual.* **2000**, *29*, 5–66. [CrossRef]
9. Ottman, M.J.; Pope, N.V. Nitrogen fertilizer movement in the soil as influenced by nitrogen rate and timing in irrigated wheat. *Soil Sci. Soc. Am. J.* **2000**, *64*. [CrossRef]
10. Yan, L.; Zhang, Z.-D.; Zhang, J.-J.; Gao, Q.; Feng, G.-Z.; Abelrahman, A.M.; Chen, Y. Effects of improving nitrogen management on nitrogen utilization, nitrogen balance, and reactive nitrogen losses in a Mollisol with maize monoculture in Northeast China. *Environ. Sci. Pollut. Res.* **2016**, *23*, 4576–4584. [CrossRef] [PubMed]
11. Zhao, R.F.; Chen, X.P.; Zhang, F.S.; Zhang, H.L. Fertilization and nitrogen balance in a wheat maize rotation system in north China. *Agron. J.* **2006**, *98*, 938–945. [CrossRef]
12. Roth, G.W.; Fox, R.H. Soil nitrate accumulations following nitrogen-fertilized corn in Pennsylvania. *J. Environ. Qual.* **1990**, *19*, 243–248. [CrossRef]

13. Liang, B.C.; Remillard, M.; MacKenzie, A.F. Influence of fertilizer, irrigation, and non-growing season precipitation on soil nitrate-nitrogen under corn. *J. Environ. Qual.* **1991**, *20*, 123–128. [CrossRef]

14. Paramasivam, S.; Alva, A.K. Leaching of nitrogen forms from controlled release nitrogen fertilizers 1. *Commun. Soil Sci. Plant Anal.* **1997**, *28*, 1663–1674. [CrossRef]

15. Toth, J.D.; Fox, R.H. Nitrate losses from a corn-alfalfa rotation: Lysimeter measurement of nitrate leaching. *J. Environ. Qual.* **1998**, *27*, 1027–1033. [CrossRef]

16. Sogbedji, J.M.; van Es, H.M.; Yang, C.L.; Geohring, L.D.; Magdoff, F.R. Nitrate leaching and nitrogen budget as affected by maize nitrogen rate and soil type. *J. Environ. Qual.* **2000**, *29*, 1813–1820. [CrossRef]

17. Chen, X.P. Optimization of the N Fertilizer Management of a Winter Wheat/Summer Maize Rotation System in the Northern China Plain. Ph.D. Thesis, Universität Hohenheim, Stuttgart, Germany, 2003.

18. Owens, L.B. Nitrate leaching losses from monolith lysimeters as influenced by nitrapyrin. *J. Environ. Qual.* **1987**, *16*, 34–38. [CrossRef]

19. Chen, S.F.; Wu, W.L.; Hu, K.L.; Du, Z.D.; Chu, Z.H. Characteristics of nitrate leaching in high yield farmland under different irrigation and fertilization managements in North China Plain. *Trans. Chin. Soc. Agric. Eng.* **2011**, *27*, 65–73. (In Chinese)

20. Darcy, H. *Les Fontaines Publiques de la Ville de Dijon*; Dalmont: Paris, France, 1856; p. 647.

21. Van Genuchten, M.T. A closed-form equation for predicting the hydraulic conductivity of unsaturated soils1. *Soil Sci. Soc. Am. J.* **1980**, *44*, 892–898. [CrossRef]

22. Kengni, L.; Vachaud, G.; Thony, J.L.; Laty, R.; Garino, B.; Casabianca, H.; Jame, P.; Viscogliosi, R. Field measurements of water and nitrogen losses under irrigated maize. *J. Hydrol.* **1994**, *162*, 23–46. [CrossRef]

23. Reynolds, W.D.; Elrick, D.E. Saturated and field-saturated water flow parameters. In *Methods of Soil Analysis. Part 4. Physical Methods*; Dane, J.H., Topp, G.C., Eds.; Soil Science Society of America: Madison, WI, USA, 2002; pp. 804–808.

24. Duncan, D.B. Multiple range and multiple F tests. *Biometrics* **1955**, *11*, 1–42. [CrossRef]

25. Zhang, Y.M.; Hu, C.S.; Zhang, J.B.; Chen, D.L.; Li, X.X. Nitrate leaching in an irrigated wheat-maize rotation field in the North China Plain. *Pedosphere* **2005**, *15*, 196–203.

26. Campbel, C.A.; de Jong, R.; Zentner, R.P. Effects of cropping summer fallow and fertilization nitrogen on nitrate-nitrogen by leaching on a brown chernozemic loam. *Can. J. Soil Sci.* **1984**, *64*, 64–74. [CrossRef]

27. Strong, W.M. Nitrogen fertilization of upland crops. In *Nitrogen Fertilization in the Environment*; Bacon, P.E., Ed.; Marcel Dekker: New York, NY, USA, 1995; pp. 129–170.

28. Zhu, A.; Zhang, J.; Zhao, B.; Cheng, Z.; Li, L. Water balance and nitrate leaching losses under intensive crop production with Ochric Aquic Cambosols in North China Plain. *Environ. Int.* **2005**, *31*, 904–912. [CrossRef] [PubMed]

29. Jin, M.; Zhang, R.; Sun, L.; Gao, Y. Temporal and spatial soil water management: A case study in the Heilonggang Region, PR China. *Agric. Water Manag.* **1999**, *42*, 173–187. [CrossRef]

30. Quemada, M.; Baranski, M.; Nobel-de Lange, M.N.J.; Vallejo, A.; Cooper, J.M. Meta-analysis of strategies to control nitrate leaching in irrigated agricultural systems and their effects on crop yield. *Agric. Ecosyst. Environ.* **2013**, *174*, 1–10. [CrossRef]

water

MDPI

Article

Assessing the Water-Resources Potential of Istanbul by Using a Soil and Water Assessment Tool (SWAT) Hydrological Model

Gokhan Cuceloglu [1,2,*], Karim C. Abbaspour [2] and Izzet Ozturk [1]

[1] Environmental Engineering Department, Istanbul Technical University, 34469 Maslak, Istanbul, Turkey; ozturkiz@itu.edu.tr
[2] EAWAG, Swiss Federal Institute of Aquatic Science and Technology, 8600 Dübendorf, Switzerland; karim.abbaspour@eawag.ch
* Correspondence: cuceloglu@itu.edu.tr; Tel.: +90-212-285-3776

Received: 15 September 2017; Accepted: 20 October 2017; Published: 24 October 2017

Abstract: Uncertainties due to climate change and population growth have created a critical situation for many megacities. Investigating spatio-temporal variability of water resources is, therefore, a critical initial step for water-resource management. This paper is a first study on the evaluation of water-budget components of water resources in Istanbul using a high-resolution hydrological model. In this work, the water resources of Istanbul and surrounding watersheds were modeled using the Soil and Water Assessment Tool (SWAT), which is a continuous-time, semi-distributed, process-based model. The SWAT-CUP program was used for calibration/validation of the model with uncertainty analysis using the SUFI-2 algorithm over the period 1977–2013 at 25 gauge stations. The results reveal that the annual blue-water potential of Istanbul is 3.5 billion m^3, whereas the green-water flow and storage are 2.9 billion m^3 and 0.7 billion m^3, respectively. Watersheds located on the Asian side of the Istanbul megacity yield more blue-water resources compared to the European side, and constitute 75% of the total potential water resources. The model highlights the water potential of the city under current circumstances and gives an insight into its spatial distribution over the region. This study provides a strong basis for forthcoming studies concerning better water-resources management practices, climate change and water-quality studies, as well as other socio-economic scenario analyses in the region.

Keywords: hydrological modeling; SWAT; SWAT-CUP; water-resources modeling; water availability; water potential; Istanbul

---

1. Introduction

As demand for water increases across the world, the availability of freshwater in many regions is likely to decrease due to population growth, industrialization, land use and climate change. Climate change due to the greenhouse effect plays a vital role in the availability of freshwater and is just one of the pressures facing water resources today. Nearly all of the climate change projections indicate substantial decreases in water availability in the Mediterranean region in the future [1–5]. On the other hand, rapid population increase, urbanization, and industrialization in this region have had a significant effect on the regional hydrological cycle. As the population increases, the provision of clean water in the megacities of developing countries becomes increasingly more complex [6]. Continuing urbanization poses a major challenge to providing adequate water services to the metropolis [7].

Quantifying the water resources of a megacity is essential for providing the strategic information needed for long-term planning of the city's water security. Conventional water-resource planning and management have mostly been based on "blue-water" resources, which serve the needs of engineers who are responsible for coping with infrastructure projects for water supply [8]. Blue water is generally defined as "the sum of river discharge and deep groundwater recharge". "Green water", however, is differentiated by Falkenmark and Rockström [8] between green-water resources and green-water flows. According to their definition, "green-water resource is the moisture in the soil". This is the renewable part that can potentially generate economic returns and the source of rainfed agriculture. Green-water flow, however, is the actual evaporation (the non-productive part) and the actual transpiration (the productive part), commonly referred to together as the actual evapotranspiration [9]. Thus, it is vital to evaluate the blue- and green-water potential for human activities. This water paradigm is successfully used to evaluate water resources and its availability throughout the world using Soil and Water Assessment Tool (SWAT) models at continent, country, or basin scales [9–15].

In 2016, Istanbul had a population of 15 million within the city proper. It is not only the most populous city in Turkey, but one of the biggest conurbations in the world whose population is still increasing due to a high level of migration by approximately 250,000 people every year [16] from all over the country. The annual population growth rate is 2.8%, which is almost twice the overall rate of Turkey [17]. Furthermore, due to the Syrian war, Istanbul hosts 600,000 Syrian refugees per year. Demand for water in the city, which is supplied mainly (about 98%) from surface water resources in 15 watersheds, is about 3 million m^3 day^{-1}, including domestic and industrial consumption as well as non-reserve water (NRW) or water losses. Water supply has always been a great challenge throughout the history of Istanbul, from ancient times to the present.

One of the most comprehensive studies on İstanbul's water resources is reported by the Istanbul Master Plan Consortium (IMC) [18] for the planning of water-supply, wastewater and stormwater investments in the Istanbul Metropolitan Area. Akkoyunlu et al. [19], Eroglu et al. [20], Yuksel et al. [21], Altinbilek [22], Saatci [23], Ozturk and Altay [24] and van Leeuwen and Sjerps [17], reported on current water-management strategies, challenges in water supply, and future directions regarding water administration in the city mostly based on the IMC report. Few studies can be found on water budgets of the watersheds of Istanbul covering individual sub-basins [25,26]. Kara and Yucel [27] and the Istanbul Water and Sewerage Administration (ISKI) [28] used trend analysis to study the impact of climate change on extreme flows in Istanbul. They concluded that a significant impact could be expected on the frequency and amount of hydrological extremes such as floods and droughts. Cuceloglu and Ozturk [29,30] quantified the water-budget components of major freshwater resources of Istanbul by using the SWAT model. Rouholahnejad et al. [12] studied the water resources of the European part of Turkey in the context of the Black Sea Basin.

To the best of our knowledge, no systematic modeling work has been done on the water resources of Istanbul to quantify both the potential of all watersheds in supplying potable water, and to quantify water availability in terms of various components of the water budget with a high resolution. This study has been conducted to fill this gap, and will be the first study covering all of the current and planned catchment areas of Istanbul. Also, the hydrological model built in this study will serve as the base model for forthcoming studies to evaluate climate change impacts and water quality in order to guide engineers, experts, as well as policy-makers and authorities in the region.

The overall objectives of the current study are: (1) to build a high-resolution hydrological model for all the watersheds (existing and planned) that could supply drinking water; (2) to calibrate and validate the model and perform sensitivity and uncertainty tests upon its outputs; and (3) to reveal the current water budget in terms of blue- and green-water resources. For this purpose, we used SWAT to build the hydrological model and SWAT-CUP to calibrate and validate the model.

2. Materials and Methods

2.1. Study Area

Istanbul is located in the north-west of Turkey within the Marmara region, intersecting two continents, namely Asia and Europe. The surface area of Istanbul is 5540 km^2. Forest territories cover 44% of the city, mainly in the northern part. A large majority of the settlements (with a 2700 capita/km^2 population density) are concentrated on the south coast of the city (Figure 1). The city has a transitional climate, impacted by the Black Sea to the north, and the Marmara Sea and the Aegean Sea to the south. The northern parts of the city, where forested areas mostly lie, is affected by northerly colder air masses of maritime and continental origins, whereas the southern part shows the general characteristics of the Mediterranean climate [31]. The average temperature in winter months is between 2 °C and 9 °C, and in summer months between 18 °C and 28 °C. The city receives about 815 mm of precipitation per year as a long-term average, according to the recorded stations in the city [32].

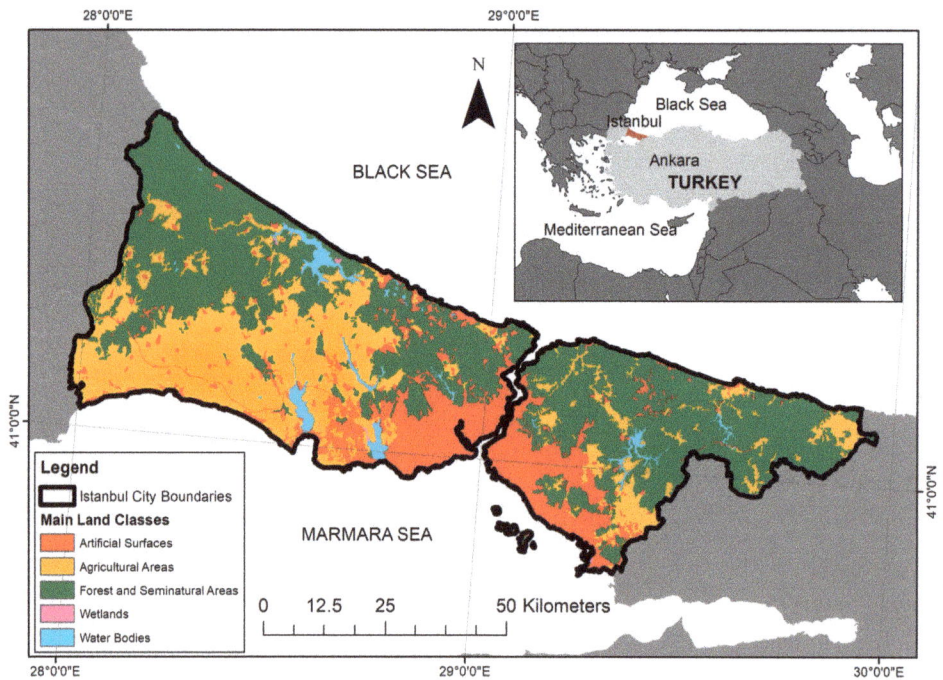

Figure 1. Istanbul city boundary and main land cover classes.

Istanbul has a population of nearly 15 million people, and it is expected that this number will grow to 21 million by 2050 [16,18]. In parallel with the increase in population, daily water consumption will grow due to changes in lifestyle, income level and eating habits. Today, gross water demand in the city is estimated to be 175 L/capita-day, and this figure is expected to reach 225 L/capita-day by 2050 including industrial usage and NRW. Figure 2 shows the historical population changes and water demand of Istanbul.

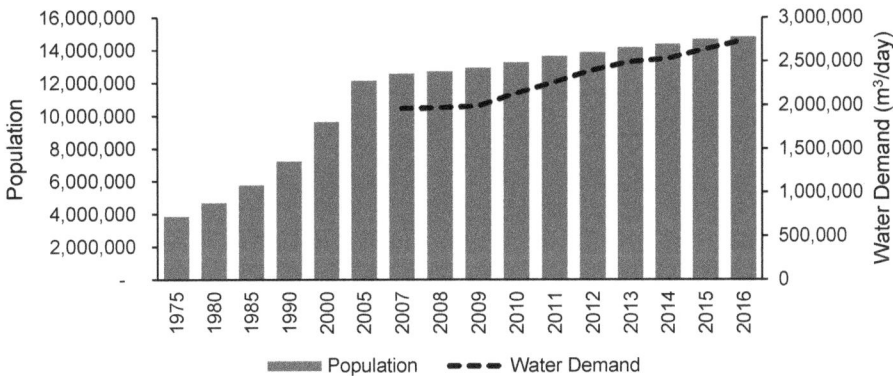

Figure 2. Population changes and annual water demand of Istanbul between 1975 and 2016.

As of today, 15 drinking-water reservoirs operate to meet the demand for potable water in Istanbul. To meet the water demand of the Istanbul metropolitan area, ISKI had to expand its service area beyond the city border (Figure 3). Currently, ISKI is responsible for the management and the protection of the water resources located in different administrative regions to supply drinking water to the Istanbul metropolitan area. There are several ongoing projects to increase the potential water capacity and protect catchments for the future. Six of the main drinking-water reservoirs and their watersheds are within the city border, namely, Terkos, Buyukcekmece, Alibey, Sazlıdere, Omerli, Elmalı and Darlık; and the rest, the Kazandere and Papucdere Reservoirs, Istranca Creeks and the Melen System, are in neighboring cities (Figure 3). Istanbul has an unbalanced distribution in terms of its water resources and population between the Asian and European sides. In numbers, the Asian side has 77% of the water resources (including the Melen System) while it hosts 35% of the population (Table 1).

Table 1. Water resources and the population distribution for Istanbul.

Water Resources	Population	Annual Water Potential (Million m³/Year)
Asian side *	5,250,000	1909 (77%)
European side	9,750,000	568 (23%)
Grand annual total	15,000,000	2477

Note: * Greater Melen Dam and Melen Stage III Transmission Line included (to be constructed in 2018).

There are plans for the Melen watershed, located in the western part of the Black Sea Region and 180 km to the east of Istanbul (Figure 3), to provide water to Istanbul in the medium and long term [33]. In order to convey water from the Asian side to the European side, a 6-m diameter and 5551-m long Bosphorous tunnel was constructed. The tunnel goes 135 m below sea level, crossing the two continents, with a capacity to transfer 3 million m³ of water daily [23].

In order to investigate the water-resources availability of Istanbul, we studied the current watersheds of the city as well as the surrounding potential catchments of Istanbul. This area is located between 40.3 to 42.1 north latitude and 27.1 to 31.7 east longitude, which includes the area next to Istanbul, the Black Sea coast of Trakya Region (Istranca Sub-region), Kapıdag Peninsula, Izmit Bay, the Sapanca and Iznik Lake watersheds, downstream of Sakarya River, and the Melen watershed in the Western Black Sea Basin in Turkey (Figure 4). The total study area is around 20,790 km². Although there are more flow stations maintained by the State Hydraulic Works (DSI) in the region, 25 stations were found to be suitable over the area for this study. Among these, 12 gauge stations are within the current watersheds of Istanbul, and 13 are located in the remaining parts of the study area (Figure 4).

Thus, we used these hydrometric stations for regionalization of the hydrological model parameters in the ungauged catchments.

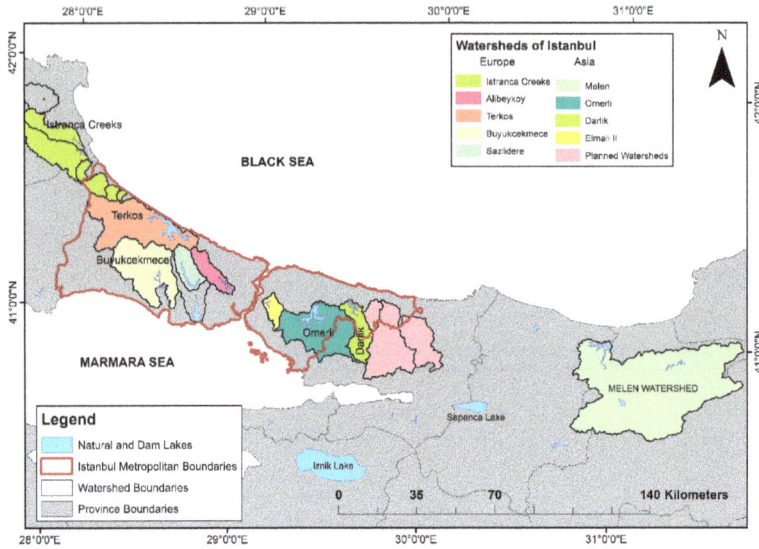

Figure 3. Watersheds used to supply drinking water located both in the Asian and European sides of Istanbul, with administrative boundaries over the region.

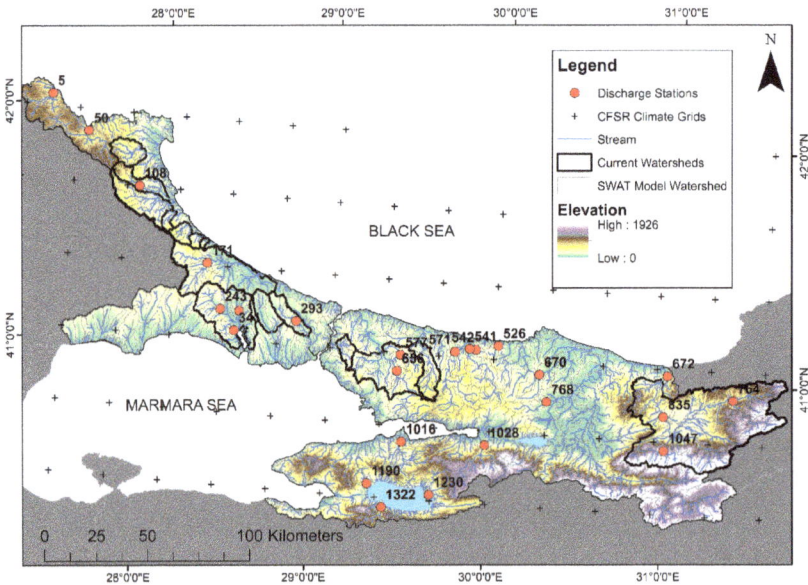

Figure 4. Location of the water resources of Istanbul, topography, rivers, climate grids and discharge stations used in the model.

2.2. SWAT Model

SWAT is a hydrological model developed by the US Department of Agriculture (USDA) Agricultural Research Service [34,35]. It is a continuous-time, semi-distributed, process-based model, developed to evaluate alternative management strategies on water resources and non-point source pollution in large river basins [36].

Water balance is the driving force behind all the processes in SWAT because it impacts plant growth and the movement of sediments, nutrients, pesticides and pathogens [36]. In SWAT, a watershed is divided into multiple sub-basins, which are then further subdivided into hydrologic response units (HRUs) based on unique combinations of land use, soil, management and topographical features. The model simulates hydrology of a watershed in two phases. In the first phase, called the land phase, the hydrological processes of a watershed are simulated at the HRU level and water balance calculated for each sub-basin. The pathways of water movement in the land phase simulated by SWAT are given as canopy storage, surface runoff, evapotranspiration, infiltration, lateral sub-surface flow, return flow, revap from shallow aquifers, and percolation to the deep aquifer. In the second phase (the routing phase), after the loadings of water, sediment, nutrients and pesticides are determined, and loadings are routed through streams and reservoirs within the watershed [37]. A schematic representation of hydrological cycle elements simulated by SWAT is given in Figure 5.

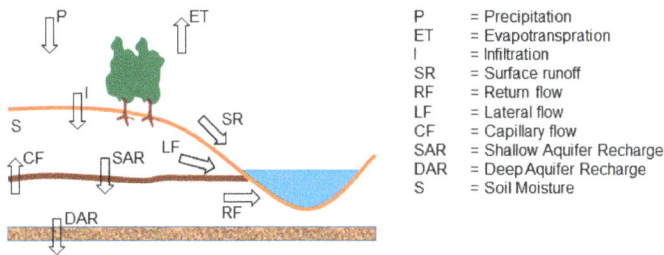

Figure 5. Schematic representation of hydrological cycle elements in SWAT.

More details and model equations can be found in the SWAT technical documentation (http://swatmodel.tamu.edu) and in Arnold et al. [34]. A general overview of SWAT model use, calibration and validation is discussed by Arnold et al. [36], and historical development, applications, and future research directions are summarized in Gasmann et al. [38] and Douglas-Mankin et al. [35].

2.3. Model Inputs and Setup

The SWAT model requires a land-use map, climate data, soil map, and topography. Due to the lack of local data to build a model, data required for this study were compiled from global datasets. River discharges and water consumption rates were obtained from local administrations (Table 2).

The soil map was produced by the Food and Agriculture Organization/United Nations Educational, Scientific and Cultural Organization (FAO–UNESCO) global soil map [39], which provides data for 5000 soil types (65 for Turkey) comprising two layers (0–30 cm and 30–100 cm depth) at a spatial resolution of 10 km. The land-use map was obtained from the CORINE 2000 Land Cover datasets (http://www.eea.europa.eu/data-and-maps/data/corine-land-cover-2000-raster-3) at a resolution of 100. A digital elevation model (DEM) was constructed from the Shuttle Radar Topography Mission (SRTM) database at a 90-m spatial resolution (http://srtm.csi.cgiar.org/). Three different climate database sources were available for the region: (1) measured data collected from the State Meteorological Service (MGM) in 17 temperature and rainfall climate stations with <15% missing data for the period 1960–2013; (2) gridded data constructed from Climate Research Units

(CRU) with a 0.5° resolution for the period 1970–2007, and 0.25° gridded data from Climate Forecast System Reanalysis (CFSR) for the period 1979–2014, amounting to 48 and 103 grid points, respectively.

Table 2. Data description, source, and resolution in the current study.

Data Type	Source	Data Resolution
Digital elevation map (DEM)	Shuttle Radar Topography Mission (SRTM) http://srtm.csi.cgiar.org/	90 m
Land use	European Environment Agency CORINE Land Cover (year 2000) http://www.eea.europa.eu/data-and-maps/data/corine-land-cover-2000-raster-3	100 m
Soil	FAO-UNESCO global soil map http://www.fao.org/nr/land/soils/digital-soil-map-of-the-world/en/	5 km
Climate data	Climate Research Unit http://www.cru.uea.ac.uk	0.5°
	Climate Forecast System Reanalysis (CFSR) http://cfs.ncep.noaa.gov/cfsr/	0.25°
	Turkish State of Meteorological Service http://www.mgm.gov.tr/	17 Stations
River discharge	Turkish State of Hydraulics http://en.dsi.gov.tr/	25 Stations Monthly
Population and water consumption rates	Turkish Statistical Institute http://www.turkstat.gov.tr/Start.do Istanbul Water and Sewage Administration www.iski.gov.tr/	Yearly

We used the ArcSWAT 2012 interface to set up the model. Despite using a high-resolution DEM in the model, in order to avoid the discrepancies, particularly during the stream network delineation, we used the burn-in feature of ArcSWAT with river data obtained from the Google Earth software. Also, to delineate coastal catchment areas more accurately, a threshold drainage area of 100 ha was chosen. Inland sub-basin outlets were manually added to represent reservoirs, gauge stations, main river channels, and other topographical features in the watershed; while coastal outlets were created automatically by the software based on the given threshold. As a result, the study area was configured with 1335 sub-basins, which were further discretized into 3315 HRUs. The model could not represent 4% of the real area as a result of the precision of basin delineation near the coastal zones. However, this missing area does not affect principal objectives of the study. The total simulated area in the current model is 19,960 km^2.

Except for two stream gauges, none of the gauges are affected by the reservoir operations. Therefore, we did not use the reservoir operation rule in the current model. The available river discharge data in the region varies between 10 years and 32 years. Five elevation bands in each sub-basin were established to adjust for orographic change in temperature ($-6.5\,^{\circ}\text{C km}^{-1}$) and rainfall ($100\ \text{mm km}^{-1}$). Potential evapotranspiration (PET) is simulated using Hargreaves method [40], actual evapotranspiration (ET) is estimated based on the methodology of Ritchie [41], and surface runoff is calculated by the Soil Conservation Service (SCS) curve number procedure [42].

According to the acquired data (Table 2), the model was simulated from 1977 to 2013 (37 years), and the first 3 years were used as a warm-up period to allow the processes simulated to reach a dynamic equilibrium and decrease the uncertainty of the initial conditions of the model. The simulation includes both dry and wet years occurring in the historical period. Figure 6 depicts the yearly cumulative precipitation in Istanbul between the years 1977 and 2014, including both drought and wet periods.

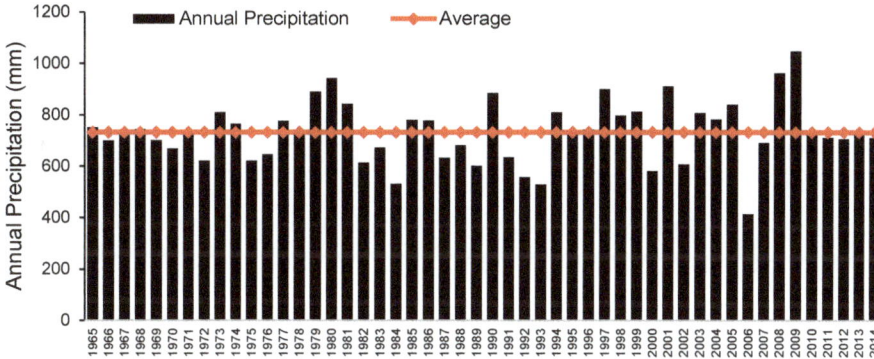

Figure 6. Annual precipitation of Istanbul including severe dry and wet periods for the city in the last 50 years.

Due to the availability of more than one climate database, we evaluated three different datasets as a preliminary analysis of the model. According to the preliminary model results during the model set-up, CFSR outperformed the local dataset and CRU captured better the streamflow dynamics and also the total rainfall distribution over the study area. Most of the watershed area evaluated in this study are protected zones without any settlements [24], thus land-use changes are negligible.

For model calibration, validation and uncertainty analysis we used the SUFI-2 algorithm [43,44] in the SWAT-CUP software. All uncertainties in the model, such as a parameter, measured data (e.g., stream flow), driving variables (e.g., rainfall), and conceptual framework, are expressed as a set of parameter ranges by SUFI-2. The algorithm tries to capture most of the measured data within the 95% band of prediction uncertainty (95PPU). The uncertainty (95PPU) is quantified at the 2.5% and 97.5% levels of the cumulative frequency distribution of an output variable obtained using the Latin hypercube sampling technique. Two indices are used to measure the goodness of calibration/uncertainty performance, the P-factor (ranges 0 to 1), which is the percentage of data captured by 95PPU band, and the R-factor (ranges 0 to ∞), which is the average thickness of the uncertainty band divided by the standard deviation of the related measured variable [43,44]. These two indices are used to judge the strength of the calibration procedure where a value of >0.7 for the P-factor and a value of around 1 for the R-factor would be satisfactory, depending on the study [11,12]. More information about the algorithm is given by Abbaspour et al. [11,44]. SUFI-2 allows the use of different objective functions such as R^2, RSR or Nash–Sutcliffe efficiency (NSE) [45]. Despite the fact that we used NSE as an objective function, percent bias (PBIAS) [46] and R^2 of the calibration/validation results were also evaluated as well as the P-factor and the R-factor in order to assess the model performance and model output uncertainty.

3. Results and Discussion

3.1. Calibration and Validation of River Discharges

Although SWAT parameterization has been done based on the parameters given in Table 3 together with initial and final ranges for the calibration, in order to improve model performance especially in high altitudes where snow processes become predominant, we changed the snow melt temperature (SMTMP), snow fall temperature (SFTMP), and maximum and minimum snowmelt rate factors (SMFMX, SMFMN) to −3.71, 3.21, 5.68 and 3.05, respectively. These numbers were obtained after several simulations in SWAT-CUP. Adjusting the parameter related to snow processes improves the model's performance, in particular at the discharge stations located in mountainous region of the study area. After fixing these parameters (to their best values obtained by the preliminary runs), the protocol given in Abbaspour et al. [11] was followed for calibration/validation as well as sensitivity and uncertainty analysis.

Table 3. List of the model calibration parameter.

SWAT Parameter	Definition	Initial Range	Final Range	*t* Value	*p* Value
r__CN2.mgt	SCS runoff curve number for moisture condition II	−0.5 to 0.5	−0.20 to 0.39	−9.742	6.73×10^{-21}
r__SOL_AWC().sol	Soil available water storage capacity (mm H_2O/mm soil)	−0.5 to 0.5	−0.06 to 0.80	0.390	0.696
r__ESCO.hru	Soil evaporation compensation factor	−0.2 to 0.2	−0.21 to 0.06	−0.572	0.567
r__GW_REVAP.gw	Groundwater revap. coefficient	−0.5 to 0.5	−0.13 to 0.60	0.033	0.973
r__GWQMN.gw	Threshold depth of water in shallow aquifer for return flow (mm)	−0.5 to 0.5	−0.52 to 0.15	−1.615	0.106
r__REVAPMN.gw	Threshold depth of water in the shallow aquifer for "revap" (mm)	−0.5 to 0.5	−0.50 to 0.16	1.695	0.090
r__ALPHA_BF.gw	Base flow alpha factor (days)	−0.5 to 0.5	0.00 to 0.97	4.497	8.28×10^{-6}
r__SOL_K().sol	Soil conductivity (mm h^{-1})	−0.5 to 0.5	−0.09 to 0.72	1.851	0.064
r__SOL_BD().sol	Soil bulk density (g cm^{-3})	−0.5 to 0.5	−0.03 to 0.89	0.011	0.990

Note: The term "r__" was used for the relative changes in the parameter between the ratio of a given range, "()" was used for all layers of soil, and the extension of the parameter states the file name of the SWAT files.

Nine SWAT parameters were selected for model calibration for all discharge stations. Half of the river discharge data were used for calibration and the remainder were used for validation. The parallel processing option of SWAT-CUP [47] considerably reduced the model simulation time. One iteration and 600 simulations with nine parameters were adequate for obtaining satisfactory calibration and validation results. There are different gridded climate datasets covering the region of our study. Using these datasets could be quite useful for evaluating the uncertainty caused by using different input data [48]. Furthermore, scientists may consider the possibility of using a stochastic hydrological rainfall-runoff model if there are not enough data of precipitations in a specific study area [49,50].

As expected, the CN2 parameter was the most sensitive parameter for outflow, followed by ALPHA_BF. Although the SOL_BD and GW_REVAP parameters seemed to be less sensitive as indicated by their *t*-stat and *p*-value (Table 3), they contributed to increased model calibration results for river discharges considerably. *t*-stat depicts parameter sensitivity: the larger the *t*-value, the more sensitive the parameter; whereas the *p*-value indicates the significance of the *t*-value: the smaller the *p*-value, the less chance of a parameter being accidentally assigned as sensitive [11].

Twenty-five gauge stations on discontinuous stream networks were parameterized and calibrated simultaneously. R^2 ranges from 0.39 to 0.82 for calibration, and 0.41 to 0.82 for validation, while NSE values vary from 0.31 to 0.81 for calibration, and 0.33 to 0.81 for validation. Model outputs for most of the discharge stations can be judged as satisfactory according to Moriasi et al. [51]. Table 4 represents the calibration and validation results and model performance criteria for the 25 gauges stations used in this study.

Except for the two gauge stations (Nos.: 1190 and 1322) our simulations captured the 60–90% (P-factor values ranges from 0.6 to 0.9) of observed data during simulation (Table 4). Poor simulation results at these stations originate from high base-flow simulations in dry periods and high peak simulations during wet periods. This is due to the CFSR climate station located in the east of the Iznik Lake (Figure 4), which overestimates the precipitation in these catchments. The larger R-factor values representing higher uncertainty for the stations (gauge station Nos.: 243, 341 and 656) located on the Istanbul city border, is most likely a result of rapid urbanization having occurred in that catchment in the last few decades. Most of the stations in the study area had R^2 and NSE of more than 0.5 both in calibration and validation (Figure 7). Poor simulation results were obtained downstream of areas with intensive water-resource development (regulators, water store operations) and urbanization (increasing settlement and urban drainage systems) because of processes that were not included the model.

Table 4. Model performance indicators of 25 gauge stations for calibration and validation.

Gauge Station No.	Calibration					Validation				
	P Factor	R Factor	R^2	NSE	PBIAS	P Factor	R Factor	R^2	NSE	PBIAS
5	0.71	1.24	0.39	0.31	−26.8	0.67	0.80	0.53	0.50	8.3
50	0.75	1.04	0.66	0.65	1.4	0.75	1.14	0.41	0.35	6.2
108	0.90	1.32	0.78	0.73	14.2	0.72	1.20	0.50	0.33	21.9
171	0.83	1.03	0.86	0.84	−1.2	0.78	1.37	0.58	0.43	−5.5
243	0.82	1.53	0.61	0.48	−23.1	0.81	1.10	0.78	0.78	1.8
252	0.62	0.83	0.82	0.81	3.8	0.49	0.86	0.68	0.67	4.4
293	0.62	0.72	0.80	0.79	15.8	0.51	1.06	0.57	0.46	−25.0
341	0.74	1.49	0.78	0.69	−5.2	0.81	1.21	0.69	0.68	9.3
526	0.69	1.10	0.67	0.64	−6.8	0.60	1.08	0.65	0.64	−5.5
541	0.80	1.00	0.52	0.50	16.6	0.81	0.93	0.80	0.79	9.2
542	0.77	1.03	0.55	0.53	14.2	0.83	0.98	0.77	0.77	5.1
571	0.69	0.97	0.68	0.68	−0.1	0.70	0.89	0.72	0.72	−7.2
577	0.72	0.97	0.79	0.77	−8.3	0.54	0.77	0.48	0.48	8.9
656	0.64	1.46	0.78	0.37	−56.6	0.57	1.32	0.81	0.55	−64.5
670	0.83	1.28	0.79	0.68	−12.3	0.67	1.63	0.80	0.41	−49.9
672	0.84	1.10	0.81	0.77	15.6	0.89	1.10	0.85	0.81	15.6
764	0.74	1.01	0.55	0.53	8.1	0.73	1.00	0.69	0.67	7.1
768	0.82	1.05	0.74	0.74	2.5	0.70	1.57	0.64	0.31	−39.1
835	0.77	1.23	0.68	0.61	−4.2	0.84	1.15	0.82	0.81	4.4
1016	0.72	1.37	0.69	0.50	−33.9	0.65	1.17	0.79	0.78	−13.1
1028	0.73	0.70	0.64	0.59	18.3	0.74	0.89	0.69	0.61	25.3
1047	0.65	0.75	0.66	0.53	32.1	0.69	0.78	0.74	0.65	28.9
1190	0.43	0.71	0.68	0.60	3.5	0.31	1.21	0.71	0.65	−31.3
1230	0.78	1.20	0.74	0.72	−10.3	0.76	1.18	0.74	0.74	−2.0
1322	0.32	0.68	0.62	0.59	−14.0	0.25	0.57	0.57	0.52	−6.2

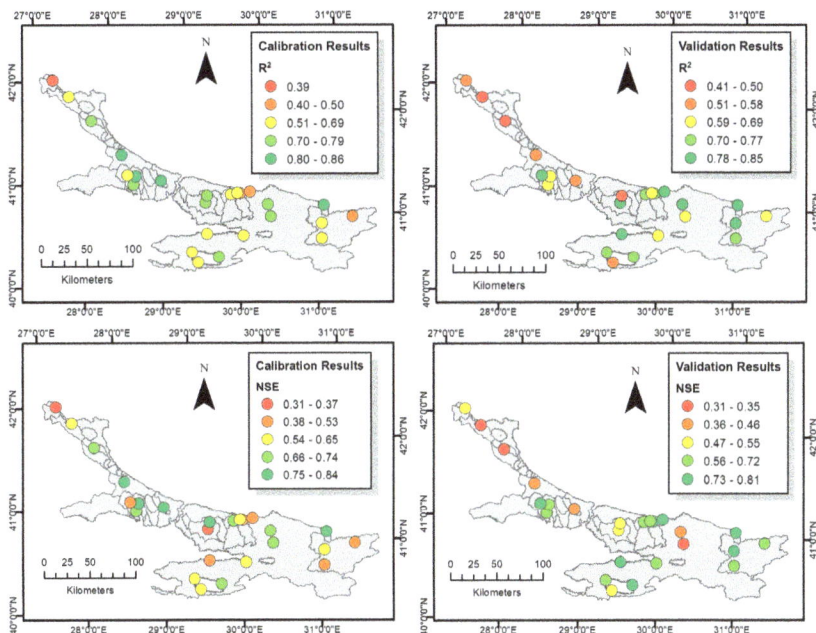

Figure 7. Comparison of simulated and observed discharge data using the coefficient of determination (R^2) and Nash–Sucthlife coefficient (NSE) for the calibration and validation period.

Average discharge rates of the rivers in the study area varies from 0.2 m^3 s^{-1} to 44 m^3 s^{-1} (Figure 8). Extreme discharge values of about 150 m^3 s^{-1} were captured by our model quite well. For the simplicity and clarity of graphs, calibration and validation periods are shown in one graph continuously (Figure 8).

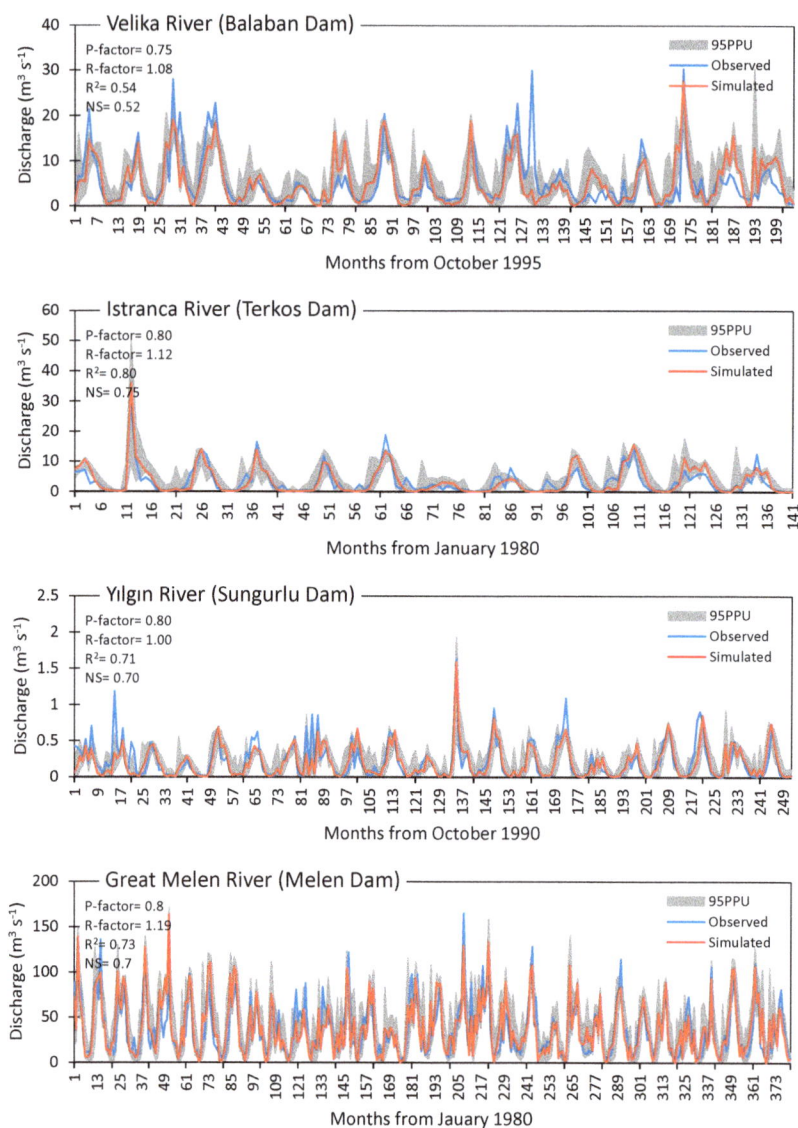

Figure 8. Illustration of the SWAT-CUP output for the simulation period depicting the observed, simulated and 95% prediction uncertainty (95PPU). These hydrographs belong to gauge station Nos. 50, 243, 542 and 835, respectively.

81

3.2. Water Availability

To show the spatial distribution of precipitation, mean annual rainfall is depicted over the study area (Figure 9a). The catchments on the Black Sea coast of the Asian side receive the highest average annual rainfall in the study area where the major water resources of the city are located. The southern part of Istanbul, with an average annual rainfall of 550–720 mm, receives the least amount of precipitation. Model results of blue and green water are represented at the 50% probability level of the the 95PPU for the period 1977–2013. These were calculated for 1335 sub-basins (Figure 9b–d). Blue-water (water yield + deep percolation) potential of the watershed on the Asian side is greater than the European side; likewise, green-water storage (soil moisture) (Figure 9b,c). The results reveal that the average blue-water potential of Istanbul is 630 mm, whereas the green-water flow and storage are 382 mm and 129 mm, respectively.

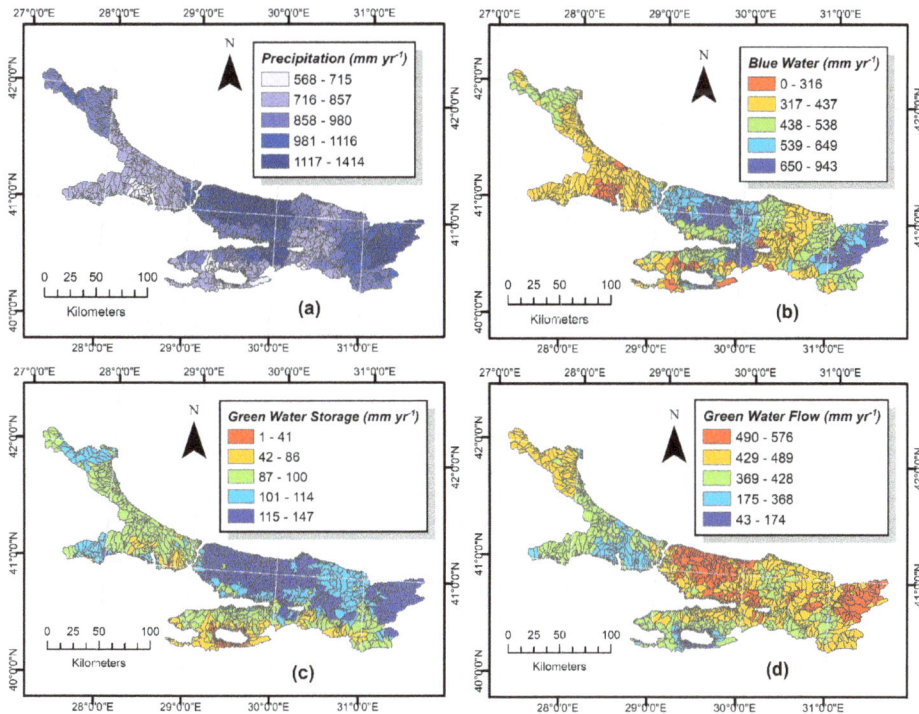

Figure 9. Spatial distribution of simulated average (1980–2013) annual (**a**) precipitation; (**b**) blue-water resources; (**c**) green-water flow and (**d**) green-water storage for the study area.

Blue-water potentials reach up to 943 mm year^{-1} in some catchments such as Omerli, Darlık and Melen which are located on the Black Sea coast. Besides, Buyukcekmece represents poor potential ranging from 0 to 316 mm year^{-1} due to less precipitation and a higher potential evapotranspiration rate as well as urbanization in that catchment. The spatial distribution of soil moisture (green-water storage) indicates higher values ranging from 100 to 150 mm year^{-1}, especially in the eastern part of the study area (Figure 9c). In the current situation, small-scale rainfed agricultural activities play an important role for local villages. Therefore, optimal management strategies are necessary to achieve a balance between supplying water to Istanbul and supporting the agricultural activities and economic growth as well as sustaining high water quality in this region. As shown in Figure 9d, due to the amount of available water for transpiration and evaporation, green-water flows in Asian-side watersheds (495 mm year^{-1}) are higher in comparison to the European side (386 mm year^{-1}).

To illustrate the reliability of water resources over the years, we calculated the coefficient of variation CV = $\sigma/\mu \times 100$ for the years 1980–2013 in each sub-basin (Figure 10), where σ is the standard deviation and μ is the mean of the variable.

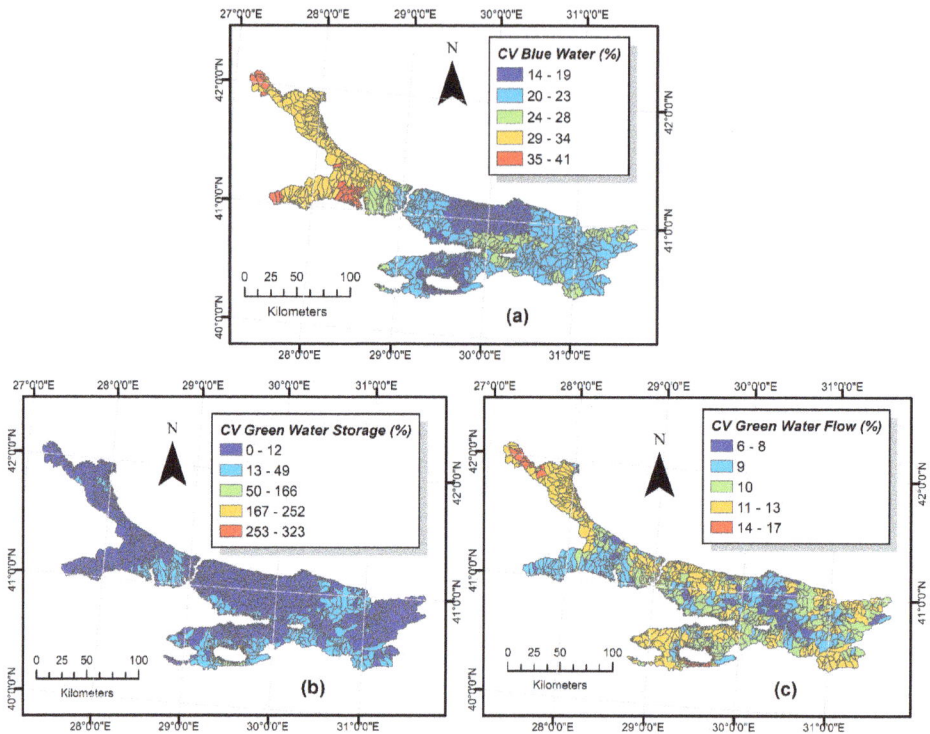

Figure 10. Coefficient of variation (CV) of the modelled annual (1980–2013) (a) blue-water resources; (b) green-water flow and (c) green-water storage.

CV is an indicator of the reliability of the water resources from year to year [9]. The smaller the CV, the smaller the year-to-year variability of a variable and the more reliable the estimates. Blue-water flows are quite important due to their contribution to the reservoir used for domestic water demand. Although the temporal variation of the blue-water resources is not very large, water resources on the Asian side seem more reliable than the European side, especially in the Darlık, Sungurlu, Isaköy and Kabakoz watersheds, which have the smallest CV values in the study area (Figure 10a). These values indicate a higher reliability of this resource over time and, hence, a less risky opportunity for a water-supplying project. Green-water storage (soil moisture) is generally relatively less variable in most of the area between the values 0–49%, except the catchments near Iznik Lake, which is located in southern part of the study area, and which has CV values between 167% and 303% (Figure 10b). This situation is most likely caused by excessive industrial consumption and agricultural irrigation in the region. Temporal variation of evapotranspiration (green-water flow) varies over the study area (Figure 10c) and larger values are obtained in the eastern part of the study area where water availability is greater (Figure 9d).

Model results in hru level aggregated to watershed level and the water potential of each reservoir include both Asian and European sources as well as the current and planned water resources together.

We calculated the water potential by multiplying the catchment area with blue-water potential (water yield + deep aquifer recharge) of each watershed, and plotted 95% prediction uncertainty band for each (Figure 11).

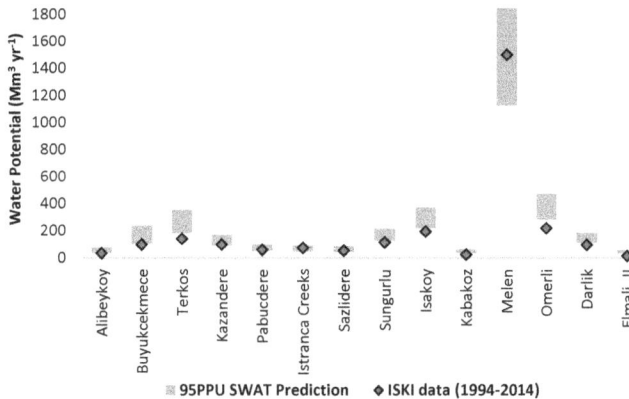

Figure 11. Comparison of the simulated 95PPU SWAT prediction (1980–2013) for water potential with the available data from the Istanbul Water and Sewerage Administration (ISKI) for the watersheds of Istanbul.

As shown in Figure 11 the ISKI estimates are within or close to the 95PPU of our model predictions. Larger differences between model results and ISKI values are observed in the watersheds that are the major reservoirs of Istanbul. This result might be due to possible water losses such as evaporation from the reservoirs, transfer operations, water usage in the watershed, and the effects of urban drainage systems, which were not included in our calculations of water potential for ISKI. Thus, our model gives the "gross" water potential of each watershed. A considerable part of water losses occurs in aged and poorly managed water distribution systems [52–54]. This phenomenon needs to be measured and calculated in detail to estimate the "net" water potential of the city.

According to the prediction of water potential, the watersheds located on the Asian side constitute almost 75% of the total water resources of Istanbul. The Melen watershed has the highest water potential with 1.5 billion m^3 year^{-1} (45% of total water resources), and is followed by Omerli, Terkos, and Buyukcekmece, which are currently in use. These three watersheds will be essential for the city considering their large water potential. The Omerli, Terkos and Buyukcekmece sources will meet 25% of the total demand in future. As these sources are located in the Istanbul administrative boundary and under pressure of urbanization, protection of these catchments will assure adequate water supply to the city next to the planned resources (Kabakoz, Isakoy and Sungurlu catchments).

Figure 11 also shows that the parameter range obtained using data from 25 discharge stations simultaneously yields quite satisfactory results for the ungauged basin in regions such as Elmali II, Darlik, Istanca Creeks, Kazandere and Papucdere.

4. Conclusions

Socio-economic developments, rapid urbanization, and population increase have put pressures on Istanbul's water resources in the last few decades. Thus, matching the water demand of the city requires effective water-management strategies and more projects to supply water from different administrative boundaries. Quantification of available water is an essential part of the management of the water resource of Istanbul. This study contributes significant insights into the water availability of the city and its vicinity, at both regional and watershed level.

Model results reveal that watersheds supplying drinking water on the Asian side are more reliable and more abundant. However, the majority of the population (thereby most of the water demand) is on the European side; planned water resources for the future are mostly located on the Asian side and are outside the city boundaries. Water transfers from these catchments are vital to meeting the water demands of Istanbul.

Water resources on the Asian side encompass almost 75% of the total potential. These catchments (except Omerli) are pristine and so far not affected by heavy urbanization. Therefore, optimal management strategies in these catchments play a significant role in balancing water supply and local activities (agriculture, energy production, recreation etc.). Also, protection of these catchments in terms of not only water quantity but also quality is vital to the city. The Melen watershed is of utmost importance as it will provide 45% of the total water demand of the city in the future. Melen and its significant water potential (1.5 billion m^3 per year) could be highly beneficial in the case of increasing population and in drought periods. In addition to increased water potential capacity, water re-use, decreasing water loss in the supply network, upgrading urban drainage system, rainwater harvesting, and the efficient use of available water will make better use of available water in Istanbul.

More detailed analysis in the study area covering quantitative assessment of each watershed for different scenarios such as drought, socio-economic change, land-use conditions, as well as climate change scenarios, could provide more information and significant knowledge of value to policy-makers, local administrators and experts. The well-established and calibrated model developed in this study provides a strong basis for forthcoming studies in the Istanbul megacity regarding reservoir operations and management, the impact of climate change, as well as water-quality and biodiversity issues in the reservoirs.

Acknowledgments: This project was funded by the Scientific and Technological Research Council of Turkey (TUBITAK) 2214 Grand Program, and Istanbul Technical University, Scientific Research Projects Fund (Project # 40227). The first author would like to thank Aysegul Tanik for her valuable comments on this manuscript.

Author Contributions: Gokhan Cuceloglu performed the data compilation, simulations, analysis, and drafted the manuscript. Karim C. Abbaspour and Izzet Ozturk made contributions in reviewing and editing the manuscript.

Conflicts of Interest: The authors declare that they have no conflicts of interest.

References

1. IPCC. *Climate Change 2013: The Physical Science Basis. Contribution of Working Group I to the Fifth Assessment Report of the Intergovernmental Panel on Climate Change*; Stocker, T.F., Qin, D., Plattner, G.-K., Tignor, M., Allen, S.K., Boschung, J., Nauels, A., Xia, Y., Bex, V., Midgley, P.M., Eds.; Cambridge University Press: Cambridge, UK; New York, NY, USA, 2013.

2. Christensen, J.H.; Kumar, K.K.; Aldria, E.; An, S.-I.; Cavalcanti, I.F.A.; De Castro, M.; Dong, W.; Goswami, P.; Hall, A.; Kanyanga, J.K.; et al. Climate Phenomena and their Relevance for Future Regional Climate Change Supplementary Material. In *Climate Change 2013: The physical Science Basis. Contribution of Working Group I to the Fifth Assessment of the Intergovernmental Panel on Climate Change*; Cambridge University Press: Cambridge, UK, 2013; pp. 1217–1308. [CrossRef]

3. Erol, A.; Randhir, T.O. Climatic change impacts on the ecohydrology of Mediterranean watersheds. *Clim. Chang.* **2012**, *114*, 319–341. [CrossRef]

4. World Water Assessment Programme. *The United Nations World Water Development Report 3: Water in a Changing World*; UNESCO: Paris, France; Earthscan: London, UK, 2009.

5. Mariotti, A.; Zeng, N.; Yoon, J.-H.; Artale, V.; Navarra, A.; Alpert, P.; Li, L.Z.X. Mediterranean water cycle changes: Transition to drier 21st century conditions in observations and CMIP3 simulations. *Environ. Res. Lett.* **2008**, *3*, 44001. [CrossRef]

6. Biswas, A.K. Water Management for Major Urban Centres. *Int. J. Water Resour. Dev.* **2006**, *22*, 183–197. [CrossRef]

7. Varis, O.; Biswas, A.K.; Tortajada, C.; Lundqvist, J. Megacities and Water Management. *Int. J. Water Resour. Dev.* **2006**, *22*, 377–394. [CrossRef]

8. Falkenmark, M.; Rockström, J. The New Blue and Green Water Paradigm: Breaking New Ground for Water Resources Planning and Management. *J. Water Resour. Plan. Manag.* **2006**, *132*, 129–132. [CrossRef]

9. Schuol, J.; Abbaspour, K.C.; Yang, H.; Srinivasan, R.; Zehnder, A.J.B. Modeling blue and green water availability in Africa. *Water Resour. Res.* **2008**, *44*. [CrossRef]

10. Veettil, A.V.; Mishra, A.K. Water security assessment using blue and green water footprint concepts. *J. Hydrol.* **2016**, *542*, 589–602. [CrossRef]

11. Abbaspour, K.C.; Rouholahnejad, E.; Vaghefi, S.; Srinivasan, R.; Yang, H.; Kløve, B. A continental-scale hydrology and water quality model for Europe: Calibration and uncertainty of a high-resolution large-scale SWAT model. *J. Hydrol.* **2015**, *524*, 733–752. [CrossRef]

12. Rouholahnejad, E.; Abbaspour, K.C.; Srinivasan, R.; Bacu, V.; Lehmann, A. Water resources of the Black Sea Basin at high spatial and temporal resolution. *Water Resour. Res.* **2014**, *50*, 5866–5885. [CrossRef]

13. Rodrigues, D.B.B.; Gupta, H.V.; Mendiondo, E.M. A blue/green water-based accounting framework for assessment of water security. *Water Resour. Res.* **2014**, *50*, 7187–7205. [CrossRef]

14. Zang, C.F.; Liu, J.; Van Der Velde, M.; Kraxner, F. Assessment of spatial and temporal patterns of green and blue water flows under natural conditions in inland river basins in Northwest China. *Hydrol. Earth Syst. Sci.* **2012**, *16*, 2859–2870. [CrossRef]

15. Faramarzi, M.; Abbaspour, K.C.; Schulin, R.; Yang, H. Modelling blue and green water resources availability in Iran. *Hydrol. Process.* **2009**, *23*, 486–501. [CrossRef]

16. TUIK Address-Based Population Registration Database System Turkish Statistical Institute Official Website. Available online: http://www.tuik.gov.tr/UstMenu.do?metod=temelist (accessed on 14 January 2016).

17. Van Leeuwen, K.; Sjerps, R. Istanbul: The challenges of integrated water resources management in Europa's megacity. *Environ. Dev. Sustain.* **2016**, *18*, 1–17. [CrossRef]

18. Istanbul Master Plan Consortium (IMC). *Istanbul Water Supply, Sewage and Drainage, Wastewater Treatment, and Disposal Master Plan*; Istanbul Master Plan Consortium: Istanbul, Turkey, 1999.

19. Akkoyunlu, A.; Yuksel, E.; Erturk, F.; Bayhan, H. Managing of watersheds of Istanbul (Turkey). In Proceedings of the Fifth Water Information Summit: Regional Perspectives on Water Information Management Systems, Ft. Lauderdale, FL, USA, 23–25 October 2002.

20. Eroglu, V.; Sarikaya, H.Z.; Ozturk, I.; Yuksel, E.; Soyer, E. Water management in Istanbul Metropolitan Area. In Proceedings of the Third International Forum, Integrated Water Management: The key to sustainable water resources, Athens, Greek, 21–22 March 2002.

21. Yuksel, E.; Eroglu, V.; Sarikaya, H.Z.; Koyuncu, I. Current and Future Strategies for Water and Wastewater Management of Istanbul City. *Environ. Manag.* **2004**, *33*, 186–195. [CrossRef] [PubMed]

22. Altinbilek, D. Water Management in Istanbul. *Int. J. Water Resour. Dev.* **2006**, *22*, 241–253. [CrossRef]

23. Saatci, A.M. Solving Water Problems of a Metropolis. *J. Water Resour. Prot.* **2013**, *5*, 7–10. [CrossRef]

24. Ozturk, I.; Altay, D.A. Water and Wastewater Management in Istanbul. In Proceedings of the UNESCO HQ International Conference on Water, Megacities and Global Change, Paris, France, 1–4 December 2015.

25. Cuceloglu, G.; Erturk, A. Model Supported Hydrological Analysis of Darlik Creek Watershed, Istanbul Turkey. *Fresenius Environ. Bull.* **2014**, *23*, 3110–3116.

26. Akiner, M.E.; Akkoyunlu, A. Modeling and forecasting river flow rate from the Melen Watershed, Turkey. *J. Hydrol.* **2012**, *456–457*, 121–129. [CrossRef]

27. Kara, F.; Yucel, I. Climate change effects on extreme flows of water supply area in Istanbul: Utility of regional climate models and downscaling method. *Environ. Monit. Assess.* **2015**, *187*. [CrossRef] [PubMed]

28. Istanbul Water and Sewerage Administrationon (ISKI). *Climate Change Impacts on Istanbul and Turkey Water Resources Project Final Report*; Istanbul Water and Sewerage Administrationon (ISKI): Istanbul, Turkey, 2010.

29. Cuceloglu, G.; Ozturk, I. Development of a Model Framework for Sustainable Water Management Practices: Case Study for the Megacity Istanbul. In Proceedings of the 9th Eastern European Young Water Professionals Conference, The International Water Association, Budapest, Hungary, 24–24 May 2017.

30. Cuceloglu, G.; Ozturk, I. Assessing the Influence of Climate Datasets for Quantification of Water Balance Components in Black Sea Catchment: Case Study for Melen Watershed in Turkey. In Proceedings of the International SWAT Conference, Warsaw, Poland, 26–30 July 2017.

31. Ezber, Y.; Sen, O.L.; Kindap, T.; Karaca, M. Climatic effects of urbanization in Istanbul: A statistical and modeling analysis. *Int. J. Climatol.* **2007**, *27*, 667–679. [CrossRef]

32. MGM Turkish State of Meteorological Service. Available online: https://www.mgm.gov.tr/veridegerlendirme/il-ve-ilceler-istatistik.aspx?k=undefined&m=ISTANBUL (accessed on 12 January 2017).

33. Ozturk, I.; Erturk, A.; Ekdal, A.; Gurel, M.; Cokgor, E.; Insel, G.; Pehlivanoglu-Mantas, E.; Ozabali, A.; Tanik, A. Integrated watershed management efforts: Case study from Melen Watershed experiencing interbasin water transfer. *Water Sci. Technol. Water Supply* **2013**, *13*, 1272–1280. [CrossRef]

34. Arnold, J.G.; Srinivasan, R.; Muttiah, R.S.; Williams, J.R. Large Area Hydrologic Modeling and Assessment Part I: Model Development. *J. Am. Water Resour. Assoc.* **1998**, *34*, 73–89. [CrossRef]

35. Douglas-Mankin, K.R.; Srinivasan, R.; Arnold, J.G. Soil and Water Assessment Tool (SWAT) Model: Current Developments and Applications. *Trans. ASABE* **2010**, *53*, 1423–1431. [CrossRef]

36. Arnold, J.G.; Moriasi, D.N.; Gassman, P.W.; Abbaspour, K.C.; White, M.J.; Srinivasan, R.; Santhi, C.; Harmel, R.D.; Van Griensven, A.; VanLiew, M.W.; et al. SWAT: Model Use, Calibration, and Validation. *Asabe* **2012**, *55*, 1491–1508. [CrossRef]

37. Neitsch, S.L.; Arnold, J.G.; Kiniry, J.R.; Williams, J.R. *Soil and Water Assessment Tool Theoretical Documentation Version 2009*; Texas Water Resources Institute Technical Report No. 406; Texas A&M University System: College Station, TX, USA, 2011.

38. Gassman, P.P.W.; Reyes, M.M.R.; Green, C.C.H.; Arnold, J.J.G. The Soil and Water Assessment Tool: Historical development, applications, and future research directions. *Trans. ASAE* **2007**, *50*, 1211–1250. [CrossRef]

39. Food and Agricultural Organization (FAO). *The Digital Soil Map of the World and Derived Soil Properties*; CD-ROM, Version 3.5; Food and Agriculture Organization of the United Nations, Land and Water Development Division: Rome, Italy, 2003.

40. Hargreaves, G.L.; Hargreaves, G.H.; Riley, J.P. Agricultural Benefits for Senegal River Basin. *J. Irrig. Drain. Eng.* **1985**, *111*, 113–124. [CrossRef]

41. Ritchie, J.T. Model for predicting evaporation from a row crop with incomplete cover. *Water Resour. Res.* **1972**, *8*, 1204–1213. [CrossRef]

42. USDA. Soil Conservation Service SCS National Engineering Handbook. Section 4, Hydrology. In *National Engineering Handbook*; USDA: Washington, DC, USA, 1972.

43. Abbaspour, K.C.; Johnson, C.A.; van Genuchten, M.T. Estimating Uncertain Flow and Transport Parameters Using a Sequential Uncertainty Fitting Procedure. *Vadose Zone J.* **2004**, *3*, 1340–1352. [CrossRef]

44. Abbaspour, K.C.; Yang, J.; Maximov, I.; Siber, R.; Bogner, K.; Mieleitner, J.; Zobrist, J.; Srinivasan, R. Modelling hydrology and water quality in the pre-alpine/alpine Thur watershed using SWAT. *J. Hydrol.* **2007**, *333*, 413–430. [CrossRef]

45. Nash, J.E.; Sutcliffe, J.V. River flow forecasting through conceptual models Part I—A discussion of principles. *J. Hydrol.* **1970**, *10*, 282–290. [CrossRef]

46. Gupta, H.V.; Sorooshian, S.; Yapo, P.O. Status of Automatic Calibration for Hydrologic Models: Comparison with Multilevel Expert Calibration. *J. Hydrol. Eng.* **1999**, *4*, 135–143. [CrossRef]

47. Rouholahnejad, E.; Abbaspour, K.C.; Vejdani, M.; Srinivasan, R.; Schulin, R.; Lehmann, A. A parallelization framework for calibration of hydrological models. *Environ. Model. Softw.* **2012**, *31*, 28–36. [CrossRef]

48. Kamali, B.; Abbaspour, C.K.; Yang, H. Assessing the Uncertainty of Multiple Input Datasets in the Prediction of Water Resource Components. *Water* **2017**, *9*. [CrossRef]

49. Aronica, G.T.; Candela, A. Derivation of flood frequency curves in poorly gauged Mediterranean catchments using a simple stochastic hydrological rainfall-runoff model. *J. Hydrol.* **2007**, *347*, 132–142. [CrossRef]

50. Morlando, F.; Cimorelli, L.; Cozzolino, L.; Mancini, G.; Pianese, D.; Garofalo, F. Shot noise modeling of daily streamflows: A hybrid spectral- and time-domain calibration approach. *Water Resour. Res.* **2016**, *52*, 4730–4744. [CrossRef]

51. Moriasi, D.N.; Arnold, J.G.; Van Liew, M.W.; Bingner, R.L.; Harmel, R.D.; Veith, T.L. Model Evaluation Guidelines for Systematic Quantification of Accuracy in Watershed Simulations. *Trans. ASABE* **2007**, *50*, 885–900. [CrossRef]

52. Covelli, C.; Cozzolino, L.; Cimorelli, L.; Della, M.R.; Pianese, D. A model to simulate leakage through joints in water distribution systems. *Water Sci. Technol. Water Supply* **2015**, *15*, 852–863. [CrossRef]

53. Covelli, C.; Cozzolino, L.; Cimorelli, L.; Della Morte, R.; Pianese, D. Optimal Location and Setting of PRVs in WDS for Leakage Minimization. *Water Resour. Manag.* **2016**, *30*, 1803–1817. [CrossRef]

54. Covelli, C.; Cimorelli, L.; Cozzolino, L.; Della, M.R.; Pianese, D. Reduction in water losses in water distribution systems using pressure reduction valves. *Water Sci. Technol. Water Supply* **2016**, *16*, 1033–1045. [CrossRef]

Article

Influence Mechanisms of Rainfall and Terrain Characteristics on Total Nitrogen Losses from Regosol

Xiaowen Ding [1,2,*], Ying Xue [1], Ming Lin [1] and Guihong Jiang [1]

[1] Key Laboratory of Regional Energy and Environmental Systems Optimization, Ministry of Education, North China Electric Power University, Beijing 102206, China; astridxuey@163.com (Y.X.); linming000000@163.com (M.L.); m15011001229@163.com (G.J.)
[2] Institute for Energy, Environment and Sustainable Communities, University of Regina, Regina, SK S4S 7H9, Canada
* Correspondence: binger2000dxw@hotmail.com; Tel.: +86-10-6177-2982

Academic Editor: Saeid Ashraf Vaghefi
Received: 25 October 2016; Accepted: 21 February 2017; Published: 3 March 2017

Abstract: The upper reach of the Yangtze River is an ecologically sensitive region where water loss, soil erosion, and nonpoint source (NPS) pollution are serious issues. In this drainage area, regosol is the most widely distributed soil type. Cultivation on regosol is extensive and total nitrogen (TN) has become a common NPS pollutant. Artificial rainfall experiments were conducted to reveal the influence mechanisms of rainfall and terrain on TN losses from regosol. The results showed that there were positive correlations between precipitations and TN loads but negative ones between precipitations and TN concentrations. Furthermore, negative correlations were more obvious on fields with slopes of $5°$ and $25°$ than on other slopes. With increasing rainfall intensity, TN loads rose simultaneously. However, TN concentration in runoff-yielding time presented a decline over time. As far as terrain was concerned, TN loads grew generally but not limitlessly when slopes increased. Similarly, TN concentrations also rose with rising slopes; upward trends were more obvious for steeper slopes. Furthermore, the initial runoff-yielding time became longer for steeper slopes and the differences under various rainfall intensity conditions diminished gradually.

Keywords: total nitrogen; artificial rainfall experiments; rainfall; terrain; regosol; influence mechanisms

---

1. Introduction

With economic and social development, water contamination has become a serious problem in many countries [1–3]. With control of point source contamination, nonpoint source (NPS) contamination has become dominant cause of water contamination due to its multi-source, wide distribution, the difficulty of controlling it and so on [4–6]. Nowadays, identifying effect factors and their mechanisms of influence on NPS pollutant exports have become research hotspots in the field of NPS pollution control [7–9].

Rainfall is known as the driving force of NPS pollution, while terrain is the main effect factor in the generation and transport of NPS pollutants [10,11]. Where NPS pollutants are concerned, total nitrogen (TN) is critical, being dissolved in runoff and adsorbed in sediment with soil and water loss [12–14].

The Yangtze River is the third longest river in the world and the longest in China. Its drainage area accounts for 18.8% of the land area in China [15–17]. The upper reach of the Yangtze River is defined from the source of the Yangtze River to the Three Gorges Dam (the largest dam in the world) and with a length of 4504 km and a drainage area of 10^6 km^2 [18]. Water pollution control for the upper reach of the Yangtze River plays a significant role in water and soil conservation, maintaining biodiversity, and ensuring water environment security in the whole river basin [19–21]. In the upper reach of the Yangtze River, regosol is the dominant soil; it is suitable for cultivation but erodible [22]. Therefore, understanding the influence mechanisms of rainfall and terrain on total nitrogen losses from regosol is essential for NPS pollution research, pollution control and the environmental protection of the Yangtze River's water [23].

In recent decades, a great deal of research concerning the influence of rainfall and terrain on NPS nitrogen pollution has been carried out worldwide [24–26]. A comparative study of the effects of rainfall events on the removal of NPS nitrogen determined the efficiency ratio indexes, load summation and load regression in the Kyeong—a stream basin, South Korea [27]. Rainfall shortly after the surface application of poultry manure was shown to have the potential to significantly increase surface runoff TN concentration in Iowa, United States [28]. A study carried out in a typical rain field of black soil in Northeast China indicated that different tilling systems led to distinctively different concentrations of losses of water-soluble nitrogen and particulate nitrogen, and loss of water-soluble nitrogen and particulate nitrogen per unit area [29]. Land use, sediment and sand content play dominant roles in affecting NPS pollutant export and contribute to the high heterogeneity of TN in regosol in the Sichuan Province in the upper reach of the Yangtze River [30]. Wilson and Weng found that concentrations of TKN, nitrate, nitrite and TN in surface water depended heavily on the spatiotemporal distribution of land use/land cover in the Lake Calumet area near Chicago, United States [31]. However, previous studies have mostly concentrated on the effects of land use and sand content on nitrogen loss in watersheds, while the influence mechanisms of rainfall and terrain with respect to nitrogen losses have not been thoroughly discussed. Therefore, these influence mechanisms need to be better understood; one method involves artificial rainfall experiments on regosol under simulated rainfall and terrain conditions.

The objective of this study is to reveal the influence mechanisms of rainfall and terrain characteristics on TN losses from regosol. This research aims to reflect how TN losses from regosol are affected by rainfall and terrain characteristics such as precipitation, rainfall intensity, rainfall duration, and slope conditions. For this study, artificial rainfall experiments were conducted in which rainfall and terrain conditions were divided into different levels. The first part of this paper shows the relationships between rainfall conditions and TN load, TN concentration, and the initial runoff-yielding time. The effects of terrain conditions on TN load, TN concentration, and the initial runoff-yielding time will also be discussed. Our goal is to give researchers a reference in understanding transport processes, establishing simulation models, and identifying parameters for modeling NPS pollution. This research may also provide support for decision making by administrators of NPS pollution control and land management.

2. Materials and Methods

2.1. Experimental Materials

Regosol soil, which accounts for 13.18% of the upper reach of the Yangtze River, has the characteristics of abundant fertility but high erodibility, resulting from hillslope in the drainage area (Figure 1) [32]. In our research, the experimental field was located in Chongqing City in the upper reach of the Yangtze River, where soil and water loss and NPS pollution are serious problems because of abundant rainfall and steep slopes [33]. The experimental field (106°43′ E, 29°53′ N) was located on a hillside of the Beibei district of Chongqing City where the soil type is regosol. Precipitation and rainfall intensity in that location were the factors representing rainfall characteristics, and the slope was chosen to represent typical terrain, one that has significant effects on TN losses.

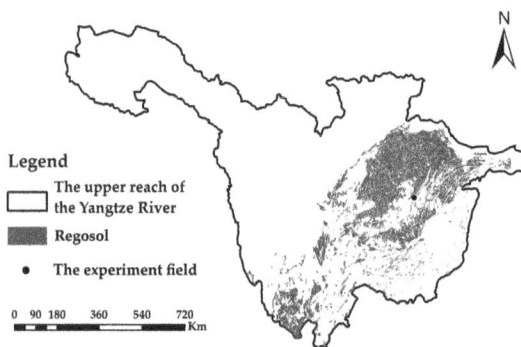

Figure 1. Distribution of regosol in the upper reach of the Yangtze River.

The 20 cm of topsoil was selected as soil samples to eliminate the spatial heterogeneity of physical and chemical properties for various soil samples. The regosol was crushed gently, passed through a 7 mm sieve to remove stones and impurities, air-dried by the oven drying method to reduce initial soil water content rate to 12.16%, and then mixed thoroughly. The physical and chemical properties of the soil samples are listed in Table 1. Regosol is thin, usually less than 50 cm; little was more than 1 m deep. Generally, regosol contains calcium carbonate, showing a neutral or slightly alkaline reaction. The sample had a clay mineral content of 7%–12% smectite, 8%–10% illite, 7%–11% chlorite, 35%–40% quartz, 27%–31% feldspar, 3% haematite, and 3% others. The microbiological component was composed of 7.0×10^6 bacteria per gram of dry soil, 6.7×10^4 actinomycetes per gram of dry soil, and 2.1×10^3 fungi per gram of dry soil. Other components also influence nitrogen retention, such as organic matter content, soil texture, and so on.

Table 1. Physical and chemical properties of soil samples.

Soil Layer (cm)	Unit Weight (g/cm^3)	Initial Soil Water Content Rate (%)	Organic Matter (g/kg)	TN (g/kg)	TP (g/kg)
0–20	1.30	12.16	8.75	0.76	0.68

Microbes have effects on the forms of nitrogen in soil. For example, the existence of nitrogen-fixing bacteria and phosphorus bacteria could add available nitrogen content in soil, denitrifying bacteria could slow the conversion of available nitrogen to ineffective nitrogen, and, in soil, some bacteria activate nitrogen and turns it into nutrients that plants can absorb [34]. Furthermore, other soil components have effects of the background value of nitrogen; for instance, the content of clay particles in soil could affect the mass fraction of TN in soil [35]. Meanwhile, those impacts mainly related to temperature [36]. However, in this study, TN was taken as the pollutant, which included various forms of nitrogen; therefore, the impacts of the conversion among various forms of nitrogen on TN concentrations and TN loads could be ignored. In addition, the background value of TN in soil and temperature were constant in the experiments. Therefore, it could be deduced that microbes and other soil component were not the key factors that caused changes of TN concentrations and TN loads in the rainfalls.

Precipitation and rainfall intensity were selected as the variables to study rainfall characteristics, and the slope was based on the typical terrain in order to study their significant effects on TN losses.

2.2. Experimental Devices

For a small-scale artificial rainfall experiment, the experimental plot is usually less than 5 m^2 with the length, width and depth of the soil box at 1–2 m, 0.5–1 m and 0.22–0.5 m, respectively [37]. In our experiments, three identical steel soil boxes, each 1.0 × 0.6 × 0.25 m and with two wheels at the base, were designed to minimize experimental error. Each soil box had apertures on the side near the wheels to allow runoff and sediment to transport freely. The bottom of the box was extended 0.1 m outward along the long side and a groove with small holes was set on the right-hand side of the extension to collect runoff and sediment. On the opposite side of the groove, a regulating screw was attached so that the slope angle of each soil box could be adjusted from 0° to 65°. The structure of the soil boxes is shown in Figure 2.

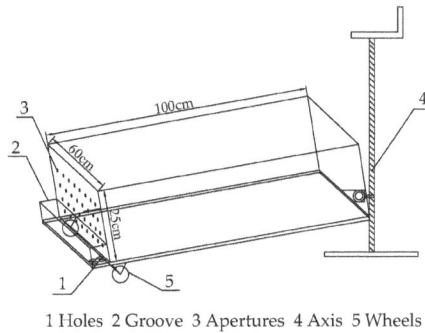

1 Holes 2 Groove 3 Apertures 4 Axis 5 Wheels

Figure 2. Structure of the soil box.

In our experiments, a Norton nozzle-type rainfall simulator was adopted, consisting of a water supply system as well as a spraying system and produced in the United States (Figure 3). The height of the nozzle was 2.5 m and the hydraulic pressure was 0.04 M Pa, which made the sizes and distribution of raindrops were similar to those in nature. The rainfall intensity could be set at different levels by changing the frequency of nozzle swings and was stable to maintain a consistent condition.

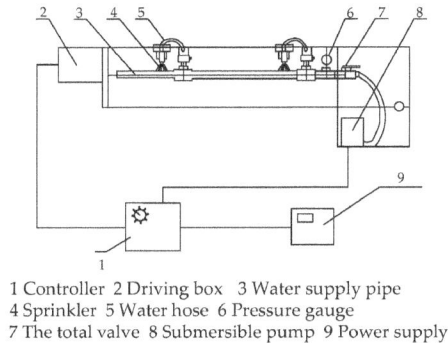

1 Controller 2 Driving box   3 Water supply pipe
4 Sprinkler 5 Water hose 6 Pressure gauge
7 The total valve 8 Submersible pump 9 Power supply

Figure 3. Structure of the Norton nozzle-type rainfall simulator.

2.3. Experimental Design

As mentioned in Section 2.1, taking soil from a hillside where the soil type was regosol and the preprocessing of the regosol included crushing, sieving, air drying, and mixing. Afterwards, five 5-cm soil layers were compacted in a soil box so that the lower soil layer was loosed before the upper one was filled, to prevent soil stratification. After the filling was completed, a cutting ring method was used to guarantee the soil bulk density was about 1.30 g/cm^3, similar to that in nature. Domestic water with a TN concentration of 1.68 mg/L was used as an artificial rainfall source to simulate rainfall. In Chongqing City, the pH of domestic water is 7.93, that of rainwater is 5.64, and difference can be found between them. Previous researches have demonstrated that pH has impacts on adsorption-desorption reactions [38,39]. In this research, the major form of TN was the dissolved one and the durations of experiments were relatively short, which led to a relatively weak adsorption-desorption effect. For those concerns, experimental error caused by such difference is regarded as acceptable.

The artificial rainfall source was assumed to have a composition similar to that of natural rainwater. According to precipitation data supplied by the National Meteorological Information Center, maximum rainfall intensity in Chongqing City in recent decades has exceeded 100 mm/h. Therefore, four rainfall intensities were adopted in this study: 30 mm/h, 60 mm/h, 90 mm/h and 120 mm/h. Normal precipitation from one rainfall would not usually exceed 80 mm; in combination with the designed rainfall intensities, we set rainfall duration at 40 min to simulate natural conditions and make the comparison and analysis more intuitive. Five slopes (5°, 10°, 15°, 20° and 25°) were designed representing regional terrain; moreover, the *Water and Soil Conservation Law of the People's Republic of China* prohibits cultivation or crop planting on fields with a slope of 25° or more. The 20 experimental scenarios with various rainfall intensities and slopes are shown in Table 2.

Table 2. Scheme of artificial rainfall experiments.

Slope (A) / Rainfall Intensity (B)	30 mm/h (B_1)	60 mm/h (B_2)	90 mm/h (B_3)	120 mm/h (B_4)
5° (A_1)	A_1B_1	A_1B_2	A_1B_3	A_1B_4
10° (A_2)	A_2B_1	A_2B_2	A_2B_3	A_2B_4
15° (A_3)	A_3B_1	A_3B_2	A_3B_3	A_3B_4
20° (A_4)	A_4B_1	A_4B_2	A_4B_3	A_4B_4
25° (A_5)	A_5B_1	A_5B_2	A_5B_3	A_5B_4

In this research, runoff-yielding times for all scenarios were 40 min; specifically, rainfall times were between 41.55 and 52.43 min. After runoff occurred on the soil surface of the soil boxes, it was collected by water butts. Specifically, the runoff occurring within 5 min was gathered in a water butt; therefore, all runoff for one scenario was collected in 8 water butts because runoff-yielding time for each scenario lasted 40 min. By this means, runoffs in all scenarios were measured. The soil was reloaded for each experiment to avoid influence caused by variations of soil water content and bulk density. The runoff and its TN concentration were measured and monitored after each simulated rainfall. As far as TN loss in regosol was concerned, dissolved nitrogen was the main form. Specifically, dissolved nitrogen was mainly nitrite, and ammonium nitrogen was the major component of the adsorbed nitrogen.

2.4. Data

In this study, the independent variables included precipitation, rainfall intensity, and slope, while the dependent ones contained TN load, TN concentration, and the initial runoff-yielding time. To ensure runoff-yielding time was 40 min for each rainfall, precipitations varied from 23.09 to 86.36 mm due to different rainfall intensities and the initial runoff-yielding times. Rainfall intensity was 30 mm/h, 60 mm/h, 90 mm/h and 120 mm/h, and slopes of 5°, 10°, 15°, 20°, and 25° were adopted to reveal the influence mechanisms of rainfall and terrain on TN in various scenarios in the study area.

The TN load measured in these experiments contained not only dissolved nitrogen in the rainfall runoff but also adsorbed nitrogen on the sediments. The general measure processes were as follows. Firstly, water samples were filtrated, by which the supernatant containing dissolved nitrogen and the sediments carrying adsorbed nitrogen were divided. Afterwards, the concentrations of dissolved nitrogen in water samples and those of adsorbed nitrogen in sediment samples were measured. Finally, TN concentration was achieved as the sum of dissolved nitrogen concentration and adsorbed one, while mg/L was adopted as the unit. The initial runoff-yielding time took into account the interval between rainfall and runoff generation, because soil water content did not reach the saturation point or generate runoff at the beginning of rainfall.

As mentioned above, the runoff occurring within one 5-min period was gathered in one water butt. In order to measure TN concentration, water was sampled in each water butt, which indicated that a measured TN concentration represented the average value in a corresponding 5 min. As far as dissolved nitrogen in the water samples was concerned, its concentration was measured by alkaline potassium persulfate digestion-UV spectrophotometric method [40]. The processes for determining concentration of dissolved nitrogen were as follows. Firstly, a calibration curve was obtained to calculate the difference between the corrected absorbance of the standard solution and that of a zero concentration solution. Secondly, a 10 mL water sample and a 10 mL pure water were measured to correct absorbance. Afterwards, the concentrations of water samples were achieved by the equation (national standard *water quality-determination of total nitrogen-alkaline potassium persulfate digestion UV spectrophotometric method HJ 636-2012*):

$$\rho = \frac{(A_r - a) \times f}{bV} \tag{1}$$

where ρ was mass concentration of TN in water sample (mg/L), A_r was the difference between the corrected absorbance of water sample and that of blank test, a was the intercept of the calibration curve, b was the slope of the calibration curve, V was the volume of water sample (mL), and f was dilution ratio of the water sample. For quality control and quality assurance, the digestion temperature was assured between 120 °C and 124 °C, and the digestion time was guaranteed as 50 min.

As for adsorbed nitrogen, its concentration was measured by semi-micro Kjeldahl method. Firstly, 1.0000 g of air-dried sediment sample was took, and its water content was tested. Secondly, the sediment sample went through heating digestion, which included ones concerned and not concerned nitrate and nitrite nitrogen. Blank tests of heating digesting were also taken. Thirdly, ammonia was distilled, distillate was titrated with a 0.005 mol/L sulfuric acid standard solution, and the volume of the acid standard solution was recorded. Lastly, the equation was adopted to calculate the concentration of adsorbed nitrogen in sediment samples (national standard *method for the determination of soil total nitrogen (semi-micro Kjeldahl method) HY/T 53-1987*):

$$\rho = \frac{(V - V_0) \times C_H \times 0.014}{m} \tag{2}$$

where ρ was the concentration of adsorbed nitrogen in the sediment sample (mg/L), V was the volume of sulfuric acid standard solution when distillate was titrated (mL), V_0 was the volume of the sulfuric acid standard solution when blank water was titrated (mL), C_H was the concentration of the sulfuric acid standard solution (mol/L), 0.014 was the millimol mass of nitrogen atom, and m was the mass of the air-dried soil sample. Meanwhile, the pH of distillate was assured as alkaline to release the ammonia completely so that accurate results would be attained. In addition, TN loads were obtained as the products of runoffs and TN concentrations, which were based on those of dissolved nitrogen and adsorbed one.

SPSS, software specialized for correlation analysis, linear, and nonlinear regression analysis [41], was used to process the experimental data. The method of correlation analysis was adopted to analyze the relationships between precipitation and TN load, precipitation and TN concentration, rainfall intensity and TN concentration, rainfall intensity and the initial runoff-yielding time, slope and TN load, slope and TN concentration, and slope and the initial runoff-yielding time. By analyzing TN loads and TN concentrations under different rainfall intensities and slopes in the artificial rainfall experiments, the effects of rainfall and terrain on TN load, TN concentration, and the initial runoff-yielding time were revealed by correlation analysis.

3. Results and Discussion

3.1. Effects of Precipitation on TN Losses

3.1.1. Effects of Precipitation on TN Load in Runoff

During the process of rainfall, dissolved nitrogen, including dissolved organic nitrogen, nitrate nitrogen, and ammonium nitrogen were carried by rainfall runoff. Meanwhile, adsorbed nitrogen was adsorbed on soil sediments and then transported by water and soil loss. In the experiments, the correlations between precipitations and TN loads under five slopes were obtained, as shown in Figure 4.

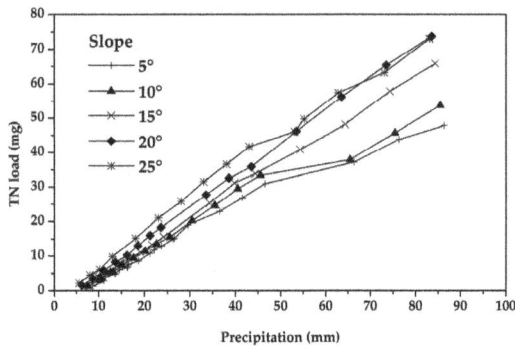

Figure 4. Correlations between precipitations and TN loads under different slopes.

Generally, we found that there were positive linear relationships between precipitation and TN loads on different slopes, i.e., TN loads rose with increased precipitation. Table 3 shows the correlations between TN loads and precipitations under different slopes.

It can be seen that runoff increased and more dissolved nitrogen was generated when precipitation increased. In addition, more sediment was transported by the runoff and adsorbed nitrogen loss became more serious. Therefore, the heavier the rainfall was, the more nitrogen the rainfall runoff and sediment generated, and the severer the TN loss became. TN load losses were slightly different for different slopes, indicating that slope is an important factor affecting TN loss, which will be discussed thoroughly in Section 3.3.1.

Table 3. Correlations of TN loads and precipitations under different slopes.

Slope	Correlation	Correlation Coefficient
5°	$y = 0.6164x - 2.1767$	0.9892
10°	$y = 0.6640x - 1.9772$	0.9880
15°	$y = 0.8357x - 4.3185$	0.9992
20°	$y = 0.9463x - 4.5229$	0.9997
25°	$y = 0.9844x + 0.0897$	0.9874

In Figure 4, the existence of the initial runoff-yielding times demonstrates that there is an interval between the beginning of rainfall and runoff generation. Rainfall merely infiltrated into the soil at the beginning of rainfall and runoff did not begin immediately, because soil water content was relatively low initially. With increased precipitation, soil water content reached the saturation point, and then runoff generation as well as TN loss began to occur from the scouring effect of rainfall runoff.

3.1.2. Effects of Precipitation on TN Concentration in Runoff

The correlations between precipitations and TN concentrations in runoff under various slopes are shown in Figure 5. Those under slope 5°, 10°, 15°, 20° and 25° are shown as Figure 5a–e, respectively. The downward sloping lines in Figure 5 show that TN concentration declines with increasing precipitation, though this is not the case for TN load.

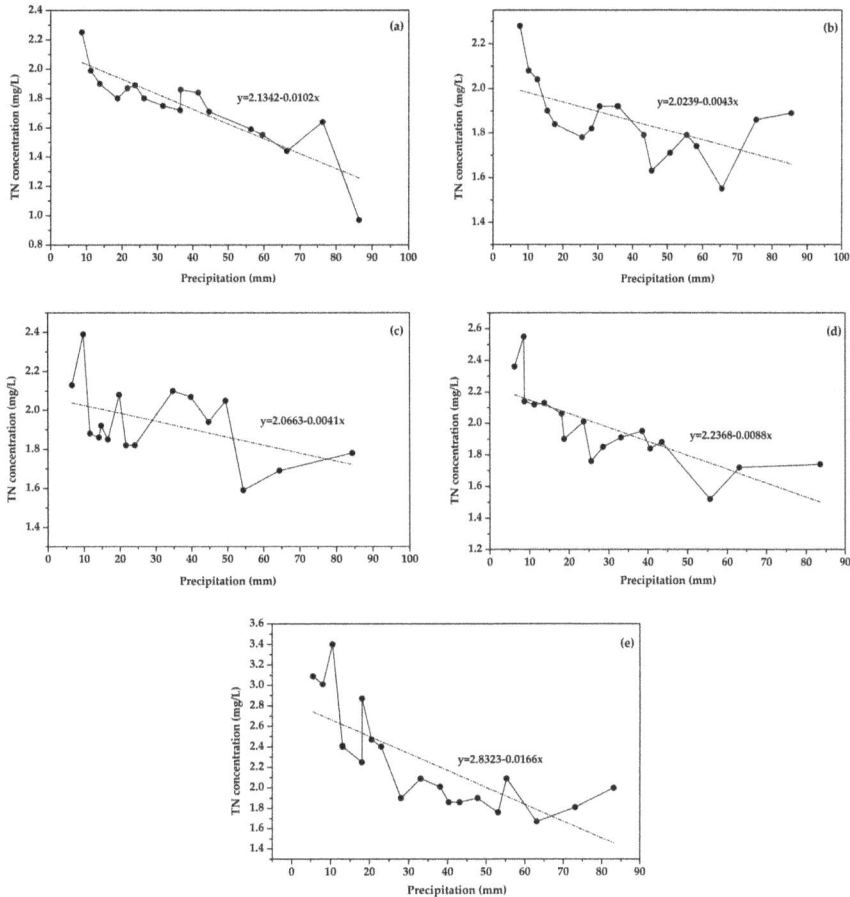

Figure 5. Correlations between precipitations and TN concentrations under slopes of (**a**) 5°; (**b**) 10°; (**c**) 15°; (**d**) 20°; (**e**) 25°.

At the start of rainfall, TN load accumulation in the soil was relatively high, and more nitrogen could dissolve in the rainfall water and adsorb on sediments. Therefore, TN went through the processes of dissolution and adsorption for several minutes before runoff began. Once soil water content reached the saturation point, the initial runoff carried the dissolved and adsorbed nitrogen, so that the TN concentration was relatively high at first. As the rainfall continued, the decline of TN concentration was caused by the decreasing background value of TN in soil, so that less nitrogen was dissolved in the runoff and adsorbed on the sediment.

Moreover, the declines on various slopes differed slightly. Compared to other slopes, TN concentrations declined most obviously on those at 5° and 25°. For a gentle slope (5°), rainfall infiltration lasted a relatively long time before runoff generation, so that nitrogen in soil dissolved sufficiently. Therefore, TN concentration on a gentle slope at the initial runoff-yielding time was relatively high. After that, the scouring effect become weaker and resulted in less dissolved and adsorbed nitrogen carried by runoff and sediment, which caused TN concentration to decrease rapidly. For a steep slope (25°), the reason of TN concentration at the initial runoff-yielding time was relatively high was that the runoff velocity was relatively small owing to gravity, the runoff infiltration rate was low, the runoff on the sloping field was large, and more TN was transported by runoff and sediment within a short time. With scouring, the background value of TN in the soil decreased, so that TN removed by runoff also diminished.

3.2. Effects of Rainfall Intensity on TN Losses

3.2.1. Effects of Rainfall Intensity on TN Concentration in Runoff

Due to the obvious relationship between precipitation and rainfall intensity, the effects of rainfall intensity on TN load were similar to those of precipitation on TN load, and therefore the effects are not discussed in detail. The correlation between rainfall intensity and TN concentration on various slopes is shown in Figure 6. It should be explained that TN concentrations on the five slopes under the same rainfall intensity were averaged as one value to eliminate the influence of slope on TN concentration.

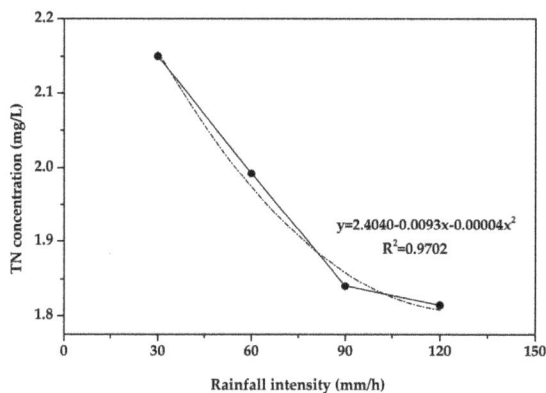

Figure 6. Correlation between rainfall intensity and TN concentration.

The effects of slope on TN loss will be discussed in Section 3.3.1. Rainfall intensity and TN concentration are negatively correlated, which can be formulated by a quadratic polynomial, and the square of the correlation coefficient (R^2) is 0.9702. It can be seen that more runoff is generated under the condition of higher rainfall intensity. Meanwhile, the background value of TN in soil was consistent throughout our experiments; hence TN concentration declined as rainfall intensity increased.

3.2.2. Effects of Rainfall Intensity on the Initial Runoff-Yielding Time

Correlation analysis on the relationship between precipitation and the initial runoff-yielding time showed that the former had no obvious effects on the latter. In contrast, there was a negative correlation between rainfall intensity and the initial runoff-yielding time (Figure 7). It can be explained that as rainfall intensity increases, more raindrops fall on a slope over a certain duration and soil water content increases rapidly. On the other hand, heavy rainfall causes broken soil particles to fill the gap and form water film, which causes the infiltration capacity of the runoff to decrease [37]. Those factors generated rapid runoff even before rainfall infiltrated thoroughly into deep soil, which caused the initial runoff-yielding time to be short.

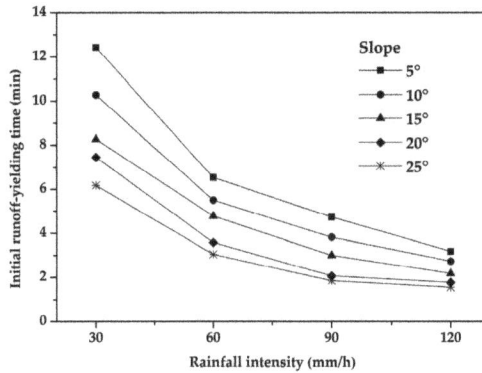

Figure 7. Correlations between rainfall intensities and the initial rainfall-yielding times.

In Figure 7, the initial runoff time shortens when the slope rises, since increasing slope makes rainfall infiltration weaker and leads to easier runoff generation. It can be also seen that the effects of slope on the initial runoff time become indistinct as rainfall intensity increases to a certain degree. The difference among the initial runoff-yielding times on different slopes was obvious under a rainfall intensity of 30 mm/h: the longest time was 12.43 min on the 5° slope and the shortest was 6.17 min on the 25° slope. With increased rainfall intensity, the difference gradually diminishes. At a rainfall intensity of 120 mm/h, the difference between the maximum and the minimum is reduced from 6.26 min under rainfall intensity of 30 mm/h to 1.63 min. It can be seen that slope is a key factor affecting runoff generation when rainfall intensity is relatively small, and the initial runoff-yielding time decreases significantly as the slope increases. With rising rainfall intensity, the effect of rainfall intensity on runoff generation became more and more important, and the differences of the initial runoff-yielding times on different slopes lessen.

3.3. Effects of Slope on TN Losses

3.3.1. Effects of slope on TN Load in Runoff

Figure 8a reflects TN loads in runoff on different slopes under different rainfall intensities. In Figure 8b, various rainfall intensities on certain slopes were averaged to eliminate the influence of rainfall intensity on TN load. The correlation between slope and TN load can be expressed as by a quadratic polynomial, and the square of correlation coefficient (R^2) is 0.9745, as shown in Figure 8b.

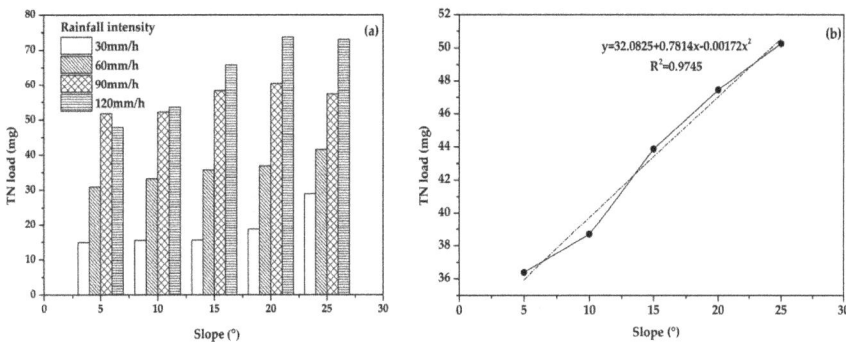

Figure 8. (**a**) Effects of slopes on TN loads under different rainfall intensities; (**b**) Correlation between slope and TN load.

Generally, TN load rises as slope increases, under various rainfall intensity conditions. The influence of slope on TN load can be explained by the following aspects. For steeper slopes, the scouring effect is stronger, a larger amount of runoff is generated, more sediment is transported, and more serious TN loss occurs. Previous studies have shown that there is a critical value for TN loss on slopes, so that soil erosion and TN loss do not increase limitlessly with greater slope [42]. In our experiments, slight drops in Figure 8a also supported the idea that TN load might reach a peak with the rise of slope. TN load decreased slightly as slope increased from 20° to 25° under rainfall intensities of 90 mm/h and 120 mm/h.

3.3.2. Effects of Slope on TN Concentration in Runoff

Figure 9 shows the correlation between slope and TN concentration, which can be expressed as a quadratic polynomial where the square of correlation coefficient (R^2) is 0.9218. In Figure 9, four rainfall intensities on varying slope were averaged in order to eliminate the influence of rainfall intensity on TN concentration.

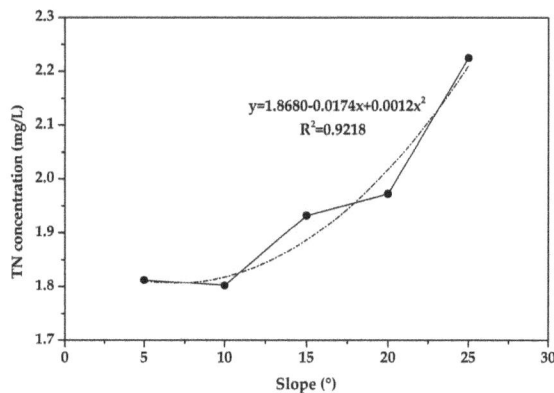

$$y=1.8680-0.0174x+0.0012x^2$$
$$R^2=0.9218$$

Figure 9. Correlation between slope and TN concentration.

TN concentration showed a positive correlation with slope, and increases were more obvious for steeper slopes. It can be explained that the steeper the slope was, the faster the velocity of water flow became, the stronger the scouring effect, the worse the soil erosion, so that nitrogen dissolution and adsorbed nitrogen loss are greater. However, runoff generated from different slopes does not vary very much with rainfall intensity; therefore, TN concentration rises with the increase of slope. Moreover, the increases were more obvious for steeper slopes, which also indicated that the increases of TN loads were more significant than those of runoff for greater slopes.

3.3.3. Effects of Slope on the Initial Runoff-Yielding Time

In our experiments, the relationships between slope and the initial runoff-yielding times showed negative correlations for all rainfall intensities (Figure 10). For a steeper slope, the contact time between rainfall and sloping field was shorter, the gravity of rainfall drops in slope surface direction became larger, rainfall infiltration was weakened, and runoff formed more easily; hence the initial runoff-yielding time was shorter. Figure 10 also shows that the differences among initial runoff-yielding times under various rainfall intensities diminish gradually with increasing of slope. Therefore, it can be deduced that slope plays a more important role in runoff generation as it increases, due to the increasing impacts of various factors.

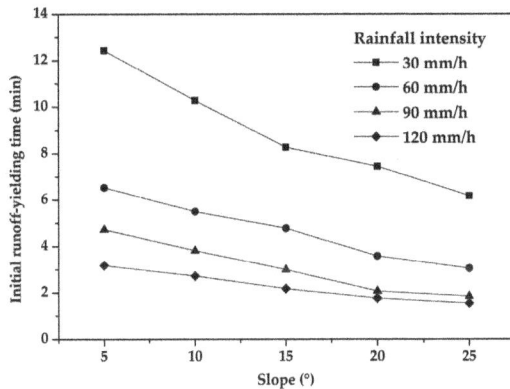

Figure 10. Correlations between slopes and the initial rainfall-yielding times.

4. Conclusions

In this study, the influence mechanisms of rainfall and terrain characteristics on total nitrogen (TN) losses from regosol were revealed by a series of artificial rainfall experiments. Natural rainfall and terrain conditions were simulated in 20 designed scenarios, which consisted of four rainfall intensities (30 mm/h, 60 mm/h, 90 mm/h, and 120 mm/h) and five slopes (5°, 10°, 15°, 20°, and 25°), with runoff-yielding time lasting 40 min.

The results showed that there were positive linear relationships between precipitations and TN loads under different slopes. There were delays before runoff generation, which demonstrated that there was an interval between the beginning of rainfall and runoff generation. In contrast, negative linear relationships represented the correlations between precipitations and TN concentrations, and the decreases of TN concentrations with increased precipitation were obvious on sloping fields at 5° and 25° of slope. The effects of rainfall on TN loss indicated that total precipitation was an important influencing factor and driving force for TN loss. Due to the obvious relationship between precipitation and rainfall intensity, the effects of rainfall intensity on TN load were similar to those of precipitation on TN load. As far as the effects of rainfall intensity on TN concentration were considered, they presented as a negative correlation. Similarly, the initial runoff-yielding time also showed a negative correlation with increased rainfall intensity. The initial runoff-yielding time was short for a steep slope, and the differences among the initial runoff-yielding times under designed slopes decreased as the rainfall intensity increased. TN load generally increased with an increase of slope, but not limitlessly, which can be deduced by slight drops of TN load on slopes with 25° comparing to TN loads on slopes with 20° under rainfall intensities of 90 mm/h and 120 mm/h. As for the relationship between slope and TN concentration, results showed that it was a positive one, with upward trends of TN concentrations with increasing slope more obvious for steeper slopes. Initial runoff-yielding times exhibited downward trends with increasing of slope, and the differences under various rainfall intensities diminished gradually with increasing of slope. It could be concluded that the steeper the slope was, the more serious the TN loss was.

Other research has shown that TN losses increased with increased rainfall intensity and TN concentration in runoff did not limitlessly increase, and those results were confirmed by this study [42,43]. The runoff volume increased with the increasing rainfall intensity and the increasing soil moisture content, and decreased with the increasing vegetation cover. These factors also significantly affected the losses of N. In other words, a longer and more intense rainfall resulted in a higher loss of N [23]. The N concentration in runoff was high at an early stage, and then reduced with increasing time [44]. All of this research presented conclusions similar to those of this study. This study reveals the effects of precipitation, rainfall intensity, and slope on TN losses from regosol with quantitative correlation analysis, which has not been discussed thoroughly in previous studies. This study provides theoretical support for NPS pollution simulation and pollution control in reference to these influential factors and their influence mechanisms. The effects of rainfall and terrain on dissolved nitrogen in runoff and adsorbed nitrogen on sediments are also shown here. In the future, we hope to determine the critical slope value of TN losses from regosol and the influence mechanisms of natural rainwater and underlying surfaces on TN loss under multiple rainfalls.

Acknowledgments: This research work was funded by the National Natural Science Foundation of China (51309097), the National Key Scientific and Technological Projects of the PRC (2014ZX07104-005), and the Fundamental Research Funds for the Central Universities of PRC. The authors gratefully acknowledge the financial support of the programs and agencies.

Author Contributions: Xiaowen Ding conceived and designed the experiments; Ming Lin performed the experiments; Guihong Jiang contributed to data measurement; Ying Xue analyzed the data; Xiaowen Ding and Ying Xue wrote the paper.

Conflicts of Interest: The authors declare no conflicts of interest.

References

1. Kalavrouziotis, I.K.; Arambatzis, C.; Kalfountzos, D.; Varnavas, S.P. Wastewater Reuse Planning in Agriculture: The Case of Aitoloakarnania, Western Greece. *Water* **2011**, *3*, 988–1004. [CrossRef]
2. Sui, J.; Wang, J.; Gong, S.; Xu, D.; Zhang, Y. Effect of Nitrogen and Irrigation Application on Water Movement and Nitrogen Transport for a Wheat Crop under Drip Irrigation in the North China Plain. *Water* **2015**, *7*, 6651–6672. [CrossRef]
3. Ahmed, W.; Hughes, B.; Harwood, V.J. Current Status of Marker Genes of Bacteroides and Related Taxa for Identifying Sewage Pollution in Environmental Waters. *Water* **2016**, *8*, 231. [CrossRef]
4. Park, Y.S.; Engel, B.A.; Harbor, J. A Web-Based Model to Estimate the Impact of Best Management Practices. *Water* **2014**, *6*, 455–471. [CrossRef]
5. Ahn, S.R.; Kim, S.J. The Effect of Rice Straw Mulching and No-Tillage Practice in Upland Crop Areas on Nonpoint-Source Pollution Loads Based on HSPF. *Water* **2016**, *8*, 106. [CrossRef]
6. Alvarez, S.; Asci, S.; Vorotnikova, E. Valuing the Potential Benefits of Water Quality Improvements in Watersheds Affected by Non-Point Source Pollution. *Water* **2016**, *8*, 112. [CrossRef]
7. Blumstein, M.; Thompson, J.R. Land-use impacts on the quantity and configuration of ecosystem service provisioning in Massachusetts, USA. *J. Appl. Ecol.* **2015**, *52*, 32–47. [CrossRef]
8. Datri, C.W.; Pray, C.L.; Zhang, Y.X.; Nowlin, W.H. Nutrient enrichment scarcely affects ecosystem impacts of a non-native herbivore in a spring-fed river. *Freshw. Biol.* **2015**, *60*, 551–562. [CrossRef]
9. Xing, W.M.; Yang, P.L.; Ren, S.M.; Ao, C.; Li, X.; Gao, W.H. Slope length effects on processes of total nitrogen loss under simulated rainfall. *Catena* **2016**, *139*, 73–81. [CrossRef]
10. Shen, Z.Y.; Chen, L.; Hong, Q.; Qiu, J.L.; Xie, H.; Liu, R.M. Assessment of nitrogen and phosphorus loads and casual factors from different land use and soil types in the Three Gorges Reservoir Area. *Sci. Total Environ.* **2013**, *454–455*, 383–392. [CrossRef] [PubMed]
11. Recanatesi, F.; Ripa, M.N.; Leone, A.; Luigi, P.; Luca, S. Erratum to: Land use, climate and transport of nutrients: Evidence emerging from the Lake Vico case study. *Environ. Manag.* **2013**, *52*, 503–513. [CrossRef] [PubMed]
12. Gaddamwar, A.G.; Rajput, P.R. Analytical study of Bembala damp water for fishery capacity, portability and suitability for agricultural purposes. *Int. J. Environ. Sci.* **2012**, *2*, 1278–1283.
13. Laine-Kaulio, H.; Koivusalo, H.; Komarov, A.S.; Lappalainen, M.; Launiainenc, S.; Laurénc, A. Extending the ROMUL model to simulate the dynamics of dissolved and sorbed C and N compounds in decomposing boreal mor. *Ecol. Model.* **2014**, *272*, 277–292. [CrossRef]
14. Darmawi, S.; Burkhardt, S.; Leichtweiss, T.; Weber, D.A.; Wenzel, S.; Janek, J.; Elm, M.T.; Klar, P.J. Correlation of electrochromic properties and oxidation states in nanocrystalline tungsten trioxide. *Phys. Chem. Chem. Phys.* **2015**, *17*, 15903–15911. [CrossRef] [PubMed]
15. Hu, B.Q.; Wang, H.J.; Yang, Z.S.; Sun, X.X. Temporal and spatial variations of sediment rating curves in the Changjiang (Yangtze River) basin and their implications. *Quat. Int.* **2011**, *230*, 34–43. [CrossRef]

16. Hu, G.Y.; Dong, Z.B.; Lu, J.F.; Yan, C.Z. Driving forces responsible for aeolian desertification in the source region of the Yangtze River from 1975 to 2005. *Environ. Earth Sci.* **2012**, *66*, 257–263. [CrossRef]

17. Pan, B.Z.; Wang, H.Z.; Ban, X.; Yin, X.A. An exploratory analysis of ecological water requirements of macroinvertebrates in the Wuhan branch of the Yangtze River. *Quat. Int.* **2015**, *380*, 256–261. [CrossRef]

18. Liu, B.; Hu, Q.; Wang, W.P.; Zeng, X.F.; Zhai, J.Q. Variation of actual evapotranspiration and its impact on regional water resources in the Upper Reaches of the Yangtze River. *Quat. Int.* **2011**, *244*, 185–193.

19. Ding, X.W.; Shen, Z.Y.; Hong, Q.; Yang, Z.F.; Wu, X.; Liu, R.M. Development and test of the Export Coefficient Model in the upper reach of the Yangtze River. *J. Hydrol.* **2010**, *383*, 233–244. [CrossRef]

20. Zhang, N.; He, H.M.; Zhang, S.F.; Jiang, X.H.; Xia, Z.Q.; Huang, F. Influence of reservoir operation in the upper reaches of the Yangtze River (China) on the inflow and outflow regime of the TGR-based on the Improved SWAT Model. *Water Resour. Manag.* **2012**, *26*, 691–705. [CrossRef]

21. Li, C.L.; Zhou, J.Z.; Ouyang, S.; Wang, C.; Liu, Y. Water Resources Optimal Allocation Based on Large-scale Reservoirs in the Upper Reaches of Yangtze River. *Water Resour. Manag.* **2015**, *29*, 2171–2187. [CrossRef]

22. Zhao, X.L.; Jiang, T.; Du, B. Effect of organic matter and calcium carbonate on behaviors of cadmium adsorption-desorption on/from purple paddy soils. *Chemosphere* **2014**, *99*, 41–48. [CrossRef] [PubMed]

23. Liu, R.M.; Wang, J.W.; Shi, J.H.; Chen, Y.X.; Sun, C.C.; Zhang, P.P.; Shen, Z.Y. Runoff characteristics and nutrient loss mechanism from plain farmland under simulated rainfall conditions. *Sci. Total Environ.* **2014**, *468*, 1069–1077. [CrossRef] [PubMed]

24. Patil, R.H.; Laegdsmand, M.; Olesen, J.E.; Porter, J.R. Effect of soil warming and rainfall patterns on soil n cycling in northern europe. *Agric. Ecosyst. Environ.* **2010**, *139*, 195–205. [CrossRef]

25. Ding, X.W.; Shen, Z.Y.; Liu, R.M.; Chen, L.; Lin, M. Effects of ecological factors and human activities on nonpoint source pollution in the upper reach of the Yangtze River and its management strategies. *Hydrol. Earth Syst. Sci. Discuss.* **2013**, *11*, 691–721. [CrossRef]

26. Kakuturu, S.; Chopra, M.; Hardin, M.; Wanielista, M. Total nitrogen losses from fertilized turfs on simulated highway slopes in Florida. *J. Environ. Eng.* **2013**, *139*, 829–837. [CrossRef]

27. Shin, J.; Gil, K. Effect of rainfall characteristics on removal efficiency evaluation in vegetative filter strips. *Environ. Earth Sci.* **2014**, *72*, 601–607. [CrossRef]

28. Diaz, D.A.R.; Sawyer, J.E.; Barker, D.W.; Mallarino, A.P. Runoff Nitrogen Loss with Simulated Rainfall Immediately Following Poultry Manure Application for Corn Production. *Soil Sci. Soc. Am. J.* **2010**, *74*, 221–230. [CrossRef]

29. Hao, C.L.; Yan, D.H.; Xiao, W.H.; Shi, M.; He, D.W.; Sun, Z.X. Impacts of typical rainfall processes on nitrogen in typical rainfield of black soil region in northeast china. *Arab. J. Geosci.* **2015**, *8*, 1–13. [CrossRef]

30. Wang, H.J.; Shi, X.Z.; Yu, D.S.; Weindorf, D.C.; Huang, B.; Sun, W.X.; Ritsema, C.J.; Milne, E. Factors determining soil nutrient distribution in a small-scaled watershed in the purple soil region of Sichuan Province, China. *Soil Tillage Res.* **2009**, *105*, 35–44. [CrossRef]

31. Wilson, C.; Weng, Q. Assessing surface water quality and its relation with urban land cover changes in the Lake Calumet area, Greater Chicago. *Environ. Manag.* **2010**, *45*, 1096–1111. [CrossRef] [PubMed]

32. Chen, X.; Huang, Y.; Zhao, Y.; Mo, B.; Mi, H. Comparison of loess and purple rill erosions measured with volume replacement method. *J. Hydrol.* **2015**, *530*, 476–483. [CrossRef]

33. Shen, Z.Y.; Chen, L.; Ding, X.W.; Hong, Q.; Liu, R.M. Long-term variation (1960–2003) and causal factors of non-point-source nitrogen and phosphorus in the upper reach of the Yangtze River. *J. Hazard. Mater.* **2013**, *252*, 45–56. [CrossRef] [PubMed]

34. Philippot, L.; Spor, A.; Hénault, C.; Bru, D.; Bizouard, F.; Jones, C.M.; Sarr, A.; Maron, P.A. Loss in microbial diversity affects nitrogen cycling in soil. *ISME J.* **2013**, *7*, 1609–1619. [CrossRef] [PubMed]

35. Hassink, J. The capacity of soils to preserve organic c and n by their association with clay and silt particles. *Plant Soil* **1997**, *191*, 77–87. [CrossRef]

36. Marcarelli, A.M.; Wurtsbaugh, W.A. Temperature and nutrient supply interact to control nitrogen fixation in oligotrophic streams: An experimental examination. *Limnol. Oceanogr.* **2006**, *51*, 2278–2289. [CrossRef]

37. Huang, J.; Wu, P.; Zhao, X.N. Effects of rainfall intensity, underlying surface and slope gradient on Soil infiltration under simulated rainfall experiments. *Catena* **2013**, *104*, 93–102. [CrossRef]

38. Hong, J.; Li, T.; Xuan, H.; Yang, X.; He, Z. Effects of pH and low molecular weight organic acids on competitive adsorption and desorption of cadmium and lead in paddy soils. *Environ. Monit. Assess.* **2012**, *184*, 6325–6335.

39. Gondar, D.; López, R.; Antelo, J.; Fiol, S.; Arce, F. Effect of organic matter and pH on the adsorption of metalaxyl and penconazole by soils. *J. Hazard. Mater.* **2013**, *260*, 627–633. [CrossRef] [PubMed]
40. Gross, A.; Boyd, C.E.; Seo, J. Evaluation of the Ultraviolet Spectrophotometric Method for the Measurement of Total Nitrogen in Water. *J. World Aquac. Soc.* **1999**, *30*, 388–393. [CrossRef]
41. Jiang, F.; Zhou, K.; Deng, H.; Li, X.; Zhong, Y. Study on Enterprise's Employees' Safety Training Based on SPSS. In Proceedings of the 2009 International Conference on Computational Intelligence and Software Engineering, Wuhan, China, 11–13 December 2009; IEEE: New York, NY, USA, 2009; pp. 1–4.
42. Shao, X.J.; Wang, H.; Hu, H.W. Experimental and modeling approach to the study of the critical slope for the initiation of rill flow erosion. *Water Resour. Res.* **2005**, *41*, W12405. [CrossRef]
43. Schwenke, G.D.; Haigh, B.M. The interaction of seasonal rainfall and nitrogen fertiliser rate on soil N_2O emission, total N loss and crop yield of dryland sorghum and sunflower grown on sub-tropical Vertosols. *Soil Res.* **2016**, *54*, 604–618. [CrossRef]
44. Qian, J.; Zhang, L.P.; Wang, W.Y.; Liu, Q. Effects of vegetation cover and slope length on nitrogen and phosphorus loss from a sloping land under simulated rainfall. *Pol. J. Environ. Stud.* **2014**, *23*, 835–843.

water

MDPI

Article

Using Modeling Tools to Better Understand Permafrost Hydrology

Clément Fabre [1,*], Sabine Sauvage [1], Nikita Tananaev [2,3], Raghavan Srinivasan [4], Roman Teisserenc [1] and José Miguel Sánchez Pérez [1]

[1]   ECOLAB, Université de Toulouse, CNRS, INPT, UPS, 31055 Toulouse, France;
     sabine.sauvage@univ-tlse3.fr (S.S.); roman.teisserenc@ensat.fr (R.T.);
     jose-miguel.sanchez-perez@univ-tlse3.fr (J.M.S.P.)
[2]   P.I. Melnikov Permafrost Institute, SB RAS, Merzlotnaya Str. 36, 677010 Yakutsk, Sakha Republic, Russia;
     nikita.tananaev@gmail.com
[3]   Ugra Research Institute of Information Technologies, Mira Str. 151, 628011 Khanty-Mansiysk, Russia
[4]   Spatial Science Laboratory in the Department of Ecosystem Science and Management,
     Texas A&M University, College Station, TX 77845, USA; r-srinivasan@tamu.edu
*   Correspondence: clement.fabre21@gmail.com

Academic Editor: Karim Abbaspour
Received: 27 April 2017; Accepted: 26 May 2017; Published: 10 June 2017

Abstract: Modification of the hydrological cycle and, subsequently, of other global cycles is expected in Arctic watersheds owing to global change. Future climate scenarios imply widespread permafrost degradation caused by an increase in air temperature, and the expected effect on permafrost hydrology is immense. This study aims at analyzing, and quantifying the daily water transfer in the largest Arctic river system, the Yenisei River in central Siberia, Russia, partially underlain by permafrost. The semi-distributed SWAT (Soil and Water Assessment Tool) hydrological model has been calibrated and validated at a daily time step in historical discharge simulations for the 2003–2014 period. The model parameters have been adjusted to embrace the hydrological features of permafrost. SWAT is shown capable to estimate water fluxes at a daily time step, especially during unfrozen periods, once are considered specific climatic and soils conditions adapted to a permafrost watershed. The model simulates average annual contribution to runoff of 263 millimeters per year (mm yr^{-1}) distributed as 152 mm yr^{-1} (58%) of surface runoff, 103 mm yr^{-1} (39%) of lateral flow and 8 mm yr^{-1} (3%) of return flow from the aquifer. These results are integrated on a reduced basin area downstream from large dams and are closer to observations than previous modeling exercises.

Keywords: permafrost; modeling; hydrology; water; Yenisei River; SWAT

1. Introduction

Ongoing climate change became consensual through a plethora of studies [1]. This global warming is particularly important at high latitudes because of the Arctic amplification effect [2]. Significant alteration of the hydrological cycle and, subsequently, in other global cycles is expected in Arctic watersheds [3–6]. Arctic hydrology is poorly understood, and largely understudied, compared to lower and mid–latitudes [7,8]. Arctic catchments are genuinely remote areas, where data acquisition is complicated by natural conditions, logistics and societal issues. Field studies are scarce and these data, particularly for the Russian territory, are virtually unexposed to a wider international audience [9]. Most of the largest Arctic rivers are followed by an extremely limited number of gauging stations, which is steadily declining throughout last decades [10].

The complexity of Arctic hydrology is also a challenge for the modelers. Insolation seasonality affects the energy state of Arctic watersheds and an enormous difference in energy input between the seasons. Hydrological processes follow this pattern, and are virtually stagnant in winter while highly dynamic in transition seasons, i.e., during spring freshet [11]. Water in the Arctic alternates between ice and liquid phases on a seasonal basis, both in surface and subsurface compartments. This regular phase transition implies sound modifications of runoff storage and pathways with the presence of permafrost, controlled by the active layer depth and its intra-annual fluctuations [12].

Permafrost hydrology emerged as a distinct branch of hydrological sciences in the U.S. in early 1980s, when sufficient data on the water balance and runoff regime of the High Arctic Rivers became available [13]. It has been widely acknowledged since then that the interaction of water fluxes and soil freeze-thaw processes has by far the most important hydrological effect in permafrost environment. The seasonally-thawed (active) layer accommodates the totality of hydrogeochemical activities in the continuous permafrost areas, the statement which became a 'mantra' since this keystone publication by M.-K. Woo saw the light. Water transport in the soil is only possible when the active layer is thawed, since frozen soil effectively acts as an aquitard [14]. Depending on soil properties and regional climate, the active layer can attain the thickness from first tens of centimeters to more than 2 m in the continuous permafrost zone [15,16]. Thicker active layer limits the occurrence of surface flow, but promotes deeper percolation of water, participation of deeper soil layers in water transfer in pores or unfrozen corridors [17].

The seasonality of freeze-thaw processes affects the changes in hydraulic properties of permafrost soils, and the watershed hydrology in general [13]. Early in winter the active layer is completely frozen, and flow is interrupted even in major Arctic catchments, e.g., the Yana (basin area $A = 90,000$ km^2) and the Anabar ($A = 107,000$ km^2) Rivers. During the spring, solar radiation penetrates the snow cover, starting the annual cycle of active layer development. By late autumn, the active layer reaches its deeper limit and water can travel freely in the upper meters of the soil profile. During winter, the active layer freezes again from the bottom and from the top and snow starts to accumulate.

Hence, while the soil is part-time frozen, precipitation mostly follows the surface and shallow subsurface pathways. Deep subsurface flow enters the hydrological stage seasonally, mostly in discontinuous permafrost. Freeze-thaw processes affect moisture storage in soils by limiting the infiltration and partitioning of water fluxes between surface and subsurface compartments [9]. The average flow partitioning between these compartments at the outlet of a permafrost watershed should be completely different from those found in other latitudes. Hydrogeological regime is insufficiently studied in permafrost areas [7,18], though several efforts have been made in recent years both to compile existing observations and to model permafrost–groundwater interactions [12,19–21].

Previous experiments in the Arctic domain permitted the establishment of an estimation of this precipitations and flows distribution. They found a ratio close to 50/50 between rainfall and snow for the Yenisei watershed [22]. Only one study tested the flows distribution for Arctic watersheds and they established a 60/40 ratio for surface and subsurface flows [17]. Snowpack contribution to the water balance has an important impact on the behavior of the river throughout the year. The release of a large part, up to 50% in some regions [9], of precipitation by snowmelt is responsible of a spring freshet in the Arctic rivers, occurring in May and June [23]. Arctic river ice breaks up between April and June. One-third to roughly half of the annual discharge delivered to the Arctic Ocean occurs from May to July [23].

The hydrological effects of the frozen ground are best detectable in the regions completely underlain by permafrost, i.e., in continuous permafrost zone with 90—100% areal coverage. In discontinuous permafrost, with 50–90% coverage, unfrozen areas represent significant pathways for both shallow and deep subsurface waters [12]. Farther south, in the sporadic (10–50%) and isolated (less than 10%) permafrost zones, hydrological significance of frozen ground is negligible, and is perceivable only locally. These four permafrost types constitute respectively 54%, 16%, 14% and 16% of permafrost soils in the Northern Circumpolar Region [24].

Coupled heat and water fluxes calculation represents the major concern in permafrost hydrology and this issue is far from being resolved [13,25]. Permafrost models perform predominantly at the large scale and rarely downscale for hydrological processes [26], or, being downscaled and adjusted for hydraulic effects, are not designed to reproduce water fluxes at the catchment scale [27]. Hydrological models, in their turn, largely oversimplify soil heat transfer and phase transitions in the subsurface compartment [28], computationally heavy [29], require over-calibration [30], or are explicitly incapable to upscale point, or stand-scale, permafrost features to the whole catchment volume [31]. Promising results have been obtained recently using a modular Cold Regions Hydrological Model (CRHM) for two permafrost watersheds in western China [32]. Introduction of water-permafrost interactions to the surface runoff module of a global land surface model (JULES) showed unexpectedly poor performance of the snowmelt water routing module, making JULES incapable to reproduce spring flood peak on the Lena River [33]. Snowmelt representation in other hydrological models deems to be imprecise [28]. The effect of permafrost continuity, in a spatial context, is rarely taken into account explicitly. Hülsmann et al. [34] diagnosed major modeling issues in a relevant study.

This work aims to analyze, to understand and to quantify water fluxes dynamics for a big permafrost watershed, the Yenisei watershed scale (2,540,000 km^2 [23]) using the hydrological modeling approach coupled to discharge data at daily time scale at the outlet of the watershed. The objectives of the study are:

- to evaluate the role of permafrost soils in water transfer,
- to identify the hydrologically relevant features for each permafrost class, and the runoff routing through a large Arctic watershed of the Yenisei River,
- to characterize and quantify the different hydrological pathways,
- to perform hydrological modeling of the Yenisei River at daily time step, accounting for the hydrological functions of permafrost, in order to allow predictions under non-stationary conditions.

2. Results

2.1. Hydrological Response

2.1.1. Daily Modeled Discharge

Daily discharge is correctly predicted during the validation period (Figure 1). Some peaks are underestimated by the model (e.g., May 2005) while others are overestimated (e.g., May 2008) but the global behavior of the modeling is good with a good detection of the high flow periods. After the freshets, our modeling underestimates sometimes the recession (e.g., 2006) and the low flows (e.g., 2009 and 2010). The statistical performance is satisfactory with a Nash and Sutcliffe Efficiency (NSE) and a coefficient of determination (R^2) above 0.75 in the calibration and in the validation period and with reasonable percent bias (PBIAS) and root mean square error-observations standard deviation ratio (RSR) (Figure 1). For a daily time step modeling, our results are considered as very good. Table 1 details the goodness of indices for low and high flow periods. The discharges above 30,000 m^3 s^{-1} represent a small part of the observed discharges and are not as well predicted as the low flow discharges which is especially seeable with the R^2. The PBIAS confirms this statement by revealing an underestimation of high flows by the model.

Table 1. Goodness of indices discretized by flow periods. Here, the discharge is considered high when it exceeds 30,000 m^3 s^{-1}.

	Number of Values		Daily Time Step Modeling		Monthly Time Step Modeling	
			Calibration	Validation	Calibration	Validation
Low flow periods	1175	NSE	0.70	0.66	0.87	0.74
		R^2	0.55	0.50	0.57	0.80
		PBIAS	2.0	−1.0	7.6	2.0
		RSR	0.43	0.67	0.36	0.51
High flow periods	189	NSE	0.75	0.78	0.71	0.86
		R^2	0.37	0.44	0.08	0.56
		PBIAS	25.7	22.0	30.5	19.2
		RSR	0.52	0.45	0.54	0.37

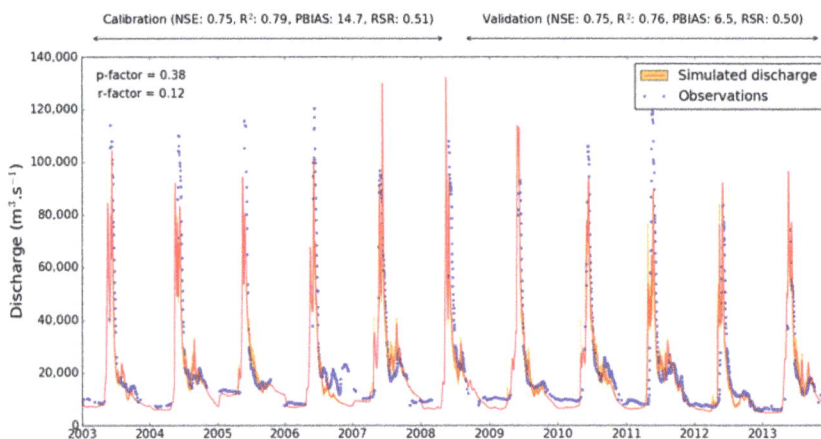

Figure 1. Daily simulated hydrograph compared to daily observations at the Yenisei outlet with goodness of indices. In orange is represented the 95PPU zone resulting from the last SWAT-CUP run. We observe a good dynamic and a good representation of spring floods.

2.1.2. Comparison with the Modeling at a Monthly Time Step

Compared to the daily modeled discharge, the monthly time step modeling underestimates considerably the freshets with peaks not exceeding 70,000 m^3 s^{-1} (Figure 2). But, low flow periods are as well represented with this time step than with daily time step and the same uncertainties remain concerning under and overestimations by the model. However, our modeling at a monthly time step predict correctly the average monthly discharge during the validation period. In details, the monthly predictions are often higher than the daily predictions during the increase in discharge (Figure 2). On the contrary, the recession and the low flow periods are higher in the daily time step modeling. The change in the time step has a low impact in the statistical performance which is in the same range than the one in Figure 1. The NSE and the R^2 increase by less than 0.1. The rising is due to the loss of information by changing the time step. The crushing of the predictions at a monthly time step are underlined by a lower R^2 during high flow periods (Table 1). Nevertheless, the statistical analysis during low flow periods is close to the one for daily time step modeling.

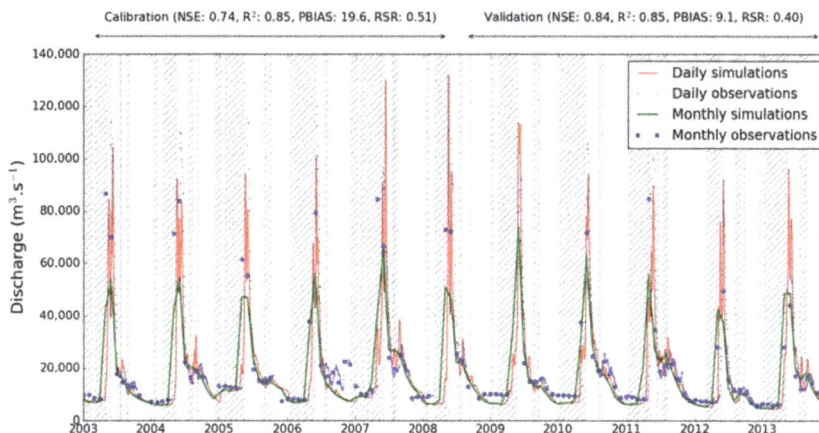

Figure 2. Simulation at a monthly time step at the Yenisei outlet compared to the daily simulation and observations shown in Figure 1 with goodness of indices for the monthly simulation. The peaks are not as well represented but the dynamic low flows are still respected. The hatched zones represents the periods where the monthly modeled discharge is higher than the daily modeled discharge.

2.2. Modeled Water Balance

By integrating the reservoirs and the total area of the watershed in the global water balance, the modeled mean annual discharge at the outlet is 237.8 millimeters per year (mm yr^{-1}) with a standard deviation of 38.7 mm yr^{-1}. The yearly average predicted water balance is close to the observed one with a lack of 3.8 mm yr^{-1} of water (Table 2).

As inputs flows, the model returns a ratio of 56/44 for rainfall/snowfall distribution with amounts of respectively 265.7 and 206.3 mm yr^{-1} resulting in an average annual precipitation amount of 472 mm yr^{-1}. Evapotranspiration is returned as 199 mm yr^{-1}, or 42% of total precipitation, with a potential evapotranspiration of 364 mm yr^{-1}. Sublimation reaches 10.5 mm yr^{-1}, or 5.5% of the annual snowfall. Percolation is as low as 11 mm yr^{-1}.

Concerning the flows entering the river, the reservoirs are excluded from the average water balance. Only are considered here fluxes flowing on the slopes of the watershed. The following results are only integrating the modeled basin surface, that is to say 1,383,398 km^2 as mentioned before, or approximately 58% of the total watershed area at the Igarka gauging station (2,440,000 km^2). Simulated average annual water flow is 263 mm yr^{-1}. Surface runoff contributes to the streamflow for 152 mm yr^{-1}, lateral runoff represents 103 mm yr^{-1} of the streamflow and the return flow from the shallow aquifer participates for 8 mm yr^{-1} (which corresponds to 58%, 39% and 3% of the contribution to the streamflow). The surface runoff is dominant during the discharge peak after snowmelt (Figure 3). The surface runoff is still dominant after the increase in discharge at the beginning of the third period (see Materials and Methods) because permafrost has not yet thawed. The lateral runoff is mostly present during the third period after the recession but explains most of the discharge during the fourth period when permafrost is freezing again from the top and the bottom and when snow starts to accumulate (see Materials and Methods). The groundwater flow is low regardless the season but follow the same trend as the lateral runoff being present after permafrost unfreezing.

Regarding the respect of the water flows distribution in the basin, the hypotheses on the permafrost properties which have been done upper (see Materials and Methods) seem good. By comparing average daily precipitations and average air temperature in the whole basin and the average flows distribution, we underline the water stock which is not restituted to the outlet before snowmelt (Figure 3). Temperature is the main vector in the distribution of water by melting snow and unfreezing permafrost. At the beginning of the snowmelt, the discharge at the outlet comes from surface runoff while during the recession, because the active layer is larger, lateral runoff explains a bigger part of the discharge.

Table 2. Interannual mean of water fluxes at the Yenisei outlet compared to previous studies. The discharges and specific discharges are calculated with the watershed area at the outlet: 2,440,000 km².

Source	ArcticGRO Dataset	This Study	Finney et al. (2012)	Ducharne et al. (2003)	Alkama et al. (2006)	Nohara et al. (2006)	Arora (2001)	Yang et al. (2015)
Years	2003–2013	2003–2013	1989–1999	1980–1988			1980–1994	1901–2010
Data	Observed	Simulated	Simulated	Simulated	Simulated	Simulated	Simulated	Simulated
Model		SWAT	TOPMODEL	RiTHM	LMDZ	TRIP	AMIP2	SDGVM
Runoff (mm yr^{-1})	241.6 ± 31.3	237.8 ± 38.7	140	140.6	151.7 ± 44.3	179	189	273.8
Discharge (10^9 m^3 yr^{-1})	589.5 ± 76.4	580.2 ± 94.4	341.6	343.1	370.1 ± 108.1	436.8	461.2	668.1
Difference with observed data (%)		1.6	42.1	41.8	37.2	26.0	21.8	13.3

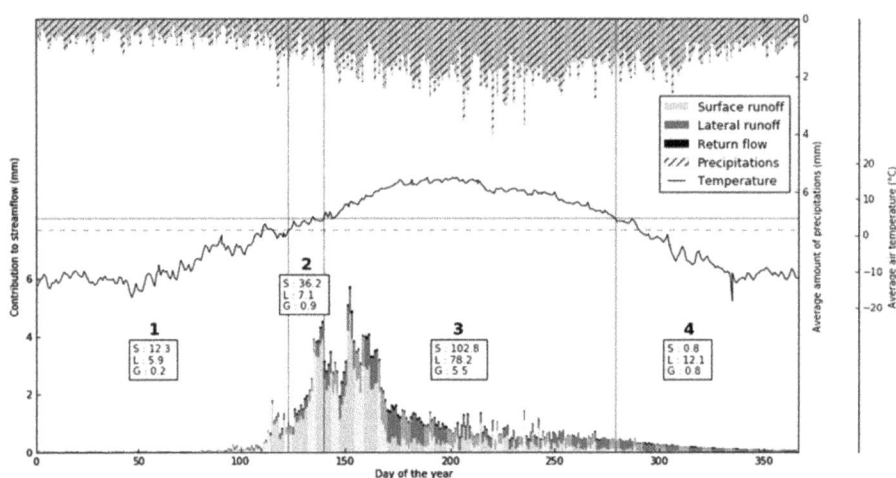

Figure 3. Interannual spatial average contribution to the streamflow per day for all the HRUs. We highlight 4 periods with 4 different hydrological behaviors corresponding to the conceptual model described in Materials and methods. In the boxes are noted the accumulated contribution on each period for surface runoff (S), lateral runoff (L) and groundwater (G). The average amount of precipitations per day and the average air temperature in the whole basin are also represented. The horizontal lines correspond to the modeled limits of snowfall (dashed line, 1.52 °C) and snowmelt (solid line, 4.75 °C) temperature (see Table 2). The vertical lines which separates the periods represents the date when temperature reaches the limits defined.

2.3. Spatial Water Transfer

A dynamic study of the spatial distribution of annual mean fluxes of surface, subsurface and groundwater runoffs has been performed in order to follow the water flows in the watershed. We have focused on the year 2013 because it yields the best results compared to the rest of the study period (NSE: 0.79; R²: 0.87). Figure 4 shows cartographies of water flows contribution to streamflow at points of interest highlighted in Figure 5. The increase in discharge is mainly explained by surface runoff in the North of the basin and by lateral runoff in the South of the basin. This date underlines the unfreezing of the active layer which occurs firstly in the South and allows the lateral runoff. At the peak of discharge, the active layer has unfrozen and surface and lateral runoff are possible in a large part of the basin. Then, the decrease in discharge is mostly sustained by subsurface runoff and local precipitation events which completes the analysis done with the Figure 3. The groundwater contribution is only significant in the southern parts of the watershed where the permafrost is not present and seems to not impact significantly the watershed functioning.

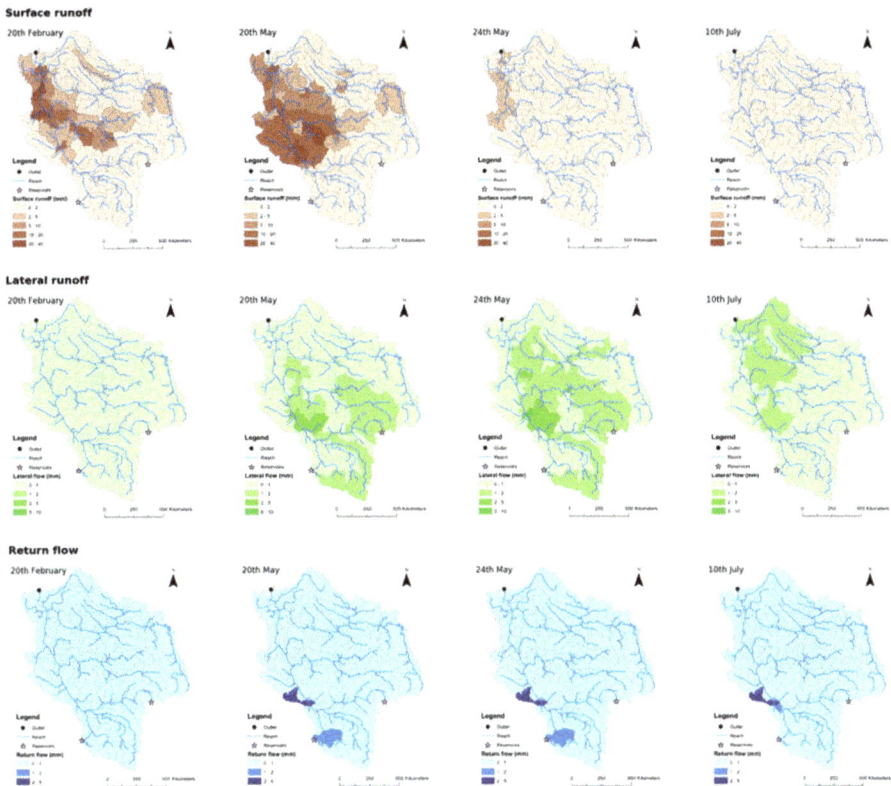

Figure 4. Spatial water flows dynamics at different key periods of the year 2013 in the Yenisei watershed (frozen period, unfreezing, peak of discharge and recession; see Figure 5). The strong disturbance in the surface and lateral runoffs during snowmelt and permafrost unfreezing is clear.

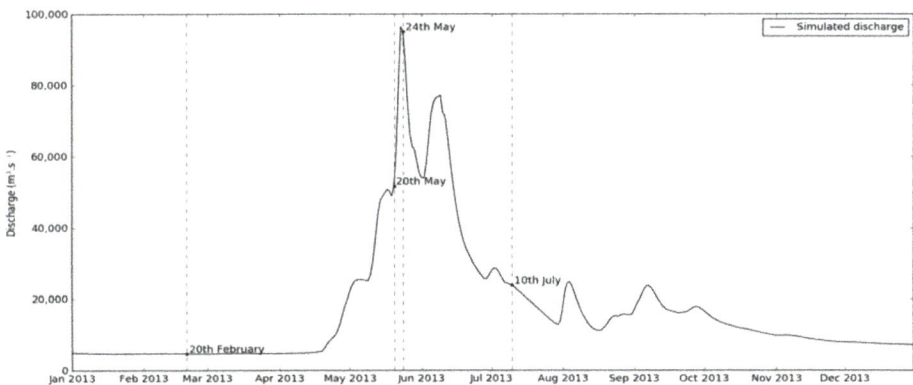

Figure 5. Representation of the modeled discharge at the Yenisei outlet.

The water contribution to streamflow are mostly dependent on the presence or not of permafrost (Figure 6). The surface runoff is doubled in sporadic zones which could be linked to a rapid increase in temperature and a rapid snowmelt in these regions. The lateral flow remain in the same range in each type of permafrost. But, in the permafrost free zones, the flow is lowered as water can be transferred in the aquifer and as expected, the groundwater flow is only significant in the zones where no permafrost is found. The absence of permafrost implies a recharge of the aquifer which provokes a return flow. Nevertheless, we observe a decrease from sporadic to isolated permafrost zones in both surface and lateral runoff. As isolated permafrost zones present higher temperature, this decrease could be linked to a highest evapotranspiration.

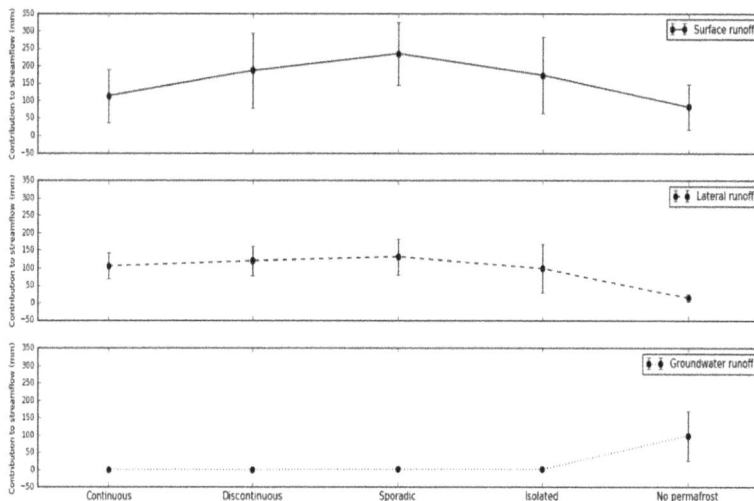

Figure 6. Average contribution to streamflow for each flow depending on the type of permafrost.

3. Discussion

3.1. Conceptualization of Permafrost Hydrology in the Yenisei Watershed and Outputs

Our conceptualization seems good regarding the results obtained in this study. Indeed, the modeled water cycle is consistent with the observed values. The resulting average flows between the compartments match observations made in other works on the field [17]. Concerning precipitation, the modeled total amount and the rain/snow ratio is in the observed range. The modeled evapotranspiration and sublimation are in the range of numerous observations made from Arctic watersheds in Russia, Canada and Alaska [22,30,33–36].

The simulated surface/subsurface ratio, which is 58/39 in this study, is close to observations made in Canadian watersheds with a ratio of 60/40 [17]. The low contribution (3%) of deep groundwater to the annual modeled streamflow is as expected for this hydrological system [19]. The average percolation rate is low, as it is naturally restricted in permafrost landscapes by an impermeable boundary of the active layer bottom.

The automatic calibration has allowed a sensitivity analysis to bring to light the most sensitive parameters calibrated. This automatic adjustment calibrates all the parameters together by creating a batch of parameters for a run and it returns the best combination of parameters after all runs. The strong influence of the snow behavior is clear in this model regarding the sensitivity analysis results (Table 4; see Materials and Methods). Snowmelt and meltwater routing are the main primers of the Yenisei River hydrology. Automatic calibration have not included the parameters DEP_IMP and

SLSOIL (depth to impermeable layer and slope length for lateral subsurface flow). They have been found to be extremely influential and they have implied strong changes in the hydrological regime, which pulled the results too far away from the reality.

Several parameters have been manually calibrated according to literature while others have been calibrated in order to return values in the hydrological cycle close to observations in literature. Ranges for snowfall and snowmelt parameters (SFTMP, SMTMP, SMFMX, SMFMN, SNO50COV and TIMP; see Materials and Methods) have been established by expertise and the final values have been kept for representing well the flows peaks and the distribution between snowfall and rainfall. The minimum snow water content that corresponds to 100% snow cover in millimeters (SNOCOVMX) has been approached with literature [37]. The DEP_IMP parameter, representing the maximum depth of the active layer, has been established with the work of Zhang et al. (2005) [15] as mentioned before. The SLSOIL parameter has been fixed for representing correctly the distribution between surface and subsurface runoff according to Carey and Woo [17] which seems to be a good representation of the flows distribution in permafrost soils. Finally, the three last parameters (CANMX, LAT_TTIME and ESCO) have been adjusted using the work of Hülsmann et al. [38].

The system definition could be discussed. By taking the reservoirs out of the conceptualization, we avoid strong controllers of the water and other elements transfer. Considering the two last reservoirs as water inlets is a good first approach in order to reproduce the behavior of water in these areas but can be improved. By integrating in the modeling the reservoirs managements, and by this way the whole upstream part of the watershed, the model would be enhanced making the refinement of our predictions possible. On the other hand, the discharge has been checked by a quick comparison between observations and predictions at the exit of the Boguchany reservoir on the Angara River and at the exits of the Nizhnyaya Tunguska and the Podkamennaya Tunguska tributaries with observed data available on the 2008–2013 period, which confirms our assumptions in different zones of the modeled watershed (see Materials and Methods). The introduction of an active layer in the modeling by using the parameter DEP_IMP allows a good representation of permafrost characteristics and thus a water flows distribution representative of Arctic watersheds.

Simulated water fluxes are closer to observations than in other previous studies (Table 2). It could be explained by our good representation of high flow periods while previous studies did not succeed in representing the snowmelt contribution to streamflow [33]. By integrating snowmelt and permafrost unfreezing events in the modeling, we can argue that this study returns more precise predictions than past researches. Nevertheless, approximately 3.8 mm yr^{-1} are still missing in the annual average discharge but the difference in the water balance is the lowest compared to past researches and is less than 10% of the total runoff (Table 2). A focus on the high flows during the whole period and on low flows which are underestimated (Table 1) during some summers (e.g., 2006, 2007 and 2008) could be done to allow a first good improvement of our modeling. These missing 3.8 mm yr^{-1} could also come from the low number of the real meteorological stations used as inputs. With data from only 9 stations, we could have missed precipitation events after the spring freshet in sub-catchments, as observed in Figure 3. Other precipitation data are available in the Yenisei watershed but not easily accessible. By collecting all these data, new stations could be implemented in the model and an improvement in the capacity of the model to represent low flow periods could be reached.

Finally, we have selected a small number of discharge data during freshets because the confidence level accorded to the rating curve method is low in Arctic watersheds due to the quick increase in discharge. It implies difficulties to study the hydrology in the watershed. Our calibration and statistical analysis performances are reduced by this strict selection of data in the ArcticGRO dataset during these particular periods.

3.2. Future Modeling Improvements for the Arctic Rivers

This study includes new strengths in modeling permafrost hydrology. We have shown the importance of modeling at a daily time step in Arctic watersheds to collect more information on the hydrological behavior of those basins. Modeling at a monthly time step limits the understanding of the freshet and of the water pathways. The transition from low flow to high flow periods occurs in few days and a daily time step modeling allows more precise predictions. Monthly time step modeling approaches the permafrost hydrology observed in the daily time step modeling especially for low flow periods, but the peaks are strongly reduced and do not match the observations due to the low observed data available during high-flows periods as shown in Table 1 which has definitely repercussions on the spatial discretization of water pathways. Precisely, daily time step modeling permits a spatial study of water pathways on days of interest which is important to characterize the origin of the water and the snowmelt intensity (Figure 4).

By integrating high flow periods, by conceptualizing and implementing the active layer in a model and by including each type of permafrost in the study, the obtained results are better than that from previous efforts. This study is the first study following spatially water flows from each compartment during the year. On the other hand, this research reveals some weaknesses. Unlike the work of Zhang et al. (2005) [15], the active layer is not implemented variable temporally and not at the same scale. Each sub-basin has received a value of the depth to the impervious layer depending on the type of permafrost according to the paper of Zhang et al. (2005) [15] and to expertise. A reconsideration of this parameters at a smaller scale could increase the goodness of the returned results or an integration of the active layer model by Zhang et al. (2005) [15] to refine the spatial representation of the active layer in our modeling could be a good improvement. In a same time, the snow cover extent could be checked as it is already studied by remote sensing [37].

However, improvements in the assumptions made in this modeling could be introduced. The representation of the active layer, which has the biggest influence on the distribution of flows, allows a good representation of the peaks and the recessions. But, the hypothesis used to represent the active layer with only spatial variations is weak because in reality, the active layer thickness varies also temporally as shown before. As a first perspective, the conceptualization could be improved by implementing the temporal fluctuation of the active layer and to better represent the dynamic of soils conditions. In order to achieve this goal, models available for big watersheds scale should integrate other equations and parameters adapted to permafrost soils. The most straight and evident way is the development of a separate SWAT module for soil physics and heat transfer. This implies also the development of a dedicated open database of permafrost soil properties, including heat transmissivity and the like. This module will make the model computationally more demanding, but this is the only way of providing relevant hydrological forecasts based on future climate scenarios.

3.3. Limit of the Model for Permafrost Soils

The SWAT model allows spatial and temporal predictions of hydrological fluxes in a large Arctic watershed. Some other models could have returned more precise results but with a need of larger number of measured variables which, unlike the variables used in SWAT, are quite difficult to collect. It could be interesting to use the Hydrograph model on the Yenisei because it has already been successfully used on another big arctic watershed, the Lena River [39], and then compare the model performance on catchments presenting permafrost soils. Explicit permafrost description through heat fluxes and variable active layer depth is not available, though essential for permafrost catchment modeling. The SWAT model does not currently include soil heat transfer module, as it is an overkill for more temperate regions.

4. Materials and Methods

4.1. Study Area

The Yenisei River has the seventh largest watershed worldwide, and the largest in the Arctic domain, with a basin area of 2,540,000 km^2 [23]. The Yenisei is the fifth longest river in the world (4803 km [23]) with the sixth biggest discharge at the outlet (17,700–19,900 m^3 s^{-1} [23,40–42]). The main stream comes from Western Sayan Mountains (Southern Siberia), crosses the Central Siberia in south to north direction, and drains into the Kara Sea (Figure 7). Its largest tributary, the Angara River, comes from Mongolia and is fed by the Lake Baikal, the biggest freshwater reserve on Earth [43]. Its average discharge is 4500 m^3 s^{-1} and it sustained water flow during the frozen period with an average discharge of 3000 m^3 s^{-1}. The mean annual discharge at the Yenisei outlet approaches 20,000 m^3 s^{-1} with peaks exceeding 100,000 m^3 s^{-1} during the highest freshets; low flow discharge around 6000 m^3 s^{-1} during winter is mainly sustained by water releases by dams [42].

The Yenisei watershed embodies three geographically distinct regions: mountainous headwater area of the Southern Siberia at the southern limit of the watershed, a relatively plain area of boreal forest in its central and northern parts, and Central Siberian Plateau in the northern part of its northernmost large tributary basin, the Nizhnyaya Tunguska River (Figure 7). The mean elevation is 670 m and the average basin slope is 0.2% [23].

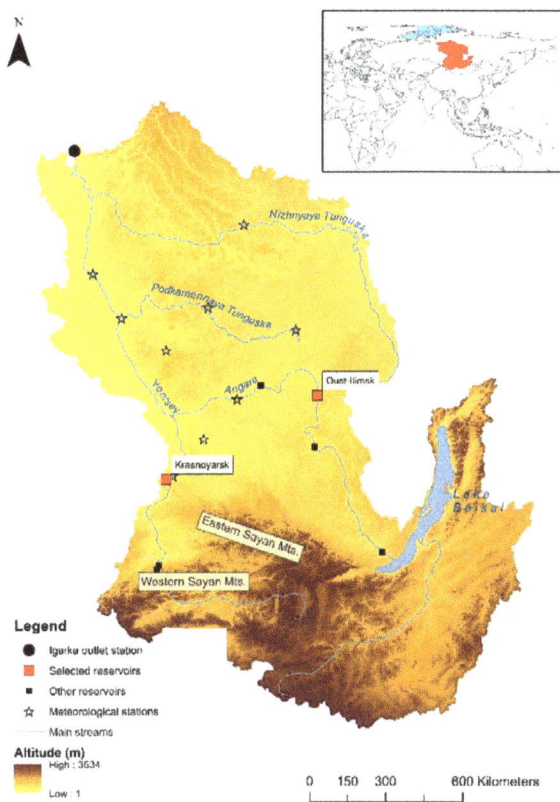

Figure 7. Topography and mains streams of the Yenisei watershed from its sources to the Igarka gauging station using a digital elevation model from de Ferranti and Hormann.

The southern and central parts of the Yenisei watershed are dominated by Podzoluvisols, Cambisols and Podzols while Cryosols and Gleysols cover the largest area in the Northern parts (Figure 8a). The taiga (boreal coniferous forest) is dominant in this watershed, but the tundra is also present in the northernmost part of the watershed, and steppe is a typical landcover class on the southern basin margin (Figure 8b).

Permafrost soils overlay 90% of the Yenisei watershed and are distributed as followed: 34% of continuous permafrost, 11% of discontinuous permafrost and 45% of sporadic and isolated permafrost (Figure 8c [44]). It may influence hydrology, as mentioned before, and a lag time for snowmelt should be taken into account in the conceptual model.

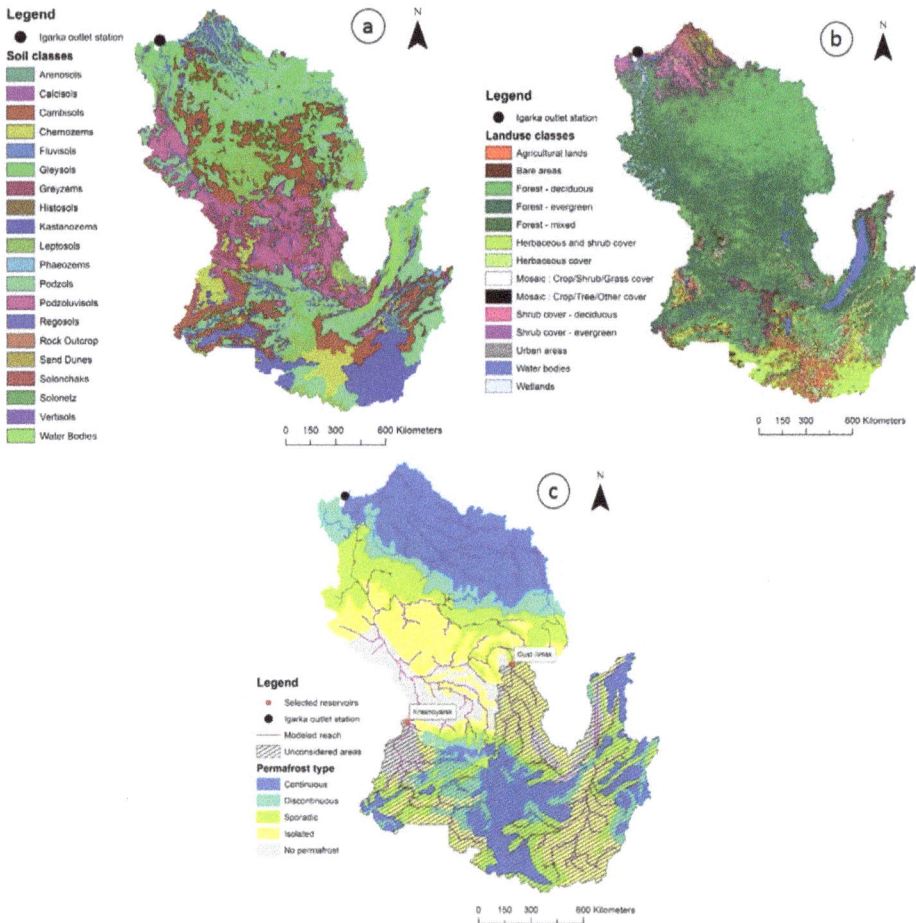

Figure 8. Soils, land use and permafrost distribution along the Yenisei watershed. (**a**) Soils distribution from the Harmonized World Soil Database at a 1 km resolution. We see a large diversity of soils among the whole watershed; (**b**) Land use distribution from the Global Land Cover 2000 Database at a 1 km resolution. The forest classes correspond to the tundra distribution. The shrub cover classes correspond to the taiga; (**c**) Model adaptation of the Yenisei watershed and permafrost extent. The GIS map classifies permafrost types as the percentage of extent: continuous (90–100%), discontinuous (50–90%), sporadic (10–50%) and isolated (0–10%). Source: National Snow and Ice Data Center (NSIDC), based on Brown et al. (1998) [45].

The watershed outlet has been established near Igarka (67°27′55″ N, 86°36′09″ E), since the river flow downstream from this settlement is affected by the marine influence, causing perturbations in discharge and a salinity increase which disturbs water chemistry. Discharge data are available at the outlet (see Section 3.2 for description).

Several physiographical objections complicate our modeling exercise. Firstly, uneven permafrost distribution across the watershed forces to implement different modeling strategies for the sub-basins, or even their parts, void of permafrost and for those which are perennially frozen. Secondly, the SWAT model does not provide a separate module for soil physics and heat transfer calculations, so the active layer presence and development are to be accounted for, using only the means, currently available in the model interface. Finally, hydropower generation is an important activity in the Yenisei basin, related to metallurgy which is highly energy-consumptive. Thus, seven large dams have been constructed on the Yenisei and the Angara rivers in the 1960s and 1970s, actually maintaining a minimum daily flow ca. 6000 m^3 s^{-1} at the outlet throughout the low flows season.

4.2. Observed Data

Observed discharge data are available at the basin outlet at a daily timestep, originating from daily water stage observations at the Igarka gauging station [46]. Water stage values are recalculated to daily flow values using a rating curve, which is not openly available. Daily flow values at this station were measured regularly from 1930s to late 1980s at a cross-section around 3 km downstream from the water stage gauge, using the standard velocity-area method [47]. Stream velocity was measured at several verticals from a boat using a propeller device at five depth points. To our knowledge, the most recent direct flow measurement at the Igarka gauge dates back to 2003.

Daily flow data used in this paper for calibration and verification purposes for the period between 1999 and 2014 comes from the Arctic Great Rivers Observatory (ArcticGRO) dataset (Table 3). Daily discharge data at the large dams exits and at the tributaries outlets are available from the Roshydromet online database (Table 3) for the 2008–2013 period. During the freshet, the discharge is not measured because of field difficulties but estimated. The confidence accorded to these data is low so we exclude them of the dataset for this study. Since the reservoirs management practices are not publicly available, we exclude from consideration the basin areas upstream from the Krasnoyarsk hydropower station (HPS) on the Yenisei River, and upstream from the Ust'-Ilimsk HPS on the Angara River which reduce the area of the basin to 1,383,398 km^2 (Figure 8c). The Ust'-Ilimsk HPS is downstream the Lake Baikal and then integrates its water delivery. On the Angara, the most downstream reservoir is currently the Boguchany HPS, but it has been under construction until 2013 and is thus considered as having no significant impact on flow redistribution in the preceding years. These two mentioned HPSs have been considered as inlets in the model, delivering an average daily flow of 3000 m^3 s^{-1} each.

Table 3. SWAT data inputs and observations datasets.

Data Type	Observations	Resolution	Source
Digital Elevation (DEM)	-	500 m	Digital Elevation Data (http://www.viewfinderpanoramas.org/dem3.html)
Soil dataset	-	1 km	Harmonized World Soil Database v 1.1 (http://webarchive.iiasa.ac.at/Research/LUC/External-World-soil-database/HTML/index.html?sb=1)
Land use dataset	-	1 km	Global Land Cover 2000 Database (http://forobs.jrc.ec.europa.eu/products/glc2000/products.php)
River network dataset	-	-	Natural Earth (http://www.naturalearthdata.com)
River discharge	2003–2013	-	Arctic Great Rivers Observatory (http://arcticgreatrivers.org/)
Reservoirs deliveries	2008–2014	-	Roshydromet (https://gmvo.skniivh.ru/)
Meteorological dataset	1999–2014	-	Observed: Global center for meteorological data, VNIIGMI-MCD, Region of Moscou (http://aisori.meteo.ru/ClimateR) Simulated: Climate Forecast System Reanalysis (CFSR) Model (http://globalweather.tamu.edu/)

4.3. Model Choice

Different models have been used in past researches to simulate permafrost hydrology. The TOPMODEL used by Stieglitz et al. (1999) [48] on the Imnavait Creek watershed in Alaska (2 km^2) and by Finney et al. (2012) [33] on the biggest Arctic watersheds is a surface runoff model which show its limits for Arctic watersheds by omitting lateral and return flows from the modeled water cycle. The Topoflow model used by Schramm et al. (2007) [28] is a spatially distributed, process-based hydrological model, designed primarily for permafrost catchments. This model is able to correctly reproduce the hydrological processes in Arctic systems but seems not easy to use for large catchments because it requires strong calculations and does not take into account soil properties at different depths. Though, modeling at large scale have been performed. The Hydrograph model [39] has been used at small and very large scale (the Lena basin) in numerous studies. This model estimates the heat fluxes and permafrost hydrology with good accuracy at a stand-scale [30], but its applicability is limited by a virtually random spatial distribution of model parameters, expulsion of lateral flow from model equations, and oversimplistic channel routing description. The RiTHM model is an adaptation of the MODCOU model, which is a regional spatially-distributed model and can estimate surface runoff, infiltration and return flow from groundwater to the streamflow. Again, by not taking into account lateral flows in model equations, permafrost hydrology is not estimated with accuracy [49]. Nohara et al. (2006) [50] have performed a simulation on different catchments in the world using TRIP, a model that considers the water transport in the watershed as a displacement of water depending on the velocity of water. This model does not handle human implications in the water cycle (e.g., dams) and does not include evaporation and sublimation module, which are essential in Arctic systems. The LMDZ model has been applied to some of the northernmost basins in the world including the Yenisei River basin [51]. This model handles snowmelt but does not consider percolation processes, which is useless in order to represent the active layer dynamics. In a same way, the SDGVM tool has been used in a multi-model analysis on various watersheds on a large period of study (1901–2010 [52]). This model is adapted to plant growth but provides runoff outputs. Future improvements in this model will allow handling of permafrost soils and snow behavior but is still unable to correctly represent permafrost hydrology.

The Soil and Water Assessment Tool (SWAT) is a hydro-agro-climatological model developed by USDA Agricultural Research Service (USDA-ARS; Temple, TX, USA) and Texas A&M AgriLife Research (College Station, TX, USA) [53]. Its performance has already been tested at multiple catchments of different sizes and in various physiographical settings ([54–58] and references therein). It is a semi-distributed model which has been firstly designed to predict impacts of human activities on water management in ungauged catchments. Importantly, both the SWAT model and the ArcSWAT interface are open–source and free software, allowing reproducibility of the results once the input data are well–documented and openly available [59].

This study uses the SWAT model to simulate the hydrology of a permafrost watershed including HPS. SWAT uses small calculating units, called Hydrological Response Units (HRUs), homogeneous in terms of land use and soil properties [60]. The SWAT system coupled with a geographical information system (GIS) engine integrates various spatial environmental data, including soil, land cover, climate and topographical features. The SWAT model manage soils types and properties. It decomposes the water cycle and returns the water pathways and other information on the water cycle in the studied watershed. Theory and details of hydrological processes integrated in SWAT model are available online in the SWAT documentation [61]. The SWAT model has already been tested in permafrost watersheds [38,39].

4.4. Modeling Data Inputs

The ArcSWAT software has been used in the ArcGIS 10.2.2 interface [62] to compile the SWAT input files. All the inputs data used in the study are detailed in Table 2. The DEM resolution has been chosen coarse because of the watershed size. The soil map comprises 6000 categories but has been simplified to 36 categories and soil properties have been adjusted by expertise. Global soils have been aggregated into categories representative with average properties. The soils have been aggregated on their common structural properties and average soils categories result. Observed meteorological data have been extracted from 9 stations located in the reduced watershed (Figure 7) for precipitations. The other variables (temperature, average wind speed, solar radiation and relative humidity) have been extracted from a global climatic model [63]. Observed variables have been compared to the predictions by the CFSR model and a good correlation has been observed for all of them, except precipitation. In order to have a larger number of inputs data, simulated data have been preferred for all the variables except for the precipitations where observations have been used as inputs. Daily discharge data at the outlet of two reservoirs have been used: one on the Yenisei at Krasnoyarsk and one on the Angara at Ust'-Illimsk (Figure 7 [64]).

The catchment has been firstly discretized automatically by ArcSWAT into 250 sub-basins. In order to take into account the effect of the big reservoirs in the upper part of the watershed, we have introduced 2 inlets in the modeling at the reservoirs localizations, which have been fed by the observed data at the reservoirs exits. This step has reduced the modeled basin to an area of 1,383,398 km^2, resulting in 140 sub-basins. These sub-basins have been further divided into 1884 HRUs which are a combination of 14 land use classes, 13 soil classes and three slope classes (0–1, 1–2, and >2%).

4.5. Conceptualization

The conceptual model used in this study is based on the snow and the soil behavior depending on the season (Figure 9). While few studies have been done on the subject, in permafrost affected areas the groundwater flow is considered low or null regardless the season. Indeed, as the soil is always frozen in the deepest layers, water is trapped as ice which inhibits a return flow from aquifers. As a first period, during winter, the stability of the snowpack and the soil freezing sustain low flows. Discharge in the main river is maintained almost exclusively by dams [42]. The second period corresponds to the spring freshet. An increase in temperature induces a rapid snowmelt which is the main contributor to the surface runoff during few days. A subsurface flow accompanies the surface flow which is a result of a first unfreezing of the superficial layers. In a third time, during the recession, the active layer reaches its maximal depth and surface and subsurface flows are at in the same range. The last period shows the active layer freezing from the bottom and the top and the snow starts to accumulate again. Only lateral flow is possible with a piston effect then cease with the permafrost freezing.

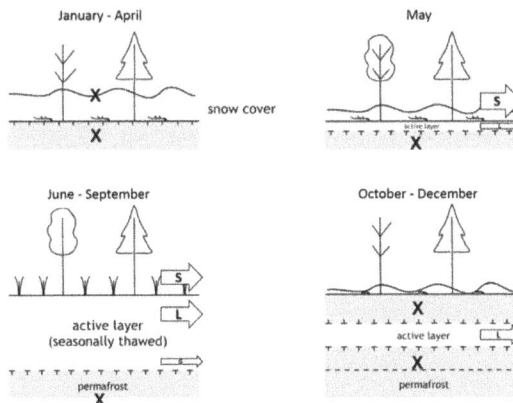

Figure 9. Conceptual model for Arctic watersheds hydrology. The cross corresponds to low interactions. The arrows represent the distribution of surface (S), lateral (L) and groundwater (G) flows depending on the snowmelt and the freeze-unfreeze processes of permafrost. The average ratio is 60/40 according to Carey and Woo (1999). Adapted from Tananaev (2015) [65] and Hülsmann et al. (2015) [38].

4.5.1. Climate Approach

Concerning the climate approach, an attention to the behavior of snow that regulates most of the spring flood in these regions has been devoted. Indeed, the annual average ratio rainfall/snowfall is assumed to be 60/40 regarding the work of Su et al. (2005) [22]. Indeed, the snow pack must melt enough before flow could be detectable.

Parameters controlled by temperature have been calibrated with attention to separate rainfall and snowfall and to contain the snowpack on the lands before the massive snowmelt. Evapotranspiration and sublimation have also been followed with attention because of their significance in these ecosystems, ranging respectively between 36 and 50% of precipitation [22,30,33–36] and between 4 and 20% of snowfall [66,67].

4.5.2. Distribution of the Flows Returning to the River

A main step of the conceptualization has been to consider a low or null groundwater contribution to streamflow which is linked to the soils conceptualization. So far, groundwater flows have not been studied well in Arctic watersheds [7] but are assumed to be extremely low due to soil permafrost conditions. The surface/subsurface flows ratio, which is assumed to be close to Canadian watersheds i.e., 60/40 [17], has been followed with attention in the manual calibration.

4.5.3. Permafrost Approach

Permafrost behavior in the watershed is accounted for in the model following an approach from Hülsmann et al. (2015) [38]. An impermeable boundary has been set within the soil profile, which corresponds roughly to the active layer bottom ultimately limiting percolation in permafrost environments [19]. Impermeable boundary depth has been assigned as a function of permafrost extent class in each sub-basin, based on the active layer depth estimates from Zhang et al. (2005) [15]. By remote sensing, they have estimated with the annual thawing index an average depth of the active layer. The conceptualized maximal depth of the active layer has been set to 800 mm, 1500 mm, 1750 mm and 2000 mm respectively for continuous, discontinuous, sporadic and isolated permafrost. This approach neglects the temporal development of the active layer, its gradual thawing through summer, and subsurface runoff inhibition in winter.

4.6. Model Calibration and Validation

The simulation has been performed from January 2003 to July 2014 (excluding a 4-year spin-up from 1999 to 2002). As a first step, the calibration has been done manually based on literature and expertise by comparison to observed data. The discharge has been calibrated at a daily time step from January 2003 to December 2008 and validated from January 2009 to July 2014. Discharge at reservoirs exits and tributaries outlets have been checked in order to supervise the good displacement of water in the basin.

In a second time, the calibration has been done automatically with 3 iterations of 500 simulations using the Sequential Uncertainty Fitting analysis routine (SUFI-2 [68,69]) of the SWAT Calibration and Uncertainty Procedures (SWAT-CUP) software [70] to select the best value for all parameters in ranges outlined by the manual calibration, and to perform sensitivity and uncertainty analysis. Each simulation has selected a list of parameters taken in the ranges defined before and the objective function has been calculated after running the model with this set of parameters on the study period. The algorithm is designed to capture the measured data in the 95% prediction uncertainty (95PPU) of the model in an iterative process with an objective function [57]. In our case, the objective function considered has been to increase the Nash and Sutcliffe efficiency (NSE; developed below). Then, Latin hypercube samplings have been performed to obtain the cumulative distribution of the output variables. The 95PPU has been calculated by integrating the cumulative distributions between 2.5% and 97.5%.

Table 4 gives calibrated and validated parameters values ranked by sensitivity.

Table 4. Calibrated values of SWAT parameters. The SLSOIL and the DEP_IMP parameters have not been integrated in the SWAT CUP runs because of their high sensitivity. Because their modification disrupt completely the water flows distribution, we have decided to keep them fixed.

Parameter	Name	Input File	Literature Range	Calibrated Value	Sensitivity Rank
SMTMP	Snow melt base temperature (°C)	.bsn	−5–5	4.75	1
SMFMN	Melt factor for snow on December 21 (mm H_2O/°C-day)	.bsn	0–10	0.25	2
TIMP	Snow pack temperature lag factor	.bsn	0–1	0.42	3
SMFMX	Melt factor for snow on June 21 (mm H_2O/°C-day)	.bsn	0–10	8.26	4
SNO50COV	Fraction of SNOCOVMX that corresponds to 50% snow cover	.bsn	0–1	0.57	5
LAT_TTIME	Lateral flow travel time (days)	.hru	0–180	9.06	6
SFTMP	Snowfall temperature (°C)	.bsn	−5–5	1.52	7
SNOCOVMX	Minimum snow water content that corresponds to 100% snow cover (mm H_2O)	.bsn	0–500	67.73	8
ESCO	Soil evaporation compensation factor	.bsn	0–1	0.86	9
CANMX	Maximum canopy storage (mm H_2O)	.hru	0–100	1.90	10
SLSOIL	Slope length for lateral subsurface flow (m)	.hru	0–150	3	X
DEP_IMP	Depth to impervious layer in soil profile (mm)	.hru	0–6000	800–2000	X

4.7. Model Evaluation

The performance of the model is evaluated using 4 indices recommended for hydrological modeling studies [65]: the NSE, the R^2, the PBIAS and the RSR. The NSE is a normalized statistic, usually used in hydrological modeling, which determines the relative magnitude of the residual variance ("noise") compared to the measured data variance ("information") [71,72].

$$NSE = 1 - \frac{\Sigma(obs - sim)}{\Sigma\left(obs - \overline{obs}\right)}$$

where *obs* and *sim* represents observed and simulated data while $\overline{obs}$ is the observed data mean. NSE ranges from $-\infty$ to 1. If NSE = 1, there is a perfect match between simulated and observed data. If NSE = 0, it indicates that model predictions are as accurate as the mean of the observed data. If NSE < 0, the mean of the observations is a better predictor than the model. The NSE is usually used because it is easy to interpret. Indeed, the more the NSE is close to 1, the more accurate the model is. Modeling at a daily step are generally considered satisfactory if NSE > 0.5 [71] and are considered really good when NSE exceed 0.75.

R^2 describes the degree of collinearity between simulated and measured data [71]. R^2 represents the proportion of the variance in measured data explained by the model and ranges from 0 to 1, with higher values indicating less error variance. As the NSE, values greater than 0.5 are typically considered good and excellent when R^2 is higher than 0.75.

$$R^2 = \frac{\Sigma\left(obs - \overline{obs}\right)(sim - \overline{sim})}{\left(\Sigma\left(obs - \overline{obs}\right)^2\right)^{0.5}\left(\Sigma(sim - \overline{sim})^2\right)^{0.5}}$$

The PBIAS measures the average tendency of the simulated data to be larger or smaller than their observed counterparts [71]. It expresses the percentage of deviation between simulations and observations and the optimal value is 0. PBIAS can be positive or negative which reveals respectively a model underestimation or overestimation bias [71].

$$PBIAS = \frac{\Sigma(obs - sim) \times 100}{\Sigma(obs)}$$

The RSR is calculated as the ratio of the RMSE and standard deviation of measured data [65]. The RSR ranges from the optimal value of 0 to $+\infty$.

$$RSR = \frac{\sqrt{\Sigma(obs - sim)}}{\sqrt{\Sigma\left(obs - \overline{obs}\right)}}$$

Two other indices, the *p*-factor and the R-factor, are used in automatic calibration [68]. The *p*-factor corresponds to the percentage of observed data included in the 95PPU. If we consider the model uncertainty, the more the *p*-factor is close to 1, the more the model is perfect. For simulations at a monthly time step, the *p*-factor is adequate if it is higher than 0.7 [57]. The R-factor is calculated by dividing the average width of the 95PPU band by the standard deviation of the considered variable. The R-factor should be lower than 1.5 to be considered adequate [57].

5. Conclusions

The objectives of this paper have been to better analyze water fluxes and pathways in Arctic rivers and produce a performing model adapted to permafrost soils. This study allows a quantification of these fluxes at the outlet of the biggest Arctic basin, the Yenisei, with a spatial variability by quantifying the fluxes in each sub-basin. This work offers a discretization of the flows distribution in a permafrost affected watershed. This is the first study trying to model water displacement in an Arctic watershed presenting different types of permafrost in these scales of time and space. Concerning modeling, this study is the first one trying to conceptualize the active layer and to implement it in the model. Another advantage of this study in front of prior researches on water discharge in Arctic rivers is the integration of high flow periods in the daily time step modeling while previous studies do not succeed to represent peaks due to snowmelt and approach the observed discharge at the outlet. The model has returned average annual water flows to the river of 152, 103 and 8 mm yr^{-1} attributed respectively to surface runoff, lateral runoff and return flow from the shallow aquifer for a yearly water inflow of 263 mm yr^{-1} for the modeled watershed. The permafrost plays a temporal role in the distribution of water. Surface runoff explains most of the peak of discharge while the recession is sustained by lateral runoff. By integrating the reservoirs and the whole area of the watershed, the simulated discharge at the outlet reaches 237.8 mm yr^{-1}, a result closer to observations than previous modeling. Daily time step modeling seems the better way to predict water flows in Arctic watersheds regarding the speed of changing between high and low flow periods. This study is still a first step in hydrological modeling of Arctic systems and need other improvements to return more trustworthy results. However, it could be used as an interesting tool to do predictions on the Yenisei hydrological cycle disturbances due to climate change and their impacts on the Arctic Ocean functioning, as shown in Kuzin et al. (2010) [6] and to follow biogeochemical flows such as organic carbon exports from permafrost soils which is a main issue in Arctic areas and a consequent threat at a global scale.

Acknowledgments: This project beneficed of funding from the TOMCAR-Permafrost Marie Curie International Reintegration Grant FP7-PEOPLE-2010-RG (project reference: 277059) within the Seventh European Community Framework Programme awarded to Roman Teisserenc (http://www.tomcar.fr). Travel and living expenses were also funded thanks to GDRI Car-Wet-Sib II and INP-Toulouse SMI program. We also thank the Arctic Great Rivers Observatory (NSF-1107774) and Roshydromet for their data.

Author Contributions: Clément Fabre, Sabine Sauvage, Nikita Tananaev, Roman Teisserenc and José Miguel Sánchez Pérez conceptualized the model for permafrost catchments. Raghavan Srinivasan set up the climatic inputs for the SWAT model and managed all the strongest issues encountered with SWAT. Clément Fabre, Sabine Sauvage and José Miguel Sánchez Pérez performed the modeling. Clément Fabre wrote the paper. Sabine Sauvage, José Miguel Sánchez Pérez, Nikita Tananaev and Roman Teisserenc supervised the paper writing.

Conflicts of Interest: The authors declare no conflict of interest.

References

1. Stocker, T.F.; Qin, D.; Plattner, G.-K.; Tignor, M.; Allen, S.K.; Boschung, J.; Nauels, A.; Xia, Y.; Bex, V.; Midgley, P.M. *Climate Change 2013: The Physical Science Basis. Contribution of Working Group I to the Fifth Assessment Report of the Intergovernmental Panel on Climate Change*, 1st ed.; Cambridge University Press: Cambridge, United Kingdom and New York, NY, USA, 2013; ISBN 978-1-107-66182-0.

2. Serreze, M.; Barry, R. Processes and impacts of Arctic amplification: A research synthesis. *Glob. Planet. Chang.* **2011**, *77*, 85–96. [CrossRef]

3. Stocker, T.F.; Raible, C.C. Climate change: Water cycle shifts gear. *Nature* **2005**, *434*, 830–833. [CrossRef] [PubMed]

4. Francis, J.; White, D.; Cassano, J.; Gutowski, W.; Hinzman, L.; Holland, M.; Steele, M.; Vörösmarty, C. An arctic hydrologic system in transition: Feedbacks and impacts on terrestrial, marine, and human life. *J. Geophys. Res.* **2005**, *114*. [CrossRef]

5. Frey, K.; McClelland, J. Impacts of permafrost degradation on arctic river biogeochemistry. *Hydrol. Process.* **2009**, *23*, 169–182. [CrossRef]

6. Kuzin, V.I.; Platov, G.A.; Golubeva, E.N. Influence that interannual variations in Siberian river discharge have on redistribution of freshwater fluxes in Arctic Ocean and North Atlantic. *Izv. Atmos. Ocean. Phys.* **2010**, *46*, 770–783. [CrossRef]

7. Woo, M.-K.; Kane, D.L.; Carey, S.K.; Yang, D. Progress in permafrost hydrology in the new millennium. *Permafr. Periglac. Process.* **2008**, *19*, 237–254. [CrossRef]

8. Briggs, M.A.; Campbell, S.; Nolan, J.; Walvoord, M.; Ntarlagiannis, D.; Day-Lewis, F.; Lane, J. Surface geophysical methods for characterizing frozen ground in transitional permafrost landscapes. *Permafr. Periglac. Process.* **2016**, *28*, 52–65. [CrossRef]

9. Tetzlaff, D.; Buttle, J.; Carey, S.K.; McGuire, K.; Laudon, H.; Soulsby, C. Tracer-based assessment of flow paths, storage and runoff generation in northern catchments: A review: Tracers in Northern catchments. *Hydrol. Process.* **2015**, *29*, 3475–3490. [CrossRef]

10. McClelland, J.W.; Tank, S.E.; Spencer, R.G.M.; Shiklomanov, A.I. Coordination and sustainability of river observing activities in the Arctic. *Arctic* **2015**, *68*, 59–68. [CrossRef]

11. Beltaos, S.; Prowse, T.D. River-ice hydrology in a shrinking cryosphere. *Hydrol. Process.* **2009**, *23*, 122–144. [CrossRef]

12. Streletskiy, D.A.; Tananaev, N.I.; Opel, T.; Shiklomanov, N.I.; Nyland, K.E.; Streletskaya, I.D.; Tokarev, I.; Shiklomanov, A.I. Permafrost hydrology in changing climatic conditions: Seasonal variability of stable isotope composition in rivers in discontinuous permafrost. *Environ. Res. Lett.* **2015**, *10*, 95003. [CrossRef]

13. Woo, M. Permafrost hydrology in North America. *Atmos. Ocean.* **1986**, *24*, 201–234. [CrossRef]

14. Burt, T.P.; Williams, P.J. Hydraulic conductivity in frozen soils. *Earth Surf. Process.* **1976**, *1*, 349–360. [CrossRef]

15. Zhang, T.; Frauenfeld, O.; Serreze, M.; Etringer, A. Spatial and temporal variability in active layer thickness over the Russian Arctic drainage basin. *J. Geophys. Res.* **2005**, *110*, D16101. [CrossRef]

16. Schuur, E.A.G.; Bockheim, J.; Canadell, J.G.; Euskirchen, E.; Field, C.B.; Goryachkin, S.V.; Hagemann, S.; Kuhry, P.; Lafleur, P.M.; Lee, H.; et al. Vulnerability of Permafrost Carbon to Climate Change: Implications for the Global Carbon Cycle. *BioScience* **2008**, *58*, 701–714. [CrossRef]

17. Carey, S.K.; Woo, M.-K. Hydrology of two slopes in subarctic Yukon, Canada. *Hydrol. Process.* **1999**, *13*, 2549–2562. [CrossRef]

18. Bense, V.F.; Kooi, H.; Ferguson, G.; Read, T. Permafrost degradation as a control on hydrogeological regime shifts in a warming climate: Groundwater and degrading permafrost. *J. Geophys. Res.* **2012**, *117*, F03036. [CrossRef]

19. Woo, M. *Permafrost Hydrology*; Springer: Heidelberg, Germany, 2012; ISBN 9783642234620.

20. McKenzie, J.M.; Voss, C.I. Permafrost thaw in a nested groundwater-flow system. *Hydrogeol. J.* **2013**, *21*, 299–316. [CrossRef]

21. Walvoord, M.A.; Kurylyk, B.L. Hydrologic Impacts of Thawing Permafrost—A Review. *Vadose Zone J.* **2016**, *15*. [CrossRef]

22. Su, F.; Adam, J.C.; Trenberth, K.E.; Lettenmaier, D.P. Evaluation of surface water fluxes of the pan-Arctic land region with a land surface model and ERA-40 reanalysis. *J. Geophys. Res.* **2006**, *111*, D05110. [CrossRef]

23. Amon, R.M.W.; Rinehart, A.J.; Duan, S.; Louchouarn, P.; Prokushkin, A.; Guggenberger, G.; Bauch, D.; Stedmon, C.; Raymond, P.A.; Holmes, R.M.; et al. Dissolved organic matter sources in large Arctic rivers. *Geochim. Cosmochim. Acta* **2012**, *94*, 217–237. [CrossRef]

24. Tarnocai, C.; Canadell, J.G.; Schuur, E.A.G.; Kuhry, P.; Mazhitova, G.; Zimov, S. Soil organic carbon pools in the northern circumpolar permafrost region: Soil organic carbon pools. *Glob. Biogeochem. Cycles* **2009**, *23*, GB2023. [CrossRef]

25. Boike, J.; Roth, K.; Overduin, P.P. Thermal and hydrologic dynamics of the active layer at a continuous permafrost site (Taymyr Peninsula, Siberia). *Water Resour. Res.* **1998**, *34*, 355–363. [CrossRef]

26. Jafarov, E.E.; Marchenko, S.S.; Romanovsky, V.E. Numerical modeling of permafrost dynamics in Alaska using a high spatial resolution dataset. *Cryosphere* **2012**, *6*, 613–624. [CrossRef]

27. Weismüller, J.; Wollschläger, U.; Boike, J.; Pan, X.; Yu, Q.; Roth, K. Modeling the thermal dynamics of the active layer at two contrasting permafrost sites on Svalbard and on the Tibetan Plateau. *Cryosphere* **2011**, *5*, 741–757. [CrossRef]

28. Schramm, I.; Boike, J.; Bolton, W.R.; Hinzman, L.D. Application of TopoFlow, a spatially distributed hydrological model, to the Imnavait Creek watershed, Alaska. *J. Geophys. Res.* **2007**, *112*, G04S46. [CrossRef]

29. Rigon, R.; Bertoldi, G.; Over, T.M. GEOtop: A Distributed Hydrological Model with Coupled Water and Energy Budgets. *J. Hydrometeorol.* **2006**, *7*, 371–388. [CrossRef]

30. Gusev, E.M.; Nasonova, O.N.; Dzhogan, L.Y. Reproduction of Pechora runoff hydrographs with the help of a model of heat and water exchange between the land surface and the atmosphere (SWAP). *Water Resour.* **2010**, *37*, 182–193. [CrossRef]

31. Semenova, O.; Vinogradov, Y.; Vinogradova, T.; Lebedeva, L. Simulation of soil profile heat dynamics and their integration into hydrologic modelling in a permafrost zone. *Permafr. Periglac. Process.* **2013**, *25*, 257–269. [CrossRef]

32. Zhou, J.; Kinzelbach, W.; Cheng, G.; Zhang, W.; He, X.; Ye, B. Monitoring and modeling the influence of snow pack and organic soil on a permafrost active layer, Qinghai–Tibetan Plateau of China. *Cold Reg. Sci. Technol.* **2013**, *90–91*, 38–52. [CrossRef]

33. Finney, D.L.; Blyth, E.; Ellis, R. Improved modelling of Siberian river flow through the use of an alternative frozen soil hydrology scheme in a land surface model. *Cryosphere* **2012**, *6*, 859–870. [CrossRef]

34. Arora, V.K. Streamflow simulations for continental scale river basins in a global atmospheric general circulation model. *Adv. Water Resour.* **2001**, *24*, 775–791. [CrossRef]

35. Fukutomi, Y.; Igarashi, H.; Masuda, K.; Yasunari, T. Interannual variability of summer water balance components in three major river basins of Northern Eurasia. *J. Hydrometeorol.* **2003**, *4*, 283–296. [CrossRef]

36. Nakai, T.; Kim, Y.; Busey, R.C.; Suzuki, R.; Nagai, S.; Kobayashi, H.; Park, H.; Sugiura, K.; Ito, A. Characteristics of evapotranspiration from a permafrost black spruce forest in interior Alaska. *Polar Sci.* **2013**, *7*, 136–148. [CrossRef]

37. Yang, D.; Zhao, Y.; Armstrong, R.; Robinson, D.; Brodzik, M.-J. Streamflow response to seasonal snow cover mass changes over large Siberian watersheds. *J. Geophys. Res.* **2007**, *112*, F02S22. [CrossRef]

38. Hülsmann, L.; Geyer, T.; Schweitzer, C.; Priess, J.; Karthe, D. The effect of subarctic conditions on water resources: Initial results and limitations of the SWAT model applied to the Kharaa River Basin in Northern Mongolia. *Environ. Earth Sci.* **2015**, *73*, 581–592. [CrossRef]

39. Vinogradov, Y.B.; Semenova, O.M.; Vinogradova, T.A. An approach to the scaling problem in hydrological modelling: The deterministic modelling hydrological system. *Hydrol. Process.* **2011**, *25*, 1055–1073. [CrossRef]

40. Ludwig, W.; Probst, J.-L.; Kempe, S. Predicting the oceanic input of organic carbon by continental erosion. *Glob. Biogeochem. Cycle* **1996**, *10*, 23–41. [CrossRef]

41. Raymond, P.A.; McClelland, J.W.; Holmes, R.M.; Zhulidov, A.V.; Mull, K.; Peterson, B.J.; Striegl, R.G.; Aiken, G.R.; Gurtovaya, T.Y. Flux and age of dissolved organic carbon exported to the Arctic Ocean: A carbon isotopic study of the five largest arctic rivers. *Glob. Biogeochem. Cycles* **2007**, *21*, GB4011. [CrossRef]

42. Holmes, R.M.; McClelland, J.W.; Peterson, B.J.; Tank, S.E.; Bulygina, E.; Eglinton, T.I.; Gordeev, V.V.; Gurtovaya, T.Y.; Raymond, P.A.; Repeta, D.J.; et al. Seasonal and Annual Fluxes of Nutrients and Organic Matter from Large Rivers to the Arctic Ocean and Surrounding Seas. *Estuar. Coasts* **2012**, *35*, 369–382. [CrossRef]

43. Lake Baikal. Available online: http://whc.unesco.org/en/list/754 (accessed on 23 May 2017).

44. McClelland, J.W. Increasing river discharge in the Eurasian Arctic: Consideration of dams, permafrost thaw, and fires as potential agents of change. *J. Geophys. Res.* **2004**, *109*, D18102. [CrossRef]

45. Brown, J.; Ferrians, J.A.; Melnikov, E. *Circum.-Arctic Map of Permafrost and Ground-Ice Conditions*, Version 2; National Snow and Ice Data Center (NSIDC): Boulder, CO, USA, 2002.

46. Roshydromet, Russian Federal Service for Hydrometeorology and Environmental Monitoring. Available online: https://gmvo.skniivh.ru (accessed on 6 June 2017).

47. Herschy, R. The velocity-area method. *Flow Meas. Instrum.* **1993**, *4*, 7–10. [CrossRef]

48. Stieglitz, M.; Hobbie, J.; Giblin, A.; Kling, G. Hydrologic modeling of an arctic tundra watershed: Toward Pan-Arctic predictions. *J. Geophys. Res.* **1999**, *104*, 27507–27518. [CrossRef]

49. Ducharne, A.; Golaz, C.; Leblois, E.; Laval, K.; Polcher, J.; Ledoux, E.; de Marsily, G. Development of a high resolution runoff routing model, calibration and application to assess runoff from the LMD GCM. *J. Hydrol.* **2003**, *280*, 207–228. [CrossRef]

50. Nohara, D.; Kitoh, A.; Hosaka, M.; Oki, T. Impact of Climate Change on River Discharge Projected by Multimodel Ensemble. *J. Hydrometeorol.* **2006**, *7*, 1076–1089. [CrossRef]

51. Alkama, R.; Kageyama, M.; Ramstein, G. Freshwater discharges in a simulation of the Last Glacial Maximum climate using improved river routing. *Geophys. Res. Lett.* **2006**, *33*, L21709. [CrossRef]

52. Yang, H.; Piao, S.; Zeng, Z.; Ciais, P.; Yin, Y.; Friedlingstein, P.; Sitch, S.; Ahlström, A.; Guimberteau, M.; Huntingford, C.; et al. Multicriteria evaluation of discharge simulation in Dynamic Global Vegetation Models: Evaluation on simulation of discharge. *J. Geophys. Res. Atmos.* **2015**, *120*, 7488–7505. [CrossRef]

53. Arnold, J.G.; Srinivasan, R.; Muttiah, R.S.; Williams, J.R. Large area hydrologic modeling and assessment part 1: Model development. *J. Am. Water Resour. Assoc.* **1998**, *34*, 73–89. [CrossRef]

54. Schuol, J.; Abbaspour, K.C.; Yang, H.; Srinivasan, R.; Zehnder, A.J.B. Modeling blue and green water availability in Africa: Modeling blue and green water availability in Africa. *Water Resour. Res.* **2008**, *44*, W07406. [CrossRef]

55. Douglas-Mankin, K.R.; Srinivasan, R.; Arnold, J.G. Soil and Water Assessment Tool (SWAT) Model: Current Developments and Applications. *Trans. Am. Soc. Agric. Biol. Eng.* **2010**, *53*, 1423–1431. [CrossRef]

56. Faramarzi, M.; Abbaspour, K.C.; Ashraf Vaghefi, S.; Farzaneh, M.R.; Zehnder, A.J.B.; Srinivasan, R.; Yang, H. Modeling impacts of climate change on freshwater availability in Africa. *J. Hydrol.* **2013**, *480*, 85–101. [CrossRef]

57. Abbaspour, K.C.; Rouholahnejad, E.; Vaghefi, S.; Srinivasan, R.; Yang, H.; Kløve, B. A continental-scale hydrology and water quality model for Europe: Calibration and uncertainty of a high-resolution large-scale SWAT model. *J. Hydrol.* **2015**, *524*, 733–752. [CrossRef]

58. Krysanova, V.; White, M. Advances in water resources assessment with SWAT—An overview. *Hydrol. Sci. J.* **2015**, *60*, 771–783. [CrossRef]

59. Hutton, C.; Wagener, T.; Freer, J.; Han, D.; Duffy, C.; Arheimer, B. Most computational hydrology is not reproducible, so is it really science? *Water Resour. Res.* **2016**, *52*, 7548–7555. [CrossRef]

60. Flügel, W.-A. Delineating hydrological response units by geographical information system analyses for regional hydrological modelling using PRMS/MMS in the drainage basin of the River Bröl, Germany. *Hydrol. Process.* **1995**, *9*, 423–436. [CrossRef]

61. Soil and Water Assessment Tool. Available online: http://swatmodel.tamu.edu/ (accessed on 6 June 2017).

62. ESRI. Available online: http://www.esri.com/ (accessed on 6 June 2017).

63. Climate Forecast System Reanalysis. Available online: http://rda.ucar.edu/pub/cfsr.html (accessed on 6 June 2017).

64. Information System of Russian Water Surveys. Available online: https://gmvo.skniivh.ru/ (accessed on 6 June 2017).

65. Tananaev, N. *Permafrost Hydrology under Changing Climate*; P.I. Melnikov Permafrost Institute: Yakutsk, Sakha Republic, Russia, 2015.

66. Suzuki, K. Estimation of Snowmelt Infiltration into Frozen Ground and Snowmelt Runoff in the Mogot Experimental Watershed in East Siberia. *Int. J. Geosci.* **2013**, *4*, 1346–1354. [CrossRef]

67. Suzuki, K.; Liston, G.E.; Matsuo, K. Estimation of Continental-Basin-Scale Sublimation in the Lena River Basin, Siberia. *Adv. Meteorol.* **2015**, *2015*, 286206. [CrossRef]

68. Abbaspour, K.C.; Johnson, C.A.; van Genuchten, M.T. Estimating Uncertain Flow and Transport Parameters Using a Sequential Uncertainty Fitting Procedure. *Vadose Zone J.* **2004**, *3*, 1340. [CrossRef]

69. Abbaspour, K.C.; Yang, J.; Maximov, I.; Siber, R.; Bogner, K.; Mieleitner, J.; Zobrist, J.; Srinivasan, R. Modelling hydrology and water quality in the pre-alpine/alpine Thur watershed using SWAT. *J. Hydrol.* **2007**, *333*, 413–430. [CrossRef]

70. Abbaspour, K.C. *Swat-Cup2: SWAT Calibration and Uncertainty Programs Manual Version 2, Department of Systems Analysis, Integrated Assessment and Modelling (SIAM), Eawag*; Swiss Federal Institute of Aquatic Science and Technology: Duebendorf, Switzerland, 2011.

71. Moriasi, D.N.; Arnold, J.G.; Liew, M.W.V.; Bingner, R.L.; Harmel, R.D.; Veith, T.L. Model Evaluation Guidelines for Systematic Quantification of Accuracy in Watershed Simulations. *Trans. Am. Soc. Agric. Biol. Eng.* **2007**, *50*, 885–900. [CrossRef]

72. Nash, J.E.; Sutcliffe, J.V. River flow forecasting through conceptual models part I—A discussion of principles. *J. Hydrol.* **1970**, *10*, 282–290. [CrossRef]

water

MDPI

Article

Climate Change Impacts on US Water Quality Using Two Models: HAWQS and US Basins

Charles Fant [1,*], Raghavan Srinivasan [2], Brent Boehlert [1,3], Lisa Rennels [1], Steven C. Chapra [4], Kenneth M. Strzepek [3], Joel Corona [5], Ashley Allen [5] and Jeremy Martinich [5]

[1] Industrial Economics, Inc., Cambridge, MA 02140, USA; BBoehlert@indecon.com (B.B.); lrennels@indecon.com (L.R.)
[2] Department of Ecosystem Science and Management and Biological and Agricultural Engineering, Texas A&M University, College Station, TX 77843, USA; r-srinivasan@tamu.edu
[3] Joint Program on the Science and Policy of Global Change, Massachusetts Institute of Technology, Cambridge, MA 02139, USA; strzepek@mit.edu
[4] Civil and Environmental Engineering, Tufts University, Medford, MA 02155, USA; steven.chapra@tufts.edu
[5] U.S. Environmental Protection Agency (EPA), Washington, DC 20460, USA; Corona.Joel@epa.gov (J.C.); Allen.Ashley@epa.gov (A.A.); Martinich.Jeremy@epa.gov (J.M.)
* Correspondence: cfant@indecon.com; Tel.: +1-617-354-0074

Academic Editor: Lei Chen
Received: 22 November 2016; Accepted: 7 February 2017; Published: 14 February 2017

Abstract: Climate change and freshwater quality are well-linked. Changes in climate result in changes in streamflow and rising water temperatures, which impact biochemical reaction rates and increase stratification in lakes and reservoirs. Using two water quality modeling systems (the Hydrologic and Water Quality System; HAWQS and US Basins), five climate models, and two greenhouse gas (GHG) mitigation policies, we assess future water quality in the continental U.S. to 2100 considering four water quality parameters: water temperature, dissolved oxygen, total nitrogen, and total phosphorus. Once these parameters are aggregated into a water quality index, we find that, while the water quality models differ under the baseline, there is more agreement between future projections. In addition, we find that the difference in national-scale economic benefits across climate models is generally larger than the difference between the two water quality models. Both water quality models find that water quality will more likely worsen in the East than in the West. Under the business-as-usual emissions scenario, we find that climate change is likely to cause economic impacts ranging from 1.2 to 2.3 (2005 billion USD/year) in 2050 and 2.7 to 4.8 in 2090 across all climate and water quality models.

Keywords: water quality; climate change; economic valuation; mitigation; greenhouse gases; model comparison

1. Introduction

Climate change is projected to have widespread effects on freshwater quality due to increasing temperatures and changes in patterns of river runoff and extreme events [1] (pp. 69–112). Rising water temperatures, reduced lake mixing, and increased biotic consumption of dissolved oxygen each reduce water quality [2] (pp. 445–456). Evidence of rising river and lake temperatures [3,4] (pp. 1–5), and decreased mixing of lakes and reservoirs (i.e., increased stratification) [5,6] have already been observed. There is an economic value associated with these changes in water quality, measured in terms of changes in the quality of recreation opportunities and commercial activity. A variety of studies have examined the impact of water quality on activities such as river and lake visits, boating, and swimming and fishing in a number of geographic contexts. The authors of [7] provide an example of this by translating biophysical modeling estimates of water quality into human preferences and find households in Virginia are willing to pay $184 million per year (in 2010 dollars) to improve water quality.

Recent studies have investigated the impacts of climate change on water quality, and one in particular focused on the resulting economic impacts. Boehlert [8] (pp. 1326–1338) used a parsimonious water quality model to analyze how climate change impacts in the contiguous United States (CONUS) translate to economic benefits of climate change mitigation. The authors find that at a national level, annual economic impacts of a high emission future scenario on water quality of $1.4 billion in 2050 and $4 billion in 2100 for the CONUS, using a water quality index approach and a willingness-to-pay valuation. Although this study employed multiple climate models to show the effect of climate uncertainty, the analysis relied on only a single water quality model, begging the question of whether the findings would hold if a different water quality models were used.

Differences in General Circulation Model (GCM) projections have been a focus of many studies, and have highlighted model bias among well-trusted and complex climate models. The findings of these studies have developed a common-practice of using many GCMs to produce a range of impacts, and therefore produce an ensemble of risks. Recently, climate change biophysical impact analyses have begun to take a similar strategy by comparing results across various biophysical models, e.g., the Agricultural Model Intercomparison and Improvement Project (AgMIP) part of the Inter-Sectoral Impact Model Intercomparison Project (ISI-MIP, https://www.isimip.org/). In addition, Schewe [9] uses a large ensemble of global hydrologic models to assess global water scarcity under climate change. The authors of [10] take the multi-model assessment further by evaluating the impacts of climate change using regional scale models on three large-scale river basins with three hydrologic models. However, to the knowledge of the authors, no existing study has used multiple water quality models to assess the impacts of climate change on water quality.

In this study we project future water quality in CONUS using two water quality models: Hydrologic and Water Quality System (HAWQS) and the model system used in [8] (pp. 1326–1338), which we refer to as "US Basins" for the remainder of the study for simplicity. HAWQS builds off of the widely accepted Soil and Water Assessment Tool (SWAT) by advancing functionality, primarily through minimizing the necessary initialization time. This improves the ease of application to national scale analyses [11] (p. 164). Although prior analyses specifically using HAWQS for water quality analyses are limited (see [11] (p. 164) for an example), the underlying SWAT model is widely used in water quality modeling ([12] (pp. 16–29), [13] (pp. 228–244)). US Basins is a linked water systems and water quality model designed to evaluate the impacts of climate change on water quantity and quality outcomes. In [8] (pp. 1326–1338), the authors use US Basins to estimate the impacts of climate change and global greenhouse gas (GHG) mitigation effects on U.S. water quality.

We present projections of future water quality parameters in CONUS—namely, river flow, water temperature, dissolved oxygen, total nitrogen, and total phosphorus—for both HAWQS and US Basins. These are projected for five climate models and two emissions scenarios, with total water quality impacts shown through a Climate-oriented Water Quality Index (CWQI) and estimates of resulting changes in economic value (willingness-to-pay; WTP). The goal of this study is not to compare the two water quality models, resulting in a recommendation of which model is more accurate. A study of that nature would be more effective either at a smaller spatial scale (e.g., a single basin) or focused on individual model components (e.g., water temperature or stream flow). Instead, this study aims to make use of both models to better understand the impacts of climate change on water quality in CONUS, which is analogous to the common use of multiple GCMs in climate change impact studies to address the uncertainty of future climate projections.

The remainder of this document presents the modeling and valuation approaches, a presentation of model results, and a discussion of findings and future research recommendations.

2. Methodological Approach

We produce biophysical outputs from two water quality models, process those into a water quality index and changes in economic outcomes, and compare these three outputs across a common set of climate scenarios. Below we describe these common climate scenarios, each of the models, the loading inputs, and the valuation approach.

2.1. Forcing Scenarios and Climate Projections

This multi-model water quality impacts modeling exercise is contributing to the Climate Change Impacts and Risk Analysis (CIRA; [14]) project, an effort to quantify the physical and economic impacts of climate change in futures with varying assumptions about global greenhouse gas (GHG) emissions. The CIRA analytic framework uses a consistent set of climate forcing, climate projection, and socioeconomic scenarios to enable comparisons of impacts across space, time, and sectors. As such, the climate scenarios and projections used in this article to estimate changes in water quality are consistent with those of the broader CIRA project.

The emissions and climate scenarios are based on those generated for the Intergovernmental Panel on Climate Change's Fifth Assessment Report (AR5). For the emissions, two Representative Concentration Pathways (RCPs) are used: RCP4.5 and RCP8.5. RCP8.5 represents a warmer global future caused by higher GHG emissions, which results in a total change in radiative forcing by 2100 (compared to 1750) of 8.5 W/m^2. RCP4.5 provides a future with additional mitigation on GHG emissions and results in a change in total radiative forcing of 4.5 W/m^2. Of the many GCMs generated for the AR5, five were selected for this study. These were selected based on multiple objectives, related to the full scope of CIRA2 studies, and capture much of the temperature and precipitation change projected for the CONUS across the broader set of CMIP5 GCMs. The selected GCMs are the CanESM2 (from Canadian Centre for Climate Modeling and Analysis), CCSM4 (Community Climate System Model version 4), GISS-E2-R (from the Goddard Institute for Space Studies), HadGEM2-ES (from Met Office Hadley Centre), and MIROC5 (Model for Interdisciplinary Research on Climate). To select these, the points of mean change in temperature and precipitation across CONUS were plotted using scatter plots (i.e., precipitation change on the X-axis and temperature on the Y-axis). The GCMs selected best represent the variability, or "scatter", of the full set. More detail is provided in Supplementary Material. These projections were downscaled using a statistical process that uses a multi-scale spatial matching scheme to select analog days from observations across CONUS [15,16]. This dataset, LOCA (Localized Constructed Analogs; [17]), results in a 1/16 degree resolution for daily maximum temperature, daily minimum temperature, and daily precipitation. Additional climate variables—solar radiation, wind speed, humidity, minimum and maximum daily air temperature, air pressure —required were developed using a binning approach, sourcing the historical values from the Princeton Land Surface Hydrology Group [18] (pp. 3088–3111). More detail on the climate scenario selection and processing is provided in the Supplementary Material. These variables were aggregated, for this analysis, to the USGS 8-digit Hydrologic Unit Code (HUC-8) scale (more detail on the HUC can be found in [19]). Furthermore, each climate projection through 2099 is split into two 20-year "eras": 2050 (2040–2059) and 2090 (2080–2099). These eras are compared to a 20-year baseline climate of 1986–2005. Average changes in temperature and precipitation across the LOCA scenarios and over time are displayed in Figure 1.

Note that in the results section (Section 3), we often focus on two GCMs (rather than the full set of 5 GCMs) for simplicity of presentation—namely, GISS-E2-R and MIROC5. These GCMs represent two extremes as pertains to water quality and exhibit different spatial patterns of changes in climate. The GISS-E2-R climate model projects less increases in air temperature than the others, with wetter conditions in the East. Alternatively, the MIROC5 climate model projects high increases in air temperature and considerable drying, especially in the central region of CONUS. The full set of the results for the five GCMs is provided in the Supplementary Material.

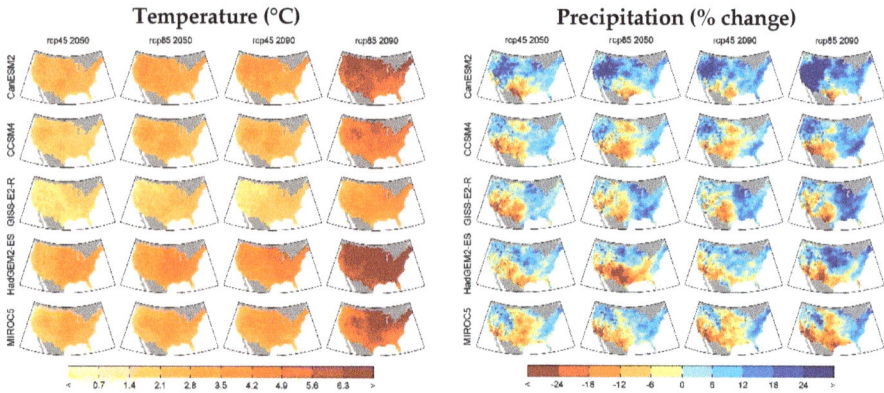

Figure 1. Mean projected changes in temperature (°C; **left**) and precipitation (%; **right**) for the five climate models, two emissions scenarios, and the 2050 and 2090 eras. Changes are between the average of the 20-year projected era and the 20-year baseline.

2.2. Description of HAWQS

The Hydrologic and Water Quality System (HAWQS; [11]) is a web-based Decision Support System developed at the Texas A&M University Spatial Sciences Laboratory and funded by the United States Environmental Protection Agency (EPA) Office of Water. HAWQS is an advanced, total water quantity and quality modeling system with databases, interfaces and models that evaluates the impacts of proposed regulations, water quality management actions and scenarios of climate and land use change on the quality and quantity of the Nation's streams and rivers.

The core engine of HAWQS is the watershed water quality and quantity simulation model, Soil and Water Assessment Tool (SWAT). Originally developed by the U.S. Department of Agriculture (USDA), SWAT has been the core simulation tool for numerous US national and international assessments of soil and water resources. SWAT is a physically-based, computationally efficient model that continuously simulates a large array of watershed processes for a defined period of record. Details of the SWAT modelling methods are described in the Theoretical Documentation [20].

HAWQS is designed to support national-scale economic benefit assessments of potential water quality management strategies (including policy scenarios and best management practices), and is capable of supporting a wide variety of national- and regional-scale economic and policy analyses by simulating baseline and alternative water quality conditions for sediments, pathogens, nutrients, biological oxygen demand, dissolved oxygen, pesticides, and other characteristics. The model follows a broad modeling sequence: (1) the landscape phase, where the primary processes are climate, soil water balance, nutrient and sediment transport and fate, land cover, plant growth, farm management; and (2) the main channel phase, where the main processes are river routing, sediment and nutrient transport through the rivers and reservoirs. While HAWQS is capable of modeling CONUS at the spatial scale of the 10- and 12-digit HUC, the 8-digit HUC is used in this study.

2.2.1. Landscape

In HAWQS, runoff is modeled using the Soil Conservation Service (SCS) curve number procedure on a daily basis, adjusting for antecedent soil moisture, canopy interception (thus effective rainfall), land cover, slope and soil type. The parameters for this calculation are collected from United States Department of Agriculture—Natural Resource Conservation Service, State Soil Geographic [21] and topography from [22]. The simulated buildup and transport of nutrients and Biological Oxygen Demand (BOD) in the landscape are modeled in HAWQS in response to agricultural management, municipal point-sources, and atmospheric deposition [23]. Agricultural land use data were derived from the National Agricultural Statistics Service [24]. In the landscape, solar radiation, relative humidity, minimum and maximum daily temperature, and wind speed, and leaf area, are used to estimate crop growth and runoff.

2.2.2. Main Reach and Reservoirs

The daily estimates of runoff, lateral and ground water flow including any contribution from tiled land surfaces are added to the main routing stream after routing through the tributaries for channel losses. Once the water is added to the main routing, the water is routed using variable storage coefficients [25] (pp. 100–103) through each 8-digit HUC reach with point sources added in each reach based on the contribution of nitrogen and phosphorus by population. If reservoirs are present, the SWAT routes water, nutrients and sediment through the reservoirs based on their characteristics. Reservoir information is sourced from the National Inventory of Dams [26]. River flows, derived from the United States Geological Survey [27], were used for calibration and validation of the flow at selected locations across CONUS. Water consumptive use data were sourced from [28] for surface and groundwater uses.

Within each 8-digit HUC, the main river reach and reservoirs are assumed to be well-mixed. Water temperature is calculated based on dampened changes in daily air temperature developed by [29]. The transport of nutrients, dissolved oxygen, and sediment in streams is modeled in the stream by keeping track of changes in mass on a daily basis. All details of SWAT calculations are well-documented in [20].

2.3. Description of US Basins

The version of the US Basins model used here is described in [8] (pp. 1326–1338). Precipitation and temperature from each climate scenario are inputs into: (a) a rainfall-runoff model (CLIRUN-II), which is used to simulate monthly runoff; and (b) a water demand model, which projects the water requirements of the municipal and industrial (M&I) and agriculture sectors. With these runoff and demand projections, a water resources systems model produces a time series of reservoir storage, release, and allocation to the various demands in the system, which include M&I, agriculture, transboundary flows, and hydropower. The water quality model is driven by QUALIDAD [30], which uses managed flows and reservoir states to simulate a number of water quality constituents in rivers and reservoirs. Since US Basins does not include a representation of loading transport through the landscape, loading into the main river reaches is exogenous. For this study, nonpoint agricultural loadings from the HAWQS landscape (phosphorus, nitrogen and BOD) are used directly in US Basins, equally distributed across each segment within the HUC-8. Due to the computational intensity of US Basins, one year of mean climatology is used for the baseline period and future eras.

2.3.1. Runoff and Water Demand

The climate projections for each emission scenario were used to develop monthly runoff estimates. Runoff modeling converts the climate shifts into changes in surface water availability important for the water resource systems model. Surface water runoff was modeled with the rainfall-runoff model CLIRUN-II (see [31,32]), the latest available application in a family of hydrologic models developed specifically for the analysis of the impact of climate change on runoff, first proposed by [33] (pp. 1–16). Water demands are the other side of the water balance, and are developed using 2005 data from the U.S. Geological Survey on annual water withdrawals and consumptive use in a range of sectors including irrigation, M&I use, mining, thermal cooling, and several other sectors [34] (p. 52). These data are available at the 3109 counties of CONUS and spatially averaged to the 8-digit HUC resolution using the same approach taken by the U.S. Forest Service in their development of the Water Supply Stress Index (WaSSI; [35]).

2.3.2. Water Resources Planning Model

Reservoir management and routing in US Basins is simulated using a water resource systems scheme, where the simulated runoff—used as surface water supply—and projected water demands are used to optimize water allocation based on a prescribed set of priorities. Three demand types, or nodes, are modeled throughout the system, which are in competition for water dependent on the sequence (upstream/downstream). The node types are municipal and industrial (M&I) water use, hydropower generation, and irrigation withdrawal. The hydrologic boundaries used to define the basins are the 2119 8-digit HUCs of CONUS. The structure of each basin is generic, prescribed with input characteristics that are unique to each HUC. Reservoir data, such as locations, hydropower capabilities, and the information needed to calculate surface area and volume are all retrieved from the Army Corps of Engineers [26]. Hydropower production is calculated and calibrated to the National Renewable Energy Laboratory (NREL) Regional Energy Deployment System (ReEDS) model [36] (pp. 275–3000). For each of the basins, the priorities of the various water users are assumed to be in the following order: (1) minimum flows driven by environmental and trans-boundary concerns; (2) M&I water demands (including mining and thermal cooling); (3) irrigation demands; and (4) hydropower production. More detail on the calibration and verification of US Basins can be found in [37].

2.3.3. Water Quality Model Description

Using the managed flows and reservoir storage and volume from the water planning model as well as climate parameters, we use the QUALIDAD model [30] to track several water quality constituents for each 8-digit HUC, including temperature, dissolved oxygen (DO), three nitrogen species, two phosphorus species, a generic metal, and salt. QUALIDAD is a parsimonious water quality model that is designed to model daily water quality dynamics at the basin scale. The mathematical representations of these processes are detailed in [30]. The mass balance equation is solved numerically using the Matlab Ordinary Differential Equation (ODE) 15 s [38] (pp. 1–22). All variables in this model have a daily time step except temperature, which is hourly. For simplicity, monthly flows are transformed to daily using a spline interpolation, which is certainly a limitation the US Basins approach and plans are in place to address this in the future. To track water quality constituents within the CONUS framework, each 8-digit HUC is divided into a number of segments based on the [39], which is a dataset built upon the EPA's digital record of over 60,000 river reaches in the U.S., intended for national water-quality modeling. For each river segment, the data set contains corresponding parameters such as flow, velocity, segment length, and the sequence of segments. Based on these parameters, the main river channel is found within each 8-digit HUC, and then separated into segments based on travel time estimates optimized to reduce numerical dispersion. Each constituent is modeled separately in each segment, and upstream to downstream mass transfer is governed using numerical methods documented by [30,31]. More detail on the CONUS routing framework is provided in [40].

Temperature is tracked within QUALIDAD using a heat budget model approach [41], that simulates the surface heat exchange of a body of water as well as water sources/sinks through inflows from upstream basins, outflows downstream, small tributaries, and groundwater. Strzepek [40] includes more detail on this approach. We assume that each riverbed is parabolic, following [42], which helps to derive a relationship of flow with surface area and velocity. Wind speed, relative humidity, daily temperature range, solar radiation, and air pressure are used in addition to precipitation and temperature to calculate water temperature.

In the summer, as temperature warms and solar radiation increases, stratification in temperate reservoirs occurs. Temperature during the season of stratification is modeled differently for reservoirs, where a two-layer model is used, representing both the epilimnion (top) and the hypolimnion (bottom) layers. For example, if the reservoir is bottom-releasing (i.e., outflow is occurring in the hypolimnion) then the water state (e.g., dissolved oxygen levels) in the bottom layer flows downstream.

A detailed sensitivity analysis of this process-based mass balance approach that governs the water quality calculations of US Basins can be found in [8] (pp. 1326–1338).

2.4. Summary of Key Differences between US Basins and HAWQS

There are several key differences between the HAWQS framework and the US Basins framework (summarized in Table 1). To start, US Basins is computationally intensive, due to the ODE solver, and for this study mean climate conditions are used for each era and scenario, as discussed. Alternatively, HAWQS runs the full set of transient years, resulting in 20 years for each era and scenario. The process of converting climate into flow is estimated differently. Although HAWQS is not fully calibrated across CONUS, HAWQS has been calibrated and verified across large areas of CONUS [43,44]. Then, runoff, routing, and water withdrawal is translated into flows based on a pre-calibrated scheme. On the other hand, US Basins uses estimates of naturalized runoff to calibrate the runoff model on a long-term mean monthly basis across CONUS. These runoff estimates are applied to a prioritization scheme of water uses to estimate flow and reservoir levels. Furthermore, US Basins focuses on the water quality of the main reach, lacking many of the landscape and tributary processes applied in HAWQS, which includes 225,000 landscape units distributed across CONUS. The loadings between the two models are also different, discussed in detail in Section 2.5.

Table 1. Summary of key differences between US Basins and Hydrologic and Water Quality System (HAWQS).

Parameter/Characteristic	US Basins	HAWQS
Number of years per "era"	One year	20-years
Runoff model	CLIRUN, calibrated, monthly	SWAT, daily
Landscape	Simple	Complex, includes land management, fertilizer application, more complex crop model, among others
Water quality model	QUALIDAD, daily	SWAT, daily
Reservoirs	Stratified, 2-layer	Well mixed with seasonal settling and decay rates
Main Rivers	Multiple segments per HUC8	Well mixed
Water Allocation/management	Priority scheme, hydropower, monthly	Based on a calibration of flows, daily
Water Temperature	Energy balance, hourly	Dampened air temperature, daily
DO saturation	Based on temperature and elevation	Based on temperature

Although the water quality models both use many of the same mass balance equations in-stream as outlined in [41], the application is different. HAWQS solves these mass balances on a daily basis assuming the main reach and reservoirs are well-mixed and the water temperature equation is estimated based on air temperature. US Basins splits the main reach into segments, as described previously, and the reservoirs into two vertical layers to account for summer stratification. In addition, these mass balance equations in the reach and reservoirs are solved numerically using ODE solvers. The water temperature is also solved numerically using an energy balance equation. Additionally, US Basins accounts for the effect of elevation on the DO saturation level while HAWQS DO saturation is based on water temperature only, without the elevation effect.

Many of the specific implications of these differences are discussed in the Results section that follows.

2.5. Loading Inputs to US Basins and HAWQS

In each model, loadings enter the system as point and nonpoint sources. Agricultural nonpoint source loadings were developed in HAWQS using data available from the Spatially Referenced Regressions on Watershed Attributes (SPARROW) model (see [45]). These included total annual nitrogen and phosphorus from fertilizer application, as well as BOD outputs from livestock. Non-point loadings were modeled using HAWQS to estimate the transport of nonpoint loadings to the main river reaches and reservoirs. These loadings vary with runoff, as SWAT includes crop, tree, and plant nitrogen, phosphorus, and carbon cycles. The longer the nutrients stay in the soil, the more they are consumed by vegetation. As climate change affects runoff, each climate scenario has a unique set of nitrogen and phosphorus loadings from agriculture (see Figure 2). Note the similarities between the inverse of changes in precipitation (Figure 1), and the changes in loadings presented below.

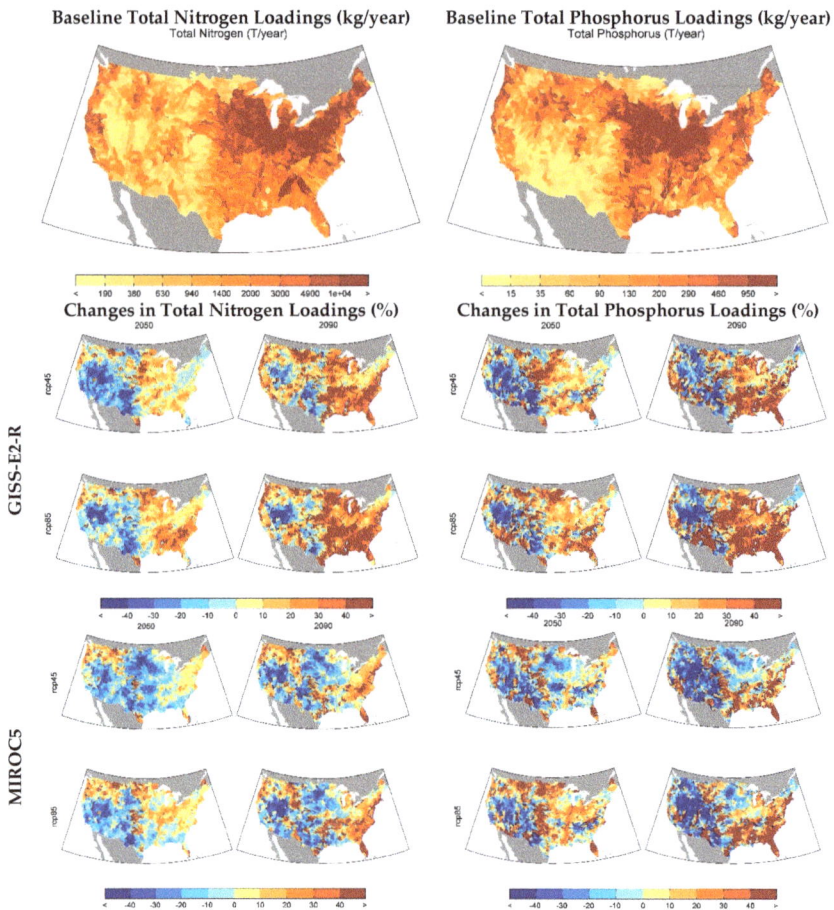

Figure 2. Non-point source nitrogen and phosphorus loadings under the baseline (**top**) and climate change (**bottom**) derived from HAWQS outputs. Variability in loading patterns across climate scenarios, emissions scenarios, and time is driven by the response of the landscape model to changes in river runoff under climate change.

In both HAWQS and US Basins, loadings from municipalities were estimated using export coefficients for both nitrogen (2 kg/person/year) and phosphorus (0.3 kg/person/year) derived from [46]. BOD loadings from municipalities are excluded. These annual per capita loadings were scaled to kilograms based on U.S. population projections developed using the Integrated Climate and Land Use Scenarios (ICLUS, version 2 [46] (pp. 20887–20892); [47]) model. Using the UN Median Variant projection for the U.S. [47], ICLUSv2 was applied to generate county-level population projections at five-year time steps between 2000 and 2100, which were then spatially averaged to the 8-digit HUCs. This population projection is consistent across both GHG mitigation scenarios. These point source loadings rose proportionately to projected population through 2100. In addition, HAWQS includes loadings from atmospheric deposition, which is based on wet and dry deposition from historical observation stations on a monthly basis by 8 digit HUCs [23]. These loadings from atmospheric deposition were excluded in US Basins because US Basins lacks a model of the landscape as discussed previously.

2.6. Valuation of Water Quality

In this study, the economic impacts of changes in water quality measures are estimated using a valuation of changes in a water quality index. Many water quality indices have been developed over the past 50 years. The National Sanitation Foundation (NSF) (explained in detail in [48]) built on previous work by incorporating expert judgement and provides a template for many water quality indices developed since (e.g., [49]). In this study, we use a water quality index following a similar approach outlined by [49], which follows three steps: (1) obtain measurements on water quality constituents, obtained directly from the water quality model previously described; (2) convert each measurement into a subindex using water quality curves and (3) aggregate the subindex values into the WQI. McClelland [48] provides water quality curves (step 2) and aggregation weights (step 3) for nine water quality parameters. Using this approach, we develop a "Climate WQI" (CWQI) similar to the one used in [8] (pp. 1326–1338), which uses four subindex calculations: water temperature, as well as the concentrations of DO, total phosphorus, and total nitrogen. In this study, we use an updated form of the subindex calculations for DO, total nitrogen and total phosphorus from [50]. The subindex curves vary for total nitrogen and total phosphorus across the U.S. by Level III Ecoregion and are based on a fitted exponential function. In contrast, the DO subindex curve, based on an exponential relationship below saturation and a second order polynomial above saturation, is the same across CONUS. The temperature subindex calculation [48] is based on deviations from mean water temperature and described in more detail in [8] (pp. 1326–1338). More detail on the subindex calculations can be found in Supplementary Material.

Similar to [8] (pp. 1326–1338), the relationship between changes in CWQI and changes in WTP—used here as an indicator of economic costs and/or benefits—is developed from the full linear meta-regression transfer function from [7], using a piecewise linear function. We use state-level data from the [50] on persons per household to convert WTP per household to WTP per person to develop a national WTP across scenarios and eras. Van Houtven [7] also distinguished WTP by users and non-users. We use state-level boating survey data [51] to weight each 8-digit HUC by fraction of users and non-users. Although "users" include a broader group than boaters, information on other categories was not available at the national level. Both the users/non-users and persons per household are scaled using the population projections discussed previously. The four water quality parameters in each HUC-8 are aggregated, weighting by total HUC-8area, to the Level-III Ecoregions. Because boaters are a subset of all users, and because users have a higher WTP per person than non-users, our approach likely underestimates aggregate WTP for improvements in the CWQI.

3. Results and Discussion

The following section outlines the water quality parameters from the models, starting with the baseline followed by the future projections. As mentioned, these focus on flow, water temperature, total nitrogen, total phosphorus, dissolved oxygen, and CWQI, followed by WTP. The section continues with a more quantitative comparison of the two water quality models and a short discussion.

3.1. Baseline Model Outputs

In this section, we describe the main differences in the baseline water quality parameters from both models and identify the main causes for these differences. Figure 3 shows the mean baseline managed flow, water temperature, and DO concentrations. Overall, the spatial patterns of flow between HAWQS and US basins are quite similar, although the magnitudes of flow differ, particularly in the western US where HAWQS flows are higher. These differences are not unexpected, as the US Basins model relies on a calibrated rainfall-runoff model (CLIRUN-II), whereas the HAWQS simulates runoff using partially calibrated curve numbers in SWAT. The lower western river flows in US Basins result in higher nitrogen and phosphorus concentrations in these rivers than in HAWQS. The situation is comparable with DO, where the spatial patterns between the two models are quite similar, although the magnitude of concentrations is generally higher in HAWQS. As discussed previously, SWAT does not adjust for elevation in the DO saturation equation, resulting in higher DO values in high elevation regions in CONUS as compared to US Basins. As for water temperature, US Basins tends to estimate higher water temperatures than HAWQS across CONUS. This is not unexpected since the water temperature in HAWQS (also in SWAT) does not take into account solar radiation, relative humidity, and water depth, while US Basins includes these effects. Also note that US Basins is using a single mean climate year, and does not include inter-annual variability, which also accounts for many of these differences. This also applies to the remainder of the results.

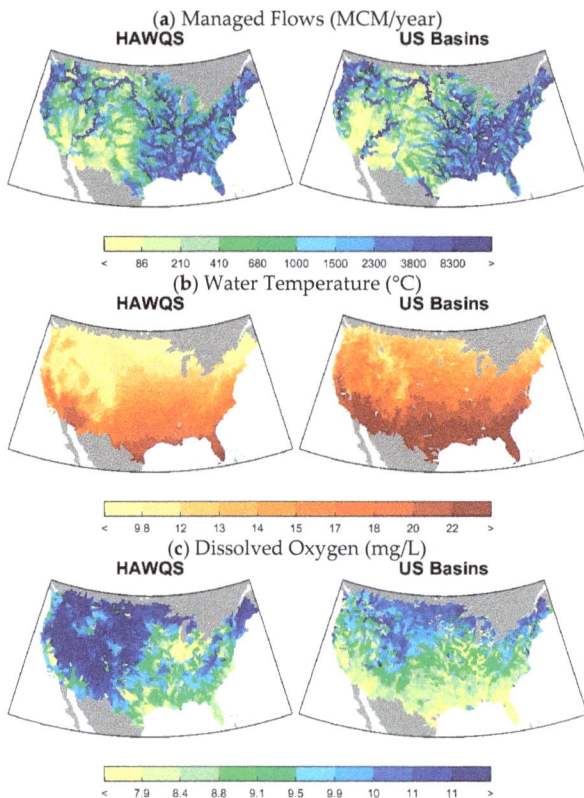

Figure 3. (a) Mean baseline flow (MCM/year); (b) water temperature and (c) dissolved oxygen (mg/L), 1986–2005, at the Hydrologic Unit Code (HUC)-8 watershed scale for US Basins and HAWQS.

Figure 4 shows the median total phosphorus and total nitrogen concentrations in the main reach in each HUC-8. Both the spatial patterns and magnitudes of each constituent are similar across the two models. As noted above, concentrations in US Basins are higher than HAWQS in the western US because of lower river flows. Also, US Basins uses monthly flows, which cannot account for intra-monthly variations in flow. This also results in differences in concentrations. In addition, US Basins excludes atmospheric deposition while it is included in HAWQS. Atmospheric deposition primarily influences nutrient concentrations in areas with higher precipitation, especially in eastern CONUS.

Figure 5 shows the scatter plot of the baseline CWQI across the 85 Level III Ecoregions as well as a map of the differences, where positive values indicate that HAWQS WQI is higher than US Basins. There is a clear difference between the east and west, where HAWQS WQI is higher in the west and US Basins is higher in the east. These differences illustrate two of the major differences in the models.

The lower CWQI values in the west in US Basins is primarily driven by lower managed flows in the west than in HAWQS, where lower flows results in higher concentrations of nitrogen and phosphorus and lower DO. In the east, these differences are primarily driven by the additional loadings in HAWQS through atmospheric deposition.

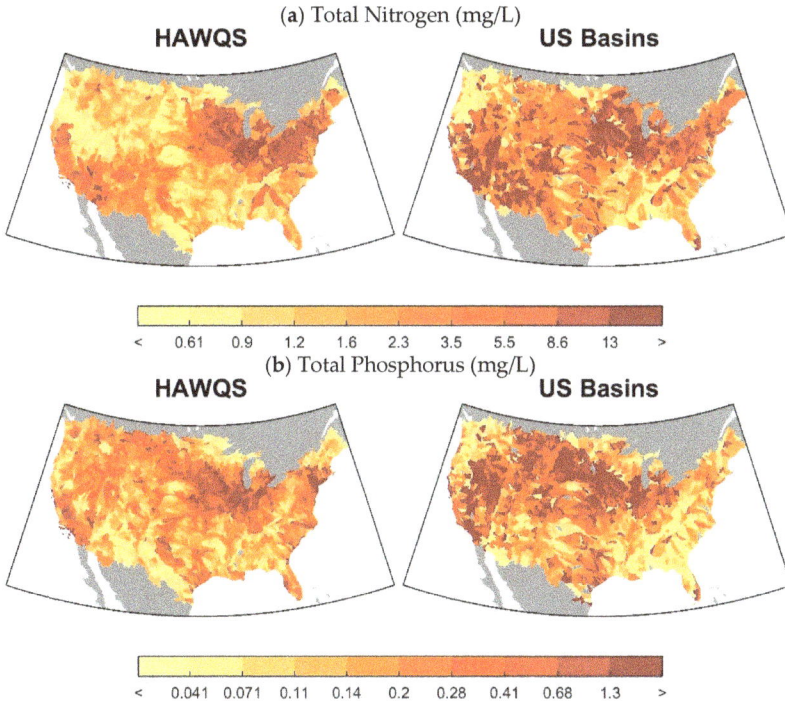

(a) Total Nitrogen (mg/L)

HAWQS   US Basins

| < | 0.61 | 0.9 | 1.2 | 1.6 | 2.3 | 3.5 | 5.5 | 8.6 | 13 | > |

(b) Total Phosphorus (mg/L)

HAWQS   US Basins

| < | 0.041 | 0.071 | 0.11 | 0.14 | 0.2 | 0.28 | 0.41 | 0.68 | 1.3 | > |

Figure 4. Mean baseline nitrogen (**a**) and phosphorus (**b**) concentrations (mg/L), 1986–2005, at the HUC-8 watershed scale for US Basins and HAWQS3.2. Climate change impacts and effect of greenhouse gas (GHG) mitigation on water quality.

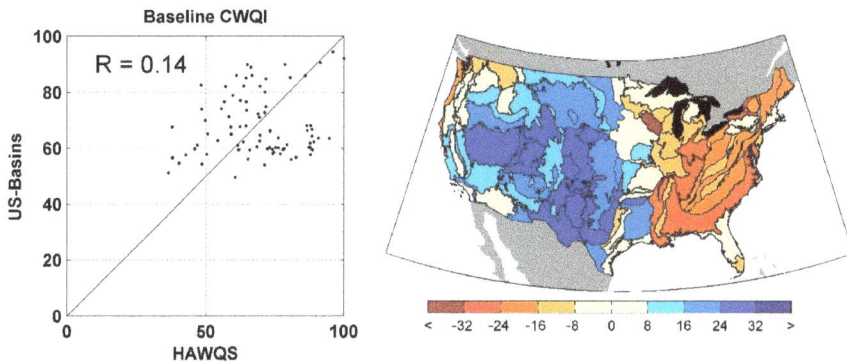

Baseline CWQI

R = 0.14

US-Basins

HAWQS

| < | -32 | -24 | -16 | -8 | 0 | 8 | 16 | 24 | 32 | > |

Figure 5. Scatter plot of baseline Water Quality Index (WQI) for HAWQS and US-Basins as well as correlation coefficient (R) and 1:1 line (**left**) and map of difference in WQI between HAWQS and US Basins (**right**), where positive values indicate that HAWQS WQI is higher than US Basins. These are both shown across the 85 Level III ecoregions in contiguous United States (CONUS).

3.2. Projected Model Outputs

Next we focus on the projections of these water quality parameters under climate change. As noted in Section 2.1, we focus on two GCMs for simplicity—GISS-E2-R and MIROC5. Figure 6 shows the change in managed flows for both water quality models for 2050 and 2090. Both water quality models generally project larger changes in flow in 2090 than 2050. There are also a number of differences between flow projections. For GISS-E2-R, HAWQS projects larger decreases in the flow in the southwest than US Basins, while US Basins projects larger decreases in flow in the central portion of the country. The increase in flow in the east is consistent across both HAWQS and US Basins. For MIROC5, both HAWQS and US Basins project decreases in flow in the central portion of CONUS with less change in the east and slight wetting in the northwest. The most striking difference between HAWQS and US Basins in MIROC5 is the southwest, where HAWQS projects large decreases. However, this portion of the country is dry, so the relative change (as shown in percent) is large but the total change in flow smaller than other portions of the country. These differences between the water quality model projections of managed flow represent a complex interaction between changes in climate, modeled runoff, and the model assumptions about reservoir management.

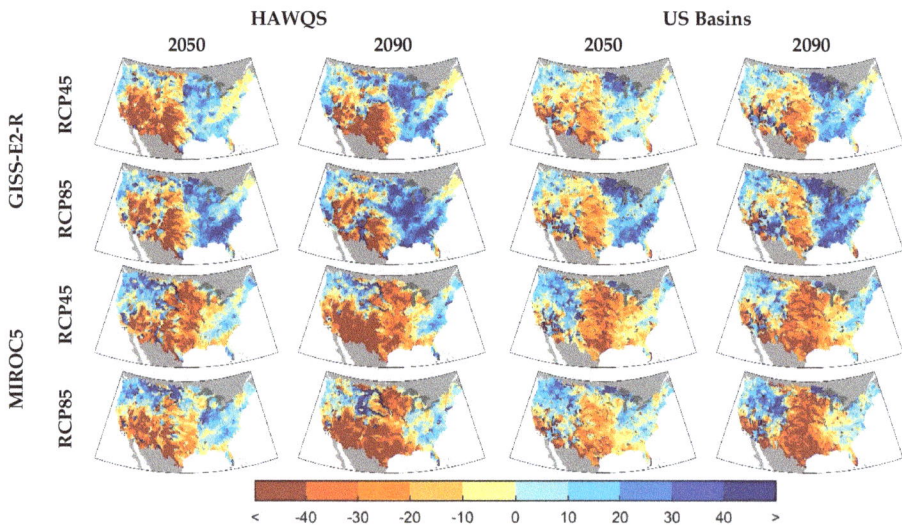

Figure 6. Percentage changes in mean projected (**left**) HAWQS and (**right**) US Basins river flow at the HUC-8 watershed scale for both the GISS-E2-R and MIROC5 climate models, two emissions scenarios, and two eras.

Figure 7 shows the changes in water temperature for both climate models and water quality models in 2050 and 2090 and both RCPs. Since water temperature is primarily driven by changes in air temperature, this water quality parameter shows the most similarity between the two water quality models. However, the two models estimate water temperature using completely different equations, as previously discussed. These, along with the differences in changes in runoff and flow account for the different spatial patterns shown for the two models. The 2050 results show moderate increase in water temperature, while the results in 2090 are more extreme with increases above 4.5 °C in MIROC5, RCP85.

Figure 8 shows the percent change in total nitrogen concentrations. Both models show larger changes in 2090 than in 2050. Both water quality models agree that the south-central U.S. and the area around the Great Lakes are likely to see increases in total nitrogen. HAWQS shows large increases in total nitrogen in the southwest for all GMS and RCPs in 2090, albeit more pronounced in MIROC5 than

GISS-E2-R, while US Basins show decreases in this region. Figure 9 shows changes in total phosphorus concentrations across CONUS. HAWQS consistently projects increase in phosphorus levels along the southwest coast and Texas. In contrast, US Basins projects larger increases in the central U.S., especially in 2090, as well as in the east. Differences in nitrogen and phosphorus concentration changes can be explained primarily by the differences in flow changes for these two models. Also notice the differences in the change in concentrations in the drier areas—namely, the southwest—where the models differ in sign. Since flows are low in these areas, the resulting concentrations are sensitive to flow changes.

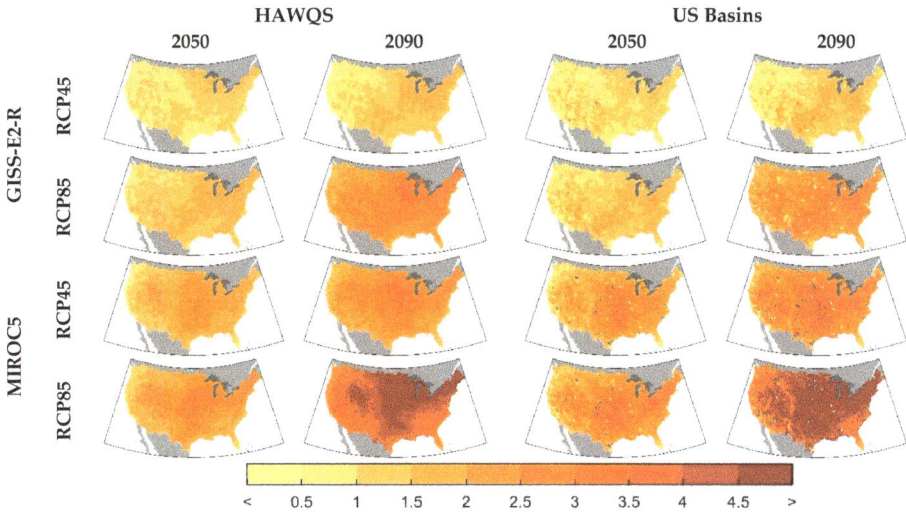

Figure 7. Changes (°C) in mean projected (**left**) HAWQS and (**right**) US Basins water temperature at the HUC-8 watershed scale for both the GISS-E2-R and MIROC5 climate models, two emissions scenarios, and two eras.

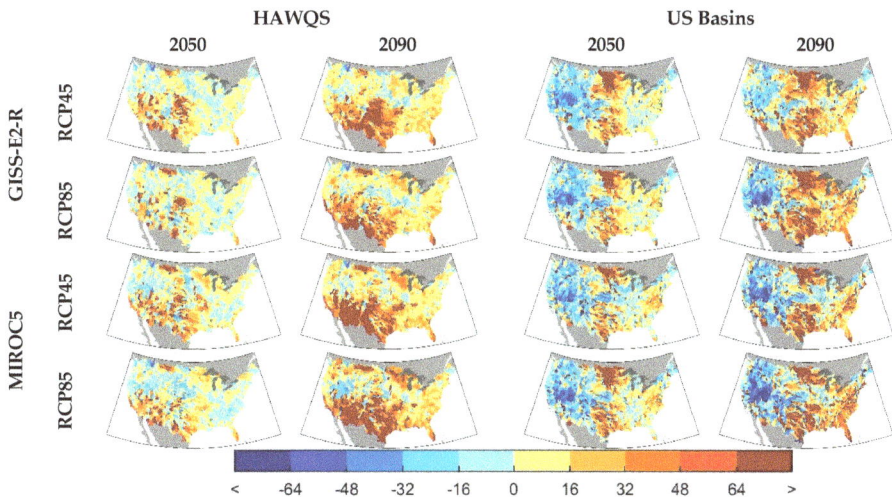

Figure 8. Percentage changes in mean (**left**) HAWQS and (**right**) US Basins nitrogen concentrations at the HUC-8 watershed scale for both the GISS-E2-R and MIROC5 climate models, two emissions scenarios, and two eras.

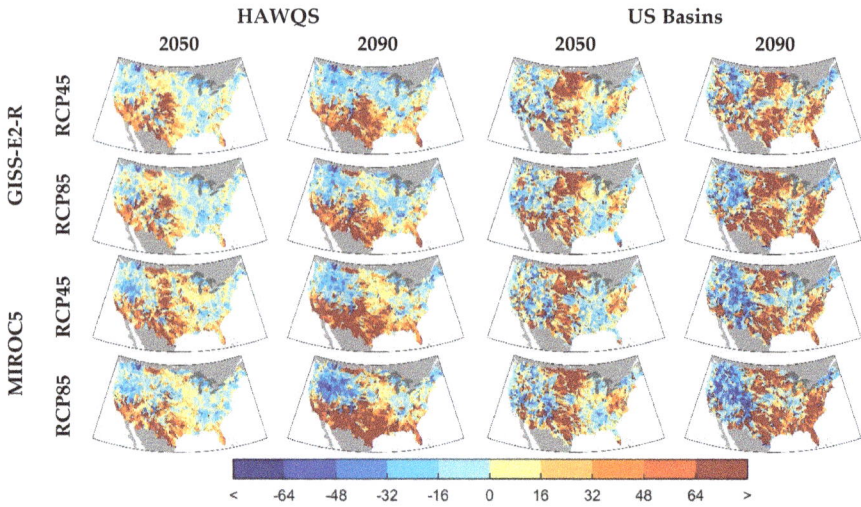

Figure 9. Percentage changes in mean (**left**) HAWQS and (**right**) US Basins phosphorus concentrations at the HUC-8 watershed scale for both the GISS-E2-R and MIROC5 climate models, two emissions scenarios, and two eras.

Figure 10 shows percent changes in DO for the same eras and scenarios. In both water quality models, there are consistent decreases in DO in the east. HAWQS shows large decreases around Texas, areas in the southwest, and along the East coast, with areas of increases in the western mountainous regions. US Basins shows the largest decrease in DO around the Great Lakes. Since DO is largely influenced by temperature through levels of DO saturation (i.e., higher temperatures reduce DO saturation levels, thereby reducing DO aeration), DO generally decreases in the future. However, DO is also influenced by changes in nitrogen, phosphorus, and BOD loadings, as well as changes in flow. In both models, changes in DO are largest for 2090 as compared to 2050 and larger for RCP85 than RCP45.

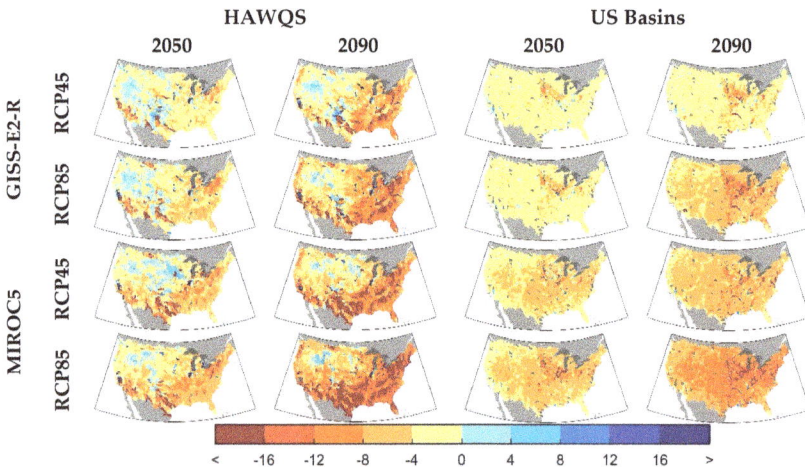

Figure 10. Percentage changes in mean (**left**) HAWQS and (**right**) US Basins dissolved oxygen for, at the HUC-8 watershed scale, both the GISS-E2-R and MIROC5 climate models, two emissions scenarios, and two eras.

Changes in CWQI are shown in Figure 11. For both water quality models, changes in CWQI are more pronounced in the east than in the west from increases in loadings, causing higher concentrations of total nitrogen and total phosphorus, and temperature being higher in this area. For HAWQS, changes in CWQI are largest along the east coast, although this pattern also shows in US Basins in MIROC5 RCP85. US Basins tends to show larger increases in CWQI in the central U.S. and around the Great Lakes. Since this is an aggregation of the changes in water quality parameters previously discussed, many of these differences between projected CWQI changes can be explained by the model differences already discussed. For example, HAWQS shows larger decreases in CWQI in the west relative to US Basins, for GISS-E2-R in particular. These can be explained by increases in nitrogen and phosphorus in the west as compared to US Basins projections of nitrogen and phosphorus.

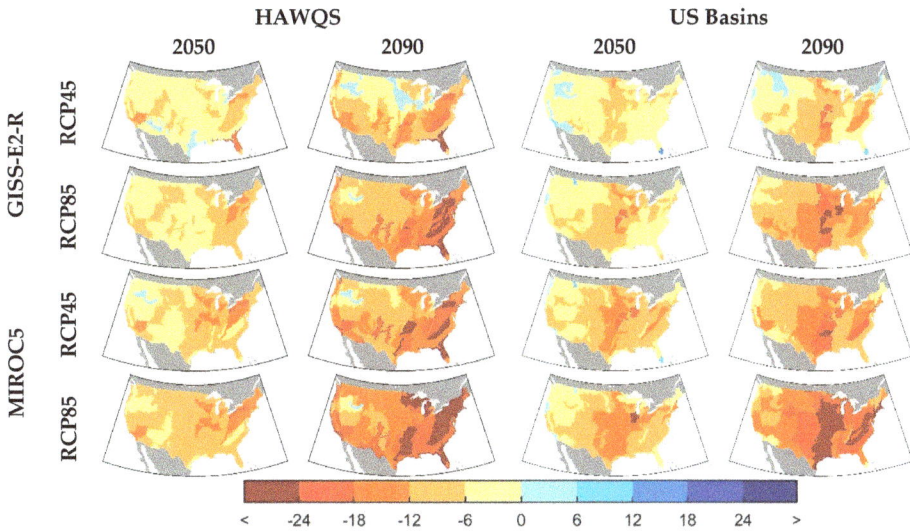

Figure 11. Changes in mean (**left**) HAWQS and (**right**) US Basins levels of the Climate-Water Quality Index at the Level-III Ecoregion scale for both the GISS-E2-R and MIROC5 climate models, two emissions scenarios, and two eras.

Tables 2 and 3 show correlations of changes in CWQI and WTP across the Level III Ecoregions for the five GCMs, two emissions scenarios, and two future eras. The least agreement between the water quality models is found in the GISS-E2-R GCM in 2050 for RCP45, which represents the climate with the least change, of the ones shown, in both temperature and precipitation. However, there is more agreement between the water quality models for future climates with higher radiative forcing, either by mitigation policy or era, and larger projected changes in climate as seen in MIROC5.

Table 2. Correlation coefficients of changes in CWQI across the Ecoregions between HAWQS and US Basins.

		CanESM2	CCSM4	GISS-E2-R	HadGEM2-ES	MIROC5
RCP45	**2050**	0.08	0.02	0.05	0.31	0.47
	2090	0.21	0.39	0.45	0.33	0.43
RCP85	**2050**	0.05	0.01	0.21	0.29	0.35
	2090	0.36	0.36	0.44	0.37	0.46

Table 3. Correlation coefficients of changes in WTP across the Ecoregions between HAWQS and US Basins.

		CanESM2	CCSM4	GISS-E2-R	HadGEM2-ES	MIROC5
RCP45	2050	0.32	0.14	0.27	0.50	0.59
	2090	0.34	0.38	0.40	0.44	0.42
RCP85	2050	0.33	0.21	0.35	0.51	0.53
	2090	0.50	0.40	0.48	0.48	0.54

We find that while the models present a different picture of baseline water quality, the projections of aggregate water quality as expressed through CWQI tend to exhibit stronger levels of agreement and that as changes in climate become more drastic—i.e., in projections of climate with higher solar forcings either in time or global GHG mitigation policy—agreement, across the two models is highest. As discussed, differences in these water quality projections from the two models point to dissimilarities in the model structure and inherent bias of each water quantity and quality model. As these are complex systems modeled over large geographic areas, inconsistencies in the outcomes of the two models are expected. However, we find greater levels of agreement across the water quality models in the direction and magnitude of CONUS-wide climate change impacts.

3.3. Valuation

The following section focuses on the valuation results, in terms of changes in WTP, for both water quality models at the Level III ecoregions. Figure 12 shows these WTP changes, in 2005 USD/year/person for the two climate models, two eras, and two mitigation policies. Note that decreases in WTP reflect the willingness to pay in order to avoid the given water quality scenario relative to the baseline. Since these values are directly related to the changes in WQI, these maps resemble the maps in Figure 11. Although there are differences in WTP changes across the two models, both models consistently show decreases in WTP in the future, with few increases across ecoregions. Decreases under RCP45 are smaller, generally, than under RCP85, with this pattern more pronounced in 2090 than 2050. WTP decreases are more pronounced particularly in the east on all counts than in the west. HAWQS continues to show more impacts on the west coast than US Basins, while US Basins shows more consistent increases in WTP in the Midwest, and the lower Mississippi basin.

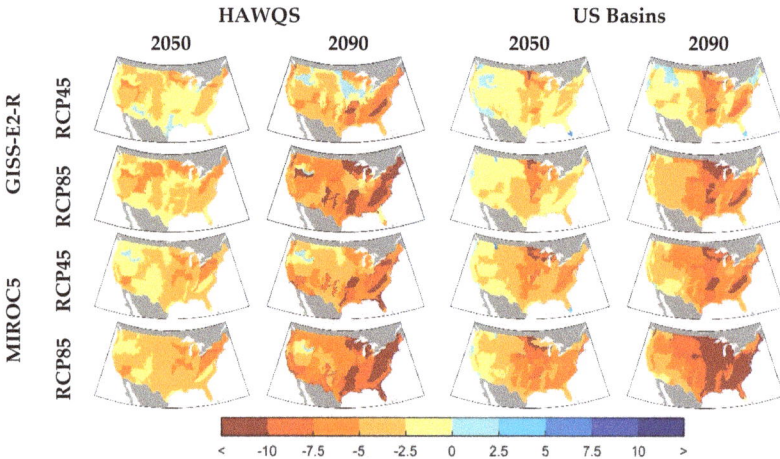

Figure 12. Changes in mean (**left**) HAWQS and (**right**) US Basins Willingness to Pay per person (USD/year) at the Level-III Ecoregion scale for both the GISS-E2-R and MIROC5 climate models, two emissions scenarios, and two eras.

National level WTP per person is shown in Figure 13 and total WTP is shown in in Table 4 for all five GCMs, RCPs, eras, and the two water quality models. In all GCMs and water quality models, WTP decreases most for RCP85 compared to RCP45. The largest changes in WTP are shown in the HadGEM2-ES GCM with the least projected for GISS-E2-R. Differences between these two GCM projections for both water quality models are 3.9 USD/person/year for RCP85 in 2090. In contrast, the largest difference between the two water quality models for RCP85 in 2090 is 1.8 USD/person/year for CanESM2. Since CanESM2 shows the largest changes in precipitation, this difference in WTP between the water quality models can be explained in part by the absence of atmospheric deposition in US Basins.

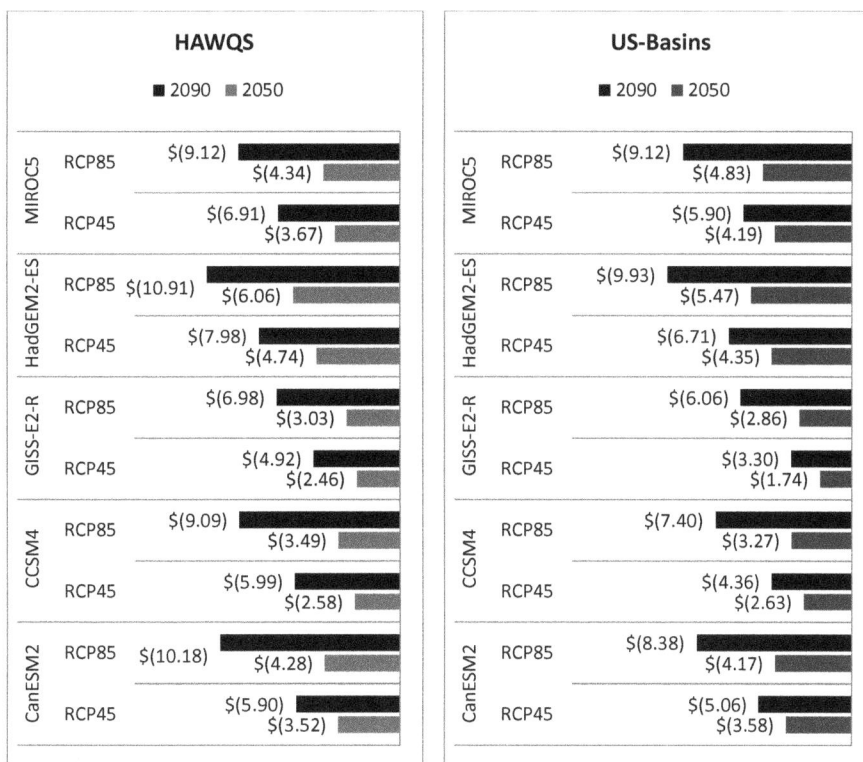

Figure 13. National willingness-to-pay (WTP) (USD/person/year) for HAWQS (**left**) and US-Basins (**right**) for all five General Circulation Models (GCMs).

Table 4. Total WTP (in billion 2005 USD/year) for all five GCMs and global GHG mitigation scenarios.

			CanESM2	CCSM4	GISS-E2-R	HadGEM2-ES	MIROC5
RCP45	2050	HAWQS	−$1.36	−$1.00	−$0.95	−$1.83	−$1.42
		US-Basins	−$1.38	−$1.01	−$0.67	−$1.68	−$1.62
	2090	HAWQS	−$2.59	−$2.62	−$2.16	−$3.50	−$3.03
		US-Basins	−$2.22	−$1.91	−$1.45	−$2.94	−$2.59
RCP85	2050	HAWQS	−$1.66	−$1.35	−$1.17	−$2.34	−$1.68
		US-Basins	−$1.61	−$1.26	−$1.10	−$2.11	−$1.87
	2090	HAWQS	−$4.46	−$3.98	−$3.06	−$4.78	−$4.00
		US-Basins	−$3.67	−$3.24	−$2.66	−$4.35	−$4.00

4. Conclusions

We find that, as the end-goal of this study is about national scale economic benefits, differences between the increases in WTP for the two water quality models is less than differences between the five GCMs. Decreases in total national WTP in RCP85 range from 1.2 to 2.3 (2005 billion USD/year) in 2050 and 2.7 to 4.8 in 2090 across all climate and water quality models. Converted to a net present value, discounting at 3% and using the mean WTP across GCMs, results in a total decline in WTP of $28.9 and $26.3 billion for RCP45 for HAWQS and US Basins, respectively. For RCP85 this value is $38.2 and $35.8 billion for HAWQS and US Basins, respectively. The overall benefit of GHG mitigation is substantial, at a present value of $9.3 (HAWQS) and $9.5 (US Basins) billion for RCP4.5 compared to RCP8.5. These results are similar to the total impacts found in [8] (pp. 1326–1338) using only one GCM, where the net present value of GHG mitigation benefit, using a policy similar to RCP4.5, was found to be $10.7 billion. Note that these WTP estimates are based on recreational value, which is only a portion of the economy likely to be affected by decreases in water quality.

As both HAWQS and US Basins represent complex hydrologic, biochemical, and heavily managed systems over a broad spatial area, there are certainly limitations to the models and data. In general, both models take a parsimonious approach to modeling the system, so instead of rigorous calibration and validation that is typically performed on detailed models of "project-scale" studies, these models use a process-based, mass balance approach in order to assess general behavior and response to a changing climate. The hydrology in both models is at least partially calibrated, as previously discussed. The water quality modelling, on the other hand, is generally uncalibrated and relies on mass balance and commonly used parameters. For this reason, we do not present results at detailed scales in either time or space and rely on large-scale changes from the baseline water quality projection for the purpose of informing policy rather than individual project construction or design. This study is not the first to use either model in this way. SWAT, the basis of HAWQS, has often been used in ungauged basins (e.g., [52–54]) and US Basins was also designed for this type of analysis. However, this is a limitation, and detailed analysis should be performed on a case-by-case basis when needed. Another limitation is that the WTP values used in this study are based on recent estimates and would likely change in the future. Also, the population projections used in this analysis do not vary by RCP, although the populations are likely to change under different forcing levels. However, this decision was made intentionally, in order to isolate the effects of climate change from changes related to population change. In addition, US Basins is the use of only one "median" year to represent the baseline and future eras.

In this study, we have compared only two water quality models. This work can be expanded by comparing more water quality models to understand how other developed methods for projecting future water quality can result in alternative conclusions. Also, as in all climate change impact studies, we are limited by the spatial scale and uncertainty of the GCMs. Climate-related uncertainties were also not fully addressed here, in part by using a subset of the CMIP-5 models, but also by not including uncertainties related to initial conditions (addressing the chaos in the system), as in [37]. In addition, over the last several decades, large improvements have been made in agricultural and urban water conservation, agricultural soil conservation, farming technologies, urban stormwater and wastewater treatment, and protection of critical natural areas. In recent years, large amounts of agricultural land have been removed from soil conservation programs in order to produce more corn for ethanol, and growth of urban areas has removed large amounts of agricultural land from production. Future modeling efforts can address many of the impacts of land use changes, new agricultural and urban water management technologies, and environmental policies. These and other research endeavors are an important part of understanding the future of water quality in the U.S. and the effect of Climate Change.

Supplementary Materials: The following are available online at www.mdpi.com/2073-4441/9/2/118/s1.

Acknowledgments: We acknowledge the financial support of the U.S. Environmental Protection Agency's (EPA's) Climate Change Division (Contract #EP-BPA-12-H-0024), and Office of Water (Contract #EP-G15H-01113), and access to reservoir datasets from the U.S. Army Corps of Engineers. Arndt Gossel, Karen Metchis, Michael Trombley, Pravin Rana, and Isabella Morin provided support on the scope and direction of the study and technical contributions provided by Jacqueline Willwerth. Data used to produce the results of this paper can be made available through the corresponding author, Charles Fant (cfant@indecon.com). The views expressed in this article are solely those of the authors, and do not necessarily reflect the views of their organizations.

Author Contributions: Brent Boehlert, Jeremy Martinich, Charles Fant and Raghavan Srinivasan, conceived and designed the experiments; Charles Fant and Raghavan Srinivasan performed the experiments; Charles Fant, Brent Boehlert and Lisa Rennels analyzed the data; Jeremy Martinich and Brent Boelhert managed the project; Charles Fant wrote the paper; Brent Boelhert, Raghavan Srinivasan, Joel Corona, Ashley Allen, and Jeremy Martinich advised on conceptual and technical, and contributed to the writing and strategy; Stephen C. Chapra, Kenneth M. Strzepek advised on technical matters.

Conflicts of Interest: The authors declare no conflict of interest.

References

1. Georgakakos, A.; Fleming, P.; Dettinger, M.; Peters-Lidard, C.; Richmond, T.C.; Reckhow, K.; White, K.; Yates, D. Chapter 3: Water resources. In *Climate Change Impacts in the United States: The Third National Climate Assessment*; Melillo, J.M., Richmond, T.C., Yohe, G.W., Eds.; U.S. Global Change Research Program: Washington, DC, USA, 2014; pp. 69–112.

2. Sahoo, G.B.; Schladow, S.G.; Reuter, J.E.; Coats, R. Effects of climate change on thermal properties of lakes and reservoirs, and possible implications. *Stoch. Environ. Res. Risk Assess.* **2011**, *25*, 445–456. [CrossRef]

3. Kaushal, S.S.; Likens, G.E.; Jaworski, N.A.; Pace, M.L.; Sides, A.M.; Seekell, D.; Belt, K.T.; Secor, D.H.; Wingate, R.L. Rising stream and river temperatures in the United States. *Front. Ecol. Environ.* **2010**. [CrossRef]

4. Schneider, P.; Hook, S.J. Space observations of inland water bodies show rapid surface warming since 1985. *Geophys. Res. Lett.* **2010**, *37*, 1–5. [CrossRef]

5. Sahoo, G.B.; Schladow, S.G.; Reuter, J.E.; Coats, R.; Dettinger, M.; Riverson, J.; Wolfe, B.; Costa-Cabral, M. The response of Lake Tahoe to climate change. *Clim. Chang.* **2012**. [CrossRef]

6. Sahoo, G.B.; Schladow, S.G. Impacts of climate change on lakes and reservoirs dynamics and restoration policies. *Sustain. Sci.* **2008**, *3*, 189–199. [CrossRef]

7. Van Houtven, G.; Powers, J.; Pattanayak, S.K. Valuing water quality improvements in the United States using meta-analysis: Is the glass half-full or half-empty for national policy analysis? *Resour. Energy Econ.* **2007**, *29*, 206–228. [CrossRef]

8. Boehlert, B.; Strzepek, K.M.; Chapra, S.C.; Fant, C.; Gebretsadik, Y.; Lickley, M.; Swanson, R.; McCluskey, A.; Neumann, J.; Martinich, J. Climate change impacts and greenhouse gas mitigation effects on US water quality. *J. Adv. Model. Earth Syst.* **2015**, *7*, 1326–1338. [CrossRef]

9. Schewe, J.; Heinke, J.; Gerten, D.; Haddeland, I.; Arneli, N.W.; Douglas, C.B.; Dankers, R.; Eisner, S.; Fekete, B.M.; Colon-Gonzales, F.J.; et al. Multimodel assessment of water scarcity under climate change. *Proc. Natl. Acad. Sci. USA* **2014**, *111*, 3245–3250. [CrossRef] [PubMed]

10. Vetter, T.; Huang, S.; Aich, V.; Yang, T.; Wang, X.; Krysanova, V.; Hattermann, F. Multi-model climate impact assessment and intercomparison for three large-scale river basins on three continents. *Earth Syst. Dyn.* **2015**, *6*, 17–43. [CrossRef]

11. Yen, H.; Daggupati, P.; White, M.J.; Srinivasan, R.; Gossel, A.; Wells, D.; Arnold, J.G. Application of large-scale, multi-resolution watershed modeling framework using the Hydrologic and Water Quality System (HAWQS). *Water* **2016**, *8*, 164. [CrossRef]

12. Ficklin, D.L.; Luo, Y.; Luedeling, E.; Zhang, M. Climate change sensitivity assessment of a highly agricultural watershed using SWAT. *J. Hydrol.* **2009**, *374*, 16–29. [CrossRef]

13. Varanou, E.; Gkouvatsou, E.; Baltas, E.; Mimikou, M. Quantity and quality integrated catchment modeling under climate change with use of soil and water assessment tool model. *J. Hydrol. Eng.* **2002**, *7*, 228–244. [CrossRef]

14. U.S. Environmental Protection Agency (U.S. EPA). *Climate Change in the United States: Benefits of Global Action*; EPA 420-R-15-001; EPA Office of Atmospheric Programs: Washington, DC, USA, 2015.

15. Pierce, D.W.; Cayan, D.R.; Thrasher, B.L. Statistical downscaling using localized constructed analogs (LOCA). *J. Hydrometeorol.* **2014**, *15*, 2558–2585. [CrossRef]

16. Pierce, D.W.; Cayan, D.R.; Maurer, E.P.; Abatzoglou, J.T.; Hegewisch, K.C. Improved bias correction techniques for hydrological simulations of climate change. *J. Hydrometeorol.* **2015**, *16*, 2421–2442. [CrossRef]

17. U.S. Bureau of Reclamation. Downscaled CMIP3 and CMIP5 Climate Projections—Addendum Release of Downscaled CMIP5 Climate Projections (LOCA) and Comparison with Preceding Information. Available online: http://gdo-dcp.ucllnl.org/downscaled_cmip_projections/ (accessed on 1 September 2016).

18. Sheffield, J.; Goteti, G.; Wood, E.F. Development of a 50-year high-resolution global dataset of meteorological forcings for land surface modeling. *J. Clim.* **2006**, *19*, 3088–3111. [CrossRef]

19. U.S. Geological Survey. Hydrologic Unit Maps. Available online: https://water.usgs.gov/GIS/huc.html (accessed on 19 January 2017).

20. Neitsch, S.L.; Arnold, J.G.; Kiniry, J.R.; Srinivasan, R.; Williams, J.R. *Soil and Water Assessment Tool, User Manual, Version 2000*; Grassland, Soil and Water Research Laboratory: Temple, TX, USA, 2002.

21. Soil Survey Staff, Natural Resources Conservation Service, United States Department of Agriculture. U.S. General Soil Map (STATSGO2). Available online: http://sdmdataaccess.nrcs.usda.gov/ (accessed on 1 October 2010).

22. United States Geological Survey. National Elevation Dataset-NED. Available online: http://nationalmap.gov/elevation.html (accessed on 1 October 2010).

23. University of Illinois at Urbana-Champaign. Atmosphere Deposition—National Atmospheric Deposition Program (NADP). Available online: http://nadp.sws.uiuc.edu/ (accessed on 1 October 2010).

24. U. S. Department of Agriculture (USDA). *Farm and Ranch Irrigation Survey (2008)*; Special Studies, Part 1; U.S. Department of Agriculture: Washington, DC, USA, 2010; Volume 3.

25. Willams, J.R. Flood routing with variable travel time or variable storage coefficients. *Trans. ASAE* **1969**, *12*, 100–103. [CrossRef]

26. United States Corps of Engineers. Reservoirs—National Inventory of Dams (NID). Available online: http://geo.usace.army.mil/pgis/f?p=397:12 (accessed on 15 June 2013).

27. United States Geological Survey. River Discharge Data. Available online: http://waterdata.usgs.gov/nwis (accessed on 19 February 2016).

28. United States Geological Survey. Global Data Explorer. Available online: http://gdex.cr.usgs.gov/gdex/ (accessed on 19 February 2016).

29. Stefan, H.G.; Prued'homme, E.B. Steam temperature estimation from air temperature. *Water Resour. Bull.* **1993**, *29*, 27–45. [CrossRef]

30. Chapra, S.C. *QUALIDAD: A parsimonious Modeling Framework for Simulating River Basin Water Quality, Version 1.1, Documentation and Users Manual*; Civil and Environmental Engineering Department, Tufts USDA: Medford, MA, USA, 2014.

31. Strzepek, K.; McCluskey, A.; Boehlert, B.; Jacobsen, M.; Fant, C. *Climate Variability and Change: A Basin Scale Indicator Approach to Understanding the Risk to Water Resources Development and Management*; Water Papers; World Bank: Washington, DC, USA, 2011.

32. Strzepek, K.; Jacobsen, M.; Boehlert, B.; Neumann, J. Toward evaluating the effect of climate change on investments in the water resources sector: Insights from the forecast and analysis of hydrological indicators in developing countries. *Environ. Res. Lett.* **2013**, *8*, 044014. [CrossRef]

33. Kaczmarek, Z. Water balance model for climate impact analysis. *ACTA Geophys. Pol.* **1993**, *41*, 1–16.

34. Kenny, J.F.; Barber, N.L.; Hutson, S.S.; Linsey, K.S.; Lovelace, J.K.; Maupin, M.A. Estimated use of water in the United States in 2005, U.S. *Geol. Surv. Circ.* **2009**, *1344*, 52.

35. U.S. Forest Service (USFS). Water Supply Stress Index (WaSSI). Available online: http://www.forestthreats.org/research/tools/WaSSI (accessed on 20 September 2014).

36. Short, W.; Sullivan, P.; Mai, T.; Mowers, M.; Uriarte, C.; Blair, N. Regional Energy Deployment System (ReEDS). *Contract* **2011**, *303*, 275–3000.

37. Boehlert, B.; Solomon, S.; Strzepek, K.M. Water under a changing and uncertain climate: Lessons from climate model ensembles. *J. Clim.* **2015**, *28*, 9561–9582. [CrossRef]

38. Shampine, L.F.; Reichelt, M.W. The MATLAB ODE Suite. *SIAM J. Sci. Comput.* **1997**, *18*, 1–22. [CrossRef]

39. U.S. Geological Survey. ERF1—Enhanced River Reach File 1.2. Available online: http://water.usgs.gov/GIS/metadata/usgswrd/XML/erf1.xml (accessed on 12 August 2013).

40. Strzepek, K.M.; Lickley, M.; Gebretsadik, Y.; Schlosser, C.A.; Chapra, S.; Fant, C.; Boehlert, B.; Adams, E.; Strzepek, A. River temperature model for assessing climate impacts on the energy system. *J. Adv. Model. Eath Syst.* **2017**, in preparation.

41. Chapra, S.C. *Surface Water-Quality Modeling*; McGraw-Hill: New York, NY, USA, 1997.

42. Leopold, L.B.; Maddock, T. *The Hydraulic Geometry Channels and Some Physiographic Implications*; Geological Survey Professional Paper 252; United States Government Printing Office: Washington, DC, USA, 1953.

43. Arnold, J.G.; Moriasi, D.N.; Gassman, P.W.; Abbaspour, K.C.; White, M.J.; Srinivasan, R.; Santhi, C.; Harmel, R.D.; van Griensven, A.; van Liew, M.W.; et al. SWAT: Model use, calibration, and validation. *Trans. ASABE* **2012**, *55*, 1491–1508. [CrossRef]

44. Daggupati, P.; Yen, H.; White, M.J.; Srinivasan, R.; Arnold, J.G.; Keitzer, C.S.; Sowa, S.P. Impact of model development, calibration and validation decisions on hydrological simulations in West Lake Erie Basin. *Hydrol. Process.* **2015**. [CrossRef]

45. Schwarz, G.E.; Hoos, A.B.; Alexander, R.B.; Smith, R.A. The SPARROW Surface Waterquality Model: Theory, Application and User Documentation. In *U.S. Geological Survey Techniques and Methods*; Book 6, Section B, Chapter 3; U.S. Geological Survey: Reston, VA, USA, 2006; p. 248. Available online: http://pubs.usgs.gov/tm/2006/tm6b3/ (accessed on 1 January 2015).

46. Henze, M.; Comeau, Y. *Biological Wastewater Treatment: Principles Modelling and Design*; Henze, M., van Loosdrecht, M.C.M., Ekama, G.A., Brdjanovic, D., Eds.; IWA Publishing: London, UK, 2008.

47. United Nations. *World Population Prospects: The 2015 Revision*; Department of Economic and Social Affairs, Population Division: New York, NY, USA, 2015.

48. McClelland, N.I. *Water Quality Index Application in the Kansas River Basin*; U.S. Environmental Protection Agency: Washington, DC, USA, 1974.

49. U. S. Environmental Protection Agency. *Environmental Impact and Benefits Assessment for Final Effluent Guidelines and Standards for the Construction and Development Category*; Office of Water; U.S. Environmental Protection Agency: Washington, DC, USA, 2009.

50. United States Environmental Protection Agency Office of Water. Benefit and Cost Analysis for the Effluent Limitation Guidelines and Standards for the Steam Electric Power Generating Point Source Category. EPA-821-R-15-005, 2015. Available online: https://www.epa.gov/sites/production/files/2015-10/documents/steam-electric_benefit-cost-analysis_09-29-2015.pdf (accessed on 15 November 2016).

51. U.S. Coast Guard Office of Auxiliary and Boating Safety. National Recreational Boating Survey (NRBS). Available online: http://www.uscgboating.org/assets/1/workflow_staging/AssetManager/821. PDF (accessed on 17 September 2014).

52. Srinivasan, R.; Zhang, X.; Arnold, J. SWAT ungauged: Hydrological budget and crop yield predictions in the Upper Mississippi River Basin. *Trans. ASABE* **2010**, *53*, 1533–1546. [CrossRef]

53. Fuka, D.R.; Walter, M.T.; MacAlister, C.; Degaetano, A.T.; Steenhuis, T.S.; Easton, Z.M. Using the climate forecast system reanalysis as weather input data for watershed models. *Hydrol. Process.* **2014**, *28*, 5613–5620. [CrossRef]

54. Tram, V.N.Q.; Liem, N.D.; Loi, N.K. Assessing Water Availability in PoKo Catchment using SWAT model. *Khon Kaen Agric. J.* **2014**, *42*, 73–84.

water

MDPI

Article

Assessment of the Combined Effects of Threshold Selection and Parameter Estimation of Generalized Pareto Distribution with Applications to Flood Frequency Analysis

Amr Gharib [1,*] **, Evan G. R. Davies** [1] **, Greg G. Goss** [2] **and Monireh Faramarzi** [3]

[1] Department of Civil and Environmental Engineering, University of Alberta, Edmonton, AB T6G 1H9, Canada; evan.davies@ualberta.ca
[2] Department of Biological Sciences, University of Alberta, Edmonton, AB T6G 2E9, Canada; ggoss@ualberta.ca
[3] Department of Earth and Atmospheric Sciences, University of Alberta, Edmonton, AB T6G 2E3, Canada; faramarz@ualberta.ca
* Correspondence: gharib@ualberta.ca; Tel.: +1-587-937-2454

Received: 29 July 2017; Accepted: 7 September 2017; Published: 10 September 2017

Abstract: Floods are costly natural disasters that are projected to increase in severity and frequency into the future. Exceedances over a high threshold and analysis of their distributions, as determined through the Peak Over Threshold (POT) method and approximated by a Generalized Pareto Distribution (GPD), respectively, are widely used for flood frequency analysis. This study investigates the combined effects of threshold selection and GPD parameter estimation on the accuracy of flood quantile estimates, and develops a new, widely-applicable framework that significantly improves the accuracy of flood quantile estimations. First, the performance of several parameter estimators (i.e., Maximum Likelihood; Probability Weighted Moments; Maximum Goodness of Fit; Likelihood Moment; Modified Likelihood Moment; and Nonlinear Weighted Least Square Error) for the GPD was compared through Monte Carlo simulation. Then, a calibrated Soil and Water Assessment Tool (SWAT) model for the province of Alberta, Canada, was used to reproduce daily streamflow series for 47 watersheds distributed across the province, and the POT was applied to each. The Goodness of Fit for the resulting flood frequency models was measured by the upper tail Anderson-Darling (AD) test and the root-mean-square error (RMSE) and demonstrated improvements for more than one-third of stations by averages of 65% (AD) and 47% (RMSE), respectively.

Keywords: peak over threshold (POT); extreme value analysis; flood; extreme hydrological events; flood quantiles

1. Introduction

Floods are considered the most destructive and wide-spread natural disaster, and account annually for about 50 percent of all natural disasters world-wide [1,2]. Further, intensification of the hydrological cycle with climate change may lead to larger and more frequent floods [3–6], and land use change may increase flood risk through expansion of urban areas, which typically limit soil permeability [7], or through development in the flood plain. Examples of recent extreme flood events include flooding in the Calgary, Alberta, area in 2013, which created the costliest natural disaster in Canadian history, with damage of approximately $6 billion [8], and the largest number of flood events recorded in a single year in the United States in 2016, the most catastrophic of which occurred in Louisiana [9]. In Europe, the May–June 2013 flooding of the Elbe, Oder, and Danube rivers in Germany produced the third "flood of the century" since 1997 [10].

To mitigate flood risks, understanding the relationship between flood events and their probability of occurrence is a critical first step from both economic and environmental points of view [1,11]. A systematic assessment of flood risks is therefore vital for sustainable watershed management and planning. More specifically, a rigorous analysis of flood events helps to identify the risk of flooding at various locations and aids in selection of appropriate design periods for structures such as dams and levees [12–14]. Thus, reliable predictions of flood magnitude will improve both the sustainability and economic efficiency of water resources systems.

Flood Frequency Analysis (FFA) is a standard method used to detect relationships between flood magnitudes and the corresponding frequency of occurrence. FFA has largely been applied in four categories: on-site FFA, climate/weather-informed on-site FFA, historical and paleo-flood analyses, and regional FFA [13]. In this paper, the focus is on-site FFA, which is performed by fitting a chosen probability distribution to flood events sampled from streamflow records at the site of interest [15]. Three different approaches are available for flood event sampling from a streamflow record: the annual maxima (AM), the partial duration series, and the peaks over a threshold (POT) [16–21]. Although the AM is the most common because of its simplicity, it samples only one event per year [14,17,19]; therefore, it may result in a loss of information, if the second or third peak within a year—which can be greater than the flood peak in other years—is ignored. This situation is particularly problematic in regions where the record of historical streamflow is short. In contrast, the POT includes all peaks above a certain flow value (the threshold), which provides flexibility in controlling the number of events included in the analysis. Comparisons between AM and POT series have found that POT offers a smaller uncertainty of estimated values, because of the larger quantity of data involved in the analysis [18,22,23]. More specifically, Cunnane [24] conducted a statistical analysis that showed greater statistical efficiency, or more precise estimation of the parameters, for POT where the average number of peaks per year included in the POT series was greater than 1.65.

The standard practice in POT fits a distribution to exceedances above a selected threshold. In selecting this threshold, it must be high enough to identify the distribution underlying the excess series and to maintain the independent and identically distributed (IID) flood variables assumption. It also should not be so high as to increase the variance by reducing the number of events needed for reliable inferences [13,25]. Therefore, the optimal threshold detection has been of interest in earlier studies [19,25–35], which have suggested various approaches based on graphical and/or analytical methods. An example of the graphical approach is the Mean Residual Life plot (MRL), while analytical approaches include the Square Error method (SE), the Multiple Threshold Method (MTM), and the Likelihood Ratio Test method (LRT). Numerous studies have stated that the performance of FFA with POT is dependent upon the threshold selection method applied but its effect was not studied (e.g., [36–38]). Zoglat et al. [25] noted the subjectivity of the graphical methods and therefore compared different analytical methods towards threshold selection, with the LRT method found to be the best and SE second-best, with satisfactory performance. However, it is difficult to generalize this approach because the comparison was performed for only one hydrometric station.

After sampling the flood peaks, a distribution is chosen for the FFA model. There are several probability distribution methods used to model extreme events. Generalized Pareto Distribution (GPD) is widely used to model extreme floods over a threshold. It has been used successfully to estimate return values of flood events in conjunction with a POT method. Pickands [39] first introduced the GPD as a two-parameter family of distributions for exceedances over a threshold and showed that it is a stable distribution for excesses over thresholds. To determine appropriate parameter values, numerous estimators for the GPD have been proposed in the literature, with the Maximum Likelihood Estimator (MLE) [39], Method of Moments (MOM) [40], and the Probability Weighted Moments (PWM) used most widely. Additional methods have also been proposed, including Likelihood Moment Estimations (LME) [41], the modified Likelihood Moment Estimator (NEWLME) [42], and the Nonlinear Weighted Least Squares estimator (NWLS) [43]. For an extensive discussion of the various methods see de Zea Bermudez and Kotz [44]. Estimator performance has been found to vary considerably with both the flood-event sample sizes and the value of the GPD shape-parameter, such that estimators that perform well in some situations may perform poorly in others. For example, Hosking and Wallis [40] introduced the PWM estimators for the GPD and compared them to the MOM and MLE estimators. They showed that the MOM and PWM estimators have lower bias and variance than MLE estimators for sample sizes less than 500; however, both are sensitive to threshold choice and sometimes result in infeasible estimates [29]. Many other simulation studies have provided a quantitative comparison of the performance of different estimators (e.g., [36–38,45]). These simulation studies pointed out that estimator performance depends on the threshold selection method without a quantitative measure of this effect.

Overall, there are numerous factors influencing the accuracy of the modeled return period of an extreme flood. Among them are the length and accuracy of flow records, the criteria used to identify independent flood peaks, the threshold selection method, and the estimator of the selected probability distribution. Moreover, the performance of various parameter and quantile estimators can vary greatly in terms of their bias, variance, and sensitivity to threshold choice; and consequently affect the accuracy of the estimated return values [37,46]. Given the key role of threshold selection in reducing uncertainty prediction, the POT method is still under-employed, particularly for studying the impacts of climate change. In climate change studies, application of the POT method can be beneficial as it provides useful information about the time of floods along with their magnitudes. Accordingly, it is important to measure the performance of the GPD estimator quantitatively along with the effect of threshold selection (POT) on its performance.

The purpose of this study is to propose a novel systematic procedure for assessing the combined effect of the threshold selection method and GPD parameter estimator on the accuracy of flood frequency distribution calculations. First, a comparison between older and more recent methods proposed for the GPD over a wide range of flood event sample sizes and shape parameters was undertaken. Further, POT was applied to a large number of streamflow series using different combinations of threshold selection methods and parameter estimators and the goodness of fit of the flood frequency distributions were assessed. We hypothesize that more accurate flood frequency models can be obtained by improving threshold selection and GPD parameter estimator selection.

This paper is organized as follows. Section 2 describes the study area, data used for the analysis, and the methods employed in this study. Section 3 presents the results from applying POT on simulated streamflow records from SWAT model along with the analysis and discussion of these results. Section 4 includes the summary and conclusions of the study.

2. Materials and Methods

2.1. Study Area, Hydrologic Model, and Data Overview

Given the large spatial extent of Alberta, the variability of the hydro-meteorological conditions, and the availability of a calibrated Soil and Water Assessment Tool (SWAT) model [47] for the province of Alberta, Canada for the period 1986–2007 that allows us to obtain streamflow records across the entire province, Alberta presented an ideal study region. One of the three Canadian Prairie Provinces, Alberta has an area of about 661,000 km^2 that encompasses 17 river basins principally originating from the east slopes of the Canadian Rockies, and that drain eastward to Hudson Bay through the provinces of Saskatchewan and Manitoba or northward to the Arctic Ocean [48]. Alberta has a relatively dry climate. In the north, the average annual precipitation ranges from 400 mm in the northeast to over 500 mm in the northwest, while in the south, it ranges from 450 mm to less than 350 mm in the southeast. Northern Alberta is dominated by Boreal forests and was wet for the 100 years between 1900 and 2000 [49]. Southern Alberta is dominated by grassland that has experienced progressively decreasing precipitation for the period 1960–2000 [49]. Alberta has cold winters and warm summers with a mean annual temperature ranging from 3.6 °C to 4.4 °C.

Despite its relatively low average precipitation, Alberta is experiencing increasing flooding trends mainly because of high variability in precipitation [50]. For example, flooding in the city of Calgary in 2013 is considered to have been the costliest natural disaster in Canadian history [8]. Thus, the province makes an interesting case study because of its heterogeneous hydro-climatic conditions, diverse land management practices, and the occurrence of extreme hydrologic events (e.g., floods). Alberta forms the basis of an assessment of our research objectives and the development of a comprehensive practical guideline for the most suitable methods for threshold selection and parameter estimation methods under diverse hydro-climatic conditions.

The stream flow data used for the study are daily average discharges as simulated by the SWAT model [48]. Input variables for the SWAT hydrologic model of Alberta have been carefully selected to represent the actual physical processes related to natural and anthropogenic features (e.g., snow, potholes, glaciers, reservoirs, dams, and irrigated agriculture), and to minimize input data uncertainties [48]. The model has been extensively calibrated and validated using the discharge data of about 130 hydrometric stations for the 1983–2007 period in the province [47] (See Table S1), and can simulate stream flow and other water resource components in 2255 sub-basins across the province. In this study we used a subset of 47 stations/outlets spread across the province (see Figure 1), based on efficiency criteria that were used to assess the calibrated performance of the model based on simulated versus observed discharges e.g., bR^2, NSE, and R^2 [48]. Only the stations with bR^2, NSE, and R^2 greater than 0.6 were considered, because these values provide both reliable streamflow records and a sufficient number of stations to cover various river basins that encompass a diverse range of topographic, hydrologic, and climatic conditions.

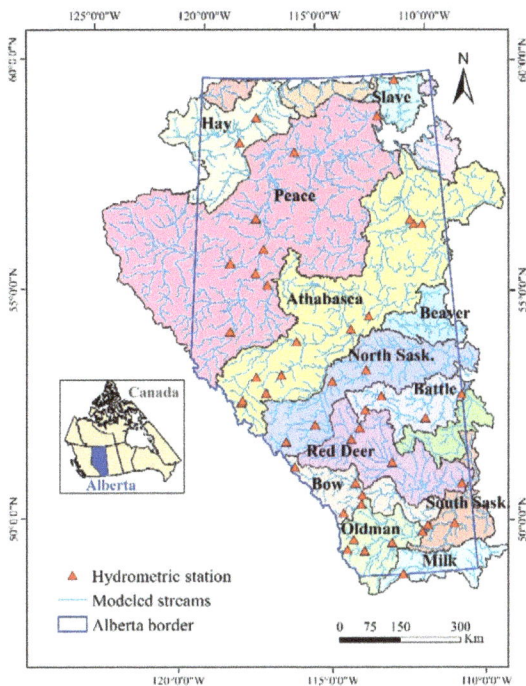

Figure 1. Map of the study area presenting geographic distribution of the 17 main river basins (background colors), and 47 hydrometric stations used in the study.

2.2. Probability Modeling

In flood probability modeling, a distribution is fitted to excesses above a high threshold (POT) for a selected streamflow sample. For flood peak flows, extreme value theory shows that the distribution of exceedances of a high-enough threshold will tend to follow a Generalized Pareto distribution [37,51].

The theoretical development of the GPD can be found in Coles [17]. The cumulative distribution frequency (CDF) for the GPD is derived from the following equation:

$$F_u(x) = Pr(X - u < x | X > u) = \begin{cases} 1 - \left(1 + \frac{\xi(x-u)}{\sigma}\right)^{\frac{-1}{\xi}} & \xi \neq 0 \\ 1 - e^{(-\frac{x-u}{\sigma})} & \xi = 0 \end{cases}, \tag{1}$$

where x is the flood peak flow in m^3/s, ξ is the shape parameter, u is the location parameter, also known as the threshold, and σ is the scale parameter. In special cases, $\xi = 0$ and $\xi = 1$ yield an exponential distribution and a uniform distribution, respectively, while a Pareto distribution is obtained when $\xi < 0$. Note that selection of the threshold, u, reduces the three-parameter GPD to a two-parameter distribution. To evaluate the Goodness of Fit, a comparison between the modeled CDF and the empirical distribution function (EDF) was performed. The EDF is calculated from,

$$\tilde{F}\left(x_{(i)}\right) = \frac{i}{n+1}, \tag{2}$$

where i is the rank of the flood event and n is the sample size [17].

In this study, the data were sampled from 47 streamflow gauges using two threshold selection methods (see Section 2.2.3). The sampled data were then fitted to GPD using each of the parameter estimators considered in this study (see Section 2.2.1), and the fitted distributions were compared with the empirical distribution. To assess the Goodness of Fit, the root-mean-square error (RMSE) and p-value of the Anderson-Darling test (AD) were computed. The Anderson Darling statistic allots extra weight to floods in the tail of the distribution [30], which are of importance when estimating floods with high return periods.

2.2.1. GPD Parameter Estimators

Several techniques are available for the estimation of GPD parameters, including MLE [39], PWM [40], LME [41], NEWLME [42], MGFAD [27], and NWLS [43].

1   Maximum Likelihood Estimator (MLE): MLE is the most efficient method of parameter estimation, particularly for large streamflow sample sizes. It maximizes the likelihood function (L) for the sampled independent flood peaks (x), and is derived from,

$$L(\xi, \sigma) = \prod_{j=1}^{n} f(x_j, \xi, \sigma), \tag{3}$$

where $f = \frac{dF}{dx}$. The estimated parameters are the values that maximize Equation (3). Although the algorithms used to compute MLE estimated parameters run into convergence problems, Ashkar and Tatsambon [36] argue that this behavior is simply due to incorrect choice of the numerical algorithm.

2   Probability Weighted Moments Estimator (PWM): The PWM estimator has a lower bias and variance than the MLE estimator for sample sizes less than 500 [40]. The PWM of a sampled flood peak (x) with distribution function F is derived from the following equation:

$$M_{p,q,r} = E\left(x^P F^q (1-F)^r\right) = \int_0^1 [x(F)]^P F^q (1-F)^r dF, \tag{4}$$

PWM always exists, is computationally straightforward, and is a function of the plotting position, which makes it more stable. However, PWM estimates are sensitive to the threshold choice [37,40].

3   The Likelihood Moment Estimator (LME): LME is a hybrid between likelihood and moment estimators, and was proposed by Zhang [41]. The LME for a streamflow sample size of n is derived from the following equation:

$$\frac{1}{n} \sum_{j=1}^{n} (1 - \theta X_i)^P - \frac{1}{1-r} = 0, \theta < X_{(n)}^{-1}, \tag{5}$$

where $\theta = \xi/\sigma$ and $P = \frac{-rn}{\sum_{j=1}^{n} \log(1-\theta x_j)}$. In this equation the parameter $r < 1$ is chosen before the estimation [42].

Zhang [41] shows that when $r = \xi$, the LME and MLE asymptotic variances are the same. The LME is simple to compute and less sensitive than other parameter estimators to threshold choice [37]. In this study, an initial guess of the value of ξ was made by the PWM as recommended by Mackay et al. [37].

4   The Modified Likelihood Moment Estimator (NEWLME): The NEWLME method was proposed by Zhang and Stephens [42] to address complexities in the MLE numerical solution and to avoid computational problems. The method is similar to the Bayesian methods [37] and its solution is calculated as follows:

$$\hat{\theta} = \frac{\int \theta \cdot \pi(\theta) L(\theta) d\theta}{\int \pi(\theta) L(\theta) d\theta},\tag{6}$$

where $L(\theta) = n \left[\log\left(\frac{\theta}{\xi}\right) \right] + \xi - 1$ with $\xi = -n^{-1} \sum_{i=1}^{n} \log(1 - \theta x_i)$ is the profile log-likelihood function, and $\pi(\theta)$ is a data-driven density function for θ [42]. This allows the parameters to be computed very efficiently [37].

5   The Maximum Goodness of Fit–Anderson-Darling Estimator (MGFAD): Moharram et al. [52] proposed least-square type estimators, which are found by minimizing the sum of squared difference between the empirical and the model quantiles. Luceño [27] proposed an estimator with a similar approach, in which the estimates are obtained by minimizing the square differences between the empirical and the model distribution functions using various Goodness of Fit statistics. Luceño [27] included the Cramer-von Mises [30], the Anderson-Darling [30], and the right-tail weighted Anderson-Darling statistics [53]. The different statistics considered by Luceño [27] were found to have strong positive bias and high root-mean-square error (RMSE) in estimating high quantiles for small sample sizes [37]. Only the Maximum Goodness of Fit estimator with Anderson-Darling statistic (MGFAD) was considered in this study.

6   The Nonlinear Weighted Least Squares Estimator (NWLS): A new estimator based on the nonlinear weighted least squares estimator (NWLS) was recently proposed by Song and Song [38], and revised and improved by Park and Kim [43]. The calculation of the NWLS estimator is a two-step procedure that is calculated using Equations (7) and (8) as follows:

$$(\hat{\xi}_1, \hat{\sigma}_1) = \operatorname{argmin}_{(\xi,\sigma)} \sum_{i=1}^{n} \left[\log \frac{1 - F_n(x_i)}{1 - F_n(u)} - \log\left(1 - G_{\xi, u, \sigma}(x_i)\right) \right]^2,\tag{7}$$

$$(\hat{\xi}_2, \hat{\sigma}_2) = \operatorname{argmin}_{(\xi,\sigma)} \sum_{i=1}^{n} \left[\frac{i(n-i+1)}{(n+1)^2(n+2)} \right]^{-1} \left[\frac{F_n(x_i) - F_n(u)}{1 - F_n(u)} - G_{\xi, u, \sigma}(x_i) \right]^2,\tag{8}$$

Park and Kim [43] compared the performance of the NWLS with some other estimators, and it was found to be highly competitive for heavy-tailed data ($\xi > 0$). In this study, the performance of the NWLS was tested for the case $\xi < 0$.

2.2.2. Performance Analyses of the GPD Estimators and the Monte Carlo (MC) Sampling Experiments

Performance of the GPD parameter estimators depends on both the sample size, n, and the value of the GPD shape parameter, ξ [37]. Since the streamflow records did not provide a sufficient range of events to yield reasonable estimates for the entire range of possible values in a given GPD model, a Monte Carlo sampling technique [54] was used to generate different samples in order to evaluate the performance of the six GPD parameter estimators described in Section 2.2.1. A total of 10,000 random samples of different n, ξ, and σ combinations was generated. Samples of size $n = 40, 50, ..., 200, 500$, and 1000 were used for the parameter estimation, with the ξ values ranging from -0.5 to 0.5 ($\xi = -0.5, -0.4, -0.3, -0.2, -0.1, 0, 0.1, 0.2, 0.3, 0.4, 0.5$), and of the scale parameter σ ranging from 20 to 1000. Then, from each sample, the shape and scale parameters were computed by the parameter estimator and the 0.99 quantile was obtained from the resulting distribution. To evaluate the performance of the six parameter estimation methods, four statistics were computed: the average biases in estimating the shape and scale parameters, the 0.99 quantile (upper quantile), and the average root-mean-square error (RMSE). The comparison between different estimators for each threshold selection method focused on the upper tail Anderson-Darling test (AD), as it evaluates the ability of the model to estimate high quantiles. The other statistics (e.g., RMSE) were used to evaluate the model performance further, if the AD test values were similar.

In the application of flood frequency analysis, the value of the scale parameter can vary significantly, from 100 to 1000 or more; therefore, we examined the effect of varying scale parameter values on the accuracy of estimating the 0.99 quantile.

2.2.3. Threshold Selection Methods

The choice of a threshold is an important practical problem with a solution that is primarily based on a compromise between estimator bias and variance. In other words, it affects the performance of the different GPD parameter estimators. To investigate further the effect of the threshold selection method on the performance of the GPD parameter estimators, data were sampled from 47 streamflow gauges using the threshold selection methods. Various methods to select thresholds have been proposed in the literature. For the application, one option includes graphical methods, which are relatively easy to apply. However, they have some shortcomings. First, the interpretation of these plots is unclear in practice [17], and it is clearly difficult to determine which portion of the curve is completely linear. Second, graphical techniques cannot be automated and the uncertainty associated with threshold selection cannot be estimated [55]. Figure 2 shows a range of potential optimum threshold values, making the use of a single value from the range in the analysis subjective. Compared to graphical methods, analytical methods are less subjective and the computations can easily be programmed. Zoglat et al. [25] compared various threshold selection techniques. Two analytical methods outperformed the other techniques and were employed in our study: The Square Error Method (SE) and Likelihood Ratio Test method (LRT). A brief description of the two methods is provided as follows:

The SE optimal threshold value is the value for which the mean square error of the EDF and CDF is the minimum for an estimator. Zoglat et al. [25] suggested an algorithm based on the work of Beirlant et al. [51]. The main steps of the algorithms are as follows (Adapted from Zoglat et al. [25]):

1. Let $u_1, \ldots, u_m$ be m equally spaced increasing threshold candidates (u_m is the threshold corresponding to the minimum number of exceedances—the number of years in the record multiplied by 1.65 [24]). For $j = 1, \ldots, m$, let $\hat{\xi}_{uj}, \hat{\sigma}_{uj}$ be the estimates from any of the six parameter estimators employed in the study of the scale and shape parameters of the GPD underlying the exceedances over the threshold u_j.
2. Find N_{uj}, the number of exceedances over u_j.
3. Simulate v independent samples of size N_{uj} from GPD with parameter $\hat{\xi}_{uj}, \hat{\sigma}_{uj}$.
4. For each $\alpha \in A = \{0.05, 0.1, 0.15, \ldots, 0.95\}$, and each $i = 1, \ldots, v$, calculate the quantile $q^i_{(\alpha, u_j)}$ of the ith simulated sample compute Equation (9).

$$q^{sim}_{(\alpha, u_j)} = \frac{1}{v} \sum_{i=1}^{v} q^i_{(\alpha, u_j)}, \tag{9}$$

5. For $j = 1, \ldots, m$ calculate the square error, $SE_{u_j} = \sum_{\alpha \in A} \left(\frac{q^{sim}_{(\alpha, u_j)}}{q^{obs}_{(\alpha, u_j)}} \right)^2$, where $q^{obs}_{(\alpha, u_j)}$ is the observed quantile corresponding to the simulated quantile.
6. The optimal threshold value is the value of minimum SE_{u_j}.

LRT was proposed by Zoglat e al. [25]. It is based on using the likelihood ration test statistic to test the hypothesis $H_0 : u_i$ is the optimal threshold against $H_1 : u_k$ is the optimal threshold. The LRT statistic is given by:

$$LR(u_i, u_k) = -2 \log \frac{f(\hat{\xi}_{u_i}, \hat{\sigma}_{u_i}, X_{u_i})}{f(\hat{\xi}_{u_k}, \hat{\sigma}_{u_k}, X_{u_i})}, \tag{10}$$

where LR is the likelihood ratio, $f(\ldots)$ is the likelihood function, X_{u_i} is the vector of observations exceeding u_i, and X_{u_k} is the vector of observations exceeding u_k. For LRT application, the maximum threshold is selected to correspond to 36 exceedances—the number of years in the record multiplied by 1.65 [24]. For more details on LRT method see Zoglat et al. [25].

Further, to meet the requirements for the POT method, an independence test is applied to ensure that peaks identified with the POT method correspond to different flood events. The test is conditional for the set of sampled peaks and is a prerequisite to any statistical frequency analysis and to the Poisson process assumption [17,19]. To assess the degree of dependence between flood peaks, we examined autocorrelation coefficients. Specifically, when the absolute values of autocorrelation coefficients for different lag times, in time series with n observations, are less than or equal to the critical values, the flood peaks can be regarded as being independent from each other [22,56]. The critical values are equal to $\pm 1.96/\sqrt{n}$, corresponding to the 0.05 significance level, where n is the sample size. If more than two values exceed the critical values, this means the flood peaks are dependent; therefore, a higher threshold level without a restriction on the duration between flood peaks must be tried, as recommended by Ashkar and Rouselle [57]. In addition, the number of the sampled peaks was compared to 1.65 times the number of years in the record to ensure that it exceeds this lower bound and that the POT is therefore more effective than the AM series, as recommended by Cunnane [24]. The peaks were sampled automatically by an algorithm we developed in R [58]. Then the autocorrelation values of peaks were calculated and the figures were plotted using the "forecast" package in R [59] and visually checked.

In application, it is more convenient to interpret the POT flood frequency model in terms of quantiles or return levels [17,22]. Flood estimates can be made based on Poisson model, negative binomial, and binomial models. It is not necessary to prefer one model to the other models. Regardless of accepting or rejecting the Poisson hypothesis any of the three models can be used [60,61]. For more details on modeling the quantiles and return levels please refer to Cunnane [60] and Coles [17].

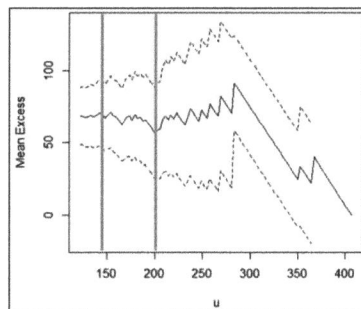

Figure 2. Mean Residual Life plot for the sampled flood peaks in (m^3/s) with approximate 95% confidence intervals represented by the dotted lines. The horizontal (u) and the vertical axes represent the threshold value and the mean excess, respectively. The dotted lines illustrate the approximate 95% confidence intervals.

3. Results and Discussion

Quantile estimates and shape parameter estimates (ξ) are presented here, with a focus on the accuracy of the predicted high quantiles (floods with high return periods) rather than on the estimated parameters of the distribution.

3.1. Performance Analyses Results of the GPD Estimators Using MC Sampling Data

The average biases of the 99% quantile and shape parameter (ξ) for sample sizes ranging from n = 40 to 500 for different parameter estimators are presented in Figures 3 and 4, respectively. The results can be summarized as follows (also see Table 1):

- Generally, the relative bias of estimated parameter values and the 99% quantile decreased with increasing sample size (n). However, for short tails ($\xi < 0$) the bias was still high in the case of MLE and LME, even if the sample size was increased to 300.
- PWM: Very consistent and among all methods the least sensitive to sample size (n). However, it had a medium sensitivity to the sample size and medium bias for heavy tails with respect to estimating the shape parameter.
- LME: For heavy tails, LME was among the parameter estimators that had the lowest bias and sensitivity to the sample size. However, for short tails the biases in estimating the 99% quantile and shape parameter were very high.
- MLE: Very accurate in estimating the 99% quantile. However, it had a high bias in estimating the shape parameter and was very sensitive to the sample size for heavy and short tailed distributions.
- NEWLME: Average performance in estimating the 99% quantile for heavy tails and the shape parameter for short tails. In contrast, NEWLME excelled when estimating the shape parameter for heavy tails and estimating the 99% quantile for short tails.
- MGFAD: The most accurate and the least sensitive to sample size in estimating the shape parameter. However, it had a very high bias and sensitivity to the sample size when estimating the 99% quantile.
- NWLS: No trend of decreasing bias with increasing sample size (n) in predicting the 99% quantile. However, it showed a similar trend in the case of shape parameter estimations with average bias compared to the other methods.

In the application of flood frequency analysis, the value of the scale parameter can vary significantly, from 10 to 1000 or more; therefore, we examined the effect of varying scale parameter values on the accuracy of estimation of the 0.99 quantile. Surprisingly, the LME method, often reported as one of the better methods for GPD fitting in previous comparative studies [41,42,62], was found to be sensitive to high values of the scale parameter. In contrast, the other methods did not show the same sensitivity to the scale parameter value.

Table 1. Sensitivity analyses of the parameter estimators to the sample size (n) for estimating the shape parameter (ξ) and the 99% quantile. Low represents an average bias range of 0–10% of the highest calculated bias, Medium represents the range 10–30% of the highest calculated bias, High represents values greater than 30%. Abbreviations are as follows: PWM—probability weighted moments, LME—likelihood moment estimator, MLE—maximum likelihood estimator, NEWLME—modified likelihood moment estimator, MGFAD—maximum goodness of fit-Anderson Darling estimator, NWLS—Nonlinear weighted least square estimator.

Parameter Estimator	Shape Parameter		99% Quantile	
	$\xi > 0$	$\xi < 0$	$\xi > 0$	$\xi < 0$
PWM	Medium	Low	Low	Low
LME	Low	High	Low	High
MLE	High	High	Low	Low
NEWLME	Low	Medium	Medium	Low
MGFAD	Low	Low	High	High
NWLS	High	Medium	High	High

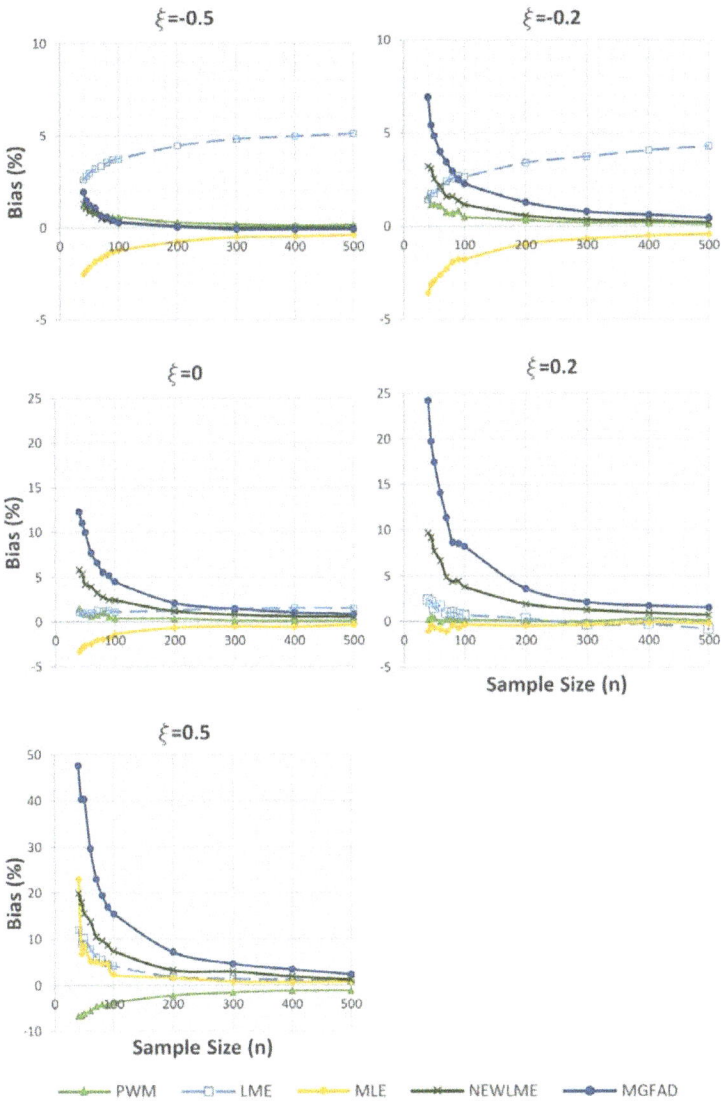

Figure 3. Bias in estimating the 99% quantile for different tails. Note that the Y axis (Bias %) scale differs between graphs. Abbreviations are as follows: PWM—probability weighted moments, LME—likelihood moment estimator, MLE—maximum likelihood estimator, NEWLME—modified likelihood moment estimator, MGFAD—maximum goodness of fit-Anderson Darling estimator.

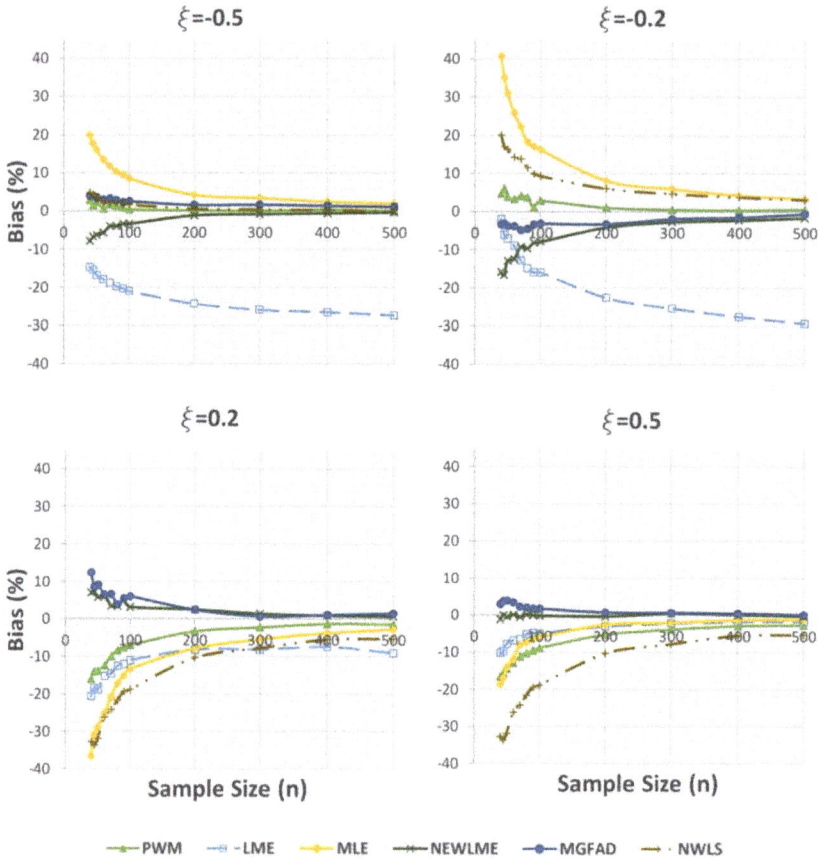

Figure 4. Bias in estimating the shape parameter (ξ) for different tails. Abbreviations are as follows: PWM—probability weighted moments, LME—likelihood moment estimator, MLE—maximum likelihood estimator, NEWLME—modified likelihood moment estimator, MGFAD—maximum goodness of fit-Anderson Darling estimator, NWLS—Nonlinear weighted least square estimator.

3.2. Application of POT to Observed Streamflows

The six parameter estimators were applied for the thresholds selected by both LRT and SE methods to the 47 stations from the study area. Accuracy of shape parameter estimation and the Goodness of Fit obtained with respect to AD and RMSE for different ranges of the shape parameter value are summarized in Table 2 for LRT and Table 3 for SE. Models developed for thresholds chosen by the SE method generally could not accurately estimate the shape parameter value. In contrast, models developed with NWLS with thresholds selected by the LRT method had high goodness of fit for heavy tails. Furthermore, models developed by MGFAD with thresholds selected by LRT method had high Goodness of Fit for short tails.

Generally, models developed with thresholds selected by the LRT had better Goodness of Fit with respect to the Anderson Darling (AD) test and higher RMSE. In contrast, models developed with thresholds selected by the SE had lower RMSE and higher AD statistics. The behavior of the parameter estimators was found to be similar within the ranges of the shape parameter identified in Tables 2 and 3.

Table 2. Goodness of Fit that can be achieved by a combination of the likelihood ratio test method (LRT) and different parameter estimators for different ranges of the shape parameter (ξ) in terms of (1) the bias in predicting the shape parameter, (2) Anderson-Darling (AD) test, and (3) the root-mean-square error (RMSE). High represents a variance ranges from 0% to 3% of the lowest possible RMSE or AD, Medium represents the range 3–10% of the lowest possible RMSE or AD, Low represents values greater than 10%. Abbreviations are as follows: AD—Anderson-Darling, RMSE—root-mean-square error, PWM—probability weighted moments, LME—likelihood moment estimator, MLE—maximum likelihood estimator, NEWLME—modified likelihood moment estimator, MGFAD—maximum goodness of fit-Anderson Darling estimator, NWLS—Nonlinear weighted least square estimator.

Parameter Estimator		PWM	LME	MLE	NEWLME	MGFAD	NWLS
Ξ	$\xi > 0$	Low	Low	Low	Low	Medium	High
	$\xi < 0$	Low	Medium	High	Medium	High	Medium
AD	$\xi > 0.4$	Low	Low	Low	Low	Low	Low
	$0.4 > \xi > 0.2$	Low	High	High	Low	Low	Low
	$0.2 > \xi > 0$	Low	Low	High	Low	Low	Low
	$0 > \xi > -0.15$	Low	Low	High	Medium	Medium	Low
	$-0.15 > \xi > -0.25$	Low	Low	Low	High	Low	Low
	$\xi < -0.25$	Medium	Low	Low	High	Medium	Medium
RMSE	$\xi > 0.4$	Low	Low	Low	Low	Low	Low
	$0.4 > \xi > 0.2$	Low	Low	Low	Low	Low	Low
	$0.2 > \xi > 0$	Low	Low	Low	Low	Low	Low
	$0 > \xi > -0.15$	Low	Low	Low	Low	Low	Low
	$-0.15 > \xi > -0.25$	Low	Low	Medium	Low	Low	Low
	$\xi < -0.25$	Low	Low	Low	Low	Low	Low

Table 3. Goodness of Fit that can be achieved by a combination of the square error method (SE) and different parameter estimators for different ranges of the shape parameter (ξ) in terms of (1) the bias in predicting the shape parameter, (2) Anderson-Darling (AD) test, and (3) the root-mean-square error (RMSE). High represents a variance ranges from 0% to 3% of the lowest possible RMSE or AD, Medium represents the range 3–10% of the lowest possible RMSE or AD, Low represents values greater than 10%.

Parameter Estimator		PWM	LME	MLE	NEWLME	MGFAD	NWLS
Ξ	$\xi > 0$	Low	Low	Low	Low	Low	Low
	$\xi < 0$	Low	Low	Low	Low	Low	Low
AD	$\xi > 0.4$	Low	Low	Low	Low	Low	High
	$0.4 > \xi > 0.2$	Low	Low	Low	Low	Low	Low
	$0.2 > \xi > 0$	Low	Low	Low	Low	Low	Low
	$0 > \xi > -0.15$	Low	Low	Low	Low	Low	Low
	$-0.15 > \xi > -0.25$	Low	Low	Low	Medium	Low	Low
	$\xi < -0.25$	High	Medium	Low	High	Medium	High
RMSE	$\xi > 0.4$	High	High	High	High	Medium	Low
	$0.4 > \xi > 0.2$	High	High	High	Medium	High	High
	$0.2 > \xi > 0$	High	High	High	Low	Low	Low
	$0 > \xi > -0.15$	High	High	High	Medium	Medium	Medium
	$-0.15 > \xi > -0.25$	High	High	High	Low	Medium	Medium
	$\xi < -0.25$	Low	High	Medium	Medium	Medium	High

We investigated different combinations of threshold selection methods and parameter estimators to develop a new framework that improves the Goodness of Fit and minimizes the AD statistic and RMSE and, consequently, obtains more accurate predictions of flood return periods. We found that the combination of LRT and MLE outperformed other combinations for shape parameter values ranging from −0.15 to 0.4, while the LRT and NEWLME combination was the best for short tails for the range of shape parameter values from −0.25 to −0.15. For short tails with shape parameter values less than −0.25 and for heavy tails with shape parameter values greater than 0.4, the SE and NWLS gave the best fit compared to the other methods. Our results demonstrate that the choice of best combination of parameter estimation and threshold selection methods depends on the range of values of the shape parameter. Hence, it is critical to have an accurate estimate of the shape parameter value prior to fitting the distribution. The combination of LRT and NWLS was found to perform best in predicting shape parameter values greater than zero, while the LRT with MGFAD was the best among other combinations to predict the shape parameter for short tails. Overall, based on the above findings, we have created a new framework for obtaining a more accurate flood frequency distribution that considers both the effect of threshold selection method and parameter estimator, as illustrated in the flow chart of Figure 5.

It could be argued that the MLE convergence problem may constrain the applicability of the resulting framework, because the MLE method for shape parameter values from −0.15 to 0.4 may produce solutions for MLE that do not converge. Although it is theoretically possible to have datasets for which no solution exists to the likelihood equations, it appears to be a very rare case in practical applications in hydrology [30].

The models for 47 hydrometric station records were fitted using the framework developed in this work (Figure 5). The resulting fitted models were compared to the models using the approach proposed by Zoglat et al. [25] (LRT-MLE). When the new framework was applied (e.g., Figure 4), a significant improvement in AD and RMSE was achieved. The Anderson-Darling (AD) test results was improved for 38% of the stations by an average of 65%. In addition, the RMSE was decreased for 35% of the stations by an average of 47% (see supplemental Table S2 for percentage improvements).

Further, the relationship between the sample size, i.e., the number of peaks, and the variance in the estimated shape and scale parameters was investigated. For the two parameters, the variance was found to increase with a decrease in the sample size, a result that agrees with those of other studies [13,25]. See supplemental Table S2 for the estimated parameters and their variance, and Figure S1 and Figure S2 for the relationship between the variance in shape and scale and the sample size. In addition, we investigated the sample size and variance of the two parameters versus the drainage area. We found that the variance in scale parameter increases with increasing drainage area, since—under the same hydrological conditions—increasing the drainage area increases the flow, which in turn increases the scale parameter value. A larger scale parameter value results in a larger variance in the scale parameter estimation (see Figure S3 for the relationship between the variance in scale and the drainage area). Further, the variance in the shape parameter decreases with increasing drainage area (see Figure S4 for the relationship between the variance in scale and the drainage area). Finally, the sample size decreases with increasing drainage area (see Figure S5 for the relationship between the sample size and drainage area).

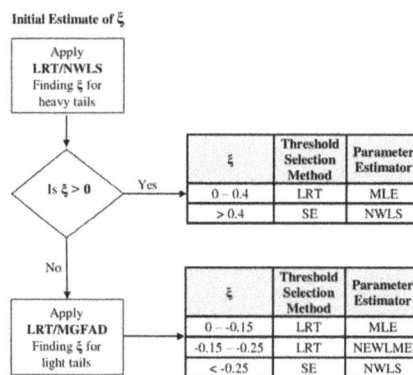

Figure 5. Flow chart of the resulted framework for the choice of accurate combination of threshold selection techniques and the parameter estimators. Abbreviations are as follow: SE—square error method, LRT—likelihood ratio test method, MLE—maximum likelihood estimator, PWM—probability weighted moments, NEWLME—modified likelihood moment estimator, NWLS—Nonlinear weighted least square estimator.

4. Conclusions

The performance of the POT method and the accuracy of estimated floods depend on both the selected threshold and the parameter estimator. In this paper, we proposed a systematic and flexible procedure to assess the combined effect of the threshold selection method and the GPD parameter estimator on the accuracy of the flood frequency distribution. The accuracy of estimating the shape parameter value and the 0.99 quantile were investigated for the MLE, PWM, LME, NEWLME, MGFAD, NWLS estimators by conducting a series of Monte Carlo simulation studies. We found that NWLS and MLE are better parameter estimation methods for heavy tails, while MLE, NEWLME, and NWLS are better parameter estimation methods for short tails. Furthermore, in addition to the tail behavior effect, the value of the scale parameter can also significantly influence the accuracy of LME.

In our study, the six parameter estimators were applied for the thresholds selected by both LRT and SE methods to each of 47 streamflow records. Our results demonstrated that the effects of variations in parameter estimators and threshold selection method must be performed a priori to allow for the most accurate estimates of flood frequency.

We tested our hypothesis by trying different combinations of threshold selection methods and parameter estimators to minimize the AD statistic and RMSE of the developed models. A new framework resulted from this minimization process. The new framework was then applied to each of 47 stations and the developed models using the new framework were compared to the models developed using MLE for parameter estimator and LRT as the threshold selection method. Using the new framework, the AD test results improved for 38% of the stations by an average of 65%. In addition, the RMSE was decreased for 35% of the stations by an average of 47%. Hence, more accurate flood frequency models and flood estimates can be obtained using the new framework, with its improved threshold selection and GPD parameter estimator selection.

The framework developed in this study is widely applicable since it is based on the analysis of data of 47 watersheds with diverse geo-hydro-climatic conditions. More importantly, given that the assumption of independent and identically distributed flood variables is maintained, it will be applicable in other catchments irrespective of their location and other hydrological conditions. Two main reasons can explain our conclusion: (i) from a statistical point of view, the sampled data covered a sufficient range of the shape parameter values that can be found in any catchment irrespective of the location of the catchment, as suggested by Hosking and Wallis [40] and later confirmed by Choulakian and Stephens [30]. In addition, the methods that were concluded to be dependent on the scale parameter value were not considered in the framework; (ii) the framework suggested in Figure 5 was generated from a combination of Monte-Carlo simulations and datasets sampled from streamflow. The results of the sampled streamflow peaks are in accordance with the findings from Monte-Carlo simulations that were based on generated datasets and covered a broader range of values of thresholds, shape parameter, and scale parameter.

The findings of this study can be applied to a wide range of situations. First, our systematic assessment procedure can be used to investigate the combined effect of new GPD parameter estimators and threshold selection methods, leading to better estimations for all POT applications in general, and specifically to flood frequency analysis. More accurate flood estimates are important for climate change adaptation requirements including insurance applications, establishment of the design flood for hydraulic and flood protection structures, and the projection of climate change impacts on flood events.

Supplementary Materials: The following are available online at www.mdpi.com/2073-4441/9/9/692, Table S1: Basic information of the 47 hydrological stations. Average daily Streamflow data are available from SWAT model, calibrated, and validated for the 1986–2007 period, Table S2: The number of peaks and values of the estimated shape (ξ) and scale (σ) parameters and their variance, along with the percentage of improvement of RMSE and AD for the stations, Figure S1: The relationship between the sample size (n) and the variance in shape parameter Var(ξ) values, Figure S2: The relationship between the sample size (n) and the variance in scale parameter Var(σ) values, Figure S3: The relationship between the variance in scale parameter Var(σ) values and the drainage area (km^2), Figure S4: The relationship between the variance in shape parameter Var(ξ) values and the drainage area (km^2), Figure S5: The relationship between the sample size (n) and drainage area (km^2).

Acknowledgments: This work was supported through Alberta Innovates grant # AI-EES 2077 (Predicting Alberta's Water Future project).

Author Contributions: All authors conceived and designed the hypotheses and experiments; Amr Gharib performed the experiments and analyzed the data; the last three authors contributed to the analysis; Amr Gharib developed the first draft and all authors revised and finalized the paper.

Conflicts of Interest: The authors declare no conflict of interest.

References

1. Salvadori, G.; De Michele, C.; Kottegoda, N.T.; Rosso, R. *Extremes in Nature An Approach Using Copulas*; Springer: Dordrecht, The Netherlands, 2007; ISBN 978-1-4020-4415-1.

2. Dahlke, H.E.; Lyon, S.W.; Stedinger, J.R.; Rosqvist, G.; Jansson, P. Contrasting trends in floods for two sub-arctic catchments in northern Sweden—Does glacier presence matter? *Hydrol. Earth Syst. Sci.* **2012**, *16*, 2123–2141. [CrossRef]

3. Mora, D.E.; Campozano, L.; Cisneros, F.; Wyseure, G.; Willems, P. Climate changes of hydrometeorological and hydrological extremes in the Paute basin, Ecuadorean Andes. *Hydrol. Earth Syst. Sci.* **2014**, *18*, 631–648. [CrossRef]

4. Huang, S.; Krysanova, V.; Hattermann, F. Projections of climate change impacts on floods and droughts in Germany using an ensemble of climate change scenarios. *Reg. Environ. Chang.* **2014**, 461–473. [CrossRef]

5. Karl, T.R.; Melillo, J.M.; Peterson, T.C. *Global Climate Change Impacts in the United States*; Cambridge University Press: Cambridge, UK, 2009; Volume 54, ISBN 9780521144070.

6. Milly, P.C.D.; Wetherald, R.T.; Dunne, K.A.; Delworth, T.L. Increasing risk of great floods in a changing climate. *Nature* **2002**, *415*, 514–517. [CrossRef] [PubMed]

7. Apollonio, C.; Balacco, G.; Novelli, A.; Tarantino, E.; Piccinni, A. Land Use Change Impact on Flooding Areas: The Case Study of Cervaro Basin (Italy). *Sustainability* **2016**, *8*, 996. [CrossRef]

8. Pomeroy, J.W.; Stewart, R.E.; Whitfield, P.H. The 2013 flood event in the South Saskatchewan and Elk River basins: Causes, assessment and damages. *Can. Water Resour. J./Rev. Can. Ressour. Hydr.* **2016**, *41*, 106–118. [CrossRef]

9. Rice, D. U.S. had more floods in 2016 than any year on record. USA Today, 4 January 2017. Available online: https://www.usatoday.com/story/weather/2017/01/04/floods-natural-disasters-2016/96120150/ (accessed on 2 February 2017).

10. In den Bäumen, H. S.; Többen, J.; Lenzen, M. Labour forced impacts and production losses due to the 2013 flood in Germany. *J. Hydrol.* **2015**, *527*, 142–150. [CrossRef]

11. Bobée, B.; Rasmussen, P.F. Recent advances in flood frequency analysis. *Rev. Geophys.* **1995**, *33*, 1111–1116. [CrossRef]

12. Mateo Lázaro, J.; Sánchez Navarro, J.Á.; García Gil, A.; Edo Romero, V. Flood Frequency Analysis (FFA) in Spanish catchments. *J. Hydrol.* **2016**, *538*, 598–608. [CrossRef]

13. Vittal, H.; Singh, J.; Kumar, P.; Karmakar, S. A framework for multivariate data-based at-site flood frequency analysis: Essentiality of the conjugal application of parametric and nonparametric approaches. *J. Hydrol.* **2015**, *525*, 658–675. [CrossRef]

14. Iacobellis, V.; Gioia, A.; Manfreda, S.; Fiorentino, M. Flood quantiles estimation based on theoretically derived distributions: Regional analysis in Southern Italy. *Nat. Hazards Earth Syst. Sci.* **2011**, *11*, 673–695. [CrossRef]

15. Mkhandi, S.; Opere, A.; Willems, P. Comparison between annual maximum and peaks over threshold models for flood frequency prediction. In Proceedings of the International Conference of UNESCO Flanders FIT FRIEND/Nile Project—'Towards a Better Cooperation', Sharm El-Sheikh, Egypt, 12–14 November 2005; Volume 1, pp. 1–15.

16. Engeland, K.; Hisdal, H.; Frigessi, A. Practical extreme value modelling of hydrological floods and droughts: A case study. *Extremes* **2005**, *7*, 5–30. [CrossRef]

17. Coles, S.G. *An Introduction to Statistical Modeling of Extreme Values*; Springer: London, UK, 2001; pp. 74–83, ISBN 1852334592.

18. Rao, A.R.; Hamed, K.H. *Flood Frequency Analysis*; CRC Press: Boca Raton, FL, USA, 1999; pp. 27–29.

19. Lang, M.; Ouarda, T.B.M.J.; Bobée, B. Towards operational guidelines for over-threshold modeling. *J. Hydrol.* **1999**, *225*, 103–117. [CrossRef]

20. Rosbjerg, D.; Madsen, H.; Rasmussen, P.F. Prediction in partial duration series with generalized pareto distributed exceedances. *Water Resour. Res.* **1992**, *28*, 3001–3010. [CrossRef]

21. Rasmussen, P.F. *The Partial Duration Series Approach to Flood Frequency Analysis*; Series Paper; Institute of Hydrodynamics and Hydraulic Engineering, Technical University of Denmark: Lyngby, Denmark, 1991.

22. Li, Z.; Wang, Y.; Zhao, W.; Xu, Z.; Li, Z. Frequency Analysis of High Flow Extremes in Northwest China. *Water* **2016**, *8*, 215. [CrossRef]

23. Madsen, H.; Pearson, C.P.; Rosbjerg, D. Comparison of annual maximum series and partial duration series methods for modeling extreme hydrologic events: 1. At site modeling. *Water Resour. Res.* **1997**, *33*, 759. [CrossRef]

24. Cunnane, C. A particular comparison of annual maxima and partial duration series methods of flood frequency prediction. *J. Hydrol.* **1973**, *18*, 257–271. [CrossRef]

25. Zoglat, A.; El Adlouni, S.; Badaoui, F.; Amar, A.; Okou, C.G. Managing Hydrological Risks with Extreme Modeling: Application of Peaks over Threshold Model to the Loukkos Watershed, Morocco. *J. Hydrol. Eng.* **2014**, *19*, 5014010. [CrossRef]

26. Smith, J.A. Estimating the Upper Tail of Flood Frequency Distributions. *Water Resour. Res.* **1987**, *23*, 1657–1666. [CrossRef]

27. Luceño, A. Fitting the generalized Pareto distribution to data using maximum goodness-of-fit estimators. *Comput. Stat. Data Anal.* **2006**, *51*, 904–917. [CrossRef]

28. Solari, S.; Losada, M.A. A unified statistical model for hydrological variables including the selection of threshold for the peak over threshold method. *Water Resour. Res.* **2012**, *48*, 1–15. [CrossRef]

29. Dupuis, D.J. Exceedances over High Thresholds: A Guide to Threshold Selection. *Extremes* **1999**, *1*, 251–261. [CrossRef]

30. Choulakian, V.; Stephens, M.A. Goodness-of-Fit Tests for the Generalized Pareto Distribution. *Technometrics* **2001**, *43*, 478–484. [CrossRef]

31. Neves, C.; Alves, M.I.F. Reiss and Thomas' automatic selection of the number of extremes. *Comput. Stat. Data Anal.* **2004**, *47*, 689–704. [CrossRef]

32. Thompson, P.; Cai, Y.; Reeve, D.; Stander, J. Automated threshold selection methods for extreme wave analysis. *Coast. Eng.* **2009**, *56*, 1013–1021. [CrossRef]

33. Wadsworth, J.L.; Tawn, J.A. Likelihood-based procedures for threshold diagnostics and uncertainty in extreme value modelling. *J. R. Stat. Soc. Ser. B Stat. Methodol.* **2012**, *74*, 543–567. [CrossRef]

34. Zhang, X.; Ge, W. A new method to choose the threshold in the POT model. In Proceedings of the 2009 1st International Conference on Information Science and Engineering (ICISE), Nanjing, China, 26–28 December 2009; pp. 750–753.

35. Davison, A.C.; Smith, R.L. Models for Exceedances over High Thresholds Published by: Wiley for the Royal Statistical Society. *J. R. Stat. Soc. Ser. B Methodol.* **1990**, *52*, 393–442.

36. Ashkar, F.; Nwentsa Tatsambon, C. Revisiting some estimation methods for the generalized Pareto distribution. *J. Hydrol.* **2007**, *346*, 136–143. [CrossRef]

37. MacKay, E.B.L.; Challenor, P.G.; Bahaj, A.S. A comparison of estimators for the generalised Pareto distribution. *Ocean Eng.* **2011**, *38*, 1338–1346. [CrossRef]

38. Song, J.; Song, S. A quantile estimation for massive data with generalized Pareto distribution. *Comput. Stat. Data Anal.* **2012**, *56*, 143–150. [CrossRef]

39. Pickands, J. Statistical Inference Using Extreme Order Statistics. *Ann. Stat.* **1975**, *3*, 119–131.

40. Hosking, J.R.M.; Wallis, J.R. Parameter and Quantile Estimation for the Generalized Pareto Distribution. *Technometrics* **1987**, *29*, 339–349. [CrossRef]

41. Zhang, J. Likelihood moment estimation for the generalized Pareto distribution. *Aust. N. Z. J. Stat.* **2007**, *49*, 69–77. [CrossRef]

42. Zhang, J.; Stephens, M.A. A New and Efficient Estimation Method for the Generalized Pareto Distribution. *Technometrics* **2009**, *51*, 316–325. [CrossRef]

43. Park, M.H.; Kim, J.H.T. Estimating extreme tail risk measures with generalized Pareto distribution. *Comput. Stat. Data Anal.* **2016**, *98*, 91–104. [CrossRef]

44. De Zea Bermudez, P.; Kotz, S. Parameter estimation of the generalized Pareto distribution-Part I. *J. Stat. Plan. Inference* **2010**, *140*, 1353–1373. [CrossRef]
45. Zhang, Y.; Cao, Y.; Dai, J. Quantification of statistical uncertainties in performing the peak over threshold method. *J. Mar. Sci. Technol.* **2015**, *23*, 717–726. [CrossRef]
46. Beguería, S. Uncertainties in partial duration series modelling of extremes related to the choice of the threshold value. *J. Hydrol.* **2005**, *303*, 215–230. [CrossRef]
47. Faramarzi, M.; Abbaspour, K.C.; Adamowicz, W.L.V.; Lu, W.; Fennell, J.; Zehnder, A.J.B.; Goss, G.G. Uncertainty based assessment of dynamic freshwater scarcity in semi-arid watersheds of Alberta, Canada. *J. Hydrol. Reg. Stud.* **2017**, *9*, 48–68. [CrossRef]
48. Faramarzi, M.; Srinivasan, R.; Iravani, M.; Bladon, K.D.; Abbaspour, K.C.; Zehnder, A.J.B.; Goss, G.G. Setting up a hydrological model of Alberta: Data discrimination analyses prior to calibration. *Environ. Model. Softw.* **2015**, *74*, 48–65. [CrossRef]
49. Jiang, R.; Gan, T.Y.; Xie, J.; Wang, N. Spatiotemporal variability of Alberta's seasonal precipitation, their teleconnection with large-scale climate anomalies and sea surface temperature. *Int. J. Climatol.* **2014**, *34*, 2899–2917. [CrossRef]
50. Jiang, R.; Gan, T.Y.; Xie, J.; Wang, N.; Kuo, C.C. Historical and potential changes of precipitation and temperature of Alberta subjected to climate change impact: 1900–2100. *Theor. Appl. Climatol.* **2015**. [CrossRef]
51. Beirlant, J.; Dierckx, G.; Guillou, A. Estimation of the extreme-value index and generalized quantile plots. *Bernoulli* **2005**, *11*, 949–970. [CrossRef]
52. Moharram, S.H.; Gosain, A.K.; Kapoor, P.N. A comparative study for the estimators of the Generalized Pareto distribution. *J. Hydrol.* **1993**, *150*, 169–185. [CrossRef]
53. Chernobai, A.; Rachev, S.T.; Fabozzi, F.J. Composite Goodness-of-Fit Tests for Left-Truncated Loss Samples. In *Handbook of Financial Econometrics and Statistics*; Lee, C.-F., Lee, J.C., Eds.; Springer: New York, NY, USA, 2015; pp. 575–596, ISBN 978-1-4614-7750-1.
54. Kroese, D.P.; Brereton, T.; Taimre, T.; Botev, Z.I. Why the Monte Carlo method is so important today. *Wiley Interdiscip. Rev. Comput. Stat.* **2014**, *6*, 386–392. [CrossRef]
55. Solari, S.; Egüen, M.; Polo, M.J.; Losada, M.A. Peaks Over Threshold (POT): Amethodology for automatic threshold estimation using goodness of fit p-value. *Water Resour. Res.* **2017**, *53*, 2833–2849, Received. [CrossRef]
56. Brockwell, P.; Davis, R. *Introduction to Time Series and Forecasting*; Springer: New York, NY, USA, 2002; pp. 94–95, ISBN 0387953515.
57. Ashkar, F.; Rousselle, J. The effect of certain restrictions imposed on the interarrival times of flood events on the Poisson distribution used for modeling flood counts. *Water Resour. Res.* **1983**, *19*, 481–485. [CrossRef]
58. R Core Team. *R: A Language and Environment for Statistical Computing*; R Foundation for Statistical Computing: Vienna, Austria, 2016.
59. Hyndman, R.J.; Khandakar, Y. Automatic time series forecasting: The forecast package for R. *J. Stat. Softw.* **2008**, *27*, 1–22. [CrossRef]
60. Cunnane, C. A note on the Poisson assumption in partial duration series models. *Water Resour. Res.* **1979**, *15*, 489. [CrossRef]
61. Önöz, B.; Bayazit, M. Effect of the occurrence process of the peaks over threshold on the flood estimates. *J. Hydrol.* **2001**, *244*, 86–96. [CrossRef]
62. Zhang, J. Improving on Estimation for the Generalized Pareto Distribution. *Technometrics* **2010**, *52*, 335–339. [CrossRef]

water

MDPI

Article

Testing the SWAT Model with Gridded Weather Data of Different Spatial Resolutions

Youen Grusson [1,2,*], **François Anctil** [1], **Sabine Sauvage** [2] and **José Miguel Sánchez Pérez** [2]

[1] Department of Civil and Water Engineering, Université Laval, Québec, QC G1V 0A6, Canada; francois.anctil@gci.ulaval.ca

[2] Laboratoire Ecologie Fonctionnelle et Environnement (EcoLab), University of Toulouse, CNRS, INPT, UPS, Avenue de l'Agrobiopole, 31326 Castanet Tolosan CEDEX, France; sabine.sauvage@univ-tlse3.fr (S.S.); jose.sanchez@univ-tlse3.fr (J.M.S.P.)

* Correspondence: youen.grusson.1@ulaval.ca

Academic Editors: Karim Abbaspour, Raghavan Srinivasan, Saeid Ashraf Vaghefi, Monireh Faramarzi and Lei Chen

Received: 23 November 2016; Accepted: 12 January 2017; Published: 17 January 2017

Abstract: This study explored the influence of the spatial resolution of a gridded weather dataset when inputted in the soil and water assessment tool (SWAT) over the Garonne River watershed. Several datasets are compared: ground-based weather stations, the 8-km SAFRAN product (Système d'Analyse Fournissant des Renseignements Adaptés à la Nivologie), the 0.5° CFSR product (Climate Forecasting System Reanalysis) and several derived SAFRAN grids upscaled to 16, 32, 64 and 128 km. The SWAT model, calibrated on weather stations, was successively run with each gridded weather dataset. Performances with SAFRAN up to 64 or 128 km were poor, due to a contraction of the spatial variance of daily precipitation. Performances with 8-km SAFRAN are similar to that of the aggregated 16- and 32-km SAFRAN grids. The ~30-km CFSR product was found to perform well at some sites, while in others, its performance was considerably inferior because of grid points where precipitation was overestimated. The same problem was found in the calibration, where data at some weather stations did not appear to be representative of the subwatershed in which they are used to compute hydrology. These results suggest that the difference in the representation of the climate was more influential than its spatial resolution, an analysis that was confirmed by similar performances obtained with the SWAT model calibrated on the 16- and 32-km SAFRAN grids. However, the better performances obtained from these two weather datasets than from the ground-based stations' dataset confirmed the advantage of using the SAFRAN product in SWAT modelling.

Keywords: SWAT; SAFRAN; weather data input; spatial resolution

1. Introduction

Semi-distributed hydrological models, such as the soil and water assessment tool (SWAT [1–3]), are becoming increasingly popular for water management at the watershed scale [4–6]. One of the main challenges in achieving their maximum potential is accessing proper data with which to establish them. Over the years, distributed soil and land cover data have become more reliable and accessible, mainly because of advancements in remote sensing and a relatively slow rate of evolution. On the other hand, climate data are often problematic. Indeed, climate networks are prone to being irregularly spaced and operated over non-uniform periods. This problem may be circumvented by using gridded climate products constructed from weather reanalysis systems, such as CFSR (Climate Forecasting System Reanalysis) from NOAA's National Centers for Environmental Prediction [7], ERAs from the European Centre for Medium-Range Weather Forecasts (ECMWF) [8], SAFRAN (Système d'Analyse

Fournissant des Renseignements Adaptés à la Nivologie) from Météo-France, the French weather agency [9–11] or the L15 dataset covering North America and described in Livneh, et al. [12].

The use of gridded climate products within an SWAT setup has recently been investigated. For instance, Fuka, et al. [13] found that the CFSR product provides stream discharge simulations that are as good as or better than models forced by using traditional weather stations. A similar conclusion was reached by Auerbach, et al. [14]. Dile and Srinivasan [15] highlighted the benefit of using the CFSR product in sparsely-monitored regions, where CFSR and conventional databases led to minor differences, except for one watershed for which CFSR gave much higher average annual rainfall. Monteiro, et al. [16] demonstrated the superiority of one ERA-derived product (WFDEI–WATCH Forcing Data ERA Interim [17]) over CFSR. Finally, in a comparison of several weather input datasets, de Almeida Bressiani, et al. [18] concluded that the best option for hydrological simulation is the CFSR product used with ground-based climate data.

Others studies explored the influence of weather data density based on a single type of data. For instance, Chaplot, et al. [19] used 1–15 precipitation gauges in two different watershed (51 and 918 km^2) and showed the benefit of a higher data resolution. Such a benefit is however more substantial when larger watersheds are considered and remains limited on small ones [20]. Based on this conclusion, subsequent studies aimed at increasing the spatial resolution of ground-based weather data, in order to improve simulation. Different methods were tested to interpolate weather data to better fit model requirements (from the nearest neighbour method to the Thiessen polygon method) [21–23]. In all of these studies, the main concern was the common lack of density when using ground-based weather station data, which could be offset using interpolated data.

Working with gridded data, the opposite question may be considered. The resolution of the gridded climate products has indeed continued to improve over the years, to the point that operational hydrologists have to question the need for more detailed information when a model such as SWAT has a user-defined areal discretisation that influences the way in which climate data are manipulated within the model. As the model uses a single weather chronicle per subbasin, a too high spatial discretisation of the weather data may lead to a loss of information to the hydrological model.

This study uses an SWAT setup on the Garonne River, a large alpine watershed in southwest France (55,000 km^2), to explore the relationship between the resolution of gridded climate data and SWAT internal discretisation using: (i) available ground-based data; (ii) the native 8-km SAFRAN product; (iii) the native ~30-km CFSR product; and (iv) several aggregated, upscaled SAFRAN-derived databases that may better suit the SWAT model discretisation and could avoid a loss of information.

2. Materials and Methods

2.1. Study Site

The 525-km Garonne River is an important French fluvial system that flows into the Atlantic Ocean after draining a watershed extending over an area of 55,000 km^2 across three distinct geographic entities: the Pyrenees to the south, with peaks exceeding 3000 m, the plateau of the Massif Central to the northeast that reaches up to 1700 m in altitude and the plain in between whose elevation is typically less than a few hundred metres (Figure 1). The actual SWAT implementation, however, is limited to the 50,000-km^2 area upstream of Tonneins, where tides cease to influence the discharge.

The Garonne watershed offers diversified topography and land cover, good data availability, good prior knowledge of the hydrological system [24] and some successful SWAT setups built around available ground-based climate data [25–28].

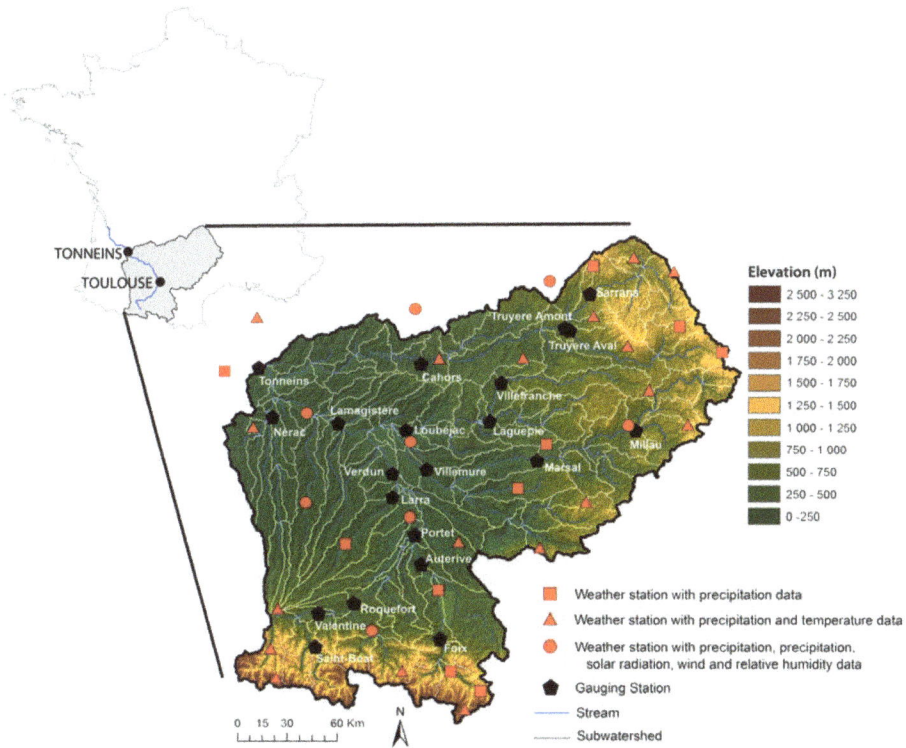

Figure 1. Location and elevation of the Garonne River watershed at Tonneins, with a 150-subwatershed areal discretisation monitored by 21 gauging stations and several types of ground-based climate stations.

2.2. The SWAT Model

SWAT [1–3] is an agro-hydrological semi-distributed model that requires an areal discretisation process that consists of dividing the watershed into subwatersheds based on the river network and topography. SWAT then identifies hydrological response units (HRUs) within each subwatershed, based on soil, land cover and slope information. HRUs are then used to compute a water balance articulated around four reservoirs: snow, soil, shallow aquifer and deep aquifer. The main hydrological processes include infiltration, runoff, evapotranspiration, lateral flow and percolation. Computation is performed at the HRU level, aggregated at the subwatershed level, and flows are routed toward reaches to the catchment outlet.

It is important to stress that SWAT uses only one climate data source per subwatershed to compute its water balance, thus opening up the issue of optimal climate data spatial resolution. Nonetheless, it has been successfully implemented in many locations worldwide to simulate a large range of water components of the hydrological cycle [4–6].

ArcSWAT 2012, a GIS-based graphical interface [29], was used to identify the subwatersheds and HRUs and to generate their associated input files. It should be noted that the number of subwatersheds within SWAT is directly influenced by the resolution of the topography and by a user-defined threshold that defines the minimum drainage area required to form the beginning of a stream, since every river confluence corresponds to a potential subwatershed outlet. Extensive SWAT and ArcSWAT documentation, including theoretical and technical manuals, can be consulted on the SWAT website [30].

2.3. Data Availability

Data sources for this study are presented in Table 1. River discharge data are available from 21 gauging stations spread over the watershed and covering the period from 2000 to 2010 (Figure 1). Two climate datasets from Météo-France were compared: standard stations providing precipitation at 36 sites, temperature at 28 sites, and solar radiation, relative humidity and wind speed at eight sites (Figure 1) and the 8-km SAFRAN product providing all of the required variables at each grid point [9–11]. SAFRAN uses the optimal interpolation (OI) method [31], which corrects background values against all nearby observed data, applying a linear regression in which observations are weighted by distance and associated error. In SAFRAN, the background values originate from the ARPEGE meteorological model from the French meteorological agency [32] or the ECMWF operational archives, while the observations are from a large range of datasets, such as ground-based climate stations, snow monitoring networks, weather balloons and dropsondes.

Table 1. Data sources. CFSR, Climate Forecasting System Reanalysis.

Data Type	Data Source	Scale
DEM	NASA/METI [33]	Grid cell 90 m × 90 m
Land cover	CORINE Land Cover [34]	1:100,000
Soil	European Soil Database [35]	1:1,000,000
Climate (stations/SAFRAN)	Météo-France [36]	8 × 8 km (SAFRAN)
Climate CFSR	NOAA [7,37]	0.5° × 0.5°
River discharge	Banque Hydro [38]	

The SAFRAN grid is constructed in three stages: interpolation of all atmospheric parameters to a 300-m vertical resolution, horizontal interpolation of the surface parameter and temporal interpolation. A more comprehensive description of this process is reported in Durand, Brun, Merindol, Guyomarch, Lesaffre and Martin [9] and Quintana-Segui, Le Moigne, Durand, Martin, Habets, Baillon, Canellas, Franchisteguy and Morel [10].

A second gridded climate product, the CFSR grid, was used in this comparison [7]. Free access is now provided to SWAT users via the Texas A&M University spatial sciences website [37] which automatically creates SWAT-formatted input files. The CFSR has latitudinal and longitudinal resolutions of 0.5°, which over the Garonne watershed correspond to a resolution of ~35-km in latitude and ~25-km in longitude. The CFSR was built around coupled atmospheric, oceanic and surface modelling components, corrected with satellite, aircraft, radiosonde, pibal and in situ data from both land and ocean [7]. Like SAFRAN, the data are interpolated according to the OI method, as described in Xie, et al. [39] for land surfaces and Reynolds, et al. [40] for ocean surfaces.

2.4. Watershed Discretisation

As mentioned above, the SWAT model only uses one source of climate information per subwatershed: the one nearest to the centroid. The number of information points used by the model is therefore directly linked to the areal discretisation defined by the user during the SWAT implementation phase. However, the delineation of the subwatersheds was initially based on the need for a fair representation of all of the hydrological processes prevailing on the watershed and on computing time allocation. In this project, data presented in Table 1 have been used to set up the SWAT model. The watershed was divided into 150 subbasins to allow the representation of the different hydrological behaviours highlighted by Probst [24]. HRUs were defined on soil, land use and slope, retaining only information covering more than 10% of the subbasin area, as proposed by Srinivasan [41].

2.5. Aggregation of the SAFRAN Product

In order to evaluate the spatial appropriateness of the SAFRAN 8-km product against the SWAT areal discretisation of the Garonne River watershed, other resolutions of the former were computed, aggregating the gridded information to 16, 32, 64 and 128 km, respectively.

All climate datasets, including information from the ground stations, were then used in turn to simulate the hydrology of the Garonne River. In the first step, a reference calibration was undertaken based on the ground stations (Figure 1). SWAT was then run with the native SAFRAN product, the native CFSR product and all SAFRAN-aggregated grids. During the second stage, new calibrations were performed based on the native 8-km SAFRAN product and the SAFRAN product aggregated to 32 km (see Section 3.1). In all instances, performance values were computed and compared for the 2000–2010 period at a monthly time step.

2.6. Sensitivity Analysis and Calibration Process

Sensitivity analysis and calibration were undertaken within SWAT-Cup [42] using the SUFI-2 (Sequential Uncertainty Fitting 2) algorithm [43]. SWAT-Cup is an external software tool that allows SWAT users to perform automatic calibrations [44]. They are then given the option of several calibration algorithms, of which SUFI-2 is known to identify an appropriate parameter set in a limited number of iterations [45].

A sensitivity analysis was performed, following the one-at-a-time procedure proposed by Abbaspour [42]. Thirty two parameters were considered in the analysis (Table 2). Five runs were performed over the 10-year period from 2000 to 2010, preceded by a three-year warming period (1997–2000).

Table 2. Parameters considered in the sensitivity analysis. HRU, hydrological response unit. For more details on parameters name see [46].

Parameters	Description	Min.	Max.	Default
EPCO	Plant uptake compensation factor	1	0	1
SURLAG	Surface runoff lag time	0.5	1	4
GW_Delay	Groundwater delay	0	500	31
GW_Revap	Groundwater "revap" coefficient	0.02	0.2	0.02
GWQMN	Threshold in the shallow aquifer for return flow to occur	0	5000	1000
GWHT	Initial groundwater height	0	25	1
GW_SPYLD	Specific yield of the shallow aquifer	0	0.4	0.003
SHALLST	Initial depth of water in the shallow aquifer	0	50,000	500
DEEPST	Initial depth of water in the deep aquifer	0	50,000	1000
ALPHA_BF	Base flow alpha factor (days)	0	1	0.048
REVAPMN	Threshold in the shallow aquifer for "revap" to occur	0	500	0
RCHRG_DP	Deep aquifer percolation fraction	0	1	0.05
ESCO	Soil evaporation compensation factor	0	1	0.95
CN2 (relative test)	Soil conservation Services (SCS) runoff curve number	−0.2	0.2	HRU
CANMX	Maximum canopy storage	0	100	HRU
OV_N	Manning's "n" value for overland flow	0.01	30	HRU
SOL_AWC (relative test)	Available water capacity of the soil layer	−0.5	0.5	soil layer
SOL_K (relative test)	Saturated hydraulic conductivity	−10	10	soil layer
SOL_Z (relative test)	Depth from soil surface to bottom of layer	−500	500	soil layer
EVRCH	Reach evaporation adjustment factor	0.5	1	1
EVLAI	LAI at which no evaporation occurs from the water surface	0	10	3
SFTMP	Snowfall temperature	−10	10	4.5
SMTMP	Snowmelt base temperature	−10	10	4.5
TIMP	Snowpack temperature lag factor	0	1	1
SMFMX	Maximum melt rate for snow during the year (summer solstice)	0	20	1
SMFMN	Minimum melt rate for snow during the year (winter solstice)	0	20	0.5
SNOW50COV	Snow water equivalent that corresponds to 50% snow cover	0	1	0.5
SNOWCOVMX	Snow water content that corresponds to 100% snow cover	0	100	1
SNO_SUB	Initial snow water content	0	300	0
TLAPS	Temperature lapse rate	−10	10	−6
PLAPS	Precipitation lapse rate	−100	500	0
SNOEB	Initial snow water content in elevation bands	0	300	0

Once the most sensitive parameters were identified (see Supplementary Materials for more detail), 1500-run calibrations were performed as recommended by Yang, Reichert, Abbaspour,

Xia and Yang [45]. The SWAT-Cup calibration was achieved sequentially from upstream to downstream, one gauging station at the time, using the Nash–Sutcliffe efficiency criterion (NSe) [47] as the objective function. NSe is a normalised metric allowing a comparison between the variance of the observed dataset and the existing variance of residual errors between this same observed dataset and the simulated one. It ranges from $-\infty$ to 1 and is sensitive to large errors. It equals 0 when the model is as accurate as the mean of the observed dataset and equals 1 when the model offers a perfect fit.

After calibration, the performance was evaluated using the same criterion, but calculated on the root square of discharge values (NSeSqrt) in order to diminish the influence of large errors on the metric. Indeed, the NSeSqrt is influenced more greatly by common flows and the error on the global simulated volume [48,49].

3. Results and Discussion

3.1. Climate Data of Different Resolutions

Seven climate datasets were compared for the Garonne River watershed: the network of standard ground stations, the 8-km SAFRAN product, SAFRAN aggregated to 16, 32, 64 and 128 km and the 30-km CFSR product. SAFRAN and CFSR come from a combination of meteorological simulations and observations and may therefore possess a different spatial variance than the network of ground stations. SAFRAN aggregation also results in a reduction in variance or, in other words, reduces its ability to describe irregular, non-uniform precipitation patterns [50].

Ranges in the spatial variance of the daily precipitation are compared in Figure 2 for the watershed of the Garonne River at Tonneins. As expected, aggregation smoothed out spatial irregularities, which is evident in Figure 2 in the significant loss of large variance events when reducing the resolution from the initial 8 km to the aggregated 128 km. On the other hand, not all precipitation events are non-uniform, so it was mostly higher values of the variance distributions that were affected by the aggregation process.

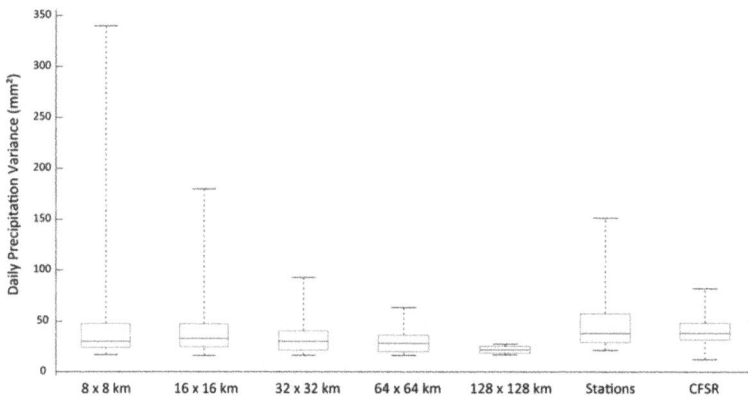

Figure 2. Ranges of the spatial variance of daily precipitation for the seven climate datasets. The box represents the 25th and 75th percentiles of the distribution; the line is the median; and the whiskers extend to the most extreme values.

Figure 2 also illustrates the spatial variance of the precipitation reported by the network of ground stations. The 8-km SAFRAN had a narrower and lower distribution and a lower median value than the network of stations, but the use of information from many different sources and an 8-km resolution allowed SAFRAN to report more irregular events. In practice, the variance of the network between the 16- and 32-km aggregated SAFRAN grids in terms of spatial variance was an indication of the factual resolution of the irregularly-spaced climate stations. Finally, the CFSR product offered yet another distribution of the spatial variance of daily precipitation, which lay between the aggregated 32-km and 64-km SAFRAN grids.

The fact that the choice was made to break down the watershed into 150 subwatersheds limited the number of local climate information points to 150 and raised the question of the optimal resolution of climate data. Indeed, too high a resolution forced SWAT to disregard much of the information, and too low a resolution forced SWAT to use the same information for many adjacent subwatersheds, whereas the resolution of the climate grid had a direct influence on its ability to describe non-uniformities in the precipitation patterns, as mentioned above. This issue is illustrated in Figure 3, which shows points of local climate information used by SWAT. It is evident that SWAT was not able to make use of all of the 8-km information; it used most of the 16-km information and used nearly all of the information from a resolution of 32-km and more. From this perspective, the 32-km SAFRAN grid was the closest to the resolution of the station network, as was the 30-km CFSR product.

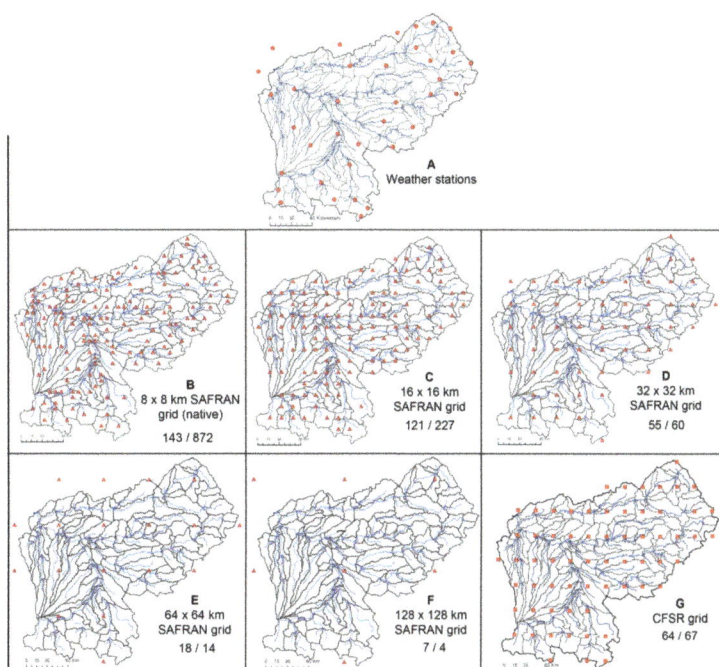

Figure 3. Weather data used by the SWAT model when fed with each different dataset. The notation 143/872 indicates the number of grid points used by the model out of the grid points located over the watershed (a higher first number indicates that the model used a grid point located outside the watershed).

3.2. Hydrological Performance

Figures 4 and 5 compare the performance values (NSe and NSeSqrt) of the SWAT model calibrated with the network of ground stations and run with each SAFRAN grid and the CFSR grid (values of the calibrated parameters can be found in the Supplementary Materials). Using the data from SAFRAN did not improve on the network of ground stations, except at two sites, Larra and Villefranche, where the initial performance was unsatisfactory. Excessive aggregation was detrimental to SWAT performance, as depicted by the much lower SWAT performance when operated with the 64- or 128-km SAFRAN grids, while the other three resolutions were closer to one another in terms of performance.

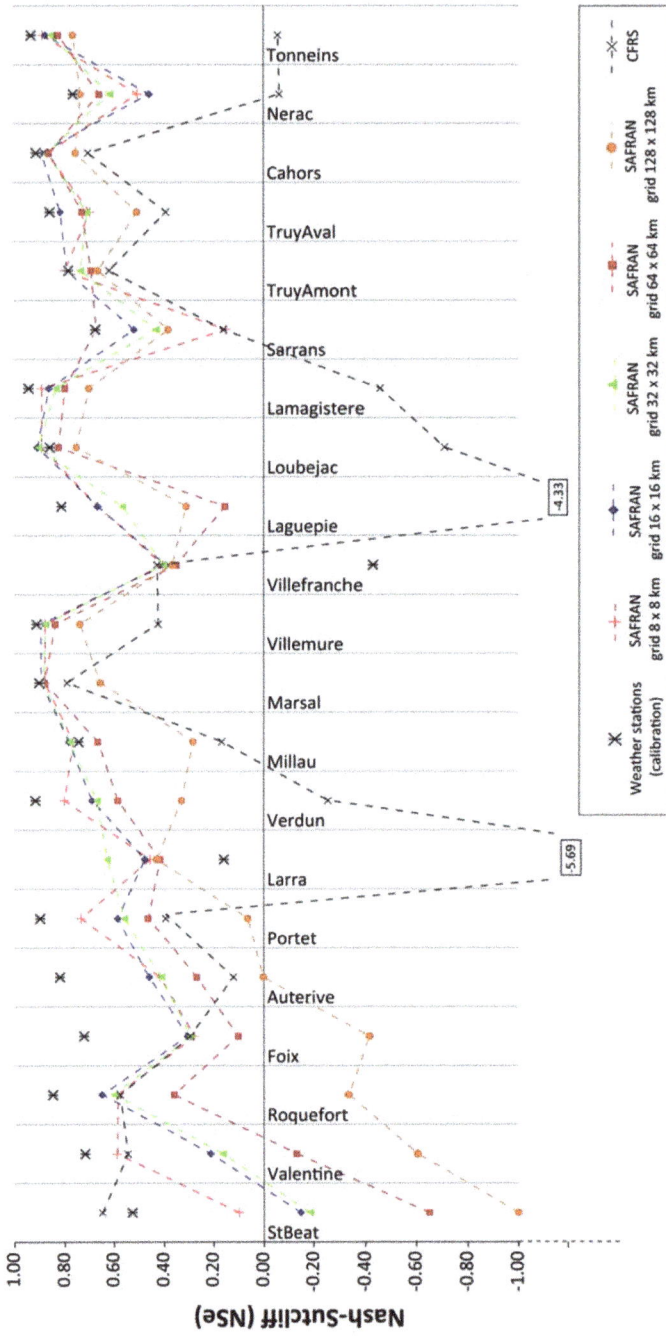

Figure 4. Nash–Sutcliff efficiency (NSe) criterion of the SWAT model calibrated with the weather stations and run with several SAFRAN grid resolutions. The Y-axis displays gauging stations from upstream to downstream (see also Figure 1).

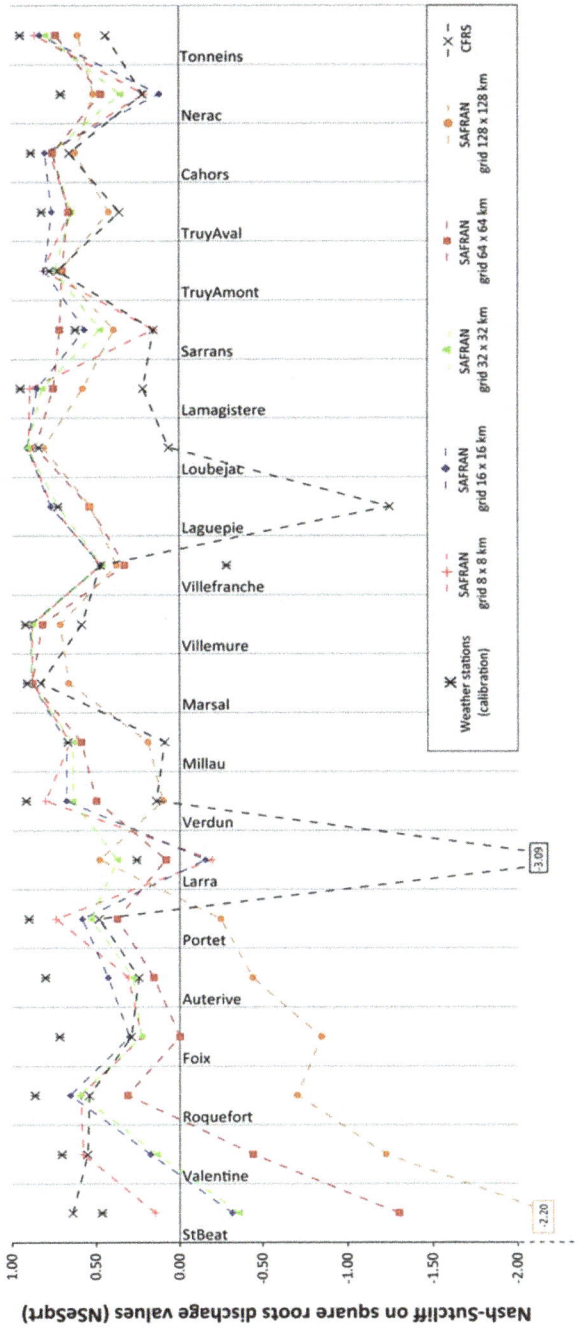

Figure 5. Nash–Sutcliff criterion calculated on the square root discharge value of the SWAT model calibrated with the weather stations and run with several SAFRAN grid resolutions. The Y-axis displays gauging stations from upstream to downstream (see also Figure 1).

CFSR data produced heterogeneous performances, mostly similar to those obtained with the 8-, 16- and 32-km SAFRAN grids, and was even better than in the calibration for the St Béat station (see Figure 1), but also for stations where the SWAT model completely failed to simulate the discharge. When comparing different products, the fact that 8-, 16- and 32-km SAFRAN grids produced similar performances, which were worse than the weather station data and for several stations, very different from the CFSR performance values, indicated that it was not so much an issue of spatial resolution than one of a different representation of the climate.

It is noteworthy that the performance of the Saint Béat and Valentine gauging stations declined as the aggregation increased. Both stations are located in the Pyrenean part of the watershed, where the model is quite sensitive to the snow-relative calibration [28].

As for the Larra and Villefranche sites, each site drains a single subwatershed, where no merging with other subwatersheds can compensate for any errors in a climate series. These poor performance values may also be a representation of a widespread problem when using in situ gauging stations: weather data used to compute hydrological processes in a subwatershed could originate from a distant gauging station and may not be representative of the subwatershed [13]. Indeed, when the weather stations' datasets were used in the SWAT model, computations in subwatersheds upstream of the Villefranche and Truyère-Amont sites were performed using the same weather station, located in the subwatershed upstream of the Truyère-Amont station (Figure 1). Unlike the Villefranche site, no important loss in performance at Truyère-Amont was seen, suggesting that even though the weather station was located close to the Villefranche watershed, it was not representative of that particular system.

As for the CFSR grid, it proposed non-representative precipitation data in some areas. Indeed, a detailed analysis of the hydrographs (not illustrated) revealed that CFSR forced SWAT to overestimate discharges greatly at Larra and Laguepie, a problem that was then transmitted downstream to Verdun, Loubéjac, Lamagistère and Tonneins, while discharge was underestimated at Millau and Sarrans. When comparing hydrographs and hyetographs, it appeared that overestimations were mainly caused by grid points where precipitation was overestimated when compared to the rest of the grid. This was consistent with the findings of Dile and Srinivasan [15], who encountered the same situation over the Blue Nile watershed.

In order to document the issue of SAFRAN and the network of stations not reporting exactly the same climatology, it was decided to recalibrate SWAT alternately using the 8- and 32-km SAFRAN grids, since the 8-km resolution described spatial non-uniformity better, and the 32-km resolution was the closest to the network's (values of calibrated parameters are provided as Supplementary Materials).

Here, Figures 6 and 7 show poorer SWAT performance values when calibrated with the network of climate stations than with any of the two SAFRAN grids, which was consistent with previous studies [13,18]. Better performance values at the Larra and Villefranche sites when using SAFRAN tended to confirm that, at least for these two sites, climate stations were either non-representative or included errors. Moreover, the performance at Tonneins, which integrates climate data across the entire watershed, confirmed the superiority of the SAFRAN product. This highlights the benefit of using a gridded dataset developed from the interpolation and cross-checking of weather data, which avoids temporal gaps, is less influenced by very local events and guarantees space-time consistency for all meteorological variables [11].

The performance of the 8- and 32-km SAFRAN grids was mostly similar, indicating that the grid resolution did not have a considerable influence on the calibration. The largest gain was at Saint Béat in the Pyrenees, where the 32-km resolution proved to be quite beneficial, as it was for the CFSR dataset.

In this case, the Pyrenean gauging stations do not behave differently from the other ones, in opposition to the previous modelling step when running SWAT (reference calibration) with the various upscale grids, which led to a loss in performance values at those sites. Moreover, for some of them (St Beat, Roquefort and Portet), a gain in performance is obtained. These results confirmed the capacity of the calibration to take full advantage of the available information, especially in mountainous regions.

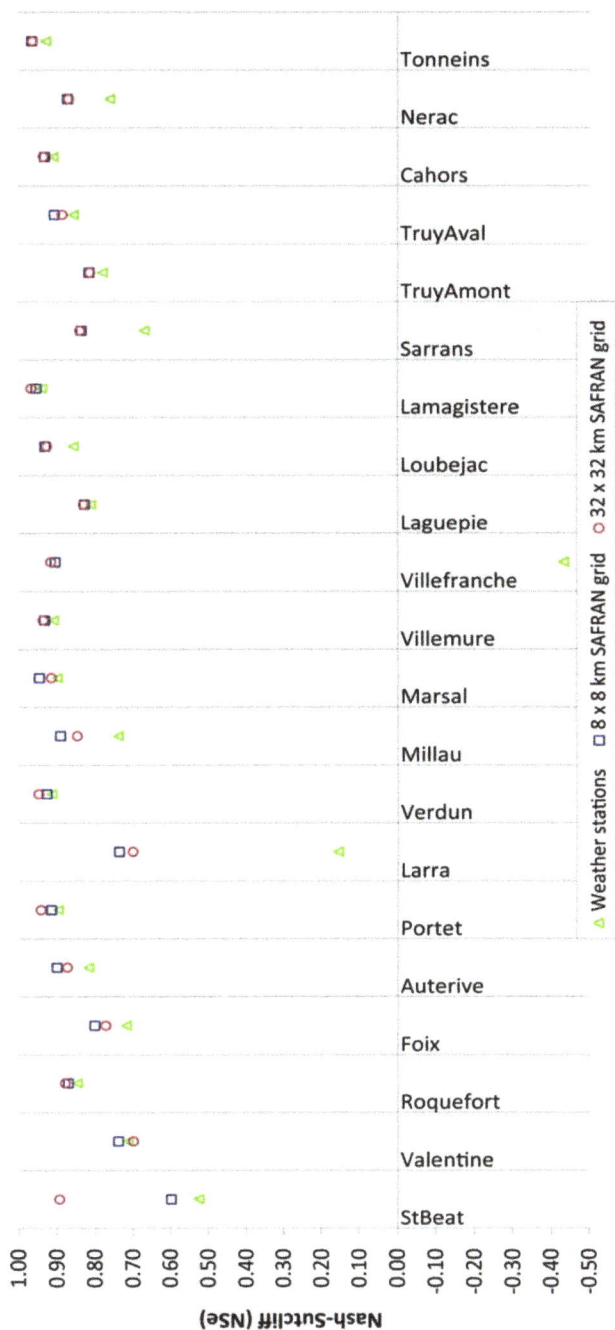

Figure 6. Nash–Sutcliff criterion of the SWAT model calibrated with the weather stations, the native SAFRAN grid and the 32-km SAFRAN grid. The Y-axis displays gauging stations from upstream to downstream (see also Figure 1). It can globally be noted that the upscaling process of the SAFRAN product has various impacts from one site to the other. For instance, upscale grids in the Pyrenean part of the watershed (St Beat, Valentine, Roquefort, Foix and Portet) lead to the largest loss of performance. This is due to the important climate inhomogeneity over mountainous regions that are highly reduced by the upscaling process (Figure 2).

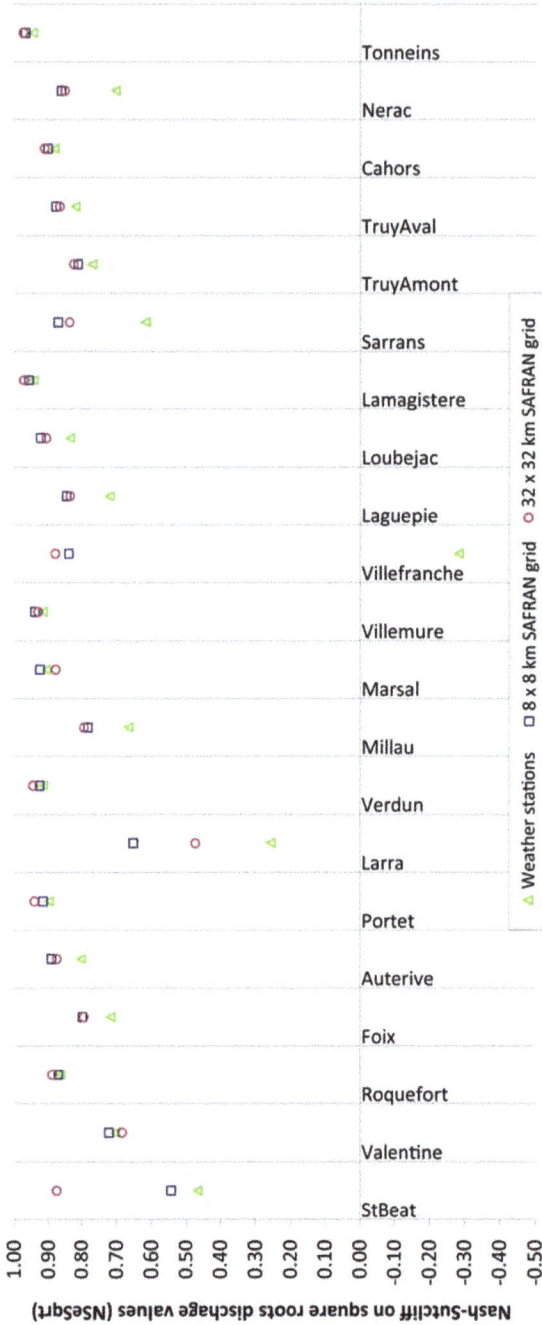

Figure 7. Nash–Sutcliff criterion calculated on the square root (NSeSqrt) discharge value calibrated with the weather stations, the native SAFRAN grid and the 32-km SAFRAN grid. The Y-axis displays the gauging station from upstream to downstream (see also Figure 1).

4. Conclusions

The implementation of a semi-distributed hydrological model generally involves breaking down the watershed space into homogeneous units that are compatible with the dynamic computation of the hydrological processes. In the case of the SWAT model, only one climate grid point per subwatershed, the nearest one, was used for the calculations, raising the question of the optimal resolution of the climate data. In this study, the 45,000-km^2 Garonne River watershed at Tonneins, which drains a substantial part of southwest France to the Atlantic Ocean, was used to compare three sources of climate information in different formats and resolutions, namely the available climate station network, the 8-km SAFRAN product and the ~30-km CFSR product. SAFRAN grids aggregated to 16, 32, 64 and 128 km were also explored.

A spatial breakdown of the Garonne watershed into 150 subwatersheds was deemed as optimal to represent the hydrological functioning of the catchment, as well as adequate regarding to computing costs. From this breakdown results also the limitation of the number of local climate grid points to 150. A higher climate data resolution would therefore force SWAT to disregard much of the information, and a much lower resolution would force SWAT to use the same information for many adjacent subwatersheds, whereas the resolution of the climate grid had a direct influence on its ability to describe non-uniformities in the precipitation patterns.

The results showed that aggregating SAFRAN up to 64 or 128 km was detrimental to the description of non-uniform precipitation events, to the point of leading to a much poorer performance when used with the SWAT implementation calibrated on the available network of climate stations. The native 8-km SAFRAN product offered the variability of daily precipitation events, while the aggregated 16- and 32-km SAFRAN grids and the ~30-km CFSR product ranges of spatial variance were similar to that of the network of climate stations.

Running the SWAT model calibrated on the network of climate stations with the 8-, 16- and 32-km SAFRAN grids led to very similar performance values at most sites, but lower than those previously obtained in calibration. These results suggest that the difference in the representation of the climate was more influential than its spatial resolution in simulating the hydrological processes within the SWAT model. Using CFSR with the same framework provided similar overall performance values as SAFRAN for some sites, but for others, CFSR provided precipitation rates that were judged to be unrealistic when compared to the other climate databases.

The great importance of the quality of the climate product over its resolution was confirmed when calibrating SWAT with the native 8-km SAFRAN and its aggregated 32-km counterpart, since both of these datasets led to very similar performances that were better overall than the calibration performance values previously obtained by calibrating SWAT with the network of climate stations.

Results obtained in this study are consistent with previous works revealing the benefit of using a gridded dataset. However, the choice of the dataset is influenced by (1) the resolution range, which must be adapted to the model definition, even if certain resolution seems to lead to similar performance values, and (2) the representation of the climate by the dataset. It is thus quite important to select a dataset that is suitable to the model and to the climatological characteristics of the watershed.

The present study was based on monthly time step performance computations, compatible with water planning needs, where the influences of extreme precipitation are limited. An equivalent study calculating performances at a daily time step could generate additional findings that could have flood warning applications.

Supplementary Materials: The following are available online at www.mdpi.com/2073-4441/9/1/54/s1.

Acknowledgments: The authors acknowledge the financial support given by the Natural Sciences and Engineering Research Council of Canada and the Institut Hydro-Québec en environnement, développement et société. This research was carried out as a part of the ADAPT'EAU project (Adaptation aux variations des régimes hydrologiques-ANR-11-CEPL-008), a project supported by the French National Research Agency (ANR—Agence Nationale de la Recherche) within the framework of the Global Environmental Changes and Societies (GEC&S) programme. This work was also part of the REGARD project (modélisation des REssources en eau sur le bassin de la GAronne: interaction entre les composantes naturelles et anthropiques et apport de la téléDétection) funded by the RTRA-STAE (Réseau Thématique de Recherche Avancées-Sciences et Technologies pour l'Aéronautique et l'Espace)—2014–2017. We sincerely thank Météo-France for providing meteorological data and AEAG (Agence de l'Eau Adour-Garonne) for providing hydrological discharge data.

Author Contributions: Youen Grusson conceived of, designed, performed the experiments and wrote the paper as part of his Ph.D. thesis. François Anctil and José Miguel Sánchez Pérez supervised the Ph.D. candidate. Sabine Sauvage also provided help throughout the work. All coauthors have collaborated on the redaction of the manuscript.

Conflicts of Interest: The authors declare no conflict of interest.

References

1. Arnold, J.G.; Allen, P.M.; Bernhardt, G. A comprehensive surface-groundwater flow model. *J. Hydrol.* **1993**, *142*, 47–69. [CrossRef]

2. Srinivasan, R.; Ramanarayanan, T.S.; Arnold, J.G.; Bednarz, S.T. Large area hydrologic modeling and assessment part II: Model application. *J. Am. Water Resour. Assoc.* **1998**, *34*, 91–101. [CrossRef]

3. Arnold, J.G.; Srinivasan, R.; Muttiah, R.S.; Williams, J.R. Large area hydrologic modeling and assessment part I: Model development. *J. Am. Water Resour. Assoc.* **1998**, *34*, 73–89. [CrossRef]

4. Gassman, P.W.; Reyes, M.R.; Green, C.H.; Arnold, J.G. The soil and water assessment tool: Historical development, applications, and future research directions. *Trans. ASABE* **2007**, *50*, 1211–1250. [CrossRef]

5. Douglas-Mankin, K.R.; Srinivasan, R.; Arnold, J.G. Soil and water assessment tool (SWAT) model: Current developments and applications. *Trans. ASABE* **2010**, *53*, 1423–1431. [CrossRef]

6. Gassman, P.W.; Sadeghi, A.M.; Srinivasan, R. Applications of the SWAT model special section: Overview and insights. *J. Environ. Qual.* **2014**, *43*, 1–8. [CrossRef] [PubMed]

7. Saha, S.; Moorthi, S.; Pan, H.-L.; Wu, X.; Wang, J.; Nadiga, S.; Tripp, P.; Kistler, R.; Woollen, J.; Behringer, D.; et al. The NCEP climate forecast system reanalysis. *Bull. Am. Meteorol. Soc.* **2010**, *91*, 1015–1057. [CrossRef]

8. Uppala, S.M.; Kallberg, P.W.; Simmons, A.J.; Andrae, U.; Bechtold, V.D.C.; Fiorino, M.; Gibson, J.K.; Haseler, J.; Hernandez, A.; Kelly, G.A.; et al. The ERA-40 re-analysis. *Q. J. R. Meteorol. Soc.* **2005**, *131*, 2961–3012. [CrossRef]

9. Durand, Y.; Brun, E.; Merindol, L.; Guyomarch, G.; Lesaffre, B.; Martin, E. A meteorological estimation of relevant parameters for snow models. *Ann. Glaciol.* **1993**, *18*, 65–71.

10. Quintana-Segui, P.; Le Moigne, P.; Durand, Y.; Martin, E.; Habets, F.; Baillon, M.; Canellas, C.; Franchisteguy, L.; Morel, S. Analysis of near-surface atmospheric variables: Validation of the SAFRAN analysis over France. *J. Appl. Meteorol. Clim.* **2008**, *47*, 92–107. [CrossRef]

11. Vidal, J.P.; Martin, E.; Franchisteguy, L.; Baillon, M.; Soubeyroux, J.M. A 50-year high-resolution atmospheric reanalysis over France with the SAFRAN system. *Int. J. Climatol.* **2010**, *30*, 1627–1644. [CrossRef]

12. Livneh, B.; Bohn, T.J.; Pierce, D.W.; Munoz-Arriola, F.; Nijssen, B.; Vose, R.; Cayan, D.R.; Brekke, L. A spatially comprehensive, hydrometeorological data set for Mexico, the U.S., and Southern Canada 1950–2013. *Sci. Data* **2015**, *2*. [CrossRef] [PubMed]

13. Fuka, D.R.; Walter, M.T.; MacAlister, C.; Degaetano, A.T.; Steenhuis, T.S.; Easton, Z.M. Using the climate forecast system reanalysis as weather input data for watershed models. *Hydrol. Process.* **2014**, *28*, 5613–5623. [CrossRef]

14. Auerbach, D.A.; Easton, Z.M.; Walter, M.T.; Flecker, A.S.; Fuka, D.R. Evaluating weather observations and the climate forecast system reanalysis as inputs for hydrologic modelling in the tropics. *Hydrol. Process.* **2016**, *30*, 3466–3477. [CrossRef]

15. Dile, Y.T.; Srinivasan, R. Evaluation of CFSR climate data for hydrologic prediction in data-scarce watersheds: An application in the Blue Nile River Basin. *J. Am. Water Resour. Assoc.* **2014**, *50*, 1226–1241. [CrossRef]

16. Monteiro, J.A.F.; Strauch, M.; Srinivasan, R.; Abbaspour, K.; Gücker, B. Accuracy of grid precipitation data for Brazil: Application in river discharge modelling of the Tocantins catchment. *Hydrol. Process.* **2016**, *30*, 1419–1430. [CrossRef]

17. Weedon, G.P.; Balsamo, G.; Bellouin, N.; Gomes, S.; Best, M.J.; Viterbo, P. The WFDEI meteorological forcing data set: Watch forcing data methodology applied to ERA-interim reanalysis data. *Water Resour. Res.* **2014**, *50*, 7505–7514. [CrossRef]

18. de Almeida Bressiani, D.; Srinivasan, R.; Jones, C.A.; Mendiondo, E.M. Effects of different spatial and temporal weather data resolutions on the streamflow modeling of a semi-arid basin, Northeast Brazil. *Int. J. Agric. Biol. Eng.* **2015**, *8*, 125–139.

19. Chaplot, V.; Saleh, A.; Jaynes, D.B. Effect of the accuracy of spatial rainfall information on the modeling of water, sediment, and NO_3-N loads at the watershed level. *J. Hydrol.* **2005**, *312*, 223–234. [CrossRef]

20. Cho, H.D.; Olivera, F. Effect of the spatial variability of land use, soil type, and precipitation on streamflows in small watersheds. *J. Am. Water Resour. Assoc.* **2009**, *45*, 673–686. [CrossRef]

21. Masih, I.; Maskey, S.; Uhlenbrook, S.; Smakhtin, V. Assessing the impact of areal precipitation input on streamflow simulations using the SWAT model. *J. Am. Water Resour. Assoc.* **2011**, *47*, 179–195. [CrossRef]

22. Shope, C.L.; Maharjan, G.R. Modeling spatiotemporal precipitation: Effects of density, interpolation, and land use distribution. *Adv. Meteorol.* **2015**. [CrossRef]

23. Cho, J.; Bosch, D.; Lowrance, R.; Strickland, T.; Vellidis, G. Effect of spatial distribution of rainfall on temporal and spatial uncertainty of SWAT output. *Trans. ASABE* **2009**, *52*, 1545–1555. [CrossRef]

24. Probst, J.L. *Hydrologie du Bassin de la Garonne: Modèles de Mélange, Bilan de l'Erosion, Exportation des Nitrates et des Phosphates*; University of Toulouse: Toulouse, France, 1983. (In French)

25. Ferrant, S.; Oehler, F.; Durand, P.; Ruiz, L.; Salmon-Monviola, J.; Justes, E.; Dugast, P.; Probst, A.; Probst, J.L.; Sanchez-Perez, J.M. Understanding nitrogen transfer dynamics in a small agricultural catchment: Comparison of a distributed (TNT2) and a semi distributed (SWAT) modeling approaches. *J. Hydrol.* **2011**, *406*, 1–15. [CrossRef]

26. Boithias, L.; Sauvage, S.; Taghavi, L.; Merlina, G.; Probst, J.L.; Perez, J.M.S. Occurrence of metolachlor and trifluralin losses in the save river agricultural catchment during floods. *J. Hazard. Mater.* **2011**, *196*, 210–219. [CrossRef] [PubMed]

27. Oeurng, C.; Sauvage, S.; Sanchez-Perez, J.M. Assessment of hydrology, sediment and particulate organic carbon yield in a large agricultural catchment using the SWAT model. *J. Hydrol.* **2011**, *401*, 145–153. [CrossRef]

28. Grusson, Y.; Sun, X.; Gascoin, S.; Sauvage, S.; Raghavan, S.; Anctil, F.; Sanchez Pérez, J.M. Assessing the capability of the SWAT model to simulate snow, snow melt and streamflow dynamics over an alpine watershed. *J. Hydrol.* **2015**, *531*, 574–588. [CrossRef]

29. Olivera, F.; Valenzuela, M.; Srinivasan, R.; Choi, J.; Cho, H.; Koka, S.; Agrawal, A. Arcgis-SWAT: A geodata model and GIS interface for SWAT. *J. Am. Water Resour. Assoc.* **2006**, *42*, 295–309. [CrossRef]

30. SWAT Documentation Web Page. Available online: http://swat.tamu.edu/documentation/ (accessed on 10 December 2016).

31. Gandin, L.S. *Objective Analysis of Meteorological Fields*; Gidrometeorologicheskoe Izdatelstvo: Leningrad, Russian, 1963. (In Russian)

32. Courtier, P.; Freydier, C.; Geleyn, J.; Rabier, F.; Rochas, M. The Arpege Project at Meteo-France. Available online: http://www.ecmwf.int/sites/default/files/elibrary/1991/8798-arpege-project-meteo-france.png (accessed on 13 January 2017).

33. NASA; JPL. ASTER—Global Digital Elevation Model v2 90 × 90 m. 2011. Available online: https://asterweb.jpl.nasa.gov/gdem.asp (accessed on 15 January 2017).

34. *Corine Land Cover*; European Union: Brussels, Belgium, 2006. Available online: http://land.copernicus.eu/pan-european/corine-land-cover (accessed on 15 January 2017).

35. *European Soil Data Base v2.0, 1 km × 1 km "Dominant Value and Dominant Stu" Rasters EEA*; European Union: Brussels, Belgium, 2006. Available online: https://eusoils.jrc.ec.europa.eu (accessed on 15 January 2017).

36. Météo France Data Services. Available online: https://donneespubliques.meteofrance.fr/ (accessed on 22 November 2014).

37. Global Weather Data for SWAT. Available online: http://globalweather.tamu.edu/ (accessed on 10 December 2015).

38. Banque Hydro. Available online: http://www.hydro.eaufrance.fr/ (accessed on 13 November 2014).

39. Xie, P.; Chen, M.; Yang, S.; Yatagai, A.; Hayasaka, T.; Fukushima, Y.; Liu, C. A gauge-based analysis of daily precipitation over East Asia. *J. Hydrometeorol.* **2007**, *8*, 607–626. [CrossRef]

40. Reynolds, R.W.; Smith, T.M.; Liu, C.; Chelton, D.B.; Casey, K.S.; Schlax, M.G. Daily high-resolution-blended analyses for sea surface temperature. *J. Clim.* **2007**, *20*, 5473–5496. [CrossRef]

41. Srinivasan, R. Soil and Water Assessment Tool—Introductory Manual—Version 2012. Available online: http://swat.tamu.edu/documentation/ (accessed on 10 September 2014).

42. Abbaspour, K.C. *SWAT-CUP 2012: SWAT Calibration and Uncertainty Programs—A User Manual*; EAWAG: Dübendof, Switzerland, 2013; p. 103.

43. Abbaspour, K.C.; Johnson, C.A.; van Genuchten, M.T. Estimating uncertain flow and transport parameters using a sequential uncertainty fitting procedure. *Vadose Zone J.* **2004**, *3*, 1340–1352. [CrossRef]

44. Arnold, J.G.; Moriasi, D.N.; Gassman, P.W.; Abbaspour, K.C.; White, M.J.; Srinivasan, R.; Santhi, C.; Harmel, R.D.; van Griensven, A.; van Liew, M.W.; et al. SWAT: Model use, calibration, and validation. *Trans. ASABE* **2012**, *55*, 1491–1508. [CrossRef]

45. Yang, J.; Reichert, P.; Abbaspour, K.C.; Xia, J.; Yang, H. Comparing uncertainty analysis techniques for a SWAT application to the Chaohe Basin in China. *J. Hydrol.* **2008**, *358*, 1–23. [CrossRef]

46. Arnold, J.G.; Kiniry, J.R.; Srinivasan, R.; Williams, J.R.; Haney, E.B.; Neitsch, S.L. Soil and Water Assessment Tool—Input/Output Documentation—Version 2012. Available online: http://swat.tamu.edu/documentation/ (accessed on 16 January 2016).

47. Nash, J.E.; Sutcliffe, J.V. River flow forecasting through conceptual models part I—A discussion of principles. *J. Hydrol.* **1970**, *10*, 282–290. [CrossRef]

48. Oudin, L.; Andréassian, V.; Mathevet, T.; Perrin, C.; Michel, C. Dynamic averaging of rainfall-runoff model simulations from complementary model parameterizations. *Water Resour. Res.* **2006**. [CrossRef]

49. Seiller, G.; Anctil, F.; Perrin, C. Multimodel evaluation of twenty lumped hydrological models under contrasted climate conditions. *Hydrol. Earth Syst. Sci.* **2012**, *16*, 1171–1189. [CrossRef]

50. Gaborit, É.; Anctil, F.; Fortin, V.; Pelletier, G. On the reliability of spatially disaggregated global ensemble rainfall forecasts. *Hydrol. Process.* **2013**, *27*, 45–56. [CrossRef]

water

MDPI

Article

Sensitivity of Calibrated Parameters and Water Resource Estimates on Different Objective Functions and Optimization Algorithms

Delaram Houshmand Kouchi [1], Kazem Esmaili [1,*], Alireza Faridhosseini [1], Seyed Hossein Sanaeinejad [1], Davar Khalili [2] and Karim C. Abbaspour [3]

[1] Water Science and Engineering Department, Ferdowsi University of Mashhad, 9177948974 Mashhad, Iran; delaram.houshmandkouchi@mail.um.ac.ir (D.H.K.); farid-h@um.ac.ir (A.F.); sanaei@um.ac.ir (S.H.S.)

[2] Water Engineering Department, Shiraz University, 7144165186 Shiraz, Iran; dkhalili@shirazu.ac.ir

[3] Eawag, Swiss Federal Institute of Aquatic Science and Technology, 8600 Dübendorf, Switzerland; karim.abbaspour@eawag.ch

* Correspondence: esmaili@um.ac.ir; Tel.: +98-915-182-5327

Academic Editor: Karl-Erich Lindenschmidt
Received: 31 March 2017; Accepted: 24 May 2017; Published: 30 May 2017

Abstract: The successful application of hydrological models relies on careful calibration and uncertainty analysis. However, there are many different calibration/uncertainty analysis algorithms, and each could be run with different objective functions. In this paper, we highlight the fact that each combination of optimization algorithm-objective functions may lead to a different set of optimum parameters, while having the same performance; this makes the interpretation of dominant hydrological processes in a watershed highly uncertain. We used three different optimization algorithms (SUFI-2, GLUE, and PSO), and eight different objective functions (R^2, bR^2, NSE, MNS, RSR, $SSQR$, KGE, and $PBIAS$) in a SWAT model to calibrate the monthly discharges in two watersheds in Iran. The results show that all three algorithms, using the same objective function, produced acceptable calibration results; however, with significantly different parameter ranges. Similarly, an algorithm using different objective functions also produced acceptable calibration results, but with different parameter ranges. The different calibrated parameter ranges consequently resulted in significantly different water resource estimates. Hence, the parameters and the outputs that they produce in a calibrated model are "conditioned" on the choices of the optimization algorithm and objective function. This adds another level of non-negligible uncertainty to watershed models, calling for more attention and investigation in this area.

Keywords: calibration; uncertainty analysis; conditional parameters; SUFI-2; GLUE; PSO

1. Introduction

Distributed hydrologic models are useful tools for the simulation of hydrologic processes, planning and management of water resources, investigation of water quality, and prediction of the impact of climate and landuse changes worldwide [1–5]. The successful application of hydrologic models, however, depends on proper calibration/validation and uncertainty analysis [6].

Process-based distributed hydrologic models are generally characterized by a large number of parameters, which are often not measurable and must be calibrated. Calibration is performed by carefully selecting the values for model input parameters (within their respective uncertainty ranges) and by comparing model simulation (outputs) for a given set of assumed conditions with observed data for the same conditions [7].

Hydrological model predictions are affected by four sources of error, leading to uncertainties in the results of the model. These are: 1- input errors (e.g., errors in rainfall, landuse map, pollutant source inputs); 2- model structure/model hypothesis errors (e.g., errors and simplifications in the description of physical processes); 3- errors in the observations used to calibrate/validate the model (e.g., errors in measured discharge and sediment); and 4- errors in the parameters, which arise from a lack of knowledge of the parameters at the scale of interest (e.g., hydraulic conductivity, Soil Conservation Service (SCS) curve number). These sources of error are commonly acknowledged in many studies (e.g., Montanari et al. [8]).

Over the years, a variety of optimization algorithms have been developed for calibration and uncertainty analysis, such as the Generalized Likelihood Uncertainty Estimation method (GLUE) [9], the Sequential Uncertainty Fitting procedure (SUFI-2) [10], Parameter Solution (ParaSol) [11], and Particle Swarm Optimization (PSO) [12,13]. Although these algorithms differ in their search strategies, their goal is to find a set of the best parameter ranges, satisfying a desired threshold assigned to an objective function. Furthermore, many objective functions have also been developed and are in common usage, such as Nash-Sutcliffe efficiency (*NSE*) [14], the root mean square error (*RMSE*), the observations standard deviation ratio (*RSR*) [15], and Kling-Gupta efficiency (*KGE*) [16], to name just a few.

A comparison of the performance of hydrological models under different optimization algorithms [17–20] and objective functions [21,22] has been the subject of some scrutiny in the literature. Examples of this are the work of Arsenault et al. [19], who compared ten optimization algorithms in terms of the method performance with respect to model complexity, basin type, convergence speed, and computing power for three hydrological models. Wu and Chen [20] compared three calibration methods (SUFI-2, GLUE, and ParaSol) within the same modeling framework and showed that SUFI-2 was able to provide more reasonable and balanced predictive results than GLUE and ParaSol. Wu and Liu [21] examined four potential objective functions and suggested SAR as a reasonable choice. In a more comprehensive study, Muleta [22] examined the sensitivity of model performance to nine widely used objective functions in an automated calibration procedure. Less attention, however, has been paid to the optimized parameter values obtained under different optimization algorithms and objective functions, in addition to their impact on the interpretation of hydrological processes in the studied watersheds.

In this study, we examine the sensitivity of optimized model parameters to different optimization algorithms and objective functions, as well as their impacts on the calculation of water resources in two different watersheds in Iran. The current paper focuses on the GLUE, SUFI-2, and PSO algorithms and the objective functions R^2, bR^2, *NSE*, *MNS*, *RSR*, *SSQR*, *KGE*, and *PBIAS* (see Table 1 for a definition of the function). To achieve our objectives, we used the Soil and Water Assessment Tool (SWAT) [23] in the Salman Dam Basin (SDB) and Karkheh River Basin (KRB). For model calibration, we used SWAT-CUP [24], which couples five optimization algorithms to SWAT, allowing the use of different objective functions for SUFI-2 and PSO algorithms.

Table 1. Formulation of the objective functions.

Objective Function	Reference	Formulation *
Modified Coefficient of determination (bR^2)	[25]	$bR^2 = \begin{cases} \|b\|.R^2 \text{ for } b \leq 1 \\ \|b\|^{-1}.R^2 \text{ for } b > 1 \end{cases}$
Coefficient of determination (R^2)	-	$R^2 = \frac{\left[\sum_i (Q_{i,o} - \bar{Q}_o)(Q_{i,s} - \bar{Q}_s)\right]^2}{\sum_i (Q_{i,o} - \bar{Q}_o)^2 \sum_i (Q_{i,s} - \bar{Q}_s)^2}$
Nash-Sutcliffe efficiency (*NSE*)	[14]	$NSE = 1 - \left[\frac{\sum_{i=1}^{n} (Q_{i,o} - Q_{i,s})^2}{\sum_{i=1}^{n} (Q_{i,o} - \bar{Q}_o)^2}\right]$
Modified Nash-Sutcliffe efficiency (*MNS*)	[25]	$MNS = 1 - \frac{\sum_{i=1}^{n} \|Q_{i,o} - Q_{i,s}\|^j}{\sum_{i=1}^{n} \|Q_{i,o} - \bar{Q}_o\|^j}$ with $j \in N$
Ratio of standard deviation of observations to root mean square error (*RSR*)	[15]	$RSR = \frac{RMSE}{STDEV_o} = \frac{\left[\sqrt{\sum_{i=1}^{n} (Q_{i,o} - Q_{i,s})^2}\right]}{\left[\sqrt{\sum_{i=1}^{n} (Q_{i,o} - Q_m)^2}\right]}$

Table 1. *Cont.*

Ranked sum of squares (*SSQR*)	[26]	$SSQR = \frac{1}{n}\sum_{i=1}^{n}[Q_{i,o} - Q_{i,s}]^2$
Kling-Gupta efficiency (*KGE*)	[16]	$KGE = 1 - \sqrt{(R-1)^2 + (\alpha-1)^2 + (\beta-1)^2}$
Percent bias (*PBIAS*)	[27]	$PBIAS = 100 * \left[\frac{\sum_{i=1}^{n}(Q_{i,o} - Q_{i,s})}{\sum_{i=1}^{n}Q_{i,o}}\right]$

* R is the correlation coefficient between observed and simulated data; b is the slope of regression line between observed an simulated data; $Q_{i,o}$ and $Q_{i,s}$ are the *i*th observed and simulated values, respectively; $\overline{Q}_o$, $\overline{Q}_s$ are the mean observed and simulated values, respectively; n is total number of observations; $\alpha = \sigma_s/\sigma_m$ and $\beta = \mu_s/\mu_m$ where σ_m and σ_s are the standard deviation of the observed and simulated data, respectively, and μ_m and μ_s are the mean of observed and simulated data, respectively.

2. Materials and Methods

2.1. Hydrologic Model SWAT

SWAT is a process-based, spatially distributed, and time continuous model. The program is open source and is commonly applied to quantify the impact of landuse and climate change, as well as the impact of different watershed management activities on hydrology, sediment movement, and water quality. SWAT operates by spatially dividing the watershed into multiple sub-basins using digital elevation data. Each sub-basin is further discretized into hydrologic response units (HRUs), which consist of uniform soil, landuse, management, and topographical classes. More information on SWAT can be found in Arnold et al. [7] and Neitsch et al. [28].

2.2. Calibration/Uncertainty Analysis Programs

SUFI-2 is a semi-automated approach used for calibration, validation, and sensitivity and uncertainty analysis. In SUFI-2, all sources of parameter uncertainties are assigned to parameters. The uncertainty in the input parameters are described as uniform distributions, while model output uncertainty is quantified by the 95% prediction uncertainty (95PPU) determined at the 2.5% and 97.5% levels of the cumulative distribution of output variables obtained through Latin hypercube sampling.

Two indices determine the model's goodness-of-fit and uncertainty: *p-factor* and *d-factor*. The *p-factor* is the percentage of observed data bracketed by the 95% prediction uncertainty (95PPU), while the *d-factor* is the average thickness of the 95PPU band divided by the standard deviation of the observed data. In the ideal situation, where the simulation exactly matches the observed data, the *p-factor* and *d-factor* tend to be 100% and 0, respectively, but these values cannot be achieved for real cases due to the errors from different sources. A wide *d-factor* can lead to a large *p-factor*, but SUFI-2 searches to bracket most of the measured data with the smallest possible uncertainty band (*d-factor*) [24].

GLUE relies on the output of numerous Monte Carlo simulations in which a global optimum parameter set is sought and any assessment of parameter uncertainty is made with respect to that global optimum. In GLUE, all sources of uncertainty (i.e., input uncertainty, structural uncertainty, and response uncertainty) are also accounted for by parameter uncertainty. This method is based on the concept of non-uniqueness, which means that different parameter sets can produce equally good and acceptable performances of model predictions due to the interactions of different parameters. This concept rejects the idea of a unique global optimum parameter set. The objective of GLUE is to identify a set of behavioral models within the universe of possible model/parameter combinations. The term "behavioral" is used to signify models that are judged to be "acceptable" on the basis of the available data. In this method, a large number of model runs are performed with different randomly chosen parameter values selected from prior parameter distributions. To quantify how well the parameter combination simulates the real system, a likelihood value is assigned to each set of parameter values by comparing the predicted simulation and observed data. Then, this value is compared to a cutoff threshold value that is selected arbitrarily. Each parameter set that leads to a likelihood value less than the threshold value is discarded from future consideration [29].

PSO is a population- (swarm-) based statistical optimization technique inspired by the social behavior of bird flocking or fish schooling. PSO is initialized with a group of random particles (solutions) moving through the search space for optima. During the optimization process, PSO generates the positions of particles (coordinate in the parameter space) and their velocities (step change in space), and then updates the velocity of each particle using the information from the best solution it has achieved so far and a global best solution obtained by all the other particles. The new position of each particle is calculated by updating the current position using the velocity vector [12].

2.3. Objective Function

To assess the impact of the objective functions on the calibration results and final sensitive parameter ranges, we used the following popularly used functions: the coefficient of determination (R^2) and modified R^2 (bR^2), Nash-Sutcliffe efficiency (*NSE*), Modified Nash-Sutcliffe efficiency (*MNS*), ratio of the standard deviation of observations to root mean square error (*RSR*), ranked sum of squares (*SSQR*), Kling-Gupta efficiency (*KGE*), and percent bias (*PBIAS*). The formulation of these eight objective functions is presented in Table 1.

2.4. Case Studies

The first case study is the Salman Dam Basin (SDB) located in the arid regions of south-central Iran. This region includes the watershed upstream of the Salman Farsi Dam (Figure 1a). The area of the SDB is approximately 13,000 km^2, with geographic coordinates of 28°26' N to 29°47' N and 51°55' E to 54°19' E. The elevation of the basin ranges from less than 800 m above sea level in the southern areas to more than 3100 m in the northern areas of the basin. The main river in the SDB is Ghareh-Aghaj, with an annual average discharge of 18 m^3·s^{-1}. The average annual precipitation is less than 250 mm·year^{-1} in the central and southern part of the watershed, and >750 mm·year^{-1} in the northwest regions.

The second case study is the Karkheh River Basin (KRB), which is the highly studied basin of the Challenge Program in Water and Food [30], located in western Iran. The KRB covers an area of 51,000 km^2 and lies between 30° N to 35° N and 46° E to 49° E geographic coordinates, with an elevation ranging from less than the mean sea level to more than 3600 m (Figure 1b). The Karkheh river is the third longest river in Iran, with an annual average discharge of 188 m^3·s^{-1} [31]. The climate is semi-arid in the uplands (north) and arid in the lowlands (south). The precipitation exhibits large spatial and temporal variability. The mean annual precipitation is about 450 mm·year^{-1}, ranging from 150 mm·year^{-1} in the lower arid plains to 750 mm·year^{-1} in the upper mountainous parts [31]. A large multi-purpose earthen embankment dam, Karkheh, was built on the river and has been utilized since 2001 in order to supply irrigation water in the Khuzestan plains (in the lower Karkheh region), and hydropower generation and flood control. Management information relating to Karkheh reservoir operation (i.e., the minimum and maximum daily outflow, reservoir surface area, and spillway conditions) was considered in the SWAT model of KRB [3].

2.5. Model Setup

2.5.1. SDB and KRB Models

For this study, we used ArcSWAT 2012 with ArcGIS (ESRI-version 10.2.2). The input data and their sources are listed in Table 2. The SDB and KRB watersheds were discretized into 184 and 333 sub-basins, respectively. The sub-basins were further subdivided into 1115 and 3002 homogeneous hydrological response units (HRUs), respectively, by fixing a threshold value of 5% for landuse and 10% for soil type. By using these thresholds, soils and landuses with smaller areas than their respective thresholds were integrated into larger soil and landuses, respectively, by an area weighted scheme.

Table 2. Data description and sources used in the SWAT projects.

Data Type	Source
Digital Elevation Maps (DEM) (resolution 90 m)	The Shuttle Radar Topography Mission (SRTM by NASA) [32]
Soil data (resolution 10 km)	The Food and Agriculture Organization of the United Nations [33]
Landuse data	Satellite images (IRS-P6 LISS-IV and IRS-P5-Pan satellite images, ETM + 2001 Landsat)
Weather data (minimum and maximum daily air temperature and daily precipitation)	Iranian ministry of Energy, The Iranian Meteorological Organization, and WFDEI_CRU data (0.5° × 0.5°)
River discharge	Iranian ministry of Energy
Digital river network and geological position of reservoirs and dams	Iranian ministry of Energy
Management information of Karkheh reservoir operation (i.e., minimum and maximum daily outflow, reservoir surface area, and spillway conditions)	Iranian ministry of Energy

The Hargreaves method [34] was used to simulate the potential evapotranspiration (PET). The maximum transpiration and soil evaporation values were then calculated in SWAT using an approach similar to Ritchie [35], where soil evaporation is estimated by using the exponential functions of soil depth and water content based on PET and a soil cover index based on aboveground biomass. Plant transpiration is simulated as a linear function of PET and the leaf area index, root depth, and soil water content [36]. The modified SCS curve number method was used to calculate surface runoff, and the variable storage routing method was used for flow routing.

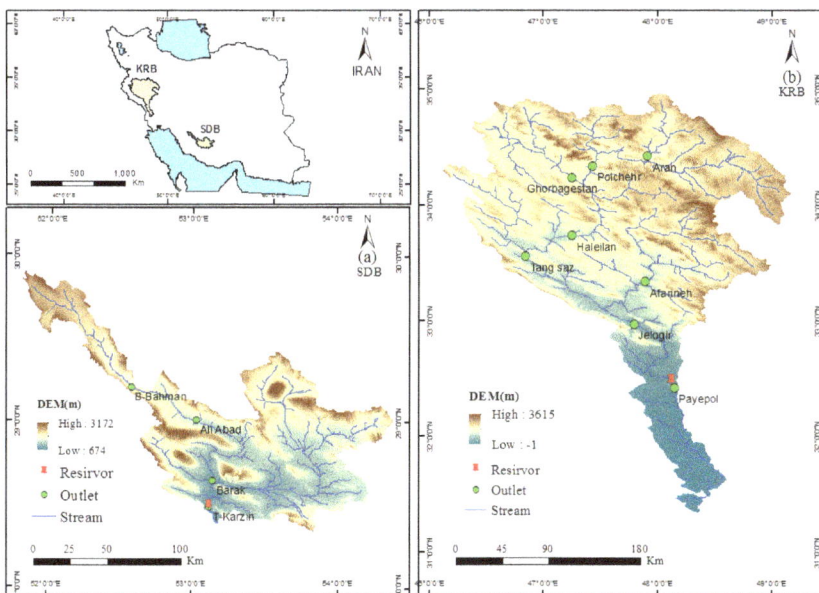

Figure 1. Location of the study areas: (**a**) Salman Dam Basin (SDB) and (**b**) Karkheh River Basin (KRB).

Monthly discharges in the SDB model were calibrated from 1990 to 2008 and validated from 1977 to 1989, and in the KRB, the same periods were 1988–2012 and 1980–1987, respectively, using daily observed flows from four and eight river discharge stations, respectively, (Figure 1). A three-year duration was considered as the warm up period, in order to account for the initial conditions.

To calibrate the model, we initially selected 15 parameters for SDB based on a preliminary one-at-a-time sensitivity analysis and nine parameters for KRB based on the previous research works [3,37]. Ashraf Vaghefi et al. [3] used SUFI-2 to model KRB and identified nine parameters based on their global sensitivity analysis (Table 3). As one-at-a-time analysis is quite limited due to the large interactions between parameters, we initially used a large number of parameters for further analysis. To evaluate the impact of various optimization algorithms and objective functions on the final calibrated parameter ranges, we calibrated each model using the same initial parameter ranges (Table 3) and followed the calibration protocol presented by Abbaspour et al. [6]. Initially, the snow parameters were fitted separately and their values were fixed to avoid identifiability problems with other parameters. Similar to rainfall, snow melt is also a driving variable and its parameters should not be calibrated simultaneously with other model parameters.

Table 3. Initial ranges and descriptions of the parameters used for calibrating the SWAT models in the Salman Dam Basin (SDB) and Karkheh River Basin (KRB).

Parameter	Description	Parameter Range			
		SDB		KRB	
		min	max	min	max
* r_CN2.mgt	SCS runoff curve number	−0.2	0.2	−0.3	0.3
** v_CH_N2.rte	Manning's "n" value for main channel	0	0.3	0	0.3
v_ALPHA_BF.gw	Baseflow alpha factor (days)	0	0.3	0	1
r_SOL_BD.sol	Moist bulk density	−0.5	0.5	−0.5	0.5
v_GW_DELAY.gw	Groundwater delay (days)	30	450	0	500
v_SMFMX.bsn	Max. melt rate for snow during year	0	10	0	20
v_SFTMP.bsn	Snowfall temperature	−5	5		
v_SMTMP.bsn	Snow melt base temperature	−5	5		
v_SMFMN.bsn	Minimum melt rate for snow during year	0	10		
v_TIMP.bsn	Snow pack temperature lag factor	0	1		
v_ESCO.hru	Soil evaporation compensation factor	0.7	1		
v_CH_K2.rte	Effective hydraulic conductivity in channel	5	130		
r_SOL_AWC.sol	Available water capacity	−0.4	0.4		
r_SOL_K.sol	Saturated hydraulic conductivity	−0.8	0.8		
v_ALPHA_BNK.rte	Baseflow alpha factor for bank storage	0	1		
v_GWQMN.gw	Threshold depth of water in the shallow aquifer			0	5000
r_OV_N.hru	Manning's "n" value for overland flow			−1	1
v_GW_REVAP.gw	Groundwater "revap" coefficient			0	0.2

* r_ refers to a relative change in the parameters were their current values are multiplied by (1 plus a factor in the given range); ** v_ refers to the substitution of a parameter value by another value in the given range [24]).

Monthly discharges in both watersheds were calibrated separately using the eight different efficiency criteria in Table 1. Then, the goodness of calibration results were compared for different objective functions using the criteria in Table 4. For bR^2 and *MNS*, we introduced measures based on the results of similar studies [22,38,39], where the satisfactory threshold values for bR^2 and *MNS* were considered greater than or equal to 0.4. No such threshold could be specified for *SSQR* as the measured and simulated variables are independently ranked and its value depends on the magnitude of the variables being investigated.

Table 4. General performance ratings for a monthly time step [15,40].

Performance Rating	R^2	NSE	RSR	PBIAS	KGE
Very good	$0.75 < R^2 \leq 1$	$0.75 < NSE \leq 1$	$0 \leq RSR \leq 0.5$	$PBIAS < \pm 10$	$0.9 \leq KGE \leq 1$
Good	$0.65 < R^2 \leq 0.75$	$0.65 < NSE \leq 0.75$	$0.5 < RSR \leq 0.6$	$\pm 10 \leq PBIAS < \pm 15$	$0.75 \leq KGE < 0.9$
Satisfactory	$0.5 < R^2 \leq 0.65$	$0.5 < NSE \leq 0.65$	$0.6 < RSR \leq 0.7$	$\pm 15 \leq PBIAS < \pm 25$	$0.5 \leq KGE < 0.75$
Unsatisfactory	$R^2 \leq 0.5$	$NSE \leq 0.5$	$RSR > 0.7$	$PBIAS \geq \pm 25$	$KGE < 0.5$

2.5.2. Optimization Algorithms

To compare the parameters obtained in each optimization method, we used similar conditions in terms of behavioral and non-behavioral parameter values, objective function types, calibration parameters and their prior ranges, number of runs, and statistical criteria. The *NSE* was selected as the common objective function for all three optimization algorithms. The behavioral threshold was set at $NSE \geq 0.5$. The criteria NSE, *p-factor*, and *d-factor* were used to evaluate model performance. In SDB, we used three iterations with 480 simulation runs (totally 1440 simulation runs) for SUFI-2 and 1440 simulation runs for GLUE and PSO. In KRB, we used five iterations with 480 simulation runs (totally 2400 simulation runs) for SUFI-2 and 2400 simulation runs for GLUE and PSO. The parallel processing option of SWAT-CUP [41] was used to run SUFI-2 with different objective functions. GLUE and PSO usually require a large number of simulations. However, because of our relatively good initial parameter values and reasonable parameter ranges (based on the previous works mentioned above), fewer runs were needed to produce satisfactory results.

2.5.3. Statistical Analysis

We used the non-parametric Kruskal-Wallis test to assess whether the ranges of sensitive parameters obtained by different objective functions were significantly different from each other. The test is based on an analysis of variance using the ranks of the data values, not the data values themselves. If the Kruskal-Wallis test was significant, we used Tukey's post-hoc test to determine which objective functions produced similar or different parameters.

3. Results

3.1. Sensitivity of Model Performance to the Objective Functions Used in SUFI-2 Algorithm

Based on the criteria of Table 4, all objective functions performed better than satisfactory, except for *PBIAS* in the calibration stage (Table 5). In the validation, the Barak station did not have satisfactory results for six of the objective functions. This could be due to extensive water management and human activities in the upstream of Barak during the validation period.

Table 5. Calibration and validation (in parentheses) results by eight different objective functions using the SUFI-2 optimization algorithm.

Station	bR^2	R^2	NSE	MNS	RSR	$SSQR$	KGE	$PBIAS$
-				Salman Dam Basin (SDB)				
B.Bahman	0.57 (0.52)	0.64 (0.57)	0.57 (0.48)	0.46 (0.38)	0.66 (0.7)	6.3 (3)	0.76 (0.7)	37.8 (34.6)
Ali abad	0.62 (0.56)	0.79 (0.71)	0.65 (0.7)	0.53 (0.5)	0.6 (0.53)	11 (3.8)	0.76 (0.75)	9.6 (−19.4)
Barak	0.57 (0.15)	0.67 (0.36)	0.64 (0.14)	0.41 (0.13)	0.61 (0.88)	0.93 (2.1)	0.65 (0.05)	−6.9 (−5.5)
T.karzin	0.62 (0.61)	0.76 (0.61)	0.74 (0.57)	0.53 (0.42)	0.52 (0.65)	13 (17)	0.84 (0.62)	−40.5 (−9)
-				Karkheh River Basin (KRB)				
Aran	0.51 (0.57)	0.61 (0.57)	0.73 (0.51)	0.49 (0.49)	0.81 (0.7)	8.1 (11)	0.48 (0.49)	−43.00 (58.5)
Polchehr	0.59 (0.54)	0.62 (0.45)	0.55 (0.5)	0.47 (0.42)	0.66 (0.76)	38 (110)	0.75 (0.74)	−7.30 (−9.5)
Ghorbaghestan	0.68 (0.71)	0.69 (0.71)	0.67 (0.66)	0.53 (0.49)	0.56 (0.6)	14 (85)	0.82 (0.72)	4.70 (10.6)
Haleilan	0.66 (0.69)	0.71 (0.58)	0.65 (0.62)	0.52 (0.5)	0.58 (0.64)	170 (130)	0.79 (0.8)	3.10 (0.2)
Tang saz	0.65 (0.73)	0.72 (0.69)	0.66 (0.54)	0.53 (0.45)	0.58 (0.66)	250 (240)	0.82 (0.8)	−1.40 (−4.5)
Afarineh	0.51 (0.37)	0.67 (0.54)	0.56 (0.42)	0.41 (0.49)	0.65 (0.78)	180 (740)	0.67 (0.48)	22.20 (32.4)
Jelogir	0.67 (0.67)	0.71 (0.62)	0.66 (0.59)	0.50 (0.5)	0.57 (0.64)	480 (980)	0.83 (0.81)	4.80 (7.2)
Payepol	0.39 (0.56)	0.43 (0.6)	0.13 (0.27)	0.16 (0.4)	0.94 (0.85)	730 (4400)	0.56 (0.67)	14.30 (14.6)

For an illustration example, we plotted the best and the worst calibration results for the T.Karzin sub-basin in SDB in Figure 2. The discharges based on *NSE* were quite similar and close to the observation, while *PBIAS* showed a systematic delay in the recession leg of the discharge.

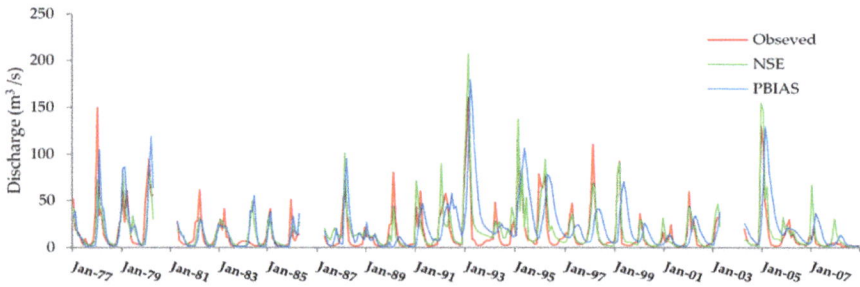

Figure 2. Calibration (1990–2008) and validation (1977–1988) results of monthly simulated discharges showing performance of the best (*NSE*) and the worst (*PBIAS*) objective functions for the T.Karzin station in Salman Dam Basin (SDB).

In KRB, all objective functions performed better than satisfactory for all sub-basins except in Payepol (Table 5). For KRB, we obtained similar results to Ashraf Vaghefi et al. [3], who also modeled this watershed with SWAT. They reported larger uncertainties in the southern parts of the Karkheh Dam (i.e., Payepol station), because of higher water management activities. While in the northern part of the Dam (i.e., Afarine and Jologir stations), the uncertainties were smaller, and in general, model performance was better [3].

At the Payepol station, the validation results were better than the calibration results because the Karkheh Dam was constructed after the validation period. However, the results from some objective functions like *NSE* and *RSR* were still unsatisfactory in Payepol.

To compare the closeness of the final discharges in all objective functions, we calculated the correlation coefficient table (Table 6). The high correlation coefficients among the best simulated discharges in KRB show that most objective functions led to similar results. As in SDB, in KRB, *PBIAS* displayed the worst correlation with the other methods.

Table 6. Correlation coefficients of the objective functions based on the best simulation in the calibration period using the SUFI-2 algorithm.

Case Study	Objective Function	bR^2	R^2	NSE	MNS	RSR	SSQR	KGE	PBIAS
	bR^2	1.00	0.88	0.96	0.93	0.96	0.94	0.95	0.58
	R^2	-	1.00	0.93	0.93	0.95	0.83	0.87	0.61
	NSE	-	-	1.00	0.98	0.98	0.96	0.98	0.56
Salman Dam	MNS	-	-	-	1.00	0.96	0.95	0.97	0.54
Basin (SDB)	RSR	-	-	-	-	1.00	0.93	0.94	0.55
	SSQR	-	-	-	-	-	1.00	0.99	0.46
	KGE	-	-	-	-	-	-	1.00	0.50
	PBIAS	-	-	-	-	-	-	-	1.00
	bR^2	1.00	0.97	0.95	0.96	0.96	0.84	0.97	0.79
	R^2	-	1.00	0.98	0.99	0.98	0.81	0.98	0.74
	NSE	-	-	1.00	0.98	1.00	0.78	1.00	0.70
Karkheh River	MNS	-	-	-	1.00	0.98	0.83	0.98	0.76
Basin (KRB)	RSR	-	-	-	-	1.00	0.78	1.00	0.69
	SSQR	-	-	-	-	-	1.00	0.79	0.95
	KGE	-	-	-	-	-	-	1.00	0.71
	PBIAS	-	-	-	-	-	-	-	1.00

We conclude here that the final results of the monthly discharges in our two case studies are not very sensitive to the objective functions in the SUFI-2 algorithm. In our two case studies, except *PBIAS*, other objective functions produce equally acceptable simulation results. However, this is not a general conclusion because in other regions, where, for example, snow melt is dominant, a certain objective function that targets a specific feature of the discharge may perform better and be more desirable.

3.2. Sensitivity of Model Parameters to Objective Functions

In SUFI-2, parameters are always expressed as distributions, beginning with a wider distribution and ending up with a narrower distribution after calibration. In this study, we used a uniform distribution to express the parameter uncertainty. The parameters obtained by each objective function in the SDB and KRB study sites showed significantly different ranges (Figure 3), even though the simulated discharges were not significantly different. This illustrates the concept of parameter "non-uniqueness" and the concept of "conditionality" of the calibrated parameters. An unconditional parameter range is a parameter range that is independent of the objective function used in calibration. By this definition, the unconditional parameter range of CN2 for B.Bahman would be the range indicated by the broken line in Figure 3. However, this translates into a very large parameter uncertainty. This indicates that there is a significant uncertainty associated with the choice of objective functions with respect to parameter ranges.

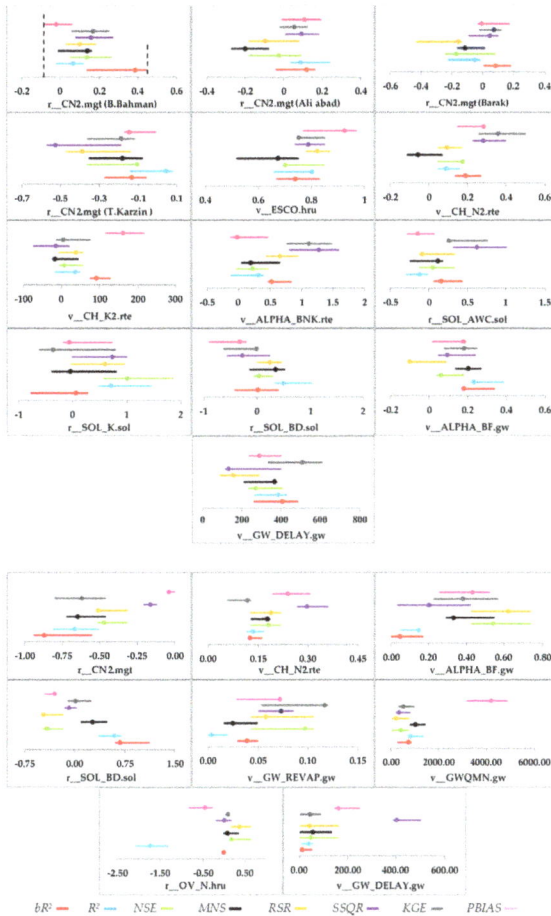

Figure 3. Uncertainty ranges of calibrated parameters using different objective functions in (Top) SDB and (bottom) KRB. The points in each line show the best value of parameters, r_ refers to a relative change where the current values are multiplied by (one plus a factor from the given parameter range), and v_ refers to the substitution by a value from the given parameter range [24]).

Using the Kruskal-Wallis test, we determined which parameter ranges were significantly different from the others (Table 7). As an example, the parameter CN2 for the upstream sub-basins of the B.Bahman outlet were not significantly different for *NSE*, *SSQR*, and *KGE*, while they were significantly different for all other objective functions. A careful analysis of the results in Table 7 reveals that there is no clear pattern of similarity or differences between the objective functions. However, it is clearly indicated that the *NSE* method has the most common parameters with other objective functions, followed by *RSR* and *KGE*.

Table 7. Results of Tukey's post-hoc test to determine if parameters obtained by different objective functions were statistically different or similar.

Case Study	Parameter	bR^2	R^2	NSE	MNS	RSR	SSQR	KGE	PBIAS
Salman Dam Basin (SDB)	r_CN2.mgt (B.Bahman)	-	-	A1	-	-	A1	A1	-
	r_CN2.mgt (Ali abad)	B1	-	B2	-	B2	B3	B1	B3
	r_CN2.mgt (barak)	-	-	C1	C1	-	-	C2	C2
	r_CN2.mgt (T.Karzin)	-	-	D1	D1	-	-	D1	-
	v_ESCO.hru	-	-	-	-	-	E1	E1	-
	v_ALPHA_BNK.rte	F1	F2	-	-	F1	F3	F3	F2
	r_SOL_BD.sol	-	-	-	G1	G1	-	-	-
	v_GW_DELAY.gw	-	-	H1	-	-	-	-	H1
	v_CH_K2.rte	-	I1	I2	I1	I2	-	I2	-
	v_CH_N2.rte	J1	J2	J2	-	J2	-	-	J1
	r_SOL_AWC.sol	-	-	K1	-	K1	-	-	-
	r_SOL_K.sol	-	-	-	L1	L2	L2	L1	l1
	v_ALPHA_BF.gw	-	-	M1	-	-	-	-	M1
Karkheh River Basin (KRB)	r_CN2.mgt	-	A1	A2	-	A2	-	A1	-
	v_CH_N2.rte	B1	B1	B2	-	B2	-	-	-
	v_ALPHA_BF.gw	C1	C1	C2	C3	C2	-	C3	C3
	r_SOL_BD.sol	-	-	D1	-	D1	-	-	D1
	v_GW_REVAP.gw	-	-	E1	-	E1	E1	-	-
	v_GWQMN.gw	-	F1	F2	F1	F2	-	F2	-
	v_GW_DELAY.gw	-	-	G1	G1	G1	-	-	-
	r_OV_N.hru	H1	-	H2	-	H2	H1	H2	-

3.3. Sensitivity of Water Resources Components to the Objective Functions

Next, we calculated the water resource components for parameters obtained by different objective functions. To show this, we calculated the actual evapotranspiration (AET), soil water (SW), and water yield (WYLD) (Figure 4). The long-term annual averages of these variables in SDB, based on the best parameter values given by different objective functions, show significant differences. Furthermore, it is seen that the regional water resources maps of AET, SW, and WYLD exhibit significant differences in their spatial distributions (Figure 5).

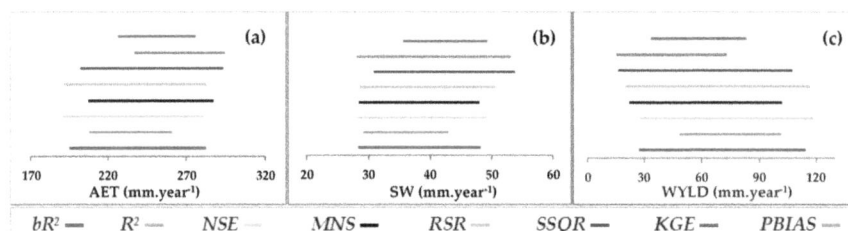

Figure 4. Uncertainty ranges of annual average (**a**) actual evapotranspiration (mm·year^{-1}); (**b**) soil water (mm·year^{-1}); and (**c**) water yield (mm·year^{-1}) derived by different objective functions in Salman Dam Basin (SDB).

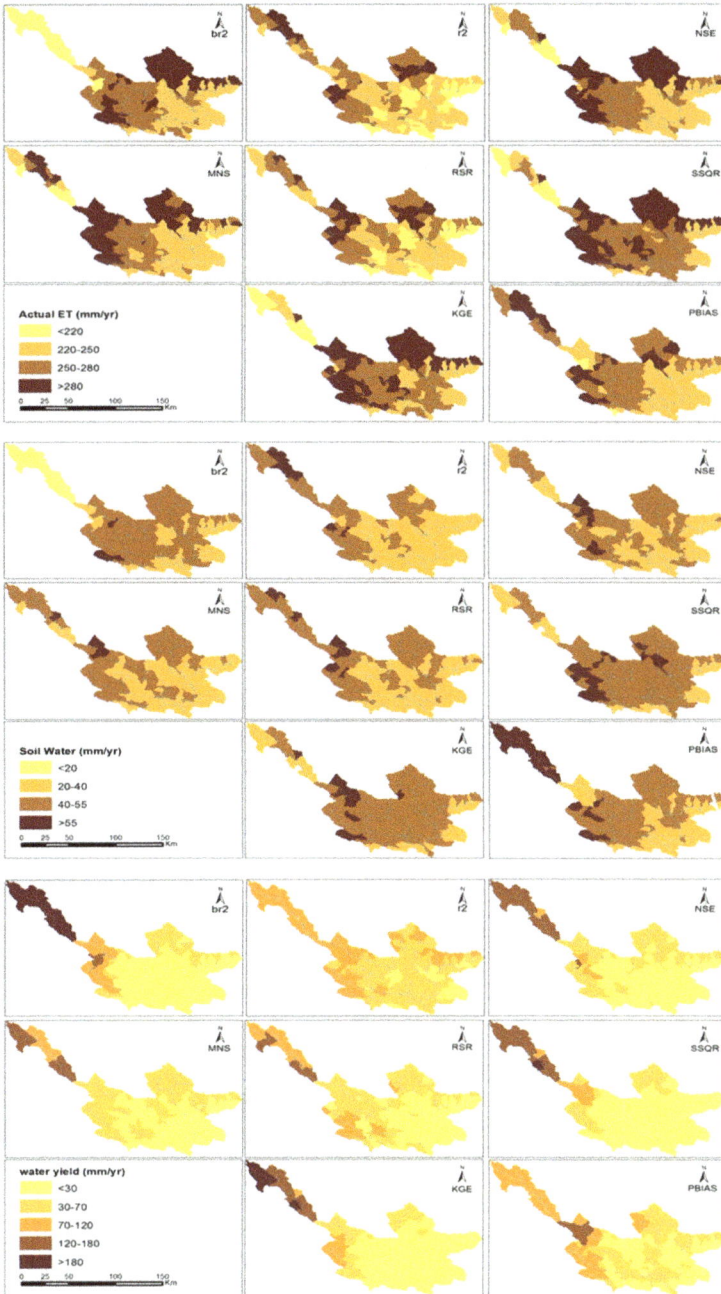

Figure 5. Distribution of annual average (top) actual evapotranspiration (mm·year^{-1}), (middle) soil water (mm·year^{-1}), and water yield (mm·year^{-1}) modeled by SWAT using different objective functions in calibration in Salman Dam Basin (SDB).

Faramarzi et al. [2] reported a range of 120–300 mm·year^{-1} in their national model for AET for the same region. In the current study, the minimum and maximum values of the annual average AET were determined by *RSR* and *KGE* as being 191 and 295 mm·year^{-1}, respectively (Figure 4a). These values are within the uncertainty ranges reported by Faramarzi et al. [2]. The results of SW and WYLD in SDB (Figure 4b,c) also corresponded well with the values reported by Faramarzi et al. [2].

3.4. Sensitivity of Calibration Performance and Model Parameters to Optimization Algorithms Using NSE

In SDB, the maximum *NSE* values in all three optimization techniques were higher than 0.6; hence, they all achieved satisfactory results (Table 8). The *p-factor* values verify that most of the observed discharges were bracketed by the 95PPU of simulations by SUFI-2, followed by GLUE and PSO during the calibration and validation periods. Using a threshold value of *NSE* ≥ 0.5, the SUFI-2 algorithm found 214 behavioral solutions in 480 simulations, while PSO and GLUE achieved 477 and 283 behavioral solutions in 1440 simulations, respectively. Although PSO and GLUE used a larger number of simulations, the *p-factor* and *d-factor* of SUFI-2 show a better performance than GLUE, followed by PSO. This would probably be expected as the latter two algorithms were not allowed to fully exploit the parameter spaces due to the limited number of runs. However, in this study, we used relatively good initial parameter values and uncertainty ranges, and all of the methods obtained quite similar and satisfactory results.

Table 8. Performance of the optimization algorithms and the number of behavioral parameter ranges for the calibration and/validation periods in Salman Dam Basin (SDB) and Karkheh River Basin (KRB).

Case Study	Performance	SUFI-2		GLUE		PSO	
		Cal.	Val.	Cal.	Val.	Cal.	Val.
SDB	*P-factor*	0.84	0.88	0.65	0.68	0.57	0.58
	d-factor	1.22	1.83	0.93	1.44	0.61	0.88
	Best *NSE* value	0.65	0.47	0.7	0.45	0.62	0.45
	No. of behavioral parameter sets	214/480	-	283/1440	-	477/1440	-
KRB	*P-factor*	0.55	0.6	0.15	-	-	-
	d-factor	0.67	0.78	0.22	-	-	-
	Best *NSE* value	0.53	0.51	0.5	-	0.47	-
	No. of behavioral parameter sets	103/480	-	3/2400	-	0/2400	-

In KRB, GLUE and PSO were not successful in calibrating the SWAT model based on the defined conditions (i.e., initial parameter ranges, number of simulation runs, and behavioral threshold value), as there were no behavioral parameter sets. The SUFI-2 algorithm achieved satisfactory simulations of discharge, with *NSE* = 0.53 and *NSE* = 0.51 for the calibration and validation periods, respectively. The *p-factor* was 55% and the *d-factor* was around 1, indicating a reasonable uncertainty in the calibration and verification results (Table 8). More than 100 behavioral solutions in 480 simulations were found with *NSE* ≥ 0.5, while only three behavioral solutions were found by GLUE and no behavioral solution was found by PSO in the 2400 simulation (Table 8). Yang et al. [17] calibrated the Chaohe Basin in China and showed that the application of SUFI-2 based on the Nash–Sutcliffe coefficient used the smallest number of model runs to achieve similar prediction results to GLUE. Additionally, in the current study, in both watersheds, the SUFI-2 algorithm used the smallest number of runs to achieve similar results to GLUE and PSO. As already mentioned, GLUE and PSO in KRB were not allowed to fully explore the parameter spaces, which is the reason for their relatively poor performances here.

Although all three algorithms underestimated the monthly discharge at SDB, they obtained similarly good results based on the performance criteria given by Moriasi et al. [15] (Figure 6) and (Table 9). The calibrated parameters estimated by the three algorithms have larger overlaps than those by different objective functions (Figure 7). PSO provided the widest ranges of parameter uncertainty, followed by GLUE and SUFI-2.

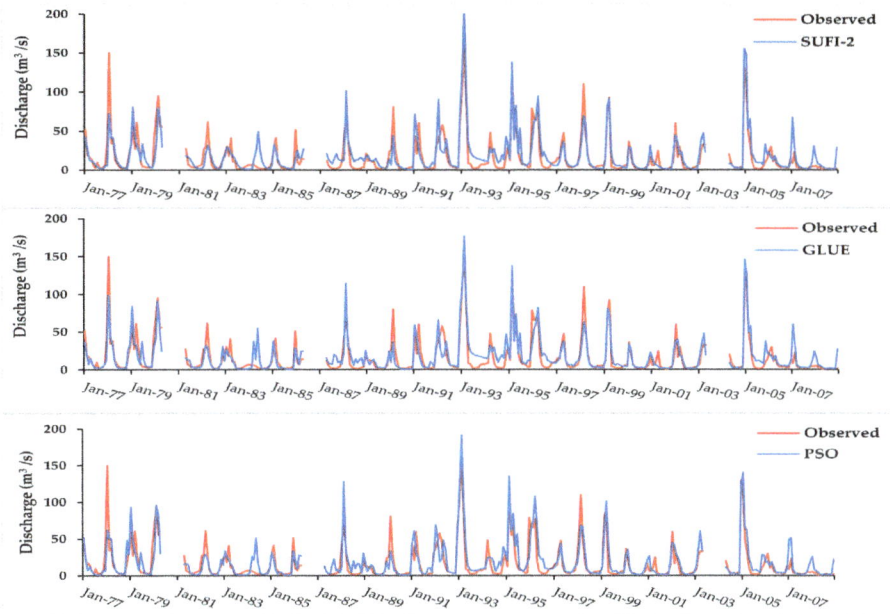

Figure 6. Calibration (1990–2008) and validation (1977–1989) results of the monthly simulated discharges using the three optimization algorithms (SUFI-2, GLUE, and PSO), with *NSE* as the objective function in the T.Karzin station in Salman Dam Basin (SDB).

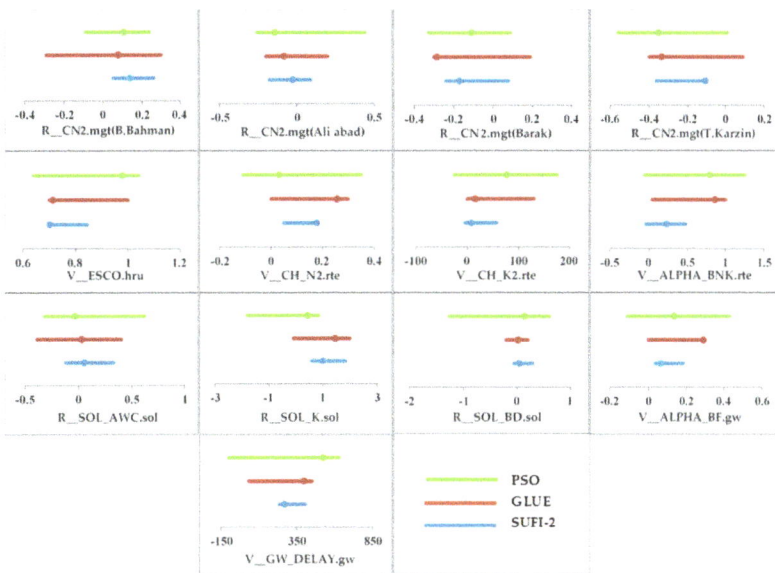

Figure 7. Uncertainty ranges of the parameters based on all three methods applied in Salman Dam Basin (SDB). The points in each line show the best value of the parameters, r_ refers to a relative change where the current values are multiplied by one plus a factor from the given parameter range, and v_ refers to the substitution by a value from the given parameter range [24]).

Table 9. Correlation coefficient among the best simulation of discharges obtained by all optimization techniques in all stations at Salman Dam Basin (SDB).

Optimization Technique	SUFI-2	GLUE	PSO
SUFI-2	1.00	0.99	0.98
GLUE	-	1.00	0.98
PSO	-	-	1.00

Based on multiple comparison tests, half of the calibrated parameter ranges obtained by SUFI-2, GLUE, and PSO were significantly different in SDB. Between GLUE-PSO, SUFI2-GLUE, and SUFI2-PSO, five, four, and four parameters out of 18 were found not to be significantly different from each other, respectively. Overall, the sensitivity of the parameters to different objective functions was found to be larger than the sensitivity to optimization algorithms. This is expected because objective functions solve different problems, while calibration methods basically solve the same problem.

4. Conclusions

We investigated the sensitivity of parameters, model calibration performance, and water resource components to different objective functions (R^2, bR^2, NSE, MNS, RSR, $SSQR$, KGE, and $PBIAS$) and optimization algorithms (e.g., SUFI-2, GLUE, and PSO) using SWAT in two watersheds. The following conclusions could be drawn:

1) In most cases, different objective functions with one optimization algorithm (in this case SUFI-2) led to satisfactory calibration/validation results for river discharges in both case studies. However, the calibrated parameters were significantly different in each case, leading to different water resource estimates.

2) Different optimization algorithms with one objective function (in this case NSE) also produced satisfactory calibration/validation results for river discharges in both case studies. However, the calibrated parameters were significantly different in each case, resulting in significantly different water resources estimates.

Finally, the important message of this work is that the calibration/validation performance may not be sensitive to the choice of optimization algorithm and objective function, but the parameters obtained may be significantly different. As parameters represent processes, the choice of calibration algorithm and objective function may be critical in interpreting the model results in terms of important watershed processes.

Acknowledgments: The authors would like to thank the Eawag Partnership Program (EPP) for the opportunity of a collaboration. We wish to express our sincere thanks to Bahareh Kamali and Mohammad Javad Khordadi for their kind assistance in providing data, useful comments, and suggestions.

Author Contributions: Delaram Houshmand Kouchi and Karim C. Abbaspour developed the methodological framework. Delaram Houshmand Kouchi prepared and calibrated the SWAT models, analysed the data and wrote the paper with assistance from Karim C. Abbaspour. Karim C. Abbaspour advised on conceptual and technical, and contributed to the strategy. Alireza Faridhosseini, Davar Khalili revised the manuscript. This study is a part of Delaram Houshmand Kouchi's Ph.D. project. Kazem Esmaili, Alireza Faridhosseini are project supervisors and Davar Khalili, Seyed Hossein Sanaeinejad and Karim C. Abbaspour are project advisor.

Conflicts of Interest: The authors declare no conflict of interest.

References

1. Tong, S.T.Y.; Chen, W. Modeling the relationship between land use and surface water quality. *J. Environ. Manag.* **2002**, *66*, 377–393. [CrossRef]
2. Faramarzi, M.; Abbaspour, K.C.; Schulin, R.; Yang, H. Modelling blue and green water resources availability in Iran. *Hydrol. Process.* **2009**, *23*, 486–501. [CrossRef]

3. Ashraf Vaghefi, S.; Mousavi, S.J.; Abbaspour, K.C.; Srinivasan, R.; Yang, H. Analyses of the impact of climate change on water resources components, drought and wheat yield in semiarid regions: Karkheh River Basin in Iran. *Hydrol. Process.* **2014**, *28*, 2018–2032. [CrossRef]

4. Rouholahnejad, E.; Abbaspour, K.C.; Srinivasan, R.; Bacu, V.; Lehmann, A. Water resources of the black sea basin at high spatial and temporal resolution. *Water. Resour. Res.* **2014**, *50*, 5866–5885. [CrossRef]

5. Thanapakpawin, P.; Richey, J.; Thomas, D.; Rodda, S.; Campbell, B.; Logsdon, M. Effects of landuse change on the hydrologic regime of the Mae Chaem River Basin, NW Thailand. *J. Hydrol.* **2007**, *334*, 215–230. [CrossRef]

6. Abbaspour, K.C.; Rouholahnejad, E.; Vaghefi, S.; Srinivasan, R.; Yang, H.; Kløve, B. A continental-scale hydrology and water quality model for Europe: Calibration and uncertainty of a high-resolution large-scale SWAT model. *J. Hydrol.* **2015**, *524*, 733–752. [CrossRef]

7. Arnold, J.G.; Moriasi, D.N.; Gassman, P.W.; Abbaspour, K.C.; White, M.J.; Srinivasan, R.; Santhi, C.; Harmel, R.; Van Griensven, A.; Van Liew, M.W. SWAT: Model use, calibration, and validation. *Trans. ASABE* **2012**, *55*, 1491–1508. [CrossRef]

8. Montanari, A.; Shoemaker, C.A.; van de Giesen, N. Introduction to special section on uncertainty assessment in surface and subsurface hydrology: An overview of issues and challenges. *Water. Resour. Res.* **2009**, *45*. [CrossRef]

9. Beven, K.; Binley, A. The future of distributed models: Model calibration and uncertainty prediction. *Hydrol. Process.* **1992**, *6*, 279–298. [CrossRef]

10. Abbaspour, K.C.; Johnson, C.A.; van Genuchten, M.T. Estimating uncertain flow and transport parameters using a sequential uncertainty fitting procedure. *Vadose Zone J.* **2004**, *3*, 1340–1352. [CrossRef]

11. Van Griensven, A.; Meixner, T. Methods to quantify and identify the sources of uncertainty for river basin water quality models. *Water. Sci. Technol.* **2006**, *53*, 51–59. [CrossRef] [PubMed]

12. Eberhart, R.; Kennedy, J. A new optimizer using particle swarm theory. In Proceedings of the Sixth International Symposium on Micro Machine and Human Science, Nagoya, Japan, 4–6 October 1995; pp. 39–43.

13. Kennedy, J.; Eberhart, R. Particle swarm optimization. In Proceedings of the 1995 IEEE International Conference on Neural Networks, Perth, Australia, 27 November–1 December 1995; Volume 1944, pp. 1942–1948.

14. Nash, J.E.; Sutcliffe, J.V. River flow forecasting through conceptual models part I—A discussion of principles. *J. Hydrol.* **1970**, *10*, 282–290. [CrossRef]

15. Moriasi, D.N.; Arnold, J.G.; Van Liew, M.W.; Bingner, R.L.; Harmel, R.D.; Veith, T.L. Model evaluation guidelines for systematic quantification of accuracy in watershed simulations. *Trans. ASABE* **2007**, *50*, 885–900. [CrossRef]

16. Gupta, H.V.; Kling, H.; Yilmaz, K.K.; Martinez, G.F. Decomposition of the mean squared error and NSE performance criteria: Implications for improving hydrological modelling. *J. Hydrol.* **2009**, *377*, 80–91. [CrossRef]

17. Yang, J.; Reichert, P.; Abbaspour, K.C.; Xia, J.; Yang, H. Comparing uncertainty analysis techniques for a SWAT application to the Chaohe Basin in China. *J. Hydrol.* **2008**, *358*, 1–23. [CrossRef]

18. Uniyal, B.; Jha, M.K.; Verma, A.K. Parameter identification and uncertainty analysis for simulating streamflow in a river basin of eastern India. *Hydrol. Process.* **2015**, *29*, 3744–3766. [CrossRef]

19. Arsenault, R.; Poulin, A.; Côté, P.; Brissette, F. Comparison of stochastic optimization algorithms in hydrological model calibration. *J. Hydrol. Eng.* **2014**, *19*, 1374–1384. [CrossRef]

20. Wu, H.; Chen, B. Evaluating uncertainty estimates in distributed hydrological modeling for the Wenjing River watershed in China by GLUE, SUFI-2, and ParaSol methods. *Ecol. Eng.* **2015**, *76*, 110–121. [CrossRef]

21. Wu, Y.; Liu, S. A suggestion for computing objective function in model calibration. *Ecol. Inform.* **2014**, *24*, 107–111. [CrossRef]

22. Muleta, M.K. Model performance sensitivity to objective function during automated calibrations. *J. Hydrol. Eng.* **2012**, *17*, 756–767. [CrossRef]

23. Arnold, J.G.; Srinivasan, R.; Muttiah, R.S.; Williams, J.R. Large area hydrologic modeling and assessment part 1: Model development. *J. Am. Water Resour. Assoc.* **1998**, *34*, 73–89. [CrossRef]

24. Abbaspour, K.C.; Yang, J.; Maximov, I.; Siber, R.; Bogner, K.; Mieleitner, J.; Zobrist, J.; Srinivasan, R. Modelling hydrology and water quality in the pre-alpine/alpine Thur watershed using SWAT. *J. Hydrol.* **2007**, *333*, 413–430. [CrossRef]

25. Krause, P.; Boyle, D.; Bäse, F. Comparison of different efficiency criteria for hydrological model assessment. *Adv. Geosci.* **2005**, *5*, 89–97. [CrossRef]

26. van Griensven, A.; Bauwens, W. Multiobjective autocalibration for semidistributed water quality models. *Water Resour. Res.* **2003**. [CrossRef]

27. Yapo, P.O.; Gupta, H.V.; Sorooshian, S. Automatic calibration of conceptual rainfall-runoff models: Sensitivity to calibration data. *J. Hydrol.* **1996**, *181*, 23–48. [CrossRef]

28. Neitsch, S.L.; Arnold, J.G.; Kiniry, J.R.; Williams, J.R. *Soil and Water Assessment Tool*; Theoretical Documentation: Version 2009; Texas Water Resource Institute: College Station, TX, USA, 2011.

29. Blasone, R.S.; Vrugt, J.A.; Madsen, H.; Rosbjerg, D.; Robinson, B.A.; Zyvoloski, G.A. Generalized likelihood uncertainty estimation (GLUE) using adaptive Markov Chain Monte Carlo sampling. *Adv. Water Resour.* **2008**, *31*, 630–648. [CrossRef]

30. Oweis, T.; Siadat, H.; Abbasi, F. *Improving On-Farm Agricultural Water Productivity in the Karkheh River Basin*; Technical Report; CGIAR Challenge Program on Water and Food: Colombo, Sri Lanka, 2009.

31. Etemad-Shahidi, A.; Afshar, A.; Alikia, H.; Moshfeghi, H. Total dissolved solid modeling; Karkheh reservoir case example. *Int. J. Environ. Res.* **2010**, *3*, 671–680.

32. Reuter, H.I.; Nelson, A.; Jarvis, A. An evaluation of void-filling interpolation methods for SRTM data. *Int. J. Geogr. Inf. Sci.* **2007**, *21*, 983–1008. [CrossRef]

33. Food and Agriculture Organization (FAO). *The Digital Soil Map of the World and Derived Soil Properties*; FAO: Rome, Italy, 1995.

34. Hargreaves, G.L.; Hargreaves, G.H.; Riley, J.P. Agricultural benefits for Senegal River Basin. *J. Irrig. Drain. Eng.* **1985**, *111*, 113–124. [CrossRef]

35. Ritchie, J.T. Model for predicting evaporation from a row crop with incomplete cover. *Water Resour. Res.* **1972**, *8*, 1204–1213. [CrossRef]

36. Wang, X.; Melesse, A.; Yang, W. Influences of potential evapotranspiration estimation methods on SWAT's hydrologic simulation in a northwestern Minnesota watershed. *Trans. ASAE* **2006**, *49*, 1755–1771. [CrossRef]

37. Ghobadi, Y.; Pradhan, B.; Sayyad, G.A.; Kabiri, K.; Falamarzi, Y. Simulation of hydrological processes and effects of engineering projects on the Karkheh River Basin and its wetland using SWAT 2009. *Quat. Int.* **2015**, *374*, 144–153. [CrossRef]

38. Mehdi, B.; Ludwig, R.; Lehner, B. Evaluating the impacts of climate change and crop land use change on streamflow, nitrates and phosphorus: A modeling study in Bavaria. *J. Hydrol.* **2015**, *4*, 60–90. [CrossRef]

39. Akhavan, S.; Abedi-Koupai, J.; Mousavi, S.F.; Afyuni, M.; Eslamian, S.S.; Abbaspour, K.C. Application of SWAT model to investigate nitrate leaching in Hamadan–Bahar watershed, Iran. *Agric. Ecosyst. Environ.* **2010**, *139*, 675–688. [CrossRef]

40. Thiemig, V.; Rojas, R.; Zambrano-Bigiarini, M.; De Roo, A. Hydrological evaluation of satellite-based rainfall estimates over the Volta and Baro-Akobo Basin. *J. Hydrol.* **2013**, *499*, 324–338. [CrossRef]

41. Rouholahnejad, E.; Abbaspour, K.C.; Vejdani, M.; Srinivasan, R.; Schulin, R.; Lehmann, A. A parallelization framework for calibration of hydrological models. *Environ. Model. Softw.* **2012**, *31*, 28–36. [CrossRef]

water

Article

MDPI

Assessing the Uncertainty of Multiple Input Datasets in the Prediction of Water Resource Components

Bahareh Kamali [1,2,*], Karim C. Abbaspour [1] and Hong Yang [1,3]

[1] Eawag, Swiss Federal Institute of Aquatic Science and Technology, 8600 Dubendorf, Switzerland; karim.abbaspour@eawag.ch (K.C.A.); hong.yang@eawag.ch (H.Y.)
[2] ETH Zürich, Department of Environmental Systems Science, Universitätstr. 16, 8092 Zürich, Switzerland
[3] Department of Environmental Sciences, University of Basel, 4003 Basel, Switzerland
* Correspondence: bahareh.kamali@eawag.ch; Tel.: +41-587655358

Received: 27 July 2017; Accepted: 7 September 2017; Published: 16 September 2017

Abstract: A large number of local and global databases for soil, land use, crops, and climate are now available from different sources, which often differ, even when addressing the same spatial and temporal resolutions. As the correct database is unknown, their impact on estimating water resource components (WRC) has mostly been ignored. Here, we study the uncertainty stemming from the use of multiple databases and their impacts on WRC estimates such as blue water and soil water for the Karkheh River Basin (KRB) in Iran. Four climate databases and two land use maps were used to build multiple configurations of the KRB model using the soil and water assessment tool (SWAT), which were similarly calibrated against monthly river discharges. We classified the configurations based on their calibration performances and estimated WRC for each one. The results showed significant differences in WRC estimates, even in models of the same class i.e., with similar performance after calibration. We concluded that a non-negligible level of uncertainty stems from the availability of different sources of input data. As the use of any one database among several produces questionable outputs, it is prudent for modelers to pay more attention to the selection of input data.

Keywords: input data uncertainty; multiple data sets; calibration; modeling; SWAT; SUFI-2

1. Introduction

The successful application of hydrological models depends on their performance during calibration/validation and the degree of model uncertainty. However, the process of calibration is difficult and subjective [1]. This is partly as a result of modeling errors stemming from different sources such as: correctness and adequacy of the input data [2,3], the model's lack of accounting of relevant physical processes in the watershed [4,5], and also the experience of the modeler in manual calibration [6,7].

In the past decade, there has been a major push towards data collection on for example climate, soil, and land use by different agencies such as government ministries (at the local and national levels), educational institutions, local companies, aeronautic industries (e.g., NASA (National Aeronautics and Space Administration) and the University of East Anglia, UK) as well as global organizations such as the FAO (Food and Agriculture Organization). These data, from a hydrological point of interest, include elevation, climate, soil, land use, and river water quantity and quality. A challenging trend that could impact model uncertainty is the availability of multiple datasets of varying and mostly unknown quality for a given region. Selection of only one dataset from among many could have a significant impact on the model calibration and output results. In general, neglecting the uncertainty stemming from different sources of input data during calibration might produce outputs that are not appropriate or representative of real situations [8]. In other words, inappropriate input data (e.g., climate data with errors or incomplete values) can result in unrealistic model parameters [9], which will in turn produce unrealistic model outputs. Therefore, no matter how the model is used, it is always good to know how it performs based on different datasets [10].

Several studies have attempted to explore the sensitivity of hydrological models to land use [11,12], climate [9,13,14], or digital elevation models (DEM) [12,15]. Some studies conducted initial tests on the available data prior to calibration, then chose the data that appeared to perform the best based on certain model efficiency criteria [16,17]. Others studied the sensitivity of model outputs to precipitation ensembles [18,19], and their effects on water resource components of non-calibrated models [14,15,20]. Although all these schemes are important and necessary, in all of them, the prediction uncertainty was based on only one dataset. However, in this work, we are concerned with the uncertainty arising from multiple datasets, where each may have its own uncertainties.

In this work, using the soil and water assessment tool (SWAT), we built eight different models based on four different climate databases and two different land use maps. These models were calibrated using nine measured discharge stations (hereafter referred to as outlets) at the Karkhe River Basin (KRB) in Iran. We then calibrated these models and (i) compare their performances and parameters; and (ii) compare their outputs in terms of water yield (WY) (total amount of water entering the main channel in each time step), blue water (BW) (water yield plus deep aquifer discharge), evapotranspiration (ET), and soil water content (SW).

2. Materials and Methods

2.1. Study Area

The Karkheh River Basin (KRB) is the third largest river basin in Iran (Figure 1). The basin is a benchmark watershed studied in the CGIAR (Consultative Group on International Agricultural Research) challenge program on water and food [21]. It is located in the western part of Iran with a total area of about 50,800 km^2 and stretches from the Zagros Mountains to the Hoor-Al-Azim Swamp (a trans-boundary wetland located at the Iran–Iraq border). The amount of yearly precipitation varies from 250 mm year^{-1} in the southern part up to 750 mm year^{-1} in the northern part of the basin [22]. The elevation of KRB varies from 3 m a.s.l in the south to over 3000 m a.s.l in the north (Figure 1). Nearly 60% of the basin is between 1000–2000 m a.s.l and 20% of the region is below 1000 m a.s.l [23,24]. The highest peak in the region is 3645 m a.s.l. In the northern regions with high elevation, the temperature decreases to below 0 and therefore snowmelt contributes to runoff. A study performed by Saghafian et al. [25] showed that the snow water equivalent is about 17% of long-term annual precipitation in the region.

2.2. Hydrological Simulation

SWAT [26] is a semi-distributed, time continuous watershed simulator operating on daily and sub-daily time steps. The model has been developed to quantify the impact of land management practices in large and complex watersheds coupling land- and routing-phases in the hydrological cycle. Spatial parameterization of SWAT is performed through dividing the watershed into subbasins and further into hydrological response units (HRU) by overlaying soil, landuse, and slope. Hydrological processes include surface runoff, percolation, lateral flow, flow to shallow and deep aquifers, return flow to streams, potential evapotranspiration, snow melt, and transmission loss. A more detailed description of SWAT is given in Neitsch et al. [27]. In this study, we used ArcSWAT 2012.10.1 (Revision 591), where the ArcGIS version 10.3.1 environment was used for project development.

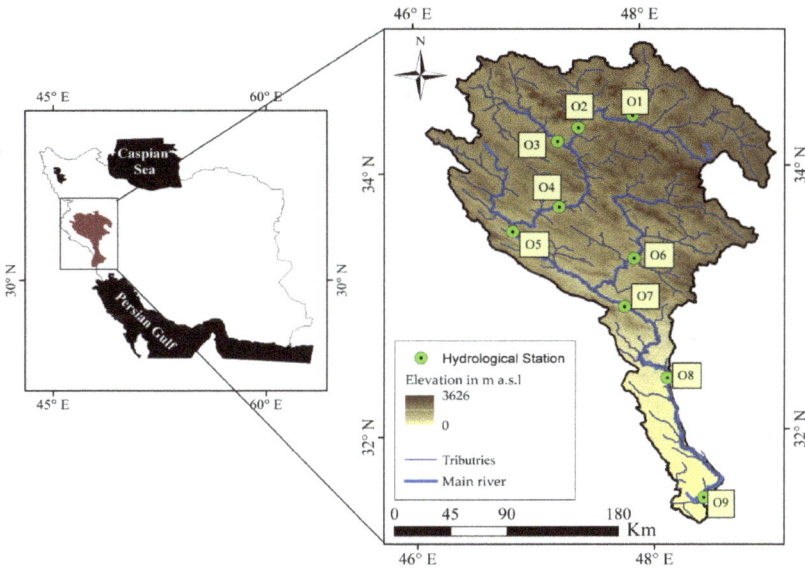

Figure 1. Left: the Karkheh River Basin and its location on a map of Iran. **Right**: the figure shows elevation, main river, and nine outlets (O1–O9) used in calibration.

2.3. Model Calibration and Parameterization

The SWAT model was calibrated using the SUFI-2 algorithm in the SWAT-CUP (SWAT calibration uncertainty procedures) software [28]. SWAT-CUP can be used for sensitivity analysis, multi-site calibration, and uncertainty analysis. SUFI-2 is an iterative algorithm. It maps all model uncertainties on the parameter ranges. The overall uncertainty in the output is quantified by the 95% predictive uncertainty (95PPU) calculated at the 2.5% and 97.5% levels of cumulative distribution of an output variable obtained through Latin hypercube sampling. In this study, we used bR^2 as the efficiency criterion (g) for comparing the simulated and observed discharge values defined as [29]:

$$g = \begin{cases} |b|R^2 \ for \ |b| \leq 1 \\ |b|^{-1}R^2 \ for \ |b| > 1 \end{cases} \tag{1}$$

where R^2 is the coefficient of determination and b is the slope of the regression line between the simulated and measured data. For multiple outlets, the objective function Θ is formulated as:

$$\Theta = \frac{1}{\sum_{i=1}^{n} w_i} \sum_{i=1}^{n} w_i g_i \tag{2}$$

where n is the number of discharge outlets; and w_i is the weight for station i which is set to 1 for all stations. The goodness-of-fit and the degree to which the calibrated model accounts for the uncertainty are assessed by *r-factor* and *p-factor*. The *p-factor* is a fraction of measured data bracketed by the 95PPU band and varies from 0 to 1, and the *r-factor* is the average width of the 95PPU band divided by the standard deviation of the measured variable. A value around 1 is targeted for this parameter [28]. These two indices can be used to judge the strength of the calibration and prediction uncertainty. A larger *p-factor* can be achieved at the expense of a larger *r-factor*. Hence, often a balance must be reached between the two. When acceptable values of *r-factor* and *p-factor* are reached, then the parameter ranges are considered to be the calibrated parameter ranges. Abbaspour et al. [17] mentioned that *p-factors* larger than 0.7 and *r-factors* smaller than 1.5 are adequate. However, this also depends on the scale of the project and the adequacy of the input. The literature shows that a *p-factor* larger than 0.5 is still acceptable [16,30].

2.4. Data and Model Setup

For the study area, a DEM map was obtained from the Shuttle Radar Topography Mission with a spatial resolution of 90 m [23] (Figure 1). The soil map was obtained from the global soil map of the Food and Agricultural Organization (FAO) of the United Nations, which provides data for 5000 soil types comprising two layers (0–30 cm and 30–100 cm depth) at the spatial resolution of 10 km. Of these, 17 soil types were used in our study area. Other soil variables such as hydralic conductivity and bulk density were obtained from the works of Schuol et al. [31]. Four sets of daily climate data (C1, C2, C3, C4) and two landuse maps (L1, L2) were obtained from different sources as described in Table 1.

Table 1. Description of climate and land use data for the Karkheh River Basin (KRB).

Data		Sources	Description
Daily climate data (1977–2004)	C1	Iranian Ministry of Energy database; local observation data based on ground level measurement [32]	Variables used are daily precipitation, maximum and minimum temperature
	C2	Iranian Meteorological Organization database; local observation data based on ground level measurement (http://www.irimo.ir/eng/index.php)	
	C3	Modeling grid cell centroids data obtained from GFDL-ESM2M (Geophysical Fluid Dynamics Laboratory of national oceanic and atmospheric administration—Earth System Model) General Circulation Model (GCM) climate model with $0.5° \times 0.5°$ resolution—Global level [33]	
	C4	Merged from selected stations in C1 and C2 based on their performance in discharge simulation—Details illustrated in Section 3.1 and Figure 2	
Landuse	L1	United States Geological Survey (USGS) Global Land Cover Characterization (GLCC) database [34] with 90m resolution for year 1997	Classification according to Figure 2e and Table 2
	L2	Created from Indian Remote Sensing-Linear P6 (IRS-P6) satellite with Linear Imaging and Self Scanning (LISS-IV) sensor, IRS-P5 satellite with panchromatic cameras, Enhanced Thematic Mapper+2001 (ETM+2001) Landsat, and 3300 field sampling points [35] with 90m resolution for year 2009_ENREF_34	Classification according to Figure 2f and Table 2

C1, C2, and C4 are based on observation data and C3 is from a GCM (General Circulation Model) model (Figure 2). For C3, the daily rainfall was bias corrected using the nearest locally measured stations from C1 and C2. We used a simple ratio method, in which for each month, we divided the average GCM data by the observed data and then divided the daily GCM data by this factor to obtain the daily rainfall data.

The locations and the numbers of climate stations (or grids) within the study region differ from one dataset to the other (Figure 2a–d). The seasonal precipitation depicts some spatial difference among the four databases (Figure S1) mostly in the upper parts of KRB compared to the lower regions. C1 shows the lowest amount of winter precipitation in the western KRB compared to the other databases where the amount of precipitation inceases to above 80 mm month^{-1}. The spring precipitation shows approximately similar distribution in all databases except in C3, where slightly high precipitation occurs in northern KRB (Figure S1). Less differences are noticed in the summer and fall precipitations. Despite approximately similar spatial distribution for the seasonal percipitation, the yearly temporal variations are not the same for all four sources of climate data (Figure S2). For example, relatively high precipitation is noticed during 2000–2004, as well as in year 1987 for C3. The average temperature shows similar values in four sources in the southern KRB in all seasons with the exception of summer temperatures in C3 (Figure S3). In the upper KRB, C3 shows slightly higher temperatures mostly in the western side (Figure S3).

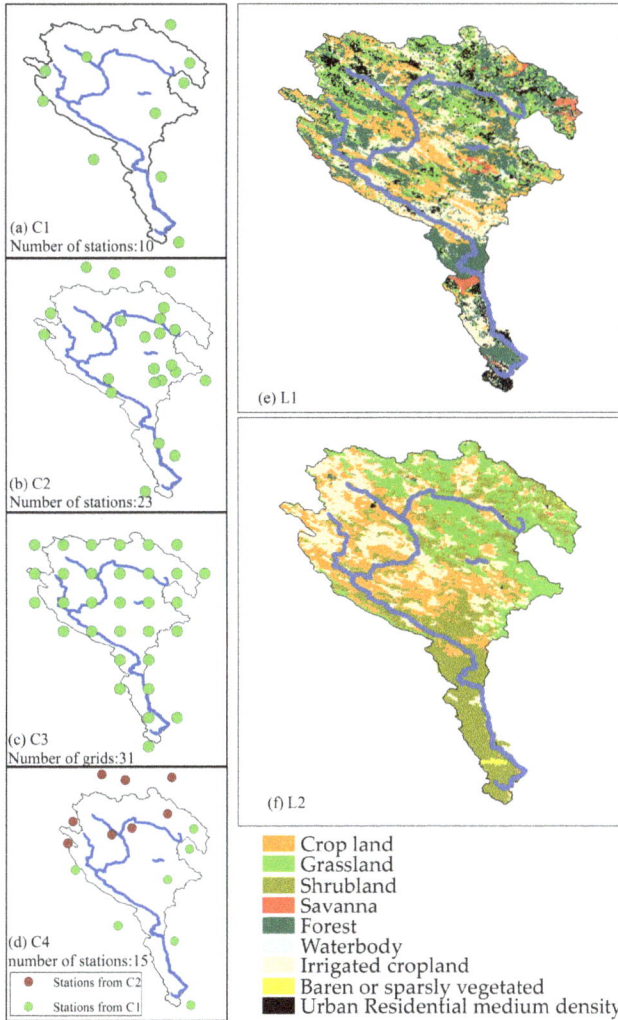

Figure 2. (**a–d**) The location of climate station in the four sources of climate data C1, C2, C3, and C4. (**e,f**) The land use classifications in the L1 and L2 maps.

L1 and L2 were produced with two different approaches in two different years, 2009 and 1997, respectively. L2 was produced locally for the region of study, whereas L1 was obtained from the USGS (United States Geological Survey) global land use map. Table 2 lists different classes of each map corresponding to the SWAT landuse database and the percentages of each land use type. For example, L1 has 25.8% forest lands, while L2 has only 0.2%. On the other hand, L2 has 32.7% shrubland, which is only 1.4% in L1. Both maps show approximately the same percentage of crop and irrigated crop lands, but with different spatial distributions.

Table 2. Percentage of area in each category of two land use maps after being fed into Soil and Water Assessment Tool (SWAT).

Land Use Categories	L1 (%)	L2 (%)
All forest types	25.8	0.2
Grassland	18.3	20.5
Crop land	19.2	22.4
Irrigated crop land	23.1	23.5
Barren and sparsely vegetated	0.0	0.5
Urban residential medium density	8.8	0.1
Shrub land	1.4	32.7
Savanna	2.0	0.1
Water bodies	1.4	0.0

The four climate databases and two land use maps were designated as C1L1, C2L1, C3L1, C4L1, C1L2, C2L2, C3L2 and C4L2, from which eight SWAT models (i.e., eight configurations) were constructed. Considering 8000 ha as the minimum drainage area, a total of 333 subbasins were created for the study area. We used three slope classes (0–2%; 2–4%; and 4–99.99%). The threshold for land use, soil, and slope were all set to 15%, which produced 1520 HRUs for L1 and 1450 HRUs for L2. Potential evapotranspiration was calculated using the Hargreaves method.

For the calibration of all configurations, we used monthly values for the nine outlets (O1–O9 in Figure 1) recorded by IWPCO (Iran Water and Power Resources Development Company, Tehran, Iran) [36]. We calibrated the models using parameters sensitive to discharge, selected based on the initial model simulation, the guidelines suggested by Abbaspour et al. [17], and the experience gained from previous work in the same river basin [37,38], as explained in Table 3. The snow parameter i.e., "maximum snow melt rate" was set to 5 mm C^{-1} day^{-1} based on the work of Vaghefi et al. [37] in all eight configurations.

All analyses were conducted for the years 1977–2004 considering the first three years as a warm-up period, 1988–2004 as calibration, and 1980–1987 as validation periods. We calibrated each model using five iterations with 480 simulations in each iteration. After an iteration, the objective function, the 95PPU band for all nine outlets, and the new ranges of parameters were calculated [17]. The best

Table 3. List of parameters included in the calibration of the eight different configurations and their description.

Parameter	Definition	Initial Values
r_CN2.mgt	SCS (Soil Conservation Service) runoff curve number for moisture condition II	Spatially variable
r_SOL_AWC.sol	Soil available water storage capacity (mm H_2O/mm soil)	Spatially variable
v_ESCO.hru	Soil evaporation compensation factor	0.95
r_OV_N.hru	Manning's *n* value for overland flow	Spatially variable
v_ALPHA_BF.gw	Base flow alpha factor (days)	0.048
v_GW_DELAY.gw	Groundwater delay time (days)	31
v_GW_REVAP.gw	Capillary flow from groundwater into root zone	0.02
r_REVAPMN.gw	Threshold depth of water in the shallow aquifer (mm)	750
v_GWQMN.gw	Threshold depth of water in the shallow aquifer required for return flow to occur (mm)	1000

parameter of the current iteration was used to calculate the new range of parameters and modify the previous ranges. The procedure continued until satisfactory *p-factor* and *r-factor* values were reached or no further improvements were seen in the objective function.

2.5. Statistical Analysis: Multiple Comparison Test

We used the non-parametric Kruskal–Wallis test to compare if the bR^2 values in the nine outlets obtained by different models significantly differed from each other or not. The test is based on an analysis of variance using the ranks of the data values, not the data values themselves. The *p-value* is the criteria used to estimate probability of rejecting the null hypothesis (H_0: all models are statistically similar) of a study when that hypothesis is true. The conventional value for the *p-value* is set at 0.05. The threshold shows that any *p-value* lower than 0.05 results in a statistically significant difference, while values above 0.05 present statistically insignificant differences. More detail is given in Zar [39].

2.6. Analytical Framework

To analyze the differences in the performances of the eight model configurations, we used the following general approach:

(1) Run each configuration before calibration and calculate the model efficiency criterion, bR^2, [28] for the nine discharge outlets. Examining model performance based on default parameters (Table 3) is important in determining how the model should be calibrated and which parameters should be adjusted [17]. Harmel et al. [40] also defined "initial evaluation of model performance" as the first step to make the best judgment to guide model refinement. If important processes or key input information are neglected, then the model should not be calibrated, because wrong and meaningless parameters will be obtained. Furthermore, comparison of the pre- and post-calibrated parameter ranges (uncertainties) indicates the information content of the variable(s) used to calibrate the model. If we achieve a large reduction in the parameter uncertainties, then the variable(s) used to calibrate the model (as they appear in the objective function) have high information content.

(2) Calibrate each configuration in the same way against the monthly observed river discharges. Then compare the efficiency criteria from after calibration with those from before.

(3) Perform a multiple comparison significance test [39] on non-calibrated and calibrated configurations to identify configurations that are significantly different or similar to each other in terms of bR^2 efficiency criteria and classify them into three classes (Class1 with high performance, Class2 with medium performance, Class3 with low performance). The selection of the number of classes and the classification were based on the null hypothesis and pair-wise comparison of the configurations. We started with C1L1 and made a pairwise comparison with the remaining seven models. Those that were significantly different from C1L1 were taken out of this class. Now, all other members of C1L1 except C1L1 were compared with each other pairwise. The set that was similar with C1L1, but was different from the others was also taken out of this class. We continued this until all members of a group were not significantly different in a pair-wise comparison. We repeated this process for the configurations that were not in the first class.

(4) Calculate and compare the annual WY, BW, SW, and ET for each model using calibrated parameter ranges obtained in the 480 simulations at the sub-basin level. The components were then aggregated to the entire watershed level using the weighted area average method.

(5) Calculate and quantify the uncertainties of the water resource components WY, BW, SW, and ET resulting from the different configurations using the coefficient of variation (%CV).

3. Results and Discussion

3.1. Model Performance and Parameters

Initial evaluation of the different configurations based on the initial parameter values (Table 3) showed significant differences in their performance compared to each other (Figure 3a). The efficiency criteria (bR^2) of all the configurations except C3L1 and C3L2 indicated that they could be improved by calibration. Looking at the hydrographs and the bR^2 values of the nine outlets and also at the climate stations that furnished the rainfall in their respective sub-basins, we noticed that C2L1 had higher bR^2 in outlets O1–O3 (0.23, 0.11, and 0.31 respectively) than C1L1 (0.12, 0.08, and 0.15 respectively). We saw the same patterns when we compared C1L2 with C2L2. We therefore constructed C4 (Figure 2d) by combining the better performing climate stations from C1 and C2 (Figure 2d). To statistically compare the eight configurations, a significance test was performed and the configurations were classified into Class1, Class2, and Class3 based on the average bR^2 of nine outlets. For the pre-calibration runs, C4L1 and C4L2 fell in Class1, C3L1 and C3L2 in Class3, and the other four in Class2 (Figure 3b). While bR^2 was used to calculate the objective function and model classification, we also computed the average Nash–Sutcliffe efficiency (NS) values of nine outlets [41] as a supplemental reference for the evaluation of all configurations which also showed relatively low values (Table 4).

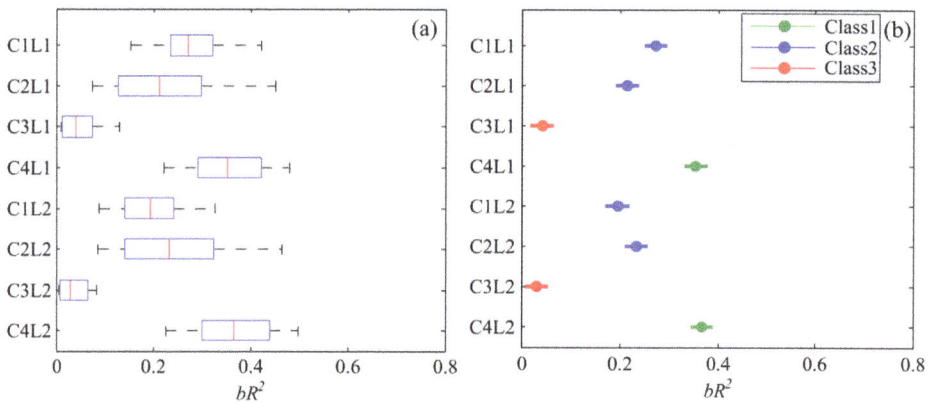

Figure 3. (**a**) The performance of the eight configurations in simulating discharge before calibration (single model run). Red lines show the average bR^2 obtained from the nine outlets and the boxes show the 25th and 75th percentiles and the whiskers show the maximum and minimum. (**b**) The three performance classes obtained from the multiple comparison significance test before calibration. The dots show the average bR^2 and the ranges indicate the standard error.

After calibration, the eight configurations showed significant improvement as indicated by bR^2 (Figure 4a) compared to pre-calibration results (Figure 3a). Similar to pre-calibration, configurations of the same climate datasets in the two different land use maps fell in the same class after calibration (Figure 4b). This indicates the insensitivity of land use to discharge in our case, which also corroborates the conclusion of Yen et al. [11] who found the same level of performance with different land uses after calibration. In our region, it could also mean that the land use maps were not too different from each other for most classes except shrub land and forest, which comprised about 30% of KRB, and urban areas (with about 8.5%), and that their influence on discharge were not significant.

Overall, C1L1 and C1L2 showed the best performance. One can see that C1 with the fewest number of climate stations (Figure 2a) performed better in combination with both land uses. SWAT assigns to each sub-basin climate data from the nearest station. The C1 climate stations better represented the entire basin albeit with fewer stations (Figures S1 and S2). Hence, in this example, the number of climate stations did not seem to have as important impact as the quality of the data in them.

C3L1 and C3L2 did not have satisfactory performance before and after calibration. This indicates the poor quality of the C3 climate database, which was generated with the bias-corrected GFDL-ESM2M (Geophysical Fluid Dynamics Laboratory of national oceanic and atmospheric administration—Earth System Model) GCM model on 0.5° grid resolution for KRB. This suggests that the measured climate data at a river basin level, which usually suffer from missing values and other quality problems, still performed better for this region than the estimated global gridded data. Looking at the spatial distribution of seasonal precipitation and temperature (Figures S1 and S3), one could see that there is no significant difference among different climate datasets, however the temporal variability (yearly precipitation) is noticeable (Figure S2). Many studies have assessed the impacts of gridded data for simulating runoff [14,42,43]. The results showed that their quality vary significantly from one region to the other. For example, Vu et al. [42] showed poor performance of the PERSIANN (precipitation estimation from remotely sensed information using artificial neural networks) and TRMM (tropical rainfall measuring mission) rainfall data compared to the station data.

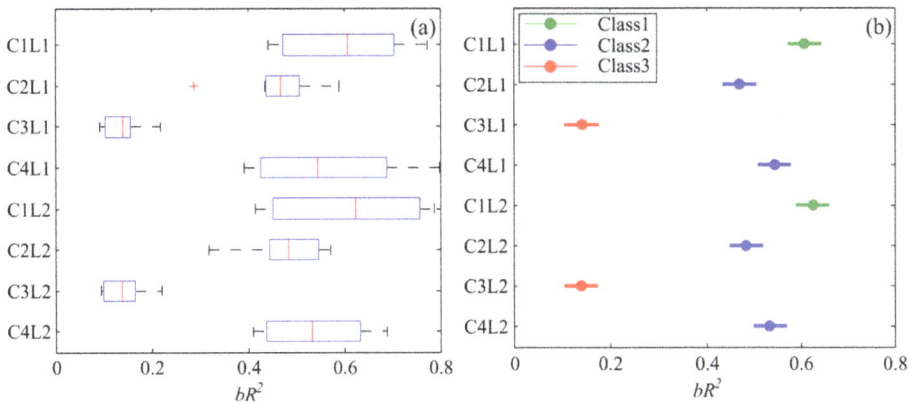

Figure 4. (**a**) The performance of the eight configurations in simulating discharge based on the best simulation after calibration. The red lines show the average bR^2 obtained from the nine outlets, the boxes show the 25th and 75th percentiles and the whiskers show the maximum and minimum. (**b**) The three performance classes obtained by the multiple comparison significance test after calibration. The dots show the average bR^2 values and the ranges indicate the standard error.

C4L1 and C4L2, which performed best before calibration, did not improve as significantly as C1L1 and C1L2 after calibration and fell in Class2. This indicates that selection of the best performing climate stations based on checking their performance prior to calibration might not work after calibration. We noticed that the initial performance of O1–O3 was low in C1L1 and C1L2 compared to C2L1 and C2L2. However, apparently this was related to the inaccuracy of the initial parameter values. After parameter adjustment, they outperformed other configurations. The average *NS* values of the nine outlets for calibrated configurations in Class1 are above 0.60, indicating good model performance (Figure 4b). Configuration models in Class2 have slightly lower *NS*, especially in C2L2. The reason is *NS* varies between $-\infty$ to 1, hence, one outlet with rather lower *NS* can lower the average *NS* of the basin. Configurations C3L1 and C3L2 had negative *NS* values, indicating very poor model performance.

The discharge hydrographs of the different configurations are shown in Figure 5 for the outlet O7 as an example, with the other outlets shown in the supplementary material (Figures S4–S11). As shown, more than 50% of the observed discharges are within the 95PPU bands depicted with green shades in all configurations, except C3L1 and C3L2, where significant overestimations can be noticed, especially after 2000. A significant decrease is recorded in the observed values in the region at the end of the period due to severe droughts occurring after 2000 [44]. This can also be noticed in the temporal variation of rainfall for these datasets (Figure S2). All configurations except C3L1 and C3L2 could capture such extreme situations.

The *p-factor* values for the calibration and validation periods were larger than 0.50 for configurations of Class1 and Class2, indicating that more than 50% of the observed data were bracketed by the 95PPU bands (Table 4). For these two classes, the *r-factor* was smaller than 1.5 during calibration and validation, indicating reasonable prediction uncertainties.

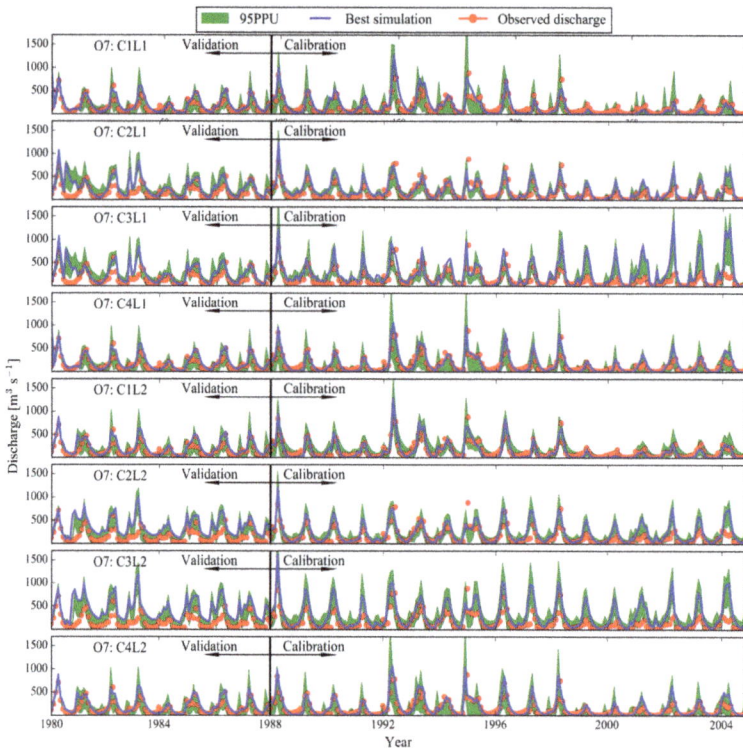

Figure 5. Comparison of simulated and observed discharge values in the O7 outlet (Figure 1) during the calibration and validation periods. The green shaded region is the 95PPU band. The best simulation (i.e., the simulation with the highest bR^2) is shown by the blue line.

Table 4. The performance of the eight configurations during the calibration and validation periods.

Configuration	Calibration Period 1988–2004			Validation Period 1980–1987		
	NS	p-factor	r-factor	NS	p-factor	r-factor
C1L1 (Class1)	0.60	0.68	1.19	0.61	0.59	1.32
C2L1 (Class2)	0.51	0.54	1.23	0.50	0.52	1.39
C3L1 (Class3)	−3.5	0.41	1.77	−0.5	0.25	1.05
C4L1 (Class2)	0.49	0.64	1.12	0.51	0.58	1.36
C1L2 (Class1)	0.62	0.71	1.37	0.60	0.67	1.50
C2L2 (Class2)	0.46	0.54	1.47	0.48	0.50	1.27
C3L2 (Class3)	−1.69	0.37	0.60	−1.75	0.38	1.32
C4L2 (Class2)	0.51	0.65	1.15	0.53	0.60	1.27

After calibration, each parameter attained a different range (Figure 6). Yang et al. [45] showed that different optimization algorithms lead to differently-calibrated parameter ranges. Here, it is seen that different existing input datasets also lead to differently-calibrated parameter ranges for the same region. This highlights the problem of the "conditionality" of calibrated models which is caused by the multimodality of the response surface of the objective function as discussed by Abbaspour [46]. Overall, the ranges of CN2 are relatively similar in all configurations (except C1L2). Similar patterns for CN2 were found in the study of Strauch et al. [19] in the Pipiripau River in Central Brazil where the fitted values of CN2 were relatively similar for all rainfall input models.

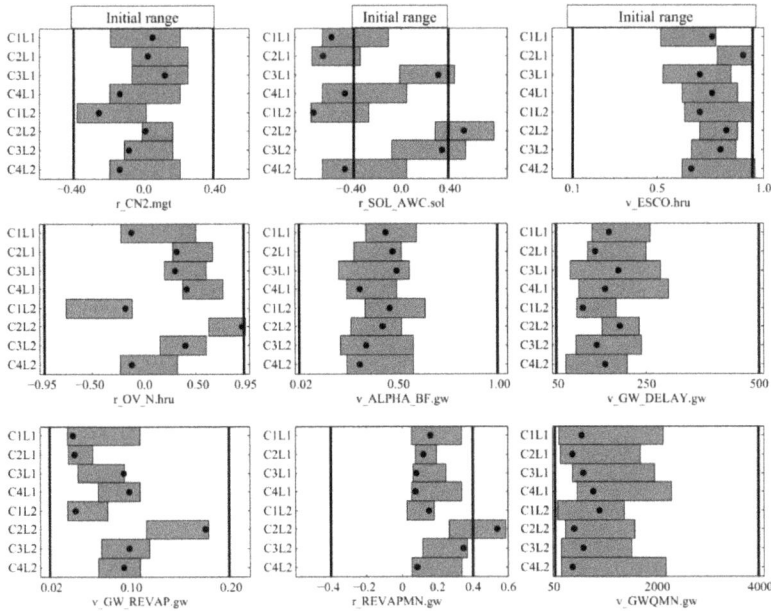

Figure 6. The initial (vertical lines) and final ranges (grey bars) of the parameters considered in the calibration. The dots show the best parameter sets based on the best value of the objective function. "v_" indicates an absolute change where the initial parameter value is replaced by another value. "r_" indicates a relative change where the initial parameters are multiplied by (1 + a given value).

The C1L1 and C1L2 configurations have statistically the same calibration results. However, it is important to note that the CN2 of these configurations have different ranges from each other, indicative of different hydrological processes in the region which are explained by the parameters. For example, while the "best" relative value of CN2 (e.g., the value of CN2 where the objective function is maximum) for C1L2 was −0.25, it was 0.05 for C1L1. The actual CN2 values for sub-basin #35 (obtained from average CN2 of all HRU in this sub-basin) as an example were 53 and 78, respectively. The latter represents a surface-runoff-dominated system, while the former is an infiltration-dominated system. Generally, non-uniqueness in the domain of parameters is an important problem in the calibration of distributed models [47]. This can partly be resolved by better understanding of the watershed hydrology leading to constraining ranges of parameters in the objective function.

Other parameters showed larger variations among different configurations in terms of both ranges and best fitted values (Figure 6). For example, significant variability is found among the ranges and best fitted values of GW_REVAP (groundwater parameters), indicating that different configurations attempt to fit differently. Our objective function is based on comparing observed and simulated discharge values and contains no measured variables that directly explain the status of groundwater processes. Therefore, a high degree of uncertainty remained in the model relative to groundwater parameters. Overall, we found that there is a high degree of parameter uncertainty, which would not be apparent if only a single dataset was used. Use of different data sources adds a new dimension to the existing category of "input data uncertainty" which mostly stem from potential errors in data collection or incomplete data.

3.2. Estimation of Water Resource Variables

The water resource components simulated with the calibrated model configurations show quite different behaviors for WY, BW, SW, and ET (Figure 7a–d) due to differences in the parameters of each configuration. Large ranges of values were obtained for different variables. For example, WY varies between 80–270 mm year^{-1} and SW varies between 30–58 mm year^{-1}. We do not have observed records for variables such as SW and ET to judge their reliability, but we can comment on their differences in different configurations. For example, C1L1 from Class1 and C3L1 from Class3 have approximately the same SW values, whereas SW based on C4L1 from Class1 is more similar to the models of Class2.

BW and WY had similar patterns of classification i.e., models with similar WY also showed similar BW. This similarity between BW and WY is related to the "deep aquifer percolation fraction (DAP)" parameter assumed to be 5% for arid regions like KRB. BW is obtained from the summation of WY and DAP. In this paper, DAP was a constant fraction which was significantly smaller than WY. Therefore, BW and WY showed similar patterns. ET is mostly influenced by temperature. Therefore, it shows similar values in models of the same climate datasets.

Another observation is that in all models and for all water resource components except SW, land use seems to have no significant impact e.g., C4L1 and C4L2 produced approximately the same results for WY, BW, and ET. Likewise, WY and BW of C1L1 and C2L1 were slightly smaller than C1L2 and C2L2, respectively, and ET of C1L1 and C2L were slightly larger than C2L1 and C2L2. Models of Class2 also showed similar results with respect to different land uses, indicating that in this work, land use is not an important factor in water resource components, with the exception of the soil water. This might be partially due to approximately similar percentage of areas allocated to each land use class in the L1 and L2 maps for most classes, except shrub land and forest, which were less than 30% different (Table 2), resulting in variability mostly in SW. Besides, while the spatial distribution of each class indicates some differences (Figure 2e,f), the percentage remained similar.

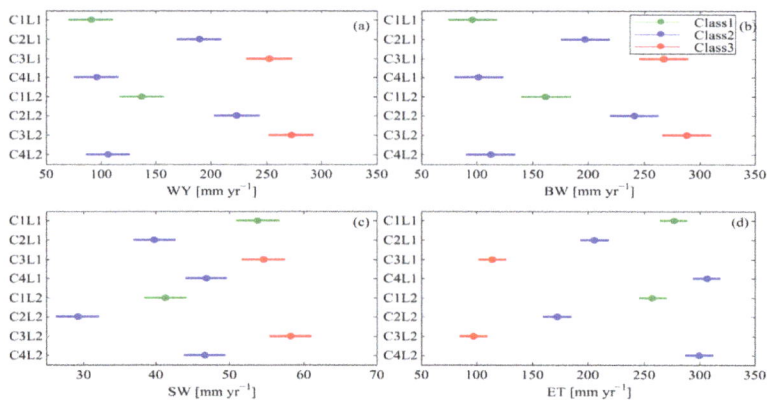

Figure 7. Range of four water resources components; (**a**) WY = water yield; (**b**) BW = blue water; (**c**) SW = soil water; (**d**) ET = evapotranspiration obtained from eight calibrated configurations during the studied period. The three colors identify configurations with high (Class1: green), medium (Class2: blue), and low (Class3: red) performance in simulating discharge values as displayed in Figure 4b.

3.3. Uncertainty in Water Resource Variables

Next, we investigated the uncertainty in the water resource components by calculating the coefficient of variation (%CV) using outputs from the eight calibrated configurations in 480 simulations. As illustrated in Figure 8, the uncertainty due to multiple input datasets is larger for ET and SW than for BW and WY. The median value of CV is 45% for SW and approximately 46% for ET. For WY and BW, this is about 31.5% and 32%, respectively. WY is directly related to the river discharge used in the objective function definition, but SW and ET are not directly adjusted based on observed data, therefore their estimates contain larger uncertainties.

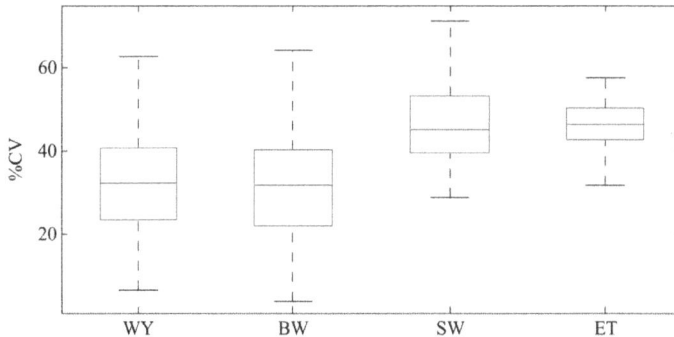

Figure 8. Comparison of the uncertainty in the water resource components stemming from the use of eight different input datasets. The boxplot shows the 25th and 75th percentiles of coefficient of variation (%CV) obtained from 480 simulations and the whiskers show the maximum and minimum %CV.

4. Conclusions

Different input datasets usually exist for modeling the hydrology of a watershed. As analysts usually consider only one database in their analysis, the uncertainty due to multiple existing databases goes unnoticed. Our findings here are based on model configurations built with different climate data and land use maps and calibrated against nine outlets using bR^2 as the objective function. All calibrated models were compared to each other in terms of simulating different components of water resources. The following points were highlighted in this research:

(i)   Multiple model configurations built for a region with datasets coming from different sources produce significantly different parameter sets after calibration, albeit with similar calibration results.

(ii)   Subsequently, water resource components are significantly different for different configurations, resulting in large model output uncertainties.

(iii) Discharge prediction seems to be less sensitive to different land uses, which is the same conclusion made by Yen et al. [11]. Additionally, the present study pointed to the impact of both land use and climate data on different components of water resources, such as SW and ET.

(iv)   The uncertainty is larger for SW and ET compared to WY. Decreasing uncertainty for these components relies on observed records data.

Our findings, therefore, highlight a significant level of uncertainty in modeling results stemming from uncertain data inputs (used in models) for a region. Ajami et al. [8] state that neglecting different aspects of uncertainty during the calibration of hydrological models may result in inconsistent outputs. We hence emphasize that it may be prudent for modelers to pay more attention to the existence of uncertainty from multiple sources of data (especially climate data) in combination with other sources of uncertainty such as spatial data resolution [48], objective functions, or optimization algorithms [38]. We also suggest that the calibration of models against more observed variables such as evapotranspiration or soil moisture may help to select better models. It is worthy to note that local decision makers and engineers should compromise between the expected accuracy of the model and the time and resources invested in data collection and assimilation.

Supplementary Materials: The following are available online at www.mdpi.com/2073-4441/9/9/709/s1.

Acknowledgments: The authors express their sincere thanks to Eawag Partnership Program (EPP) for the opportunity of a collaboration.

Author Contributions: Bahareh Kamali and Karim C. Abbaspour prepared the SWAT model. Bahareh Kamali designed the methodology framework under supervision of Karim C. Abbaspour and Hong Yang. Karim C. Abbaspour, Hong Yang advised on conceptual and technical, and contributed to the strategy.

Conflicts of Interest: The authors declare no conflict of interest.

References

1. Kalantari, Z.; Lyon, S.W.; Jansson, P.E.; Stolte, J.; French, H.K.; Folkeson, L.; Sassner, M. Modeller subjectivity and calibration impacts on hydrological model applications: An event-based comparison for a road-adjacent catchment in South-East Norway. *Sci. Total Environ.* **2015**, *502*, 315–329. [CrossRef] [PubMed]
2. Shrestha, R.; Tachikawa, Y.; Takara, K. Input data resolution analysis for distributed hydrological modeling. *J. Hydrol.* **2006**, *319*, 36–50. [CrossRef]
3. Montanari, A.; Di Baldassarre, G. Data errors and hydrological modelling: The role of model structure to propagate observation uncertainty. *Adv. Water Resour.* **2013**, *51*, 498–504. [CrossRef]
4. Orth, R.; Staudinger, M.; Seneviratne, S.I.; Seibert, J.; Zappa, M. Does model performance improve with complexity? A case study with three hydrological models. *J. Hydrol.* **2015**, *523*, 147–159.
5. Martina, M.L.V.; Todini, E. Watershed hydrological modeling: Toward physically meaningful processes representation. *Water Sci. Technol. Libr.* **2008**, *63*, 229–241.
6. Confesor, R.B.; Whittaker, G.W. Automatic calibration of hydrologic models with multi-objective evolutionary algorithm and Pareto optimization. *J. Am. Water Resour. Assoc.* **2007**, *43*, 981–989. [CrossRef]
7. Kim, S.M.; Benham, B.L.; Brannan, K.M.; Zeckoski, R.W.; Doherty, J. Comparison of hydrologic calibration of HSPF using automatic and manual methods. *Water Resour. Res.* **2007**, *43*. [CrossRef]
8. Ajami, N.K.; Duan, Q.Y.; Sorooshian, S. An integrated hydrologic Bayesian multimodel combination framework: Confronting input, parameter, and model structural uncertainty in hydrologic prediction. *Water Resour. Res.* **2007**, *43*. [CrossRef]
9. Faramarzi, M.; Srinivasan, R.; Iravani, M.; Bladon, K.D.; Abbaspour, K.C.; Zehnder, A.J.B.; Goss, G.G. Setting up a hydrological model of Alberta: Data discrimination analyses prior to calibration. *Environ. Model. Softw.* **2015**, *74*, 48–65. [CrossRef]
10. Bennett, N.D.; Croke, B.F.W.; Guariso, G.; Guillaume, J.H.A.; Hamilton, S.H.; Jakeman, A.J.; Marsili-Libelli, S.; Newham, L.T.H.; Norton, J.P.; Perrin, C.; et al. Characterising performance of environmental models. *Environ. Model. Softw.* **2013**, *40*, 1–20. [CrossRef]
11. Yen, H.; Sharifi, A.; Kalin, L.; Mirhosseini, G.; Arnold, J.G. Assessment of model predictions and parameter transferability by alternative land use data on watershed modeling. *J. Hydrol.* **2015**, *527*, 458–470. [CrossRef]
12. Cotter, A.S.; Chaubey, I.; Costello, T.A.; Soerens, T.S.; Nelson, M.A. Water quality model output uncertainty as affected by spatial resolution of input data. *J. Am. Water Resour. Assoc.* **2003**, *39*, 977–986. [CrossRef]
13. Yen, H.; Su, Y.W.; Wolfe, J.E.; Chen, S.T.; Hsu, Y.C.; Tseng, W.H.; Brady, D.M.; Jeong, J.; Arnold, J.G. Assessment of input uncertainty by seasonally categorized latent variables using SWAT. *J. Hydrol.* **2015**, *531*, 685–695. [CrossRef]
14. Biemans, H.; Hutjes, R.W.A.; Kabat, P.; Strengers, B.J.; Gerten, D.; Rost, S. Effects of precipitation uncertainty on discharge calculations for main river basins. *J. Hydrometeorol. JHM* **2009**, *10*, 1011–1025. [CrossRef]
15. Zhang, P.P.; Liu, R.M.; Bao, Y.M.; Wang, J.W.; Yu, W.W.; Shen, Z.Y. Uncertainty of SWAT model at different DEM resolutions in a large mountainous watershed. *Water Res.* **2014**, *53*, 132–144. [CrossRef] [PubMed]
16. Rouholahnejad, E.; Abbaspour, K.C.; Srinivasan, R.; Bacu, V.; Lehmann, A. Water resources of the Black Sea Basin at high spatial and temporal resolution. *Water Resour. Res.* **2014**, *50*, 5866–5885. [CrossRef]

17. Abbaspour, K.C.; Rouholahnejad, E.; Vaghefi, S.; Srinivasan, R.; Yang, H.; Klove, B. A continental-scale hydrology and water quality model for Europe: Calibration and uncertainty of a high-resolution large-scale SWAT model. *J. Hydrol.* **2015**, *524*, 733–752. [CrossRef]

18. Monteiro, J.A.F.; Strauch, M.; Srinivasan, R.; Abbaspour, K.; Gücker, B. Accuracy of grid precipitation data for Brazil: Application in river discharge modelling of the Tocantins catchment. *Hydrol. Process.* **2015**, *30*, 1419–1430. [CrossRef]

19. Strauch, M.; Bernhofer, C.; Koide, S.; Volk, M.; Lorz, C.; Makeschin, F. Using precipitation data ensemble for uncertainty analysis in SWAT streamflow simulation. *J. Hydrol.* **2012**, *414*, 413–424. [CrossRef]

20. Getirana, A.C.V.; Espinoza, J.C.V.; Ronchail, J.; Rotunno, O.C. Assessment of different precipitation datasets and their impacts on the water balance of the Negro River basin. *J. Hydrol.* **2011**, *404*, 304–322. [CrossRef]

21. Oweis, T.; Siadat, H.; Abbasi, F. *Improving On-Farm Agricultural Water Productivity in the Karkheh River Basin (KRB)*; CPWF Project Report-Project Number 08: CGIAR Challenge Program on Water and Food; Department for International Development: Chatham, UK, 2009.

22. Ahmad, M.U.D.; Giordano, M. The Karkheh river basin: The food basket of Iran under pressure. *Water Int.* **2010**, *35*, 522–544. [CrossRef]

23. Jarvis, A.; Reuter, H.; Nelson, A.; Guevara, E. Hole-Filled SRTM for the Globe Version 4. CGIAR-CSI SRTM 90 m Database. 2008. Available online: http://srtm.csi.cgiar.org (accessed on 15 January 2008).

24. Masih, I. *Understanding Hydrological Variability for Improved Water Management in the Semi-Arid Karkheh Basin, Iran*; CRC Press/Balkema: Leiden, The Netherlands, 2011.

25. Saghafian, B.; Davtalab, R. Mapping snow characteristics based on snow observation probability. *Int. J. Climatol.* **2007**, *27*, 1277–1286. [CrossRef]

26. Arnold, J.G.; Srinivasan, R.; Muttiah, R.S.; Williams, J.R. Large area hydrologic modeling and assessment—Part 1: Model development. *J. Am. Water Resour. Assoc.* **1998**, *34*, 73–89. [CrossRef]

27. Neitsch, S.L.; Arnold, J.G.; Kiniry, J.R.; Williams, J.R.; King, K.W. *Soil and Water Assessment Tool. Theoretical Documentation: Version 2009*; Texas Water Resources Institute: College Station, TX, USA, 2005.

28. Abbaspour, K.C.; Yang, J.; Maximov, I.; Siber, R.; Bogner, K.; Mieleitner, J.; Zobrist, J.; Srinivasan, R. Modelling hydrology and water quality in the pre-ailpine/alpine Thur watershed using SWAT. *J. Hydrol.* **2007**, *333*, 413–430. [CrossRef]

29. Krause, P.; Boyle, D.P.; Bäse, F. Comparison of different efficiency criteria for hydrological model assessment. *Adv. Geosci.* **2005**, *5*, 89–97. [CrossRef]

30. Monteiro, J.A.F.; Kamali, B.; Srinivasan, R.; Abbaspour, K.C.; Gücker, B. Modelling the effect of riparian vegetation restoration on sediment transport in a human-impacted Brazilian catchment. *Ecohydrology* **2016**, *9*, 1289–1303. [CrossRef]

31. Schuol, J.; Abbaspour, K.C.; Srinivasan, R.; Yang, H. Estimation of freshwater availability in the West African sub-continent using the SWAT hydrologic model. *J. Hydrol.* **2008**, *352*, 30–49. [CrossRef]

32. Ministry of Energy of Iran. *An Overview of National Water Planning of Iran*; Ministry of Energy of Iran: Tehran, Iran, 1998. (In Persian)

33. Taylor, K.E.; Stouffer, R.J.; Meehl, G.A. An overview of CMIO5 and the experiment design. *Bull. Am. Meteorol. Soc.* **2012**, *93*, 485–498. [CrossRef]

34. USGS. *Global Land Cover Characterization*; United State Geological Survey: Washington, DC, USA, 1997.

35. Iran Water and Power Resources Development Company (IWPCO). *Systematic Planning of Karkheh Watershed; Land Use Studies*; Iranian Ministry of Energy: Tehran, Iran, 2009. (In Persian)

36. Iran Water and Power Resources Development Company (IWPCO). *Systematic Studies of Karkheh River Basin*; Iranian Ministry of Energy: Tehran, Iran, 2010. (In Persian)

37. Vaghefi, S.A.; Mousavi, S.J.; Abbaspour, K.C.; Srinivasan, R.; Yang, H. Analyses of the impact of climate change on water resources components, drought and wheat yield in semiarid regions: Karkheh river basin in Iran. *Hydrol. Process.* **2014**, *28*, 2018–2032. [CrossRef]

38. Kouchi, D.H.; Esmaili, K.; Faridhosseini, A.; Sanaeinejad, S.H.; Khalili, D.; Abbaspour, K.C. Sensitivity of calibrated parameters and water resource estimates on different objective functions and optimization algorithms. *Water* **2017**, *9*, 384. [CrossRef]

39. Zar, J.H. Statistical procedures for biological-research-a citation classic commentary on biostatistical analysis. *Agric. Biol. Environ. Sci.* **1989**, *6*, 1–20.

40. Harmel, R.D.; Smith, P.K.; Migliaccio, K.W.; Chaubey, I.; Douglas-Mankin, K.R.; Benham, B.; Shukla, S.; Munoz-Carpena, R.; Robson, B.J. Evaluating, interpreting, and communicating performance of hydrologic/water quality models considering intended use: A review and recommendations. *Environ. Model. Softw.* **2014**, *57*, 40–51. [CrossRef]

41. Nash, J.E.; Sutcliffe, J.V. River flow forecasting through conceptual models part I—A discussion of principles. *J. Hydrol.* **1970**, *10*, 282–290. [CrossRef]

42. Vu, M.T.; Raghavan, S.V.; Liong, S.Y. SWAT use of gridded observations for simulating runoff—A Vietnam river basin study. *Hydrol. Earth Syst. Sci.* **2012**, *16*, 2801–2811. [CrossRef]

43. Thom, V.T.; Khoi, D.N.; Linh, D.Q. Using gridded rainfall products in simulating streamflow in a tropical catchment—A case study of the Srepok River Catchment, Vietnam. *J. Hydrol. Hydromech.* **2017**, *65*, 18–25. [CrossRef]

44. Kamali, B.; Houshmand Kouchi, D.; Yang, H.; Abbaspour, K.C. Multilevel drought hazard assessment under climate change scenarios in semi-srid regions—A case study of the Karkheh River Basin in Iran. *Water* **2017**, *9*, 241. [CrossRef]

45. Yang, J.; Reichert, P.; Abbaspour, K.C.; Xia, J.; Yang, H. Comparing uncertainty analysis techniques for a SWAT application to the Chaohe Basin in China. *J. Hydrol.* **2008**, *358*, 1–23. [CrossRef]

46. Abbaspour, K.C. Calibration of hydrologic models: When is a model calibrated? In Proceedings of the Modsim 2005: International Congress on Modelling and Simulation: Advances and Applications for Management and Decision Making, Melbourne, Australia, 12–15 December 2005; pp. 2449–2455.

47. Bardossy, A. Calibration of hydrological model parameters for ungauged catchments. *Hydrol. Earth Syst. Sci.* **2007**, *11*, 703–710. [CrossRef]

48. Chapiot, V. Impact of spatial input data resolution on hydrological and erosion modeling: Recommendations from a global assessment. *Phys. Chem. Earth* **2014**, *67–69*, 23–35. [CrossRef]

water

MDPI

Article

Multilevel Drought Hazard Assessment under Climate Change Scenarios in Semi-Arid Regions—A Case Study of the Karkheh River Basin in Iran

Bahareh Kamali [1,2,*], Delaram Houshmand Kouchi [1,3], Hong Yang [1,4] and Karim C. Abbaspour [1]

[1]   Eawag, Swiss Federal Institute of Aquatic Science and Technology, 8600 Duebendorf, Switzerland;
      delaram.houshmandkouchi@mail.um.ac.ir (D.H.K.); hong.yang@eawag.ch (H.Y.);
      karim.abbaspour@eawag.ch (K.C.A.)
[2]   ETH Zürich, Department of Environmental Systems Science, Universitätstr. 16, 8092 Zürich, Switzerland
[3]   Department of Water Engineering, Ferdowsi University of Mashhad, Mashhad 9177948974 AzadiSq,
      Khorasan Razavi, Iran
[4]   Department of Environmental Sciences, University of Basel, 4003 Basel, Switzerland
*   Correspondence: bahareh.kamali@eawag.ch; Tel.: +41-58-765-5358

Academic Editor: Athanasios Loukas
Received: 17 February 2017; Accepted: 24 March 2017; Published: 30 March 2017

Abstract: Studies using Drought Hazard Indices (*DHIs*) have been performed at various scales, but few studies associated *DHIs* of different drought types with climate change scenarios. To highlight the regional differences in droughts at meteorological, hydrological, and agricultural levels, we utilized historic and future *DHIs* derived from the Standardized Precipitation Index (*SPI*), Standardized Runoff Index (*SRI*), and Standardized Soil Water Index (*SSWI*), respectively. To calculate *SPI*, *SRI*, and *SSWI*, we used a calibrated Soil and Water Assessment Tool (SWAT) for the Karkheh River Basin (KRB) in Iran. Five bias-corrected Global Circulation Models (GCMs) under two Intergovernmental Panel on Climate Change (IPCC) scenarios projected future climate. For each drought type, we aggregated drought severity and occurrence probability rate of each index into a unique *DHI*. Five historic droughts were identified with different characteristics in each type. Future projections indicated a higher probability of severe and extreme drought intensities for all three types. The duration and frequency of droughts were predicted to decrease in precipitation-based *SPI*. However, due to the impact of rising temperature, the duration and frequency of *SRI* and *SSWI* were predicted to intensify. The *DHI* maps of KRB illustrated the highest agricultural drought exposures. Our analyses provide a comprehensive way to monitor multilevel droughts complementing the existing approaches.

Keywords: SWAT; drought hazard index; future drought projection

1. Introduction

Drought is a natural hazard with adverse impacts on water resources, agriculture, and the environment [1–3]. In the literature, it is defined as a recurring prolonged dry period, which affects different components of the hydrological process [4]. Drought is a complex phenomenon that is difficult to quantify. This is because its characterization relies on different components of the water cycle; drought impacts evolve over time, so it is time-dependent. Climate change is likely to shift the patterns of drought and exacerbate the frequency and intensity of drought events in the foreseeable future. Therefore, a more comprehensive insight to drought should simultaneously take into account: (1) different components of the hydrological cycle and their interactions; (2) drought features in spatial and temporal domains using aggregation methods; and (3) future changes of components under projected climate change scenarios. Existing literatures mostly look at only one or two of the abovementioned aspects. Despite its significance for effective regional drought management, considering all perspectives together using a standardized procedure has not been well documented so far.

Depending on the scope, drought has been classified into meteorological, agricultural, hydrological, and socioeconomic categories [5,6]. The first three types of droughts reflect the physical characteristics of a drought phenomenon (namely physical drought). Socioeconomic drought is concerned with the water shortfall whose impact ripples through socioeconomic systems [7]. Although all types of droughts originate from a deficiency of precipitation [5], hydrological drought is usually out of phase with or lags behind the occurrence of a meteorological drought [8]. This is mainly because it takes some time before precipitation shortfall emerges in different subsurface components of the hydrological system, such as soil moisture, groundwater, and streams [8].

In order to alleviate the expected impacts of droughts, decision makers need to monitor drought using timely and reliable indices on both spatial and temporal scales. A common measurement tool used for this purpose is drought indices, which are believed to be more functional than raw precipitation or runoff variables for evaluating spatial and temporal characteristics of drought [9]. The Standardized Precipitation Index (*SPI*) [10] is broadly applied to monitor meteorological droughts [11–13]. Meteorological drought indices have been evaluated together with hydrological and agricultural indices to gain a broader understanding of drought propagation through the hydrological cycle (here called multilevel drought assessment). Hisdal et al. [14] assessed meteorological and hydrological droughts in Denmark on a regional scale and found that hydrological drought is less frequent, more persistent, and less homogeneous compared to meteorological droughts. Liu et al. [15] characterized drought propagation in groundwater systems using a standardized groundwater level index and *SPI*, showing that groundwater drought lasts longer with higher intensity. Tallaksen et al. [16] explored drought propagation in hydrology by looking at precipitation, groundwater recharge, hydraulic head, and river discharge in a groundwater-fed catchment in UK. Tadesse et al. [17], Vidal et al. [1], Tokarczyk et al. [18], and Duan et al. [19] found that drought impacts can be seen differently in each type, and more importantly, in the different affected regions. As such, their findings explain the reason for developing a comprehensive drought monitoring model for different types of droughts to give decision-makers detailed information on drought characteristics.

Drought has been inevitably interwoven with climate change impacts. Central to this concern is whether drought will become more frequent, severe, and widespread in the coming decades or not [20–22]. Water resource management to mitigate drought risks relies on understanding future characteristics such as the degree of severity, probability of occurrence, frequency, and duration of expected droughts [23–25]. Many researchers have projected occurrences of droughts under future climate scenarios by using Global Circulation Models (GCMs) [2]. Lee et al. [26] analyzed climate change impacts on different characteristics of drought in the Seoul region using four GCMs and reported a decrease in mild drought frequency, but an increase in the frequency of severe and extreme droughts._ENREF_18 Leng et al. [27] assessed the climate change impact on biophysical droughts using daily climate projections under five GCMs with the RCP8.5 (Representative Concentration Pathways) scenario in China. Their findings confirmed that meteorological, agricultural, and hydrological droughts will variably occur on different temporal and spatial scales. Liu et al. [28] used *SPI*, Standardized Runoff Index (*SRI*), and Palmer Drought Severity Index (*PDSI*) to construct historical and future projection of drought patterns for the Blue River Basin in Oklahoma. Their results predicted more drought events in the future (2010–2099). They also recommended *PDSI* and *SRI* as the most functional indices for drought risk assessment.

Drought hazard is usually defined as an aggregation of the frequency, intensity, duration, and spatial extent of occurrences [29]. Despite the extensive research on multilevel drought identification using drought indices under historic and future conditions, fewer studies have focused on associating climate change scenarios with composite drought hazard indices of different drought types. This level of analysis has received even less attention in Iran's river basins with semi-arid climate. To fulfill this research demand, we examine the historic and future drought hazard using an ensemble of climate scenarios in the Karkheh River Basin (KRB) of Iran.

KRB is one of the nine watersheds studied in the CGIAR (Consultative Group on International Agricultural Research) Challenge Program on Water and Food (CPWF) [30]. The basin is one of the most agriculturally important areas in Iran, which produces about 10% of the country's wheat [31]. It is also an example of a dryland system with a wide spectrum of bio-physical and socio-economic conditions as well as complex agricultural problems. While the properties of drylands around the world can widely vary [32,33], lessons learned from the drought assessment of such a complex system can be useful in other catchments in terms of methodology and providing detailed insights on key elements required for assessing different aspects of drought. The standardized and holistic drought hazard assessment implemented in this study can be conducted in other basins to identify regions exposed to drought.

Most of the research studies conducted in KRB have concentrated on water resource allocation [31,34], variability assessment in one or two components of water cycle [35–37], historic meteorological and agricultural droughts [38], or future projection in one drought type [39]. None of these research studies have looked at drought hazard indices of three different types. There is also an apparent lack of implementation of hazard analyses considering historic and future perspectives. Such detailed analyses are an essential step toward evaluating drought vulnerability of agricultural and water resources sectors and help policymakers recognize threats to different sectors.

The current study was carried out in order to analyze characteristics and relationships among meteorological, agricultural, and hydrological droughts using Drought Hazard Index (*DHI*) derived from a Soil and Water Assessment Tool (SWAT) hydrologic model. In the sections that follow, we analyze drought characteristics such as severity, frequency, and duration using *SPI*, *SRI*, and Standardized Soil Water Index (*SSWI*) for historical (1980–2012) and near future (2020–2052) periods to identify drought hotspots in the region.

2. Materials and Methods

2.1. Study Area

KRB covers an area of 51,000 km^2. It is the third largest basin in Iran and the food basket of the country [40]. The basin is divided into three catchments: Northern Karkheh (NKRB), Central Karkheh (CKRB), and Southern Karkheh (SKRB) (Figure 1). The climate of KRB is mainly semi-arid with annual precipitation ranging from 150 mm in SKRB to 750 mm in NKRB [40]. A number of dams were built or have been proposed for construction for irrigation and hydropower purposes [41]. The Karkheh dam located in the most downstream part of the basin, was constructed in 2002 to provide irrigation to the dry and lowland plains and is the largest reservoir in the basin (Figure 1) [41]. The Seymareh dam, the second most notably multipurpose dam, is under construction and is expected to be completed by 2025 [41].

KRB uses a rainfed production system in areas upstream of the Karkheh dam. The upper basin is dominated by pasture and scattered and sparse forest, which has been converted into rainfed and partially irrigated agriculture [41]. In recent years, groundwater has been excessively used for irrigation purposes. In contrast, SKRB is mostly under an irrigated production system (71% irrigated and 29% rainfed), but the amount of precipitation does not fulfill crop water requirements [30,41]. Wheat is the dominant crop, especially in rainfed condition. Other cultivated crops are chickpea, barley, and maize [30].

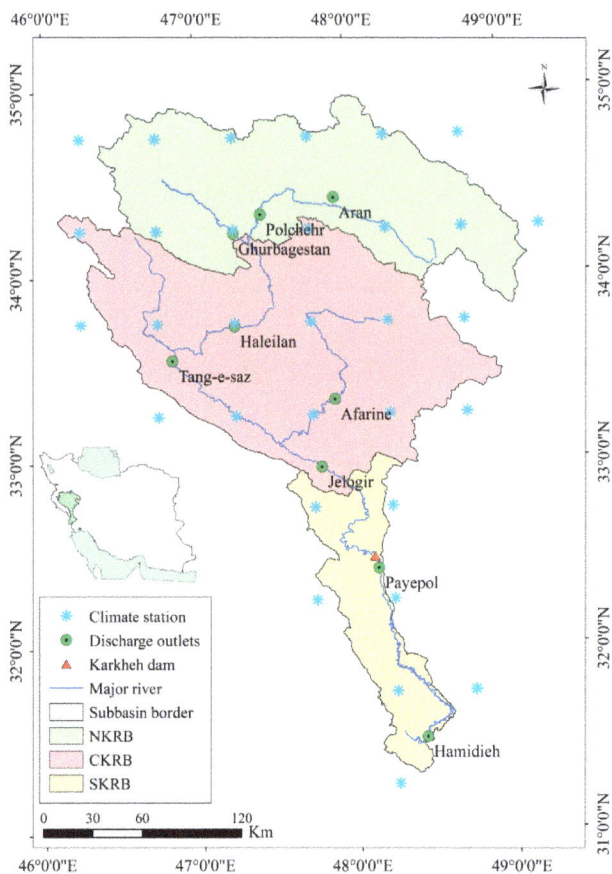

Figure 1. The Karkheh River Basin (KRB) and the three major catchments (Northern Karkheh (NKRB), Central Karkheh (CKRB), and Southern Karkheh (SKRB)). The figure shows the main river, Karkheh dam, 31 climate stations, and 9 observed discharge outlets used for calibration.

2.2. Agro-Hydrological Simulation and Model Calibration

SWAT [42] is a process-based, semi-distributed, continuous-time model, used to estimate water budget components in many studies. Hydrologic modeling in SWAT is based on a soil water balance equation. The primary components estimated in the model include surface water flow, evapotranspiration, soil infiltration, and percolation to shallow and deep aquifers. The model estimates surface water flow using the modified SCS-CN (Soil Conservation Service-Curve Number) method, which estimates the amount of infiltration and runoff from rainfall excess based on land use, hydrologic soil group, and antecedent moisture condition. According to the SCS-CN method, the total rainfall is divided into initial abstraction, continuous abstraction, and excess rainfall [43]. Daily precipitation, land use characteristics, and soil profile features are used as input for calculations. A detailed description of all hydrological processes in the model is provided by Neitsch et al. [44].

The Sequential Uncertainty Fitting Procedure (SUFI-2) is used for model calibration [45]. SUFI-2 quantifies prediction uncertainty using a 95% prediction uncertainty (95PPU) band calculated by expressing a range for parameters to map all sources of uncertainties. Two indices are used to measure the goodness-of-fit of the calibrated model: *p-factor* and *r-factor* [46]. The *p-factor* is the percentage of measured data bracketed by the 95PPU band. It varies between 0 and 1, where 1 indicates an ideal case, meaning that 100% of the measured data are inside the 95PPU band. The *r-factor* is the relative width of the uncertainty band divided by the standard deviation of the observed variable. More details are given by Abbaspour et al. [47]. The bR^2 criterion (the weighted version of coefficient of determination R^2) [46] and the Nash Sutcliffe (*NS*) [48] were used as objective functions to measure the degree of match between simulated and observed discharge values.

2.3. Model Set-Up and Data

For the study area, a digital elevation map (DEM) was obtained from NASA's Shuttle Radar Topography Mission (SRTM) with a spatial resolution of 90 m [49]. A soil map, containing information such as maximum rooting depth of soil profile, soil porosity, and bulk density, was obtained from the global soil map of Food and Agricultural Organization (FAO). The database provided over 5000 soil types from which 17 were in our study area. Each soil type comprised two layers (0–30 and 30–100 cm) at the spatial resolution of 10 km and other soil variables calculated by Schuol et al. [50]. Daily climate data including precipitation and temperature at 31 stations (Figure 1) were obtained from WATCH (Water and Global Change) Forcing Data methodology applied to ERA-Interim (a re-analysis of meteorological observations produced by the European Centre for Medium-Range Weather Forecasts) data-Climate Research Unit (WFDEI-CRU) [51] at $0.5° \times 0.5°$ resolution for 1980–2012. The land use map was created from the Indian Remote Sensing-Linear P6 (IRS-P6) satellite with Linear Imaging and Self Scanning (LISS-IV) sensor, IRS-P5 satellite with panchromatic cameras, Enhanced Thematic Mapper+2001 (ETM+2001) Landsat, and from 3300 field sampling points collected by IWPCO (Iran Water and Power Resources Development Company, Tehran, Iran) [52]. The monthly discharge values at nine observed discharge outlets (Figure 1) from IWPCO [53] were used for model calibration (1988–2012) and validation (1980–1987).

We obtained future daily climate data, including precipitation and minimum and maximum temperatures, from the Inter-Sectoral Impact Model Inter-comparison Project (ISI-MIP) for five GCMs based on Coupled Model Intercomparison Project (CMIP5) data [54] driven by RCP scenarios of the Intergovernmental Panel on Climate Change (IPCC) fifth assessment report [55] at a $0.5° \times 0.5°$ spatial resolution. Details of the five GCMs (HADGEMES, GFDL, IPSL, MIROC, and NORESM) are summarized in Table 1.

Table 1. Description of the five Global Circulation Models (GCMs) used in this study obtained from Coupled Model Intercomparison Project (CMIP5).

GCM Name	Institute Full Name
HadGEM2-ES	Met Office Hadley Centre (additional HadGEM2-ES realizations contributed by Instituto Nacional de Pesquisas Espaciais)
IPSL-CM5A-LR	Institute Pierre-Simon Laplace
GFDL-ESM2M	NOAA Geophysical Fluid Dynamics Laboratory-Earth System Model
MIROC-ESM-CHEM	Japan Agency for Marine-Earth Science and Technology, Atmosphere and Ocean Research Institute (The University of Tokyo) and National Institute for Environmental Studies
NorESM1-M	Norwegian Climate Centre-Earth System Model

The daily rainfall and temperature data from the five GCMs were bias corrected using the nearest local measured stations. For rainfall, we used a simple ratio method, in which for each month, we divided the average GCM data by the observed data and divided the daily GCM data by this factor to obtain future daily rainfall data. For the temperature, we tested linear and nonlinear models as described by Wilby et al. [56] and chose a fourth-degree regression model. In general, the results of the first-degree linear and fourth-degree nonlinear models were similar except for very small and very large temperature values, where the nonlinear model performed systematically better, as was also reported by Abbaspour et al. [47].

We used ArcSWAT 2012 with Esri's ArcGIS version 10.2. A total of 333 subbasins and 1507 HRUs (Hydrologic Response Units) were created. The model was calibrated in five iterations with 480 simulations in each iteration. The time required for one single 33-year simulation was about 13 min. Considering 480 simulations in each iteration of calibration, we needed 100 h. In this paper, we used the parallel processing features of SWAT-CUP (a calibration/uncertainty or sensitivity program interface for SWAT) [57], where simulations were distributed over 24 CPUs, decreasing the required time to approximately 4.5 h. After calibration, model outputs including soil water, discharge, and precipitation at subbasin level were used as input variables for drought analysis.

2.4. Drought Analysis Methods

The commonly used *SPI* [10] was selected to monitor meteorological drought. It is computed by fitting a suitable probability distribution function (f_x) to the frequency distribution of precipitation. We chose a 2-parameter gamma distribution as the probability density function [58,59]. The cumulative distribution function (F_x) is then the integral over f_x as:

$$F_x = \int_0^x f_x(x)\, dx \quad x : precipitation \tag{1}$$

To obtain the *SPI*, we transformed F_x using an inverse normal transformation function with mean 0 and standard deviation 1. Six *SPI* classes were defined as: extreme wet, wet, mild, moderate, severe, and extreme drought [59] (Table 2).

Table 2. Six drought classes and weight and rate assigned to each drought class based on drought severity and drought occurrence probability, respectively. SPI, Standardized Precipitation Index; SRI, Standardized Runoff Index; SSWI, Standardized Soil Water Index.

Class	SPI, SRI, SSWI Values	Weight	Rates Based on % of Occurrence Probability (Pr)
Extreme wet	Larger than 1	0	-
Wet	0 to 0.99	0	-
Mild	−0.99 to 0	$W_1 = 1$	If $(17.9 < \text{Pr} \le 25.7) \to R_1 = 1$ If $(25.7 < \text{Pr} \le 30.4) \to R_1 = 2$ If $(30.4 < \text{Pr} \le 34.6) \to R_1 = 3$
Moderate	−1.49 to −1	$W_2 = 2$	If $(5.9 < \text{Pr} \le 8.3) \to R_2 = 1$ If $(8.3 < \text{Pr} \le 10.3) \to R_2 = 2$ If $(10.3 < \text{Pr} \le 13) \to R_2 = 3$
Severe	−1.99 to −1.5	$W_3 = 3$	If $(1.5 < \text{Pr} \le 3.7) \to R_3 = 1$ If $(3.7 < \text{Pr} \le 5.6) \to R_3 = 2$ If $(5.6 < \text{Pr} \le 8.3) \to R_3 = 3$
Extreme	Smaller than −2	$W_4 = 4$	If $(0.7 < \text{Pr} \le 2.2) \to R_4 = 1$ If $(2.2 < \text{Pr} \le 3.4) \to R_4 = 2$ If $(3.4 < \text{Pr} \le 7.6) \to R_4 = 3$

SPI-X could be defined over different time scales (*X* = 1, 3, 6, 12, and 24-month). *SPI-X* at each month is obtained from total precipitation over the last *X* months. For example, *SPI-3* at the end of February compares the December–January–February precipitation total in that particular year with the December–January–February precipitation totals of all other years. The *SPI* method can also be applied to soil moisture and discharge variables [10,60,61] as indicators of hydrological and agricultural droughts, respectively. In this study, we used the same method to calculate *SRI* based on discharge and *SSWI* based on soil water content.

2.5. Drought Hazard Index

To aggregate the severity and occurrence probability features of each index into one unique index for the entire study period, we calculated the Drought Hazard Index (*DHI*) using the methodology proposed by Shahid et al. [62] and later by Rajsekhar et al. [63]. In this method, each of the four drought classes is given a particular weight from 1 to 4, which represent mild (W_1), moderate (W_2), severe (W_3), and extreme droughts (W_4), respectively (Table 2). Furthermore, each class i receives a rate R_i from 1 to 3, based on its probability of occurrence obtained from the Jenks natural break method [64] (Table 2). The final *DHI* is aggregated as:

$$DHI = (W_1 \times R_1) + (W_2 \times R_2) + (W_3 \times R_3) + (W_4 \times R_4) \tag{2}$$

As a result, three degrees of hazard intensity, namely low (*DHI* < 18), medium (21 < *DHI* < 18), and high (*DHI* > 21), are defined using Jenks natural break classification method.

3. Results

3.1. Performance of the KRB Hydrologic Model

The KRB hydrologic model provided reasonable accuracy after calibration. The *p-factor* for calibration (1988–2012) and validation (1980–1987) periods were larger than 0.55, indicating that more than 55% of the observed data were bracketed by the 95PPU band (Table 3). The *r-factor* values were mostly around 1 for all discharge stations, indicating reasonable prediction uncertainties in both calibration and validation periods. The average values of bR^2 were 0.53 and 0.60 for calibration and validation periods, respectively. The *NS* efficiency values were larger than 0.5 in most discharge outlets, which are satisfactory results.

Table 3. Calibration and validatation performances of simulated discharge in SWAT. NS, Nash Sutcliffe.

Outlet Names	Calibration Period (1988–2012)				Validation Period (1980–1987)			
	p-Factor	*r-Factor*	bR^2	*NS*	*p-Factor*	*r-Factor*	bR^2	*NS*
Akan	0.56	1.04	0.51	0.37	0.66	1.17	0.57	0.51
Polchehr	0.57	0.82	0.49	0.55	0.70	0.92	0.54	0.50
Ghurbagestan	0.74	0.83	0.59	0.67	0.73	0.93	0.71	0.66
Haleilan	0.67	0.82	0.62	0.65	0.73	0.95	0.68	0.62
Tangsaz	0.75	0.93	0.64	0.66	0.65	1.09	0.73	0.54
Afrine	0.53	0.66	0.46	0.56	0.63	0.58	0.37	0.42
Jelogir	0.64	0.89	0.67	0.66	0.71	1.12	0.67	0.59
Payepol	0.52	1.04	0.38	0.13	0.64	1.08	0.56	0.27
Hamidieh	0.55	1.08	0.43	0.18	0.65	1.23	0.51	0.17

3.2. Temporal Propagation of Droughts in Historic Period

To calculate *SPI*, *SRI*, and *SSWI*, monthly values of precipitation, river discharge, and soil water (1980–2012) from 333 subbasins were aggregated into the NKRB, CKRB, and SKRB catchments levels (Figure 1) using weighted areal averages. The *SPI* evolution over 1, 3, 6, 9, 12, and 24-month time scales (Figure S1) showed higher drought frequency for shorter time scales. Moreover, less persistency was noticed at time scales shorter than six months. On the other hand, although in *SPI*-24, the severe drought period of 2000–2002 was identified, the extreme drought years of 1992 and 2008 were less obvious. This shows that 6, 9, 12-month time scales are more representative of drought periods. In this paper, *SPI*-12 was selected as the time scale of interest, as was also suggested by Lloyd-Hughes et al. [59], Gocic et al. [65], and Raziei et al. [66].

The historic time series of *SPI*-12, *SRI*-12, and *SSWI*-12 in the three catchments (Figure 2) show that the basin experienced most severe drought conditions after 1999 and most extreme wet conditions during 1993–1996. Overall, five drought events (D1–D5) with different meteorological, hydrological, and agricultural drought characteristics were identified between 1980 and 2012 (Figure 2a–c). In the meteorological sector (Figure 2a), the first drought event (D1) started in late 1983 with mostly moderate severity and lasted until late 1984. The event, however, persisted until early 1986 in SKRB. Meteorological drought D2 started in 1989 with mild severity in the three catchments and lasted until late 1991 with severe intensity in NKRB and CKRB. The subsequent event (meteorological D3) in 1997 had a short duration with mostly mild to moderate severity in all catchments. Meteorological event D4 started in mid-1999 with extreme severity. It lasted until 2001 in NKRB and CKRB and until 2004 in SKRB. The basin experienced another extreme event D5 from 2007 to 2010 with higher severity at the beginning and in SKRB at the end of the period.

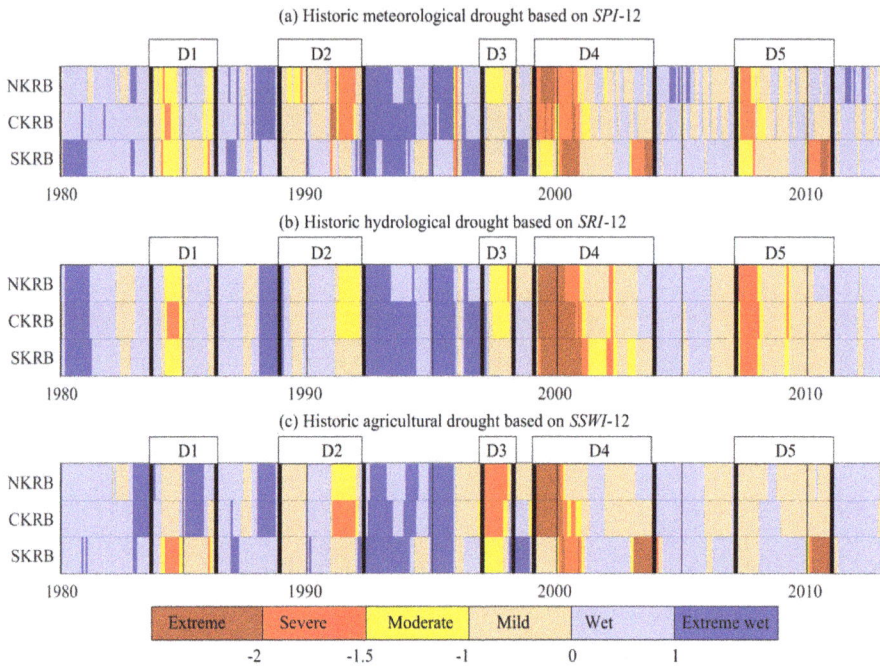

Figure 2. The historic patterns of (**a**) *SPI*-12 for meteorological; (**b**) *SRI*-12 for hydrological; and (**c**) *SSWI*-12 for agricultural droughts in three major catchments of KRB.

Not all meteorological droughts had hydrological (Figure 2b) and agricultural (Figure 2c) signatures. Meteorological drought D1 registered as severe hydrological drought only in late 1984 in CKRB and as agricultural drought in late 1984 in SKRB. The D2 event had mostly a mild to moderate effect on hydrological drought, while producing severe agricultural drought in CKRB in 1991. D3 had almost the same pattern in hydrological sectors, whereas severe agricultural droughts were identified in 1997 in NKRB and CKRB. The major reason for higher severity of D3 in the agricultural sector is probably related to two months of extreme drought in early 1996, resulting in mild agricultural drought in 1996. From this time until the start of event D3, there was not enough time for soil moisture to replenish itself. In SKRB, the extreme wet conditions after 1996 accelerated replenishment of soil moisture and this caused a less severe agricultural drought during the D3 event. Meteorological drought D4 resulted in extreme hydrological and agricultural droughts. The meteorological event D5 resulted in a similar pattern for the hydrological sector, except for the extreme case in 2010, which showed up as an agricultural drought in SKRB.

Comparison of the correlation coefficient in droughts of different sectors in the six time scales (*SPI-*, *SRI-1*, 3, 6, 9, 12, 24, and *SSWI-1*, 3, 6, 9, 12, 24 months) and in the three catchments (Table 4) showed that the meteorological droughts are better correlated with hydrological and agricultural droughts of longer time periods. For example, *SPI-1* is mostly correlated with *SRI-3* (0.83 in NKRB, 0.79 in CKRB, and 0.60 in SKRB). *SPI-3* and *SPI-6* are mostly correlated with *SRI-6* and *SRI-9*, respectively (highlighted box in Table 4). Similarly, *SPI-3* shows the highest correlation with *SSWI-9* (0.79 in NKRB, 0.78 in CKRB, and 0.76 in SKRB). This most likely suggests a 3-month lag of hydrological and agricultural responses to meteorological drought.

Table 4. Correlation coefficient of *SPI* with *SRI* and *SSWI* in different time scales and in three catchments; the highlighted boxes show the highest correlation values of *SPIs* with *SRIs* and *SSWIs*.

Catchment		SPI-1	SPI-3	SPI-6	SPI-9	SPI-12	SPI-24
NKRB	SRI-1	0.55	0.27	0.12	0.05	0.07	0.09
	SRI-3	0.83	0.71	0.38	0.23	0.16	0.16
	SRI-6	0.76	0.87	0.77	0.53	0.40	0.35
	SRI-9	0.64	0.74	0.85	0.77	0.59	0.47
	SRI-12	0.58	0.65	0.75	0.81	0.76	0.56
	SRI-24	0.42	0.50	0.60	0.64	0.68	0.83
CKRB	SRI-1	0.49	0.24	0.10	0.05	0.07	0.10
	SRI-3	0.79	0.66	0.37	0.23	0.18	0.19
	SRI-6	0.72	0.83	0.76	0.53	0.42	0.40
	SRI-9	0.61	0.73	0.85	0.78	0.62	0.54
	SRI-12	0.54	0.64	0.75	0.82	0.80	0.64
	SRI-24	0.39	0.50	0.62	0.66	0.71	0.87
SKRB	SRI-1	0.36	0.20	0.11	0.07	0.08	0.12
	SRI-3	0.60	0.51	0.32	0.22	0.18	0.25
	SRI-6	0.46	0.56	0.57	0.43	0.35	0.41
	SRI-9	0.41	0.49	0.64	0.58	0.48	0.49
	SRI-12	0.37	0.48	0.61	0.67	0.64	0.58
	SRI-24	0.26	0.36	0.47	0.51	0.56	0.79
NKRB	SSWI-1	0.63	0.30	0.16	0.09	0.10	0.10
	SSWI-3	0.70	0.65	0.37	0.27	0.23	0.19
	SSWI-6	0.67	0.79	0.74	0.56	0.49	0.41
	SSWI-9	0.57	0.71	0.81	0.78	0.67	0.56
	SSWI-12	0.47	0.59	0.72	0.78	0.79	0.64
	SSWI-24	0.32	0.42	0.50	0.52	0.56	0.82
CKRB	SSWI-1	0.68	0.32	0.15	0.09	0.09	0.09
	SSWI-3	0.71	0.68	0.37	0.26	0.23	0.20
	SSWI-6	0.63	0.78	0.74	0.55	0.47	0.42
	SSWI-9	0.51	0.67	0.77	0.75	0.63	0.55
	SSWI-12	0.41	0.56	0.67	0.73	0.74	0.62
	SSWI-24	0.32	0.41	0.48	0.50	0.52	0.75
SKRB	SSWI-1	0.71	0.45	0.28	0.19	0.10	0.14
	SSWI-3	0.62	0.74	0.50	0.43	0.28	0.30
	SSWI-6	0.56	0.76	0.81	0.61	0.48	0.43
	SSWI-9	0.49	0.69	0.85	0.83	0.65	0.53
	SSWI-12	0.39	0.59	0.77	0.85	0.84	0.63
	SSWI-24	0.32	0.43	0.53	0.53	0.56	0.88

3.3. Future Characteristics of Droughts (Severity-Frequency-Duration)

The temporal variation of the three types of droughts under the RCP2.6 (Figure S2) and RCP8.5 (Figure S3) scenarios in five GCMs (2020–2052) shows that KRB will likely be more susceptible to droughts in the future. However, drought periods and their severities are different among GCM models and for different types of droughts. Overall, *SPI-12* patterns (Figure S2a) show more severe meteorological droughts after 2045 under RCP2.6. In this scenario, severe hydrological (Figure S2b) droughts and extreme agricultural (Figure S2c) droughts are observed after 2035. In RCP8.5 (Figure S3), severe and extreme droughts for all three types also mostly appear in the same time period.

We compared historic and future droughts by using Probability Density Functions (PDFs) of *SPI*-12, *SRI*-12, and *SSWI*-12 for different severities (Figure 3a–r). The uncertainty bands stemming from the differences in the five GCMs are wider for mild and moderate meteorological and hydrological drought classes (*SPI*-12 and *SRI*-12 between −1.5 and 0) as compared with other classes, indicating lesser agreement between different GCMs. For agricultural drought, larger uncertainty is noticed for wet conditions (*SSWI*-12 between 0 and 1).

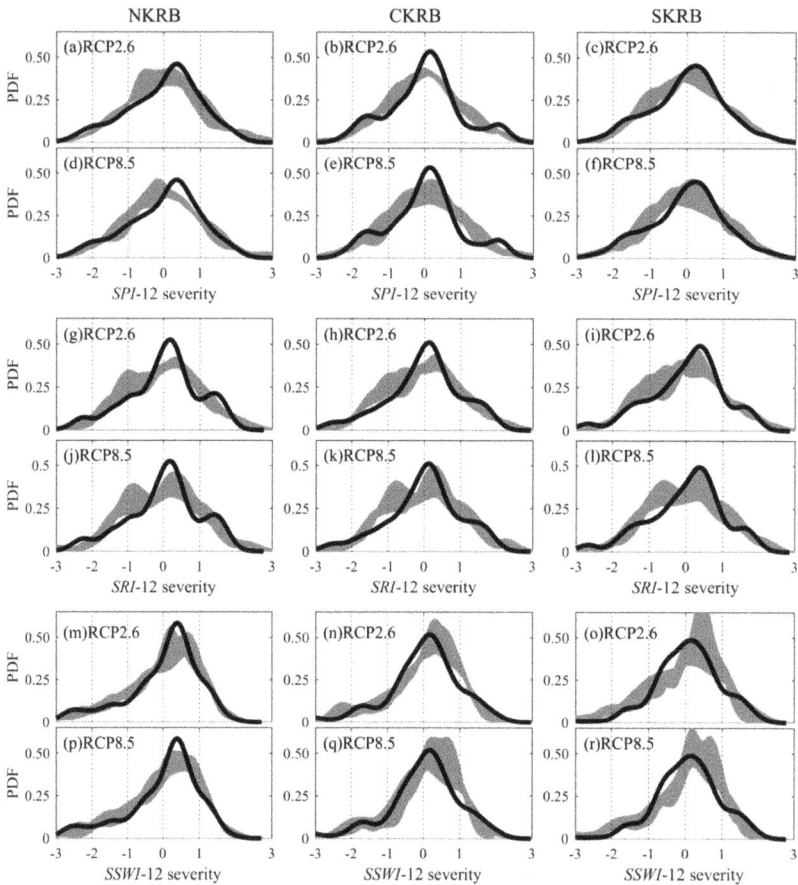

Figure 3. The probability density function (PDF) of different severities of *SPI*-12 (**a–f**), *SRI*-12 (**g–l**), and *SSWI*-12 (**m–r**) in RCP2.6 and RCP8.5 scenarios. The grey bands are extracted from maximum and minimum values in the five GCMs and the black lines indicate the historic PDFs.

The resulting PDFs in the entire region, with the exception of agricultural drought in CKRB and SKRB, show a shift to left in the grey band, especially in the left leg of graphs for both RCP2.6 and RCP8.5, indicating higher probability of droughts (especially mild and moderate droughts) in the future. The left shift is slightly larger in RCP8.5 compared to RCP2.6 in all catchments. Agricultural drought in CKRB and NKRB, however, shows a tendency to shift to the right for most GCM models, indicating smaller probability of mild to moderate droughts. No significant change is observed in the probability of extreme meteorological droughts in the three catchments (the left tails of the distributions). The wider bands of *SRI*-12 compared to *SPI*-12 indicate larger uncertainties in the hydrological drought predictions by GCMs. The agricultural drought index (Figure 3m–r) shows a shift to the right, especially in the right tail of graphs for both the RCP2.6 and RCP8.5 scenarios, indicating higher probability of wet conditions. On the other hand, the agricultural sector is more exposed by extreme and severe droughts in CKRB and SKRB, as their probabilities are higher. The wide uncertainty band during wet conditions shows less agreement among the five GCMs.

To compare frequency and duration of historic droughts, we defined a drought event as having *SPI*-12, *SRI*-12, or *SSWI*-12 < 0 for at least two months. The historic frequency shows there were on the average 15 meteorological droughts, 5 hydrological droughts, and 10 agricultural droughts (Figure 4a–c). These droughts had durations of 8 months, 3.5 months, and 6 months, respectively (Figure 4d–f). During the historic period, meteorological droughts were more frequent with longer duration. The reason is that we considered periods of longer than two months as a drought event. So, some of the very short and mild meteorological events did not register signatures in other sectors (Figure 2). Besides, *SRI*-12 and *SSWI*-12 are influenced by precipitation as well as temperature, whereas *SPI*-12 depends only on precipitation.

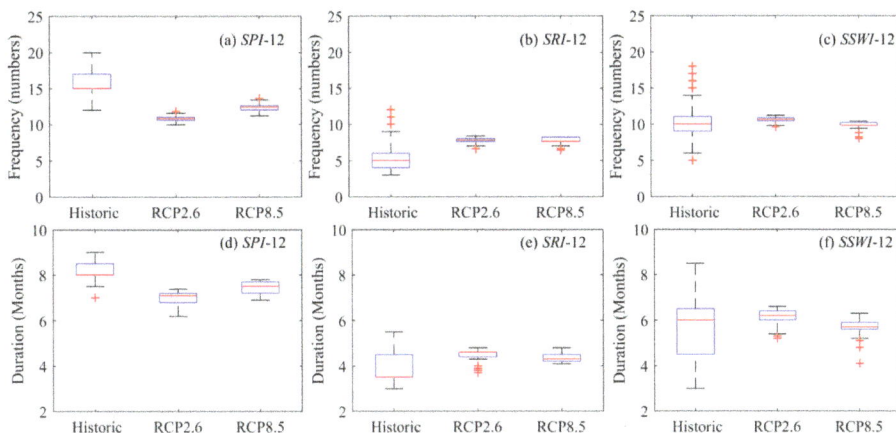

Figure 4. Comparison of the frequency and duration of historic and future of (**a,d**) *SPI*-12, (**b,e**) *SRI*-12, and (**c,f**) *SSWI*-12 droughts based on an ensemble of five GCMs. The red lines inside the boxes show the median, the boxplots show the 25 and 75 percentiles, and the whiskers show 5 and 95 percentiles from 333 subbasins in KRB.

Similarly, future projections show the highest frequency of meteorological drought. However, compared to the historic period, there are fewer differences between frequency and duration of the three drought types. Moreover, the duration and frequency of future hydrological and agricultural droughts are predicted to increase compared to historic period. It is interesting to note that future prediction of meteorological droughts is smaller in frequency and shorter in duration (Figure 4a,d) compared to historic meteorological drought. Drought frequency is expected to decrease from a median value of 15 to 10 and 12 in RCP2.6 and RCP8.5, respectively. This pattern is mostly caused by an unusually large number of droughts in the KRB during 2000 to 2010, which resulted in a high historic drought frequency. Hydrological droughts are, however, more frequent in the future with longer duration. There does not seem to be a large difference in the historic and future agricultural droughts. This is mainly due to the impact of both precipitation and temperature variables in the calculation of hydrologic and agricultural indices. Only slight differences between RCP2.6 and RCP8.5 are observed in all cases.

3.4. Composite Droughts Index, DHI

Spatial distributions of the composite meteorological *DHI* under future climate change conditions show that KRB would probably experience a higher degree of meteorological *DHI* compared to the historic period. Most of KRB will probably be exposed to medium meteorological *DHI* except eastern sides of CKRB where high meteorological *DHI* is predicted. Hydrological *DHI* responded differently for both historic and future conditions (Figure 5d–f). Generally, KRB is predicted to be more exposed to high hydrological *DHI* in western NKRB and CKRB. SKRB will probably be less exposed to high hydrological *DHI*. High agricultural *DHI* during the historic period was limited to the western part of KRB (Figure 5a,d), however, both RCP2.6 and RCP8.5 predict higher agricultural *DHI* (Figure 5g–i) in all catchments. In fact, RCP8.5 puts most of KRB under high agricultural *DHI*.

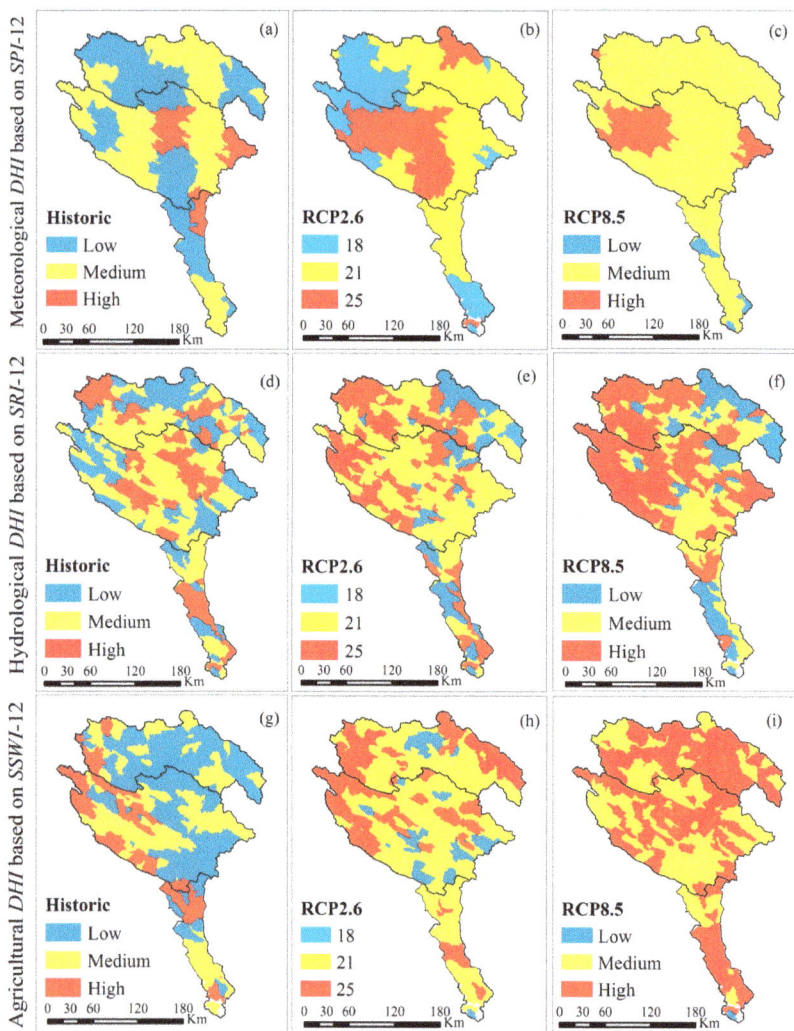

Figure 5. Spatial variations of meteorological *DHI* (**a–c**), hydrological *DHI* (**d–f**), and agricultural *DHI* (**g–i**) comparing the historic and future variations in RCP2.6 and RCP8.5. The results are obtained from an ensemble of five GCMs.

4. Discussion and Conclusions

SPI, *SRI*, and *SSWI* with the aid of a SWAT model captured past drought periods. The selected indices were found to be suitable for drought monitoring, since the severe and extreme periods agreed with historical records over the past 33 years of the study period reported previously [37,41]. We found a 3-month lag between incipient meteorological droughts and the time that hydrological droughts were observed. While occurrence of a lag between meteorological and hydrological droughts is reported in many studies [8,67], the length of lag time varies by study. For example, Liu et al. [28] found that hydrological drought was not observed until 2 months after meteorological drought, and Stefan et al. [68] reported a 2–3-month lag between the precipitation and river discharge anomalies during winter and a 0–1 month delay in summer. Generally, not only precipitation, but also factors such as rainfall interception, temperature, evapotranspiration, and a basin's morphological conditions contribute to discharge formation. In our study, the lag time between meteorological and hydrological droughts might be related to different reasons in the upper to lower catchments. In NKRB and CKRB, the lag might be mostly due to the mountainous characteristic of catchments. The flow that contributes to rivers in these catchments are mostly from snow melt of the mountainous areas, which occurs at a later time than the actual precipitation. However, lagged response in SKRB is most probably associated with a mixed-flow regime. Within this catchment, not only precipitation, but also the discharge from NKRB and SKRB contributes to flow, however, with some lag due to the varying time of concentration. Therefore, the timing of hydrological drought differs from meteorological drought.

We chose soil moisture to quantify *SSWI* to monitor agricultural drought because it is more relevant compared to evapotranspiration in basins with semi-arid climates. In the semi-arid regions, the rate of potential evapotranspiration (atmospheric demand) is substantially larger than actual evapotranspiration (soil's ability to supply water), causing soil moisture to be at the wilting point for most of the year [69]. In our study, *SSWI*-12 showed a 3-month delay with *SPI*-12, as temperature also influenced soil moisture content. With increasing temperature in the summer season, evaporation increases, causing a depletion of soil water content.

For KRB as a whole, the future climate is likely to increase the probability of severe and extreme droughts. Comparison of the results with future projection of the Köppen-Geiger climate classification [70] also confirms a shift of climate zone from warm to arid in SKRB. The frequency and duration of future droughts will probably increase based on *SRI*-12 and *SSWI*-12, but decrease for *SPI*-12. The reason is that *SPI*-12 is computed based on precipitation alone, while indices of hydrological and agricultural droughts depend on both temperature and precipitation. The spatial extent of high agricultural *DHI* is predicted to be much larger in the future, especially in SKRB. This shows the complexity in translating meteorological droughts to agricultural and hydrological sectors, as drought propagation into two latter types depends on the climate of the region as well as the responses of the hydrological cycles and differs depending on physiographic characteristics of the regions such as permeability, topography, and land use. Higher exposure of the agricultural sector to drought poses additional challenges to agricultural production, as KRB has already experienced serious water shortages in the last two major droughts (D4 and D5) and irrigated agriculture had to rely heavily on the exploitation of groundwater.

In conclusion, one of the strengths of our applied approach is the use of a standardized definition of drought indices, which made our analysis consistent for comparing different drought types irrespective of the climatic conditions and the regions. The paper also made some contributions toward exploring behaviors of drought propagation in hydrological systems and identifying regions that will be more exposed to drought risks in the future. The distributed agro-hydrological model SWAT was used to estimate soil water content and runoff at a fine spatial resolution. Comparison of multiple drought indices of different aspects allows for a better monitoring of space-time drought characteristics.

Similar analyses and sets of selected indices could be applied to other basins with different scales for a better understanding of drought effects. The high spatial resolution obtained from applying a physically based model can be aggregated to district, farm, and provincial levels, as the findings from multiple scales are usually complementary to each other [71]. The standardized procedure facilitates linking drought indices with socioeconomic factors to broaden the knowledge on physical and social vulnerability. For example, by linking the agricultural *DHI* with crop yields, one can quantify crop drought vulnerability and risks, which are essential for food security purposes. Similarly, hydrological *DHI* is an appropriate candidate to measure drought indices that assess the status of water resources vulnerability. Such joint interpretations help decision makers with proposing better allocation of resources.

Supplementary Materials: The following are available online at www.mdpi.com/2073-4441/9/4/241/s1, Figure S1: Evolution of SPI over different time scales in three catchments of KRB; Figure S2: The future heat map of (a) meteorological, (b) hydrological, and (c) agricultural droughts in RCP2.6 scenario in three catchments of KRB; Figure S3: The future heat map of (a) meteorological, (b) hydrological, and (c) agricultural droughts in RCP8.5 scenario in three catchments of KRB.

Acknowledgments: The first and second authors express their sincere thanks to Eawag Partnership Program (EPP) for the opportunity of a collaboration. The authors also thanks the financial support of Swiss National Science Foundation (SNF No.: CR21I3_146430. Dec.2013-Nov.2016).

Author Contributions: Bahareh Kamali and Delaram Houshmand Kouchi prepared the SWAT model and wrote the paper with the assistance from Karim C. Abbaspour. Bahareh Kamali and Delaram Houshmand Kouchi designed the methodology framework under supervision of Karim C. Abbaspour and Hong Yang. Karim C. Abbaspour, Hong Yang advised on conceptual and technical, and contributed to the strategy.

Conflicts of Interest: The authors declare no conflict of interest.

References

1. Vidal, J.P.; Martin, E.; Franchisteguy, L.; Habets, F.; Soubeyroux, J.M.; Blanchard, M.; Baillon, M. Multilevel and multiscale drought reanalysis over France with the Safran-Isba-Modcou hydrometeorological suite. *Hydrol. Earth Syst. Sci.* **2010**, *14*, 459–478. [CrossRef]
2. Dai, A.G. Drought under global warming: A review. *WIREs Clim. Chang.* **2011**, *2*, 45–65. [CrossRef]
3. Vicente-Serrano, S.M.; Begueria, S.; Lorenzo-Lacruz, J.; Camarero, J.J.; Lopez-Moreno, J.I.; Azorin-Molina, C.; Revuelto, J.; Moran-Tejeda, E.; Sanchez-Lorenzo, A. Performance of drought indices for ecological, agricultural, and hydrological applications. *Earth Interact.* **2012**, *16*, 1–27. [CrossRef]
4. Peters, E.; Bier, G.; van Lanen, H.A.J.; Torfs, P.J.J.F. Propagation and spatial distribution of drought in a groundwater catchment. *J. Hydrol.* **2006**, *321*, 257–275. [CrossRef]
5. Wilhite, D.A.; Glantz, M.H. Understanding the drought phenomenon: The role of definitions. *Water Int.* **1985**, *10*, 111–120. [CrossRef]
6. Orville, H.D. American meteorological society statement on meteorological drought. *Bull. Am. Meteorol. Soc.* **1990**, *71*, 1021–1023.
7. Wang, D.B.; Hejazi, M.; Cai, X.M.; Valocchi, A.J. Climate change impact on meteorological, agricultural, and hydrological drought in central Illinois. *Water Resour. Res.* **2011**, *47*. [CrossRef]
8. Lglesias, L.; Garrote, L.; Cancelliere, A.; Cubillo, F.; Wilhite, D. *Coping with Drought Risk in Agriculture and Water Supply Systems, Drought Management and Policy Development in the Mediterranean*, 1st ed.; Springer: Dordrecht, The Netherlands, 2009; p. 320.
9. Koutroulis, A.G.; Vrohidou, A.E.K.; Tsanis, I.K. Spatiotemporal characteristics of meteorological drought for the Island of Crete. *J. Hydrometeorol.* **2011**, *12*, 206–226. [CrossRef]
10. McKee, T.B.; Doesken, N.J.; Kleist, J. The relationship of drought frequency and duration to time scales. In Proceedings of the 8th Conference on Applied Climatology, Anaheim, CA, USA, 17–22 January 1993; pp. 179–184.
11. Stagge, J.H.; Kohn, I.; Tallaksen, L.M.; Stahl, K. Modeling drought impact occurrence based on meteorological drought indices in Europe. *J. Hydrol.* **2015**, *530*, 37–50. [CrossRef]
12. Belayneh, A.; Adamowski, J.; Khalil, B.; Ozga-Zielinski, B. Long-term SPI drought forecasting in the Awash River Basin in Ethiopia using wavelet neural network and wavelet support vector regression models. *J. Hydrol.* **2014**, *508*, 418–429. [CrossRef]
13. Moreira, E.E.; Pires, C.L.; Pereira, L.S. SPI drought class predictions driven by the North Atlantic Oscillation index using log-linear modeling. *Water* **2016**, *8*, 1–18. [CrossRef]

14. Hisdal, H.; Tallaksen, L.M. Estimation of regional meteorological and hydrological drought characteristics: A case study for Denmark. *J. Hydrol.* **2003**, *281*, 230–247. [CrossRef]

15. Liu, B.; Zhou, X.; Li, W.; Lu, C.; Shu, L. Spatiotemporal characteristics of groundwater drought and its response to meteorological drought in Jiangsu Province, China. *Water* **2016**, *8*, 1–21. [CrossRef]

16. Tallaksen, L.M.; Hisdal, H.; Van Lanen, H.A.J. Space-time modelling of catchment scale drought characteristics. *J. Hydrol.* **2009**, *375*, 363–372. [CrossRef]

17. Tadesse, T.; Brown, J.F.; Hayes, M.J. A new approach for predicting drought-related vegetation stress: Integrating satellite, climate, and biophysical data over the US central plains. *ISPRS J. Photogramm.* **2005**, *59*, 244–253. [CrossRef]

18. Tokarczyk, T.; Szalinska, W. Combined analysis of precipitation and water deficit for drought hazard assessment. *Hydrol. Sci. J.* **2014**, *59*, 1675–1689. [CrossRef]

19. Duan, K.; Mei, Y.D. Comparison of meteorological, hydrological and agricultural drought responses to climate change and uncertainty assessment. *Water Resour. Manag.* **2014**, *28*, 5039–5054. [CrossRef]

20. Trenberth, K.E.; Dai, A.G.; van der Schrier, G.; Jones, P.D.; Barichivich, J.; Briffa, K.R.; Sheffield, J. Global warming and changes in drought. *Nat. Clim. Chang.* **2014**, *4*, 17–22. [CrossRef]

21. Rummukainen, M. Changes in climate and weather extremes in the 21st century. *WIREs Clim. Chang.* **2012**, *3*, 115–129. [CrossRef]

22. Touma, D.; Ashfaq, M.; Nayak, M.A.; Kao, S.C.; Diffenbaugh, N.S. A multi-model and multi-index evaluation of drought characteristics in the 21st century. *J. Hydrol.* **2015**, *526*, 196–207. [CrossRef]

23. Wilhite, D.A.; Sivakumar, M.V.K.; Pulwarty, R. Managing drought risk in a changing climate: The role of national drought policy. *Weather Clim. Extremes* **2014**, *3*, 4–13. [CrossRef]

24. Awal, R.; Bayabil, H.K.; Fares, A. Analysis of potential future climate and climate Extremes in the Brazos Headwaters basin, Texas. *Water* **2016**, *8*, 1–18. [CrossRef]

25. Lu, G.; Wu, H.; Xiao, H.; He, H.; Wu, Z. Impact of climate change on drought in the upstream Yangtze river region. *Water* **2016**, *8*, 1–20. [CrossRef]

26. Lee, J.H.; Kim, C.J. A multimodel assessment of the climate change effect on the drought severity-duration-frequency relationship. *Hydrol. Processes* **2013**, *27*, 2800–2813. [CrossRef]

27. Leng, G.Y.; Tang, Q.H.; Rayburg, S. Climate change impacts on meteorological, agricultural and hydrological droughts in China. *Glob. Planet Chang.* **2015**, *126*, 23–34. [CrossRef]

28. Liu, L.; Hong, Y.; Bednarczyk, C.N.; Yong, B.; Shafer, M.A.; Riley, R.; Hocker, J.E. Hydro-climatological drought analyses and projections using meteorological and hydrological drought indices: A case study in Blue river basin, Oklahoma. *Water Resour. Manag.* **2012**, *26*, 2761–2779. [CrossRef]

29. Han, L.Y.; Zhang, Q.; Ma, P.L.; Jia, J.Y.; Wang, J.S. The spatial distribution characteristics of a comprehensive drought risk index in southwestern China and underlying causes. *Theor. Appl. Climatol.* **2016**, *124*, 517–528. [CrossRef]

30. Oweis, T.; Siadat, H.; Abbasi, F. *Improving On-Farm Agricultural Water Productivity in the Karkheh River Basin (KRB)*; CPWF Project Report-Project Number 08: CGIAR Challenge Program on Water and Food; Department for International Development: Chatham, UK, 2009.

31. Vaghefi, S.A.; Abbaspour, K.C.; Faramarzi, M.; Srinivasan, R.; Arnold, J.G. Modeling crop water productivity using a coupled SWAT–MODSIM model. *Water* **2017**, *9*, 1–15.

32. Sietz, D.; Feola, G. Resilience in the rural Andes: Critical dynamics, constraints and emerging opportunities. *Reg. Environ. Chang.* **2016**, *16*, 2163–2169. [CrossRef]

33. Millennium Ecosystem Assessment. *Ecosystems and Human Well-Being: Synthesis*; Island Press: Washington, DC, USA, 2005.

34. Vaghefi, S.A.; Mousavi, S.J.; Abbaspour, K.C.; Srinivasan, R.; Arnold, J.R. Integration of hydrologic and water allocation models in basin-scale water resources management considering crop pattern and climate change: Karkheh River Basin in Iran. *Reg. Environ. Chang.* **2015**, *15*, 475–484. [CrossRef]

35. Masih, I.; Ahmad, M.U.D.; Uhlenbrook, S.; Turral, H.; Karimi, P. Analysing streamflow variability and water allocation for sustainable management of water resources in the semi-arid Karkheh river basin, Iran. *Phys. Chem. Earth* **2009**, *34*, 329–340. [CrossRef]

36. Jamali, S.; Abrishamchi, A.; Marino, M.A.; Abbasnia, A. Climate change impact assessment on hydrology of Karkheh Basin, Iran. *Proc. Inst. Civ. Eng.-Water Manag.* **2013**, *166*, 93–104. [CrossRef]

37. Zamani, R.; Tabari, H.; Willems, P. Extreme streamflow drought in the Karkheh river basin (Iran): Probabilistic and regional analyses. *Nat. Hazards* **2015**, *76*, 327–346. [CrossRef]

38. Golian, S.; Mazdiyasni, O.; AghaKouchak, A. Trends in meteorological and agricultural droughts in Iran. *Theor. Appl. Climatol.* **2015**, *119*, 679–688. [CrossRef]

39. Vaghefi, S.A.; Mousavi, S.J.; Abbaspour, K.C.; Srinivasan, R.; Yang, H. Analyses of the impact of climate change on water resources components, drought and wheat yield in semiarid regions: Karkheh river basin in Iran. *Hydrol. Processes* **2014**, *28*, 2018–2032. [CrossRef]

40. Ahmad, M.U.D.; Giordano, M. The Karkheh river basin: The food basket of Iran under pressure. *Water Int.* **2010**, *35*, 522–544. [CrossRef]

41. Marjanizadeh, S.; Qureshi, A.S.; Turral, H.; Talebzadeh, P. *From Mesopotamia to the Third Millennium: The Historical Trajectory of Water Development and Use in the Karkheh River Basin, Iran*; IWMI Working Paper 135; International Water Management Institute: Colombo, Sri Lanka, 2009.

42. Arnold, J.G.; Srinivasan, R.; Muttiah, R.S.; Williams, J.R. Large area hydrologic modeling and assessment—Part 1: Model development. *J. Am. Water Resour. Assoc.* **1998**, *34*, 73–89. [CrossRef]

43. Hjelmfelt, A.T. Investigation of Curve Number Procedure. *J. Hydraul. Eng.—ASCE* **1991**, *117*, 725–737. [CrossRef]

44. Neitsch, S.L.; Arnold, J.G.; Kiniry, J.R.; Williams, J.R.; King, K.W. *Soil and Water Assessment Tool; Theoretical Documentation: Version 2009*; Texas Water Resources Institute: College Station, TX, USA, 2005.

45. Abbaspour, K.C.; Yang, J.; Maximov, I.; Siber, R.; Bogner, K.; Mieleitner, J.; Zobrist, J.; Srinivasan, R. Modelling hydrology and water quality in the pre-ailpine/alpine Thur watershed using SWAT. *J. Hydrol.* **2007**, *333*, 413–430. [CrossRef]

46. Krause, P.; Boyle, D.P.; Bäse, F. Comparison of different efficiency criteria for hydrological model assessment. *Adv. Geosci.* **2005**, *5*, 89–97. [CrossRef]

47. Abbaspour, K.C.; Faramarzi, M.; Ghasemi, S.S.; Yang, H. Assessing the impact of climate change on water resources in Iran. *Water Resour. Res.* **2009**, *45*, 1–16. [CrossRef]

48. Nash, J.E.; Sutcliffe, J.V. River flow forecasting through conceptual models part I—A discussion of principles. *J. Hydrol.* **1970**, *10*, 282–290. [CrossRef]

49. Jarvis, A.; Reuter, H.; Nelson, A.; Guevara, E. Hole-Filled SRTM for the Globe Version 4, the CGIAR-CSI SRTM 90m Database. Available online: http://srtm.csi.cgiar.org (accessed on January 2008).

50. Schuol, J.; Abbaspour, K.C.; Yang, H.; Srinivasan, R.; Zehnder, A.J.B. Modeling blue and green water availability in Africa. *Water Resour. Res.* **2008**, *44*, 1–18. [CrossRef]

51. Weedon, G.P.; Gomes, S.; Viterbo, P.; Shuttleworth, W.J.; Blyth, E.; Osterle, H.; Adam, J.C.; Bellouin, N.; Boucher, O.; Best, M. Creation of the WATCH forcing data and its use to assess global and regional reference crop evaporation over land during the twentieth century. *J. Hydrometeorol.* **2011**, *12*, 823–848. [CrossRef]

52. IWPCO (Iran Water and Power Resources Development Company). *Systematic Planning of Karkheh Watershed; Land Use Studies (Available in Persian)*; Iran Water and Power Resources Development Company: Iranian Ministry of Energy, Tehran, Iran, 2009.

53. IWPCO (Iran Water and Power Resources Development Company). *Systematic Studies of Karkheh River Basin (Available in Persian)*; Iran Water and Power Resources Development Company: Iranian Ministry of Energy, Tehran, Iran, 2010.

54. Hagemann, S.; Chen, C.; Haerter, J.O.; Heinke, J.; Gerten, D.; Piani, C. Impact of a statistical bias correction on the projected hydrological changes obtained from three GCMs and two hydrology models. *J. Hydrometeorol.* **2011**, *12*, 556–578. [CrossRef]

55. Hempel, S.; Frieler, K.; Warszawski, L.; Schewe, J.; Piontek, F. A trend-preserving bias correction—The ISI-MIP approach. *Earth Syst. Dyn.* **2013**, *4*, 219–236. [CrossRef]

56. Wilby, R.L.; Wigley, T.M.L. Downscaling general circulation model output: A review of methods and limitations. *Prog. Phys. Geogr.* **1997**, *21*, 530–548. [CrossRef]

57. Rouholahnejad, E.; Abbaspour, K.C.; Vejdani, M.; Srinivasan, R.; Schulin, R.; Lehmann, A. A parallelization framework for calibration of hydrological models. *Environ. Model. Softw.* **2012**, *31*, 28–36. [CrossRef]

58. Bordi, I.; Frigio, S.; Parenti, P.; Speranza, A.; Sutera, A. The analysis of the Standardized Precipitation Index in the Mediterranean area: Large-scale patterns. *Ann. Geophys.* **2001**, *44*, 965–978.

59. Lloyd-Hughes, B.; Saunders, M.A. A drought climatology for Europe. *Int. J. Climatol.* **2002**, *22*, 1571–1592. [CrossRef]

60. Hao, Z.; AghaKouchak, A.; Nakhjiri, N.; Farahmand, A. Global integrated drought monitoring and prediction system. *Sci. Data* **2014**, 140001. [CrossRef] [PubMed]

61. Mo, K.C. Model-based drought indices over the United States. *J. Hydrometeorol.* **2008**, *9*, 1212–1230. [CrossRef]

62. Shahid, S.; Behrawan, H. Drought risk assessment in the western part of Bangladesh. *Nat. Hazards* **2008**, *46*, 391–413. [CrossRef]

63. Rajsekhar, D.; Singh, V.P.; Mishra, A.K. Integrated drought causality, hazard, and vulnerability assessment for future socioeconomic scenarios: An information theory perspective. *J. Geophys. Res.-Atmos.* **2015**, *120*, 6346–6378. [CrossRef]

64. Jenks, G.F. The Data Model Concept in Statistical Mapping. *Int. Yearbook Cartogr.* **1967**, *7*, 186–190.

65. Gocic, M.; Trajkovic, S. Analysis of precipitation and drought data in Serbia over the period 1980–2010. *J. Hydrol.* **2013**, *494*, 32–42. [CrossRef]

66. Raziei, T.; Saghafian, B.; Paulo, A.A.; Pereira, L.S.; Bordi, I. Spatial patterns and temporal variability of drought in western Iran. *Water Resour. Manag.* **2009**, *23*, 439–455. [CrossRef]

67. Wilhite, D.A. *Drought Assessment, Management, and Planning: Theory and Case Studies*; Springer: New York, NY, USA, 1993; Volume 2.

68. Stefan, S.; Ghioca, M.; Rimbu, N.; Boroneant, C. Study of meteorological and hydrological drought in southern Romania from observational data. *Int. J. Climatol.* **2004**, *24*, 871–881. [CrossRef]

69. Tallaksen, L.M.; van Lanen, H.A.J. *Hydrological Drought: Processes and Estimation Methods for Streamflow and Groundwater*; Developments in Water Science; Elsevier: Amsterdam, The Netherlands, 2004; p. 579.

70. Rubel, F.; Kottek, M. Observed and projected climate shifts 1901–2100 depicted by world maps of the Koppen-Geiger climate classification. *Meteorol. Z.* **2010**, *19*, 135–141.

71. Sietz, D. Regionalisation of global insights into dryland vulnerability: Better reflecting smallholders' vulnerability in Northeast Brazil. *Glob. Environ. Chang.* **2014**, *25*, 173–185. [CrossRef]

water

MDPI

Article

Assessment of Flood Frequency Alteration by Dam Construction via SWAT Simulation

Jeong Eun Lee [1,2], Jun-Haeng Heo [2,*], Jeongwoo Lee [1] and Nam Won Kim [1]

[1] Hydro Science and Engineering Research Institute, Korea Institute of Civil Engineering and Building Technology (KICT), Goyang 10223, Korea; jeus22@kict.re.kr (J.E.L); ljw2961@kict.re.kr (J.L.); nwkim@kict.re.kr (N.W.K.)

[2] School of Civil and Environmental Engineering, Yonsei University, Seoul 03722, Korea

* Correspondence: jhheo@yonsei.ac.kr; Tel.: +82-2-2123-2805; Fax: +82-2-364-5300

Academic Editor: Ataur Rahman
Received: 29 December 2016; Accepted: 5 April 2017; Published: 8 April 2017

Abstract: The purpose of this study is to evaluate the impacts of the upstream Soyanggang and Chungju multi-purpose dams on the frequency of downstream floods in the Han River basin, South Korea. A continuous hydrological model, SWAT (Soil and Water Assessment Tool), was used to individually simulate regulated and unregulated daily streamflows entering the Paldang Dam, which is located at the outlet of the basin of interest. The simulation of the regulated flows by the Soyanggang and Chungju dams was calibrated with observed inflow data to the Paldang Dam. The estimated daily flood peaks were used for a frequency analysis, using the extreme Type-I distribution, for which the parameters were estimated via the L-moment method. This novel approach was applied to the study area to assess the effects of the dams on downstream floods. From the results, the two upstream dams were found to be able to reduce downstream floods by approximately 31% compared to naturally occurring floods without dam regulation. Furthermore, an approach to estimate the flood frequency based on the hourly extreme peak flow data, obtained by combining SWAT simulation and Sangal's method, was proposed and then verified by comparison with the observation-based results. The increased percentage of floods estimated with hourly simulated data for the three scenarios of dam regulation ranged from 16.1% to 44.1%. The reduced percentages were a little higher than those for the daily-based flood frequency estimates. The developed approach allowed for better understanding of flood frequency, as influenced by dam regulation on a relatively large watershed scale.

Keywords: SWAT; regulated and unregulated streamflows; flood frequency analysis; sangal's method

1. Introduction

Growing industrial, municipal, and agricultural demands for water have increased the need for dam construction [1]. Numerous dams have been built in the majority of the world's rivers to provide water supply, electricity production, and the mitigation of flood risk. River hydrology, geomorphology, and ecology have been substantially altered through dam construction. In particular, dams have major impacts on river hydrology, primarily through changes in the timing, magnitude, and frequency of high and low flows, ultimately producing a hydrologic regime that differs significantly from the natural pre-dam conditions [2–4].

Various studies dealing with downstream hydrograph alterations caused by dams have been undertaken. For example, Gregory and Park [5] examined the changed pattern of river discharge and channel capacity below a dam in the basin of the River Tone, England. Their results showed that the downstream peak discharge was reduced by 40% after the dam construction, and the reduction of channel capacity downstream from the reservoir persisted for a distance of 11 km. Page [6] demonstrated that the construction of the Burrinjuck Dam increased the return period of bankfull discharge in Wagga Wagga, Australia. Galat and Lipkin [7] studied the hydrological alterations of the Missouri River flows using the Index of Hydrological Alteration (IHA), indicating that the river flows were heavily influenced by the reservoirs, but they dissipated below tributary junctions. Maingi and Marsh [8] analyzed pre- and post-dam daily discharge data for the Tana River, Kenya, using flood frequency analysis and the computation of various IHAs, demonstrating a significant modification in the hydrologic regime after dam construction. They also estimated the frequency and duration of floods for 71 vegetation sample plots by simulating the hydrologic water profiles, showing an increase in days flooded from the pre- to the post-dam period. Magilligan et al. [9] assessed hydrologic changes at 21 dams of various sizes scattered across the U.S., and found that on average the 2-year flow decreased by 60% after dam installation [10]. Pegg et al. [11] used a time series approach to examine the daily mean flow for 10 Missouri River locations with data from the pre- and post-alteration periods. The results suggest that human alterations on the Missouri River, particularly in the middle portion that is most strongly affected by impoundments and channelization, have resulted in changes to the natural flow regime. Batalla et al. [12] investigated how hydrograph alteration from dams varies though the river network, and determined that the influence of dams on mean monthly flow and flood magnitude diminished with the distance downstream because of increasing drainage area. Magilligan and Nislow [13] presented an analysis of pre- and post-dam hydrological changes from dams that cover the spectrum of hydrologic and climatic regimes across the U.S. They applied the IHA to assess the type, magnitude, and direction of hydrologic shifts for 21 gauge stations due to dam installation, and found that 1-day through 90-day minimum flows increased significantly, while 1-day though 7-day maximum flows decreased significantly following impoundment. Singer [14] reported basinwide patterns of hydrograph alteration via statistical and graphical analysis from a network of long-term streamflow gauges in the Sacramento River basin in California, U.S. In this study, pre- and post-dam flows were compared with respect to the annual flood peak, annual flow trough (lowest value), annual flood volume, time to peak, flood drawdown time, and interarrival time. Yang et al. [1] investigated the spatial patterns of the hydrologic alterations caused by dam construction in the middle and lower Yellow River, China during the most recent five decades using the range of variability approach (RVA) method in which 33 hydrologic parameters were used to assess the hydrologic alterations in terms of streamflow magnitude, timing, frequency, duration, and rate of change. Romano et al. [15] evaluated the pre-dam versus post-dam differences in flood frequency and duration below the Carlyle Dam, lower Kaskaskia River, U.S. The results indicated a decrease in flood duration and frequency, and a decrease in annual flood frequency variation at a near site below the dam. However, their results also showed that hydrological alterations were related to climate rather than dam effects at a far distance downstream from the dam, emphasizing that the distance downstream from the dam and downstream tributary and watershed characteristics should be considered. These works mentioned above have indicated that dams appear to decrease flood peaks and increase low flows. However, these have all focused only on comparing pre- and post-dam flows with respect to the hydrograph characteristics of frequency, magnitude, and shape; they do not focus on naturalized flow conditions that would have existed before dam construction without the effect of regulation.

Some investigations have compared historical regulated flow data with natural flow data generated by rainfall-runoff modeling to assess the impacts of dams on flood flows. The U.S. Geological Survey (USGS) [16,17] has published reports evaluating the effects of flood detention reservoirs on flood frequencies, as well as peak discharges along downstream reaches for several small urban drainage basins with areas of 0.1–0.37 square miles in Georgia, U.S. In the study, the Distributed Routing Rainfall-Runoff Model (DR3M), developed by the USGS, was used to simulate the long-term peak discharges for conditions with and without the flood detention reservoirs. The relationships of flood-frequency, between those with and without reservoirs, were developed based on these long-term peak discharges. Peters and Prowse [18] quantified the effects of regulation on the Peace River, Canada, based on comparisons between pre-regulated, regulated, and naturalized flow regimes.

In their work, a modified Streamflow Synthesis and Reservoir Regulation (SSARR) model was used to naturalize the river flows. Montaldo et al. [19] analyzed the flood hydrograph attenuation induced by a reservoir system for the Toce River basin, Italy, using a Flash-flood Event-based Spatially distributed rainfall-runoff Transformation, including Reservoir System (FEST98RS). Sayama et al. [20] developed a rainfall-runoff prediction system for the Yodo River basin, Japan, and demonstrated that the system was able to simulate complex dam operations, such as preliminary release, peak attenuation, and cooperative operations by multiple dams. Gross and Moglen [21] developed regression equations, both natural and with the dam, for predicting the influence distance of flow regulation downstream of a dam using the Geographic Information System (GIS), the HEC (Hydrologic Engineering Center)-1 model, and the stage-storage-discharge curves for 34 dams in Maryland, U.S. All these studies relied on rainfall-runoff modeling to assess the post-dam effects on flood flows. From these reviews, it can be seen that previous studies on the impact of dams and reservoirs have predominantly focused on alterations in the hydrograph. However, none have quantified the effects of regulation on flood frequency, except those conducted by the USGS [16,17]. A possible reason for the lack of studies assessing the effects of dams on flood frequencies during the post-dam period is that generating flood data over a sufficient period of time for a flood frequency analysis is difficult.

Traditionally, flood frequency estimations have been based on statistical analyses using available gauged flood peaks. In cases where long records of measured streamflow data are not available, an estimate of flood frequency must be based on event simulation or continuous simulation modeling. The availability of powerful computers has made it possible to move towards techniques for flood estimation based on continuous simulation modeling [22]. Since the late 1990s, the estimation of flood frequency was approached using the continuous simulation of stream flow time series [23–31]. The main advantages of this approach are that the streamflow can be considered as a single term, without explicit prior separation into direct runoff and baseflow, and the problem associated with the antecedent wetness condition, which is very important in the consideration of event simulation, is largely removed because this condition is an integral part of the modeling procedure [23]. Furthermore, other merits of using a continuous simulation model to simulate streamflow include: the use of longer data series for the frequency analysis, the ability to simulate streamflow under future land use/climate conditions, and no assumption of a direct transformation of a design rainfall to a design flood [25]. These advantages of continuous simulation modeling make it possible to more elaborately explore various hydrological problems.

Therefore, the objective of this study is to investigate the potential of a continuous simulation model for assessing the effects of a dam on the characteristics of downstream flood frequencies. A continuous simulation model, SWAT (Soil and Water Assessment Tool), was used to predict the stream flow time series, both with and without dams, which were then analyzed to produce flood frequency curves. The SWAT model was developed by the U.S. Department of Agriculture (USDA) Agricultural Research Service and is a quasi-distributed model used for simulating runoff at the watershed scale [32,33]. The reasons for selecting the SWAT model include its wide global use and good performance for predicting daily runoff hydrographs. Moreover, the model is an open source code program; thus, it can be modified to improve its performance. This paper outlines the overall procedure of the SWAT-simulation-based flood frequency estimation and discusses the results of the impacts of the upstream Soyanggang and Chungju multi-purpose dams on the flood frequency of downstream flows in the Han River basin, South Korea.

2. Methodology

2.1. SWAT Model Description

The SWAT model was developed as a watershed scale hydrological model for assessing the effects of land use and management on the long-term runoff, sediment, and non-point source pollution loads to surface water bodies on a daily basis. The major components of SWAT include: weather, hydrology, erosion, plant growth, nutrients, pesticides, land management, and streamflow routing. The methods for estimating the hydrological components are briefly described below. The model allows for simulation with high level spatial detail by dividing the watershed into multiple sub-watersheds, which are then partitioned into additional areas, called Hydrologic Response Units (HRUs). The water in each HRU is stored in or released from four storage volumes: snow, soil profile, shallow aquifer, and deep aquifer. Snow melts on days when the maximum temperature exceeds a preselected threshold value. Melted snow is treated as rainfall for estimating the amount of runoff and percolation. The soil profile is subdivided into multiple layers for soil water processes, such as infiltration, evaporation, plant uptake, lateral flow, and percolation. Surface runoff or infiltration from daily rainfall is calculated using a modified Soil Conservation Service (SCS) Curve Number (CN) method, or is otherwise calculated using the Green-Ampt method. The model individually computes the evaporation from soils and transpiration from plants. Potential evapotranspiration (PET) can be calculated using the Penman–Monteith, Priestley–Taylor, or Hargreaves methods. Potential soil water evaporation is estimated using a function of the PET and leaf area index. Actual soil evaporation is estimated using exponential functions for the soil depth and water content. Plant water uptake is simulated by a linear function of the PET, leaf area index, and root depth, and is limited by the soil water content. The soil percolation component is estimated using a water storage capacity technique, in which downward flow occurs when the field capacity of a soil layer is exceeded. The shallow aquifer is recharged by the water from the bottom of the soil profile by the percolation process. Simultaneous to percolation, the lateral sub-surface flow in the soil profile is calculated using the kinematic storage model, which is a function of the saturation hydraulic conductivity, slope length, and slope. The recharge and discharge of groundwater were estimated using exponential attenuation weighting functions. These hydrological components for each HRU are summed over a sub-watershed. The calculated flow obtained for each sub-watershed is then routed through the river channels. Finally channel routing is simulated using the variable storage or Muskingum method. More details can be found in the work of [34].

2.2. Study Area

The seven dams located upstream of the Paldang Dam are shown in Figure 1, including: the Soyanggang and Chungju multi-purpose dams, and the Hwacheon, Chuncheon, Uiam, Cheongpyeong, and Goesan dams for electricity generation in the Han River basin. In this study, the impacts of the two major multi-purpose dams, Soyanggang and Chungju, on the downstream flood frequency were assessed at the Paldang Dam. According to [35], the dams for electricity generation (Hwacheon, Chuncheon, Uiam, Cheongpyeong, and Goesan) have little effect on flood control and were therefore not included in this assessment. The Chungju Dam, which was built lastly among the dams considered in this study, was constructed in 1986, therefore, the data period used in this work starts from that year.

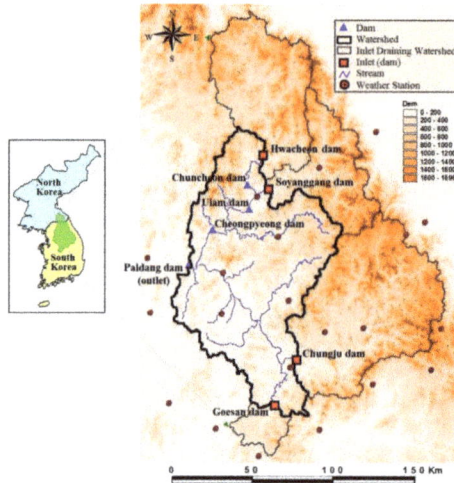

Figure 1. Study area.

2.3. Approach for Flood Frequency Assessment

Current inflows to the Paldang Dam are regulated by the upstream multi-purpose dams, Chungju and Soyanggang. Data without the influence of the dams are not available; therefore, a frequency analysis was not possible. The SWAT model allows for daily streamflow estimation, without the influence of the dam, by designating upstream dams as inlet conditions of the model and plugging upstream dam inflows into the inlets. The historical dam inflows can be regarded as unregulated flows below the dams. Therefore, in this study, SWAT simulations were performed to generate streamflows for the conditions without dams; then the simulated streamflow data was used for the flood frequency analysis.

The effects of dams on the peak inflows and flood frequencies at the Paldang Dam were determined by simulating the annual peak flows with and without dams. Annual peak flows for the period 1986–2015 were simulated by the SWAT model. Several scenarios were built based on the presence of dams, and the resulting flood frequencies were compared to each other to characterize the effects of dam regulation on floods. The first simulation was for existing conditions, i.e., with dams in place. Subsequent simulations involved eliminating individual dams to determine the effect of a single dam on the peak flows for the entire basin. The model was also used to simulate conditions without any dams for a basin to examine the cumulative effects of upstream dams on the peak flow at the outlet.

SWAT simulation is based on a daily time step; as such, the SWAT model cannot represent a small time scale, such as hours or minutes, for simulating sharp peaks. To overcome this, Sangal's method was used to estimate the instantaneous peak flow from the mean daily flow. Sangal [36] suggested a practical formula, shown in Equations (1) and (2), based on the assumption of a triangular hydrograph, in which the instantaneous peak flow is estimated from the mean daily flow for three consecutive days, including the peak day.

$$Q_{max} = \frac{Q_1 + Q_3}{2} + \frac{2Q_2 - Q_1 - Q_3}{K} \tag{1}$$

$$K = \frac{4Q_2 - 2Q_1 - 2Q_3}{2Q_{peak} - Q_1 - Q_3} \tag{2}$$

where Q_{max} is the predicted instantaneous peak flow (m^3/s), Q_2 is the mean daily flow (m^3/s) of the day that includes the peak flow, Q_1 and Q_3 are the mean daily flow (m^3/s) for the days before and after the peak flow day, respectively, K is the base factor calculated from the historical mean daily and peak flows, and Q_{peak} is the actual peak flow (m^3/s). Sangal [37] applied this method to streams with drainage areas that varied from less than 1 km^2 to more than 100,000 km^2 in Ontario, Canada, using the data of 3946 peak flows collected from a total of 387 stations and achieved results with reasonable accuracies. Also, the base factor K is an important parameter and can be estimated from historical flows. In this study, a reasonable estimate of $K(= 0.96)$ is calculated from historical flows at the Paldang Dam.

3. SWAT Modeling

3.1. Input Data and Model Preparation

The SWAT model is applicable for assessing the effects of dam regulation on downstream floods. Accordingly, the model was used in this study to assess the changes in downstream floods regulated by the upstream multi-purpose dams in the Paldang Dam watershed. As shown in Figure 1, the Soyanggang, Chungju, Hwacheon, and Goesan Dams comprised the model inlets from their respective upstream watersheds. Instead of simulating runoff from the entire watershed, the observed discharges from the Soyanggang and Chungju dams were directly plugged into the model, with runoffs from the remaining watersheds then simulated to determine the flood inflows to the Paldang Dam. This was intended to eliminate errors associated with the runoff simulation for upstream watersheds. In addition, the Chuncheon, Uiam, and Cheongpyeong electricity generation dams that are located in the middle of the study watershed were not considered in the modeling because they are incapable of flood control and discharge all inflows nearly without detention for electricity generation. Therefore, the inflows to the Paldang Dam can be simulated reasonably even though the electricity generation dams were not considered. Table 1 shows the inlet conditions at the upper dams to be implemented. The inflow and outflow for the Soyanggang and Chungju dams in 1990 is illustrated in Figure 2.

Table 1. Inlet Conditions at the Upper Dams.

Scenarios	Soyanggang	Chungju	Hwacheon	Goesan
Current state	Outflow	Outflow	Outflow	Outflow
Scenario 1	Inflow	Outflow	Outflow	Outflow
Scenario 2	Outflow	Inflow	Outflow	Outflow
Scenario 3	Inflow	Inflow	Outflow	Outflow

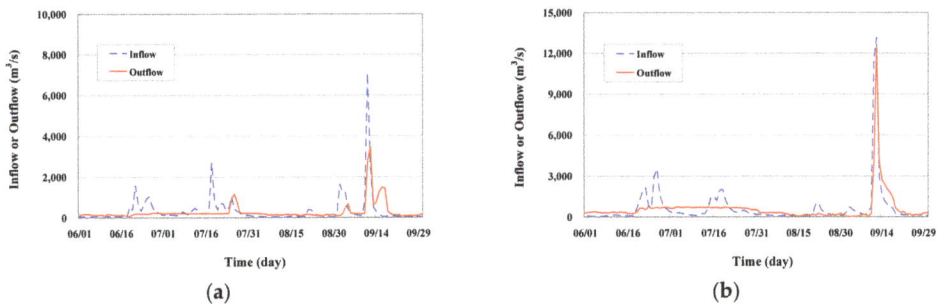

Figure 2. Inflow and Outflow in 1990. (a) Soyanggang Dam; (b) Chungju Dam.

Daily climatic data of precipitation, solar radiation, maximum/minimum temperature, wind speed, and relative humidity from 1986 to 2015 were used for the SWAT model simulation. A Digital Elevation Model (DEM), with a 100 m mesh size, mid-classified land cover from the Ministry of Environment, and soil map from the National Institute of Agricultural Sciences were utilized as the spatial input data for the model. The entire study watershed was divided into 21 sub-watersheds. The two methods of the temporally weighted average CN method [37] and non-linear storage routing technique [38] were incorporated into the SWAT model in this study for better simulation of the surface runoff and streamflow routing, respectively. Evapotranspiration was estimated using the Penman-Monteith equation.

3.2. Model Calibration and Validation

The parameters of the SWAT model were calibrated using the observed daily inflow data collected from 1986 to 1995 at the Paldang Dam. The parameters were manually adjusted to fit the daily simulations as close as possible to the observations. The statistics of the coefficient of determination (R^2) and Nash–Sutcliffe efficiency (NSE) [39] were used to evaluate model performances. For a brief description of the calibration procedure, firstly, the soil evaporation compensation coefficient (ESCO) parameter was calibrated to match the total amount of runoff. The other internal parameters were then adjusted for better simulation of the hydrograph, especially for the flows during recession and at peak time. We increased the CN value with the Antecedent Moisture Condition (AMC)-II condition (CN2) during the flood season (June–September) by 20% from the default value to increase the peak flow. In addition, we calculated the CN value of each day using the temporally weighted average CN method [37]. The CH_N (Manning's n) parameter related to the channel routing was estimated as 0.030 using the non-linear storage routing technique [38], which is higher than the default value, in consideration of the natural channel condition and reliability of the value estimated. The two parameters of average slope length for the subbasin (SLSUBBSN) and adjustment factor for the lateral flow (ADJF), which are related to the interflow generation, were then calibrated. The delay of groundwater recharge (GW_DELAY) parameter was adjusted to generate a reasonable shape of the recession hydrograph. Table 2 shows a list of the calibration parameters and their optimal values.

Table 2. Calibrated model parameters.

Parameter	Description	Calibrated Value
ESCO	Soil evaporation compensation factor	0.65
CN2	CN value with the AMC-II	+20%
CH_N	Manning's n	0.03
ADJF	Adjustment factor for lateral flow (default = 1)	2
SLSUBBSN	Average slope length for subbasin (m)	Max (10, default)
GW_DELAY	Groundwater delay (days)	150

Figure 3a shows a good agreement between the observed and simulated daily runoff for the calibration period of 1986–1995. The model performance statistics of R^2 and NSE using the optimal parameter values for the calibration period were satisfactory, at more than 0.85. The calibrated parameter values were then applied to the SWAT model to simulate the flow time series of the validation period of 1996–2015 to evaluate the model prediction ability. Figure 3b shows a good model performance with R^2 and NSE values greater than 0.85. It reveals that the hydrologic processes were reasonably modeled by the SWAT model.

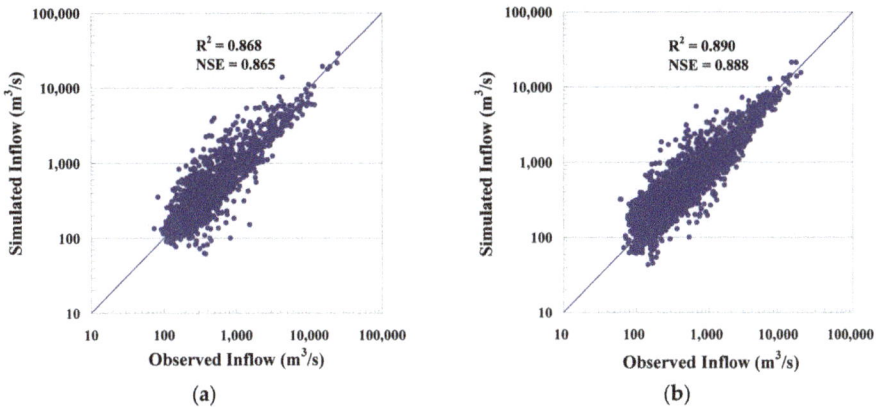

Figure 3. Comparison of the observed and simulated inflows to the Paldang Dam for (a) calibration period (1986–1995) and (b) validation period (1996–2015).

3.3. Model Results

The runoff ratio, flow duration curve, and daily streamflow discharge simulated by SWAT were analyzed at the Paldang Dam. The runoff ratio is defined as the runoff volume divided by the rainfall volume that occurred during the flood season (June–September) of each year. Table 4 contains the observed and simulated runoff ratios from 1986 to 2015, as well as the statistical values of R^2 between the simulated and observed daily flows. The results are highly acceptable.

Table 3. Runoff ratio and Determination coefficient.

Year	Rainfall (mm)	Runoff (mm)		Runoff Ratio (%)		Determination Coefficient (R^2)
		Observed	Simulated	Observed	Simulated	
1986	788.9	378.0	357.9	47.9	45.4	0.795
1987	1206.4	940.1	865.5	77.9	71.7	0.867
1988	712.8	374.6	398.5	52.5	55.9	0.849
1989	850.7	334.8	372.1	39.4	43.7	0.733
1990	1545.5	1128.7	1040.7	73.0	67.3	0.856
1991	925.4	421.9	429.0	45.6	46.4	0.722
1992	719.7	249.7	277.1	34.7	38.5	0.709
1993	797.8	508.1	481.1	63.7	60.3	0.745
1994	687.7	220.9	261.8	32.1	38.1	0.721
1995	1236.1	822.6	750.4	66.5	60.7	0.904
1996	715.7	370.3	365.0	51.7	51.0	0.854
1997	730.4	458.6	386.2	62.8	52.9	0.729
1998	1276.0	726.0	702.9	56.9	55.1	0.877
1999	1108.3	580.2	610.5	52.4	55.1	0.854
2000	926.1	424.3	449.5	45.8	48.5	0.770
2001	753.5	285.7	311.6	37.9	41.4	0.798
2002	926.0	465.0	537.0	50.2	58.0	0.934
2003	1277.0	724.2	717.5	56.7	56.2	0.842
2004	1051.5	614.4	628.4	58.4	59.8	0.864
2005	1209.1	527.0	501.0	43.6	41.4	0.749
2006	1242.3	732.2	756.6	58.9	60.9	0.971
2007	1027.9	589.7	525.8	57.4	51.1	0.960
2008	909.5	374.3	373.2	41.2	41.0	0.941
2009	1013.4	487.4	459.0	48.1	45.3	0.925
2010	1074.0	477.4	436.2	44.4	40.6	0.914

Table 4. Runoff ratio and Determination coefficient.

Year	Rainfall (mm)	Runoff (mm)		Runoff Ratio (%)		Determination Coefficient (R^2)
		Observed	Simulated	Observed	Simulated	
2011	1568.8	919.7	903.4	58.6	57.6	0.974
2012	990.0	405.9	382.6	41.0	38.6	0.870
2013	1050.5	548.9	496.5	52.3	47.3	0.964
2014	494.9	125.7	144.0	25.4	29.1	0.914
2015	395.8	86.5	91.0	21.8	23.0	0.980
Average	973.7	510.1	500.4	50.0	49.4	0.828

From the simulated daily flows, the exceedance probability of the flow, referred to as the flow duration curve, was plotted with that based on the observed data, as shown in Figure 4. In can be seen that all sections of the high, intermediate, and low flows were simulated with reasonable accuracy.

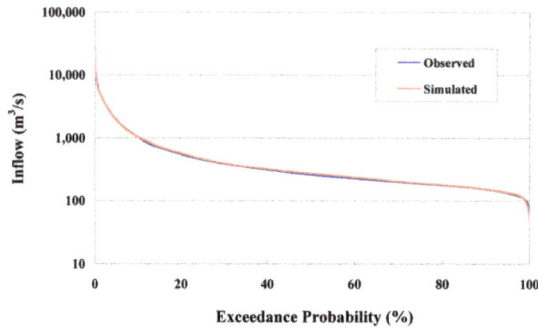

Figure 4. Observed and Simulated Flow Duration Curve at the Paldang Dam.

Figure 5 compares the hydrographs of the observed and simulated daily inflows to the Paldang Dam for large, mid, and low flood years between 1986 and 2015. The simulated inflows matched well with the observed inflows (average R^2 = 0.828 in Table 4).

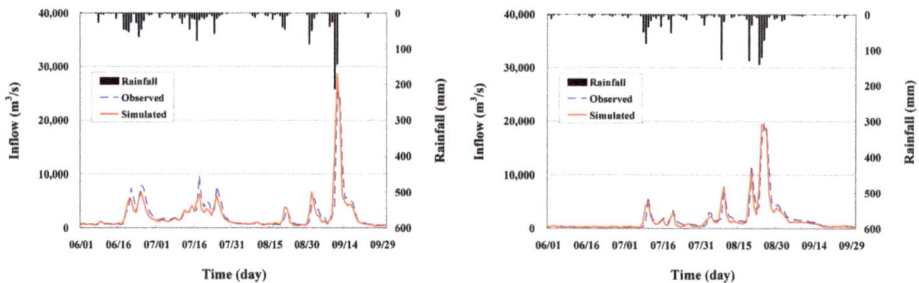

(a)

Figure 5. *Cont.*

(b)

(c)

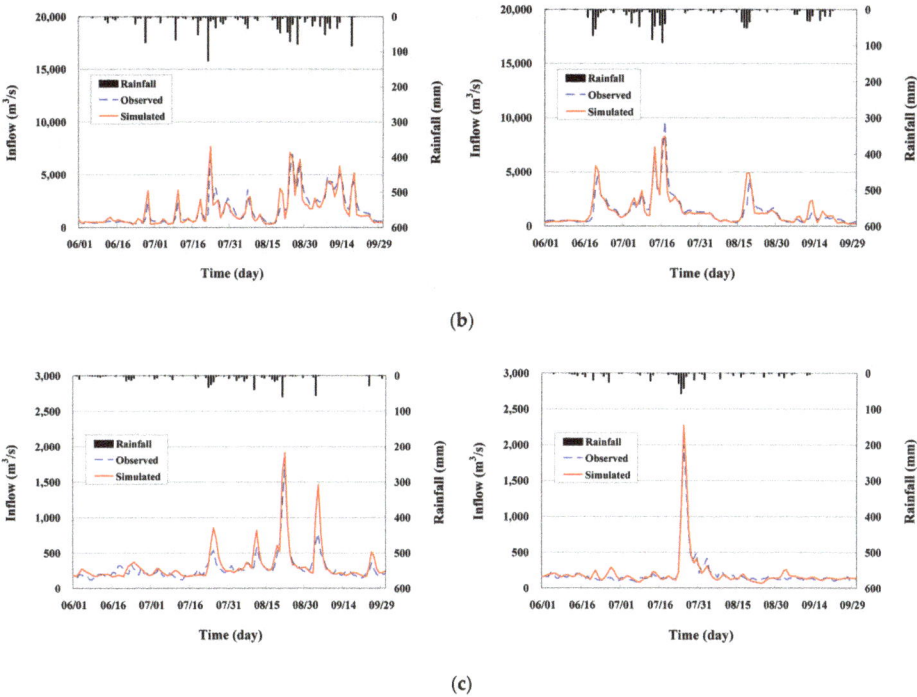

Figure 5. Observed and Simulated Daily Inflow at the Paldang Dam. (**a**) Large flood years (left: 1990, right: 1995); (**b**) Mid flood years (left: 2003, right: 2004); (**c**) Low flood years (left: 2014, right: 2015).

4. Results and Discussion

4.1. Flood Frequency Analysis of the Observed and Simulated Daily Inflows

The annual maximum series of simulated and observed inflows at the Paldang Dam were extracted from the simulated and observed hydrographs. Comparisons between the annual maximum series of simulated and observed inflows for the study period (Figure 6) showed the robustness of the SWAT simulation (R^2 = 0.903, Root Mean Square Error (RMSE) = 0.187 m^3/s).

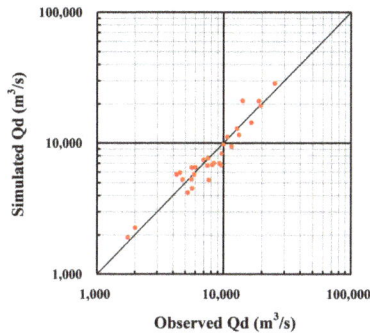

Figure 6. Annual Maximum Series of Observed and Simulated Daily Inflow.

A flood frequency analysis was performed on the annual maximum series of simulated and observed daily peak inflows at the Paldang Dam. The data used is shown in Figure 6. In this study, the probability distribution of Extreme Value Type-I was selected for the flood frequency analysis, with the relevant parameters estimated using the method of L-moments, as described by [40]. A comparison of flood estimates and frequencies computed from both the simulated and observed data is shown in Figure 7. Comparing the observed and simulated flood estimates, the flood estimates based on the simulated streamflow were slightly overestimated compared to the observed data. The overall errors ranged from −1.84% to 3.18%. This small error suggests satisfactory correspondence between the simulated and observed flood frequencies.

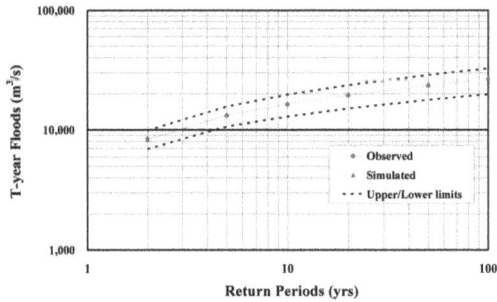

Figure 7. Observed and Simulated T-year Floods at the Paldang Dam (Daily).

4.2. Influence of Regulated Flow by the Dam on Daily Flood Frequency

The inflows to the Paldang Dam, under the regulation of the upstream dams, were evaluated in the previous section. To assess the dam effects on the flood frequency, flood flows into the Paldang Dam without upstream dam regulation need to be estimated and compared with regulated floods. Three scenarios were constructed for the unregulated flood simulation based on the existence of an individual dam: Scenario 1 (Chungju dam only), Scenario 2 (Soyanggang dam only), and Scenario 3 (no dam), as shown in Table 1. The runoff from all watersheds without any dam (Scenario 3) was simulated to estimate the unregulated floods to the Paldang Dam. Similar to the regulated flow simulation, the measured outflows or inflows of each dam were used as inlets to the model. The measured outflow or inflows were subjected to flow directly to the downstream channel of each dam for the watersheds with a dam, i.e., the Soyanggang and Chungju dams, for Scenarios 1 and 2.

Figure 8 shows a comparison of the hydrographs for the current state with the Chungju and Soyanggang dams (w/CJ, SY) and the three scenarios for 1990, 1995, 1999, and 2006, where high flood peaks occurred. It was obvious that regulation resulted in a marked reduction in discharges. For example, between Scenario 3 (natural flow condition; w/o CJ, SY) and the current state, the peaks were reduced by an average of 30.9% for the years studied.

Figure 9 depicts the simulated annual maximum series of inflows for the three scenarios at the Paldang Dam from 1986 to 2015. As expected, the modeled peaks for the three scenarios were higher than those found in the historical records for the current state with the effects of the Soyanggang and Chungju multi-purpose dams. In Figure 9, the slopes indicate the degree of increase in the peaks. Disregarding the effects of the dams on the peak attenuation would lead to considerably larger design flood estimates.

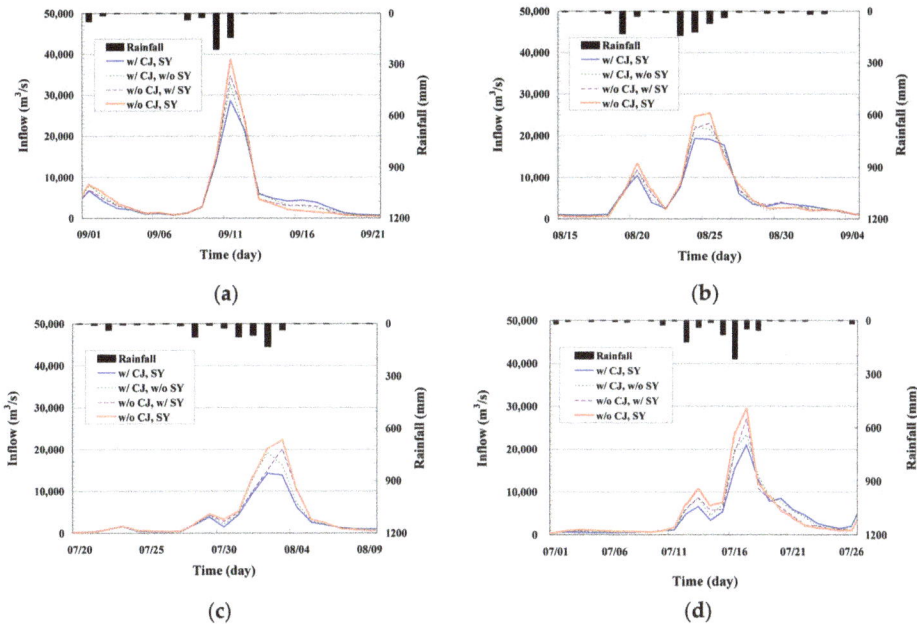

Figure 8. Hydrographs according to the Scenarios at the Paldang Dam. (**a**) 1990; (**b**) 1995; (**c**) 1999; (**d**) 2006.

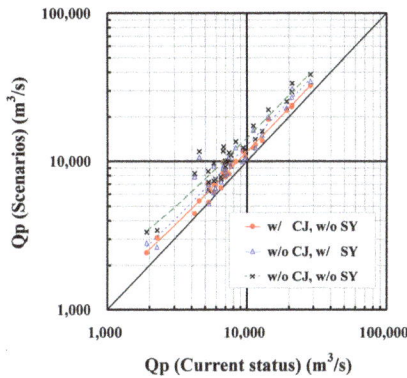

Figure 9. Annual Maximum Series according to the Scenarios (Daily).

Figure 10 presents the flood frequency for each scenario. Natural (Scenario 3) and regulated (Scenario 1, 2, current) flood frequency curves indicated similar patterns for all scenarios. The discrepancy between the flood with and without the dams indicates the effect of regulation by the dam. Assuming Scenario 3 represents the flood frequency under natural conditions, the regulated flood frequencies estimated from the current state and Scenarios 1 and 2 were 70.1%, 81.2%, and 90.1% of the natural flood, respectively. The 100-year flood under the current state (27,050 m^3/s) appeared to be equivalent to the natural flood, with 10 to 20-year return periods, which implies the substantial effect of the dam on flood regulation. The percentage increases expected in the flood estimates for Scenarios 1, 2, and 3 for the return periods of 2, 5, 10, 20, 50, and 100 years without dams are listed in Table 5. The data presented in Table 5 reveal that the removal of the Soyanggang Dam from the basin would increase flood peaks by 15.6% to 16.5%, the removal of the Chungju Dam would increase flood peaks by 28.5% to 28.7%, and the removal of both dams would increase floods by 41.5% to 45.4%. The effect of the Chungju Dam on the flood peak attenuation was about twice that of the Soyanggang Dam. This difference in the effects of dams on flood peaks can be attributed to the dam size and stage-outflow in relation to each dam.

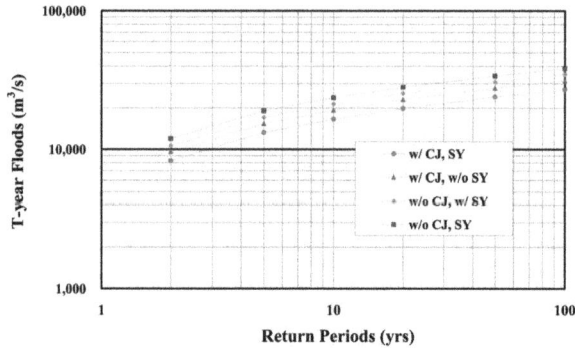

Figure 10. Flood Change for different Return Periods according to the Scenarios (Daily). T-year represents the different return period.

Table 5. Percent Increase in T-year Floods according to the Scenarios (Daily) (%).

Scenarios	Return Periods (Years)						
	2	5	10	20	50	100	Ave.
Scenario 1	16.5	16.0	15.9	15.8	15.7	15.6	15.9
Scenario 2	28.5	28.6	28.6	28.6	28.6	28.7	28.6
Scenario 3	45.4	43.3	42.6	42.1	41.7	41.5	42.7

4.3. Influence of Regulated Flow by the Dam on the Hourly Flood Frequency

Knowledge of instantaneous peak flows is often required to estimate or assess the design flood for hydraulic structures, such as dams and levees, because there may be significant streamflow variations within hours, especially for small basins. For the basin of interest in this study, the annual maximum hourly and daily peak inflow data observed at the Paldang Dam from 1986 to 2015 are plotted in Figure 11. The mean ratio of annual maximum hourly to daily peaks was 1.47, which indicates that the use of mean daily data may cause underestimation of the design flood.

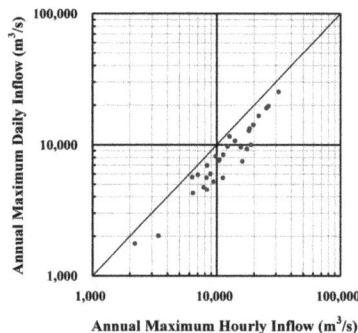

Figure 11. Annual Maximum Series of Hourly and Daily Observed Inflow at the Paldang Dam.

Figure 12 shows the comparison of flood frequency curves based on the observed daily and hourly annual extreme data. The ratios of hourly to daily floods were calculated for return periods of 2, 5, 10, 20, 50 and 100 years, which resulted in an average of approximately 1.39. It is obvious that there was a significant difference between the hourly and daily peaks; therefore, it is necessary to use small time scale peak flow data for a precise flood frequency assessment.

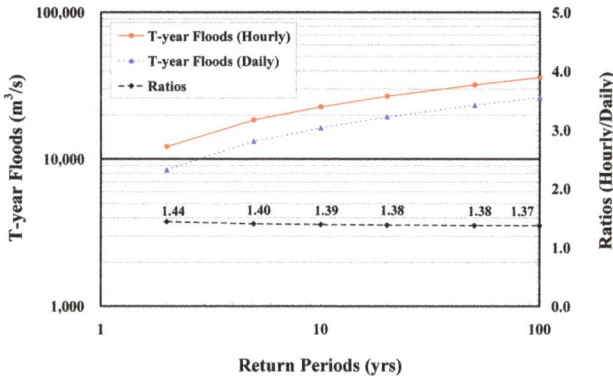

Figure 12. Comparison of the Hourly and Daily Frequency Curve at the Paldang Dam.

To test the validity of Sangal's method for predicting the instantaneous peak flow at the Paldang Dam, a comparison between the observed and estimated hourly peak flows was conducted, the results of which are shown in Figure 13. The estimated values of the hourly peak flow presented a RMSE of less than 0.170. The R^2 value between the observed and estimated data was 0.882, which represents a high correlation. From these statistical values, it was concluded that Sangal's method is reasonable for estimating the hourly peak flows at the Paldang Dam, the outlet of the study area.

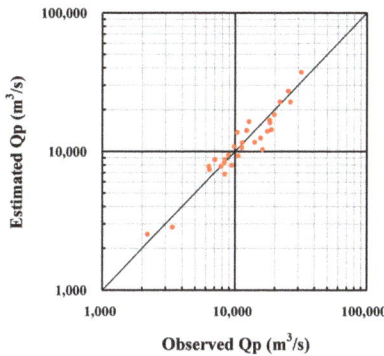

Figure 13. Annual Maximum Series of Observed and Estimated Peak Inflow at the Paldang Dam.

The daily flow peaks simulated by SWAT were converted to hourly flow peaks using Sangal's method, and a series of annual maximum hourly peak flows was then constructed for a flood frequency analysis (Figure 14).

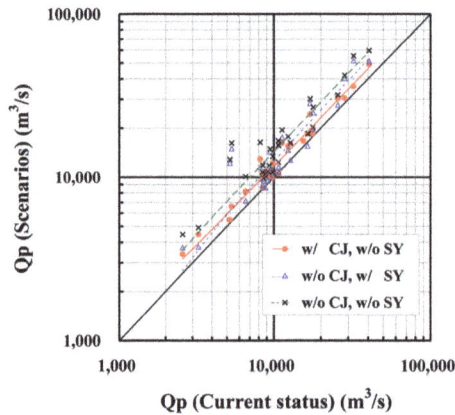

Figure 14. Annual Maximum Series according to the Scenarios (Hourly).

The flood frequency estimates based on hourly data, obtained by combining the SWAT simulation and Sangal's method, were compared with the observation-based results, as shown in Figure 15. Flood frequency estimates based on the simulated data were slightly overestimated, but the errors for return periods of 2, 5, 10, 20, 50, and 100 years were less than 8% (ranged from −2.49% to 7.13%). Therefore, the combination of the SWAT simulation and Sangal's method is suitable for estimating small time scale flood peaks.

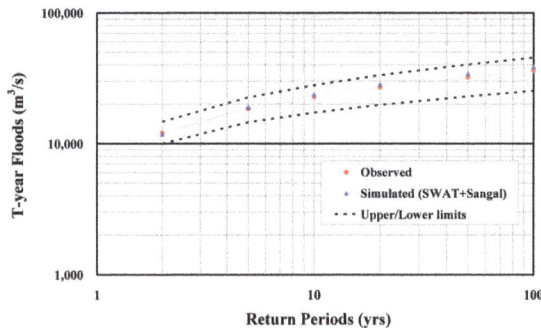

Figure 15. Observed and Simulated Floods for different Return Periods at the Paldang Dam (Hourly). T-year represents the different Return period.

For the same scenarios described in Section 4.2, the effects of dams on peak inflows and flood frequencies on an hourly basis at the Paldang Dam were assessed for conditions with and without dams (Figure 16 and Table 6). The hourly peak flow data were simulated by the procedure mentioned above. A comparison of the flood frequency estimates at the Paldang Dam for the current condition (with the Soyanggang and Chungju dams) and for the condition representing the removal of the Soyanggang Dam (Scenario 1), indicated that the peak inflows increased from 16.0% to 16.1% for the 2, 5, 10, 20, 50 and 100 year return periods. The removal of the Chungju Dam (Scenario 2) caused the peak inflows to increase from 26.0% to 32.1%. Therefore, Scenario 2 had a greater effect on the peak flow than Scenario 1. The peak flow increment ratio with the removal of both dams (Scenario 3) was slightly lower than the summation of the respective peak flow increment ratios for the removal of the individual dams (Scenarios 1 and 2).

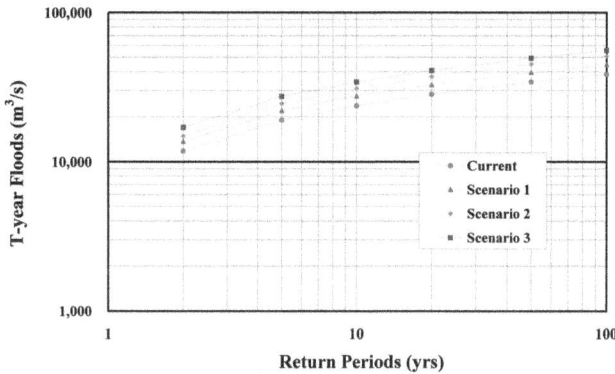

Figure 16. Flood Change for different Return periods according to the Scenarios (Hourly). T-year represents the different return period.

Table 6. Percent Increase in T-year Floods according to the Scenarios (Hourly) (%).

Scenarios	Return Periods (Years)						
	2	5	10	20	50	100	Ave.
Scenario 1	16.0	16.1	16.1	16.1	16.1	16.1	16.1
Scenario 2	26.0	29.3	30.4	31.1	31.8	32.1	30.1
Scenario 3	43.6	44.0	44.2	44.3	44.3	44.4	44.1

5. Conclusions

This study explored the downstream flood frequency alterations caused by dam construction in the Han River basin, South Korea, during 1986–2015 using streamflows simulated by a continuous model, SWAT. The model was used to simulate the long-term peak flows at the Paldang Dam for conditions both with (current state) and without (removal of the Soyanggang Dam or/and the Chungju Dam) upstream dams. Flood frequency curves were established from the simulated annual peak flows for each of these conditions by fitting a series of annual peak flow data to the Extreme Value Type-I distribution. The individual and cumulative effects of upstream dams on the downstream flood frequency were assessed by comparing the developed flood frequency curves. The main conclusions are as follows:

(1) A comparison of the simulated daily peak flows with the observed data indicated that the use of the SWAT model was suitable for estimating the flood frequency. A close correlation ($R^2 = 0.903$, RMSE = 0.187 m^3/s) between the daily flood estimates for the return periods of 2, 5, 10, 20, 50 and 100 years, computed from the observed and simulated daily inflows to the Paldang Dam, was achieved under the current condition, i.e., with the Soyanggang and Chungju dams in place.

(2) The effect of the Chungju Dam on the flood frequency at the Paldang Dam was found to be greater that of the Soyanggang Dam during the simulation periods. The removal of the Soyanggang Dam (Scenario 1; regulation by the Chungju Dam only) caused an increase in the daily flood peaks by 15.9%, while the removal of the Chungju Dam (Scenario 2; regulation by the Soyanggang Dam only) increased the daily flood peaks by 28.6%.

(3) The peak flow increment ratio by removing both Soyanggang and Chungju dams (Scenario 3) was slightly lower than the summation of the respective peak flow increment ratios from Scenarios 1 and 2.

(4) To overcome the inability of SWAT to reproduce sharp events within hours, a procedure incorporating Sangal's method for estimating instantaneous peak flow from the daily flow

into the SWAT simulation has been proposed in the present work. As a result of the flood frequency analysis on an hourly basis using this procedure, the errors in the flood estimates were less than 8%, which leads to acceptable accuracy.

(5) The increased average percentage of the hourly flood estimates for the three scenarios, relative to the current state, were 16.1%, 30.1%, and 44.1%, for the removals of the Soyanggang, Chungju, and both dams, respectively. These increased percentages were a little higher than those for the estimated daily flood frequencies.

(6) The developed approach allows for a better understanding of flood frequency alterations during the post-dam period, which will improve the applicability of continuous simulation models for the analysis of flood frequency.

Acknowledgments: This research was supported by a grant (code number: 17RDRP-B076272-04) from the Jeju Regional Infrastructure Technology Development Center funded by Ministry of Land, Infrastructure and Transport of the Korean government.

Author Contributions: Jeong Eun Lee performed SWAT modeling as well as flood frequency analysis and wrote the manuscript; Jun-Haeng Heo contributed in-depth analysis of the results and provided editing and writing support; Jeongwoo Lee provided technical advice and writing support; Nam Won Kim developed the methodology and managed the research.

Conflicts of Interest: The authors declare no conflict of interest.

References

1. Yang, T.; Zhang, Q.; Chen, Y.D.; Tao, X.; Xu, C.; Chen, X. A spatial assessment of hydrologic alteration caused by dam construction in the middle and lower Yellow River, China. *Hydrol. Process.* **2008**, *22*, 3829–3843. [CrossRef]
2. Poff, N.L.; Allan, J.D.; Bain, M.B.; Karr, J.R.; Prestegaard, K.L.; Richter, B.D.; Sparks, R.E.; Stromberg, J.C. The natural flow regime. A paradigm for river conservation and restoration. *BioScience* **1997**, *47*, 769–784. [CrossRef]
3. Richter, B.D.; Baumgartner, J.V.; Powell, J.; Braun, D.P. A method for assessing hydrologic alteration within ecosystems. *Conserv. Biol.* **1996**, *10*, 1163–1174. [CrossRef]
4. Walker, K.F.; Sheldon, F.; Puckridge, J.T. A perspective on dryland river ecosystems. *Regul. Rivers Res. Manag.* **1995**, *11*, 85–104. [CrossRef]
5. Gregory, K.J.; Park, C. Adjustment of river channel capacity downstream from a reservoir adjustment of river channel capacity downstream from a reservoir. *Water Resour. Res.* **1974**, *10*, 870–873. [CrossRef]
6. Page, K.J. Bankfull Discharge Frequency for the Murrumbidgee River, New South Wales. In *Fluvial Geomorphology of Australia*; Warner, R.F., Ed.; Academic Press: Sydney, Australia, 1988; pp. 267–281.
7. Galat, D.L.; Lipkin, R. Restoring ecological integrity of great rivers: Historical hydrographs aid in defining reference conditions for the Missouri River Restoring ecological integrity of great rivers: Historical hydrographs aid in defining reference conditions for the Missouri River. *Hydrobiologia* **2000**, *422/423*, 29–48.
8. Maingi, J.K.; Marsh, S.E. Quantifying hydrologic impacts following dam construction along the Tana River, Kenya. *J. Arid Environ.* **2002**, *50*, 53–79. [CrossRef]
9. Magilligan, F.J.; Nislow, K.H.; Graber, B.E. Scale-independent assessment of discharge reduction and riparian disconnectivity following flow regulation by dams. *Geology* **2003**, *31*, 569–572. [CrossRef]
10. Graf, W.L. Downstream hydrologic and geomorphic effects of large dams on American rivers. *Geomorphology* **2006**, *79*, 336–360. [CrossRef]
11. Pegg, M.A.; Pierce, C.L.; Roy, A. Hydrological alteration along the Missouri River Basin: A time series approach. *Aquat. Sci.* **2003**, *65*, 63–72. [CrossRef]
12. Batalla, R.J.; Gomez, C.M.; Kondolf, G.M. Reservoir-induced hydrological changes in the Ebro River basin, NE Spain. *J. Hydrol.* **2004**, *290*, 117–136. [CrossRef]
13. Magilligan, F.J.; Nislow, K.H. Changes in hydrologic regime by dams. *Geomorphology* **2005**, *71*, 61–78. [CrossRef]
14. Singer, M.B. The influence of major dams on hydrology through the drainage network of the Sacramento River Basin, California. *River Res. Appl.* **2007**, *23*, 55–72. [CrossRef]

15. Romano, S.P.; Baer, S.G.; Zaczek, J.J.; Williard, K.W.J. Site modelling methods for detecting hydrologic alteration of flood frequency and flood duration in the floodplain below the carlyle dam, lower Kaskaskia River, Illinois, USA. *River Res. Appl.* **2009**, *25*, 975–984. [CrossRef]

16. Hess, G.W.; Inman, E.J. *Effects of Urban-Detention Reservoirs on Peak Discharges in Gwinnett County, Georgia*; Water-Resources Investigations Report 94-4004; U.S. Geological Survey: Atlanta, GA, USA, 1994.

17. Hess, G.W.; Inman, E.J. *Effects of Urban Flood-Detention Reservoirs on Peak Discharges and Flood Frequencies, and Simulation of Flood-Detention Reservoir Outflow Hydrographs in Two Watersheds in Albany, Georgia*; Water-Resources Investigations Report 94-4158; U.S. Geological Survey: Atlanta, GA, USA, 1994.

18. Peters, D.L.; Prowse, T.D. Regulation effects on the lower Peace River, Canada. *Hydrol. Process.* **2001**, *15*, 3181–3194. [CrossRef]

19. Montaldo, N.; Mancini, M.; Rosso, R. Flood hydrograph attenuation induced by a reservoir system: Analysis with a distributed rainfall-runoff model. *Hydrol. Process.* **2004**, *18*, 545–563. [CrossRef]

20. Sayama, T.; Tachikawa, Y.; Takara, K. Assessment of dam flood control using a distributed rainfall-runoff prediction system. In Proceedings of the International Conference on Monitoring, Prediction and Mitigation of. Water-Related Disasters (MPMD2005), Kyoto, Japan, 12–15 January 2005; pp. 59–64.

21. Gross, E.J.; Moglen, G.E. Estimating the Hydrological Influence of Maryland State Dams Using GIS and the HEC-1 Model. *J. Hydrol. Eng.* **2007**, *12*, 690–693. [CrossRef]

22. Cameron, D.S.; Beven, K.J.; Tawn, J.; Blazkova, S.; Naden, P. Flood frequency estimation by continuous simulation for a gauged upland catchment (with uncertainty). *J. Hydrol.* **1999**, *219*, 169–187. [CrossRef]

23. Calver, A.; Lamb, R. Flood frequency estimation using continuous rainfall-runoff modelling. *Phys. Chem. Earth* **1995**, *20*, 479–483. [CrossRef]

24. Blazkova, S.; Beven, K. Flood frequency prediction for data limited catchments in the Czech Republic using a stochastic rainfall model and TOPMODEL. *J. Hydrol.* **1997**, *195*, 256–278. [CrossRef]

25. Smithers, J.; Schulze, R.; Kienzle, S. *Design Flood Estimation Using a Modelling Approach: A Case Study Using the ACRU Model*; IAHS Publication 240; International Association of Hydrological Sciences: Wallingford, UK, 1997; pp. 365–376.

26. Cameron, D.S.; Beven, K.J.; Naden, P. Flood frequency estimation by continuous simulation under climate change (with uncertainty). *Hydrol. Earth Syst. Sci.* **2000**, *4*, 393–406. [CrossRef]

27. Cameron, D.S.; Beven, K.J.; Tawn, J.; Naden, P. Flood frequency estimation by continuous simulation (with likelihood based uncertainty estimation). *Hydrol. Earth Syst. Sci.* **2000**, *4*, 23–34. [CrossRef]

28. Blazkova, S.; Beven, K. Flood frequency estimation by continuous simulation of subcatchment rainfalls and discharges with the aim of improving dam safety assessment in a large basin in the Czech Republic. *J. Hydrol.* **2004**, *292*, 153–172. [CrossRef]

29. Soong, D.T.; Straub, T.D.; Murphy, E.A. *Continuous Hydrologic Simulation and Flood-Frequency, Hydraulic, and Flood-Hazard Analysis of the Blackberry Creek Watershed, Kane County, Illinois*; Scientific Investigations Report 2005-5270; U.S. Geological Survey: Reston, VA, USA, 2005.

30. Brath, A.; Montanari, A.; Moretti, G. Assessing the effect on flood frequency of land use change via hydrological simulation (with uncertainty). *J. Hydrol.* **2006**, *324*, 141–153. [CrossRef]

31. Boni, G.; Ferraris, L.; Giannoni, F.; Roth, G.; Rudari, R. Flood probability analysis for un-gauged watersheds by means of a simple distributed hydrologic model. *Adv. Water Resour.* **2007**, *30*, 2135–2144. [CrossRef]

32. Arnold, J.G.; Forher, N. SWAT2000: Current capabilities and research opportunities in applied watershed modeling. *Hydrol. Process.* **2005**, *19*, 563–572. [CrossRef]

33. Gassman, P.W.; Reyes, M.; Green, C.H.; Arnold, J.G. The Soil and Water Assessment Tool: Historical development, applications, and future directions. *Trans. ASABE* **2007**, *50*, 1211–1250. [CrossRef]

34. Neitsch, S.L.; Arnold, J.G.; Kiniry, J.R.; Williams, J.R. *Soil and Water Assessment Tool, Theoretical Documentation*, version 2009; Texas Water Resources Institute Technical Report No. 406; Texas Water Resources Institute: College Station, TX, USA, 2011.

35. Ministry of Construction and Transportation. *Investigation Report on the Water Supply Capacity of Existing Dams in Han River Basin*; Investigation Report DM-97-1; Ministry of Construction and Transportation: Gwacheon, Korea, 1997.

36. Sangal, B.P. Practical method of estimating peak flow. *J. Hydraul. Eng.* **1983**, *109*, 549–563. [CrossRef]

37. Kim, N.W.; Lee, J. Temporally weighted average curve number method for daily runoff simulation. *Hydrol. Process.* **2008**, *22*, 4936–4948. [CrossRef]

38. Kim, N.W.; Lee, J. Enhancement of the channel routing module in SWAT. *Hydrol. Process.* **2010**, *24*, 96–107. [CrossRef]
39. Nash, J.E.; Sutcliffe, J.V. River flow forecasting through conceptual models part I—A discussion of principles. *J. Hydrol.* **1970**, *10*, 282–290. [CrossRef]
40. Hosking, J.R.M. L-moments: Analysis and estimation of distributions using linear combinations of order statistics. *J. R. Stat. Soc. Ser. B (Methodological)* **1990**, *52*, 105–124.

Article

Effects of Urban Non-Point Source Pollution from Baoding City on Baiyangdian Lake, China

Chunhui Li [1,2], Xiaokang Zheng [1,3,*], Fen Zhao [1,*], Xuan Wang [1,2], Yanpeng Cai [2] and Nan Zhang [1]

[1] Ministry of Education Key Lab of Water and Sand Science, School of Environment,
    Beijing Normal University, Beijing 100875, China; chunhuili@bnu.edu.cn (C.L.);
    wangx@bnu.edu.cn (X.W.); zhangnan0227@163.com (N.Z.)
[2] State Key Laboratory of Water Environment Simulation, School of Environment, Beijing Normal University,
    Beijing 100875, China; yanpeng.cai@bnu.edu.cn
[3] Yellow River Engineering Consulting Co. Ltd., Zhengzhou 450003, China
* Correspondence: zhengxk@yrec.cn (X.Z.); 201631180021@mail.bnu.edu.cn (F.Z.); Tel.: +86-10-5880-2928 (F.Z.)

Academic Editors: Karim Abbaspour, Raghavan Srinivasan, Saeid Ashraf Vaghefi, Monireh Faramarzi and
Lei Chen
Received: 29 December 2016; Accepted: 30 March 2017; Published: 1 April 2017

Abstract: Due to the high density of buildings and low quality of the drainage pipe network in the city, urban non-point source pollution has become a serious problem encountered worldwide. This study investigated and analyzed the characteristics of non-point source pollution in Baoding City. A simulation model for non-point source pollution was developed based on the Stormwater Management Model (SWMM), and, the process of non-point source pollution was simulated for Baoding City. The data was calibrated using data from two observed rainfall events (25.6 and 25.4 mm, the total rainfall on 31 July 2008 (07312008) was 25.6 mm, the total rainfall amount on 21 August 2008 (08212008) was 25.4 mm) and validated using data from an observed rainfall event (92.6 mm, the total rainfall on 08102008 was 92.6 mm) (Our monitoring data is limited by the lack of long-term monitoring, but it can meet the requests of model calibration and validation basically). In order to analyze the effects of non-point source pollution on Baiyangdian Lake, the characteristics and development trends of water pollution were determined using a one-dimensional water quality model for Baoding City. The results showed that the pollutant loads for Pb, Zn, TN (Total Nitrogen), and TP (Total Phosphorus) accounted for about 30% of the total amount of pollutant load. Finally, applicable control measures for non-point source pollution especially for Baoding were suggested, including urban rainwater and flood resources utilization and Best Management Practices (BMPs) for urban non-point source pollution control.

Keywords: Baoding City; SWMM; non-point source (NPS) pollution; Baiyangdian Lake; rainfall-runoff

---

1. Introduction

Over the past decades, high-speed urbanization has led to increasing imperviousness in urban-underlying surface in many parts of the world [1]. An increase in imperviousness results in marked changes in water circulation patterns and may result in higher risks of flood disaster in urban areas [2–4]. Particularly, this problem is exacerbated by increases in urban dust levels due to a sharp growth of the urban population and industrial activities in developing countries such as China [5,6]. Since a large quantity of urban dust is transported into water bodies by rainfall-runoff processes, this might cause serious deterioration of urban water quality. This process has been recognized as urban non-point source pollution and has become a great threat to the urban water environment [7–9]. This pollution process is very complex because it involves diverse pollutants that originate from various non-point sources in an urban environment, such as suspended solids, organic materials, nutrients, heavy metals, and pesticide residue [10–13]. Health-related conditions for both human beings and aquatic organisms can be greatly affected by this type of pollution due to stormwater and the associated urban non-point sources [14]. This is particularly true in many cities in China, which is one of the most expedite economic and industrialized countries in the world. Therefore, urban stormwater can lead to both qualitative and quantitative problems in the receiving waters.

In recent years, the management of the quantity and quality of stormwater runoff from urban areas has become a complex task and an increasingly important environmental issue for urban communities [15–17]. To deal with this issue, computer-aided models are extremely useful for simulating and predicting the quantity and quality of urban stormwater. For example, since the 1970s, the United States and other developed countries have started to apply mathematical models to simulate the processes of urban rainfall-runoff. In addition, these models were widely used for evaluating effects of surface runoff pollution on various drainage systems and the corresponding receiving water bodies. These widely used urban stormwater models include the Source Loading and Management Model (SLAMM), the Storage, Treatment, Overflow, Runoff Model (STORM), the Model for Urban Sewers (MOUSE), the Stormwater Management Model (SWMM), and various derivative models [18–20]. Among them, SWMM is a computerized program that can assess the impacts of surface runoff pollution and evaluate the effectiveness of many mitigation strategies. It was first developed in 1971 and has undergone several major upgrades since then [18,21]. The current edition, Version 5, which runs under Windows, is a complete revision of the previous release. It is widely used throughout the world for supporting planning, analysis, and design related to stormwater runoff, and it integrates sewers, sanitary sewers, and other drainage systems in urban areas [22].

Research into the applications of the SWMM model in China has been conducted for decades. The SWMM model has been described as the classic non-point source pollution model [23,24]. The Morris screening method was used for a local sensitivity analysis of the parameters in the hydrologic and hydraulic modules of the SWMM model, in order to identify the sensitivity of the model parameters and perform an uncertainty analysis [25] (Section 3, the Electronic Supplementary Information, ESI). The results showed that the impact factors of the three most sensitive parameters were coefficients of imperviousness [26].

Baiyangdian Lake, situated centrally in the North China Plain, is the kidney of North China. The lake plays vital roles in flood reservation and environmental pollution decomposition [27]. However, Baoding City is the largest city in the upper basin, the human activities in the city have an important impact on the ecological environment of the Baiyangdian Lake. The quantity of sewage that drained into Baiyangdian Lake from Baoding City was about 25 to 33.6×10^4 t per day, and the quantity of sewage that drained into lake by secondary storm water runoff can reach about 25 to 30×10^4 t per day [28]. A large number of contaminants draining into Baiyangdian Lake is a serious threat to the Baiyangdian water ecological environment. Therefore, it is of great practical significance to research the impact of non-point source pollution of Baoding City on the water environment of Baiyangdian Lake.

Therefore, the objective of this research is to understand the process of non-point source pollution in Baoding City and the concentration of non-point source pollutants at the outfall of the catchment. Our monitoring data is limited by the lack of long-term monitoring, the use of only three sets of data, two used to calibrate and the other one to validate the model, is rather insufficient. However, it was already demonstrated by Di Modugno et al. that even a minimum amount of experimental observations may provide relevant information necessary to enhance design procedures and to improve the efficiency of systems aimed at first flush separation, storage, and treatment [29]. Our main objectives are to (1) develop a simulation model for non-point source pollution based on SWMM, and based on the model results, analyze the characteristic effects of the non-point source pollution in an urban catchment in Baoding City and (2) reveal the effects of the non-point source pollution of Baoding City on Baiyangdian Lake with a one-dimensional water quality model, the results will be useful for better control of urban non-point source pollution and water environmental recovery of Baiyangdian Lake.

2. Materials and Methods

2.1. Overview of Baoding City and Baiyangdian Lake

Baoding City is located in the mid-east of Hebei Province, and Baiyangdian Lake is located in the east, Taihang Mountain in the west, and the vast and fertile Hebei Great Plains in the north and south (Figure 1). The urban area is 312.3 km^2, with a population of 1.07 million people. The city is located in the plains, and the terrain slopes from northwest to southeast and extends to the Taihang Mountain area in the northwest with a gentle slope. The city has 12 rivers and streams, of which the Caohe and Tanghe River cross the city and flow into Baiyangdian Lake, which is called the "Pearl of North China" [28].

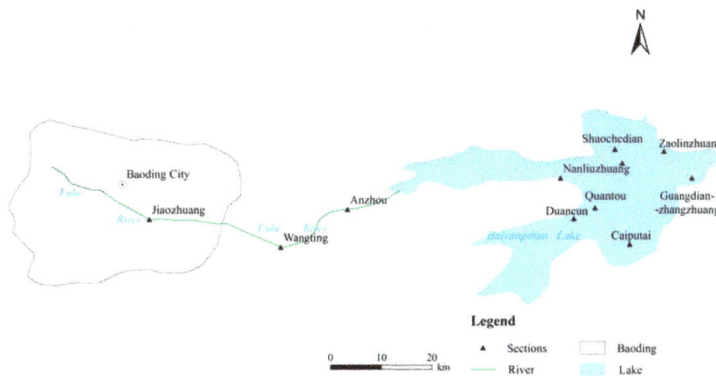

Figure 1. Monitoring sections location in Baiyangdian Lake.

With the rapid economic development, water resource shortages, water consumption, and sewage emissions in Baoding City continue to increase, and because the pollution control measures are less developed, several water-related environmental problems have become more significant. The first problem is the pollution of the flood drainage system. Sewage is discharged into the rivers, causing the destruction of the water environment.

Lake Baiyangdian, situated centrally in the North China Plain, is located 130 km south of Beijing (Figure 1). The surface area of the lake is 366 km^2, with a catchment area of 31,200 m^2. The lake depth varies according to the hydrologic conditions, but is usually less than 2.0 m [27,30,31]. The annual mean precipitation is less than 450 mm, and the annual mean ambient temperature is less than 17 °C in climate changes. With average annual runoff of 3.57×10^9 m^3, the lake plays vital roles in flood reservation, environmental pollution decomposition, etc. Moreover, the lake is a monomictic lake with only one entrance accepting pollutant emissions from Fuhe River [24] (Figure 1).

2.2. Available Data

The data of rainfall on 31 July 2008 (07312008), 21 August 2008 (08212008), and 10 August 2008 (08102008), was measured by rain gauge. Precipitation was recorded every 5 min. The total rainfall on 07312008 was 25.6 mm, but only lasted for approximately 2 h and had a high rainfall intensity. The total rainfall amount on 08212008 was 25.4 mm, lasted for approximately 7 h, but had a much lower rainfall intensity. The total rainfall on 10 August 2008 (08102008) was 92.6 mm, lasted approximately seven hours and was the strongest rainstorm that occurred from July to September 2008.

Our monitoring data is limited by the lack of long-term monitoring, the use of only three sets of data, two used to calibrate and the other one to validate the model, is rather insufficient. However, it can meet the requests of model calibration and validation at a basic level [29].

2.3. Methods

To analyze the process of non-point source pollution of Baoding City and the effects of non-point source pollution on Baiyangdian Lake, we use the Stormwater Management Model (SWMM) and the one-dimensional water quality model. In the course of the study, the technical route is as follows (Figure 2).

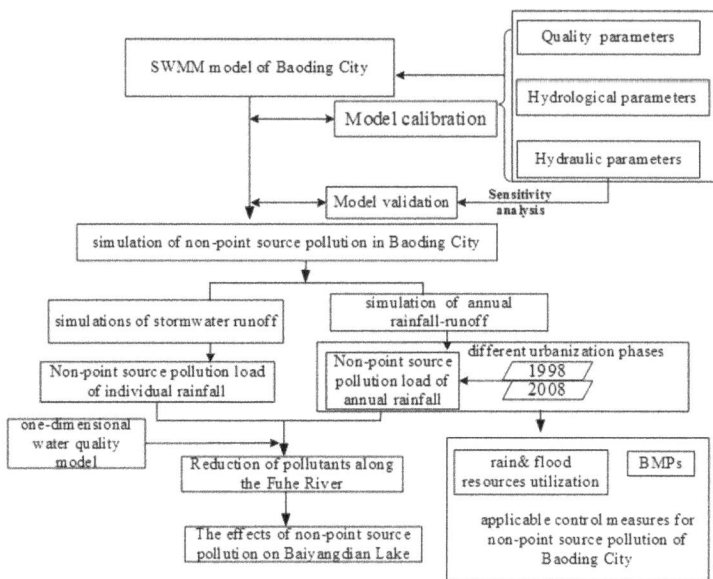

Figure 2. Diagram of the technical route.

According to the hydrological parameters, quality parameters and hydraulic parameters of the model, the sensitivity analysis of the model parameters was carried out, and the SWMM model of Baoding City was established after repeated calibration and verification. The simulation process of non-point source pollution in Baoding City is divided into typical simulation of stormwater runoff and simulation of annual rainfall-runoff. The influence of stormwater runoff and annual rainfall-runoff pollution on the Baiyangdian Lake water environment were studied by analyzing the characteristics of pollution reduction along the Fuhe River.

2.3.1. SWMM Model

SWMM, a dynamic rainfall-runoff simulation model, is used for single event or long-term (continuous) simulation of runoff quantity and quality from primary urban areas. The runoff component of SWMM operates on a collection of sub-catchment areas that receive precipitation and generate runoff and pollutant loads. The routing portion of SWMM transports this runoff through a system of pipes, channels, storage/treatment devices, pumps, and regulators. SWMM tracks the quantity and quality of runoff generated from each sub-catchment, and the flow rate, flow depth, and quality of water in each pipe and channel during a simulation period are comprised of multiple time steps. This section briefly describes the methods SWMM uses to model stormwater runoff quantity and quality through the following two processes [21].

(1)   Runoff Simulation

SWMM uses the St. Venant equations for flow simulation [32,33]. The St. Venant equations represent the principles of conservation of momentum (Equation (1)) and conservation of mass (Equation (2))

$$\frac{\partial y}{\partial x} + \frac{v}{g}\frac{\partial v}{\partial x} + \frac{1}{g}\frac{\partial v}{\partial t} = S_o - S_f \tag{1}$$

$$\frac{\partial Q}{\partial x} + \frac{\partial A}{\partial t} = 0 \tag{2}$$

where y is the water depth, m; v is the velocity, $L \cdot t^{-1}$; x is the longitudinal distance, m; t is the time, s; g is the gravitational acceleration, 9.8 m/s; S_o is channel slope, dimensionless; S_f is friction slope, dimensionless; A is the area of the flow cross-section and a function of y based on the geometry of the conduit, m^2; Q is the discharge and it is equal to $A \times v$, $m^3 \cdot s^{-1}$.

Equation (1) represents hydrostatic pressure, convective acceleration, local acceleration, and gravity and frictional forces, respectively. Representing the effects of turbulence and viscosity, the friction slope (S_f) is calculated in SWMM using Manning's equation [18]:

$$S_f = \frac{Q^2}{\frac{1}{n^2}A^2 R^{4/3}} \tag{3}$$

where n is Manning's roughness coefficient, $t \cdot L^{-1/3}$; R is hydraulic radius, m; Q and A are defined previously.

Equation (3) is substituted into Equation (1), and the resulting equation is solved for Q:

$$Q = \frac{1}{n}AR^{2/3}\left(\frac{\partial y}{\partial x} + \frac{v}{g}\frac{\partial v}{\partial x} + \frac{1}{g}\frac{\partial v}{\partial t} - S_O\right)^{1/2} \tag{4}$$

Since there is no known analytic solution, an iterative finite-difference method is applied to Equations (2) and (4) to solve the equations. For each time step, the discharge, area, and water depth (y) at the outlet of each conduit are derived.

(2)   Water quality simulation

In this paper, the exponential function is used as the surface pollutant build-up and wash-off algorithms of the SWMM model, and the build-up and wash-off algorithms used to simulate these two processes (Section 2 in the ESI).

Water quality routing within conduit links assumes that the conduit behaves as a continuously-stirred tank reactor (CSTR). Although the assumption of a plug flow reactor might be more realistic, the differences will be small if the travel time through the conduit is on the same order as the routing time step. The concentration of a constituent exiting the conduit at the end of a time step is determined by integrating the conservation of mass equation and using average values for quantities that might change over the time step, such as flow rate and conduit volume [22].

Solute transport is simulated with the assumption of complete and instantaneous mixing within each element of the sewer system. The instantaneous mixing assumption introduces artificial dispersion; however, as the number of conduit elements is increased within a system, solute transport is represented by pure advection [22]. The overall transport of solutes through the system is executed through a mass balance calculation that incorporates decay. The concentrations of solutes are determined using the finite difference form of the continuity equation [18]:

$$\frac{\partial(Vc)}{\partial t} = Q_i c_i - Q_o c_o - kcV + s \tag{5}$$

where c is the concentration in the mixed volume, $m \cdot L^{-3}$; V is the volume, L^3; t is the time, t; Q_i and Q_o are the inflow (i) and outflow (o) rate, $L^3 \cdot t^{-1}$; c_i and c_o are the concentrations of the influent and effluent, $m \cdot L^{-3}$; k is the decay constant, t^{-1}; s is the source (or sink), $m \cdot t^{-1}$.

Assuming complete mixing and applying a finite-difference scheme, Equation (5) becomes:

$$c_{j+1} = \frac{(c_j(V_j(\frac{2}{\Delta t} - (D_1 + D_2)) - Q_{o,j}) + (c_{i,j} + Q_{i,j}) + (c_{i,j+1} + Q_{i,j+1}) + D_2 \cdot S(V_j + V_{j+1}))}{(V_{j+1} \cdot (\frac{2}{\Delta t} + (D_1 + D_2)) + Q_{o,j+1})} \qquad (6)$$

where j is the time-step number; D_1 is the decay constants, t^{-1}; D_2 is the growth constant, t^{-1}; S = maximum growth, $m \cdot L^{-1}$; D_t is the time increment, t, and the other parameters were defined with Equation (5).

2.3.2. One-Dimensional Water Quality Model

Sewage and rainwater from Baoding City flow into the Fuhe River and a biochemical reaction occurs gradually in the flow process under the action of natural microorganisms. As a result, the main pollutant COD is decomposed as the flow process has been lengthened. For the Fuhe River with small ratio of width to depth, pollutants can be mixed in these sections (Jiaozhuang, Wangting, Anzhou, Nanliuzhuang) in a relatively short period of time. A one-dimensional water quality model that can simulate the migration of pollutants along the river longitudinal has been described in [34,35] and is defined as:

$$C(x) = C_0 \exp(-k \cdot \frac{x}{u}) \qquad (7)$$

where $C(x)$ is the contaminant concentration of the control section, mg/L; C_0 is the contaminant concentration of the initial section, mg/L; k is the self-purification capacity of the pollution, 1/d; x is the longitudinal distance of the control section of the downstream section of the sewage outfalls, m; u is the average flow velocity of the polluted belts along the river banks, m/s.

3. Results and Discussions

3.1. Model Calibration and Model Validation

In this study, the drainage system of Baoding City was generalized based on an analysis of the drainage system and field reconnaissance of the study area. The entire city was divided into three sub-watersheds: a middle sub-watershed, a southern sub-watershed, and a northern sub-watershed, based on the actual drainage system (Figure 3). The rainfall-runoff of the middle sub-watershed flowed into Yimuquan River, Hou River, and Qingshui River, and then flowed together into the Fuhe River. The urban rainfall-runoff of the southern sub-watershed and northern sub-watershed was also generalized to flow into flood embankments as gravity flow and then flow into the Fuhe River, without regard for processed rainfall-runoff through the sewage treatment plant.

A total of 447 rainwater pipe nodes and 447 rainwater pipes were generalized (including rainwater pipes, sewage pipes, and open channels). The diameter of rainwater pipes ranged between 400 and 1400 mm, and some pipes had a rectangular cross-section of 2000 mm × 2000 mm, while the bottom width of open channels varied between 5 and 14 m. Furthermore, three outlets were generalized, Node_556, Node_557, and Node_558. Node_557 was the Jiaozhuang section on the Fuhe River of the middle sub-watershed, Node_556 and Node_558 were the sections on the northern and southern flood embankments, respectively, and they represented the outlets of the northern and southern sub-watersheds (Figure 4, Tables S1 and S2 in the ESI). Based on the generalization of the drainage system, the boundaries of trunk roads and streets in conjunction with field reconnaissance and research, and taking into account a topographic map and the convergence characteristics of Baoding City, the whole urban catchment was divided into 450 sub-catchments, with a total area of 130.76 km^2 (Figure 5 and Tables S3 in the ESI).

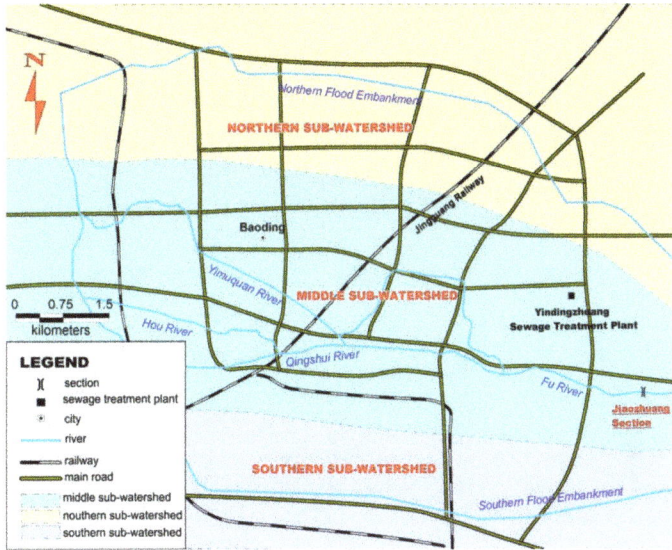

Figure 3. Drainage system diagram of Baoding City.

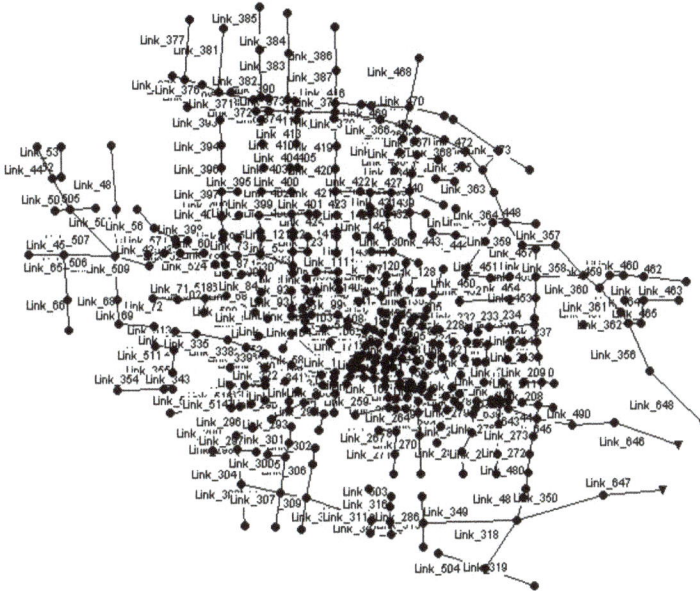

Figure 4. Drainage system generalization of Baoding City.

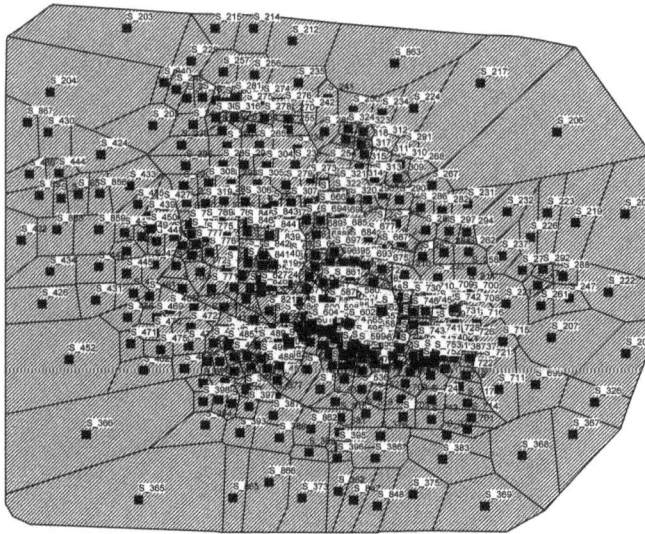

Figure 5. Sub-catchment diagram of Baoding City.

3.1.1. Model Calibration

(1)　Parameters of SWMM

SWMM parameters include hydrological, hydraulic, and water quality parameters. It is relatively easy to determine the hydraulic parameters by surveying pipes and networks. However, the hydrological parameters are relatively difficult to determine. The parameters in this study were obtained through the model handbook and field surveys. Table 1 lists the major hydrological and hydraulic parameters, their ranges, and methods for determining the parameters.

Table 1. Major hydrological, hydraulic parameters of Stormwater Management Model (SWMM) and their range and obtain methods.

NO.	Parameter	Meaning	Data Range	Data Source
1	Manning-N	Manning coefficient of Pipe	0.005~0.04	SWMM handbook
2	N-Imperv	Mannings N of impervious area	0.005~0.04	field survey, SWMM handbook
3	N-perv	Mannings N of pervious area	0.1~0.8	field survey, SWMM handbook
4	S-Imperv	Depression storage on impervious area	0.2~2 (mm)	field survey, SWMM handbook
5	S-perv	Depression storage on pervious area	2~10 (mm)	field survey, SWMM handbook
6	Pct-Zero	Percent of impervious area with no depression storage	50~80 (%)	field survey, SWMM handbook
7	MaxRate	Maximum rate on infiltration curve	3~50 (mm/h)	SWMM handbook
8	MinRate	Minimum rate on infiltration curve	1~3 (mm/h)	SWMM handbook
9	Decay	Decay constant for infiltration curve	2~7	SWMM handbook
10	Imperv (%)	Percent of impervious area	10~90 (%)	field survey, Google Earth
11	Width	Width of overland flow path	depends on the area of sub-catchment	GIS
12	Slope	Average surface slope (%)	0.1~2 (%)	GIS

The parameters for the accumulation of pollutants in water quality portion of the model were obtained by measurements, but empirical values were used for the scouring parameters. The sub-catchments in the SWMM model were divided into four types of land-use including business areas, residential areas, industrial areas, and green areas. Moreover, the main pollution factors including COD, TN, TP, Pb, and Zn, which were produced by rainwater mixing with the urban dust, were simulated in the model. The parameters for pollutant accumulation and erosion in different land-use types are shown in Tables 2 and 3, wash off exponent means the runoff exponent in wash off function, and wash off coefficient means wash off coefficient or Event Mean Concentration (EMC). The cumulative amount of surface pollution is directly related to land use condition, greening condition, traffic condition, rainfall interval and rainfall intensity. The distribution of dust accumulation on the surface is: industrial area > traffic area > residential area > green area. In this study, we regard the different land use types had the same cumulative rate constants and half-saturated accumulation times. The cumulative rate constant was 0.5 and the half-saturated accumulation time was 10 days.

Table 2. Maximum accumulation quantity of pollutants on different land use types kg/hm^2.

Area	COD	TN	TP	Pb	Zn
Commercial area	46	15.6	0.64	0.12	0.62
Residential area	58	8.6	0.24	0.056	0.22
Industrial area	43	10.3	0.38	0.12	0.22
Green area	20	14.5	0.17	0.045	0.11

Table 3. Wash off parameters of pollutants on different land use types.

Area	Parameter	COD	TN	TP	Pb	Zn
Commercial area	wash off coefficient	0.003	0.004	0.004	0.004	0.004
	wash off exponent	1.4	1.8	1.7	1.7	1.8
Residential area	wash off coefficient	0.003	0.004	0.002	0.004	0.004
	wash off exponent	1.4	1.8	1.7	1.7	1.8
Industrial area	wash off coefficient	0.003	0.004	0.004	0.004	0.004
	wash off exponent	1.4	1.8	1.7	1.7	1.8
Green area	wash off coefficient	0.003	0.002	0.001	0.001	0.001
	wash off exponent	1.2	1.4	1.2	1.2	1.2

(2) Sensitivity Analysis

A sensitivity analysis is used to study the impacts of parameters on the model output to identify the key parameters of the model. The results from a sensitivity analysis by Huang [26] and Wang [27] and the sensitivity ranking of the hydraulic and hydrological parameters in SWMM are listed in Table 4. For different output variables, the sensitivity order of each parameter is also different, for runoff coefficient, the order of parameter according to the sensitivity is Imperv(%) > S-Imperv > Pct-Zero > N-Imperv > Width.

Table 4. Sensitivity ranking of hydraulic and hydrological parameters in SWMM.

Runoff Factor	1	2	3	4	5
Runoff Coefficient	Imperv(%)	S-Imperv	Pct-Zero	N-Imperv	Width
Peak Discharge	Imperv(%)	S-Imperv	N-Imperv	Width	Pct-Zero
Peak Discharge Time	Manning-N	N-Imperv	S-Imperv	Width	Pct-Zero

(3) Calibration Results

The hydrology, hydraulic, and water quality parameters in SWMM were calibrated by using rainfall data from 31 July 2008 (07312008) and 21 August 2008 (08212008). The rainfall amount was nearly identical but had different characteristics. The total rainfall on 07312008 was 25.6 mm, but only lasted for approximately 2 h and had a high rainfall intensity. The total rainfall amount on 08212008 was 25.4 mm, lasted for approximately 7 h, but had a much lower rainfall intensity. Therefore, these two rainfall events were well suited to represent rainfall and could be used to calibrate the SWMM model for Baoding City. First, the SWMM parameters were established by using measured parameters and the empirical coefficient, and then the parameters were adjusted manually until the simulated results agreed well with the measured results. Based on the parameter sensitivity analysis, the parameter calibration mainly focused on parameters with relatively high sensitivity, while parameters with low sensitivity were adjusted roughly or the empirical coefficient was used directly. We used trial and error for adjustments and there was a good fit for the quantity and quality curves between the simulated and the measured processes. The results are shown in Figures 6 and 7. The simulated water quality process line of 07312008 rainfall and 08102008 rainstorm fit well with the measured water quality process line, but the simulated water quality process line of 08212008 did not fit well.

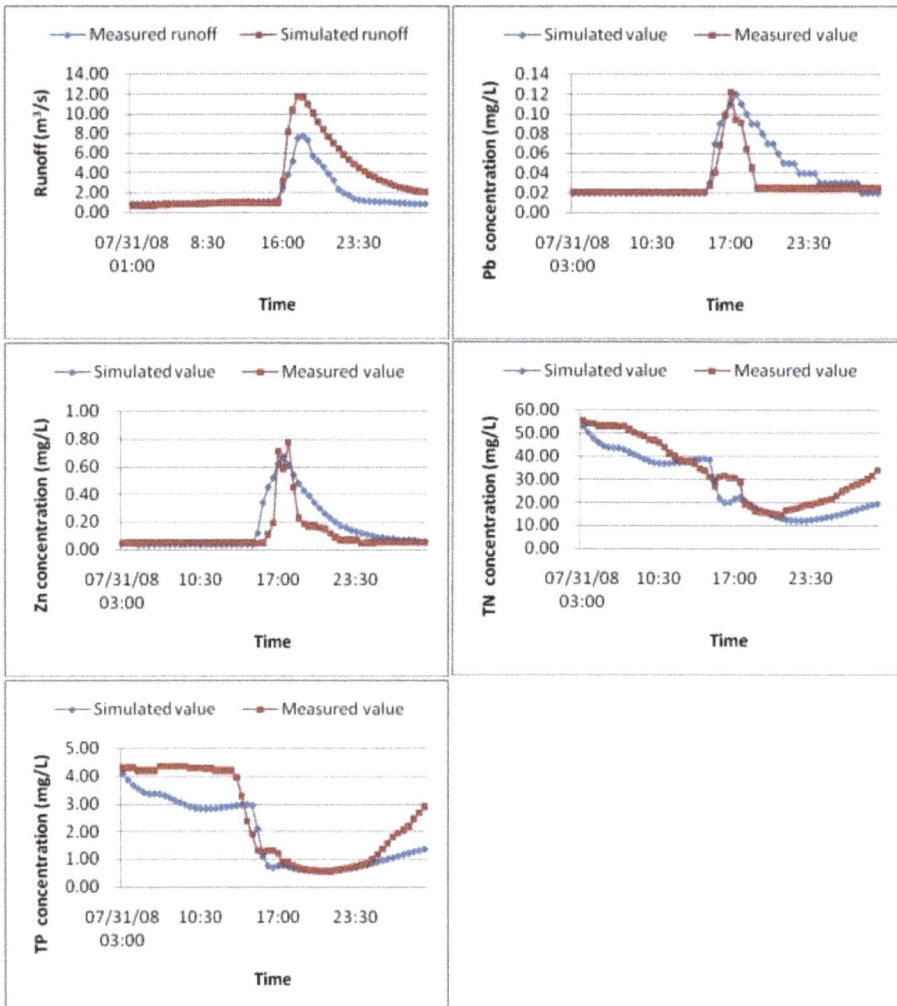

Figure 6. Measured and simulated hydrograph of rainfall-runoff and water quality of 07312008 on Jiaozhuang section.

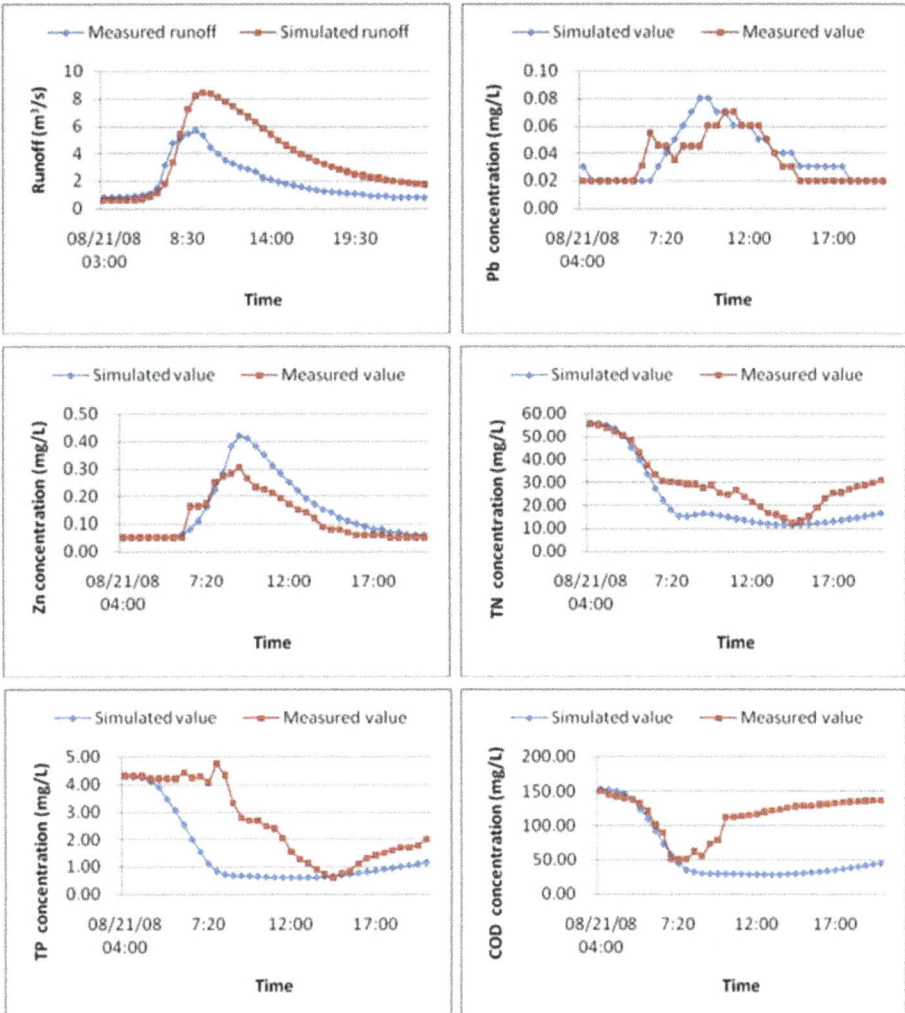

Figure 7. Measured and simulated hydrograph of rainfall-runoff and water quality of 08212008 on Jiaozhuang section.

3.1.2. Model Validation

The rainfall-runoff data from 10 August 2008 (08102008) was used to validate the simulated results of the SWMM output, which was based on the calibrated data. The total rainfall on 08102008 was 92.6 mm, lasted approximately seven hours and was the strongest rainstorm that occurred from July to September 2008. The curves of measured and simulated runoff and water quality are shown in Figure 8.

The relative error (RE) between the mean value of the measured data and the corresponding simulation data is used to test the goodness-of-fit.

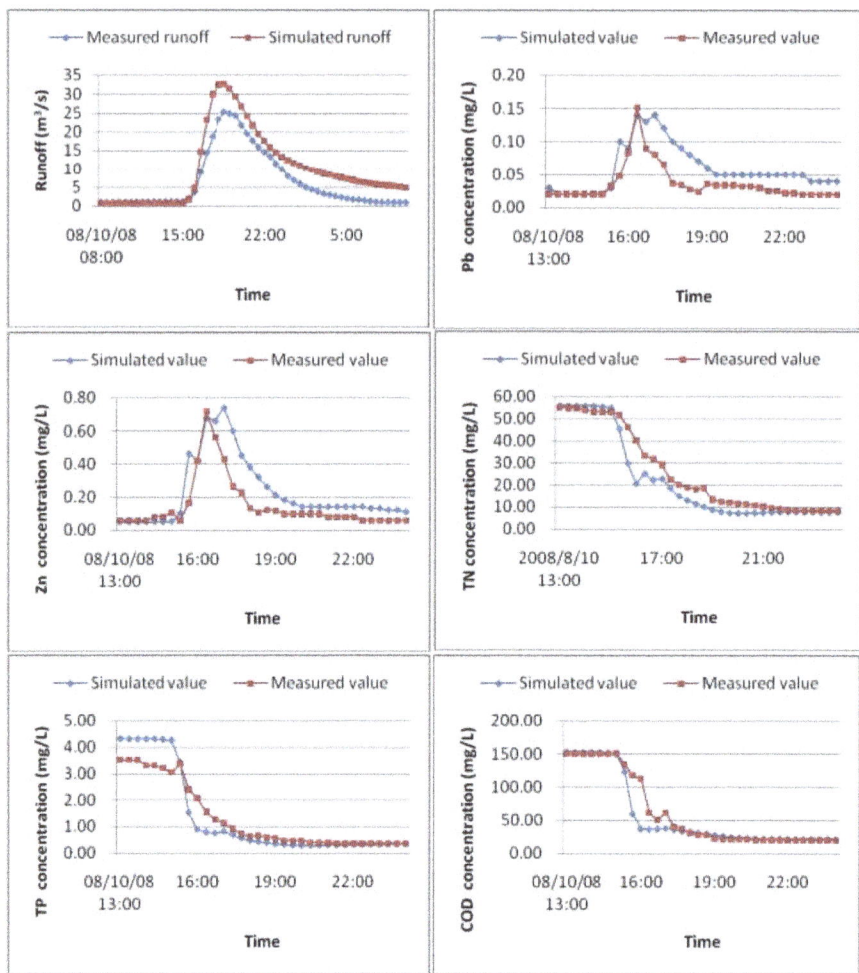

Figure 8. Measured and simulated hydrograph of rainfall-runoff and water quality of 08102008 on Jiaozhuang section.

3.2. Error Analysis

3.2.1. Runoff

We obtained a good fit for the simulated and measured hydrograph data for the Jiaozhuang section, but the simulated peak runoff volume was higher than the measured peak runoff volume by about 30%, and the simulated peak runoff occurred about 30 min prior to the measured peak runoff. However, the rate of decrease in the simulated peak runoff was also slower than the measured peak runoff. Moreover, there were certain differences between simulated and measured runoff volume (Table 5). In the middle sub-watershed of Baoding City, rainwater flowed into the Yindingzhuang sewage plant and, therefore, the volume was 80,000–100,000 cubic meters. By adding this portion of the runoff to the other runoff data, the relative error of the SWMM results was acceptable. The calibrated relative error (RE*) is shown in Table 5.

Table 5. Simulated runoff volume and measured runoff volume on Jiaozhuang section.

Analysis Period	Rain Time	Rainfall (mm)	Simulated Runoff Volume (10^4 m^3)	Measured Runoff Volume (10^4 m^3)	RE (%)	RE* (%)
Calibration Period	07312008	25.6	33.40	18.87	−77.0	−24.3 ~−15.7
	08212008	25.4	28.83	15.87	−81.6	−20.8 ~−11.4
Validation Period	08102008	92.6	94.19	60.97	−54.5	−36.6 ~−32.7

3.2.2. Rainfall Runoff Coefficient

Based on the integrated rainfall-runoff coefficient point, the rainfall-runoff coefficients on 07312008, 08212008, and 08102008 were 0.376, 0.337, and 0.653, respectively, which was consistent with the empirical rainfall-runoff coefficient values (0.3–0.7) of Baoding City (Section 4 and Table S4 in the ESI).

3.2.3. Water Quality

With regard to the water quality simulation, the results indicated that there was a better fit for the simulated and measured water quality hydrograph data for 07312008 and 08102008 than for 08212008. In the three simulation processes, the heavy metals Pb and Zn exhibited peak values, and the simulated peak values and times of occurrence were consistent with the measured values and times, indicating that the simulations for Pb and Zn were successful. However, concentrations of TN, TP, and COD did not exhibit peak values. The concentrations of the three elements decreased during the rainfall-runoff process and then gradually returned to levels seen prior to the rainfall-runoff event. The measured times of recovery for the TN, TP, and COD concentrations were earlier than the simulated times, especially for the rainfall event on 08212008. This was possibly due to the low rainfall intensity and short duration; consequently, the conditions were not ideal for simulating erosion processes, and poor simulation results were obtained. This was consistent with the results of the rainfall-runoff process that showed that the rate of decrease of the simulated runoff was longer than the measured time. This indicated that the SWMM model simulated water quality appropriately, especially for a rainfall with heavy intensity.

3.2.4. Continuity Errors

The continuity errors for the three rainfall-runoff simulations, including the calibration and validation periods, are shown in Table 6. The continuity error was lower for surface runoff and flow routing than for the simulation of water quality. Research has suggested that the continuity error should be less than 10% in SWMM [22]. The continuity error was acceptable for the simulation process in this research, except for slightly higher values for the simulation of water quality on 07312008.

Table 6. Continuity errors of simulation.

Analysis Period	Rain Time	Surface Runoff	Flow Routing	Runoff Quality
Calibration period	07312008	−0.09%	−0.15%	14.23%
	08212008	−0.01%	−0.05%	8.09%
Validation period	08102008	−0.06%	−0.07%	6.45%

3.3. Simulation Results

3.3.1. Simulation Results of 08102008 Storm Water Runoff

The results of the simulations for stormwater runoff on 08102008 are shown in Table 7 and are as follows: (1) the runoff and output of the total non-point source pollution load were higher for the northern and middle sub-watersheds than for the southern sub-watershed. This was not only due to the larger area of the northern and middle sub-watersheds compared to the southern sub-watershed, but also because the imperviousness was lower for the southern sub-watershed than for the other two sub-watersheds; (2) the overall stormwater runoff on 08102008 was 4.52×10^6 m^3, the average flow was 50.83 m^3/s, and the maximum flow for the system was 136.48 m^3/s; and 3) the output values for the non-point source pollution loads for Pb, Zn, TN, TP, and COD in the stormwater runoff on 08102008 were 145.50, 556.72, 29,412.84, 874.98 and 74,218.42 kg, respectively.

Table 7. Runoff and output of total non-point source pollution load on export section of 08102008.

Node	Average Runoff (m^3/s)	Max. Runoff (m^3/s)	Total Runoff (10^4 m^3)	Output Non-Point Source Pollution Load (kg)				
				Pb	Zn	TN	TP	COD
Node_556	22.51	61.55	196.70	60.56	238.82	11453.24	224.75	27970.87
Node_557	15.75	43.00	142.60	66.36	254.04	13425.89	595.08	35596.38
Node_558	12.57	32.97	112.53	18.58	63.87	4533.71	55.15	10651.17
System	50.83	136.48	451.83	145.50	556.72	29412.84	874.98	74218.42

3.3.2. Simulation Results of 2008 Rainfall-Runoff

The results of the simulations for rainfall-runoff in 2008 are shown in Table 8. The annual rainfall in 2008 was 564.3 mm, and the rainfall from June to September was 451 mm, which accounted for 80% of the total annual rainfall. The simulation started on 1 January 2008, and ended on 31 December 2008. Winter snowmelt runoff was not taken into account in the simulation. The infiltration losses were 103.48 mm, evaporation losses were 227.81 mm, and surface runoff was 233.39 mm. Furthermore, the rainfall-runoff coefficient was 0.414 and the continuity error was -0.068%.

Table 8. Runoff and output of total non-point source pollution load on export section of 2008.

Node	Average Runoff (m^3/s)	Max. Runoff (m^3/s)	Total Runoff (10^4 m^3)	Output Non-Point Source Pollution Load (kg)				
				Pb	Zn	TN	TP	COD
Node_556	0.38	18.48	1200.77	140.4	526.3	26127.3	523.7	91681.3
Node_557	0.99	14.97	3116.06	682.2	1747.7	1089088	82203	2997034
Node_558	0.16	8.6	496.30	29.4	88.9	7947.5	92.1	27371.1
System	1.53	41.81	4813.13	852.0	2362.8	1123163	82819	3116086

The annual rainfall-runoff in 2008 was 48.13×10^6 m^3, the average flow was 1.53 m^3/s, and the maximum flow for the system was 41.81 m^3/s. The total runoff of the middle sub-watershed was 31.16×10^6 m^3, which accounted for 60% of the total annual runoff. This was due to the location of the middle sub-watershed in the center of the city in an area of high imperviousness. For the 2008 rainfall-runoff, the annual output of the non-point source pollution loads for total Pb, Zn, TN, TP, and COD was 852.0 kg, 2362.8 kg, 1,123,163 kg, 82,819 kg, and 3,116,086 kg respectively. The pollution load of the middle sub-watershed accounted for 80% of those values. Furthermore, the pollution load for TN and TP of the middle sub-watershed accounted for more than 90% of the total due to an average imperviousness of as much as 85% for the central city, which was conducive to rainfall-runoff generation. Another reason was the growth of inflow during dry season mainly generated in the middle sub-watershed.

3.4. Influence of Non-Point Source Pollution from Baoding City on Baiyangdian Lake

3.4.1. Analysis of Non-Point Source Pollution Load in Baoding City

The total discharge and non-point source pollution loads of the rainstorm and annual rainfall processes in Baoding City are shown in Table 9. The total discharge of a single storm on 20080810 was 4.52×10^6 m^3, and the output of the non-point source pollution load of the total Pb, Zn, TN, TP, and COD was 145.5 kg, 556.7 kg, 29,412.8 kg, 875 kg, and 74,218.4 kg, respectively. Without any Best Management Practices (BMPs), the total amounts of the output pollutants Pb, Zn, TN, TP, and COD were almost doubled. With street-sweeping and other BMPs, the total output of the pollution loads for Pb and Zn increased by more than 50%, while TN, TP, and COD approximately doubled in value.

The removal efficiency of BMPs to Pb and Zn was approximately 10%, while for TN, TP, and COD, the removal efficiency was only about 1%, demonstrating that BMPs had a positive influence on the control of the non-point source pollution load in the basin.

Table 9. Total discharge and non-point source pollution load of Baoding City Basin.

Rain Time	Rainfall (mm)	Total Discharge (10^4 m^3)	BMPs	Output Non-Point Source Pollution Load (kg)				
				Pb	Zn	TN	TP	COD
08102008	92.32	451.83	N	145.5	556.7	29,412.8	875.0	74,218.4
1998	514.8	3453.31	N	554.3	1560.9	591,187.1	41,738.1	1,616,665.4
			Y	518.0	1416.9	584,777.5	41,563.7	1,595,238.3
2008	564.3	4813.13	N	852.0	2362.8	1,123,163.2	82,818.9	3,116,086.2
			Y	797.2	2148.6	1,114,198.4	82,559.8	3,081,398.4

3.4.2. Reduction of Pollutants along the Fuhe River

The sewage and stormwater discharge from Baoding City through the city's drainage network and eventually pass through the Fuhe River into Baiyangdian Lake (Figures 1 and 2), which is located approximately 45 km away. This small flow of sewage from Baoding City to Baiyangdian Lake requires about two days, while a storm flood with large velocity requires about 5 h [28]. During this time, major pollutants are degraded and reduced.

Based on a study of COD reduction along the Fuhe River [36,37] and Equation (7), the value for k is 0.5 day^{-1} and u = 22.5 km/day Section 5 in the ESI). The coefficients were verified by using the measured data shown in Table 10, and Equation (6) was well-suited to simulate the process of the longitudinal attenuation of COD.

Table 10. Pollutant concentration of the sections mg/L.

Section	TN	TP	COD
Jiaozhuang	49.33	2.34	97.75
Wangting	27.45	1.22	42.74
Anzhou	25.55	1.21	31.84
Nanliuzhuang	15.31	0.34	32.52

3.4.3. Effects on the Water Environment of Baiyangdian Lake

There were seven sample locations for water quality monitoring in Baiyangdian Lake (Figure 1), and the results are shown in Table 11.

Table 11. Water quality monitoring data of Baiyangdian Lake.

Parameters	Concentration (mg/L)							Mean (mg/L)
	Zaolinhzuang	Wangjiazhai	Guangdianzhangzhuang	Quantou	Caiputai	Duancun	Shaochedian	
Pb	0.01	0.01	0.01	0.01	0.01	0.01	0.01	0.01
TN	1.79	2.15	2.68	1.57	0.8	1.19	1.74	1.71
TP	0.07	0.14	0.12	0.09	0.05	0.09	0.06	0.09

The total discharge and the point and non-point source pollution load in the Jiaozhuang section of Baoding City are shown in Table 12. The input concentration of Pb, TN, and TP from the storm runoff on 08102008 was two times, 3.8 times, and 2.1 times higher, respectively, than the average concentration in Baiyangdian Lake.

Table 12. Total discharge and pollution load of the point source and non-point source of Baoding City.

Time	Emission Source	Total Flow of Discharging (10^4 m^3)	Pollution Load (kg)				
			Pb	Zn	TN	TP	COD
1998	Rainfall-runoff	3453.31	554.3	1560.9	591,187.1	41,738.1	1,616,665.4
	Industrial Sewage	4500.00	1350.0	2250.0	2,475,000.0	225,000.0	6,750,000.0
	Total pollution	7953.31	1904.3	3810.9	3,066,187.1	266,738.1	8,366,665.4
2008	Rainfall-runoff	4813.13	852.0	2362.8	1,123,163.2	82,818.9	3,116,086.2
	Industrial sewage	6000.00	1800.0	3000.0	3,300,000	300,000.0	9,000,000.0
	Total pollution	10,813.13	2652.0	5362.8	4,423,163.2	382,818.9	12,116,086.2

As rainfall-runoff and industrial sewage moves from Baoding City from the cross section into Baiyangdian Lake, and water losses occur along the way, including evaporation loss, river leakage loss, and loss due to agricultural irrigation. The pollutant load of rainfall-runoff input Baiyangdian Lake from Baoding City was deducted due to the losses of water along the way. The mean annual losses account for 40% of total water volume. If only river leakage losses are considered for a single rainstorm event, the leakage loss accounts for 15% of the total runoff. During the 10 years from 1998 to 2008, the growth in the non-point source pollution loads for Pb, Zn, TN, TP, and COD were 178.6 kg, 481.2 kg, 134,539.6 kg, 7866 kg, and 330,784.6 kg, respectively. The non-point source pollution load input accounted for the proportion of total pollution load input also increased.

4. Conclusions

In this paper, a simulation model for non-point source pollution of Baoding City was developed based on SWMM. Two typical measured rainfall-runoff processes on 07312008 and 08212008 were used to calibrate hydraulic and hydrological parameters of SWMM using trial and error for debugging. After the calibration of the model simulation error can be controlled within −36.6% to −11.4%, the actual process and the simulation process have achieved good fitting effect. The fit between measured and simulated processes was good, demonstrating that the model calibration was successful. The simulation results showed that a typical rainstorm on 08102008 produced a total runoff of 4.52×10^6 m^3, and the non-point source pollution loads for Pb, Zn, TN, TP, and COD were 145.50, 556.72, 29,412.84, 874.98 and 74,218.42 kg, respectively.

The annual rainfall-runoff volume in 2008 was 48.13×10^6 m^3, and the total runoff of the pipe network was 31.16×10^6 m^3, which accounted for 60% of the total annual runoff. The annual non-point source pollution loads for Pb, Zn, TN, TP, and COD were 852.0 kg, 2362.8 kg, 1,123,163 kg, 82,819 kg, and 3,116,086 kg, respectively, and the pollution load of the pipe network accounted for about 80% of those values.

The one-dimensional water quality model was applied in the research, the simulation results showed that the average concentration of TN and TP of the annual rainfall-runoff was about 10 times higher than that of Baiyangdian Lake. The input water for rainfall-runoff was 20.72×10^6 m^3 in 1998, accounting for 43% of the total amount of rain and sewage in Baoding City. The non-point source pollution loads for Pb, Zn, TN, TP, and COD were 332.6 kg, 936.5 kg, 341,866.4 kg, 19,868.7 kg, and 356,833.7 kg, respectively. The rainfall-runoff water input was 28.88×10^6 m^3 in 2008, accounting for 45% of the total amount of rain and sewage. The non-point pollution loads for Pb, Zn, TN, TP, and COD were 511.2 kg, 476,406.0 kg, 1,417.7 kg, 27,734.7 kg, and 687,618.3 kg, respectively. From 1998–2008, the total input of the non-point source pollution load for rainfall-runoff in Baoding City has increased, and the annual input accounted for about 30% of the total amount of pollutant load.

Based on the simulation results of non-point source pollution, applicable control measures for non-point source pollution especially for Baoding City would be taken, such as urban rainwater and flood resources utilization and Best Management Practices (BMPs) for urban non-point source pollution control, which including engineering and non-engineering measures. In future research, the control measures can be enhanced for floods control and pollutant reduction. In this way, the research would be more practical guidance.

Our monitoring data is limited by the lack of long-term monitoring, the use of only three sets of data, two used to calibrate and the other one to validate the model, these can meet the requests of model calibration and validation at a basic level, and the result of this study has its limitations. Further research will be needed to improve this study.

Supplementary Materials: The following are available online at http://www.mdpi.com/2073-4441/9/4/249.

Acknowledgments: This research was supported by National key research and development program (2016YFC0401302). We would like to extend special thanks to the editor and the anonymous reviewers for their valuable comments in greatly improving the quality of this paper.

Author Contributions: Fen Zhao and Xiaokang Zheng conceived and designed the experiments; Chunhui Li performed the experiments; Xuan Wang and Yanpeng Cai analyzed the data; Nan Zhang contributed reagents/materials/analysis tools; Chunhui Li wrote the paper.

Conflicts of Interest: The authors declare no conflict of interest.

References

1. Schueler, T.R. The importance of imperviousness. *Watershed Prot. Tech.* **1994**, *1*, 100–111.
2. Leopold, L.B. *Hydrology for Urban Land Planning—A Guidebook on the Hydrologic Effects of Urban Land Use*; USGS Circular: Menlo Park, CA, USA, 1968; p. 554.
3. Paul, M.J.; Meyer, J.L. Stream in the Urban Landscape. *Annu. Rev. Ecol. Syst.* **2001**, *32*, 333–365. [CrossRef]
4. Fischer, D.; Charles, E.G.; Baehr, A.L. Effects of stormwater infiltration on quality of groundwater beneath Retention Basins. *J. Environ. Eng.* **2003**, *129*, 464–471. [CrossRef]
5. Akhter, M.S.; Madany, I.M. Heavy metals in street and house dust in Bahrain. *Water Air Soil Pollut.* **1993**, *66*, 111–119. [CrossRef]
6. Kelly, J.; Thornton, I.; Simpson, P.R. Urban Geochemistry: A study of the influence of anthropogenic activity on the heavy metal content of soils in traditionally industrial and non-industrial areas of Britain. *Appl. Geochem.* **1996**, *11*, 363–370. [CrossRef]
7. Deletic, A.B.; Maksimovic, C.T. Evaluation of water quality factors in storm runoff from paved areas. *J. Environ. Eng.* **1998**, *124*, 869–879. [CrossRef]
8. Novotny, V. Urban diffuse pollution: Sources and abatement. *Water Environ. Technol.* **1991**, *12*, 60–65.
9. U.S. EPA. Meeting the Environmental Challenge. In *EPA's Review of Progress and New Directions in Environmental Protection*; United States (U.S.) Environmental Protection Agency (EPA): Washington, DC, USA, 1990; p. 26.
10. Dinius, S.H. Design of an index of water quality. *Water Resour. Bull.* **1987**, *23*, 833–843. [CrossRef]
11. Whipple, W.; Grigg, S.; Gizzard, T. *Stormwater Management in Urbanizing Areas*; Prentice-Hall: Englewood Cliffs, NJ, USA, 1983.
12. Chapman, D. *Water Quality Assessments: A Guide to the Use of Biota, Sediments and Water in Environmental Monitoring*; Chapman Hall: London, UK, 1992.
13. Han, Y.M.; Du, P.X.; Cao, J.J.; Posmentier, E.S. Multivariate analysis of heavy metal contamination in urban dusts of Xi'an, Central China. *Sci. Total Environ.* **2006**, *355*, 176–186.
14. Qi, J.Y. *Quantity Study on Non-Point Source Pollution of City*; HoHai University: Nanjing, China, 2005.
15. Choi, K.S.; Ball, J.E. Parameter estimation for urban runoff modelling. *Urban Water J.* **2002**, *4*, 31–41. [CrossRef]
16. Liu, Y.; Puripus, S.; Li, J.; Christensen, E.R. Stormwater Runoff Characterized by GIS Determined Source Areas and Runoff Volumes. *Environ. Manag.* **2011**, *47*, 201–204. [CrossRef] [PubMed]
17. Barbosa, A.E.; Fernandes, J.N.; David, L.M. Key issues for sustainable urban stormwater management. *Water Res.* **2012**, *46*, 6787–6790. [CrossRef] [PubMed]
18. Metcalf, Eddy, Inc.; University of Florida; Water Resources Engineers, Inc. *Storm Water Management Model, Version I: Final Report*; Report 11024DOC07/71(NTIS PB-203289); Environmental Protection Agency (EPA): Washington, DC, USA, 1971.

19. Hydrologic Engineering Center. *Storage, Treatment, Overflow, Runoff Model, STORM, Generalized Computer Program 723–58-L7520*; Hydrologic Engineering Center, US Corps of Engineers: Davis, CA, USA, 1977.

20. Corbett, C.W.; Matthew, W.; Dwayne, E.P. Non-point source runoff modeling: A comparison of a forested watershed and an urban watershed on the South Carolina coast. *J. Exp. Mar. Biol. Ecol.* **1997**, *213*, 133–149. [CrossRef]

21. Huber, W.C.; Dickinson, R.E. *Storm Water Management Model, Version 4: User's Manual*; EPA/600/3–88/001a; Environmental Research Laboratory, U.S. Environmental Protection Agency: Athens, GA, USA, 1992.

22. Rossman, L.A. *Storm Water Management Model User's Manual, Version 5.0*; National Risk Management Research Laboratory, Office of Research and Development, US Environmental Protection Agency: Cincinnati, OH, USA, 2010.

23. Peterson, E.W.; Wicks, C.M. Assessing the importance of conduit geometry and physical parameters in karst systems using the storm water management model (SWMM). *J. Hydrol.* **2006**, *329*, 294–305. [CrossRef]

24. Wang, H.C.; Du, P.F.; Zhao, D.Q.; Wang, H.Z.; Li, Z.Y. Global sensitivity analysis for urban rainfall-runoff model. *China Environ. Sci.* **2008**, *28*, 725–729.

25. Francos, A. Sensitivity analysis of distributed environmental simulation models: Understanding the model behavior in hydrological studies at the catchment scale. *Reliab. Eng. Syst. Saf.* **2003**, *79*, 205–206. [CrossRef]

26. Huang, J.L.; Du, P.F.; He, W.Q.; Ao, Z.D.; Wang, H.C.; Wang, Z.S. Local sensitivity analysis for urban rainfall-runoff modeling. *China Environ. Sci.* **2007**, *27*, 549–555.

27. Wang, F.; Wang, X.; Zhao, Y.; Yang, Z.F. Long-term periodic structure and seasonal-trend decomposition of water level in Lake Baiyangdian, Northern China. *Int. J. Environ. Sci. Technol.* **2014**, *11*, 327–338. [CrossRef]

28. Zheng, X.K. *Simulation of Non-Point Source Pollution in Baoding City and Its Effects on Baiyangdian Lake*; Beijing Normal University: Beijing, China, 2009.

29. Di Modugno, M.; Gioia, A.; Gorgoglione, A.; Iacobellis, V.; la Forgia, G.; Piccinni, A.F.; Ranieri, E. Build-Up/Wash-Off Monitoring and Assessment for Sustainable Management of First Flush in an Urban Area. *Sustainability* **2015**, *7*, 5050–5067. [CrossRef]

30. Zhuang, C.; Ouyang, Z.; Xu, W.; Bai, Y.; Zhou, W.; Zheng, H.; Wang, X.K. Impacts of human activities on the hydrology of Baiyangdian Lake, China. *Environ. Earth Sci.* **2011**, *62*, 1343–1350. [CrossRef]

31. Punam, P.; Michael, A.; Taylor, L. Application of market mechanisms and incentives to reduce stotmwater runoff. *Environ. Sci. Policy* **2004**, *8*, 133–144.

32. Morquecho, R.; Pitt, R. Pollutant associations with particulates in stormwater. In Proceedings of the World Water and Environmental Resources Congress, Anchorage, AK, USA, 15–19 May 2005; pp. 4973–4999.

33. Huang, G.R.; Nie, T. Characteristics and load of non-point source pollution of urban rainfall-runoff in Guangzhou, China. *J. South China Univ. Technol.* **2012**, *40*, 142–148.

34. Zhou, Q.; Ren, Y.; Xu, M.; Han, N.; Wang, H. Adaptation to urbanization impacts on drainage in the city of Hohhot, China. *Water Sci. Technol.* **2016**, *73*, 167–175. [CrossRef] [PubMed]

35. Lee, S.B.; Yoon, C.G.; Jung, K.W. Comparative evaluation of runoff and water quality using HSPF and SWMM. *Water Sci. Technol.* **2010**, *62*, 1401–1409. [CrossRef] [PubMed]

36. Li, Y.X.; Ma, J.H.; Yang, Z.F. Influence of non-point source pollution on water quality of wetland Baiyangdian, China. *Desalinat. Water Treat.* **2011**, *32*, 291–296. [CrossRef]

37. Qiu, R.Z.; Li, Y.X.; Yang, Z.F. Influence of water quality change in Fu River on Wetland Baiyangdian. *Front. Earth Sci. China* **2009**, *3*, 397–401. [CrossRef]

Article

Modeling the Fate and Transport of Malathion in the Pagsanjan-Lumban Basin, Philippines

Mayzonee Ligaray [1], Minjeong Kim [1], Sangsoo Baek [1], Jin-Sung Ra [2], Jong Ahn Chun [3], Yongeun Park [1], Laurie Boithias [4], Olivier Ribolzi [4], Kangmin Chon [5] and Kyung Hwa Cho [1,*]

[1] School of Urban and Environmental Engineering, Ulsan National Institute of Science and Technology, Ulsan 44919, Korea; mayzonee@unist.ac.kr (M.L.); paekhap0835@unist.ac.kr (M.K.); kbcqr@unist.ac.kr (S.B.); phdyongeun@gmail.com (Y.P.)
[2] Eco-Testing & Risk Assessment Center, Korea Institute of Industrial Technology, Ansan-si 426910, Korea; jinsungra@kitech.re.kr
[3] APEC Climate Center, Busan 48058, Korea; jachun@apcc21.org
[4] Géosciences Environnement Toulouse, Université de Toulouse, CNES, CNRS, IRD, UPS, Toulouse 31400, France; laurie.boithias@get.omp.eu (L.B.); olivier.ribolzi@get.omp.eu (O.R.)
[5] Department of Environmental Engineering, College of Engineering, Kangwon National University, Kangwondaehak-gil 1, Chuncheon-si, Gangwon-do 24341, Korea; kmchon@gmail.com
* Correspondence: khcho@unist.ac.kr; Tel.: +82-052-217-2829

Received: 17 April 2017; Accepted: 19 June 2017; Published: 22 June 2017

Abstract: Exposure to highly toxic pesticides could potentially cause cancer and disrupt the development of vital systems. Monitoring activities were performed to assess the level of contamination; however, these were costly, laborious, and short-term leading to insufficient monitoring data. However, the performance of the existing Soil and Water Assessment Tool (SWAT model) can be restricted by its two-phase partitioning approach, which is inadequate when it comes to simulating pesticides with limited dataset. This study developed a modified SWAT pesticide model to address these challenges. The modified model considered the three-phase partitioning model that classifies the pesticide into three forms: dissolved, particle-bound, and dissolved organic carbon (DOC)-associated pesticide. The addition of DOC-associated pesticide particles increases the scope of the pesticide model by also considering the adherence of pesticides to the organic carbon in the soil. The modified SWAT and original SWAT pesticide model was applied to the Pagsanjan-Lumban (PL) basin, a highly agricultural region. Malathion was chosen as the target pesticide since it is commonly used in the basin. The pesticide models simulated the fate and transport of malathion in the PL basin and showed the temporal pattern of selected subbasins. The sensitivity analyses revealed that application efficiency and settling velocity were the most sensitive parameters for the original and modified SWAT model, respectively. Degradation of particulate-phase malathion were also significant to both models. The rate of determination (R^2) and Nash-Sutcliffe efficiency (NSE) values showed that the modified model ($R^2 = 0.52$; NSE = 0.36) gave a slightly better performance compared to the original ($R^2 = 0.39$; NSE = 0.18). Results from this study will be able to aid the government and private agriculture sectors to have an in-depth understanding in managing pesticide usage in agricultural watersheds.

Keywords: soil and water assessment tool; pesticides; malathion; agricultural watershed; modified SWAT model

1. Introduction

Agriculture has been substantial to the Philippine economy and has contributed 10.2% to 13.2% of the country's GDP in the past decade [1]. To keep up with this demand, various kinds of pesticides were applied to different crops and vegetables to help increase food supplies and provide greater revenue for farmers. However, exposure to pesticides could potentially cause cancer and disrupt the development of vital systems (endocrine, reproductive, and immune systems) [2–4]. Pesticide contamination in soil and water also has negative effects on the diversity of the flora and fauna of local areas thereby disturbing the existing ecosystem [5–7]. Various kinds of pesticides are used in agriculture depending on the target pests. Hence, many kinds of chemicals exist and find their way into the groundwater, surface water, soils, and eventually drinking water [8–11]. Several attempts have been made to monitor and map out their potential areas of contamination in the Philippines, especially in highly agricultural areas [12–14].

Laguna de Bay is the second largest freshwater lake in Southeast Asia and the largest in the Philippines. It is located east of Metro Manila, the Philippine capital, and is part of the Laguna de Bay basin. The basin has one of the fastest economic growth among others and it is a major water resource for agriculture, fisheries, and domestic use of the surrounding communities that has an estimated population of six million people [15]. In the recent years, the lake has been threatened by waste discharges of the industrial, urban, and residential areas from the west and by intensive agricultural activities from the east [16]. The presence of pesticides and other micropollutants led to the increasing levels of toxicity and fish-kill occurrences in the lake [17,18]. Many efforts have been made to improve the water quality of the lake such as rehabilitation programs and cleanup operations within the vicinity of the basin. Monitoring activities were also performed to assess the level of contamination. However, these were short-term and limited to a few selections of pesticides [17–19]. Applying environmental models to available monitoring datasets of pesticides can help broaden the understanding of the behavior of these micropollutants in the environment. However, existing modeling studies of Laguna de Bay basin lack watershed-scale analyses of pesticides used in its agricultural activities [14].

Processes driving pesticide fate and transport are on the whole well-known and are incorporated in various pesticide models operating at plot or watershed spatial scales, such as the crop model STICS (Simulateur mulTIdiscplinaire pour les Cultures Standard), MACRO (Water and solute transport in macroporous soils), PEARL (Pesticide Emission Assessment at Regional and Local scales), PRZM (Pesticide Root Zone Model), and SWAT (Soil and Water Assessment Tool) [20,21]. Several studies have already shown that the watershed-scale SWAT model was an efficient tool to model pesticides fate and transport [22–25]. However, the SWAT model performance can be restricted by its two-phase partitioning approach, which is inadequate when it comes to simulating pesticides with limited dataset. In this study, we modified the SWAT model by incorporating the three-phase partitioning model to improve the pesticide simulations, especially for watersheds with scarce dataset that are often common in developing countries. The modified model considered the three-phase partitioning model that classifies the pesticide into three forms: dissolved, particle-bound, and dissolved organic carbon (DOC)-associated pesticide. This approach is a first for pesticides and it differs from the original SWAT model that classified pesticides into two categories: pesticide sorbed into solid phase and pesticide in solution. The addition of DOC-associated pesticide particles increases the scope of the pesticide model by also considering the adherence of pesticides to the organic carbon in the soil.

We aimed to: (1) conduct a watershed-scale analysis of the fate and transport of pesticides, specifically malathion, and increase the accuracy of the simulated malathion loading using the modified pesticide model; and (2) perform a case study by applying the original SWAT model and modified pesticide model to a catchment with limited dataset, such as PL basin, and compare their performance. Malathion is an organophosphate insecticide used in PL basin for crops and vegetables. It is preferred by farmers due to its effectiveness against a wide range of pests and short half-life. The SWAT model was used to construct the watershed model for one of the subbasins of the Laguna de Bay basin, namely the Pagsanjan-Lumban (PL) basin. SWAT is a widely-used, physically-based hydrologic model that can predict the impact of water management practices [26,27]. It can simulate the flowrate and the transport of nutrients, pesticides, and sediments in watersheds. Implementing a watershed-scale analysis of the fate and transport of the malathion in the Laguna de Bay basin using watershed models will give an insight on the dominant processes affecting pesticide loadings to the soil and water. Results from this study will aid the government and private agriculture sectors to have an in-depth understanding in managing pesticide usage in an agricultural watershed.

2. Materials and Methods

2.1. Study Area

PL basin is located at the southeastern part of Laguna de Bay basin in the Southern Tagalog Region (CALABARZON) of the Philippines. It has a catchment area of 454.45 km^2 (121°24′ E~121°37′ E, 14°37′ N~14°21′ N) that drains to Laguna de Bay. The watershed experiences two types of Philippine climate: (1) Type II; and (2) Type III. The eastern part of the basin experiences Type II climate that has no dry season with a very pronounced maximum rain period from December to February and a minimum rainfall period from March to May [28]. On the other hand, the western part has a Type III climate that has a short dry season, varying from 1 to 3 months, in December to February [28]. Areas close to Mt. Banahaw at the southernmost part of the basin have relatively uniform rainfall distribution throughout the year [29]. However, the basin in general experiences a dry period from November to April due to the rain shadow effect of the Sierra Madre mountain range while the wet period occurs for the remaining months [29]. The average annual rainfall of the basin is 2996 mm, which mostly fall during the monsoon period.

Figure 1 shows the Digital Elevation Model (DEM) of the PL basin. The areas near Mt. Banahaw at the southern region have the highest elevation, ranging from 560 m to 2170 m, while the eastern region near Sierra Madre ranges from 350 m to 560 m. The region near the outlet, including Lumban delta, has the lowest elevation, which ranges from 0 m to 200 m. Negative values can also be observed within the Lumban delta indicating that the elevations are below sea level and are often submerged in water. The outlet of the basin was set at the Lumban Station before the river branched out to the Lumban delta to exclude the possibility of water intrusion from the lake.

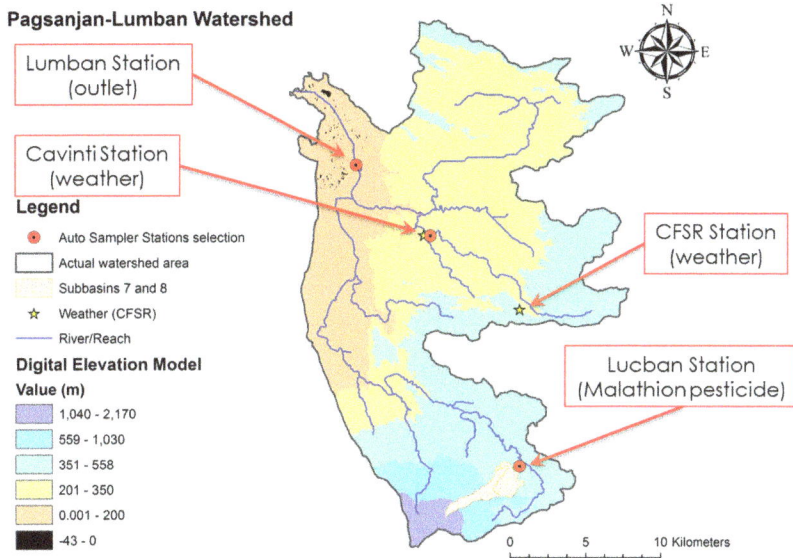

Figure 1. The digital elevation model of the Pagsanjan-Lumban watershed.

The basin has two major tributaries that branch out after the Lumban Station (outlet) as shown in Figure 1. Due to the presence of two reservoirs in the northern half of the basin, most of the discharge during dry days comes from the Pagsanjan River situated at the southern half of the PL basin. Pagsanjan River has a length of 54.1 km and a drainage area of 311.8 out of the 454.45 km^2 of PL basin, with a mean annual runoff of 53.1 m$^3 \cdot$s^{-1} [29]. The runoff pathways in the north were modified to collect the water in the reservoirs; hence, water only flows to the outlet during extreme rain events.

2.2. Monitoring Data

The locations of the monitoring stations for the flowrate, weather, and malathion are shown in Figure 1. The daily flowrate data from May 2014 to October 2016 at Lumban Station and the weather data (precipitation, temperature, humidity, and wind speed) at Cavinti Station from 2014 to 2016 were acquired from the Integrated National Watershed Research and Development Project (INWARD), the weather data from 1979 to 2014, including the solar values, were generated from the Climate Forecast System Reanalysis (CFSR) of the Global Weather Data for SWAT [30,31], and the malathion concentrations were monitored at Lucban Station by Varca [16]. A total of 26 sampling events at Lucban Station, with a frequency of at least two water samples a month, were carried out from December 2007 to November 2008 to measure the malathion concentrations. Each water sample was analyzed to measure the total concentration of malathion, which was used as comparison for the pesticide simulations in this study.

2.3. Hydrology Model

SWAT is a physically-based watershed model developed for the USDA Agricultural Research Service (ARS) to simulate the impact of land management practices on water, sediment, nutrients, and pesticide yields in large complex watersheds [32,33]. The model operates at a daily time step and uses readily available inputs such as [34]: topography (DEM with a 90 m resolution from United States Geological Survey (USGS)/National Aeronautics and Space Administration Shuttle Radar Topography Mission), hydrography, weather data (INWARD and CFSR), landuse/land cover (USGS Global Land Cover Characterization database), and soil type (Food and Agriculture Organization). The delineation threshold of the PL basin was 2.5 km^2, thus; it was divided into eight subbasins with 54 hydrological response units (HRU). Each of these HRUs is a unique combination of soil type, landuse, and slope. The threshold for soil, landuse, and slope was set to 0 to include non-agricultural areas in the basin that are less than 1% of the subbasin areas.

Table 1 summarizes the calibration and validation periods of the flowrate simulation. The calibration period was from September 2014 to May 2015 while the validation was from May 2014 to July 2014 and June 2016 to September 2016. The available flowrate dataset started from May 2014 until September 2016. We first compared the observed flowrate to the precipitation and noticed that the peaks of the flowrate did not match the precipitation for a period. This period was removed after concluding that the sensor was faulty at that time. The SWAT—Calibration and Uncertainty Program (SWAT-CUP) was then used to calibrate and validate the SWAT flowrate parameters shown in Table 2. Simulated flowrates in 2007 and 2008 were then applied to simulate malathion fate based on the available malathion dataset (December 2007 to November 2008).

Table 1. Summary of the calibration and validation periods.

Process	Period
Spinup Time (2 years)	January 2005–December 2006
Pesticide Calibration	December 2007–November 2008
Flow Calibration	September 2014–May 2015
Flow Validation	May 2014–July 2014 and June 2016–September 2016

Table 2. Soil and Water Assessment Tool (SWAT) flowrate parameters for calibration and sensitivity analysis.

Parameter	Description	Module	Method	MIN	MAX
CN2	Initial SCS runoff curve number for moisture condition II	MGT	Relative	−0.1	0.1
BIOMIX	Biological mixing coefficiency	MGT	Replace	0	1
ALPHA_BF	Baseflow alpha factor (days)	GW	Replace	0.01	1
GW_DELAY	Groundwater delay time (days)	GW	Replace	0	500
GWQMN	Threshold depth of water in the shallow aquifer required for return flow to occur (mm H_2O)	GW	Replace	0	50
REVAPMN	Threshold depth of water in the shallow aquifer for revap or percolation to the deep aquifer to occur (mm H_2O)	GW	Replace	0	750
RCHRG_DP	Deep aquifer percolation fraction	GW	Replace	0.01	0.99
GW_REVAP	Groundwater "revap" coefficient	GW	Replace	0.02	0.2
ESCO	Soil evaporation compensation factor	HRU	Replace	0.7	1
EPCO	Plant uptake compensation factor	HRU	Replace	0.7	1
SLSUBBSN	Average slope length (m)	HRU	Replace	10	150
LAT_TTIME	Lateral flow travel time (days)	HRU	Replace	0	180
OV_N	Manning's "n" value for overland flow	HRU	Replace	0.01	0.5
CANMX	Maximum canopy storage (mm H_2O)	HRU	Replace	0	100
CH_K2	Effective hydraulic conductivity in main channel alluvium (mm/h)	RTE	Replace	0.025	76
CH_N2	Manning's "n" value for the main channel	RTE	Replace	0.025	0.15
SOL_BD	Moist bulk density (Mg/m^3 or g/cm^3)	SOL	Relative	−0.1	0.1
SOL_CBN	Organic carbon content (% soil content)	SOL	Relative	−0.1	0.1
SOL_K	Saturated hydraulic conductivity (mm/h)	SOL	Relative	−0.1	0.1
SOL_AWC	Available water capacity of the soil layer (mm H_2O/mm soil)	SOL	Relative	−0.1	0.1
CH_K1	Effective hydraulic conductivity in tributary channel alluvium (mm/h)	SUB	Replace	0.025	76
CH_N1	Manning's "n" value for the tributary channel	SUB	Replace	0.025	0.15
SURLAG	Surface runoff lag coefficient (h)	BSN	Replace	0.05	10

2.4. Pesticide Modeling

The malathion loadings were then calibrated using the original SWAT and modified SWAT pesticide models. Both models need the management operation schedule to simulate the application of pesticide in the basin. In this study, malathion was applied to HRUs with "Tomato" (TOMA) as land cover. This is based on a previous study that summarized the pesticide usage of the farmers in the PL basin, which affects two HRUs from Subbasins 7 and 8 (shown in Figure 1) [35]. TOMA was assumed as a collective representative of the vegetable crops in the PL basin that used malathion during the pesticide application period. Two planting seasons were implemented for TOMA. The first season starts in January while the second season starts in June. Malathion was then applied for 12 times for 3 months after planting at a rate of 0.57 kg/ha for every application (first season: January to March; second season: June to August). Based on this schedule, the pesticide models were then run to simulate malathion loading in the HRUs with TOMA as land cover. The models are further discussed in the next subsections.

2.4.1. Original SWAT Pesticide Model

The pesticide module in the SWAT model was applied to calibrate the malathion loadings in the PL basin. Figure 2 shows that the SWAT pesticide model used the two-phase partitioning approach that classify the pesticides as: pesticide sorbed into solid phase and pesticide in solution or liquid phase. Table 3 shows the pesticide parameters that describe the reaction and transport processes of malathion starting from the application (foliar, soil surface, and subsurface). These processes include degradation, infiltration, leaching, surface runoff, volatilization, and wash off mechanisms.

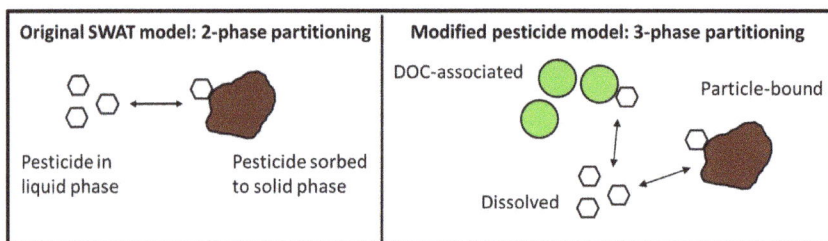

Figure 2. Original SWAT model applies the two-phase partitioning approach: pesticide in liquid phase and pesticide sorbed to the solid phase. Modified pesticide model assumes the three-phase partitioning model: dissolved pesticide, particle-bound pesticide, and DOC-associated pesticide.

Table 3. Pesticide parameters for calibration and sensitivity analysis of the original SWAT model.

Parameter	Description	Module	Method	MIN	MAX
SKOC	Soil adsorption coefficient normalized for soil organic carbon (L·kg^{-1})	PEST	Replace	1	5000
HLIFE_S	Degradation half-life of the chemical on the soil (day^{-1})	PEST	Replace	0	100
HLIFE_F	Degradation half-life of the chemical on the foliage (day^{-1})	PEST	Replace	0	100
WSOL	Solubility of the chemical in water	PEST	Replace	0	1000
WOF	Wash off fraction	PEST	Replace	0	1
AP_EF	Application efficiency	PEST	Replace	0	1
PST_DEP	Depth of pesticide incorporation in the soil (mm)	MGT	Replace	0	500
PERCOP	Pesticide percolation coefficient	BSN	Replace	0	1
CHPST_KOC	Pesticide partition coefficient between water and sediment in reach (m^3·g^{-1})	SWQ	Replace	0	0.1
CHPST_REA	Pesticide reaction coefficient in reach (day^{-1})	SWQ	Replace	0	0.1
CHPST_VOL	Pesticide volatilization coefficient in reach (m·day^{-1})	SWQ	Replace	0	10

Table 3. *Cont.*

Parameter	Description	Module	Method	MIN	MAX
CHPST_STL	Settling velocity for pesticide sorbed to sediment (m·day^{-1})	SWQ	Replace	0	10
SEDPST_REA	Pesticide reaction coefficient in reach bed sediment (day^{-1})	SWQ	Replace	0	0.1
CHPST_RSP	Resuspension velocity for pesticide sorbed to sediment (m·day^{-1})	SWQ	Replace	0	1
SEDPST_ACT	Depth of active sediment layer for pesticide (m)	SWQ	Replace	0	1
CHPST_MIX	Mixing velocity (diffusion/dispersion) for pesticide in reach (m·day^{-1})	SWQ	Replace	0	0.1
SEDPST_BRY	Pesticide burial velocity in reach bed sediment (m·day^{-1})	SWQ	Replace	0	0.1
PSTENR	Enrichment ratio for pesticide in the soil	CHM	Replace	0	5

2.4.2. Modified Pesticide Model

Figure 3 shows the schematic diagram of the fate and transport of the pesticides for the modified SWAT pesticide model. The modified model applied in this study was based on the watershed-scale model from a previous study of the same authors about modeling the fate and transport of polycyclic aromatic hydrocarbons (PAH) and linking the PAH model with SWAT [36]. This study further developed the model to include the pesticide application based on the original SWAT model and other equations related to the fate and transport of pesticides. The approach of the original 2-phase partitioning SWAT model on the fate and transport of pesticides was modified by considering the three-phase partitioning model shown in Figure 2. Figure 4 shows the diagram of using MATLAB as platform for the modified approach. Table 4 shows the parameters of the modified pesticide model.

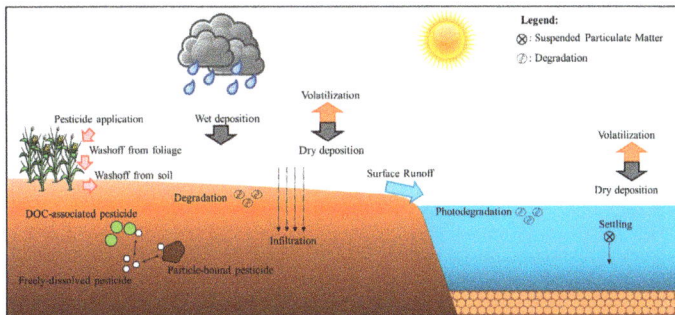

Figure 3. Fate and transport diagram of pesticides in the environment with the modified SWAT model. The three-phased partitioning model approach that we applied for the pesticide was also based on a previous study of the same authors about modeling the fate and transport of polycyclic aromatic hydrocarbons (PAH) and linking the PAH model with SWAT.

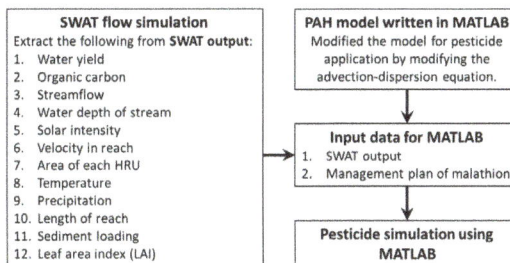

Figure 4. Flow diagram of the modified pesticide model. SWAT output and management plan of malathion were used as input in MATLAB for the pesticide simulations.

Table 4. Pesticide parameters for calibration and sensitivity analysis of the modified model.

Parameter	Description	Unit
f_{oc}	Organic carbon fraction in soil	-
ρ_{soil}	Soil density	$kg \cdot m^{-3}$
$poro$	Porosity	
f_{DOC}	Fraction of the dissolve organic carbon	-
x_1	Enrichment ratio coefficient 1	-
x_2	Enrichment ratio coefficient 2	-
v_s	Settling velocity of suspended particles in the channel	$m \cdot s^{-1}$
En	Diffusion coefficient	-
$C_{p,1}$	Wash off coefficient for particle-bound pesticide	-
$C_{fd,1}$	Wash off coefficient for dissolved pesticide	-
$C_{p,2}$	Wash off exponent for particle-bound pesticide	-
$C_{fd,2}$	Wash off exponent for dissolved pesticide	-
α	Decay coefficient in water due to solar intensity	-
$C_{DOC,1}$	Wash off coefficient for DOC-associated pesticide	-
$C_{DOC,2}$	Wash off exponent for DOC-associated pesticide	-
$\mu_{k,p}$	Degradation rate constant for the particle-bound pesticide on the soil surface	s^{-1}
$\theta_{k,p}$	Temperature adjustment factor for particle-bound pesticide	-
$\mu_{k,fd}$	Degradation rate constant for the dissolved pesticide on the soil surface	s^{-1}
$\theta_{k,fd}$	Temperature adjustment factor for dissolved pesticide	-
$\mu_{k,DOC}$	Degradation rate constant for the DOC-associated pesticide on the soil surface	s^{-1}
$\theta_{k,DOC}$	Temperature adjustment factor for DOC-associated pesticide	-

Accumulation of Pesticide

Pesticides are usually distributed to the foliage and soil surface during application. In this case, separate equations were applied to calculate the amount of pesticides on the foliage and the soil surface [37]. This approach is similar to the original SWAT model. The amount of pesticides on foliage (pst_f, kg of pesticide ha^{-1}) and on the soil surface (pst_{surf}, kg of pesticide ha^{-1}) were determined by the equations below [37]:

$$pst_f = gc \times pst' \tag{1}$$

$$pst_{surf} = (1 - gc) \times pst' \tag{2}$$

where gc is the fraction of the ground surface covered by plants (-) and pst' is the efficient amount of pesticide applied (kg of pesticide ha^{-1}). These terms were determined by the following equations [37]:

$$gc = (1.99532 - erfc \times [1.333 \times LAI - 2])/2.1 \tag{3}$$

$$pst' = ap_{ef} \times pst \tag{4}$$

where $erfc$ is the complementary error function (-), LAI is the leaf area index (-), ap_{ef} is the pesticide application efficiency (-), and pst is the original amount of pesticide applied (kg of pesticide ha^{-1}).

Pesticide on the foliage was assumed to be affected by wash off due to rain events and degradation. The amount of pesticides washed off by precipitation from the plants ($pst_{f,wsh}$, kg of pesticide ha^{-1}) were determined using the equation below [37]:

$$pst_{f,wsh} = fr_{wsh} \times pst_f \tag{5}$$

where fr_{wsh} is the wash off fraction for the pesticide on the foliage (-). Degradation of pesticide on the foliage was then calculated using the following equation [37]:

$$pst_{f,t} = pst_{f,0} \times exp[-k_{p,f} \times t] \tag{6}$$

where $pst_{f,t}$ is the amount of pesticide on the foliage at time t (kg of pesticide ha^{-1}), $pst_{f,0}$ is the initial amount of pesticide on the foliage (kg of pesticide ha^{-1}), $k_{p,f}$ is the degradation rate constant of the pesticide (day^{-1}), and t is time (day).

Pesticides on Soil

The original SWAT model assumed that the pesticide is either sorbed to the solid phase or dissolved in solution [37]. The three-phase partitioning model was used to estimate the different forms of pesticide in the soil. This method classifies the pesticide into three classes: dissolved pesticide, pesticide adsorbed on dissolved organic carbon ($[DOC]$), and pesticide adsorbed on particles. The total pesticide and dissolved pesticide concentrations in the bulk saturated soil are defined by the following equations [38]:

$$C_{bs}{}^t = C_{bs}{}^p + C_{bs}{}^d \tag{7}$$

$$C_{bs}{}^d = C_{bs}{}^{fd} + C_{bs}{}^{DoC} \tag{8}$$

where $C_{bs}{}^t$ is the total pesticide concentration in the bulk soil (kg·L^{-1} bulk soil), $C_{bs}{}^p$ is the concentration of particle-bound pesticide (kg·L^{-1} bulk soil), $C_{bs}{}^d$ is the dissolved pesticide concentrations in the bulk saturated soil (kg·L^{-1} bulk soil), $C_{bs}{}^{fd}$ is the dissolved pesticide, and $C_{bs}{}^{DOC}$ is the DOC-associated pesticide. Combining Equations (7) and (8) yields the following equation:

$$C_{bs}{}^t = C_{bs}{}^p + C_{bs}{}^{fd} + C_{bs}{}^{DoC} \tag{9}$$

The terms $C_{bs}{}^p$ and $C_{bs}{}^{DoC}$ can also be determined by multiplying $C_{bs}{}^{fd}$ by the coefficients shown in Equations (10)–(12) below [38]:

$$C_{bs}{}^p = r_{sw} \times K_{sw} \times C_{bs}{}^{fd} \tag{10}$$

$$C_{bs}{}^{DoC} = [DOC] \times K_{DOC} \times C_{bs}{}^{fd} \tag{11}$$

$$[DOC] = f_{DOC} \times OM \tag{12}$$

where r_{sw} is the soil-to-water ratio (kg·L^{-1}), K_{sw} is the soil-water distribution coefficient [L·kg^{-1}], K_{DOC} is the dissolved organic carbon-water partition coefficient (L·kg^{-1}), f_{DOC} is the fraction of DOC, and OM is the concentration of the organic matter (kg·L^{-1}). OM was calculated by dividing the mass of the soil carbon in the soil organic matter with the water yield, which were both simulated by the original SWAT model [36]. $C_{bs}{}^p$, $C_{bs}{}^{fd}$, and $C_{bs}{}^{DoC}$ from Equation (9) can also be determined by using the pesticide in bulk soil fractions [36,38]:

$$C_{bs}{}^p = f_p \times W_{cp} \times \exp(-\mu_p) \times \varepsilon_{pstsed} \tag{13}$$

$$C_{bs}{}^{fd} = f_d{}^{fd} \times W_{cf} \times \exp(-\mu_f) \tag{14}$$

$$C_{bs}{}^{DoC} = f_d{}^{DoC} \times W_{cDOC} \times \exp(-\mu_{DOC}) \tag{15}$$

where f_p (-), W_{cp} (kg·L^{-1} bulk soil), and μ_p (s^{-1}) are the fraction, wash off load, and rate constant of particle-bound pesticide, ε_{pstsed} is the enrichment ratio (-), $f_d{}^{fd}$ (-), Wcf (kg·L^{-1} bulk soil), and μ_f (s^{-1}) are the fraction, wash off load, and rate constant of dissolved pesticide, and $f_d{}^{DOC}$ (-), W_{cf} (kg·L^{-1} bulk soil), and μ_{DOC} (s^{-1}) are the fraction, wash off load, and rate constant of DOC-bound pesticide. The enrichment ratio was calculated using this equation [37] :

$$\varepsilon_{pstsed} = x_1 \times (sed/WY)^{x_2} \tag{16}$$

where x_1 and x_2 are the enrichment ratio coefficients, sed is the sediment yield (metric tons), and WY is the water yield (mm H$_2$O). Surface runoff of the suspended particles to the channel was also considered in the wash off loads and wash off fractions of the pesticide. The suspended particles affected by the wash off mechanism are described in the following equations [36]:

$$W_{cp} = C_{p1} \times q^{C_{p2}} \times C_{bs}{}^t \tag{17}$$

$$W_{cf} = C_{f1} \times q^{C_{f2}} \times C_{bs}{}^t \tag{18}$$

$$W_{cDOC} = C_{DOC1} \times q^{C_{DOC2}} \times C_{bs}{}^t \tag{19}$$

where W_{cp}, W_{cf}, and W_{cDOC} are the wash off loads of the particle-bound, dissolved, and DOC-associated PAHs exported to the river via runoff (kg·L^{-1} bulk soil); C_{p1}, C_{f1}, and C_{DOC1} are the wash off coefficients for the particle-bound, dissolved, and DOC-associated PAHs (-); q is the runoff rate per unit area; and C_{p2}, C_{f2}, and C_{DOC2} are the wash off exponents for the particle-bound, dissolved, and DOC-associated PAHs (-), respectively.

The degradation term, previously shown in Equations (13)–(15), is a temperature-dependent rate constant expressed as the first-order reaction. The rate constant for the three phases was defined as:

$$\mu = \mu_i \times \theta^{(T - 20)} \tag{20}$$

where μ_i is the initial rate constant for the pesticide [s^{-1}], θ is the temperature adjustment factor for pesticide (-), and T is the temperature (°C). μ and θ were then calibrated for the particle-bound, dissolved, and DOC-associated pesticide.

Pesticide loadings on the soil and into the water were computed by [38]:

$$C_p = (C_{bs}{}^p - C_{bs,out}{}^p)/\rho \tag{21}$$

$$C_{wfinal} = C_{bs}{}^d + C_{bs,out}{}^p \tag{22}$$

where C_p and C_w are the pesticide loading on soil [pesticide per solid mass] and in water [pesticide per fluid mass], respectively, and C_{wfinal} is the final pesticide loading in water.

Pesticides in Water

The pesticides in the channels are then subjected to the following mechanisms and processes due to water movement and reactivity of the pesticide with other components present in the water: advection, dispersion, photodegradation, and settling processes upon entering the channel [36]. The concentration of the pesticide in the waterbody was determined after considering these processes. This is expressed by the modified advection-dispersion equation:

$$(\partial C/\partial t) + u \times (\partial C/\partial x) = D_L \times (\partial^2 C/\partial x^2) - C \times [f \times (v_s/h) + aI] \tag{23}$$

where C is the concentration of pesticide in water (g·m^{-3}), t is time (day), x is distance (m), u is the velocity of the water (m·s^{-1}), D_L is the dispersion coefficient (m^2·d^{-1}), f is the fraction of the particulate pesticide in water (-), v_s is the settling velocity (m·s^{-1}), h is the depth of the channel (m), a is the photodegradation coefficient (m^2·MJ^{-1}·d^{-1}), and I is the solar intensity (MJ·m^{-2}). The photodegradation term in the equation was added to the advection-dispersion equation to fit the pesticide model.

2.5. Sensitivity Analyses

The sensitivity analyses for the flowrate and pesticide were simultaneously done with the calibration. SWAT-CUP applies the Latin-Hypercube (LH) sampling method, which is based on the Monte Carlo simulation, and set the Nash–Sutcliffe efficiency (NSE) as the objective function (Section 2.6). This is a robust method that requires a large number of simulations and computational resources [39]. LH sampling randomly assigns a value within the permitted range of each parameter to complete a parameter set. The One-factor-At-a-Time (OAT) sensitivity test was then commenced after the sampling. This method takes one parameter for each run and changes its value to determine how each parameter affects the results. Sensitive parameters were determined based on their probability values or *p*-values. Parameters with less than 0.01 *p*-value were labeled as sensitive. The LH-OAT method was applied to the SWAT hydrology and pesticide models in this study.

2.6. Evaluation Criteria

The coefficient of determination (R^2), Nash–Sutcliffe Efficiency coefficient (NSE), root mean square error (RMSE), and percent bias (PBIAS) were then calculated to evaluate the SWAT and the modified model. These statistical indices can determine whether the model performance is satisfactory or unsatisfactory [40]. R^2 and NSE were calculated for the calibration and validation periods of the flowrate (Table 1), while R^2, NSE, RMSE and PBIAS were determined for pesticide. For a daily time-step, the model is acceptable or satisfactory when the R^2 and NSE values are greater than 0.5 and when the RMSE and PBIAS values reach the optimal value of 0. A lower RMSE value is an indicator that the model has less residual variance [41]. PBIAS shows if the model has under- or overestimated the results compared to the observations, and lower absolute PBIAS values indicate more accurate model simulations [42].

SWAT-CUP also has other criteria to quantify the strength and uncertainties of the calibration analysis. Given the small dataset of this study, the P-factor and R-factor of the iteration were also noted. P-factor represents percentage of the simulated data covered by the 95% uncertainty band (95PPU) and it ranges from 0 to 1 [43]. A P-factor value greater than 0.5 is desirable since it entails that more than 50% of the simulated data are acceptable. R-factor, on the other hand, estimates strength of the calibration by dividing the average thickness of the 95PPU band with the standard deviation of the measured data [43]. The range of the R-factor starts from 0 to infinity; high and low R-factor values correspond to a thicker and thinner 95 PPU bands, respectively. A P-factor of 1.0 and an R-factor of 0 signifies a perfect simulation of the observed data.

3. Results and Discussion

3.1. Flowrate Calibration

Figure 5 shows the comparison of the observed and SWAT-simulated flowrate at Lumban Station. The model did not implement the presence of reservoirs and paddy fields due to the lack of information; hence, it was assumed that these factors would affect the flowrate calibration when it comes to water storage. The calibration process yielded an R^2 value of 0.42 and an NSE of 0.22 while the validation period has 0.10 and −2.87, respectively. These values are less than the desired value of 0.5 [40], showing that the SWAT model is slightly unfit and underperforming. The lack of information regarding the management of the two reservoirs (storage, release, and distribution) may have significantly degraded the performance of the model [44]. The SWAT model of the PL basin was limited to a small dataset, which was a major disadvantage during the calibration (227 observed flowrate) and validation (158 observed flowrate) of the flowrate. The intense rainfall events happened in the calibration period; hence, the entire basin drains to the outlet, as mentioned in Section 2.1. This is in contrast with the validation period that included the days with dry to mild rainfall, which only drains the southern part of the basin. The 95PPU band of the flowrate calibration in Figure 5 has a P-factor value greater than 0.5 and an R-factor value of less than 1. This result indicates that more than 50% of the simulated flowrate is within the acceptable uncertainty [43]. The validation period, on the other hand, has a P-factor of 0.18 and an R-factor of 0.72, indicating that only 18% of the simulated flowrate during this period is within the uncertainty range.

Nine parameters were found to have a significant effect on the flowrate (Table 5). Surface runoff lag coefficient (SURLAG) was the most sensitive parameter with a calibration value of 0.051 h. This value is negligible compared to the default value in SWAT, which is four hours. As SURLAG decreases, more water is stored in the basin, indicating that surface runoff does not go directly to the channel in the PL basin and is instead stored elsewhere [45]. The presence of two reservoirs in the northern half of the basin, mentioned in Section 2.1, may have affected the SURLAG value. A huge percentage of the runoff was immediately collected by the reservoirs instead of getting released across the basin. The water can also be stored in the foliage and paddies that are present in the basin. This is supported by the other sensitive parameters: Manning's "n" value for the tributary channels (CH_N1), second most sensitive; the initial SCS runoff curve number for moisture condition II (CN2), third; baseflow alpha factor (ALPHA_BF), fourth; and effective hydraulic conductivity in tributary channel alluvium (CH_K1), fifth.

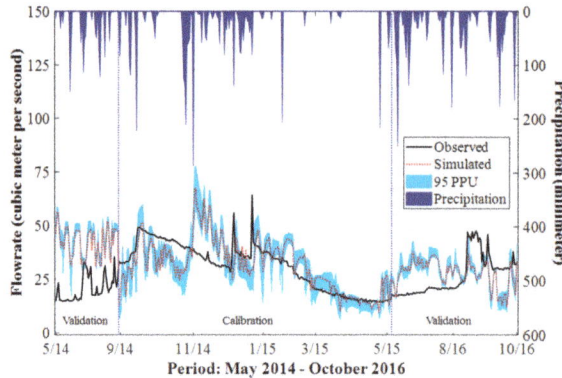

Figure 5. Observed and simulated flowrate ($m^3 \cdot s^{-1}$) at Lumban Station over the calibration (September 2014–May 2015) and the validation (May 2014–July 2014 and June 2016–September 2016) periods, together with the 95% uncertainty band (95 PPU). The flowrate is plotted against the precipitation (mm) for the whole period.

Table 5. SWAT parameters for calibration and sensitivity analysis of the original SWAT pesticide model.

Rank	Parameter	Fitted Value
1	SURLAG	0.051
2	CH_N1	0.12
3	CN2	−0.096
4	ALPHA_BF	0.46
5	CH_K1	48.4
6	RCHRG_DP	0.97
7	CH_K2	42.2
8	SLSUBBSN	75.8
9	OV_N	0.33

CN2 signifies the soil permeability of the basin [46]. Increasing CN2 values is associated with the increase of imperviousness of the basin. This can be related to how urbanized the basin is. During calibration, SWAT-CUP relatively changes the CN2 value at the HRU level since HRUs have varying CN2 values. In general, CN2 in PL basin has a calibrated value of −0.096 or −9.6%, indicating a general decrease in the CN2 values of each HRU. Hence, the basin is slightly more pervious compared to the default values suggested by SWAT. This result was expected since PL basin is highly agricultural [29]. ALPHA_BF, on the other hand, is the baseflow recession constant, a direct index of groundwater flow response and is a basin-wide parameter. It has a calibrated value of 0.46, suggesting an intermediate or average response to the change in recharge, slightly leaning towards the slow response (slow response: 0.1–0.3; rapid response: 0.9–1.0). The result can indicate that it is possible for water to be stored in the shallow aquifer that can also be associated to the CH_K1 value. CH_K1 controls the transmission losses from the surface runoff in the tributary. It is determined by the type of bed materials present in the channel bed. The calibrated result of 48.4 mm·h^{-1} falls on the moderately high loss rate that ranges from 25 mm·h^{-1} to 76 mm·h^{-1}. This value characterizes the tributary bed material as sand and gravel mixture with low silt-clay content, which makes it easier for the transmission losses of to percolate into the shallow aquifer. Another sensitive parameter that has a similar description to CH_K1 is the effective hydraulic conductivity in the main channel alluvium (CH_K2). The calibrated value of CH_K2, 42.2 mm·h^{-1}, falls within the same range as CH_K1, indicating that the main channel has similar characteristics as the tributary channel. Having a lower CH_K2 value compared to CH_K1 signifies that the bed material of the main channel is a slightly more consolidated compared to the tributary channels. The second most sensitive parameter also describes the tributary channel in the basin, CH_N1. CH_N1 has a calibrated value of 0.12, which indicates the channel is well maintained and full of weeds and brushes (excavated or dredged) or it is heavy timbered with lots of vegetation as well (natural stream). The deep aquifer percolation fraction (RCHRG_DP) was also found to be sensitive, with a calibrated value of 0.97. This result indicates that a huge fraction of percolation from the root zone recharges the deep aquifer [47]. However, the value is extremely high compared to the other studies with high RCHRG_DP values. Schuol et al. [48] estimated a range of 0.4 to 0.65 for the West Africa subcontinent, while Me et al. [47] yielded a value of 0.87 for a New Zealand catchment with mixed landuse. Though it can be assumed that the RCHRG_DP value for PL basin is also high, it should also be noted that this parameter may have been affected by the discrepancies formulated from the limited dataset (Section 2.2) and that the fraction (0.97) is too high for this basin. Lastly, the average slope length (SLSUBBSN) and Manning's value for overland flow (OV_N) have calibrated values of 75.8 m and 0.33, respectively.

3.2. Pesticide Calibration with SWAT Pesticide Model

After the flow calibration and validation, SWAT-CUP was applied to calibrate and analyze the sensitive parameters of the SWAT pesticide model. The calibration processes of the flowrate and pesticide were done separately to incorporate the same calibrated values of the flowrate parameters for the original SWAT model and the modified model. Figure 6 shows the observed and the simulated results of the malathion concentrations at Lucban Station using the original SWAT model. The calibration process yielded an R^2 of 0.39, an NSE of 0.18, a PBIAS value of 59.2%, and an RMSE of 3492.23 mg. Five parameters were found to be significant in the SWAT pesticide model (Table 6). The most sensitive parameter was the application efficiency (AP_EF) of the malathion with a calibrated value of 0.13. This indicates that only 13% of the applied malathion is deposited on the foliage and soil surface while the rest are lost in the atmosphere, which can be due to the type of pesticide application and management practices of the farmers in the basin. The pesticide partition coefficient between water and sediment in the reach (CHPST_KOC) was the second most sensitive parameter with a fitted value of 0.0012. A low value indicates that malathion is highly mobile in the water in its dissolved form, thereby increasing its potential for long-distance transport [16,49]. This parameter is followed by the degradation half-life of malathion on the soil surface (HLIFE_S), the reaction coefficient in the channel (CHPST_REA), and the soil adsorption coefficient for soil organic carbon (SKOC) that have calibrated values of 6.26 day^{-1}, 0.037 day^{-1}, and 3594.52 L·kg^{-1}, respectively. The sensitivity analysis determined that the pesticide application, degradation in the soil, and the sediment interaction and reactivity of malathion in water were the important processes that influenced the fate and transport of malathion in the basin.

Table 6. Sensitive parameters of the SWAT pesticide model.

Rank	Parameters	Fitted Value
1	AP_EF	0.13
2	CHPST_KOC	0.0012
3	HLIFE_S	6.26
4	CHPST_REA	0.037
5	SKOC	3594.52

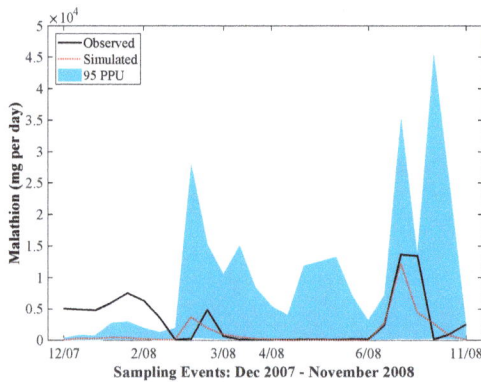

Figure 6. Observed and simulated malathion concentrations at Lucban Station over the calibration (December 2007–November 2008) period together with the 95% uncertainty band (95 PPU).

3.3. Pesticides Calibration with Modified Pesticide Model

Table 7 revealed that the settling velocity (v_s) was the most sensitive parameter, which suggest that the interaction of malathion particles in the water greatly affects the malathion transport. Degradation rate constant of the dissolved malathion particles ($\mu_{k,fd}$) was the second most sensitive, which can be attributed to how malathion readily dissolves in water compared to soil. This was followed by, in no particular order, the wash off coefficient ($C_{fd,1}$) and exponent ($C_{fd,2}$), diffusion coefficient (En), porosity ($poro$), soil density (ρ_{soil}), organic carbon fraction in soil (f_{oc}), and temperature adjustment factor of the particle-bound malathion ($\theta_{k,p}$). Malathion parameters that are associated with the particulate phase were sensitive for both models. However, the specific parameters are not exactly the same. The modified model has a more detailed formalism since it applied the three-phase partitioning model. This gave a visual understanding of the different forms of malathion that were greatly affected by wash off, which is the dissolved malathion. In this case, it can be assumed that dissolved malathion is more susceptible to wash off and most likely to end up in the channel compared to the other two malathion phases, particle-bound and DOC-associated malathion. Aside from the malathion-specific parameters, soil properties were also important such as the soil density, porosity, and organic carbon fraction of the soil.

Table 7. Sensitive parameters of the modified pesticide model.

Rank	Parameters	Fitted Value
1	v_s	0.0001
2	$\mu_{k,fd}$	4.25×10^{-5}
3	$C_{fd,1}$	1.14
4	$C_{fd,2}$	0.549
5	$poro$	0.50
6	f_{oc}	0.20
7	En	0.1006
8	$\theta_{k,p}$	1.16×10^{-5}
9	ρ_{soil}	1.06

3.4. Comparison of Observed and Simulated Malathion Loading

Figure 7 shows the observed and simulated values of the monitoring dataset. The modified model has 0.52, 0.36, 48.6%, and 3088.05 mg values for R^2, NSE, PBIAS, and RMSE, respectively. Based on these evaluation criteria, the modified model performed better compared to the SWAT model (statistics mentioned in Section 3.2).

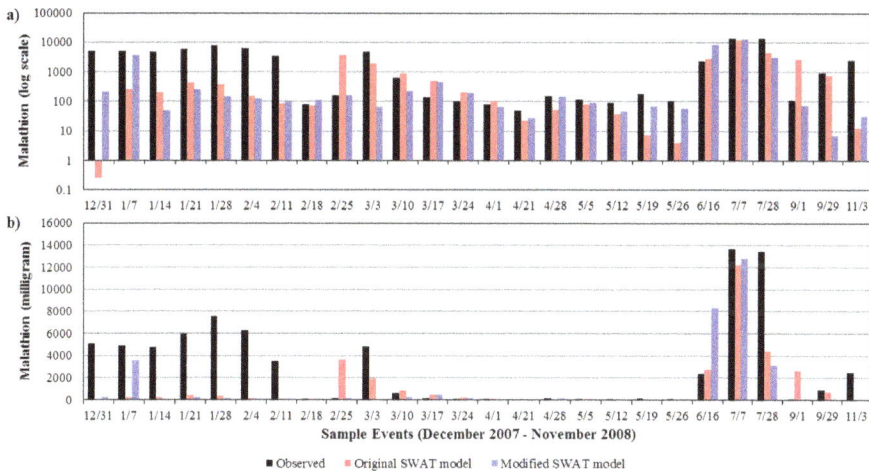

Figure 7. (**a**) Logarithmic scale (-); and (**b**) normal scale (milligram) of the observed and simulated malathion loadings at Lucban Station. This shows the comparison of the observed malathion data and simulated loading by the original SWAT and modified SWAT model.

The observed data were compared to the simulated loading by SWAT and the modified model in Figure 8. The figure includes the logarithmic (Figure 8a) and actual (Figure 8b) scale of the values, which reveals the similarities and differences between the models. Both models were able to achieve a small deviation between the observed and simulated low malathion loading. Comparing the two simulations, the modified model captured more low values compared to SWAT. However, the models were poorly able to simulate the high values as seen in the two plots (Figure 7), the results were comparable for both models. Figure 7 shows the time series of the malathion loading simulated by the SWAT and modified models compared to the SWAT-projected flowrate from January 2007 to December 2008 at the Lucban Station. The malathion simulations peaked during the duration of the pesticide application. However, the peaks of modified model showed more consistency compared to the increasing peaks of the SWAT model. Both models have similar peak levels at the fourth peak, but the SWAT model gave a more distinct pattern compared to the modified model.

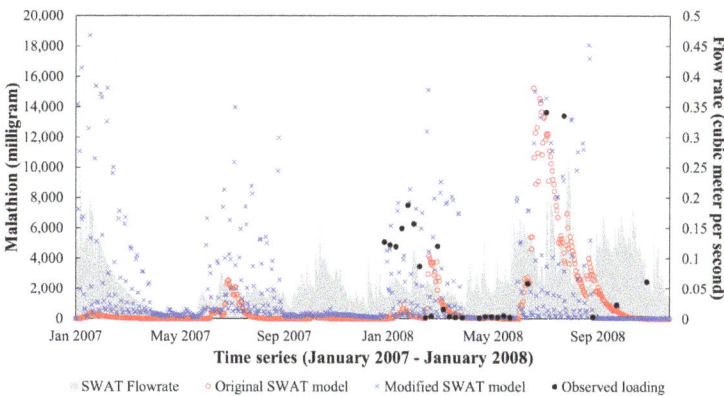

Figure 8. Time series of malathion loading simulated by the original SWAT and the modified SWAT models, observed malathion loading at the Lucban Station, and SWAT simulated flowrate.

4. Conclusions

Insufficient data were one of the limitations of this study that imply high uncertainty in both models. However, building a SWAT model for the flowrate calibration was made possible in this study. Equations that are relevant to the fate and transport of pesticides were added to the modified model prior to simulations. The objectives of this study were also met by simulating the malathion loading in the PL basin using the SWAT and modified model. Hence, the following conclusions were derived:

1. The sensitivity analysis of the hydrology model revealed that the flowrate of the PL basin is greatly influenced by the perviousness of the soil and the characteristics of the tributary channel that stores and retains the water in the basin.
2. The modified pesticide model gave a slightly better performance compared to the original SWAT model, considering the statistical analyses performed (R^2, NSE, PBIAS, and RMSE).
3. Application efficiency was the most sensitive parameters for the original SWAT model, suggesting a possible need to improve the pesticide application and management practices of farmers in the basin, while settling velocity was the most sensitive for the modified models. Parameters associated with particulate-phase malathion, especially the degradation of particle-bound malathion, were also significant to both models.
4. The temporal patterns of the target subbasin simulated by the models showed that the modified model has more consistent peaks during the duration of pesticide application compared to SWAT.

This study focused on the comparison of the outcomes of the modified model with the commonly used SWAT hydrological model. The modified model and the original SWAT model were able to identify similar sensitive parameters. However, further development of the model is needed to incorporate pesticide application scenarios and interaction of soil and water media to the atmosphere.

Acknowledgments: This study was made possible by the support of the Korea Ministry of Environment (MOE) as part of "The Chemical Accident Prevention Technology Development Project". We would like to acknowledge the Integrated Watershed Research and Development Project (INWARD) for providing the relevant information.

Author Contributions: Mayzonee Ligaray built the original SWAT model and wrote the paper; Sangsoo Baek developed the modified model; Minjeong Kim processed the data gathered; Yongeun Park processed the partial results of the SWAT model; Jin-Sung Ra and Jong Ahn Chun helped analyze the results; Laurie Boithias and Olivier Ribolzi provided their insight and expertise in improving the methods applied to the data and revising paper; Kangmin Chon made the plots and helped revised the paper; and Kyung Hwa Cho gathered the input data and designed the research plan.

Conflicts of Interest: The authors declare no conflict of interest and the founding sponsors had no role in the design of the study; in the collection, analyses, or interpretation of data; in the writing of the manuscript, and in the decision to publish the results.

References

1. The World Bank. Agriculture, Value Added (% of GDP). In *World Development Indicators*, 17 November 2016 ed.; The World Bank Group: Washington, DC, USA, 2016.
2. Bassil, K.L.; Vakil, C.; Sanborn, M.; Cole, D.C.; Kaur, J.S.; Kerr, K.J. Cancer health effects of pesticides: Systematic review. *Can. Fam. Phys.* **2007**, *53*, 1704–1711.
3. Colborn, T.; vom Saal, F.S.; Soto, A.M. Developmental effects of endocrine-disrupting chemicals in wildlife and humans. *Environ. Health Perspect.* **1993**, *101*, 378–384. [CrossRef] [PubMed]
4. Bolognesi, C. Genotoxicity of pesticides: A review of human biomonitoring studies. *Mutat. Res.* **2003**, *543*, 251–272. [CrossRef]
5. Jacobsen, C.S.; Hjelmsø, M.H. Agricultural soils, pesticides and microbial diversity. *Curr. Opin. Biotechnol.* **2014**, *27*, 15–20. [CrossRef] [PubMed]
6. Geiger, F.; Bengtsson, J.; Berendse, F.; Weisser, W.W.; Emmerson, M.; Morales, M.B.; Ceryngier, P.; Liira, J.; Tscharntke, T.; Winqvist, C.; et al. Persistent negative effects of pesticides on biodiversity and biological control potential on european farmland. *Basic Appl. Ecol.* **2010**, *11*, 97–105. [CrossRef]

7. Schäfer, R.B.; Caquet, T.; Siimes, K.; Mueller, R.; Lagadic, L.; Liess, M. Effects of pesticides on community structure and ecosystem functions in agricultural streams of three biogeographical regions in Europe. *Sci. Total Environ.* **2007**, *382*, 272–285. [CrossRef] [PubMed]
8. Wauchope, R. The pesticide content of surface water draining from agricultural fields—A review. *J. Environ. Qual.* **1978**, *7*, 459–472. [CrossRef]
9. Ritter, W. Pesticide contamination of ground water in the United States—A review. *J. Environ. Sci. Health Part B* **1990**, *25*, 1–29. [CrossRef]
10. Balinova, A.; Mondesky, M. Pesticide contamination of ground and surface water in Bulgarian Danube plain. *J. Environ. Sci. Health Part B* **1999**, *34*, 33–46. [CrossRef] [PubMed]
11. Kumari, B.; Madan, V.; Kathpal, T. Status of insecticide contamination of soil and water in Haryana, India. *Environ. Monit. Assess.* **2008**, *136*, 239–244. [CrossRef] [PubMed]
12. Snelder, D.J.; Masipiqueña, M.D.; de Snoo, G.R. Risk assessment of pesticide usage by smallholder farmers in the Cagayan Valley (Philippines). *Crop Prot.* **2008**, *27*, 747–762. [CrossRef]
13. Lapong, E.R.; Ella, V.B.; Villano, M.G.; Bato, P.M. Effects of climatic factors and land use on runoff, sediment load, and pesticide loading in upland microcatchments in Bukidnon, Philippines. *Philipp. J. Agric. Biosys. Eng.* **2008**, *6*, 49–56.
14. Senoro, D.B.; Maravillas, S.L.; Ghafari, N.; Rivera, C.C.; Quiambao, E.C.; Lorenzo, M.C.M. Modeling of the residue transport of lambda cyhalothrin, cypermethrin, malathion and endosulfan in three different environmental compartments in the Philippines. *Sustain. Environ. Res.* **2016**, *26*, 168–176. [CrossRef]
15. Santos-Borja, A.; Nepomuceno, D. Laguna de bay: Experience and lessons learned brief. *World Lake Database* **2006**, *15*, 225–258. Available online: http://wldb.ilec.or.jp/data/gef_reports/15_Laguna_de_Bay_27February2006.pdf (accessed on 20 June 2017).
16. Varca, L.M. Pesticide residues in surface waters of Pagsanjan-Lumban catchment of Laguna de Bay, Philippines. *Agric. Water Manag.* **2012**, *106*, 35–41. [CrossRef]
17. Bajet, C.M.; Kumar, A.; Calingacion, M.N.; Narvacan, T.C. Toxicological assessment of pesticides used in the Pagsanjan-Lumban catchment to selected non-target aquatic organisms in Laguna Lake, Philippines. *Agric. Water Manag.* **2012**, *106*, 42–49. [CrossRef]
18. Sanchez, P.B.; Oliver, D.P.; Castillo, H.C.; Kookana, R.S. Nutrient and sediment concentrations in the Pagsanjan-Lumban catchment of Laguna de Bay, Philippines. *Agric. Water Manag.* **2012**, *106*, 17–26. [CrossRef]
19. Hallare, A.V.; Kosmehl, T.; Schulze, T.; Hollert, H.; Köhler, H.R.; Triebskorn, R. Assessing contamination levels of Laguna Lake sediments (Philippines) using a contact assay with zebrafish (Danio rerio) embryos. *Sci. Total Environ.* **2005**, *347*, 254–271. [CrossRef] [PubMed]
20. Marín-Benito, J.M.; Pot, V.; Alletto, L.; Mamy, L.; Bedos, C.; Barriuso, E.; Benoit, P. Comparison of three pesticide fate models with respect to the leaching of two herbicides under field conditions in an irrigated maize cropping system. *Sci. Total Environ.* **2014**, *499*, 533–545. [CrossRef] [PubMed]
21. Brisson, N.; Gary, C.; Justes, E.; Roche, R.; Mary, B.; Ripoche, D.; Zimmer, D.; Sierra, J.; Bertuzzi, P.; Burger, P.; et al. An overview of the crop model stics. *Eur. J. Agron.* **2003**, *18*, 309–332. [CrossRef]
22. Boithias, L.; Sauvage, S.; Srinivasan, R.; Leccia, O.; Sánchez-Pérez, J.-M. Application date as a controlling factor of pesticide transfers to surface water during runoff events. *CATENA* **2014**, *119*, 97–103. [CrossRef]
23. Boithias, L.; Sauvage, S.; Taghavi, L.; Merlina, G.; Probst, J.-L.; Sánchez Pérez, J.M. Occurrence of metolachlor and trifluralin losses in the save river agricultural catchment during floods. *J. Hazard. Mater.* **2011**, *196*, 210–219. [CrossRef] [PubMed]
24. Fohrer, N.; Dietrich, A.; Kolychalow, O.; Ulrich, U. Assessment of the environmental fate of the herbicides flufenacet and metazachlor with the SWAT model. *J. Environ. Qual.* **2014**, *43*, 75–85. [CrossRef] [PubMed]
25. Holvoet, K.; Gevaert, V.; van Griensven, A.; Seuntjens, P.; Vanrolleghem, P.A. Modelling the effectiveness of agricultural measures to reduce the amount of pesticides entering surface waters. *Water Resour. Manag.* **2007**, *21*, 2027–2035. [CrossRef]
26. Zettam, A.; Taleb, A.; Sauvage, S.; Boithias, L.; Belaidi, N.; Sánchez-Pérez, J. Modelling hydrology and sediment transport in a semi-arid and anthropized catchment using the SWAT model: The case of the Tafna river (northwest Algeria). *Water* **2017**, *9*, 216. [CrossRef]
27. Wu, Y.; Shi, X.; Li, C.; Zhao, S.; Pen, F.; Green, T. Simulation of hydrology and nutrient transport in the Hetao Irrigation District, Inner Mongolia, China. *Water* **2017**, *9*, 169. [CrossRef]

28. Kintanar, R.L. *Climate of the Philippines*; PAGASA: Quezon City, Philippines, 1984.

29. Cruz, R.V.O.; Pillas, M.; Castillo, H.C.; Hernandez, E.C. Pagsanjan-Lumban catchment, Philippines: Summary of biophysical characteristics of the catchment, background to site selection and instrumentation. *Agric. Water Manag.* **2012**, *106*, 3–7. [CrossRef]

30. Dile, Y.T.; Srinivasan, R. Evaluation of CFSR climate data for hydrologic prediction in data-scarce watersheds: An application in the Blue Nile River Basin. *J. Am. Water Resour. Assoc.* **2014**, *50*, 1226–1241. [CrossRef]

31. Fuka, D.R.; Walter, M.T.; MacAlister, C.; Degaetano, A.T.; Steenhuis, T.S.; Easton, Z.M. Using the climate forecast system reanalysis as weather input data for watershed models. *Hydrol. Proc.* **2014**, *28*, 5613–5623. [CrossRef]

32. Holvoet, K.; van Griensven, A.; Seuntjens, P.; Vanrolleghem, P.A. Sensitivity analysis for hydrology and pesticide supply towards the river in SWAT. *Phys. Chem. Earth Parts A/B/C* **2005**, *30*, 518–526. [CrossRef]

33. Ligaray, M.; Kim, H.; Sthiannopkao, S.; Lee, S.; Cho, K.; Kim, J. Assessment on hydrologic response by climate change in the Chao Phraya River Basin, Thailand. *Water* **2015**, *7*, 6892–6909. [CrossRef]

34. Luo, Y.; Zhang, M. Management-oriented sensitivity analysis for pesticide transport in watershed-scale water quality modeling using SWAT. *Environ. Pollut.* **2009**, *157*, 3370–3378. [CrossRef] [PubMed]

35. Fabro, L.; Varca, L.M. Pesticide usage by farmers in Pagsanjan-Lumban catchment of Laguna de Bay, Philippines. *Agric. Water Manag.* **2012**, *106*, 27–34. [CrossRef]

36. Ligaray, M.; Baek, S.S.; Kwon, H.-O.; Choi, S.-D.; Cho, K.H. Watershed-scale modeling on the fate and transport of polycyclic aromatic hydrocarbons (PAHs). *J. Hazard. Mater.* **2016**, *320*, 442–457. [CrossRef] [PubMed]

37. Neitsch, S.L.; Arnold, J.G.; Kiniry, J.R.; Williams, J.R. *Soil and Water Assessment Tool Theoretical Documentation Version 2009*; Texas Water Resources Institute: College Station, TX, USA, 2011.

38. Bergknut, M.; Meijer, S.; Halsall, C.; Ågren, A.; Laudon, H.; Köhler, S.; Jones, K.C.; Tysklind, M.; Wiberg, K. Modelling the fate of hydrophobic organic contaminants in a boreal forest catchment: A cross disciplinary approach to assessing diffuse pollution to surface waters. *Environ. Pollut.* **2010**, *158*, 2964–2969. [CrossRef] [PubMed]

39. Van Griensven, A.; Meixner, T.; Grunwald, S.; Bishop, T.; Diluzio, M.; Srinivasan, R. A global sensitivity analysis tool for the parameters of multi-variable catchment models. *J. Hydrol.* **2006**, *324*, 10–23. [CrossRef]

40. Moriasi, D.N.; Gitau, M.W.; Pai, N.; Daggupati, P. Hydrologic and water quality models: Performance measures and evaluation criteria. *Trans. ASABE* **2015**, *58*, 1763–1785.

41. Roy, K.; Das, R.N.; Ambure, P.; Aher, R.B. Be aware of error measures. Further studies on validation of predictive qsar models. *Chem. Intell. Lab. Syst.* **2016**, *152*, 18–33. [CrossRef]

42. Moriasi, D.; Arnold, J.; Van Liew, M.; Bingner, R.; Harmel, R.; Veith, T. Model evaluation guidelines for systematic quantification of accuracy in watershed simulations. *Trans. ASABE* **2007**, *50*, 885–900. [CrossRef]

43. Abbaspour, K.C. *SWAT Calibration and Uncertainty Programs—A User Manual*; Eawag—Swiss Federal Institute of Aquatic Science and Technology: Dübendorf, Switzerland, 2007; Volume 103.

44. Bouraoui, F.; Benabdallah, S.; Jrad, A.; Bidoglio, G. Application of the swat model on the Medjerda river basin (Tunisia). *Phys. Chem. Earth Parts A/B/C* **2005**, *30*, 497–507. [CrossRef]

45. Li, Z.; Xu, Z.; Shao, Q.; Yang, J. Parameter estimation and uncertainty analysis of SWAT model in upper reaches of the Heihe river basin. *Hydrol. Proc.* **2009**, *23*, 2744–2753. [CrossRef]

46. Neitsch, S.; Arnold, J.G.; Kiniry, J.R.; Srinivasan, R.; Williams, J.R. *Soil and Water Assessment Tool Input/Output File Documentation Version 2009*; Texas Water Resources Institute: Forney, TX, USA, 2010; Volume 365.

47. Me, W.; Abell, J.M.; Hamilton, D.P. Effects of hydrologic conditions on SWAT model performance and parameter sensitivity for a small, mixed land use catchment in New Zealand. *Hydrol. Earth Syst. Sci.* **2015**, *19*, 4127–4147. [CrossRef]

48. Schuol, J.; Abbaspour, K.C.; Srinivasan, R.; Yang, H. Estimation of freshwater availability in the West African sub-continent using the SWAT hydrologic model. *J. Hydrol.* **2008**, *352*, 30–49. [CrossRef]

49. Newhart, K. *Environmental Fate of Malathion*; California Environmental Protection Agency: Sacramento, CA, USA, 2006.

water

MDPI

Article

The Mitigation Potential of Buffer Strips for Reservoir Sediment Yields: The Itumbiara Hydroelectric Power Plant in Brazil

Marta P. Luz [1,*], Lindsay C. Beevers [2], Alan J. S. Cuthbertson [2], Gabriela M. Medero [2], Viviane S. Dias [3] and Diego T. F. Nascimento [3]

[1] Eletrobras Furnas, Pontifical Catholic University of Goiás, BR153, km 510, Zona Rural, Aparecida de Goiânia-Goiás CEP 74923-650, Brazil

[2] School of Energy, Geoscience, Infrastructure and Society, Heriot-Watt University, Edinburgh EH14 4AS, UK; L.Beevers@hw.ac.uk (L.C.B.); A.Cuthbertson@hw.ac.uk (A.J.S.C.); G.Medero@hw.ac.uk (G.M.M.)

[3] Pontifical Catholic University of Goiás, Av. Universitária 1.440, Setor Universitário, Goiânia-Goiás CEP 74605-010, Brazil; engvivianedias@gmail.com (V.S.D.); diego.tarley@gmail.com (D.T.F.N.)

* Correspondence: martaluz@furnas.com.br; Tel.: +55-062-3239-6550

Academic Editor: Karim Abbaspour

Received: 5 August 2016; Accepted: 25 October 2016; Published: 28 October 2016

Abstract: Soil erosion and deposition mechanisms play a crucial role in the sustainability of both existing reservoirs and newly planned projects. Soil erosion is one of the most important factors influencing sediment transport yields, and, in the context of existing reservoirs, the surrounding watersheds supply both runoff and sediment yield to the receiving water body. Therefore, appropriate land management strategies are needed to minimize the influence of sediment yields on reservoir volume and, hence, the capacity of power generation. In this context, soil erosion control measures such as buffer strips may provide a practical and low-cost option for large reservoirs, but need to be tested at the catchment scale. This paper represents a study case for the Itumbiara hydroelectric power plant (HPP) in Brazil. Four different scenarios considering radially planted buffer strips of Vetivergrass with widths of 20 m, 40 m, 100 m and 200 m are analyzed. A semi-distributed hydrological model, SWAT, was used to perform the simulations. Results indicate a reduction of sediments transported to the reservoir of between 0.2% and 1.0% per year is possible with buffer strip provision, and that this reduction, over the life of Itumbiara HPP, may prove important for lengthening the productivity of the plant.

Keywords: sediments; Indian grass; reservoir; SWAT; Brazil

1. Introduction

Soil erosion and subsequent land degradation is recognized as an internationally important issue that has significant environmental and socio-economic impacts. There are direct links between land management techniques and the rate of sediment erosion, driven by wind and water processes. Focusing on water related processes at the watershed scale, eroded sediment is transported across the land and into receiving water bodies from where it is conveyed through river systems, eventually depositing within the linked fluvial-estuarine-coastal system. In fluvial systems with large anthropogenic interventions (e.g., hydropower dams and reservoirs), sediment deposition can be exacerbated by large impounds that act as sediment sinks within the watershed. With limited potential for large-scale sediment flushing, this sedimentation can build up over time, reducing available storage and decreasing the efficiency of the reservoir system (e.g., through lower attenuation of flood flows, reduced potential for hydropower production, less water for irrigation supply, or any combination of these) [1].

In agricultural watersheds, inappropriate cultivation practices often accelerate erosion rates and thereby increase sediment movement from the land surface and subsequent transport in streams and rivers. Similarly, the occurrence of large areas of exposed soils between cultivation seasons influences ground infiltration rates and overland surface flows, thus potentially increasing soil erosion rates observed during this period significantly [2,3]. When accelerated sediment erosion occurs in upstream river basins, it can result in detrimental impacts to downstream engineering infrastructure—in particular, reservoirs impoundments and associated hydropower operations [4]. Implementation of best management practices is therefore required in these critical erosion prone areas to control such losses and to protect receiving impoundments from high sediment loads [5–8]. Thus, improved insight and understanding of the interplay between soil erosion/sedimentation mechanisms within the surrounding watershed and potential land management strategies, such as buffer strip implementation, designed to mitigate these processes (and, hence, reduce sediment transport yields from the watershed), will have a crucial role in formulating "best-practice" design and management to ensure the sustainability of planned or existing reservoirs [9,10].

The government of Brazil is currently investing heavily in large hydropower plants to meet the increasing energy demands of the country. However, the loss of water storage volume within these impoundments due to sedimentation from the surrounding watersheds is recognized as a significant problem for some of these newly constructed reservoirs, impacting upon their useful operating life. In an attempt to address this issue, engineers are involved in developing better management strategies to identify critical regions within the watersheds that contribute most to these land erosion (and subsequent reservoir sedimentation) problems and to propose possible intervention measures to manage water and sediment resources more effectively.

A number of different hydro-mechanical properties can be identified to explain the protective role that vegetation has in promoting slope stabilization, reducing soil erosion risk, and filtering sediment movement through overland flow (i.e., runoff). In this context, a tight, dense cover of grass or herbaceous vegetation can provide superior protection against the impact of water (e.g., arising from precipitation) and wind erosion, whilst filtering and trapping the sediment load carried in overland flow. The deep-rooted, woody vegetation is effective in mitigating or preventing shallow mass stability slope failures. Therefore, the loss or removal of slope vegetation can result in either increased rates of erosion or a higher incidence of mass slope failure.

The use of Vetiver grass (Indian grass) as a vegetation type for delivering such potential sediment or soil erosion mitigation can provide both environmental and financial benefits. It has a root system that is resistant to changes in the water level within the reservoir and does not require frequent maintenance. Furthermore, it will grow virtually anywhere (i.e., not constrained by site conditions), which makes it unique for erosion mitigation and slope stabilization. When planted closely (approximately 10 cm apart) across the slope to form a hedge, its biological growth characteristics provide an effective dense vegetation barrier that filters out run-off sediment, dissipates hydraulic forces, and spreads out excess water evenly across the length of the hedge barrier. These properties make it an ideal vegetation type for buffer strip implementation and can stabilize slopes and filter sediment from overland flow. Figure 1 shows the schematized role of Vetiver grass in interrupting overland flow and trapping/filtering out sediments.

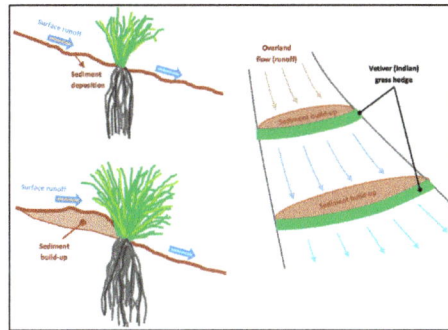

Figure 1. Schematic of the role of Vetiver grass in filtering sediment carried in overland flow.

Table 1. Existing literature overview of experimental studies applying Indian grass.

Study	Application	Scale	Results
[11]	Experimental investigation of runoff reduction and sediment removal by vegetated filter strips into cropland. Study with grass species including Vetiver grass. (Debre Mewi Basin Ethiopia)	Use of 1.5 m wide strips	Desho with the highest tiller number and density, and the second highest in root length revealed better STE than the other grass species, Vetiver (59%), Senbelet (49%), Akirma (36%) and Sebez (20%).
[12]	Experimental investigation of runoff reduction and sediment removal by vegetated filter strips into cropland. Effectiveness of tropical grass species (Elephant grass, Lemon grass, paspalum and sugarcane) as sediment filters in the riparian zone of Lake Victoria, Uganda.	At filter lengths of 2.5, 5 and 10 m.	Under natural rainfall, more than 70% of sediment was trapped in the first 5 m, and lengthening the strip to 10 m only resulted in a marginal increase in sediment trapping effectiveness.
[13]	Experimental investigation in Three Gorges Dam Area, China. Contour hedgerows have been used in this area to control soil erosion and to improve hillslope stability in the catchment of this river section.	Use of 10 m wide strips.	Measured runoffs during natural rainfall events show that all types of plant hedgerow had notable effects on reducing runoff and soil loss. The reduction in soil losses ranged from 18.4% to 70.0% and runoffs were reduced by 17.2% to 70.8%.
[14]	Experimental investigation. Reduction of runoff and soil loss over steep slopes by using Vetiver. Field experiments were conducted at Kasetsart University, Thailand.	2 m in width, 3 m in vertical height, 10.44, 8.08, and 6.71 m in length.	The study found that Vetiver hedgerows reduce runoff volume by 31%–69% and soil loss by 62%–86% on steep slopes of 30%–50%.
[15]	Experimental investigation of runoff reduction and sediment removal by vegetated filter strips into cropland. Two types of hedgerow widths (two-row and three-row) were planted for each of three species of vegetation Bahia grass, Vetiver and Daylily. (Red soil region of China)	The hedgerows were 10 m long with a spacing of 5 m between the rows.	The three selected vegetation types exhibited the similar efficiencies in filtering sediment under the experimental conditions. Generally, the soil loss from the grass hedges was controlled by the characteristics of the grass stems, regardless of the hedge widths.
[16]	Experimental investigation of runoff reduction and sediment removal by vegetated filter strips into cropland. Six different treatments: control (without any treatment), soil bund alone, and soil bund combined with tephrosia, Vetiver grass, Elephant grass and a local grass called Sembelet. (northwestern of Ethiopia)	Each of the treatments was tested on an area of 180 m².	Soil bund combined with Elephant grass had the lowest runoff (40%) and soil loss (63%) as compared to the other treatments.

International research and applications of the Vetiver eco-engineering technique have grown since the 1980s in terms of both theory and practice (detailed in Table 1). Standard sites for the application of Indian grass include river banks, reservoirs edges, slopes, and critical erosion areas, such as end zones of flow, where it can form effective buffer strip vegetation. Brazil's hydropower reservoirs have significant perimeters that make them susceptible to erosion and stabilization issues. Hence, erosion treatments that focus on reducing sediment delivery to the reservoir and do not use significant areas of land must be a priority. In the existing literature on Vetiver grass, while there are variations in the scales of experimental studies conducted to date (see Table 1 for details), the majority of studies have focused on plots between 1 m and 30 m wide. As such, there is no clear consensus in the literature about the large-scale application of Vetiver grass in erosion control and sediment trapping/filtering.

There is therefore a need to upscale these field findings to larger watershed scales in order to understand their effectiveness in reducing sediment delivery to the river basin, hence their ability to control sediment deposition within receiving reservoirs. Consequently, the objective of this study was to identify specific erosion prone areas within a case study watershed—the Itumbiara hydroelectric power plant (HPP) in Brazil—and to investigate the efficiency of the identified potential biostabilization/sediment filtering methods on sediment yields. The paper aims to upscale in situ findings to understand the efficiency of geotechnical interventions at the watershed scale using the Soil and Water Assessment Tool (SWAT), a basin scale modeling approach. Results are presented from the preliminary assessment of the role of edge vegetation in the form of a buffer strips composed of Vetiver (Indian) grasses and in mitigating sediment delivery to the Itumbiara HPP reservoir.

2. Materials and Methods

For the purposes of the study, a semi-distributed, basin-scale hydrological model, capable of simulating surface water and sediment movement was required to analyze sediment delivery (yield) to the reservoir and the potential efficiency of any proposed biostabilization and sediment filtering methods. Over the years, a number of hydrological models (e.g., MIKE SHE, AGNPS, and the Soil and Water Assessment Tool (SWAT)) have been developed [4,17–19] to simulate water flow and sediment transport at the river-basin scale. The current study uses SWAT, a process-based hydrological model, developed by the USDA, Agricultural Research Service (ARS), which can be applied to large ungauged basins [20–22]. Previous studies using SWAT have addressed a variety of watershed issues (e.g., van Griensven et al. [23]; Gassman et al. [24]; Mishra et al. [5]; Cao et al. [25]; Tuppad et al. [26]; Mukhtar et al. [27]). A detailed description of the SWAT model, and its capabilities for watershed hydrological and sediment modeling can be found in van Griensven et al. [23].

2.1. Study Site: Itumbiara HPP, Brazil

The Itumbiara Dam is an earth-fill embankment dam on the Paranaíba River near Itumbiara city in Goiás, Brazil (Figure 2), incorporating a HPP with an installed capacity of 2082 MW. The impounded reservoir has a plan area of 778 km^2 and can store 12.5 km^3 of useful water volume for power generation. The upstream watershed of the Itumbiara HPP is approximately 5685 km^2. A weather station is located adjacent to the northwest sub-basin (i.e., sub basin 82, Figure 2), which records average annual rainfall of 1638 mm. The sediment erosion characteristics in the reservoir watershed are regarded as being representative of those found around many other Brazilian reservoirs.

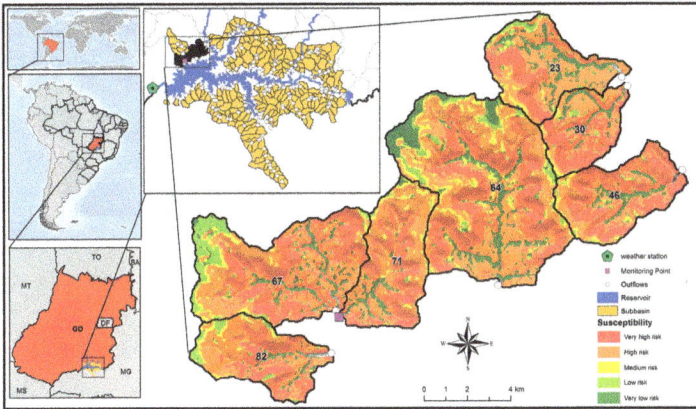

Figure 2. Itumbiara hydroelectric power plant (HPP)—Brazil: left hand figures (top to bottom): Brazil, Goias State in Brazil, location of Itumbiara HPP in Goias Sate; right hand figures (top to bottom): Map of Itumbiara HPP showing sub-basins; zoomed detail of the study area divided into 7 sub-basins (23, 30, 46, 64, 67, 71, and 82), showing laminar erosion susceptibility.

A team of researchers from the Institute of Socio-Environmental Studies at the Federal University of Goiás (IESA-UFG) developed an erosion potential map (see Figure 2) for the watershed, as part of a R&D project funded by the HPP operating company (Eletrobras Furnas). This map was created by estimating the density of erosion at the reservoir edge and associated interfluves, using a statistical tool (Kernel, ArcMap software), and combining this with other environmental factors for the watershed sub-basins such as soil classification, coverage and landuse, surface geomorphic character, slope and hypsometry. From the erosion susceptibility map (Figure 2), it is clear that, of the total reservoir watershed (approximately 5685 km^2), only about 167 km^2 reports significantly high erosion potential. The study used the erosion potential map to focus on this smaller critical area to investigate in detail the potential of Vetiver grass to reduce sediment yield to the reservoir. This detailed analysis (Figure 2) therefore focused on 7 adjacent sub-catchments with significant erosion potential that all drained directly to the reservoir. This sub-area is referred to herein as the study watersheds. Finally, Figure 3a–d show the landuse, soil classification, slope, and digital elevation model for the Itumbiara HPP study watersheds.

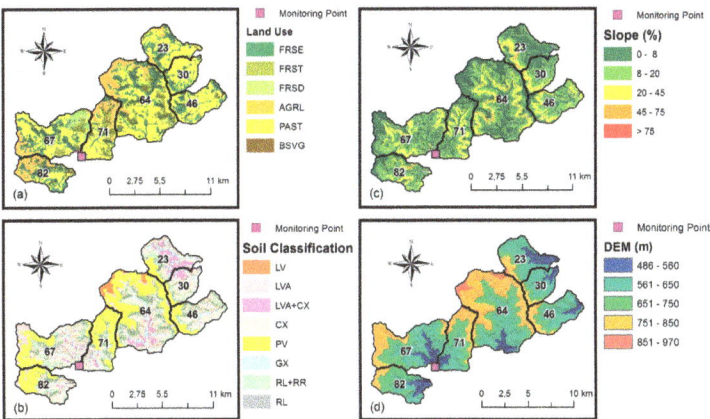

Figure 3. Study area detailed data maps (**a**) Landuse; (**b**); soils (**c**) topography; (**d**) DEM.

2.2. SWAT Model Input Data

The required input data for the SWAT model setup in the current Itumbiara HPP case study is described below:

- Landuse: The landuse map (Figure 3a) used images from INPE via Landsat-5 satellite TM sensor (Thematic Mapper) with a spatial resolution of 30 m. Complementary images from Japanese satellite ALOS were used, with 2.5 m spatial resolution. Field visits were undertaken to validate the images where necessary. The catchment is dominated by pasture—PAST (42.2%), generic agricultural land—AGRL (27.4%), forest-evergreen—FRST (21.4%), forest-mixed—FRST (6.7%), forest-deciduous—FRSD (0.8%), and barren or sparsely vegetated—BSVG (2.3%).
- Digital elevation model (DEM): A 30 m by 30 m resolution DEM was obtained from the TOPODATA project [28]. The DEM was used to delineate the upstream watershed of the Itumbiara HPP. Sub-watershed parameters such as slope gradient, length, and the stream network characteristics such as channel slope, width, and length were derived from the DEM (Figure 3d).
- Soil data: The soil data of Itumbiara HPP watershed was added to the SWAT soil database manually (at a resolution of 30 m by 30 m). The soil groups were classified using RADAM BRASIL [29] and the FAO's (Food and Agricultural Organization of United Nations) texture classification for tropical soils. The average altitude of the survey was 12 km at 690 km/h. The imaging system used GEMS (Goodyear Mapping System 1000), which operates at X-band (wavelengths close to 3 cm and often between 8 and 12.5 GHz). In addition, other methodologies were used to provide information on soil type including infrared and multispectral radar images, low altitude overflights, on-site field visits, and petrographic analysis. The soil map of the study watershed is shown in Figure 3b. For the whole catchment, the following soils are present (with percentage abundance in brackets): hapliccambisol—CX (33%), leptsol-regosol—RL + RR (16.5%), red oxisol—LV (13.6%), ultisol—PV (12.7%), gleysol—GX (10.1%), red yellow oxisol—LVA (8.2%), red yellow oxisol-haplic—LVA + CX (5.7%), leptsol—RL (0.2%).
- Hydrometeorological data: The data required for the model included rainfall, river discharge, and climate data (temperatures, solar radiation, humidity, and wind speed). Daily rainfall and climate data was available for one station inside the Itumbiara HPP (shown in Figure 2). The analyzed rain gauge provided data over the period from 1987 to 2013.

3. Results

3.1. SWAT Model Calibration

As part of the model calibration, a sensitivity analysis of the SWAT model parameters was performed, using the Latin hypercube one-factor-at-a-time (LH-OAT SWAT option) sampling procedure, to determine the most influential parameters for runoff and sediment yield [23]. The first three years were used as a warm-up period to minimize uncertain initial conditions, as the SWAT manual recommends.

Table 2 presents the results of this sensitivity analysis, showing the ranking and sensitivity level of each parameter tested. The sensitivity index (*SI*) was defined using the manual mode based on Equation (1) [30]. The higher the obtained sensitivity index value is, the higher the model sensitivity is compared to the parameter, where (a) values larger than 1 indicate high sensibility; (b) values between 1 and 0.8 indicate intermediate sensibility; and (c) values smaller than 0.8 indicate low sensibility. It is valuable to notice that values close to zero indicate that the model does not present sensibility to the parameter.

$$SI = \frac{\frac{R1-R2}{R12}}{\frac{I1-I2}{I12}} \tag{1}$$

where *SI* is the index in relation to the entry parameters; *R1* is the obtained result with the model related to the smaller entry data; *R2* is the obtained result with the model related to the largest entry data; $R12$ is the average of the obtained results with smaller and largest entry data; $I1$ is the smaller entry data; $I2$ is the largest entry data; $I12$ is the average of the entry data.

Table 2. Result of sensitivity analyses.

Parameter	Description	Rank	Sensitivity Level	Sensitivity Index (SI)
Cn2	Initial SCS runoff curve number	1	High	4.25
USLE_P	USLE equation support	2	High	2.90
Sol_Z	Soil Depth	3	High	1.31
Esco	Soil evaporation compensation	4	Intermediate	0.98
Slope	Average slope steepness	5	Intermediate	0.96
Sol_Awc	Available water capacity	6	Intermediate	0.85
Canmx	Max canopy storage	7	Intermediate	0.83
Blai	Max potential lead area index	8	Intermediate	0.82
Biomix	Biological efficiency	9	Weak	0.63
Surlag	Surface runoff lag time	10	Weak	0.55
Slsubbsn	Average slope length	11	Weak	0.28

The results indicate that the most important (Rank 1, "excessively sensitive", Table 2) parameter was Cn2 (initial SCS runoff curve number), suggesting that this parameter directly affects the soil permeability, the landuse, and the antecedent soil water conditions.

Within the study, watershed sub-basin 71, flow (velocity), and sediment (turbidity) data was measured from a sampling campaign at the main basin outlet to the reservoir, which was used for calibration of the SWAT model. The data was in the form of daily spot samples, taken at a depth of 1 m from the surface.

Next, the model was calibrated, using the top three ranked sensitive parameters in the SWAT model (Table 2), against observed field data collected in 2013 (See Figure 4). Manual calibration was undertaken, focusing on the most sensitive parameters. These were varied until reasonable agreement was achieved (Figure 4). The SWAT model performance for the calibration period (April 2013) was evaluated using Nash–Sutcliffe efficiency (NSE), and the results were 0.889 for the flow data and 0.751 for the sediment data. Figure 4 presents the results of the calibration process for both sediment yields and flow over the calibration period, while Table 3 shows the calibrated parameters—Cn2, USLE_P, and SOL_Z—and multiplying factors used. Due to the length of the measured record, validation is not possible in this case.

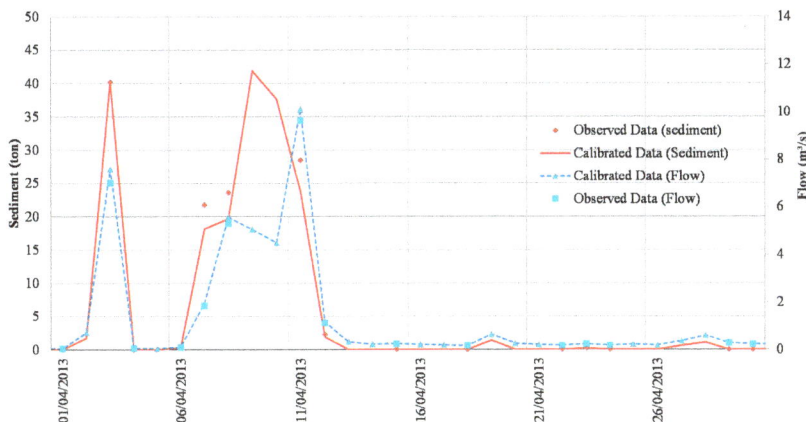

Figure 4. Calibrated (simulated) and observed sediment delivery and flow from sub-basin 71.

Table 3. Calibration parameters.

Variable	Ranking	Original	Multiplying Factor	Final Values
Cn2	1	42.50	1.19	50.58
USLE_P	2	0.29	3.07	0.89
Sol_Z (mm)	3	1.31	0.66	0.86

In cases where validation is not possible, it is standard to undertake an uncertainty or sensitivity assessment of the calibrated model. This was undertaken, and the results indicated that changing the calibrated parameters ±20% reduced the NSE value for sediment to 0.50.

3.2. Land Management Scenario: Buffer Strip Implementation

Once calibrated, the model could be used to implement different Vetiver/Indian grass buffer strip configurations to test their effectiveness in reducing sediment yields. Within the SWAT model, the hydro-mechanical processes (Figure 1) of the Vetiver grass were modeled by implementing a new landuse layer consisting of the Vetiver/Indian grass vegetation type along the perimeter of the reservoir. This captured the interruption of the overland flow process and represented appropriately the sediment filtering/trapping properties of the grass buffer strips (Figure 1). Specifically hydrological parameters were modified in order to do this. Manning's n (roughness) alongside the SCS runoff curve numbers was modified to represent the buffer strip implementation. The Vetiver/Indian grass was adopted as the treatment method along the perimeter of the reservoir as shown in Figure 5.

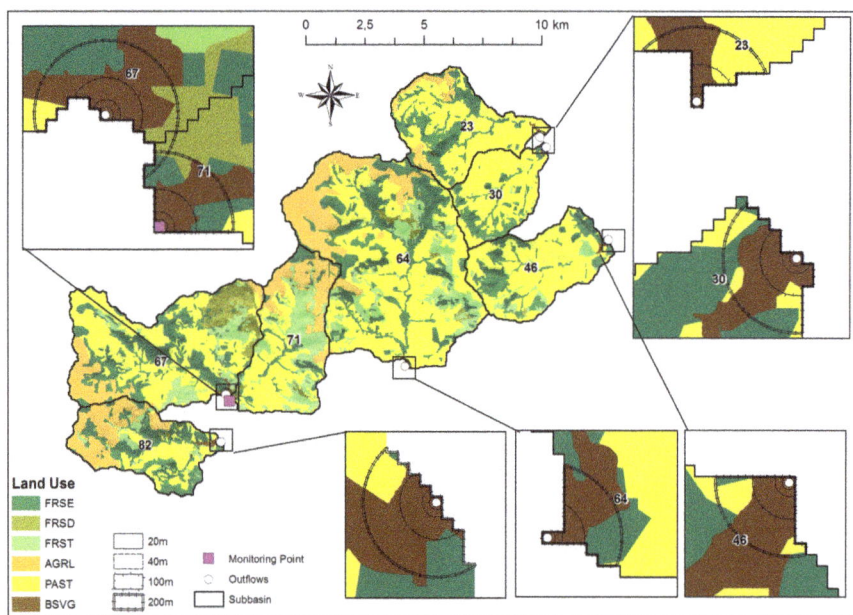

Figure 5. Indication of buffer strip implementation at the sub-basin outflows into the reservoir.

In order to upscale the In Situ findings of the influence of Vetiver/Indian grass buffer strips on the reservoir sediment yield, a number of different scenarios were considered and modeled with the SWAT model. Four different scenarios were devised to understand the potential benefits for erosion mitigation and sediment filtering/trapping by implementing this grass biostabilization measure as watershed-scale buffer strips at the Itumbiara HPP site (note: this buffer strip comprises Vetiver/Indian grass at the outflows from critical watersheds, thus avoiding existing forested areas):

- Baseline Scenario: Existing model with no additional vegetation;
- Scenario 1: incorporate a 20 m wide buffer strip around the sub-basin outflows into the reservoir;
- Scenario 2: incorporate a 40 m wide buffer strip around the sub-basin outflows into the reservoir;
- Scenario 3: incorporate a 100 m wide buffer strip around the sub-basin outflows into the reservoir.
- Scenario 4: incorporate a 200 m wide buffer strip around the sub-basin outflows into the reservoir.

These scenarios were modeled by changing the landuse adjacent to the reservoir shoreline to implement the buffer strip (Figure 5).

3.3. SWAT Model Simulations with Buffer Strips Implemented

The potential benefits to be gained from varying widths of buffer strips planted with Vetiver (Indian) grass were determined in comparison to the baseline scenario. Table 4 details the sediment production for each sub-basin in the study watershed (Figure 2) and shows the potential reduction in sediment yield obtained for each of the four scenarios (i.e., for increasing buffer radius). It is clear that this is non-linear in some instances (e.g., basins 46 and 71). In order to determine the effectiveness of the buffer strips, Figure 5 shows the actual landuse in the sub-basins in the study watershed. Each sub basin has a different landuse categorization, which is plotted alongside the reduction in sediment delivery for each sub-basin in Figure 6.

Table 4. Sediment yield results.

Sub-Basin	Area (ha)	Baseline Sed. (ton/Year)	Indian Grass Buffer							
			20 m		40 m		100 m		200 m	
			Sed. (ton/Year)	Reduction (%)	Sed. (ton/Year)	Reduction (%)	Sed. (ton/Year)	Reduction (%)	Sed. (ton/Year)	Reduction (%)
23	1766	3693	3683	0.27	3679	0.38	3663	0.81	3643	1.35
30	1154	2272	2265	0.31	2262	0.44	2251	0.92	2237	1.54
46	1719	3619	3593	0.72	3589	0.83	3541	2.16	3489	3.59
64	5167	10,004	9998	0.06	9996	0.08	9986	0.18	9974	0.30
67	3318	2386	2382	0.17	2381	0.21	2374	0.50	2366	0.84
71	1960	2468	2445	0.93	2436	1.30	2425	1.74	2385	3.36
82	1628	3682	3669	0.35	3663	0.52	3643	1.06	3605	2.09
Sum.	16,712	28,124	28,035	0.32	28,006	0.42	27,883	0.86	27,699	1.51

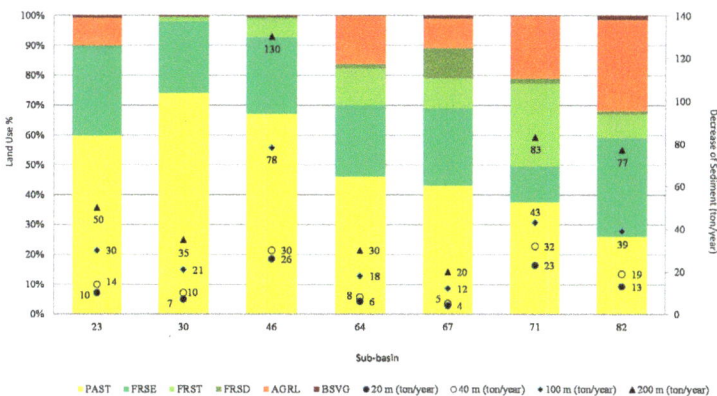

Figure 6. Decrease of sediment delivery per sub-basin (and landuse designation breakdown), demonstrating the implementation of varying buffer strip widths.

From the results, it is possible to understand the influence of replacing the current landuse by a Vetiver (Indian) grass buffer strip in each sub basin. Quantitative variations of the response of each sub basin are complicated by the heterogeneity of landuse, the soil properties, and the slope in each. There is no direct relationship between the area of the sub-basin and Vetiver (Indian) grass efficiency potential (Table 4).

4. Discussion

Each sub-basin has particular characteristics that influence its sediment transport capacity and subsequent delivery, including the distribution of landuse, the slope, the percentage of shoreline to the reservoir, and the soil type. This makes the analysis complex. Results indicate that planting Vetiver/Indian grass at the reservoir sub-basin outflows can contribute to a reduction of carried sediments. Therefore, providing buffer strips of Indian grass on the margins of the Itumbiara HPP reservoir generates a reduction in sediment transportation into the reservoir. In a conservative scenario, considering a typical tropical soil density value (2.65 ton/m^3), it is possible to obtain a decrease of around 160.4 m^3/annum of soil sediments using Vetiver; taking into consideration that the study area represents around 3% of upstream watershed of the Itumbiara HPP. This indicates that the methodology would be beneficial for the Itumbiara HPP system.

This finding demonstrates the usefulness of buffer strip implementation for reservoir management in erosion susceptible areas. Ignoring the problem of sediment erosion and subsequent sediment delivery to the reservoir could cost up to 30% of the generation capacity of a hydroelectric plant, as evidenced in China, Africa, and the United States [31]. Studies such as the one reported here can provide a useful insight into the role of land management and erosion control practices (in this case, buffer strips) in reducing sediment yield. However, any management techniques identified as successful through a modeling strategy must be accompanied with sufficient and appropriate institutional support to coordinate the correct implementation and maintenance of measures.

In the current study, relatively low percentages of sediment transport reduction at the outflow of each studied sub-basin were observed due to the introduction of the Vetiver/Indian grass in comparison to other studies. This shows some degree of contribution by the introduced methodology but also demonstrates the need for further investigation of potential alternatives of soil use and adoption of larger areas of permanent preservation.

The model has a number of limitations. For example, (i) the paucity of calibration data requires the upscaling in parameterization for the model, and (ii) the buffer strips are currently only implemented on the reservoir outflows from the 7 studied sub-basins out of the 275 hydrological sub-basins comprising the total reservoir watershed. Thus, if there were more widespread sediment management strategies imposed in the form of buffer strips along the full perimeter of each sub-basin that could result in significant reductions to the total sediment reservoir yield. Another constraint to be taken in consideration is that changes to reservoir water level are currently not considered. Added benefits could be realized with the extension of these buffers adjacent to all watercourses through the catchment. In fact, this type of intervention may be more beneficial than increasing the width of the buffer strips at the reservoir outflows. Additionally, the current study only considers the hydrological response to landuse change, but ignores the impact of climate change and human factors on hydrological factors. However, vegetation and land surface hydrology are intrinsically linked with long-term climate change [32], and water abstractions and climate change have resulted in variations in annual runoff [33]. Therefore, in future studies, forcing factors related to climate and human impacts should be introduced into the input layer of the model structure so that future landuse/cover types will be more realistically reflected.

5. Conclusions

To adequately and effectively target and implement erosion control measures to reduce reservoir sedimentation, distributed erosion modeling can be used to support decision making. However, the availability of sufficient data to calibrate and validate streamflow and sediment dynamics is crucial for the successful application of such models. The presented methodology and results indicate that distributed erosion and sediment yield modeling with SWAT, supported by sufficient data on discharge and sediment yields of different points in time and space, can provide quantitative insight into scaling up the effectiveness of site-specific erosion control measures and the subsequent benefit to downstream reservoir sedimentation. Thus, this paper presents a first step towards evaluating the role of buffer strips on sediment yield, demonstrating the benefit of scaling up alternative techniques to treat erosion in reservoirs of hydroelectric plants. However, this is an indicative, exploratory study; further, more detailed studies are required.

Acknowledgments: The work was supported by National Council of Scientific and Technological Development (CNPq) through the program Science without Borders, and National Agency of Electric Power (ANEEL), through Eletrobras Furnas Company, both from Brazil's government. The authors gratefully acknowledge IESA-UFG (Institute of Socio-environmental Studies of the Federal University of Goiás) for soil data and classifications and the making of the maps (Figures 2 and 3).

Author Contributions: Marta P. Luz and Lindsay C. Beevers wrote the paper and analyzed the data. Alan J. S. Cuthbertson designed the experiments and assisted in the paper preparation; Gabriela M. Medero contributed by analyzing data; Viviane S. Dias and Marta P. Luz performed the experiments; Diego T. F. Nascimento has developed the GIS maps.

Conflicts of Interest: The authors declare no conflict of interest.

References

1. Petkovsek, G.; Roca, M. Impact of reservoir operation on sediment deposition. *Proc. ICE-Water Manag.* **2013**, *167*, 577–584. [CrossRef]
2. Lenhart, T.; Fohrer, N.; Frede, H.-G. Effects of landuse changes on the nutrient balance in mesoscale catchments. *Phys. Chem. Earth* **2003**, *28*, 1301–1309. [CrossRef]
3. Lin, Y.-P.; Hong, N.-M.; Wu, P.-J.; Wu, C.-F.; Verburg, P.H. Impacts of landuse change scenarios on hydrology and landuse patterns in the Wu-Tu watershed in Northern Taiwan. *Landsc. Urban Plan.* **2007**, *80*, 111–126. [CrossRef]
4. Tripathi, M.P.; Panda, R.K.; Raghuwanshi, N.S. Identification and prioritization of critical sub-watersheds for soil conservation management using SWAT model. *Biosyst. Eng.* **2003**, *85*, 365–379. [CrossRef]
5. Mishra, A.; Kar, S.; Singh, V.P. Prioritizing structural management by quantifying the effect of LULC on watershed runoff and sediment yield. *Water Resour. Manag.* **2007**, *21*, 1899–1913. [CrossRef]
6. Mishra, A.; Kar, S.; Raghuwanshi, N.S. Modelling non-point source pollutant losses from a small watershed using HSPF model. *J. Environ. Eng.* **2009**, *135*, 92–100. [CrossRef]
7. Mishra, A.; Kar, S. Modelling hydrologic processes and NPS pollution in a small watershed in sub-humid subtropics using SWAT. *J. Hydrol. Eng.* **2012**, *17*, 445–454. [CrossRef]
8. Sardar, B.; Singh, A.K.; Raguwanshi, N.S.; Chatterjee, C. Hydrological modelling to identify and manage critical erosion prone areas for improving reservoir life: A case study of Barakar basin. *J. Hydrol. Eng.* **2014**, *19*, 196–204. [CrossRef]
9. Eckhardt, K.; Breuer, L.; Frede, H.-G. Parameter uncertainty and the significance of simulated landuse change effects. *J. Hydrol.* **2003**, *273*, 164–176. [CrossRef]
10. Huisman, J.A.; Breuer, L.; Frede, H.-G. Sensitivity of simulated hydrological fluxes towards changes in soil properties in response to landuse change. *Phys. Chem. Earth* **2004**, *29*, 749–758. [CrossRef]
11. Mekonnen, M.; Keesstra, S.D.; Ritsema, C.J.; Stroosnijder, L.; Baartman, J.E.M. Sediment trapping with indigenous grass species showing differences in plant traits in northwest Ethiopia. *Catena* **2015**, *147*, 755–763. [CrossRef]
12. Wanyama, J.; Herremans, K.; Maetens, W.; Isabiry, M.; Kahimba, F.; Kimaro, D.; Poesen, J.; Deckers, J. Effectiveness of tropical grass species as sediment filters in the riparian zone of Lake Victoria. *Soil Use Manag.* **2012**, *28*, 409–418. [CrossRef]
13. Bu, C.F.; Cai, Q.G.; Ng, S.L.; Chau, K.C.; Ding, S.W. Effects of hedgerows on sediment erosion in Three Gorges Dam Area, China. *Int. J. Sediment Res.* **2008**, *23*, 119–129. [CrossRef]

14. Donjadee, S.; Tingsanchali, T. Reduction of runoff and soil loss over steep slopes by using vetiver hedgerow systems. *Paddy Water Environ.* **2013**, *11*, 573–581. [CrossRef]

15. Cao, L.; Zhang, Y.; Lu, H.; Yuan, J.; Zhu, Y.; Liang, Y. Grass hedge effects on controlling soil loss from concentrated flow: A case study in the red soil region of China. *Soil Tillage Res.* **2015**, *148*, 97–105. [CrossRef]

16. Amare, T.; Zegeye, A.D.; Yitaferu, B.; Steenhuis, T.S.; Hurni, H.; Zeleke, G. Combined effect of soil bund with biological soil and water conservation measures in the northwestern Ethiopian highlands. *Ecohydrol. Hydrobiol.* **2014**, *14*, 192–199. [CrossRef]

17. Young, R.A.; Onstad, C.A.; Bosch, D.D.; Anderson, W.P. AGNPS: A nonpoint source pollution model for valuating agricultural watersheds. *J. Soil Water Conserv.* **1989**, *44*, 168–173.

18. Refsgaard, J.C.; Storm, B. MIKE SHE. In *Computer Models in Watershed Hydrology*; Singh, V.P., Ed.; Water Resources Publications: Highland Ranch, CO, USA, 1995; pp. 809–846.

19. Arnold, J.G.; Srinivasan, R.; Muttiah, R.S.; Williams, J.R. Large area hydrologic modelling and assessment Part I: Model development. *J. Am. Water Resour. Assoc.* **1998**, *34*, 73–89. [CrossRef]

20. Arnold, J.G.; Allen, P.M. Estimating hydrologic budgets for three Illinois watersheds. *J. Hydrol.* **1996**, *176*, 57–77. [CrossRef]

21. Arnold, J.G.; Fohrer, N. SWAT 2000: Current capabilities and research opportunities in applied watershed modelling. *Hydrol. Process.* **2005**, *19*, 563–572. [CrossRef]

22. Ndomba, P.; Mtalo, F.; Killingtveit, A. SWAT model application in a data scarce tropical complex catchment in Tanzania. *Phys. Chem. Earth* **2008**, *33*, 626–632. [CrossRef]

23. Van Griensven, A.; Popescu, I.; Abdelhamid, M.R.; Ndomba, P.M.; Beevers, L.; Betrie, G.D. Comparison of sediment transport computations using hydrodynamic versus hydrologic models in the Simiyu River in Tanzania. *Phys. Chem. Earth* **2013**, *61–62*, 12–21. [CrossRef]

24. Gassman, P.W.; Reyes, M.R.; Green, C.H.; Arnold, J.G. The soil and water assessment tool: Historical development, applications and future research directions. *Trans. ASABE* **2007**, *50*, 1211–1250. [CrossRef]

25. Cao, W.; Bowden, W.B.; Davie, T.; Fenemor, A. Modelling impacts of land cover change on critical water resources in the motueka river catchment, New Zealand. *Water Resour. Manag.* **2009**, *23*, 137–151. [CrossRef]

26. Tuppad, P.; Kannan, N.; Srinivasan, R.; Rossi, C.G.; Arnold, J.G. Simulation of agricultural management alternatives for watershed protection. *Water Resour. Manag.* **2010**, *24*, 3115–3144. [CrossRef]

27. Mukhtar, M.A.; Dunger, V.; Merkel, B. Assessing the impacts of climate change on hydrology of the upper reach of the spree river: Germany. *Water Resour. Manag.* **2014**, *28*, 2731–2749. [CrossRef]

28. TOPODATA. Available online: http://www.dsr.inpe.br/topodata/ (accessed on 15 January 2015).

29. RADAM BRASIL, Projeto RADAM. *Parte das Folhas SC.23 Rio São Francisco e SC.24—Aracaju: Geologia, Geomorfologia, Solos, Vegetação e uso Potencial da Terra*; Levantamento de Recursos Naturais, v. 01; Ministério de Minas e Energia—Secretaria Geral: Rio de Janeiro, Brazil, 1973.

30. Liu, Y.; Zhang, X.; Xia, D.; You, J.; Rong, Y.; Bakir, M. Impacts of landuse and climate changes on hydrologic processes in the Qingyi River watershed, China. *J. Hydrol. Eng.* **2013**, *18*, 1495–1512. [CrossRef]

31. Lelis, T.A.; Calijuri, M.L.; da Fonseca Santiago, A.; de Lima, D.C.; de Oliveira Rocha, E. Análise de sensibilidade e calibração do modelo SWAT aplicado em bacia hidrográfica da regiãosudeste do Brasil. *Rev. Bras. Cienc. Solo* **2012**, *36*, 623–634. [CrossRef]

32. Yuang, F.; Ren, L.; Yu, Z.; Zhu, Y.; Xu, J.; Fang, X. Potential natural vegetation dynamics driven by future long-term climate change and its hydrological impacts in the Hanjiang River Basin. China. *Hydrol. Res.* **2012**, *43*, 73–90. [CrossRef]

33. Lu, G.B.; Wang, J.; Wang, H. J.; Xia, Z.Q. Impacts of human activities on the flow regime of the Hanjiang River. *Res. Environ. Sci.* **2009**, *15*, 12–13.

water

MDPI

Article

Effect of Climate Change on Hydrology, Sediment and Nutrient Losses in Two Lowland Catchments in Poland

Paweł Marcinkowski [1,*], Mikołaj Piniewski [1,2], Ignacy Kardel [1], Mateusz Szcześniak [1], Rasmus Benestad [3], Raghavan Srinivasan [4], Stefan Ignar [1] and Tomasz Okruszko [1]

[1] Department of Hydraulic Engineering, Warsaw University of Life Sciences, Warsaw 02-774, Poland; m.piniewski@levis.sggw.pl (M.P.); i.kardel@levis.sggw.pl (I.K.); m.szczesniak@levis.sggw.pl (M.S.); s.ignar@levis.sggw.pl (S.I.); t.okruszko@levis.sggw.pl (T.O.)

[2] Potsdam Institute for Climate Impact Research, Potsdam 14473, Germany

[3] Norwegian Meteorological Institute, Oslo 0313, Norway; rasmus.benestad@met.no

[4] Departments of Ecosystem Science and Management and Biological and Agricultural Engineering, Texas A&M University, College Station, TX 77843, USA; r-srinivasan@tamu.edu

* Correspondence: p.marcinkowski@levis.sggw.pl; Tel.: +48-225-935-268

Academic Editor: Karim Abbaspour
Received: 30 December 2016; Accepted: 21 February 2017; Published: 23 February 2017

Abstract: Future climate change is projected to have significant impact on water resources availability and quality in many parts of the world. The objective of this paper is to assess the effect of projected climate change on water quantity and quality in two lowland catchments (the Upper Narew and the Barycz) in Poland in two future periods (near future: 2021–2050, and far future: 2071–2100). The hydrological model SWAT was driven by climate forcing data from an ensemble of nine bias-corrected General Circulation Models—Regional Climate Models (GCM-RCM) runs based on the Coordinated Downscaling Experiment—European Domain (EURO-CORDEX). Hydrological response to climate warming and wetter conditions (particularly in winter and spring) in both catchments includes: lower snowmelt, increased percolation and baseflow and higher runoff. Seasonal differences in the response between catchments can be explained by their properties (e.g., different thermal conditions and soil permeability). Projections suggest only moderate increases in sediment loss, occurring mainly in summer and winter. A sharper increase is projected in both catchments for TN losses, especially in the Barycz catchment characterized by a more intensive agriculture. The signal of change in annual TP losses is blurred by climate model uncertainty in the Barycz catchment, whereas a weak and uncertain increase is projected in the Upper Narew catchment.

Keywords: climate change effect; sediment; nutrients; SWAT; water quality

---

1. Introduction

The threat of climate change is one of the greatest challenges of the modern age and preventing it is a key strategic priority for the European Union. According to Intergovernmental Panel on Climate Change (IPCC) Synthesis Report [1], climate change will cause significant changes in the quality and availability of water resources. However, while it is a robust finding that precipitation is projected to grow in northern Europe and decrease in southern Europe [2], both annually and during the summer, changes in central and eastern Europe are more complex. There is a moderate consensus between large-scale hydrological projections driven by EURO-CORDEX that both floods and droughts might be on the rise in this region [3–5].

Although climate change is not explicitly included in the text of the European Water Framework Directive (WFD), the step-wise approach of the river basin management planning process makes it well suited to adaptively manage climate change impacts. Potentially, all elements included in the definition of WFD qualitative and quantitative status of water are sensitive to climate change. However, the present practice shows that climate change problems have not been adequately dealt with in water resources management and policy formulation in Poland and many other European countries. For example, in Poland, recent updates of the river basin management plans lacked consideration of effects of climate change on water quality and did not look beyond the upcoming horizon of 2030. The role of research within the context of international and national policies and actions to adapt to climate change is crucial. It provides the basis for: (i) understanding the causes of climate change; (ii) projecting future changes; (iii) assessing and quantifying the impacts and vulnerabilities at global and regional scale; and (iv) elaborating effective adaptation and mitigation policies and their practical implementation [1]. The great challenge for policy and decision-makers is to understand these climate change impacts and to develop policies while ensuring an optimal level of adaptation. In order to make decisions on how to best adapt, it is crucial to have access to accurate and reliable data on the possible impact of climate change.

Climate scenarios downscaled from GCMsthat use either empirical-statistical or dynamical downscaling, provide the best available information for assessing future impacts of climate change on the water quality of surface water bodies [6]. A common technique for investigating their impact at the catchment scale is to use climate forcing data (precipitation, temperature, and sometimes other variables) obtained from climate models as new input for hydrological models [7]. Modeling, with a notable use of fully-distributed physically-based or semi-distributed process-based models of intermediate complexity, is the most feasible approach to establish projections of climate change impacts on freshwater resources [6]. There are a great number of studies, which have been carried out to assess the possible effects of climate change on the water quality parameters using different hydrological models at a range of spatial scales.

Table 1 lists selected studies applying different hydrological models to assess the impact of future climate change projections. The projections are based on various emission scenarios and climate models, on water flow and water quality parameters. Most studies focus on multi-variable analysis (mostly total nitrogen (TN), total phosphorus (TP), total suspended sediment (TSS) and nitrate nitrogen (NO_3-N)), but single-variable studies can also be found. Nearly all studies have shown that climate change is likely to have a significant impact on contaminants' loads. Most indicate an overall increase in contaminants loads [8–16]. It is obvious that this increase corresponds to water flow augmentation driven by precipitation increase. The opposite results that indicate the contaminants loads are decreasing [17–19] are likewise strongly correlated with the flow pattern which is projected to decrease in these particular studies. Mixed nutrients emission response reported by Arheimer et al. [20], Records et al. [21] and Molina-Navaro et al. [22] is an effect of diverse flow changes during the projected periods. Very few studies indicate that future climate change is likely to have a negligible impact on single variables like sediment [11,23], TN [24], and NO_3-N [25].

Table 1. Selected studies assessing climate change impact on water quantity and quality. The last four columns show the dominant direction of simulated effects of climate change on different parameters (see legend below the table).

Reference	Country/ Region	Area (km²)	Hydrological Model	Climate Models (Emission Scenarios)	Future Horizons	Flow	Sediment Load	TN * Load	TP * Load
[11]	USA	248	SWAT	112(3)	2015–2034 2045–2064 2080–2099	—	—	↑	↑
[20]	Baltic Sea Basin	1,700,000	HYPE	16(4)	1971–2000 2071–2100	↓↑		↓↑	↓↑
[18]	USA	17,000	SWAT	19(4)	2046–2065 2080–2099	↓	↓		
[25]	Canada	3858	SWAT	1(1)	2025–2050	↑		— NO_3	↑PO_4

Table 1. *Cont.*

Reference	Country/ Region	Area (km²)	Hydrological Model	Climate Models (Emission Scenarios)	Future Horizons	Effect on:			
						Flow	Sediment Load	TN * Load	TP * Load
[13]	Slovenia	30	SWAT	6(1)	2001–2030 2031–2060 2061–2090	↑	↑	↑	↑
[15]	Canada	630	SWAT	6(1)	2041–2070		↑	↑	↑
[26]	Poland, Russia	20,730	SWIM	15(1)	1971–2000 2011–2040 2041–2070 2071–2098	↑		↓NO₃	↑PO₄
[12]	Finland	301,300	VEMALA	3(1)	1971–2000 2010–2039 2040–2069	↑		↑	↑
[24]	USA	7588	SWAT	3(3)	2046–2065 2080–2099	↑	↑	—	↑
[19]	USA	492,000	SWAT	1(1)	2046–2065	↓		↓NO₃	
[16]	Mongolia	447,000	WaterGAP3	1(1)	2071–2100	↑	↑		
[8]	Czech Republic	2180	SWIM	2(1)	2011–2040 2041–2070 2071–2100	↑		↑NO₃	
[23]	Canada	629	SWAT	3(1)	2041–2070	↑	—	↑	↑
[14]	Germany	980	SWAT	7(2)	2041–2070	↑		↑NO₃	↑
[9]	Baltic Sea Basin	1,700,000	HYPE/STAT	8(2)	1961–2099	↑		↑	↑
[22]	Spain	88	SWAT	11(3)	2046–2065 2081–2100	↓		↓NO₃	↓↑
[10]	Poland	482	SWAT	1(1)	2050	↑		↑NO₃	↑PO₄
[21]	USA	4000	SWAT	6(2)	2030–2059	↓↑	↓↑	↓↑	↓↑
[17]	USA	505	SWAT	1(1)	2011–2040 2041–2070 2071–2100	↓		↓NO₃	

Notes: Legend: ↑ mostly increase; ↓ mostly decrease; ↓↑ mixed pattern; — no significant changes. * Whenever NO₃ or PO₄ is given in parentheses, it means that the study dealt with either NO₃–N or PO₄–P, and not TN and TP.

To date, Poland has not been a region with intensive studies investigating climate change effects on water, sediment and nutrient losses. Two exceptions, included in Table 1, are: (1) the study of Piniewski et al. [10] conducted in a small catchment in north Poland, using only one climate scenario, and the "delta change" approach as the method of processing the climate forcing into the hydrological model; and (2) the study of Hesse et al. [26], covering mainly Russia and only a small part of coastal area in north Poland, using 15 scenarios from the ENSEMBLES project [27]. No studies were performed for the dominant type of Polish landscape, i.e., the Polish plain, a diverse region with variable levels of agricultural intensity and other pressures on water resources. In a wider context, none of the studies listed in Table 1 used the newest generation of climate model runs from CORDEX experiment (although two studies [18,21] used statistically downscaled CMIP5 projections). They are available at higher resolution than all predecessors which is an important features for hydrological modeling.

Against this background, the objective of this paper is to assess the effect of projected climate change on water quantity (annual and seasonal water balance components and discharge) and quality (sediment, TN and TP losses). The SWAT model is used in two Polish catchments and it is representative for the majority of the lowland areas of the country. The study looks into projected changes for two future time horizons within 21st century (2021–2050 and 2071–2100) under the Representative Concentration Pathway (RCP) 4.5, using an ensemble of nine EURO-CORDEX model scenarios [2].

2. Materials and Methods

2.1. Study Area

The Upper Narew (NE Poland) and the Barycz (SW Poland) catchments in which the study was conducted are the sub-catchments of two large Polish river basins (the Vistula and the Odra, respectively) (Figure 1). They drain areas of 4231 km^2 (Upper Narew, of which 27% belong to Belarus) and 5522 km^2 (Barycz). Both belong to the vast Polish Plain. According to the geographical regionalization of Kondracki (1997), the Barycz catchment belongs to the Central European Plain, while the Upper Narew catchment to Easter European Plain. These two regions were formed by glacial erosion in the Pleistocene ice age. Both catchments are within the extent of most Pleistocene glaciations, with two exceptions: the first one, Gunz, that covered only the Upper Narew catchment; and the last one, Würm, whose southern border almost touched both watersheds. Consequently, both catchments are characterized by a flat relief with an average elevation of 152 m a.s.l. in the Upper Narew and 127 m a.s.l. in the Barycz. In both, the prevailing type of soils are sands and loamy sands, whereas heavy, impervious soils are rare. However, the fraction of permeable soils in the Barycz catchment is distinctly higher (62.8% vs. 27.3%, estimates based on the input soil map and classification of Pazdro [28]). Moderate differences in land cover also can be observed. Total area of forests is slightly higher in the Upper Narew than in the Barycz catchment (43.6% vs. 38.9%). Compared to much lower values for the Barycz catchment (0% and 8%), the Upper Narew catchment has a high abundance of wetlands and grasslands (8% and 16%, respectively).

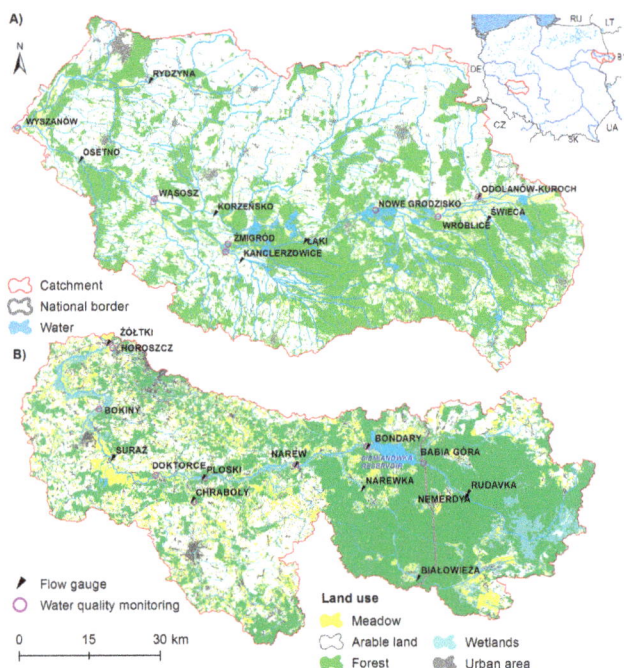

Figure 1. Location of investigated catchments: (**A**) the Barycz catchment; and (**B**) the Upper Narew catchment. Three gauges labeled with red font (Osetno and Korzeńsko for the Barycz and Żółtki for the Upper Narew) are used for showing plots of measured flow and concentrations in Figure 2.

The climate of the Upper Narew catchment is more continental, being often influenced by cold polar air masses from Russia and Scandinavia, whereas the climate of the Barycz catchment is milder, with more frequent influence of maritime air from the West. This is reflected in mean annual air temperature that equals 7.1 and 8.3 °C for the Upper Narew and the Barycz catchments, respectively (climate statistics based on [29]). The difference in mean winter temperature (−3.2 vs. −0.6 °C) is much larger than between mean summer temperature (17 vs. 17.7 °C). Mean annual precipitation total, equal to 670 mm in the Upper Narew catchment, is slightly higher compared to the Barycz catchment (632 mm). However, winter and summer total precipitation have very similar magnitude in both catchments: 127–129 mm, and 234–237 mm, respectively.

The differences in climatic and physiographic characteristics between two catchments clearly affect their hydrology. Annual total runoff coefficient equal to 0.26 in the Upper Narew catchment is much higher than the corresponding value for the Barycz catchment (0.19). However, what is important is the difference in monthly distribution of runoff (Figure 2a). A more continental climate together with less permeable soils and higher water retention capacity (wetlands, grasslands and forests) in the Upper Narew catchment lead to a higher magnitude and later occurrence of spring snow-melt floods. The magnitude of these types of floods, occurring in the Barycz catchment in March, is roughly half of the magnitude of the Narew floods. At the same, time runoff in January and February is higher in the Barycz catchment than in the Upper Narew catchment.

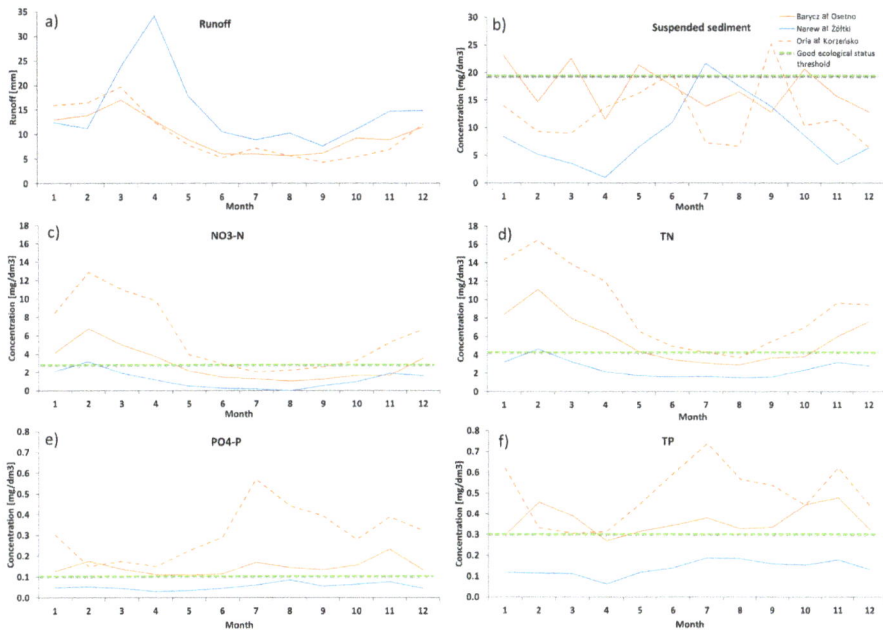

Figure 2. Mean monthly statistics of hydrological and water quality parameters for two stations (cf. Figure 1 for location) in the Barycz catchment (the Barycz river at Osetno and the Orla river at Korzeńsko) and one in the Upper Narew catchment (the Narew river at Żółtki): (**a**) runoff; (**b**) sediment concentration; (**c**) NO_3-N; (**d**) TN; (**e**) PO_4-P; and (**f**) TP. Joint period of flow data (source: Institute of Meteorology and Water Management—National Research Institute) availability (1961–1986) was selected for calculations of runoff. In the case of water quality parameters (source: General Inspectorate of Environmental Protection), the period of available data was 1992–2013, with typically one measurement per month, although many years had missing values.

Significant differences, placing the studied catchments on the extreme opposite ends, are noted in terms of the human dimension (Table 2, Figure 3):

- agriculture: its intensity, reflected by the crop structure, fertilizer rates, livestock density and the level of drainage;
- population density and its derivatives, e.g., the amount of pollution from the wastewater treatment plants (WWTPs); and
- water retention (reservoirs and ponds).

In general, waters of the Barycz catchment are subject to more intensified human pressures due to greater numbers of point sources and more intensive agriculture. In this context, the Upper Narew catchment is representative for less economically developed eastern Poland, while the Barycz catchment is more similar (although less developed) to western European countries. Additionally, it has probably the most intensive level of freshwater aquaculture (carp ponds) in Poland, with 8100 ha of ponds of the total capacity estimated as 73.1 million m^3. In contrast, the Upper Narew catchment has very little ponds and one relatively large reservoir (Siemianówka, situated in the upstream part) with a total capacity of 79.5 million m^3, which is the only important water management facility in this catchment. Mean monthly runoff of the Upper Narew shown in Figure 2 is not influenced by Siemianówka reservoir because the underlying data come from the period prior to construction year (1991). However, the effect of fish ponds on the Barycz runoff can be assessed by comparing the plots between two gauges: Osetno (influenced by the whole pond system), and Korzeńsko (under the negligible influence of ponds). Lower runoff values in January and February for Osetno reflect upstream withdrawals for filling the ponds. Higher values of runoff observed in September and October at Osetno gauge illustrate upstream discharges of pond water into the stream network.

Table 2. Comparison of selected human pressure characteristics of the Upper Narew and the Barycz catchment (sources: [30,31]).

Category	Parameter	Barycz	Upper Narew *
Agriculture	Fraction of arable land (%)	47	23
	Fraction of grassland (%)	9	18
	Mineral nitrogen fertilizer rate (kg·ha^{-1})	91	45
	Mineral phosphorus fertilizer rate (kg·ha^{-1})	17	10
	Livestock density (LSU·ha^{-1})	1.21	0.73
Urban	Population density (persons·km^{-2})	89	36
	Fraction of high density urban land cover (%)	1.2	0.45
	Number of point sources (per 1000 km^2)	7.1	3.5
	Specific wastewater discharge from WWTPs (dm^3·s^{-1}·km^{-2})	0.09	0.03
	Specific sediment load from WWTPs (Mg year^{-1}·km^{-2})	0.3	0.03
	Specific TN load from WWTPs (kg·year^{-1}·km^{-2})	47.5	36.9
	Specific TP load from WWTPs (kg·year^{-1}·km^{-2})	8.2	2.8
Water Retention	Fish ponds volume (10^3 m^3/km^2)	12.9	1.3
	Reservoir volume (10^3 m^3/km^2)	-	20

Note: * All parameters are calculated exclusively for the Polish part of the Upper Narew catchment.

The differences in human pressures between catchments are well reflected in their surface water quality characteristics, as shown in Figure 2b–f. Annual mean concentrations of five analyzed elements, total suspended solids (TSS), nitrate-nitrogen (NO$_3$-N), total nitrogen (TN), mineral phosphorus (PO$_4$-P) and total phosphorus (TP), are distinctly higher for both stations in the Barycz catchment than in station located in the Upper Narew catchment. It is noteworthy that the threshold concentrations of good ecological status are frequently exceeded in the Barycz catchment, while being rarely exceeded in the Narew catchment. With exception of TSS, pollution is much higher in the Orla tributary of the Barycz river, which can be explained by the fact that its catchment has the highest level of agricultural intensity within the Barycz catchment (cf. Figure 3). The monthly dynamic of the nitrogen and phosphorus compounds differs considerably. Both TN and NO$_3$-N have a strong correlation with runoff and achieve the highest values in winter and the lowest in summer, in all three stations. Such a seasonal pattern, related to the physics of nitrogen transport within the catchment, is typical for catchments in Poland [32,33]. This type of seasonal fluctuation is caused mainly by high mobility of nitrates not being assimilated by plants during dormancy season and contributing to streams via lateral and groundwater flow. These two transport pathways are favored, especially in winter and early spring, when evapotranspiration is low whereas infiltration can be high. During the growing season and intensive plant uptake, less mineral nitrogen particles are transported to streams. A different pattern, with highest values in the low flow period (summer and autumn), can be observed for both phosphorus forms, which is also in line with literature on P dynamics in different types of Polish rivers [33,34].

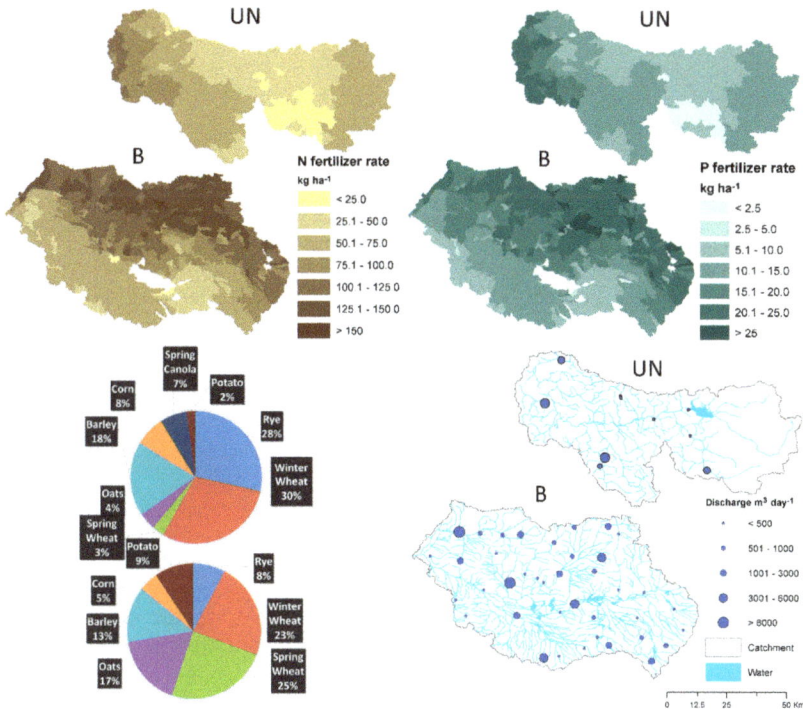

Figure 3. Spatial comparison of selected human pressure characteristics of the Upper Narew (UN) and the Barycz (B) catchment. Top panels show fertilizer rates, bottom left panel shows the crop structure and bottom right panel shows discharges from the wastewater treatment plants in both catchments.

2.2. Modelling Approach

In this study, we build upon the existing, extensively calibrated and validated SWAT models of the Barycz and the Upper Narew catchments [35]. While the full description of model setup, calibration and validation was presented in the latter study, here we provide a brief overview, important in the context of the main goal of the present paper.

2.2.1. Model Setup, Calibration and Validation

SWAT is a process-based, semi-distributed, continuous-time model which simulates the movement of water, sediment, and nutrients on a catchment scale [36]. It is a comprehensive tool suitable for investigating the interaction between climate, land use and water quantity or quality. It enables simulation of long-term impacts of land use and climate changes on water, sediment, and nutrient yields in catchments with varied topography, land use, soils and management conditions [22].

Major data items and their sources used to create the SWAT model setup of the Upper Narew and Barycz catchments are listed in Table 3. Throughout the whole process of developing the model setups, an attempt was made to use the same data sources and approaches for both catchments. Nevertheless, for the upstream part of the Upper Narew (lying in Belarus), data from various global databases, usually characterized by lower resolution had to be used.

Table 3. Data items and sources used to create the SWAT model setup of the Upper Narew and Barycz catchments.

Data Type	Source	Resolution/Scale
DEM PL	CODGiK	10 m
DEM BY	SRTM v4.1 (NASA)	Horizontal 90 m; Vertical 16 m
Rivers and lakes PL	MPHP2010 (IMGW-PIB)	1:10,000
Land Cover PL	Landsat 8 CLC 2006 (GDOS)	30 m 100 m
Land Cover BY	MODIS Landcover	500 m
Soil map PL	IUNG-PIB	1:100,000
Soil map BY	HWSD v 1.2	1:1,000,000
Climate PL/BY	CPLFD-GDPT5	5 km
Atmospheric deposition of nitrogen (dry and wet)	GIOS	1 station for the Upper Narew/3 stations for the Barycz (outside the catchment)
Agricultural statistics	GUS	Commune level

Notes: Abbreviations: BY, Belarus; CLC, Corine Land Cover; CODGiK, Central Agency for Geodetic and Cartographic Documentation; CPLFD-GDPT5, CHASE-PL Forcing Data–Gridded Daily Precipitation & Temperature Dataset–5 km [37]; DEM, Digital Elevation Model; GDOS, General Directorate of the Environmental Protection; GIOS, Chief Inspectorate of Environmental Protection; GUS, Central Statistical Office of Poland; HWSD, Harmonized World Soil Database; IMGW-PIB, Institute of Meteorology and Water Management, National Research Institute; IUNG-PIB, Institute of Soil Science and Plant Cultivation, National Research; MPHP, Hydrographic Map of Poland; NASA, National Aeronautics; PL, Poland; SRTM, Shuttle Radar Topography Mission.

Delineation of the catchment based on the 10-m resolution DEM resulted in division of the Upper Narew catchment into 243 sub-basins and 503 of the Barycz catchment. The land cover map was a combination of CORINE Land Cover (CLC) 2006 and post-processed Landsat 8. Intersection of land cover map, soil map, and slope classes resulted in creation of 4509 HRUs in the Upper Narew catchment and 8569 in the Barycz catchment. Daily precipitation and air temperature (minimum and maximum) data (1951–2013) were acquired from 5 km resolution gridded, interpolated using kriging techniques, dataset (CPLFD-GDPT5) based on meteorological observations coming from the Institute of Meteorology and Water Management (IMGW-PIB; Polish stations) [37]. The use of interpolated climate data in the SWAT model was reported to increase the model performance for a case study in Poland [38].

Parameterization of different pollution sources present in the catchment plays a critical role in water quality modeling. The following anthropogenic pollution sources were analyzed:

1. Diffuse pollution from agricultural areas: Commune-level statistical data were used to determine mineral fertilizer use and livestock population in order to impose a spatial variability of fertilizer rates in the model setup.
2. WWTPs: Defined in the model setup only when the daily average wastewater discharge exceeded 50 m^3·day^{-1}. For each WWTP, discharge and nutrient loads were expressed as constant or mean yearly values depending on the available data, usually originating from plant operators.

3. The septic systems function of SWAT was used to model the effect of pollution loads coming from population not connected to WWTPs (using cesspits or septic tanks, with or without sub-surface drainage).

4. Atmospheric deposition (dry and wet) of nitrogen (nitrate and ammonium): Defined based on one station for the Upper Narew and three stations for the Barycz as a fixed average value for the entire catchments.

Calibration phase was conducted in SWAT-CUP using the SUFI-2 algorithm (Sequential Uncertainty Fitting Procedure Version 2) where the Kling–Gupta efficiency (KGE) was used as an objective function [39]. Additionally, percent bias (PBIAS) that measures the average tendency of the modeled data to be larger or smaller than their observed counterparts, was also tracked. In the calibration and validation, ten flow gauges (data acquired from IMGW-PIB) and nine water quality monitoring stations (concentration data acquired from the General Inspectorate of Environmental Protection) were used in the Upper Narew. Likewise, in the Barycz there were seven flow gauges and eight water quality monitoring stations (Figure 1). Discharge, TSS, NO_3-N, TN, PO_4-P and TP loads were calibrated and validated in each catchment. For both catchments the calibration period for discharge was 1976–1985, and the validation period was 1986–1991, whereas for water quality variables these periods were set to 1999–2005 and 2006–2010, respectively. The inconsistency in selection of periods for discharge and water quality was because selection was optimized with respect to the abundance of observation data. Due to an objective of capturing spatial patterns of runoff and sediment/nutrient transport, a good spatial representation of gauges was crucial. About one half of flow gauging stations in both catchments were closed in 1990s, which was a reason for selecting an earlier period for discharge. In contrast, water quality monitoring by state agencies became more frequent and more abundant only in late 1990s.

Marcinkowski et al. [35] reported variable values of goodness-of-fit measures across different gauges and variables. For discharge, simulations were assessed as good (median KGE above 0.7 in both catchments). For other variables, spatial, multi-site calibration revealed problems in achieving satisfactory results for the entire set of stations taken into consideration. In consequence, there were both stations with good and satisfactory fit (KGE above 0.5), and stations with unsatisfactory behavior (PBIAS higher than 55% for sediment and higher than 75% for nutrients, cf. Moriasi et al. [40] for evaluation criteria). Among reasons for poor behavior in some stations, Marcinkowski et al. [35] reported: (1) the dominant importance of global over local parameters in calibration; (2) simultaneous calibration of different pools of water quality parameters (with different optimal parameter sets achieved for different pools); and (3) input uncertainty (e.g., differences between defined agricultural management operations and the reality). A previous study applying SWAT in Poland for modeling water quality also showed that [41], frequently, the magnitude of the highest observed loads of nutrients is captured well by the model, but there is a shift in timing by a few days (the flood peak is sometimes advanced or lagged by 1–3 days compared with the timing of the peak identified in the observed data) which has a negative effect on the objective function value.

It should be noted that even though there was a temporal inconsistency between certain input (e.g., land cover) and output (discharge) data of over 20 years, it did not affect the results much. We estimated the magnitude of land cover changes between 1990 and 2012 using CORINE Land Cover maps from the corresponding years. The analysis indicated that the patterns of change in both catchments were similar (agriculture areas converted mainly into artificial surfaces or forests). However, the rates of change were not very high, not exceeding 5% in any of the catchments. Furthermore, additional evaluation of discharge simulation in the more contemporary period (1990–2013) showed that the goodness-of-fit measures remain satisfactory.

2.2.2. Climate Change Scenarios

In this paper, SWAT is driven by climate forcing data from the CHASE-PL Climate Projections: 5-km Gridded Daily Precipitation & Temperature Dataset (CPLCP-GDPT5) [42], consisting of nine bias-corrected GCM-RCM runs (involving four different GCMs and four different RCMs) provided within the EURO-CORDEX experiment projected to the year 2100 under RCP 4.5 [43]. A quantile mapping method (QMAP) developed by the Norwegian Meteorological Institute was applied as a bias correction procedure [44]. All bias-corrected values of parameters of concern were available for the following three time slices: 1971–2000, 2021–2050, and 2071–2100. Three first years of each period were truncated, since a warm-up period of three years is used for SWAT simulations. The corresponding time horizons will be hereafter referred to as "historical period", "near future" and "far future", respectively. Future changes in simulated discharge, water balance components and water quality variables were estimated by comparing model outputs for the future periods relative to historical period.

The model runs were carried out assuming constant land use and absence of water management (reservoirs, fish ponds), in order to illustrate pure climate change effect. For the sake of map presentation, projected changes from nine ensemble members were summarized as the ensemble median change, whereas climate model uncertainty was analyzed on the level of areal mean catchment responses.

3. Results

3.1. Climatic Projections

Since within-catchment spatial variability of projected temperature and precipitation change is low in both catchments, the analysis focuses on areal mean changes. The annual and seasonal climate change signal is similar in both catchments (Figure 4). The warming is ubiquitous and accelerating in time for each individual climate model. The mean annual warming rate is slightly higher in the Upper Narew than in the Barycz catchment. Seasonal patterns are similar, with the winter increase higher than the increase projected in remaining seasons. The largest difference between two investigated catchments is projected for the minimum temperature in winter and spring in the far future: it is higher by 0.5 °C in the Upper Narew than in the Barycz catchment. The robustness (*sensu* [45]) of annual temperature increase is high in both catchments (cf. [43]). Seasonal temperature projections are more robust for the minimum temperature, T_{min}, than for the maximum temperature, T_{max}. Notably, in the near future, T_{max} projections in winter and summer are characterized by a substantial model disagreement in the Barycz catchment.

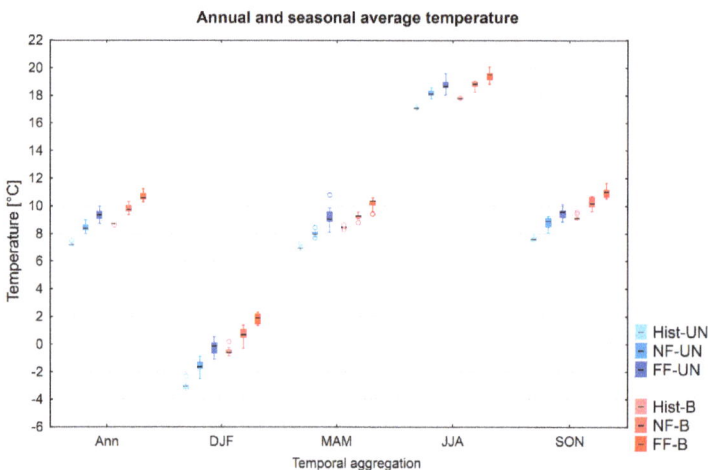

Figure 4. Multi-model ensemble projections of annual and seasonal average temperature for the near (NF) and the far (FF) future under RCP 4.5 in comparison to the historical (Hist) period. B stands for the Barycz catchment and UN stands for the Upper Narew catchment.

Annual total precipitation is projected to increase in both catchments by 5.6% in the near future and by 9.1%–9.5% in the far future. Although the spread in projections related to different RCMs is substantial (slightly higher for the Upper Narew catchment), the agreement on the direction of change is ubiquitous (Figure 5). The seasonal patterns are also similar between catchments, with a relatively high increase in winter and spring and a weaker increase or a decrease in summer and autumn. In the far future the spring precipitation increase is distinctly higher than in other seasons, exceeding 20% in both catchments. The largest difference between catchments can be observed for summer precipitation in the far future that is (i.e., the ensemble median) projected to increase by 6.5% in the Barycz catchment and only by 0.1% in the Upper Narew catchment. The uncertainty of summer precipitation is the largest among all seasons in both catchments.

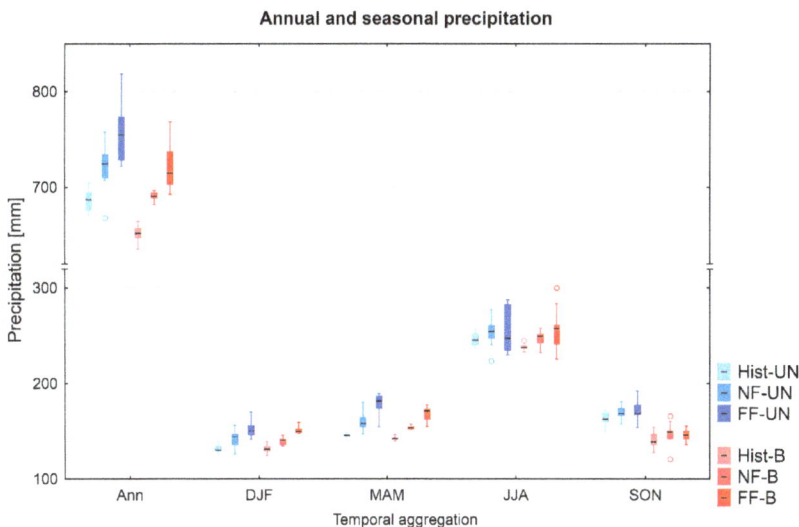

Figure 5. Multi-model ensemble projections of annual and seasonal precipitation for the near (NF) and the far (FF) future under RCP 4.5 in comparison to the historical (Hist) period. B stands for the Barycz catchment and UN stands for the Upper Narew catchment.

Annual precipitation change projections for the near future are not statistically significant according to most of the climate models in both catchments. The models agree well that the projected change is low. Despite the fact that the distance between catchments is almost 500 km, the precipitation change signal is similar. Seasonal projections of changes are significant for winter and spring, and insignificant for summer and autumn. Lack of robustness (statistically significant changes, but large disagreement about the magnitude) can be observed in the far future for both annual and spring totals. More in-depth characteristics of robustness of precipitation projections performed at a larger scale of the Vistula and Odra basins can be found in Piniewski et al. [43].

3.2. Hydrological Response to Climate Change

Hydrology of both catchments is considerably affected by projected warming and changes in precipitation patterns. As shown in Figure 2 and discussed in Section 2.1, the baseline hydrology of investigated catchments differs substantially, so it is very interesting to assess the effect of roughly similar climate change signal (cf. Figures 3 and 4) on different baseline hydrological conditions of two lowland catchments.

3.2.1. Snow Melt

Snow conditions are characterized in SWAT by the amount of melted snow [36]. The amount of water originating from snow melt is projected to substantially decrease, by 23% and 40% (ensemble median) in both catchments, in the near and far future, respectively (Figure 6). However, due to the difference in climate conditions (i.e., the frequency of temperatures falling below zero) between catchments, the response varies considerably across months. In the Barycz catchment snow melt in autumn and spring is projected to almost vanish by the end of 21st century, whereas in winter it is shown to decrease by 37%. In contrast, snow melt occurring between November and February in the Upper Narew catchment will remain almost unchanged, which can be explained by an increase in precipitation compensating an increase in temperature (cf. Figure 4). However, snow melt occurring in March and April in the Upper Narew catchment will undergo the largest change. While in the historical period a very distinct peak in snow melt occurs in March, in the near future this peak is much less apparent, and in the far future it is shifted to February. April snow melt is expected to literally vanish by the end of the century.

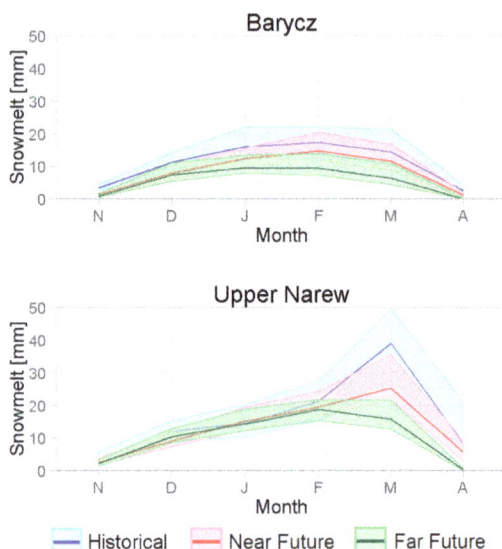

Figure 6. Multi-model ensemble projections of monthly snow melt (between November and April) for the near and the far future under RCP 4.5 in comparison to the historical period.

3.2.2. Evapotranspiration and Soil Water

Actual evapotranspiration (ET) is projected to increase in both catchments by 2.6%–3.3% in the near future and by 3.7%–6.8% in the far future (ensemble medians), in accordance with projected temperature increase (cf. Figure 4). Actual ET in the Upper Narew catchment is projected to undergo a higher increase than in the Barycz catchment, and this happens mainly due to the projected increase in spring season. Both the magnitude of change and the spread of the ET projections among all ensemble members are relative low (Figure 7). The highest relative increase, reaching 8% in the far future, is projected in winter, but since the historical value for winter is very low, this change is not very high when expressed in absolute values. It is noteworthy that projected changes in potential evapotranspiration (simulated in SWAT using Hargreaves method) are quite similar, although the magnitude of change in the far future is slightly lower.

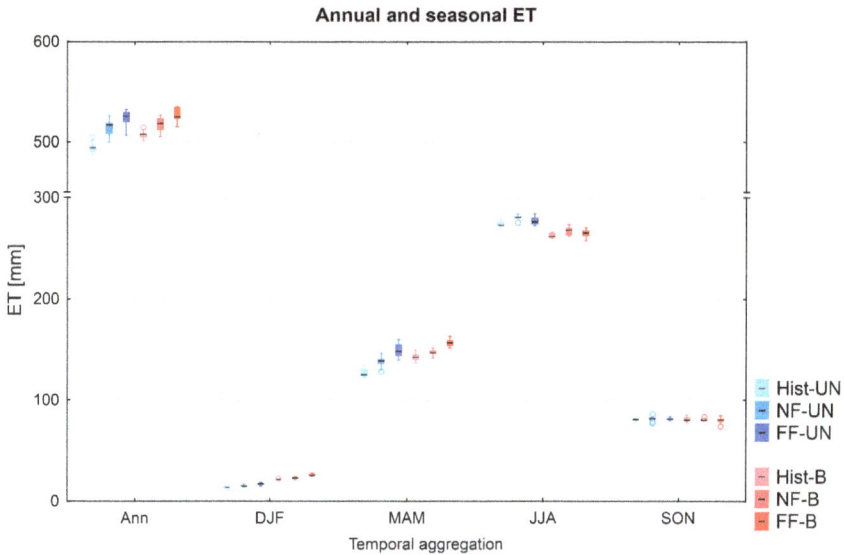

Figure 7. Multi-model ensemble projections of annual and seasonal actual evapotranspiration (ET) for the near (NF) and the far (FF) future under RCP 4.5 in comparison to the historical (Hist) period. B stands for the Barycz catchment and UN stands for the Upper Narew catchment.

According to the ensemble median, projected increase in mean annual soil water content (amount of water in the soil profile expressed in mm) is relatively low in both catchments, not exceeding 9% in the far future, and is slightly higher for the Barycz than for the Upper Narew catchment (Figure 8a). Climate model spread for the far future is more than double of the baseline period spread. Since the Upper Narew catchment is characterized by heavier soils, the mean soil water content is slightly higher there. However, seasonal patterns in both catchments are the same, with winter maximum and summer minimum. While in winter and spring soil water is projected to increase according to SWAT projections driven by the majority of RCMs in both catchments. The difference between catchments can be observed for summer and autumn: in the Barycz the increase is projected, but for the Upper Narew the direction and magnitude of projected changes are highly uncertain. This can be related to lower increases (or decreases) in summer and autumn precipitation for the latter, particularly in the far future (cf. Figure 5), but also to the differences in soil physical characteristics.

Annual percolation (movement of water past the bottom of the soil profile to the groundwater aquifers) is projected to increase by a rate at least two times higher in the Barycz catchment than in the Upper Narew catchment (Figure 8b). Due to the nature of projected changes in winter precipitation and temperatures, more rainfall is projected in winter in both catchments, which triggers a sharp increase in percolation in this season in both catchments, i.e., more than the two-fold increase for the far future. Catchments behave differently for the remaining seasons: while for the Upper Narew catchments no clear conclusion can be made, as the model spread increases, low to moderate increases are projected for the Barycz catchment.

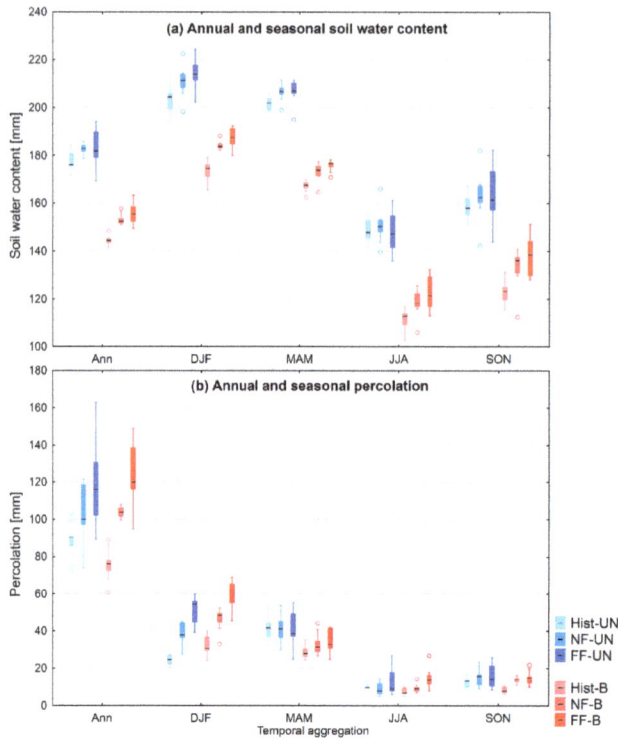

Figure 8. Multi-model ensemble projections of annual and seasonal soil water content (**a**) and percolation (**b**) for the near (NF) and the far (FF) future under RCP 4.5 in comparison to the historical (Hist) period. B stands for the Barycz catchment and UN stands for the Upper Narew catchment.

3.2.3. Water Yield, Surface Runoff and Baseflow

In SWAT, water yield is calculated as the sum of surface runoff, lateral (sub-surface) flow and baseflow, in the absence of transmission losses. Both in the Barycz and the Upper Narew catchment surface runoff and baseflow are dominating components and constitute approximately 90% of total water yield, so they are discussed in more detail below.

The median of projected changes in water yield, i.e., the portion of precipitation that reaches the stream, is significantly higher for the Barycz catchment (24% and 38% in the near and far future, respectively), than in the Upper Narew catchment (9% and 20%, respectively; Figure 9a). This large difference is partly explained by the fact that the baseline value for the latter is considerably higher, i.e., 170 mm vs. 123 mm (cf. Figure 2). Seasonal patterns of change are quite similar, with the most pronounced increase occurring in winter, which is in line with projections of other variables shown above. In three remaining seasons, the increases are either low or the uncertainty is so high that it is difficult to conclude on the direction of change.

Present differences in water yield between two investigated catchments can be to a large extent explained by differences in surface runoff, whose annual total is equal to 40 mm in the Barycz catchment, and nearly the double of it in the Upper Narew catchment (Figure 9b). Little can be concluded on projections of surface runoff on annual level, as the climate model uncertainty dominates. However, interesting patterns can be noted on seasonal level. In the Upper Narew catchment, a moderate increase in surface runoff is projected in winter and a moderate decrease in spring. In contrast, in the Barycz catchment surface runoff decreases in both seasons, although with a rather low rate. These behaviors can be well explained by projected patterns in precipitation (Figure 6) and snow melt (Figure 7). With milder and wetter winters, more (or less) melted snow forms more (or less) surface runoff, whereas more rainfall contributes to higher infiltration, as the occurrence of soil freezing is more rare. In contrast, in summer and autumn, changes in surface runoff follow to a large extent changes in rainfall. As shown in Figure 9b, overall, the uncertainty in these two seasons increases, especially in summer. Higher projected summer precipitation increase for the Barycz catchment translates into higher surface runoff change, although in absolute values the figure remains low (14 mm).

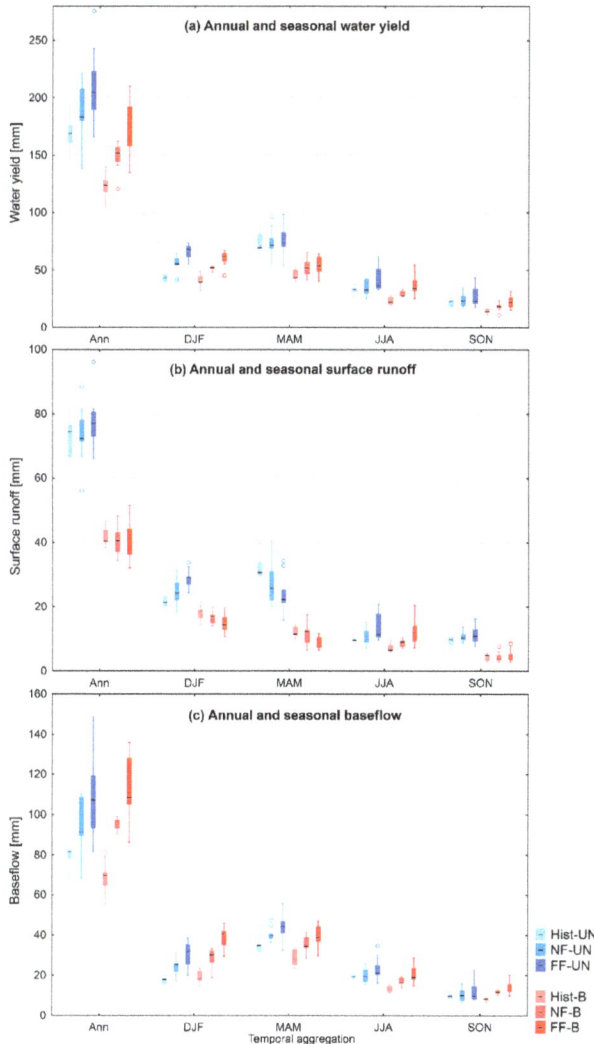

Figure 9. Multi-model ensemble projections of annual and seasonal water yield (**a**) and its two major components, surface runoff (**b**) and baseflow (**c**), for the near (NF) and the far (FF) future under RCP 4.5 in comparison to the historical (Hist) period. B stands for the Barycz catchment and UN stands for the Upper Narew catchment.

Projected changes in baseflow (Figure 9c) follow to a large extent changes in percolation (Figure 8b), although a lag in seasonal pattern can be visible (maximum values reached in spring rather than in winter). In general, both the signal of change and the uncertainty increase their magnitude in the future horizons. While in the baseline period the Upper Narew catchment has higher baseflow than the Barycz catchment, an inverse relationship occurs in the far future. Projected changes in the lateral flow component (not shown) are similar to those presented for the baseflow.

3.3. Sediment and Nutrient Transport Response to Climate Change

Sediment and nutrient transport response to climate change forcing is presented in two forms: (1) catchment-averaged sediment, TN and TP losses, i.e., the amount of sediment, TN and TP that is transported from land (sub-basins) to the river network, shown as box plots across all climate models; and (2) spatially-explicit changes in sediment, TN and TP losses presented on maps of the ensemble median. The results are presented as differences between future periods and the reference period, expressed in $kg \cdot ha^{-1}$.

Mean annual sediment losses are projected to increase in both catchments, although the baseline levels are different: roughly three-fold higher values in the Upper Narew catchment, illustrating higher fraction of erosive soils in this region (Figure 10a). Projected changes follow, to some extent, changes in surface runoff (Figure 9b), showing an increase in sediment losses in winter and summer in the Upper Narew catchment, and a decrease in winter and an increase in summer in the Barycz catchment.

Mean annual TN losses in the historical period are nearly three-fold higher in the Barycz catchment (5.6 kg/ha) than in the Upper Narew catchment (1.9 kg/ha; Figure 10b). This is presumably related to different levels of agricultural intensification of both catchments (cf. Figure 2). An increase by 35% in TN losses is projected for the Barycz catchment in the far future, whereas an increase by 45% is projected for the Upper Narew catchments according to the ensemble median. In both catchments, but notably in the Barycz catchment, most of projected increase occurs in winter, which is in line with projections of percolation (Figure 8b) and baseflow (Figure 9c). While in the present climate, spring is the season with highest TN losses in the Barycz catchment, in the far future climate it is likely to be winter rather than spring.

Intensive agriculture of the Barycz catchment is likely to explain differences in the baseline period mean annual TP losses, i.e., values that are nearly two-fold higher than in the Upper Narew catchment (Figure 10c). The SWAT model projections of climate change impacts show moderate increases for the Upper Narew catchment and high uncertainty for the Barycz catchment. However, seasonal patterns are slightly different. In the Barycz catchment, the most distinct signal is projected in summer, forced by an increase in precipitation in this season. In contrast, in winter, TN losses are projected to decrease. In the Upper Narew catchment, increases are prevailing in winter and summer, whereas small decreases occur in spring. Autumn is the season with high model spread.

Projected sediment, TN and TP losses are characterized by high spatial variability (Figures 11–13). For TN, the western part of the Upper Narew catchment (including sub-catchments of Horodnianka, Awissa and Orlanka) has the highest increase, exceeding 2 $kg \cdot ha^{-1}$ in the far future (Figure 12). In the Barycz catchment, spatial variability is even higher, and the north of the catchment, including sub-catchments of Orla, Dąbroczna and Polski Rów, has the highest increase, exceeding 5 $kg \cdot ha^{-1}$ in the far future. In both catchments, areas with the highest projected increase in TN losses coincide with areas with the most intensive agriculture (Figure 2). For both sediment and TP losses, the situation is more complex, i.e., there are areas with both increases and decreases in each catchment and projection horizon. This is presumably related to a different dominant transport pathway of sediment and TP (surface runoff), whose projected changes are also variable in space. Patchy patterns also reflect the fact that, as shown in Figure 10a,c, sediment and TP losses projections are actually highly uncertain, so within the ensemble there exist climate models for which the increases would be prevailing as well as models for which decreases would be prevailing.

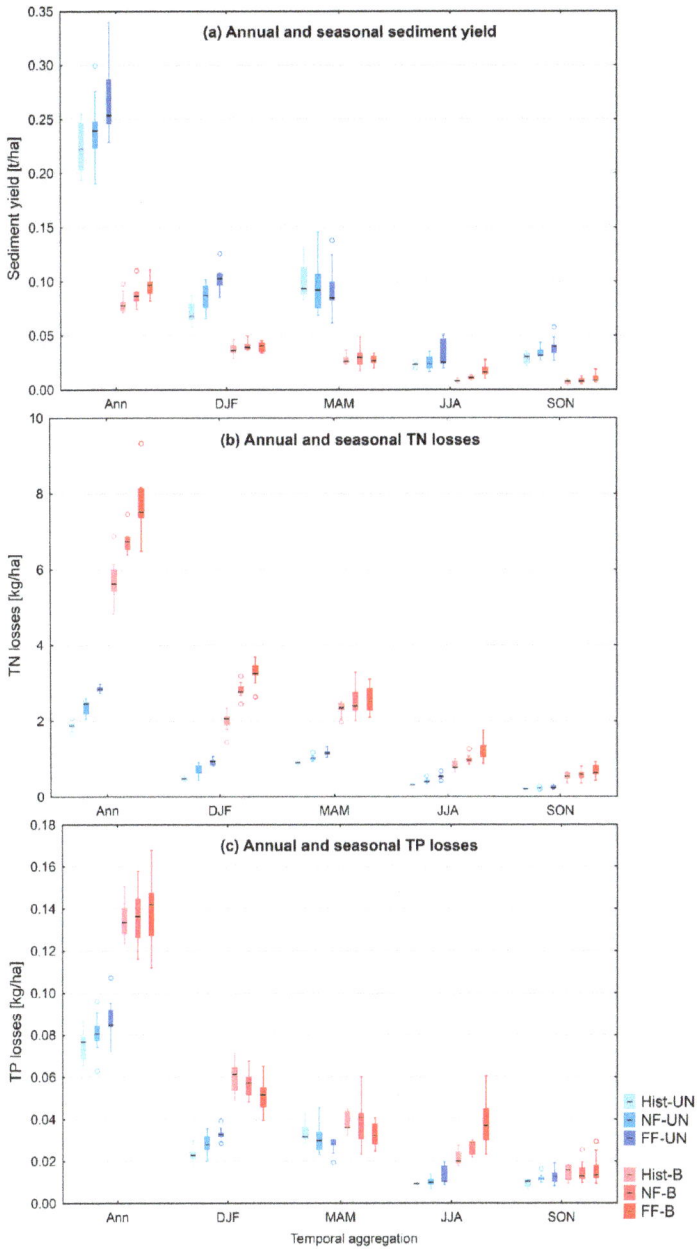

Figure 10. Multi-model ensemble projections of annual and: seasonal sediment (**a**); TN (**b**); and TP (**c**) losses, for the near (NF) and the far (FF) future under RCP 4.5 in comparison to the historical (Hist) period. B stands for the Barycz catchment and UN stands for the Upper Narew catchment.

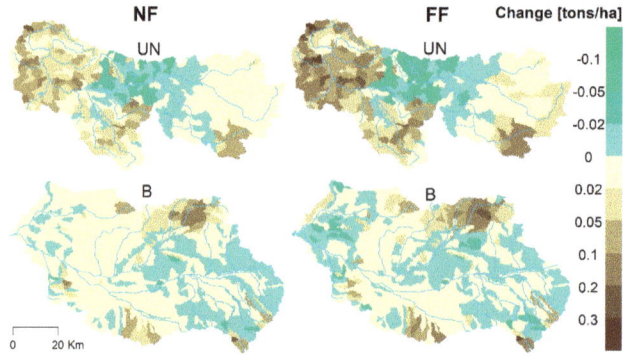

Figure 11. Projected change in sediment losses (total amount of sediment transported from sub-basins to streams) in the Upper Narew and Barycz catchments in the near and far future according to the ensemble median.

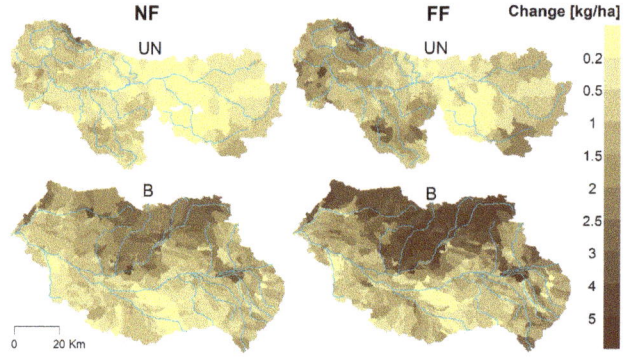

Figure 12. Projected change in TN losses (TN transported by all types of pathways from sub-basins to streams) in the Upper Narew and Barycz catchments in the near and far future according to the ensemble median.

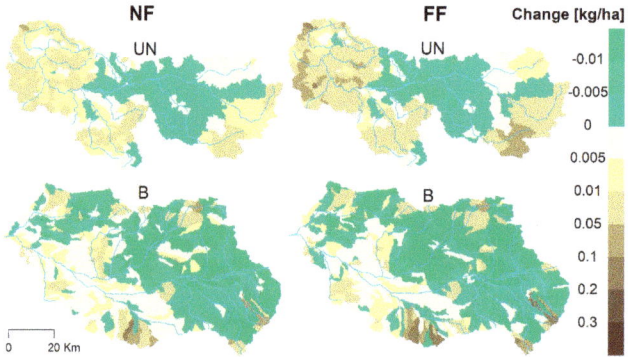

Figure 13. Projected change in TP losses (TP transported by all types of pathways form sub-basins to streams) in the Upper Narew and Barycz catchments in the near and far future according to the ensemble median.

4. Discussion

Projections of climate change derived from an ensemble of nine bias-corrected RCMs under RCP 4.5 consistently suggest an increase in temperature and precipitation over Poland [29,42]. An important feature of precipitation change for two catchments investigated in this paper, the Upper Narew located in the east and the Barycz located in the west, is that it is not seasonally constant, but is much higher for winter (by 13%–15% in the far future) and spring (by 21%–24% in the far future) than for summer and autumn (changes not statistically significant). This signal, uniform across two catchments, forces a complex response in hydrology. First, snow melt is projected to decrease considerably, but this decrease is distributed equally over winter and spring in the Barycz catchment, and occurs almost exclusively in spring in the Upper Narew catchment. Small increases (but with a low spread as well) in actual evapotranspiration are projected in both catchments. In contrast, increases in soil water content are blurred by high climate model spread, with exception of winter and spring, when the signal is stronger. Higher fraction of permeable soils in the Barycz catchment leads to a higher increase in percolation and baseflow as compared to the Upper Narew catchment. For annual surface runoff projections, the signal is overshadowed by the noise, but two different types of signals emerge in the seasonal projections: mild decreases in winter and spring and a mild increase in summer in the Barycz catchment, and an increase in winter and summer accompanied by a decrease in spring in the Upper Narew catchment.

Projected changes in sediment and nutrient losses result from a combination of reasons: climate change itself, projected changes in hydrology, as well as different soil conditions and land cover. Soil erosion was not a major problem in investigated catchments in the reference period and future projections suggest only moderate increases in sediment loss, occurring mainly in summer (both catchments) and winter (the Upper Narew, related to increased surface runoff). A sharper increase is projected in both catchments for TN losses. Here, much higher changes are projected for the Barycz catchment, which is already subject to a nearly three-fold higher TN losses than the Upper Narew catchment in the reference period. Seasonal changes in TN losses are connected to the dominant transport pathway of TN, which is sub-surface flow. The strongest increase is projected for winter season in the Barycz catchment, when percolation and baseflow are also projected to increase significantly. These results are overall consistent with the previous study carried out in Poland by Piniewski et al. [10], reporting projected increases in NO_3-N leaching to groundwater and river loads in a small coastal catchment in north Poland, according to a single, "warmer and wetter" climate scenario. In contrast, Hesse et al. [26] reported that majority of Polish and Russian rivers in the Vistula Lagoon are expected to have decreased loads of NO_3-N and NH_4-N. On the other hand, the ENSEMBLES projections used in their study were less consistent in agreement on precipitation increase than the EURO-CORDEX projections used here.

A slightly different picture occurs for TP losses: at annual level, the uncertainty dominates in the Barycz catchment, whereas a weak and uncertain increase is projected in the Upper Narew catchment. Since surface runoff is the principal transport pathway of TP, the seasonal changes in TP losses follow those of surface runoff: an increase in summer in both catchments (but stronger in the Barycz catchment) and in winter season, an increase in the Upper Narew catchment and a decrease in the Barycz catchment. Previous impact studies in Polish catchments [10,26] reported more apparent increases in phosphorus (PO_4-P) loads than in the present study.

This study has evaluated the pure effect of changing climate on water quantity and quality in two different lowland catchments in Poland, using state-of-the-art climate projections and estimating their uncertainty propagating by the hydrological model. Among several limitations of this study, one has to note that the results are based on a single RCP 4.5. It is well known that the current greenhouse gases emissions are on the RCP 8.5 trajectory, so it would be interesting to analyze the projections for this forcing as well. The same ensemble of climate models as the one used here, but driven by RCP 8.5, shows that both the rate of temperature increase and the rate of precipitation increases are expected to be higher for this RCP in both studied catchments [43]. Particularly, high increases in precipitation are projected in winter and spring seasons. Runoff change projections studied in another paper [46] demonstrate that the increases in runoff are also higher under RCP 8.5 than under RCP 4.5. This shows that the changes of precipitation are not compensated by the changes in temperature and evapotranspiration under warmer and wetter conditions. Even though water quality simulations have not been carried out under RCP 8.5 within this study, it can be expected that with a higher magnitude of increase in winter runoff, higher TN losses could be projected, whereas the results for sediment and TP losses are more uncertain. In fact, as shown in the study of Sun et al. [47], the effect of water quality parameter uncertainty on total suspended solids and total phosphorus load projections was generally greater than the effect of GCM uncertainties, particularly during high-load events.

For water resources management in Poland, the message is mixed. First, "wetter" scenarios on the Polish Plain may seem beneficial, as this region is generally known to be affected by water scarcity [48]. Particularly, in the Barycz catchment, increased water availability is likely to help sustain water-demanding fish pond systems. In the Upper Narew catchment, it may help sustain environmental flows through the wetlands of the Narew National Park [49]. Secondly, increased sub-surface runoff is expected to trigger an increase in TN losses, particularly in the Barycz catchment, characterized by a high fraction of land vulnerable to nitrate leaching. These results suggest that climate change may require additional adaptation actions on top of the "business-as-usual" actions aimed at non-point source pollution mitigation in Poland. Future studies should assess what kind of measures would help achieve the highest reduction in future TN losses, particularly in the more vulnerable Barycz catchment. An important finding of this study is that the majority of the projected increase in TN losses occurs in winter season, suggesting that maintaining vegetative cover on agricultural fields in winter could be a good solution [10,50,51].

Acknowledgments: Support of the project CHASE-PL (Climate change impact assessment for selected sectors in Poland) of the Polish-Norwegian Research Programme operated by the National Centre for Research and Development (NCBiR) under the Norwegian Financial Mechanism 2009–2014 in the frame of Project Contract No. Pol Nor/200799/90/2014 is gratefully acknowledged. The Institute of Meteorology and Water Management —State Research Institute (IMGW-PIB) is kindly acknowledged for providing the hydrometeorological data used in this work. The second author is grateful for support to the Alexander von Humboldt Foundation and the Ministry of Science and Higher Education of the Republic of Poland. Constructive comments from two anonymous reviewers that helped to improve the quality of the manuscript are highly appreciated.

Author Contributions: Mikołaj Piniewski, Ignacy Kardel, Stefan Ignar, Tomasz Okruszko, Rasmus Benestad and Raghavan Srinivasan developed the methodological framework. Paweł Marcinkowski, Mikołaj Piniewski and Ignacy Kardel developed the model setup. Paweł Marcinkowski and Mikołaj Piniewski performed model calibration. Mikołaj Piniewski, Paweł Marcinkowski, Mateusz Szcześniak, and Ignacy Kardel run the model scenarios. Paweł Marcinkowski and Mikołaj Piniewski wrote the manuscript. Paweł Marcinkowski, Mikołaj Piniewski and Ignacy Kardel created the art work.

Conflicts of Interest: The authors declare no conflict of interest.

References

1. Pachauri, R.K.; Meyer, L.A. *Climate Change 2014: Synthesis Report. Contribution of Working Groups I, II and III to the Fifth Assessment Report of the Intergovernmental Panel on Climate Change*; IPCC: Geneva, Switzerland, 2014; p. 151.
2. Jacob, D.; Petersen, J.; Eggert, B.; Alias, A.; Christensen, O.B.; Bouwer, L.M.; Braun, A.; Colette, A.; Déqué, M.; Georgievski, G.; et al. EURO-CORDEX: New high-resolution climate change projections for European impact research. *Reg. Environ. Chang.* **2014**, *14*, 563–578. [CrossRef]
3. Alfieri, L.; Burek, P.; Feyen, L.; Forzieri, G. Global warming increases the frequency of river floods in Europe. *Hydrol. Earth Syst. Sci.* **2015**, *19*, 2247–2260. [CrossRef]
4. Roudier, P.; Andersson, J.C.M.; Donnelly, C.; Feyen, L.; Greuell, W.; Ludwig, F. Projections of future floods and hydrological droughts in Europe under a +2 °C global warming. *Clim. Chang.* **2016**, *135*, 341–355. [CrossRef]

5. Papadimitriou, L.V.; Koutroulis, A.G.; Grillakis, M.G.; Tsanis, I.K. High-end climate change impact on European runoff and low flows—Exploring the effects of forcing biases. *Hydrol. Earth Syst. Sci.* **2016**, *20*, 1785–1808. [CrossRef]
6. Krysanova, V.; Kundzewicz, Z.W.; Piniewski, M. Assessment of climate change impact on water resoures. In *Handbook of Applied Hydrology*, 2nd ed.; Singh, V.P., Ed.; McGraw-Hill Education: New York, NY, USA, 2016; p. 1440.
7. Teutschbein, C.; Seibert, J. Regional Climate Models for Hydrological Impact Studies at the Catchment Scale: A Review of Recent Modeling Strategies: Regional climate models for hydrological impact studies. *Geogr. Compass* **2010**, *4*, 834–860. [CrossRef]
8. Martínková, M.; Hesse, C.; Krysanova, V.; Vetter, T.; Hanel, M. Potential impact of climate change on nitrate load from the Jizera catchment (Czech Republic). *Phys. Chem. Earth Parts ABC* **2011**, *36*, 673–683. [CrossRef]
9. Meier, H.E.M.; Müller-Karulis, B.; Andersson, H.C.; Dieterich, C.; Eilola, K.; Gustafsson, B.G.; Höglund, A.; Hordoir, R.; Kuznetsov, I.; Neumann, T.; et al. Impact of Climate Change on Ecological Quality Indicators and Biogeochemical Fluxes in the Baltic Sea: A Multi-Model Ensemble Study. *AMBIO* **2012**, *41*, 558–573. [CrossRef] [PubMed]
10. Piniewski, M.; Kardel, I.; Giełczewski, M.; Marcinkowski, P.; Okruszko, T. Climate Change and Agricultural Development: Adapting Polish Agriculture to Reduce Future Nutrient Loads in a Coastal Watershed. *AMBIO* **2014**, *43*, 644–660. [CrossRef] [PubMed]
11. Ahmadi, M.; Records, R.; Arabi, M. Impact of climate change on diffuse pollutant fluxes at the watershed scale. *Hydrol. Process.* **2014**, *28*, 1962–1972. [CrossRef]
12. Huttunen, I.; Lehtonen, H.; Huttunen, M.; Piirainen, V.; Korppoo, M.; Veijalainen, N.; Viitasalo, M.; Vehviläinen, B. Effects of climate change and agricultural adaptation on nutrient loading from Finnish catchments to the Baltic Sea. *Sci. Total Environ.* **2015**, *529*, 168–181. [CrossRef] [PubMed]
13. Glavan, M.; Ceglar, A.; Pintar, M. Assessing the impacts of climate change on water quantity and quality modelling in small Slovenian Mediterranean catchment—Lesson for policy and decision makers: Assessing the impacts of climate change on river basin modelling. *Hydrol. Process.* **2015**, *29*, 3124–3144. [CrossRef]
14. Mehdi, B.; Ludwig, R.; Lehner, B. Evaluating the impacts of climate change and crop land use change on streamflow, nitrates and phosphorus: A modeling study in Bavaria. *J. Hydrol. Reg. Stud.* **2015**, *4*, 60–90. [CrossRef]
15. Gombault, C.; Madramootoo, C.A.; Michaud, A.; Beaudin, I.; Sottile, M.-F.; Chikhaoui, M.; Ngwa, F. Impacts of climate change on nutrient losses from the Pike River watershed of southern Québec. *Can. J. Soil Sci.* **2015**, *95*, 337–358. [CrossRef]
16. Malsy, M.; Flörke, M.; Borchardt, D. What drives the water quality changes in the Selenga Basin: Climate change or socio-economic development? *Reg. Environ. Chang.* **2016**, *16*, 209–216. [CrossRef]
17. Ye, L.; Grimm, N.B. Modelling potential impacts of climate change on water and nitrate export from a mid-sized, semiarid watershed in the US Southwest. *Clim. Chang.* **2013**, *120*, 419–431. [CrossRef]
18. Cousino, L.K.; Becker, R.H.; Zmijewski, K.A. Modeling the effects of climate change on water, sediment, and nutrient yields from the Maumee River watershed. *J. Hydrol. Reg. Stud.* **2015**, *4*, 762–775. [CrossRef]
19. Jha, M.K.; Gassman, P.W.; Panagopoulos, Y. Regional changes in nitrate loadings in the Upper Mississippi River Basin under predicted mid-century climate. *Reg. Environ. Chang.* **2015**, *15*, 449–460. [CrossRef]
20. Arheimer, B.; Dahné, J.; Donnelly, C. Climate Change Impact on Riverine Nutrient Load and Land-Based Remedial Measures of the Baltic Sea Action Plan. *AMBIO* **2012**, *41*, 600–612. [CrossRef] [PubMed]
21. Records, R.M.; Arabi, M.; Fassnacht, S.R.; Duffy, W.G.; Ahmadi, M.; Hegewisch, K.C. Climate change and wetland loss impacts on a western river's water quality. *Hydrol. Earth Syst. Sci.* **2014**, *18*, 4509–4527. [CrossRef]
22. Molina-Navarro, E.; Trolle, D.; Martínez-Pérez, S.; Sastre-Merlín, A.; Jeppesen, E. Hydrological and water quality impact assessment of a Mediterranean limno-reservoir under climate change and land use management scenarios. *J. Hydrol.* **2014**, *509*, 354–366. [CrossRef]
23. Mehdi, B.; Lehner, B.; Gombault, C.; Michaud, A.; Beaudin, I.; Sottile, M.-F.; Blondlot, A. Simulated impacts of climate change and agricultural land use change on surface water quality with and without adaptation management strategies. *Agric. Ecosyst. Environ.* **2015**, *213*, 47–60. [CrossRef]
24. Jayakody, P.; Parajuli, P.B.; Cathcart, T.P. Impacts of climate variability on water quality with best management practices in sub-tropical climate of USA. *Hydrol. Process.* **2014**, *28*, 5776–5790. [CrossRef]

25. El-Khoury, A.; Seidou, O.; Lapen, D.R.; Que, Z.; Mohammadian, M.; Sunohara, M.; Bahram, D. Combined impacts of future climate and land use changes on discharge, nitrogen and phosphorus loads for a Canadian river basin. *J. Environ. Manag.* **2015**, *151*, 76–86. [CrossRef] [PubMed]

26. Hesse, C.; Krysanova, V.; Stefanova, A.; Bielecka, M.; Domnin, D.A. Assessment of climate change impacts on water quantity and quality of the multi-river Vistula Lagoon catchment. *Hydrol. Sci. J.* **2015**, 1–22. [CrossRef]

27. Van der Linden, P.; Mitchell, J.F.B. *ENSEMBLES: Climate Change and Its Impacts: Summary of Research and Results from the ENSEMBLES Project—European Environment Agency*; Met Office Hadley Centre: Exeter, UK, 2009.

28. Pazdro, Z.; Kozerski, B. *Hydrogeologia Ogólna*; Wyd. 4. uzup.; Wydaw. Geol: Warsaw, Poland, 1990. (In Polish)

29. Piniewski, M.; Szcześniak, M.; Kardel, I.; Berezowski, T.; Okruszko, T.; Srinivasan, R.; Schulerd, V.; Kundzewicz, Z.W. Hydrological modelling of the Vistula and Odra river basins using SWAT. *Hydrol. Sci. J.* **2017**, accepted.

30. Central Statistical Office Local Data Bank—Statistics for Year 2010. Available online: http://bdl.stat.gov.pl (accessed on 20 November 2016).

31. *Map of Hydrological Division of Poland in the scale 1:10 000*; General of National Water Management Authority: Warsaw, Poland, 2013.

32. Miatkowski, Z.; Smarzyńska, K. Dynamika zmian stężenia związków azotu w wodach górnej Zgłowiączki w latach 1990–2011. *Woda-Śr.-Obsz. Wiej.* **2014**, *3*, 99–111. Available online: http://www.itep.edu.pl/wydawnictwo/woda/zeszyt_47_2014/artykuly/Miatkowski%20Smarzynska.pdf (accessed on 20 November 2016).

33. Ilnicki, P.; Gorecki, K.; Lewandowski, P.; Farat, R. Long-Term Variability of Total Nitrogen and Total Phosphorus Concentration and Load in the South Part of the Baltic Sea Basin. *Fresenius Environ. Bull.* **2016**, *25*, 3923–3940.

34. Banaszuk, P.; Wysocka-Czubaszek, A. Phosphorus dynamics and fluxes in a lowland river: The Narew Anastomosing River System, NE Poland. *Ecol. Eng.* **2005**, *25*, 429–441. [CrossRef]

35. Marcinkowski, P.; Piniewski, M.; Kardel, I.; Srinivasan, R.; Okruszko, T. Challenges in modelling of water quantity and quality in two contrasting meso-scale catchments in Poland. *J. Water Land Dev.* **2016**, *31*, 97–111. [CrossRef]

36. Neitsch, S.; Arnold, J.; Kiniry, J.; Williams, J. *Soil and Water Assessment Tool Theoretical Documentation Version 2009. Technical Report TR-406*; Texas A&M University: College Station, TX, USA, 2011. Available online: http://swat.tamu.edu/media/99192/swat2009-theory.pdf (accessed on 1 November 2016).

37. Berezowski, T.; Szcześniak, M.; Kardel, I.; Michałowski, R.; Okruszko, T.; Mezghani, A.; Piniewski, M. CPLFD-GDPT5: High-resolution gridded daily precipitation and temperature data set for two largest Polish river basins. *Earth Syst. Sci. Data* **2016**, *8*, 127–139. [CrossRef]

38. Szcześniak, M.; Piniewski, M. Improvement of Hydrological Simulations by Applying Daily Precipitation Interpolation Schemes in Meso-Scale Catchments. *Water* **2015**, *7*, 747–779. [CrossRef]

39. Gupta, H.V.; Kling, H.; Yilmaz, K.K.; Martinez, G.F. Decomposition of the mean squared error and NSE performance criteria: Implications for improving hydrological modelling. *J. Hydrol.* **2009**, *377*, 80–91. [CrossRef]

40. Moriasi, D.N.; Arnold, J.G.; Van, L.; Bingner, R.L.; Harmel, R.D.; Veith, T.L. Model evaluation guidelines for systematic quantification of accuracy in watershed simulations. *Trans. ASABE* **2007**, *50*, 885–900. [CrossRef]

41. Piniewski, M.; Marcinkowski, P.; Kardel, I.; Giełczewski, M.; Izydorczyk, K.; Frątczak, W. Spatial Quantification of Non-Point Source Pollution in a Meso-Scale Catchment for an Assessment of Buffer Zones Efficiency. *Water* **2015**, *7*, 1889–1920. [CrossRef]

42. Mezghani, A.; Dobler, A.; Haugen, J.H. CHASE-PL Climate Projections: 5-km Gridded Daily Precipitation & Temperature Dataset (CPLCP-GDPT5). Available online: http://data.4tu.nl/repository/uuid:e940ec1a-71a0-449e-bbe3-29217f2ba31d (accessed on 10 November 2016).

43. Piniewski, M.; Szcześniak, M.; Mezghani, A.; Kundzewicz, Z.W. Regional projections of temperature and precipitation changes: Robustness and uncertainty aspects. *Meteorol. Z.* **2017**. accepted.

44. Gudmundsson, L.; Bremnes, J.B.; Haugen, J.E.; Engen-Skaugen, T. Technical Note: Downscaling RCM precipitation to the station scale using statistical transformations—A comparison of methods. *Hydrol. Earth Syst. Sci.* **2012**, *16*, 3383–3390. [CrossRef]

45. Knutti, R.; Sedláček, J. Robustness and uncertainties in the new CMIP5 climate model projections. *Nat. Clim. Chang.* **2013**, *3*, 369–373. [CrossRef]

46. Piniewski, M.; Szcześniak, M.; Huang, S.; Kundzewicz, Z.W. Projections of runoff in the Vistula and the Odra river basins with the help of the SWAT Model. *Hydrol. Res.* **2017**, under review.

47. Sun, N.; Yearsley, J.; Baptiste, M.; Cao, Q.; Lettenmaier, D.P.; Nijssen, B. A spatially distributed model for assessment of the effects of changing land use and climate on urban stream quality: Development of a Spatially Distributed Urban Water Quality Model. *Hydrol. Process.* **2016**, *30*, 4779–4798. [CrossRef]

48. Kundzewicz, Z.W. Water problems of central and eastern Europe—A region in transition. *Hydrol. Sci. J.* **2001**, *46*, 883–896. [CrossRef]

49. Szporak-Wasilewska, S.; Piniewski, M.; Kubrak, J.; Okruszko, T. What we can learn from a wetland water balance? *Narew National Park case study. Ecohydrol. Hydrobiol.* **2015**, *15*, 136–149. [CrossRef]

50. Thorup-Kristensen, K.; Nielsen, N.E. Modelling and measuring the effect of nitrogen catch crops on the nitrogen supply for succeeding crops. *Plant Soil* **1998**, *203*, 79–89. [CrossRef]

51. Laurent, F.; Ruelland, D. Assessing impacts of alternative land use and agricultural practices on nitrate pollution at the catchment scale. *J. Hydrol.* **2011**, *409*, 440–450. [CrossRef]

water

MDPI

Article

Assessment of Nitrogen Inputs into Hunt River by Onsite Wastewater Treatment Systems via SWAT Simulation

Supria Paul [1], Michaela A. Cashman [1], Katelyn Szura [2] and Soni M. Pradhanang [1,*]

[1] Department of Geosciences, University of Rhode Island, Kingston, RI 02881, USA;
    supria_paul@uri.edu (S.P.); mcashman@uri.edu (M.A.C.)
[2] Department of Biological Sciences, University of Rhode Island, Kingston, RI 02881, USA; kszura@my.uri.edu
*   Correspondence: spradhanang@uri.edu; Tel.: +1-401-874-5980

Received: 15 June 2017; Accepted: 8 August 2017; Published: 16 August 2017

Abstract: Nonpoint source nitrogen pollution is difficult to effectively model in groundwater systems. This study aims to elucidate anthropogenic nonpoint source pollution discharging into Potowomut Pond and ultimately Narragansett Bay. Hydrologic modeling with Soil and Water Assessment Tool (SWAT) and SWAT Calibration and Uncertainty Program (SWAT-CUP) was used to simulate streamflow and nitrogen levels in the Hunt River with and without onsite wastewater treatment systems (OWTS). The objective of this study was to determine how input of OWTS data impacts nitrogen loading into the Hunt River Watershed in Rhode Island, USA. The model was simulated from 2006 to 2014, calibrated from 2007 to 2011 and validated from 2012 to 2014. Observed streamflow data was sourced from a US Geological Survey gauge and nitrogen loading data from University of Rhode Island Watershed Watch (URIWW). From the results, adding OWTS data to the SWAT simulation produced a better calibration and validation fit for total fit (Nash–Sutcliffe Efficiency (NSE) = 0.50 calibration, 0.78 validation) when compared with SWAT simulation without OWTS data (NSE = −1.3 calibration, −6.95) validation.

Keywords: nitrogen; SWAT; OWTS; waste water; septic; watershed

1. Introduction

Eutrophication poses a severe threat to coastal waters on local, regional, and global scales. Coastal ecosystems are naturally nitrogen limited and runoff from terrestrial systems from agriculture, wastewater, and industrial practices contributes to eutrophic areas of low dissolved oxygen, known as dead zones [1]. Dead zones have been documented as doubling in occurrence every decade since the mid 1900s and have been increasing in areal extent. Factors such as rising ocean temperatures, increasing ocean acidification, sea-level rise, and changing climate variables are acting synergistically to exacerbate the eutrophication problem [2]. Often, eutrophication, leading to conditions of hypoxia and anoxia in coastal waters, is overlooked until broad scale detrimental effects are obvious [3]. The breakdown of coastal fisheries and large fish kills garner the most attention to eutrophic conditions. Fisheries located in shallow coastal waters are the most at risk and instances of fishery and shellfish closures have increased in recent years.

Narragansett Bay, located in the state of Rhode Island (RI), USA is no exception to the problem of excess nitrogen (N) and resulting eutrophication of coastal waters. Areas of the bay have experienced N inputs for over the past 200 years and impacted areas have expanded in extent as the state's population has grown and development has increased throughout the state. Rhode Island is the smallest state and is the second most densely populated with roughly one million residents [4]. Over the past several decades, the bay has ranked as one of the most heavily fertilized estuaries in the USA with the majority of nutrients originating from point sources [5,6]. The largest point source contributor is sewage treatment plants (STPs) [6] in the northern portion of the bay where the state capitol is located and the population density is highest. Through the years, the state has faced harmful algal blooms stemming from eutrophication, leading to fisheries and beach closures, most often during the peak summer tourism season. Both fisheries and tourism are vital to the state's economy. In 2003, the state experienced a severe fish kill with over one million menhaden, and other finfish and shellfish, killed from anoxic conditions within Greenwich Bay, which is a smaller offshoot of Narragansett Bay [7]. This event spurred initiative for action to clean up pollution within the bay and to combat excess N from entering the state's coastal waters in order to reduce risks to the health of the coastal ecosystems and the communities that rely upon them. Targeting the N point sources, the state set a goal of 50% reduction of N discharge into the bay from 1995 to 1996 levels, limiting the allowable discharge volumes of N from 11 wastewater treatment facilities in upper Narragansett Bay. The targeted reduction was achieved through stages by 2012 [8]. Today, as N inputs decline from point sources, nonpoint sources, such as urban runoff from septic systems, are becoming a proportionately larger contributor of N into the bay. In RI, 30% of homes are on septic systems, also known as onsite wastewater treatment systems (OWTSs) [8,9]. For coastal watersheds with households served predominantly by OWTSs, these systems can serve as significant sources of N which can contribute to detrimental eutrophication of coastal waters.

The Hunt River and two of its tributaries (Fry Brook and Scrabbletown Brook), lie within coastal watersheds served predominantly by OWTSs. These waterbodies were listed as 303 (d) Group 1, under the Clean Water Act for highest priority impacted waters in RI. The Hunt River and its tributaries have made the 1998 303 (d), under the Clean Water Act list of impaired waterbodies for several years for being impacted by fecal coliform. The majority of high bacteria counts were found to occur during wet weather conditions. The 2001 total maximum daily loads (TMDL) report determined the largest wet weather source of fecal coliform to be stormwater runoff. Fecal coliform monitoring is often used as indication of pathogen presence; however, modeling fecal coliform in freshwater systems holds a unique set of challenges. Bacteria fate and transport relies heavily on the nature of pollution events and sensitivity of fecal coliform to environmental conditions [10]. Leaking sewer and septic infrastructure greatly contributes to nonpoint nitrogen pollution to urban watersheds

In order to protect the health of any coastal community, it is important to quantify the influence of STPs and OWTSs on groundwater flow and the contributions of nutrients from these systems to coastal waters. For mitigation purposes, it is helpful to model these inputs on local watershed scales to identify potential areas of high N inputs to target for reduction through policies and infrastructure development. Few studies have been conducted modeling nutrient inputs on the watershed scale using the Soil Water Assessment Tool (SWAT) due to complexities stemming from the uncertainty of differences in subsurface hydrology, soil and water dynamics, and the often lack of detailed information about OWTS [11]. Increased densities of OWTS have been shown to increase baseflow within a watershed system [12–14], which can have the effect of increasing nutrient loading into ground and surface waters, eventually entering into coastal waters. OWTSs have a biozone layered designed to filter out nutrients, preventing their leaching into the groundwater. However, these biozone layers can become ineffective over time, resulting in an increased discharge of nutrients into sensitive coastal waters. Nutrient loading into a waterbody depends on local surrounding conditions, thus it is helpful to conduct modeling on smaller scales to identify areas to target for N mitigation.

In this study, we modeled the amount of N entering into Narragansett Bay on the watershed scale, with focus on the Hunt River Watershed. The Soil Water Assessment Tool (SWAT) was used to model N inputs into Potowomut Cove, which is fed by the Hunt River, and connects to Narragansett Bay. The effects on receiving waters due to nonpoint source loads, including OWTS could be effectively modeled using SWAT. The development and application of hydrologic models such as SWAT could therefore be used for the Hunt River Watershed to determine Best Management Practices (BMPs). We used a gauged watershed to calibrate the SWAT model, and estimate streamflow and nutrient loading into the bay. Recent development in the SWAT model allows the users to account for presence of septic systems or OWTSs. SWAT modeling also is unique in its ability to account for both watershed and in-stream processes.

2. Materials and Methods

2.1. Study Area and SWAT Model Description

The Hunt River Watershed, as shown in Figure 1, is a small-scale (59.9 km^2) watershed that cuts across central Rhode Island, USA. The 11 kilometer Hunt River winds through several towns and ultimately discharges through rural East Greenwich and into Potowomut Pond. From there, the river known as Potowomut River flows southeasterly into Potowomut Cove and then into Narragansett Bay. While the watershed is predominantly rural and undisturbed, rapid urbanization along the waterway has driven significant increases in anthropogenic pollutants. The SWAT model is hydrologic modeling software developed by the United States Department of Agriculture (USDA). The SWAT model was originally developed to predict impact of land management on water, nutrients and sediments of large ungauged basins. It was created as physically based, continuous, and computationally efficient models [15,16]. In SWAT, the catchment area is subdivided into subbasins, river reaches and hydrological response units (HRUs). Further subdivisions of HRUs are statistically performed by considering a determined percentage of subbasin area that is independent of location within the subbasin. SWAT uses a different routing scheme for runoff and chemistry through watersheds while maintaining water balance for each watershed sector. Normally, the routing for chemical transport includes sediments, erosion, plant growth, pesticides and nutrients. Modeling in-stream nutrient processes is valuable for simulating more accurate environmental conditions. A nutrient process in SWAT includes fate and transport of numerous nitrogen and phosphorus pools including organic and inorganic forms in soil and stream. The SWAT in-stream water quality algorithms incorporate nutrient interactions and relationships used in the Enhanced Stream Water Quality Model (QUAL2E) [17]. Recent advancements in SWAT modeling consider nutrient processes from OWTSs based on a biozone algorithm [18]. This algorithm is a function of the net growth of septic biomass dependent upon soil temperature and the leaching of soluble phosphorus through soil layers.

2.2. SWAT-CUP Model Description

The SWAT Calibration and Uncertainty Program (SWAT-CUP) is widely used for SWAT model calibration and uncertainty analysis. This program offers various calibration algorithms, of which, Sequential Uncertainty Fitting (SUFI2), a robust method compared to deterministic and stochastic method [19], is used for this study. The SUFI2 approach was implemented due to its comprehensive consideration of uncertainty sources in static input variables. Moreover, it gives the 95% probability distribution model output variables based on propagation of uncertainty, often referred to as 95% Prediction Uncertainty (95 PPU).

2.3. Input Data and Model Preparation

Numerous data sources were used to develop the SWAT model. Land cover data and soil data were obtained from the Rhode Island Geographic Information System (RIGIS) [20]. Precipitation data was acquired through the Parameter-elevation Regressions on Independent Slopes Model (PRISM) climate mapping system [21]. The spatial coverage representing the watershed area which included eight data stations of precipitation and temperature were downloaded from PRISM. Streamflow was calibrated using USGS Gauge 01117000 (41°38′28″, −71°26′42″), located in North Kingstown, Rhode Island, 4 kilometers upstream from the mouth of Narragansett Bay. The National Elevation Dataset was downloaded for the Hunt River Watershed at a 10 m resolution (U.S. Geological Survey). Official Soil Series Descriptions data was obtained from the Natural Resources Conservation Service (NRCS) Soil Survey Geographic Database (SSURGO) [22]. Land Use/Land Cover (LULC) data was obtained from the United States Geological Survey (USGS) 2006 National Land Cover Dataset (NLCD) [23] in 30 m resolution. Elevation data was obtained from Digital Elevation Model (DEM) obtained from United States Geological Survey National Elevation Data (NED) [24].

LULC data indicated that the land within the watershed delineation was dominated by forested and urban areas. The headstream of the Hunt River is predominantly urbanized (Figure 1).

According to the LULC classification system, the watershed area is dominantly classified as "Urban Development" (34%), and "Sewered Urban Area" (26%). Subbasins 1 and 5 are preeminently urbanized, whereas subbasins 2–4, and 6 are a mixture of urban development and undisturbed habitat. In respective order from highest to lowest, the remaining land use classifications identified in the Hunt River Watershed were "Major Parks and Open Space" (18%), "Conservation and Limited" (12%), "Reserve" (6%), and "Non-urban Development" (3%). Land classified as "Water Bodies" and "Prime Farmland" were both identified at less than 1%. OWTS information is not publicly available for the state of Rhode Island. In order to estimate the number of households on OWTS, sewer line data from RIGIS [25] was overlaid on the Hunt River Watershed area to determine the sewered areas of the delineated watershed using ArcGIS (ESRI—version 10.3.1). Urban areas served by sewers were subtracted from the map area to estimate land area serviced by OWTSs. Using this method, an estimated 94% of households (13.36 km^2) residing on urban land within the watershed do not have access to sewer, and were therefore classified as relying on OWTS for the purpose of this study. There are no industrial waste water treatment facilities within the delineated watershed. Due to absence of agricultural land use, nitrogen inputs from fertilizers, farmland, and livestock were negligible and not included within the model.

A challenging aspect of modeling nutrients in Rhode Island is lack of continuous sampling data. This project used data from the University of Rhode Island Watershed Watch (URIWW) program, collected from five monitoring stations located along the Hunt River. The URIWW is a volunteer, citizen-science program which conducts statewide monthly water quality sampling from May–October each year, beginning in 2007 for the Hunt River [26]. Total nitrogen (TN) data collected from 2007 to 2014 was used for this study.

2.4. Model Setup and HRU Definitions

For this study, six outlet subbasins were created based on data from URIWW monitoring stations of TN (Figure 1). Subbasins were divided into HRUs by assigning threshold values of land use and land cover, soil, and slope percentage (Figure 2). Development of HRUs determines the land use/soil/slope class combinations in the delineated subbasins based on predetermined threshold values. Thresholds are typically chosen based on SWAT model objectives. Because the scope of this project focuses on OWTSs in urban areas, the thresholds needed to be small enough to recognize the fragmented urban areas within each subbasin. Therefore, HRU definitions were set at 5% land use, 5% soil, and 5% slope percentage. These small-scale thresholds yielded 320 HRUs within the six subbasins. The SWAT model was run with two distinct simulations to determine the effects of OWTS on modeling streamflow and nitrogen inputs. Changing the status of septic systems was done by changing the "isep_opt" value from 0 to 1. Other OWTS input parameters were considered as default which is as shown in the Table 1. The OWTS variable in SWAT (isep = 1) converted the urban areas within subbasin 1 to active septic system users. Both scenarios were simulated for 36 years, from 1981 to 2016, following the SWAT model setup illustrated in Figure 3. The warm up period for the model was used by one year. The observed data for Total Nitrogen data was used the years 2007 to 2016 for the simulation period. The timespan for calibration and validation was confided by the availability of streamflow and TN data. The calibration was completed with 30 TN data points, and validated with 19 TN data points.

Table 1. Onsite wastewater treatment system (OWTS) input parameter.

Parameter Name	Parameter Descriptions	Value
SEP_CAP	Average number of permanent residents in a house	2.5
BZ_Area	Surface area of drain field (m^2)	100
ISEP_TFAIL	Time until failing system gets fixed (days)	70
BZ_Z	Depth to the top of biozone layer (mm)	500
BZ_THK	Thickness of biozone layer(mm)	50
SEP_STRM_DIST	Distance to the stream from the septic hydrological response unit (HRU) (km)	0.5

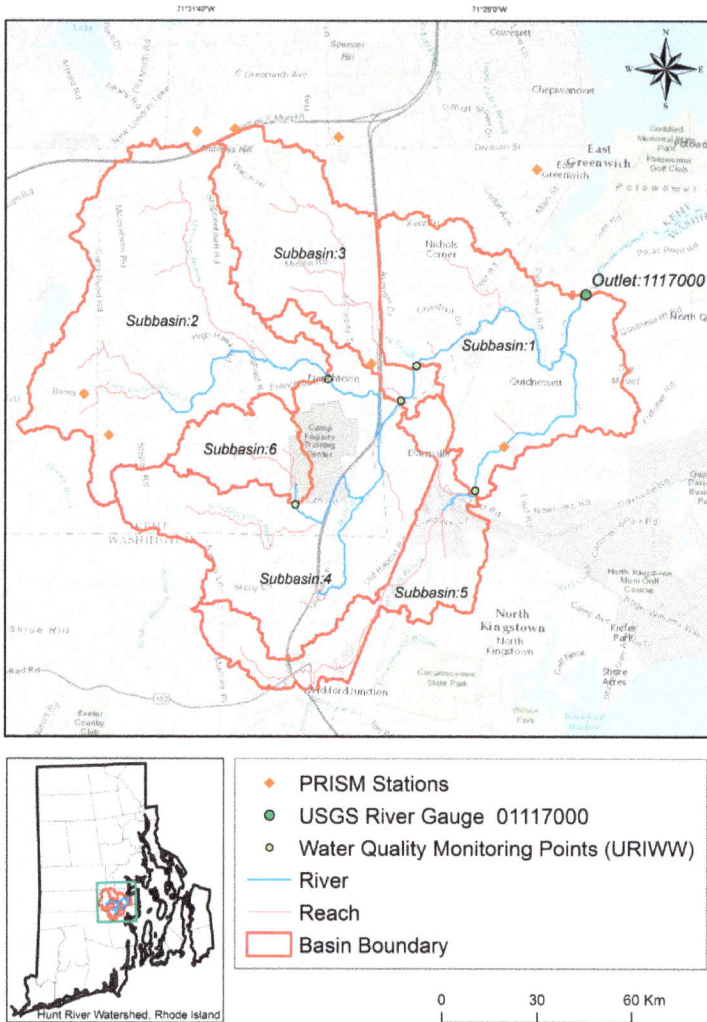

(a)

Figure 1. *Cont.*

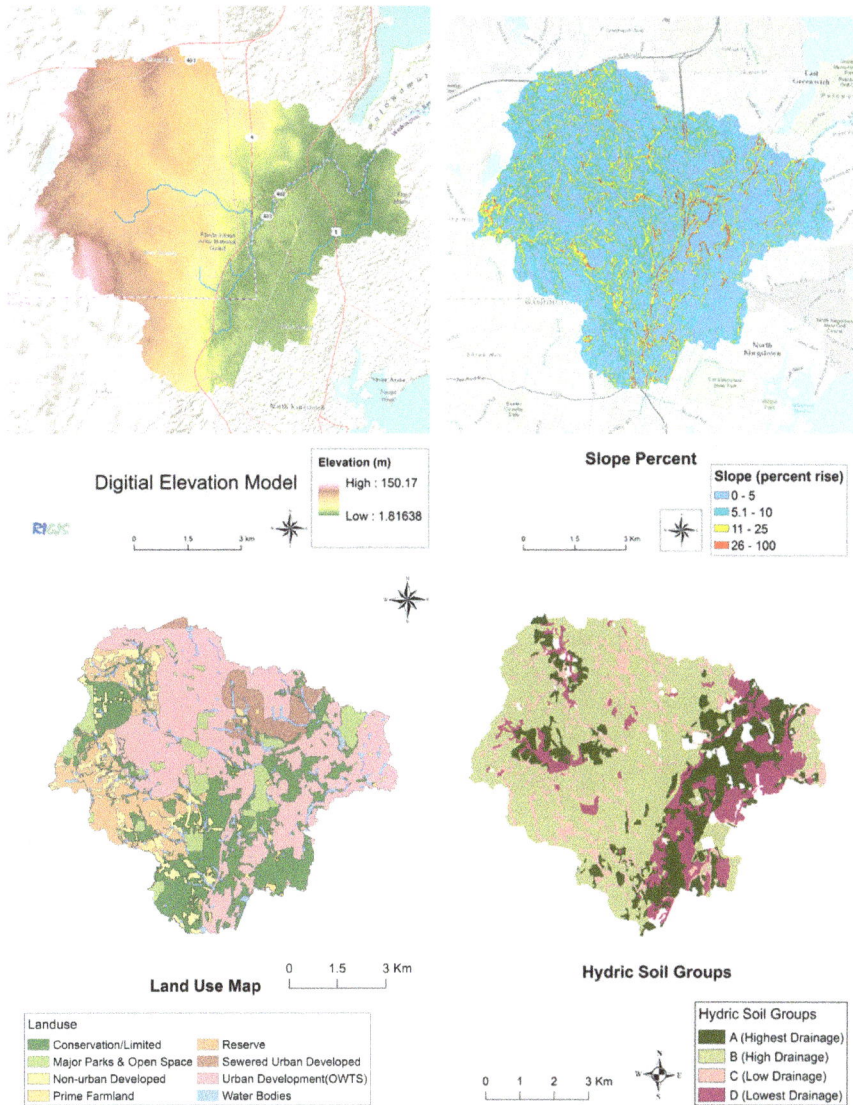

Digitial Elevation Model

Elevation (m)
High : 150.17
Low : 1.81638

0   1.5   3 km

Slope Percent

Slope (percent rise)
0 - 5
5.1 - 10
11 - 25
26 - 100

0   1.5   3 Km

Land Use Map

0   1.5   3 Km

Landuse
Conservation/Limited
Major Parks & Open Space
Non-urban Developed
Prime Farmland
Reserve
Sewered Urban Developed
Urban Development(OWTS)
Water Bodies

Hydric Soil Groups

Hydric Soil Groups
A (Highest Drainage)
B (High Drainage)
C (Low Drainage)
D (Lowest Drainage)

0   1   2   3 Km

(b)

Figure 1. Figure 1 shows the Hunt River Watershed as delineated by the Soil and Water Assessment Tool (SWAT) model: (**a**) The Hunt River Watershed delineated into six subbasins; (**b**) The Digital Elevation Model, Slope Percent, Land Use, and Hydric Soil Groups for the Hunt River Watershed.

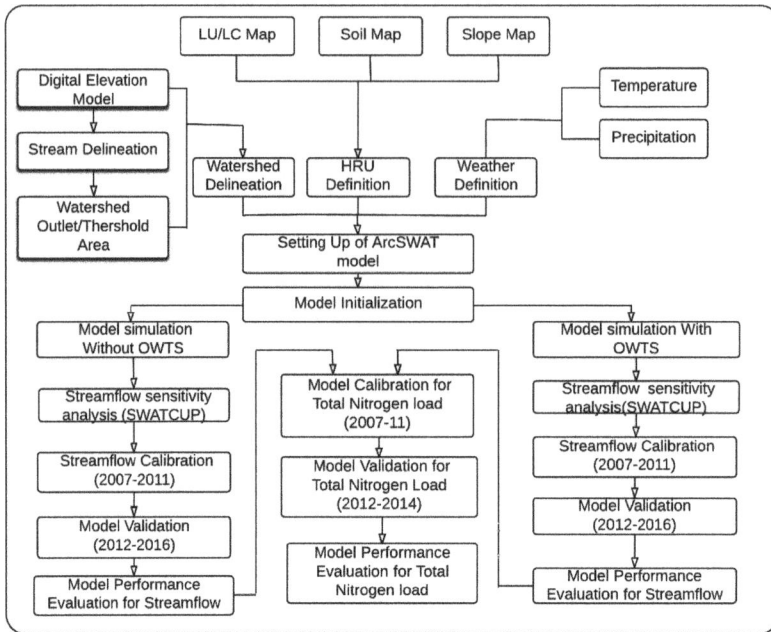

Figure 2. Flowchart of SWAT Model Development and Calibration Scheme.

2.5. Model Evaluation Performance

Model evaluation performance through statistical analysis is crucial. Evaluation recommendations prescribed by previous work [15,27] were used for this study. The three major categories of quantitative statistics were standard regression, Nash and Sutcliffe dimensionless, and error index [28,29]. Coefficient of determination (R^2), Nash and Sutcliffe Efficiency (NSE), and percent bias (PBIAS) were used for model prediction in this study [15]. Coefficient of determination (R^2) was used for standard regression analysis by describing the degree of collinearity between simulated and measured data. Values for the coefficient of determination range from −1 to 1, with 1 indicating a perfect linear relationship, and 0 indicating no linear relationship.

The NSE is a dimensionless statistical measure that indicates the magnitude of residual variance to measured variance through normalized statistics [28]. The NSE values may range between −∞ and 1, with 1 being completely inclusive. Values between 0.0 and 1.0 are generally acceptable values for SWAT models [15]. Percent bias (PBIAS) is an error index analysis that measures the average tendency of simulated data to be larger or smaller than their observed counterparts [30]. The optimal PBIAS value is 0.0, with lower values indicating accurate model simulation. Positive PBIAS values indicate model under-simulation bias, whereas negative values indicate overestimation bias [30].

3. Results

3.1. Model Sensitivity, Calibration and Validation

Nine hydrologic parameters were used in SWAT-CUP to calibrate streamflow. The parameters, fitted value, range of parameter and P-value are presented in Tables 2 and 3. In the model, 1000 iterations were used for a 6-year simulation including a one-year warm-up period, with the 48th iteration yielding the best results. The two simulations analyzed in this study were streamflow and N loadings with and without OWTSs. In both simulations, the hydrologic parameters selections were identical. Sensitivity of parameters is determined by identifying the lowest values for the parameter "*p*-Value". The "*p*-Value" serves as an indicator for sensitivity significance. By evaluating the *p*-Values, Tables 2 and 3 indicate that CN2 is sensitive both with and without OWTS simulation. Other sensitive parameters include GWQMN and REVAPMN for simulation without OWTS and GW_DELAY and GW_REVAP for simulation with OWTS. In both simulations, curve Number and groundwater parameter were common sensitive parameters and influenced more surface runoff and base flow. In this study, the BFlow separation tool was used to determine the initial Alpha_BF [31]. As shown in both Tables 2 and 3, a higher ALPHA_BF value indicates higher base flows and lower GWQMN resulted in lower base flows [11,16]. Finding optimal values for ALPHA_BF and GWQMN ensures that the streamflow is separated well to account for runoff and baseflow.

Table 2. Sensitivity analysis for the SWAT model without OWTS.

Parameter Name	Descriptions	Fitted Value	Range	*p*-Value
R__CN2	Initial Soil Conservation Service Runoff Curve	−1.11	−2–1.5	0.00
V__ALPHA_BF	Base flow alpha factor (days)	0.74	0–1	0.88
V__GW_DELAY	Groundwater delay (days)	30.62	30–450	0.37
V__GWQMN	Thresholds depth of water in the shallow aquifer required for return flow to occur (mm)	1.88	0–2	0.02
V__GW_REVAP	Groundwater "revap" coefficient	0.04	0.02–0.2	0.52
V__SURLAG	Surface lag factor	1.66	0–2	0.41
V__ESCO	Soil evaporation compensation factor	0.207	0–1	0.42
V__EPCO	Plant uptake compensation factor	0.08	0–1	0.45
V__REVAPMN	Threshold depth shallow aquifer for "revap" to occur (mm)	43.95	0–300	0.22

Note: "R" indicates that the existing parameter is added as a percentage of a given value and "V" is the existing parameter value replaced by a given value.

Table 3. Sensitivity analysis for the SWAT model with OWTS.

Parameter Name	Influence	Fitted Value	Range	*p*-Value
R__CN2	Management	−1.38	−2–1.5	0.00
V__ALPHA_BF	Groundwater	0.745	0–1	0.23
V__GW_DELAY	Groundwater	30.42	30–450	0.01
V__GWQMN	Groundwater	1.57	0–2	0.31
V__GW_REVAP	Groundwater	0.097	0.02–0.2	0.16
V__SURLAG	Basin	1.47	0–2	0.81
V__ESCO	HRUs	0.56	0–1	0.96
V__EPCO	HRUs	0.26	0–1	0.28
V__REVAPMN	Groundwater	84.9	0–300	0.79

Note: R indicates that the existing parameter is added as a percentage of a given value and "V" is the existing parameter value replaced by a given value.

Figure 3 illustrates the observed streamflow, simulated streamflow, precipitation, and 95 PPU. Table 4 shows all nitrogen parameters that were used in the total nitrogen calibration. Statistical analysis for model calibration and validation is shown in Tables 5 and 6. Model fit value is classified as "very good" [15] for stream calibration, with daily NSE of 0.75 and R^2 of 0.78 (Table 5). For the validation period, model fit was decreased with a NSE value of 0.4 and R^2 of 0.68. Overall, model performance was better for streamflow calibration in both scenarios. Meteorological forcing plays an important role in hydrologic model performance. Better spatial representation of precipitation data such as data from PRISM and temperature data ensures higher model performance compared to the NCEP dataset [32]. The average precipitation for the Hunt Watershed is 1270 mm per year. Our model's average precipitation value was overestimated by 1367 mm per year, which could be due to Rhode Island's 2010 flood event. With fitted parameters, the model was simulated again from January 2012 to August 2016 for validation. Validation indicated that simulated flow was overestimated by 4% which is indicated by smaller NSE values compared to the calibration period. This change in model performance could be attributed to the presence of a dry year in the validation period; however, the model captured peak flow and remained consistent with the precipitation pattern. According to the 95 PPU plot (Figure 3), the uncertainty band was narrow for stream flow's calibration period, ranging between 34% and 50% on a daily basis.

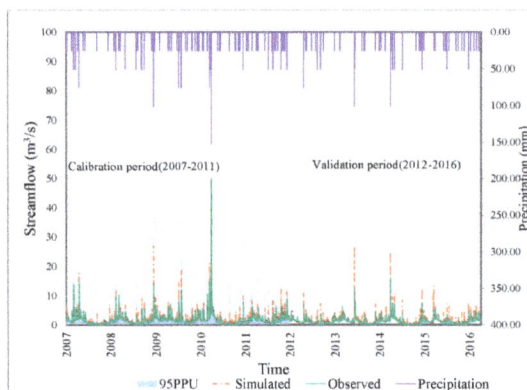

Figure 3. Flow calibration and validation in the Hunt River Watershed.

Total nitrogen was calibrated and validated after streamflow was calibrated. Nitrogen data, obtained from URIWW, was collected weekly from the months of May through October of each year. Figure 4 show the calibration and validation of total nitrogen load using SWAT model. NSE, percent bias, and coefficient of determination were calculated on a daily basis dependent on available data from URIWW. A few particular point data were overestimated; however, the overall data fit was acceptable. The model performance for total nitrogen loads for both simulations is shown in Tables 5 and 6. For the calibration period in the presence of OWTS, the model fit was satisfactory [15] with NSE of 0.5, R^2 of 0.5, and PBIAS of around 4%. For the validation period in the presence of OWTS, the model fits were very good with NSE of 0.78, R^2 of 0.81 and PBIAS of around 7%. Without the presence of OWTS, model fit was poor with R^2 of 0.14, negative NSE values, and high percentage bias. According to Moriasi, negative NSE, high percentage bias, and $R^2 < 0.5$ are considered as very poor calibration [15]. The septic tank is a better reflection of nitrogen over the watershed.

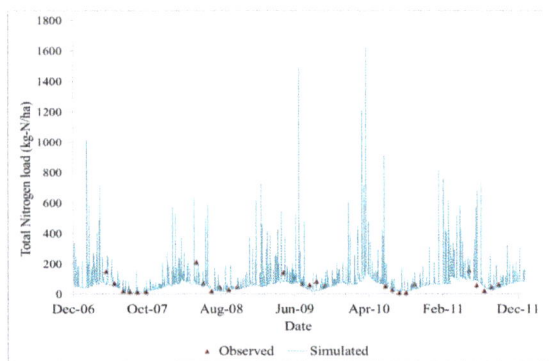

(a)

Figure 4. *Cont.*

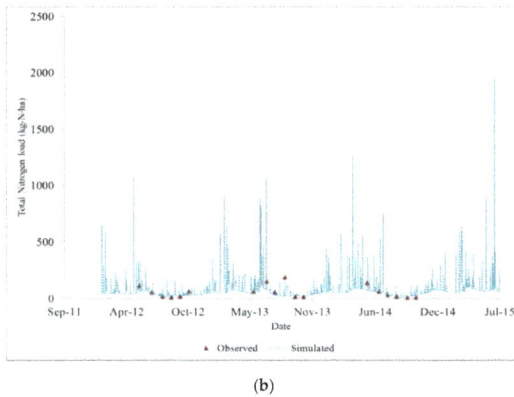

(b)

Figure 4. Figure 4 shows Total Nitrogen simulation from the SWAT model using OWTS simulations: (a) SWAT calibration from December 2006 to December 2011; (b) SWAT validation from September 2011 to July 2015.

Table 4. Total nitrogen calibration parameters.

Name	Parameters	Fitted Value
Nitrogen percolation coefficient	V__NPERCO.bsn	0.2
Denitrification rate coefficient	V__COEFF_DENITR.sep	0.32
Denitrification exponential rate coefficient	R__CDN.bsn	1.4
Denitrification threshold water content	R__SDNCO.bsn	1.1
Nitrification rate coefficient	V__COEFF_NITR.sep	1.5

Table 5. Without OWTS input.

Statistical Metric	Flow Calibration	Flow Validation	Total Nitrogen Calibration	Total Nitrogen Validation
R^2	0.79	0.7	0.14	0.15
NSE	0.78	0.5	−1.3	−6.95
PBIAS	−8	−12	72	−36

Table 6. With OWTS input.

Statistical Metric	Flow Calibration	Flow Validation	Total Nitrogen Calibration	Total Nitrogen Validation
R^2	0.78	0.68	0.5	0.81
NSE	0.75	0.4	0.5	0.78
PBIAS	−18.4	−30	3.95	6.95

3.2. Influence of OWTS on Hydrologic Cycle

Figure 5a shows the percent increase in groundwater contributing to streamflow with and without OWTS scenarios. When OWTS data was included, groundwater contribution to streamflow at the outlet was increased for each year of simulation. The annual precipitation from 2008 to 2011 was roughly 1500 mm. From 2008 to 2011, the percentage increase in groundwater contribution to streamflow was low compared to from 2012 to 2016. Therefore, groundwater contribution to streamflow due to septic systems decreased in wet years and increased during dry years. There was a 30% increase in groundwater during 2012, which was known to be a relatively dry year. Results also showed that without the presence of OWTS, groundwater contribution to streamflow decreased in each year of simulation. Therefore, it can be concluded that the influence of OWTS was high on the partition of total streamflow into its runoff and baseflow component. The study also showed that the presence of OWTS has much less influence on the groundwater during wet year than dry year. Figure 5b shows that comparison of annual total nitrogen (TN) load in each year of simulation for both scenarios. The TN loads in each year of simulation for OWTS scenarios were twice as high as scenarios run without OWTS input.

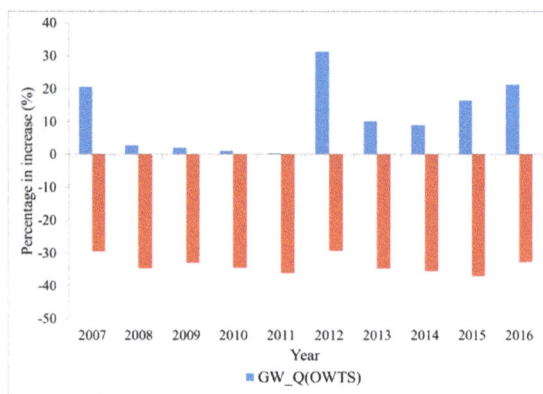

(a)

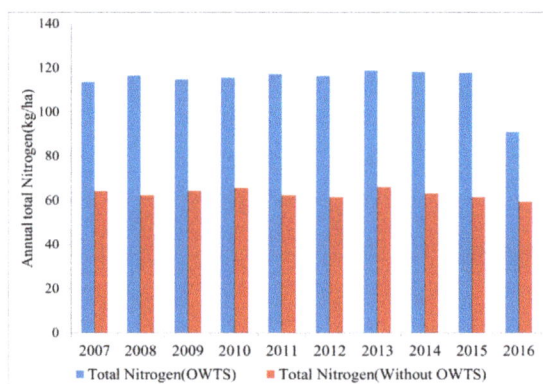

(b)

Figure 5. Plots of groundwater flow and total nitrogen inputs. Both charts are from 1 January 2007 to 31 August 2016. (**a**) Plot of annual percent increase in the groundwater contribution to streamflow between model simulation with and without OWTS at the subbasin scale; (**b**) Plot of annual total nitrogen with and without OWTS.

4. Discussion

Streamflow and TN load calibration and validation were found to be successful using the SWAT model for the Hunt River Watershed. The model effectively captured nonpoint nitrate inputs from OWTSs through hydrologic modeling of upstream and instream processes. By representing the area under the septic system, using available sewer data and TN loads data, a more accurate model of anthropogenic nitrogen inputs into the Hunt River Watershed was able to be developed. This study showed the importance of groundwater contributions to streamflow and nitrogen loading and the findings are corroborated by research conducted by Jeong et al. and Hoghooghi et al. [11,14].

According to Jeong et al. [11] and Hoghooghi et al. [14], the contribution of groundwater to streamflow is relatively high during dry years, similar to the research findings of this study.

Since the Hunt River Watershed is listed among the highest priority impacted waterbodies within the state of RI, the ability to effectively model nonpoint nitrogen pollution is essential. Accurate predictions of streamflow and nutrient simulations are important to effectively manage water quality in the receiving watershed and therefore play an important role in nitrogen mitigation and decision-making processes. The Hunt River Watershed's proximity to Narragansett Bay emphasizes the need for analyzing the impacts of nitrogen loading since this estuary has historically faced threats from eutrophication [5,6]. While the Hunt River Watershed is predominantly forested, the urbanization concentrated near the outlet amplifies the impact of OWTS into the receiving waters. Since the state has significantly reduced the input of point source nitrogen pollution from sewage treatment plants, OWTSs are proportionally becoming a larger source of nitrogen into the bay [8]. Despite reductions in point sources to the bay, eutrophication remains a significant threat to Rhode Island's fisheries and tourism industries, thus managers and policy makers need accurate models for identifying areas to target nitrogen pollution mitigation.

The use of URIWW data to successfully model nutrients entering Narragansett Bay highlights the value of volunteer water quality monitoring programs for their long-term monitoring assessments. While Narragansett Bay is no stranger to pollution, Rhode Island lacks historical nutrient data at daily and statewide monitoring levels. The lack of data on these scales inhibits the ability for modelers to make predictions based on past trends. Successful calibration of the Hunt River Watershed model could largely be attributed to the availability and use of URIWW data.

While the Nash–Sutcliffe index reflects strong calibration and validation for the model, several changes could be made to improve model fit. Limited knowledge of other nitrogen inputs into the watershed prohibited higher calibration achievement. As listed in the Hunt River TMDL, multiple point and nonpoint sources exist as major pollution contributors to the watershed. This model could be expanded upon by evaluating other potential sources of total nitrogen (TN) including lawn fertilizers, animal waste, and agricultural pollution. Another source of pollution that was not considered in this study includes backflow from the Potowomut Cove, which has the potential to be a significant contribution to the overall nitrogen budget within the area [33]. Additional nitrogen input sources need to be evaluated for higher model calibration. Another source of potential error is a lack of data on OWTS health, and how failing OWTS contribute to nitrogen loading within the watershed. Currently, little data exists on the overall health of Rhode Island's OWTSs. Additionally, RI has begun implementing and requiring advanced nitrogen removal OWTS that remove nitrogen through bacterial processes. However, there is still uncertainty on their magnitude of effectiveness and there is limited information on how many of these septic systems are currently installed. In the future, this model could be improved upon as more data becomes available on the status and health of OWTSs and other sources of pollutants.

Overall, SWAT and SWAT-CUP can be used as a tool that supports land use management decisions. Understanding how OWTS alter nitrogen and streamflow inputs into a small-scale watershed allows for analysis of base restoration efforts on numerous user end-benefits. SWAT models are largely acclaimed for their scenario modeling systems [34]. This research did not explore projected changes via modeling scenarios; however, further research efforts could include scenario modeling due to anticipated changes in climate patterns and urbanization. Overall, the model results imply that OWTS input provides a better estimate of nitrogen loading, resulting in more accurate simulations of pollutant loading into Potowomut Cove. Better prediction models are therefore important to provide informed decision-making tools for the watershed managers and regulators.

5. Conclusions

In this study, nitrogen nonpoint source pollution from OWTSs was modeled within the Hunt River Watershed using the SWAT model. Although the Hunt River Watershed has urbanization, it is served predominantly by OWTSs. Given that the waterbodies associated with the Hunt River Watershed are listed among the highest priority impacted waters in RI, it is important to quantify the impact that OWTSs have within this watershed. This study found that the presence of OWTSs increased nitrogen loading within the watershed and the model fit increased for simulating TN loading with the presence of OWTSs. The daily NSE TN load was 0.5 for the 6-year calibration period and 0.8 for the 6-year validation period. Narragansett Bay is no stranger to nitrogen pollution and often faces threats to its local tourism and fishing economies due to eutrophication stemming from nitrogen pollution. Modeling nitrogen pollution is essential to help quantify nitrogen loading into the bay to identify target areas for possible management and mitigation. This study highlights the utility of using SWAT to model nitrogen pollution from nonpoint sources. Nonpoint sources of nitrogen pollution are proportionally increasing as contributors to nutrient enrichment in coastal waters as point sources of nitrogen loading are decreasing within the state. More research is needed to monitor nonpoint nitrogen sources throughout Narragansett Bay to strengthen current models and identify priority locations for mitigation and management.

Acknowledgments: The authors would like to thank Linda Green and Elizabeth Herron of URI Watershed Watch; Jose Amador of the Department of Natural Resource Sciences; University of Rhode Island; and Anne Kuhn of the Environmental Protection Agency. The authors would also like to thank the S-1063 USDA Multistate Hatch Project for research support.

Author Contributions: Supria Paul, Michaela A. Cashman, Katelyn Szura and Soni M. Pradhanang conceived and designed the experiments; Supria Paul. and Soni M. Pradhanang designed the SWAT model; Michaela A. Cashman and Supria Paul analyzed the data; Supria Paul contributed analysis tools; Michaela A. Cashman, Katelyn Szura and Supria Paul wrote the paper.

Conflicts of Interest: The authors declare no conflict of interest.

References

1. Diaz, R.J. Overview of hypoxia around the world. *J. Environ. Qual.* **2001**, *30*, 275–281. [CrossRef] [PubMed]
2. Altieri, A.H.; Gedan, K.B. Climate change and dead zones. *Glob. Chang. Biol.* **2015**, *21*, 1395–1406. [CrossRef] [PubMed]
3. Diaz, R.J.; Rosenberg, R. Spreading dead zones and consequences for marine ecosystems. *Science* **2008**, *321*, 926–929. [CrossRef] [PubMed]
4. Blank, R.; Mesenbourg, T. *2010 Census of Population and Housing Unit Counts*; U.S. Census Bureau: Washington, DC, USA, 2010.
5. Nixon, S.W.; Pilson, M.E.Q. Nitrogen in estuarine and coastal marine ecosystems. *Nitrogen Mar. Environ.* **1983**, *565*, 648.
6. Oczkowski, A.; Nixon, S.; Henry, K.; DiMilla, P.; Pilson, M.; Granger, S.; Buckley, B.; Thornber, C.; McKinney, R.; Chaves, J. Distribution and trophic importance of anthropogenic nitrogen in Narragansett Bay: An assessment using stable isotopes. *Estuaries Coasts* **2008**, *31*, 53–69. [CrossRef]
7. DEM, R.I. *The Greenwich Bay Fish Kill–August 2003: Causes, Impacts and Responses*; Rhode Island Department of Environmental Management: Narragansett, RI, USA, 2003.
8. Amador, J.A.; Loomis, G.; Lancellotti, B.; Hoyt, K.; Avizinis, E.; Wigginton, S. *Reducing Nitrogen Inputs to Narragansett Bay: Optimizing the Performance of Existing Onsite Wastewater Treatment Technologies*; Department of Natural Resources Science, New England Onsite Wastewater Training Program, Coastal Institute University of Rhode Island: Providence, RI, USA, 2017.
9. U.S. Environmental Protection Agency. *Onsite Wastewater Treatment Systems Manual*; USEPA, Ed.; U.S. Environmental Protection Agency: Washington, DC, USA, 2002; ISBN 9781467376556.
10. Freedman, P.L.; Pendergast, J.F.; Canale, R.P. Modeling storm overflow impacts on eutrophic lake. *J. Environ. Eng. Div.* **1980**, *106*, 335–349.

11. Jeong, J.; Santhi, C.; Arnold, J.G.; Srinivasan, R.; Pradhan, S.; Flynn, K. Development of algorithms for modeling onsite wastewater systems within SWAT. *Trans. ASABE* **2011**, *54*, 1693–1704. [CrossRef]

12. Hoghooghi, N.; Radcliffe, D.E.; Habteselassie, M.Y.; Clarke, J.S. Confirmation of the Impact of Onsite Wastewater Treatment Systems on Stream Base-Flow Nitrogen Concentrations in Urban Watersheds of Metropolitan Atlanta, GA. *J. Environ. Qual.* **2016**, *45*, 1740–1748. [CrossRef] [PubMed]

13. Oliver, C.W.; Radcliffe, D.E.; Risse, L.M.; Habteselassie, M.; Mukundan, R.; Jeong, J.; Hoghooghi, N. Quantifying the contribution of on-site wastewater treatment systems to stream discharge using the SWAT model. *J. Environ. Qual.* **2014**, *43*, 539–548. [CrossRef] [PubMed]

14. Hoghooghi, N.; Radcliffe, D.E.; Habteselassie, M.Y.; Jeong, J. Modeling the Effects of Onsite Wastewater Treatment Systems on Nitrate Loads Using SWAT in an Urban Watershed of Metropolitan Atlanta. *J. Environ. Qual.* **2017**, *46*, 632–640. [CrossRef] [PubMed]

15. Moriasi, D.N.; Arnold, J.G.; Van Liew, M.W.; Binger, R.L.; Harmel, R.D.; Veith, T.L. Model evaluation guidelines for systematic quantification of accuracy in watershed simulations. *Soil Water Div. ASABE* **2007**, *50*.

16. Neitsch, S.L.; Arnold, J.G.; Kiniry, J.R.; Srinivasan, R.; Williams, J.R. *Soil and Water Assessment Tool User's Manual*; Texas Water Resources Institute, College Station, Texas A&M Universitry: College Station, TX, USA, 2002.

17. Brown, L.; Barnwell, T. *The Enhanced Stream Water Quality Models QUAL2E and QUAL2E-UNCAS*; US Environmental Protection Agency, Office of Research and Development, Environmental Research Laboratory: Washington, DC, USA, 1987.

18. Weintraub, L.H.Z.; Chen, C.W.; Goldstein, R.A.; Siegrist, R.L. WARMF: A watershed modeling tool for onsite wastewater systems. In *Proceedings of the 10th National Symposium on Individual and Small Community Sewage Systems, ASAE*; American Society of Agricultural Engineers: St. Joseph, MI, USA, 2004; pp. 636–646.

19. Abbaspour, K.C. *User Manual for SWAT-CUP, SWAT Calibration and Uncertainty Analysis Programs*; Swiss Federal Institute of Aquatic Science and Technology: Eawag, Duebendorf, Switzerland, 2007; pp. 1596–1602.

20. Rhode Island Geographic Information System (RIGIS). Available online: http://www.edc.uri.edu/rigis/ (accessed on 10 May 2017).

21. Daly, C.; Halbleib, M.; Smith, J.I.; Gibson, W.P.; Doggett, M.K.; Taylor, G.H.; Curtis, J.; Pasteris, P.P. Physiographically sensitive mapping of climatological temperature and precipitation across the conterminous United States. *Int. J. Climatol.* **2008**, *28*, 2031–2064. [CrossRef]

22. Rhode Island Geographic Information System (RIGIS) Data Distribution System SOIL_soils. Available online: http://www.rigis.org/geodata/soil/Soils16.zip (accessed on 28 September 2016).

23. Rhode Island Geographic Information System (RIGIS) Data Distribution System Land use and Land Cover. Available online: http://data.rigis.org/PLAN/rilc0304.zip (accessed on 7 October 2007).

24. U.S. Geological Survey. National Elevation Dataset (NED) 1/3 Arc-Second. Available online: https://viewer.nationalmap.gov/basic/?basemap=b1&category=ned,nedsrc&title=3DEPView (accessed on 1 January 2014).

25. Rhode Island Geographic Information System (RIGIS) Data Distribution System Sewered Areas; sewerAreas12. Available online: http://www.rigis.org (accessed on 16 October 2014).

26. Green, L.T.; Herron, E.M. University of Rhode Island Watershed Watch. Available online: www.uri.edu/watershedwatch/ (accessed on 1 January 2017).

27. Santhi, C.; Arnold, J.G.; Williams, J.R.; Dugas, W.A.; Srinivasan, R.; Hauck, L.M. Validation of the swat model on a large RWER basin with point and nonpoint sources. *JAWRA J. Am. Water Resour. Assoc.* **2001**, *37*, 1169–1188. [CrossRef]

28. Nash, J.E.; Sutcliffe, J. V River flow forecasting through conceptual models part I—A discussion of principles. *J. Hydrol.* **1970**, *10*, 282–290. [CrossRef]

29. Legates, D.R.; McCabe, G.J. Evaluating the use of "goodness-of-fit" measures in hydrologic and hydroclimatic model validation. *Water Resour. Res.* **1999**, *35*, 233–241. [CrossRef]

30. Gupta, H.V.; Sorooshian, S.; Yapo, P.O. Status of automatic calibration for hydrologic models: Comparison with multilevel expert calibration. *J. Hydrol. Eng.* **1999**, *4*, 135–143. [CrossRef]

31. Arnold, J.G.; Allen, P.M. Automated methods for estimating baseflow and ground water recharge from streamflow records. *J. Am. Water Resour. Assoc.* **1999**, *35*, 411–424. [CrossRef]

32. Radcliffe, D.E.; Mukundan, R. PRISM vs. CFSR Precipitation Data Effects on Calibration and Validation of SWAT Models. *J. Am. Water Resour. Assoc.* **2017**, *53*, 89–100. [CrossRef]

33. DiMilla, P.A.; Nixon, S.W.; Oczkowski, A.J.; Altabet, M.A.; McKinney, R.A. Some challenges of an "upside down" nitrogen budget—Science and management in Greenwich Bay, RI (USA). *Mar. Pollut. Bull.* **2011**, *62*, 672–680. [CrossRef] [PubMed]

34. Gassman, P.W.; Reyes, M.R.; Green, C.H.; Arnold, J.G. The soil and water assessment tool: Historical development, applications, and future research directions. *Trans. ASABE* **2007**, *50*, 1211–1250. [CrossRef]

water

MDPI

Article

Comparison of SWAT and GWLF Model Simulation Performance in Humid South and Semi-Arid North of China

Zuoda Qi [1], Gelin Kang [1], Chunli Chu [1,*], Yu Qiu [1], Ze Xu [2] and Yuqiu Wang [1,*]

[1]   MOE Key Laboratory of Pollution Processes and Environmental Criteria, College of Environmental Science and Engineering, Nankai University, Tianjin 300350, China; 2120140471@mail.nankai.edu.cn (Z.Q.); 2120150515@mail.nankai.edu.cn (G.K.); 2120150502@mail.nankai.edu.cn (Y.Q.)

[2]   Department of Environmental Science and Engineering, Nankai University Binhai College, Tianjin 300272, China; xuze@mail.nankai.edu.cn

*   Correspondence: chucl@nankai.edu.cn (C.C.); yqwang@nankai.edu.cn (Y.W.); Tel.: +86-22-23501117 (C.C.); +86-22-8535-8068 (Y.W.)

Received: 31 May 2017; Accepted: 27 July 2017; Published: 30 July 2017

Abstract: Watershed models have gradually been adapted to support both decision and policy making for global environmental pollution control. In this study, two watershed models with different complexity, the Soil and Water Assessment Tool (SWAT) and the Generalized Watershed Loading Function (GWLF), were applied in two catchments in data scarce China, namely the Tunxi and the Hanjiaying basins with contrasting climatic conditions (humid and semi-arid, respectively). The performances of both models were assessed via comparison between simulated and measured monthly streamflow, sediment yield, and total nitrogen. Time series plots as well as four statistical measures (the coefficient of determination (R^2), the Nash–Sutcliffe efficiency (NSE), percent bias (PBIAS), and RMSE (root mean square error)—observations standard deviation ratio (RSR)) were used to estimate the performance of both models. The results show that both models were generally able to simulate monthly streamflow, sediment, and total nitrogen loadings during the simulation period. However, SWAT performed better for detailed representations, while GWLF could produce much better average values of the observed data. Thus, GWLF offers a user-friendly prospective alternative watershed model that requires little input data and that is applicable for areas where the input data required for SWAT are not always available. SWAT is more suitable for projects that require high accuracy and offers an advantage when measured data are scarce.

Keywords: SWAT; GWLF; watershed modeling; comparison; streamflow; sediment; total nitrogen

1. Introduction

China is the biggest developing country in the world, and its rapid economic development has resulted in a large number of significant water quality issues such as eutrophication of lakes and reservoirs, deterioration of river water, and groundwater pollution [1,2]. To resolve these environmental issues, the Chinese government gradually resorted to mathematical models to provide a scientific basis for quantitative environmental management rather than exclusively depending on empirical qualitative analyses [3]. Currently, numerous watershed models with various capabilities are widely used in hydrological research and environmental resource management around the world [4,5]. These are powerful tools that enable us to understand the natural processes, as well as to find solutions for problems, while assessing the environmental conditions on a the watershed scale [6]. However, typically, there is a trade-off between model complexity, input data availability, and prediction ability in a certain application objective [7]. Butts et al. developed a hydrological modeling framework that allows for the application of different model structures by providing varying levels of model complexity. The authors reported that an increase of model complexity did not increase model performance for a number of investigated cases. Accordingly, different models with different complexities had to be selected for an exploration of the applicability of watershed models.

SWAT is a semi-distributed and physical-based hydrological model, which has evolved from multiple previous models over more than 30 years [8,9]. Considerable applications in a wide range of regions and environmental conditions have indicated SWAT to be an effective and acceptable tool both for scientific research and policy making [10]. It has been extensively implemented throughout the world, e.g., in America [11], Africa [12], and Australia [13]. In China, it has been used in the Chaohe basin in the north of China [14], the Heihe basin in the west of China [15], and the Three Gorges Reservoir Region in the south of China [16]. The primary categories to which SWAT has been applied include hydrologic assessments [17,18], pollutant assessments [19], and climate change impacts [20,21]. The GWLF is a simpler, continuous process-based model, which has been used in America [22], Ireland [23], and China [24] for various purposes. The Ministry of Environmental Protection of China has endorsed the GWLF as an alternative model to promote water quality and to meet environmental quality standards [25]. Both models were used to support the development of Total Maximum Daily Loads (TMDLs) [26]. Due to their wide applicability, acceptance by the authorities, as well as their different complexities, SWAT and GWLF were selected and compared in China for regions where monitoring networks are incomplete compared to developed countries.

There have been many studies that compared watershed models. Li et al. [27] compared the conceptual, lumped Water and Snow balance MODeling system (WASMOD) model to SWAT for the Yingluoxia watershed and found that MASMOD provided the same, or even better results than SWAT for the simulated hydrograph. Parajuli et al. [28] employed both the Annualized AGricultural Non-Point Source (AnnAGNPS) and SWAT in south-central Kansas and their study indicated SWAT as the most appropriate model for this particular watershed. Wilcox et al. [29] simulated the runoff on six uncalibrated catchments using both a simple model and a complex model. Although their results demonstrated that more complex catchment models yield more accurate results, the superiority of complex models is not immutable for all watersheds. These studies show that different models lead to different performance in different applications. A model comparison without considering the regional differences is easily one-sided. Niraula et al. [30] applied the SWAT and GWLF models in east central Alabama to identify critical source areas (CSAs) of sediment and nutrients. Both models performed well for streamflow; however, SWAT slightly outperformed GWLF for sediment, total nitrogen (TN), and total phosphorus (TP). The purpose of their study was to assess whether different model choice would lead to a variance in the locations of CSAs and the authors did not conduct a comprehensive comparison between the simulation results of SWAT and GWLF. Moreover, the authors conducted the models on one site only, suggesting limited implications. Therefore, the objective of this study was to conduct a comprehensive comparison between SWAT and GWLF and to evaluate their applicability in two catchments with different climate, landuse, and soil type for monthly stream flow, sediment, and total nitrogen in the data scarce China.

2. Materials and Methods

2.1. Study Sites

The study sites were chosen based on data availability and differences in climate, landuse and soil types (Figure 1). The Tunxi catchment is located in Anhui Province, which was selected to represent the humid south of China. It covers an area of approximately 2674 km^2 with forest covering 74%, agriculture area 15.8%, urban 4.6% and others. Red soil (55%), paddy soil (13%), and purple soil (9.8%) are the predominant soil types. The basin had a subtropical humid monsoon climate with a mean annual temperature of 15.5 °C and a mean annual precipitation of 1752 mm during the period from 1993 to 2013. The daily temperature was always above 0 °C. The 6736 km^2 of the Hanjiaying basin were selected as a representation of the semi-arid north of China. It is one of the largest subbasins of the Luan River watershed and located in Hebei Province, which is situated in the north of the Qinling Mountains-Huaihe River line. Forest (49%) and agricultural land (25%) are the major land uses within the basin. Brown soil (65%), and cinnamon soil (22%) are predominant in this watershed. The basin plays an important role for ecological servicing and water supply to the region. The climate is dominated by a temperate continental monsoon climate with a mean annual temperature of 5.62 °C and a mean annual precipitation of 446 mm from 1993 to 2013. Monthly mean temperatures range below 0 °C during the period from November to March and above 10 °C during the summer months (June–August).

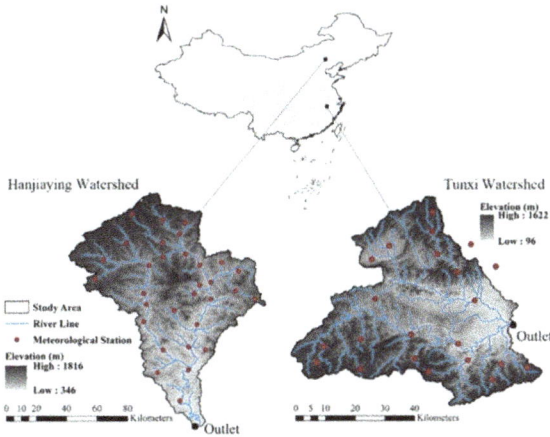

Figure 1. Location and elevation of the Hanjiaying and Tunxi watersheds.

Due to significant differences in meteorological conditions, these two sites are typical representatives of the semi-arid north and the humid south of China, respectively (Figure 2).

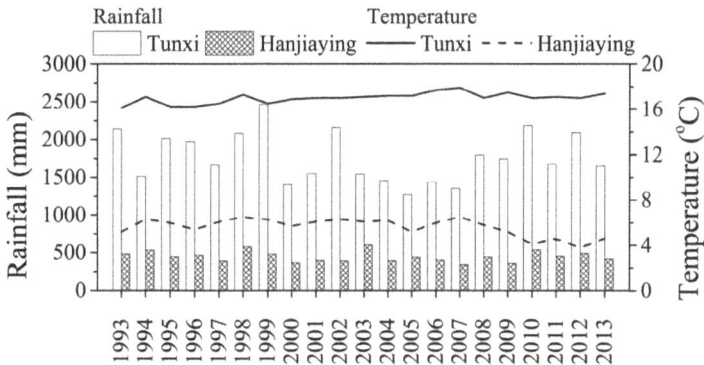

Figure 2. Mean annual rainfall and temperature of two sites between 1993 and 2013.

2.2. Watershed Models

SWAT is a distributed-parameter model, which was primarily designed by the Agricultural Research Service (ARS) of the United States Department of Agriculture (USDA) to assess the effect of land management practices on water, sediment, and agricultural chemical yields in large complex watersheds over extended periods of time [31]. GWLF is a combined distributed/lumped parameter model, which is based on a combination of simple runoff, sediment, and groundwater relationships and empirical chemical parameters [32]. Both SWAT and GWLF models are continuous, pollutant-loading models that operate with a daily time step.

Table 1 lists the major processes and related methods considered by SWAT and GWLF models. SWAT and GWLF differ greatly in the way in which they delineate the watershed. Based on the topological structure of river networks, SWAT first discretized the watershed into a number of subbasins, subsequently dividing each subbasin into hydrologic response units (HRUs) according to the unique land use, soil, and slope combinations [31]. In SWAT, each physical and chemical process is modeled at HRU scale within the subbasin and then routed along the river network toward the outlet of the watershed. However, the conception of subbasin does not exist in GWLF; therefore, it can only identify surface loading from different land covers and the results of each area are simply added into the watershed summation. In some sense, the model is distributed but lacks a spatial conception as well as a channel route component. For sub-surface modeling however, it is considered a lumped parameter model because it used uniform parameters for the entire watershed, ignoring the spatial variability of physical and chemical processes [33]. The differences in emphasis on simplifying the real environment lead to the diverse properties of various watershed models.

The hydrological process is the most important component in any watershed model as the drive force during the whole simulation. Both models simulate the hydrological component based on the water balance equation for the shallow aquifer. The SWAT model provides two methods to estimate surface runoff: the modified SCS-curve number and the Green-Ampt infiltration method. In this study, both models used different versions of SCS-CN to estimate the surface runoff volume, considering the remaining amount for infiltration [34]. The GWLF describes groundwater with the linear reservoir model, while SWAT uses empirical relationships. In addition, SWAT can calculate the lateral flow in the unsaturated zone.

Table 1. Summary of the major processes and related methods used by SWAT and GWLF.

Description	SWAT	GWLF
Model capabilities	Hydrology, sediment, nutrients, pesticides, bacteria, and other water quality factors; channel and reservoir routing, crop growth, transport in soil, management practices, and impoundment structures	Hydrology, sediment, and nutrients
Temporal resolution	Daily	Monthly
Spatial representation	Hydrologic response units	Landuse category
Interception	Water balance	-
Potential evapotranspiration (PET)	Penman–Monteith method; Priestley-Taylor method [35]; Hargreaves method [36]	Hamon method [37]
Runoff	SCS-CN [34]; Green and Ampt [38]	SCS-CN [39]
Infiltration/percolation	Water balance	Water balance
Lateral flow	Kinematic storage model [40]	-
Base flow	Empirical relations	Linear reservoir model [41]
River flow routing	Manning's equation; variable storage routing method or Muskingum river routing method	-
Sediment yield	Modified Universal Soil Loss Equation (MUSLE) [42,43]	Universal Soil Loss Equation [44]
Sediment channel routing	Simplified Bagnold model or physics based approach	-
Nitrogen forms in output	Organic nitrogen, nitrate, nitrite, ammonium, and total nitrogen	Dissolved nitrogen, solid-phase nitrogen, and total nitrogen
Atmospheric nitrogen deposition	Wet and dry deposition	Nitrogen deposition
Nitrogen cycle and transmission in land phase	Mass balance and empirical relation	-
Nitrogen cycle and transmission in routing phase	QUAL2E model [45]	-
Nitrogen load	Empirical relations	User defined concentration

In GWLF, erosion is simulated via the Universal Soil Loss Equation (USLE), which predicts the average erosion, using a function of rainfall energy [44]. Then, a sediment delivery ratio and transport capacities are applied to determine monthly sediment yield for each source area [33]. In contrast, SWAT uses a modified version of the Universal Soil Loss Equation (MUSLE), which introduces a runoff factor displacing energy factor to daily estimate erosion and sediment yield. A delivery ratio is not required and sediment yields of single storms can be calculated [42,43].

Both models are also quite different in the way they estimate nutrient loads. The GWLF simply calculates nutrient loads by multiplying N and P concentration coefficients with the runoff volume or sediment yield at a monthly scale. It uses denitrification loss fractions to calculate the denitrification amount. With the daily time step, SWAT models nutrient cycles via different pools to simulate their mineralization, decomposition, and immobilization between inorganic and organic forms within the soil. Then, the amount of mineral and organic nutrients transported in both land phase and routing phase is calculated.

In addition to these basic components, SWAT has the additional powerful ability to simulate crop growth, management, as well as the amount of pesticide, bacteria, algae, dissolved oxygen, carbonaceous biological oxygen demand (CBOD), and their routing in the channel or reservoir.

2.3. Model Inputs

Table 2 summarizes the data used for the model setup in this study. To avoid different results based on variations of model input data, we kept the input data of GWLF consistent with SWAT. The SWAT (Version 2012) and the ReNuMa (Regional Nutrient Management) (Version 2.2.2) modeling platform of GWLF (Version 2) were used. Thirty-meter resolution DEMs were used to determine the watershed and sub-watershed boundaries in SWAT and GWLF identified runoff source areas based on the same delineation. At both sites, land use data were used to obtain major cover classification information and SWAT needs extra spatial datasets. Soil datasets were only used in SWAT to partition the watershed into HRUs along with landuse and slope datasets. In SWAT, a combination of these three datasets divided the Tunxi watersheds into 40 subbasins and 307 HRUs, while it divided Hanjiaying into 33 subbasins and 258 HRUs. In GWLF, there were nine major landuse classes in Tunxi and seven in Hanjiaying. Meteorological data of each subbasin were obtained from the weather station nearest to its centroid for SWAT, while average climatic data were used for GWLF. Agriculture management information of the Tunxi watershed was referenced to [46] and obtained from the local government in Hanjiaying. Furthermore, population data were also required for the GWLF.

Table 2. Input data used in SWAT and GWLF.

Type of Data	SWAT	GWLF	Tunxi	Hanjiaying
DEM	Digital elevation map	Digital elevation map	30 m	30 m
Landuse	Grid	Proportion	1:100,000	1:100,000
Soil	Grid and properties	-	1:1,000,000	1:1,000,000
Meteorological data	Daily air temperature (maximum, minimum, average), daily precipitation, daily wind, daily solar radiation, daily relative humidity	Daily average temperature and precipitation	28 stations (2000–2013)	30 stations (2006–2014)
Flow discharge	monthly	Monthly	1 station (2000–2013)	1 station (2006–2014)
Sediment yield	monthly	Monthly	1 station (2000–2011)	1 station (2006–2014)
Nutrient load	monthly	Monthly	1 station (2002–2013)	1 station (2006–2014)

2.4. Model Calibration, Validation, and Evaluation

In the Tunxi watershed, the period from 2001 to 2008 was chosen for model calibration, and the data in 2000 were used as "warm up" to define appropriate initial conditions, and the latest five years from 2009 to 2013 were used for model validation of streamflow and total-nitrogen, while the sediment was validated from 2009 to 2011. For the Hanjiaying watershed, the periods of 2006–2011 and 2012–2014 were selected as the calibration and validation periods, respectively for flow, sediment, and total-nitrogen, while 2005 was used as the warm-up period. The simulation of SWAT and GWLF was conducted with a monthly time step and followed the calibration sequence: flow, sediment, and nitrogen.

Although multiple sets of parameters can obtain optimal fitting with the measured data, we only selected one of them as representation to facilitate comparison of both models. Tables 3 and 4 show the parameters that were chosen and defined in this study. In SWAT, a sensitivity analysis was conducted prior to model calibration and more than 20 major parameters were selected in Tunxi and Hanjiaying. Calibration was manually and automatically conducted via SUFI-2 uncertainty analysis through the SWAT-CUP program [47]. The SCS curve number (CN2) was the most critical parameter for both stations, which is directly related to the runoff yield. As the value of CN decreased, overland flow reduced, but infiltration potential increased. The base flow recession constant, αALPHA_BF, is a direct index of groundwater flow response to recharge from the vadose zone [48]. Values vary from 0.1–0.3 for land with slow response to recharge to 0.9–1.0 for land with rapid response. The SLSOIL was the key parameter, which we chose to adjust the lateral flow yield. By default, it is equal to the value of the average slope length of the subbasin (SLSUBBSN), which tends to result in a high lateral flow ratio. Therefore, we appropriately reduced its value for both sites. In Hanjiaying, two additional parameters were considered due to their influence on the snowmaking process. SMTMP defines the base temperature above which snowmelt is allowed. SNOCOVMX is the threshold depth of snow above which the basin would be completely (100%) covered with snow. The soil property parameter SOL_K was also included because the soil categories in the Hanjiaying basin are relatively coarse. Parameters related to groundwater balance and channel routing were also taken into account. Seven parameters were chosen to calibrate the sediment simulation with respect to erosion, maximal sediment amount, and routing in the channel. For nitrogen, four parameters about nitrite and one parameter about organic nitrogen were considered. Furthermore, we distributed several parameters depending on landuse, soil texture, and slope. When the calibration of one variable was completed, we retained an unchanged parameter range and began calibration of the next variable, unless results were not satisfactory [47].

For the GWLF, parameters related to watershed specific characteristics such as runoff source areas and populations were identified via GIS data analysis. Transport and nutrient parameters could be estimated using default coefficients according to [49]. In this study, we used them as initial values and manually calibrated them. A total of 11 parameters were selected for calibration. The meaning of each parameter is listed in detail in Tables 3 and 4. After model calibration, the values of input parameters remained unchanged during the validation process.

The model performance for fitting measured constituent data was qualitatively evaluated via time series plots and quantitatively evaluated via four widely used statistics in watershed model evaluation (Table 5).

The coefficient of determination (R^2) indicates the degree of linear relationship between simulated and observed data. A R^2 value close to one indicates a better performance. However, it is very sensitive to extremely high values. The Nash–Sutcliffe efficiency (NSE) is one of the most commonly used criteria [50]. This is a normalized statistic, which can be used to determine the goodness of fit. The NSE ranges from $-\infty$ to 1, with 1 indicating a perfect match. The squared difference in equation becomes the limitation of the NSE for overestimating higher values and neglecting lower values [51]. Percent bias (PBIAS) is an error index, generally used to measure the deviation of the constituent of data. It calculates the average tendency of the simulated data to be either larger or smaller than their observed counterparts with zero indicating the optimal value [52]. The RMSE (root mean square error)-observations standard deviation ratio (RSR) combines the feature of an error index RMSE and a normalization factor so that it can be applied to various constituents [53]. RSR ranges from the optimal value of 0 to infinity and the smaller the RSR, the better the simulation results will be. Model performance was judged based on statistics performance ratings as previously recommended [28,53].

Table 3. Parameters selected for the calibration for streamflow.

Model	Parameter Name	Description	Default Range	Calibrated Value	
				Tunxi Watershed	Hanjiaying Watershed
SWAT	CN2	Initial SCS Runoff curve number for moisture condition II	40–100	Varies (45–95) [1]	Varies (43–77) [1]
	GWQMN	Threshold depth of water in shallow aquifer required for the return flow to occur	0–500	364.2203	201.3421
	ALPHA_BF	Base flow alpha factor	0–1	0.7759	0.381446
	RCHRG_DP	Deep aquifer percolation factor	0–0.5	0.0493	0.042637
	ESCO	Soil evaporation compensation factor	0–1	0.6737	0.59
	CH_N2	Manning's "n" value for the main channel	0.01–0.3	0.0148	0.22892
	CH_K2	Main channel conductivity	0.01–500	102.495	407.831421
	SLSOIL	Slope length for lateral subsurface flow	10–120	Varies (21–99) [2]	Varies (145–258) [2]
	SNOCOVMX	Minimum snow water content that corresponds to 100% snow cover	0–500	-	12.482321
	SMTMP	Snowmelt base temperature	−20–20	-	−1.019328
	SOL_K	Saturated hydraulic conductivity	0–2000	-	Varies (7–66) [1]
GWLF	CN2	Initial SCS Runoff curve number for moisture condition II	0–100	Varies (45–100) [1]	Varies (23–98) [1]
	Recession coefficient	Groundwater discharge coefficient	0.1	0.25	0.0017
	Seepage coefficient	Groundwater seepage constant	0	0.044	0.0096
	Unsaturated available water	Available soil water capacity	-	8.75	17.54

[1] varied with landuse; [2] varied with slope.

Table 4. Parameters selected for the calibration for sediment.

Model	Parameter Name	Description	Default Range	Calibrated Value	
				Tunxi Watershed	Hanjiaying Watershed
SWAT	USLE_K	USLE equation soil erodibility (K) factor	0–0.65	Varied (0.12–0.60) [3]	Varied (0.09–0.54) [3]
	USLE_P	USLE equation support parameter	0–1	Varied (0.22–0.49) [2]	Varied (0.28–0.40) [2]
	PRF_BSN	Peak rate adjustment factor for sediment routing in the main channel	0–2	0.0483	0.795668
	SPEXP	Exponent parameter for calculating sediment re-entrained in channel	1–1.5	1.3454	1.171564
	SPCON	Linear parameter for calculating sediment re-entrained in channel	0.0001–0.01	0.0052	0.00255
	CH_COV1	Channel erodibility factor	−0.05–0.6	0.26	0.193029
	CH_COV2	Channel cover factor	−0.001–1	0.4776	0.26354
	CMN	Rate factor for humus mineralization of active organic nitrogen	0.0001–0.003	0.0014	0.0004
	CDN	Denitrification exponential rate coefficient	0–3	1.0039	0.004
	SDNCO	Denitrification threshold water content	0–1	0.3846	0.1541
	NPERCO	Nitrite percolation coefficient	0–1	0.3646	0.3441
	SHALLST_N	Concentration of nitrate in groundwater contribution to streamflow from subbasin	0–1000	0.735	25.5726
	ERORGN	Organic N enrichment ratio	0–5	0.0025	0.1253
GWLF	Sediment delivery ratio	Used to calculate sediment supply	-	0.039	0.1078
	Erosivity coefficient	Used to calculate rainfall erosivity	-	0.08 (November–February) [1] 0.45 (March–October)	0.1 (November–February) 0.27 (March–October)
	USLE parameter	Integrated parameter	-	Varied (0–0.14) [1]	Varied (0–0.1) [1]
	Urban N accumulation rate	-	-	0.045	0.1
	Nitrogen runoff coefficient	Rural runoff N concentration	-	Varied (0.4–2) [1]	Varied (0.5–4.5) [1]
	N (mg/L) in groundwater	N concentration in groundwater	-	0.6	7
	N (mg/kg) in sediment	N concentration in sediment	-	1000	2147

[1] varied with landuse; [2] varied with slope; [3] varied with soil texture.

Table 5. Statistics used to evaluate models.

Statistics	Excellent	Very Good	Good	Fair	Unsatisfactory
$R^2 = \dfrac{\left(\sum_{i=1}^{n}\left(Y_{obs,i}-\overline{Y}_{obs}\right)\left(Y_{sim,i}-\overline{Y}_{sim}\right)\right)^2}{\sum_{i=1}^{n}\left(Y_{obs,i}-\overline{Y}_{obs}\right)^2\sum_{i=1}^{n}\left(Y_{sim,i}-\overline{Y}_{sim}\right)^2}$	(0.90, 1]	(0.75, 0.9]	(0.65, 0.75]	(0.50, 0.65]	(0, 0.5]
$NSE = 1 - \dfrac{\sum_{i=1}^{n}\left(Y_{sim,i}-\overline{Y}_{sim}\right)^2}{\sum_{i=1}^{n}\left(Y_{obs,i}-\overline{Y}_{obs}\right)^2}$	(0.90, 1]	(0.75, 0.9]	(0.65, 0.75]	(0.50, 0.65]	$(-\infty, 0.5]$
$RSR = \dfrac{\sqrt{\sum_{i=1}^{n}\left(Y_{obs,i}-Y_{sim,i}\right)^2}}{\sqrt{\sum_{i=1}^{n}\left(Y_{obs,i}-\overline{Y}_{obs}\right)^2}}$	[0.00, 0.25)	[0.25, 0.50)	[0.50, 0.60)	[0.60, 0.70)	[0.70, +∞)
$PBIAS = \dfrac{\sum_{i=1}^{n}\left(Y_{obs,i}-\overline{Y}_{sim,i}\right)\times100}{\sum_{i=1}^{n}Y_{obs,i}}$	[0, 5) [0, 5) [0,10)	[5, 10) [5, 15) [10, 25)	[10, 15) [15, 30) [25, 40)	[15, 25) [30, 55) [40, 70)	[25, +∞) [55, +∞) [70, +∞)

3. Results and Discussion

3.1. Flow

Figure 3 illustrates a comparison between observed and simulated monthly mean streamflow series of both SWAT and GWLF models in two sites; the numerical criteria of model performance are summarized in Table 6.

In the Tunxi watershed, SWAT and GWLF almost replicated the entire trend of the discharge hydrograph with the simulated peak values and low flows consistently and perfectly matching the observed data throughout all years (Figure 3a). The high R^2 and NSE (above 0.9) values and the reasonably low RSR (below 0.25) for the calibration and validation periods indicate the excellent correlation and agreement between measured and simulated runoff for both models. Both SWAT and GWLF models underestimated streamflow by 9.69% and 4.03% during calibration, respectively, while overestimating the flow volume by 1.17% and 2.97% during the validation period. The average runoff simulated by SWAT and GWLF were both close to the average of observations. For the Hanjiaying watershed, the performance of both models degraded compared to the results for Tunxi. The shape of the monthly hydrograph was largely reproduced and relatively large fluctuations were found for the simulation of peak and low flows, contrasting with the measured data (Figure 3b). Based on the similar values of R^2, NSE, and RSR between SWAT and GWLF, both models were equally able to predicted monthly streamflow during the entire duration of the simulation. However, GWLF produced marginally better PBIAS values and slightly more accurate average monthly flow than SWAT, especially during the validation period. According to these results, both models had an almost equal ability to simulate the monthly streamflow with sufficient accuracy after adequate calibration. Furthermore, the average runoff simulated by SWAT and GWLF were both close to the average of the observations.

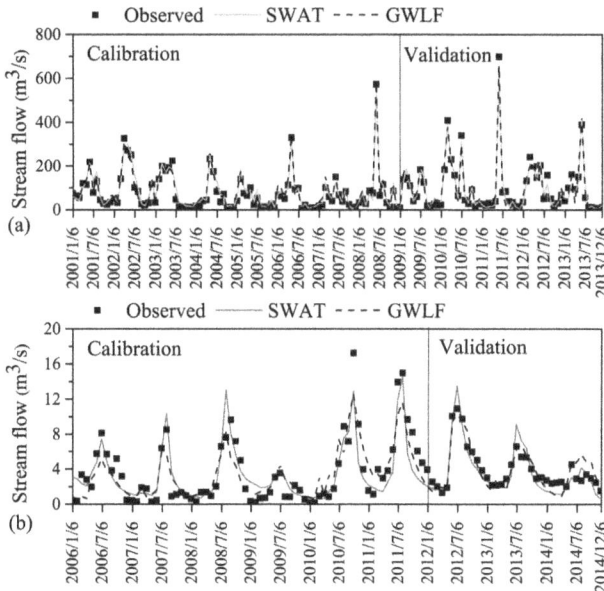

Figure 3. Simulated and observed monthly streamflow for: (a) Tunxi watershed; and (b) Hanjiaying watershed.

The critical reason for why the performance of both models was highly consistent at the same site is that the same runoff calculating method (SCS CN) was utilized in both models. Furthermore, the distinctly different behavior between the Tunxi and Hanjiaying watersheds of both models indicates that the SCS is more suitable for areas with high flow. Some previous applications in areas with less runoff yielded relatively poor statistics. Shen et al. [54] obtained a NSE of 0.711 and 0.690 during calibration and validation periods for the monthly runoff of the Three Gorges Reservoir with mean monthly observed values below 0.05 m^3/s. Parajuli et al. [28] obtained a NSE of 0.56 and a PBIAS of -95.06 in Red Rock Creek with normal flow volume below 1 m^3/s. Li et al. [27] obtained a NSEs of 0.948 and 0.923, and REs of -0.071 and -0.084 during calibration and validation periods for the Heihe River basin in China with mean monthly observed runoff above 49 m^3/s. Other publications reported that the performance of SWAT and GWLF in simulating low flows is not as useful as those of high or normal flows [22,55]. In fact, Chahinian et al. [56] compared four different infiltration-runoff models and all tested models had difficulties simulating low runoff events and even events characterized by a mild rainfall hiatus. Furthermore, the authors contributed this phenomenon to the absence of soil moisture re-distribution during flood events and to a constant value during the whole duration of the flood event.

For the calibrated parameters in SWAT, Tunxi had a higher CN2; thus, more streamflow was generated than in Hanjiaying, which is perhaps due to more abundant rainfall of Tunxi. The GWQMN is considerably higher in the Tunxi watershed than in Hanjiaying, indicating that Tunxi has more groundwater storage. The higher ALPHA_BF in Tunxi suggests a more rapid response to recharge entering the aquifers than in Hanjiaying, which was further confirmed by the higher value of recession coefficient in Tunxi of GWLF. The higher CH_K2 in Hanjiaying implies that its channel was easier to loose water via transmission when there is no groundwater contribution. As for GWLF, CN2 is also higher in Tunxi than Hanjiaying, which is consistent with SWAT. The parameter, unsaturated available water, is mainly related to soil property. Red soil and brown soil are the main soil types of Tunxi and Hanjiaying respectively. Brown soil is usually formed through eluviation and clayization processes and has thus poor water permeability and good water holding capacity [57]. Hanjiaying has a higher value of unsaturated available water than Tunxi, partially indicating that more water can be sorted in brown soil than in red soil. As a whole, the variances among these parameters of both models consistently reflect differences in hydrological processes under different catchments to some extent. However, these differences still need to be experimentally verified.

3.2. Sediment

Figure 4 shows a graphical representation of the predicted and measured sediment yield on a monthly basis. Furthermore, the numerical criteria of model performance in simulating sediment load are summarized in Table 6.

In the Tunxi watershed, both models adequately simulated the trend of monthly sediment yield, but tended to underestimate extremely high values. Furthermore, the GWLF performed worse than SWAT in tracking peak timing (Figure 4a). In summary, both models showed very good correlations and sufficient agreement between monthly measured and predicted sediment values according to statistical criteria, except for PBIAS. During the calibration time, the GWLF model performed slightly better than the SWAT model, based on the same R^2, higher NSE, lower RSR, and lower PBIAS. During the validation process, SWAT responded noticeably better than during the calibration period, while the performance of the GWLF did not show an apparent improvement. In addition, the PBIAS degrees for both models did not agree with other criteria and the values of SWAT were always higher than those of GWLF throughout entire periods, indicating a higher bias to predict sediment. In the Hanjiaying watershed, the performance of both models decreased compared to the results for Tunxi. The trend shape of the monthly sediment was roughly represented and there were large fluctuations for the simulation of peak and low flows compared to the measured data (Figure 4b). In general, both models equally predicted monthly sediment loads with reasonable accuracy during the entire simulation time based on the approximately identical values of R^2, NSE, and RSR. In addition, the performance of both models during the validation period increased compared to the calibration period. Furthermore,

SWAT performed marginally better than GWLF to some extent; however, this difference was so small that it was negligible. Furthermore, the average monthly sediment yield simulated by SWAT was much higher and closer to the observed values than for GWLF. According to the analysis above, both models were capable to predict the monthly sediment yield with adequate accuracy after sufficient calibration and SWAT was more reliable during the validation period.

The similarity of the results of both models suggests that the difference between MUSLE and USLE is not apparent in simulating monthly sediment loads, which has previously been suggested [54]. The good representation and increased performance the SWAT model during calibration and validation periods may be attributed to the distributed property assessing spatial variations of the study sites. In Tunxi catchment, the consistent performance of GWLF was partially achieved due to its capability allowing sediment delivery ratio to be calibrated during different months. It simulated the peak values of sediment between April and July during calibration period reasonably, whereas it did not capture the peak values in February 2009 and March 2010 during validation periods. This indicates that the GWLF lacks adequate flexibility in case when evident difference exists between calibration and validation observed data, mainly due to its simple sediment parameters. Furthermore, errors of manual measurement and adaption of empirical calculating equation could also affect the performance of sediment in both models.

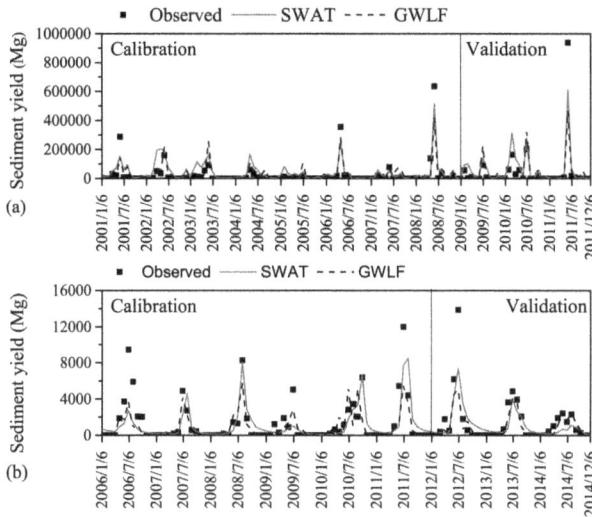

Figure 4. Simulated and observed monthly sediment yield for the: (**a**) Tunxi watershed; and (**b**) Hanjiaying watershed.

Table 6. Statistics values of model performance.

Statistics	Tunxi Watershed						Hanjiaying Watershed					
	Flow		Sediment		TN		Flow		Sediment		TN	
	SWAT	GWLF	SWAT	GWLF	SWAT	GWLF	SWAT	GWLF	SWAT	GWLF	SWAT	GWLF
Calibration Period												
Mean Observed	78.8		24394.8		248689.4		3.6		1340.3		80542.8	
Mean Simulated	86.4	82.0	43,403.3	29,324.6	137,255.9	24,1884.4	3.4	3.4	1140.1	929.8	65,820.1	85,912.1
R^2	0.95	0.96	0.75	0.75	0.89	0.88	0.78	0.80	0.57	0.59	0.81	0.79
NSE	0.94	0.95	0.68	0.74	0.65	0.87	0.78	0.77	0.57	0.54	0.77	0.77
RSR	0.24	0.21	0.56	0.51	0.58	0.35	0.47	0.48	0.65	0.67	0.48	0.47
PBIAS	−9.69	−4.03	−77.92	−20.21	39.85	−9.32	5.31	4.97	14.93	30.63	18.28	−6.67
Validation Period												
Mean Observed	100.5		46960.5		319899.5		3.9		1417.1		84859.2	
Mean Simulated	99.3	97.5	63,640.2	38,824.1	179,762.7	283,900.1	3.6	3.9	1107.2	953.8	76,385.6	76,626.8
R^2	0.96	0.96	0.84	0.74	0.85	0.88	0.87	0.82	0.79	0.76	0.70	0.35
NSE	0.96	0.96	0.80	0.67	0.60	0.86	0.77	0.78	0.68	0.61	0.72	0.57
RSR	0.21	0.21	0.44	0.56	0.63	0.36	0.47	0.46	0.56	0.61	0.52	0.65
PBIAS	1.17	2.97	−35.51	17.32	43.81	11.25	8.98	−0.50	21.86	32.69	9.99	9.70

3.3. Total Nitrogen

Time series plots and numerical criteria of simulated and measured total nitrogen loads are summarized in Figure 5 and Table 6, respectively.

For the Tunxi watershed, both the SWAT and GWLF models produced acceptable fluctuations in comparison to the observed data, while the peak values tended to be underestimated by SWAT in particular (Figure 5a). Although the R^2 value of SWAT was similar to that of GWLF, the GWLF model outperformed SWAT remarkably during both calibration and validation periods based on NSE, RSR, and PBIAS. The GWLF constantly predicted the monthly TN loadings with very good accuracy: R^2 and NSE were above 0.8, RSR was above 0.5, and PBIAS stayed within 20. Compared to GWLF, SWAT improved the results from fair to good. Furthermore, the average monthly total nitrogen yield of GWLF was closer to the observed values than SWAT. The performance of both models during the validation period did not have obvious change contrasted to that during the calibration period. In the Hanjiaying watershed, the results of both models were not as satisfactory as those for Tunxi. The SWAT model roughly represented the trend shape of the monthly TN loadings and had criteria values ranging from good to very good during both calibration and validation periods. However, the GWLF did not provide acceptable simulation results for all years, although its statistics analysis during the calibration period was very good. Especially during the validation period, the time series of the GWLF was too gentle to capture each fluctuation of the observed data (Figure 5b). In contrast to SWAT, the average monthly TN predicted via GWLF was generally nearer to the measured values during both the calibration and verification periods.

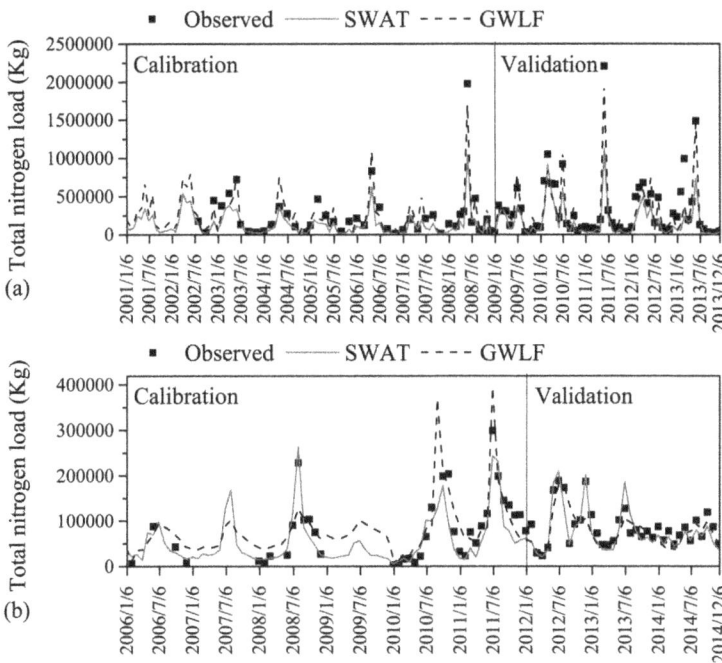

Figure 5. Simulated and observed monthly total nitrogen for the: (**a**) Tunxi watershed; and (**b**) Hanjiaying watershed.

Based on the comparison above, the SWAT model was capable of providing a reasonable and reliable prediction of monthly TN loadings especially in the Hanjiaying watershed where measured data were scarce. Several published studies verified the robustness of SWAT in representing nitrogen loadings. Stewart et al. [19] used SWAT to predict water quality changes in Texas, reporting a very good correlation (R^2 = 0.89, 0.87) and good agreement (NS = 0.71, 0.73) of monthly organic nitrogen in calibration and validation periods. Jha et al. [58] reported that SWAT performed very well on annual and monthly nutrient predictions in the Raccoon River watershed during the simulation periods with R^2 and NSE exceeding 0.7 in most cases. Gassman et al. [10] summarized more than twenty peer-reviewed articles and the values for R^2 and NSE mostly exceeded 0.5, indicating that the SWAT model is able to replicate a wide range of observed in-stream pollutant levels. This is due to SWAT considering five different chemical forms of nitrogen as well as the mutual transformation between them in the nitrogen cycle. However, GWLF only considers two different physical forms of nitrogen and does not take the conversion between them into account. Furthermore, the nitrogen concentrations remain constant during the whole model operating time. Thus, the accuracy achieved by GWLF is heavily dependent on the efficacy of calibration, which perhaps results in its poor performance in the Hanjiaying watershed where the measured data were limited. Furthermore, the value of CMD was higher in the Tunxi watershed than in Hanjiaying, indicating that microbial activity tended to be higher in this humid and warm area. In addition, the higher SHALLST_N of SWAT and nitrogen concentrations in sediment and groundwater of GWLF indicate that Hanjiaying suffers more human intervention than Tunxi. Actually, Hanjiaying has more agricultural land than Tunxi. This perhaps contributed to the relatively degraded performance of both models for the Hanjiaying watershed.

4. Conclusions

In this study, we conducted a comparison between two watershed models with different complexities and construction in two discrete sites that represent the semi-arid north and the humid south of China. According to the quantitative statistics and graphical techniques, both the SWAT and the GWLF model were capable of simulating monthly flow, sediment, and total nitrogen with adequate accuracy. They performed similarly well in terms of streamflow and sediment. Furthermore, GWLF outperformed SWAT in the Tunxi watershed, while it had opposite performance in Hanjiaying for nitrogen simulation. The main conclusions of our study are listed below.

- The performances of both models in arid areas were not as good as the performances in humid areas, indicating that climatic conditions could greatly affect the applicability of a given model.
- Due to the same adopted surface runoff calculation method (SCS CN), results of both models in monthly streamflow were quite similar, even though the complexity of the model structures was quite different.
- In contrast to GWLF, SWAT performed more dependable and robust in sediment and total nitrogen and could reproduce the fluctuations of the observed data more accurately due to its spatial property and more detailed description of reality.
- GWLF could provide similar or even better results and much closer average values to measured data than SWAT in some cases.
- Due to its simpler structure, GWLF requires fewer data to set up, less time to run, and is easier to be used than SWAT. However, it is not suitable for application in large catchments and cannot reflect spatial variations due to the absence of channel route and spatial topological relationship of land uses. Furthermore, GWLF is more dependent on the calibration process than SWAT.

Overall, the user friendly GWLF is more suitable for a basic analysis to support environmental management in data-deficient areas such as China, where the basic data required by SWAT are not always available or credible. Furthermore, SWAT has an advantage in areas where measured data are scarce and is more suitable for projects that require high accuracy.

Supplementary Materials: The following are available online at www.mdpi.com/2073-4441/9/8/567/s1, Figure S1: The interface of the SWAT model, Figure S2: The interface of the SWAT-CUP, Figure S3: The interface of GWLF modeling platform-ReNuMa. The Software is freely available for download from: http://swat.tamu.edu/ (assessed on 28 July 2017) and http://www.eeb.cornell.edu/biogeo/nanc/usda/renuma.htm (assessed on 28 July 2017), respectively.

Acknowledgments: We wish to thank the Environmental Protection Bureau and Environment Monitoring Station of Huangshan City as well as the Hai River Conservancy Commission of the Ministry of Water Resources for providing hydrology and chemistry data. We would also like to acknowledge the National Science Data Share Project—Data Sharing Infrastructure of Earth System Science (China) for the provided data support.

Author Contributions: C.C. and Y.W. conceived and designed the framework; Z.X. and Y.Q. analyzed the data; Z.Q. and G.K. wrote the paper; and C.C. and Y.W. contributed to improving the article.

Conflicts of Interest: The authors declare no conflict of interest.

References

1. Qu, J.; Fan, M. The Current State of Water Quality and Technology Development for Water Pollution Control in China. *Crit. Rev. Environ. Sci. Technol.* **2010**, *40*, 519–560. [CrossRef]

2. Ongley, E.D.; Zhang, X.; Yu, T. Current status of agricultural and rural non-point source Pollution assessment in China. *Environ. Pollut.* **2010**, *158*, 1159–1168. [CrossRef] [PubMed]

3. Sha, J.; Liu, M.; Wang, D.; Swaney, D.P.; Wang, Y. Application of the ReNuMa model in the Sha He river watershed: Tools for watershed environmental management. *J. Environ. Manag.* **2013**, *124*, 40–50. [CrossRef] [PubMed]

4. Shoemaker, L.; Dai, T.; Koenig, J.; Hantush, M. *TMDL Model Evaluation and Research Needs*; National Risk Management Research Laboratory, US Environmental Protection Agency: Washington, DC, USA, 2005.

5. Wellen, C.; Kamran-Disfani, A.-R.; Arhonditsis, G.B. Evaluation of the Current State of Distributed Watershed Nutrient Water Quality Modeling. *Environ. Sci. Technol.* **2015**, *49*, 3278–3290. [CrossRef] [PubMed]

6. Borah, D.K.; Bera, M. Watershed-scale hydrologic and nonpoint-source pollution models: Review of mathematical bases. *Trans. ASAE* **2003**, *46*, 1553–1566. [CrossRef]

7. Butts, M.B.; Payne, J.T.; Kristensen, M.; Madsen, H. An evaluation of the impact of model structure on hydrological modelling uncertainty for streamflow simulation. *J. Hydrol.* **2004**, *298*, 242–266. [CrossRef]

8. Arnold, J.G.; Srinivasan, R.; Muttiah, R.S.; Williams, J.R. Large area hydrologic modeling and assessment—Part 1: Model development. *JAWRA* **1998**, *34*, 73–89. [CrossRef]

9. Arnold, J.G.; Moriasi, D.N.; Gassman, P.W.; Abbaspour, K.C.; White, M.J.; Srinivasan, R.; Santhi, C.; Harmel, R.D.; van Griensven, A.; Van Liew, M.W.; et al. Swat: Model Use, Calibration, And Validation. *Trans. ASABE* **2012**, *55*, 1491–1508. [CrossRef]

10. Gassman, P.W.; Reyes, M.R.; Green, C.H.; Arnold, J.G. The soil and water assessment tool: Historical development, applications, and future research directions. *Trans. ASABE* **2007**, *50*, 1211–1250. [CrossRef]

11. Yang, Q.; Zhang, X. Improving SWAT for simulating water and carbon fluxes of forest ecosystems. *Sci. Total Environ.* **2016**, *569*, 1478–1488. [CrossRef] [PubMed]

12. Begou, J.C.; Jomaa, S.; Benabdallah, S.; Bazie, P.; Afouda, A.; Rode, M. Multi-Site Validation of the SWAT Model on the Bani Catchment: Model Performance and Predictive Uncertainty. *Water* **2016**, *8*, 178. [CrossRef]

13. Sun, H.; Cornish, P.S. Estimating shallow groundwater recharge in the headwaters of the Liverpool Plains using SWAT. *Hydrol. Process.* **2005**, *19*, 795–807. [CrossRef]

14. Yang, J.; Reichert, P.; Abbaspour, K.C.; Xia, J.; Yang, H. Comparing uncertainty analysis techniques for a SWAT application to the Chaohe Basin in China. *J. Hydrol.* **2008**, *358*, 1–23. [CrossRef]

15. Li, Z.; Shao, Q.; Xu, Z.; Cai, X. Analysis of parameter uncertainty in semi-distributed hydrological models using bootstrap method: A case study of SWAT model applied to Yingluoxia watershed in northwest China. *J. Hydrol.* **2010**, *385*, 76–83. [CrossRef]

16. Shen, Z.Y.; Chen, L.; Chen, T. Analysis of parameter uncertainty in hydrological and sediment modeling using GLUE method: A case study of SWAT model applied to Three Gorges Reservoir Region, China. *Hydrol. Earth Syst. Sci.* **2012**, *16*, 121–132. [CrossRef]

17. Easton, Z.M.; Fuka, D.R.; Walter, M.T.; Cowan, D.M.; Schneiderman, E.M.; Steenhuis, T.S. Re-conceptualizing the soil and water assessment tool (SWAT) model to predict runoff from variable source areas. *J. Hydrol.* **2008**, *348*, 279–291. [CrossRef]

18. Spruill, C.A.; Workman, S.R.; Taraba, J.L. Simulation of daily and monthly stream discharge from small watersheds using the SWAT model. *Trans. ASAE* **2000**, *43*, 1431–1439. [CrossRef]

19. Stewart, G.R.; Munster, C.L.; Vietor, D.M.; Arnold, J.G.; McFarland, A.M.S.; White, R.; Provin, T. Simulating water quality improvements in the Upper North Bosque River watershed due to phosphorus export through turfgrass sod. *Trans. ASABE* **2006**, *49*, 357–366. [CrossRef]

20. Dlamini, N.S.; Kamal, M.R.; Soom, M.A.B.M.; bin Mohd, M.S.F.; Abdullah, A.F.B.; Hin, L.S. Modeling Potential Impacts of Climate Change on Streamflow Using Projections of the 5th Assessment Report for the Bernam River Basin, Malaysia. *Water* **2017**, *9*, 226. [CrossRef]

21. Franczyk, J.; Chang, H. The effects of climate change and urbanization on the runoff of the Rock Creek basin in the Portland metropolitan area, Oregon, USA. *Hydrol. Process.* **2009**, *23*, 805–815. [CrossRef]

22. Schneiderman, E.M.; Pierson, D.C.; Lounsbury, D.G.; Zion, M.S. Modeling the hydrochemistry of the Cannonsville watershed with Generalized Watershed Loading Functions (GWLF). *JAWRA* **2002**, *38*, 1323–1347. [CrossRef]

23. Jennings, E.; Allott, N.; Pierson, D.C.; Schneiderman, E.M.; Lenihan, D.; Samuelsson, P.; Taylor, D. Impacts of climate change on phosphorus loading from a grassland catchment: Implications for future management. *Water Res.* **2009**, *43*, 4316–4326. [CrossRef] [PubMed]

24. Lin, C.-H.; Huang, T.-H.; Shaw, D. Applying Water Quality Modeling to Regulating Land Development in a Watershed. *Water Resour. Manag.* **2010**, *24*, 629–640. [CrossRef]

25. DPPC. *Guidelines for the Programming of Water Bodies Meeting Standards*; Department of Pollution Prevention and Control, Ministry of Environmental Protection of the People's Republic of China: Beijing, China, 2016.

26. Borah, D.K.; Yagow, G.; Saleh, A.; Barnes, P.L.; Rosenthal, W.; Krug, E.C.; Hauck, L.M. Sediment and nutrient modeling for TMDL development and implementation. *Trans. ASABE* **2006**, *49*, 967–986. [CrossRef]

27. Li, Z.; Xu, Z.; Li, Z. Performance of WASMOD and SWAT on hydrological simulation in Yingluoxia watershed in northwest of China. *Hydrol. Process.* **2011**, *25*, 2001–2008. [CrossRef]

28. Parajuli, P.B.; Nelson, N.O.; Frees, L.D.; Mankin, K.R. Comparison of AnnAGNPS and SWAT model simulation results in USDA-CEAP agricultural watersheds in south-central Kansas. *Hydrol. Process.* **2009**, *23*, 748–763. [CrossRef]

29. Wilcox, B.P.; Rawls, W.J.; Brakensiek, D.L.; Wight, J.R. Predicting runoff from Rangeland Catchments: A comparison of two models. *Water Resour. Res.* **1990**, *26*, 2401–2410. [CrossRef]

30. Niraula, R.; Kalin, L.; Srivastava, P.; Anderson, C.J. Identifying critical source areas of nonpoint source pollution with SWAT and GWLF. *Ecol. Model.* **2013**, *268*, 123–133. [CrossRef]

31. Neitsch, S.L.; Williams, J.; Arnold, J.; Kiniry, J. *Soil and Water Assessment Tool Theoretical Documentation Version 2009*; Texas Water Resources Institute: College Station, TX, USA, 2011.

32. Haith, D.A.; Shoemaker, L.L. Generalized Watershed Loading Functions for Stream Flow Nutrients. *JAWRA* **1987**, *23*, 471–478. [CrossRef]

33. Evans, B.M.; Lehning, D.W.; Corradini, K.J.; Petersen, G.W.; Nizeyimana, E.; Hamlett, J.M.; Robillard, P.D. A Comprehensive GIS-Based Modeling Approach for Predicting Nutrient Loads in Watersheds. *J. Spat. Hydrol.* **2002**, *2*, 1–18.

34. Mockus, V. *National Engineering Handbook Section 4, Hydrology*; NTIS: Alexandria, VA, USA, 1972.

35. Priestley, C.; Taylor, R. On the assessment of surface heat flux and evaporation using large-scale parameters. *Mon. Weather Rev.* **1972**, *100*, 81–92. [CrossRef]

36. Hargreaves, G.H.; Samani, Z.A. Reference crop evapotranspiration from temperature. *Appl. Eng. Agric.* **1985**, *1*, 96–99. [CrossRef]

37. Hamon, W.R. Estimating Potential Evapotranspiration. *Proc. Am. Soc. Civ. Eng.* **1961**, *87*, 107–120.

38. Green, W.H.; Ampt, G. Studies on Soil Phyics. *J. Agric. Sci.* **1911**, *4*, 1–24. [CrossRef]

39. Ogrosky, H.O.; Mockus, V. Hydrology of Agricultural Lands. In *Handbook of Applied Hydrology*; Chow, V.T., Ed.; McGraw-Hill: New York, NY, USA, 1964.

40. Sloan, P.G.; Moore, I.D. Modeling subsurface stormflow on steeply sloping forested watersheds. *Water Resour. Res.* **1984**, *20*, 1815–1822. [CrossRef]

41. Haan, C. A water yield model for small watersheds. *Water Resour. Res.* **1972**, *8*, 58–69. [CrossRef]

42. Williams, J.R. Sediment Routing for Agricultural Watersheds. *JAWRA* **1975**, *11*, 965–974. [CrossRef]

43. Williams, J.R. Sediment-yield prediction with universal equation using runoff energy factor. *Present Prospect. Technol. Predict. Sedim. Yield Sources* **1975**, *40*, 244–252.

44. Wischmeier, W.H.; Smith, D.D. Predicting rainfall erosion losses-a guide to conservation planning. In *Predicting Rainfall Erosion Losses—A Guide to Conservation Planning*; Department of Agriculture: Washington, DC, USA, 1978.

45. Brown, L.C.; Barnwell, T.O. *The Enhanced Stream Water Quality Models QUAL2E and QUAL2E-UNCAS: Documentation and User Manual*; US Environmental Protection Agency, Office of Research and Development, Environmental Research Laboratory: Washington, DC, USA, 1987.

46. Zhai, X.; Zhang, Y.; Wang, X.; Xia, J.; Liang, T. Non-point source pollution modelling using Soil and Water Assessment Tool and its parameter sensitivity analysis in Xin'anjiang catchment, China. *Hydrol. Process.* **2014**, *28*, 1627–1640. [CrossRef]

47. Abbaspour, K.C.; Yang, J.; Maximov, I.; Siber, R.; Bogner, K.; Mieleitner, J.; Zobrist, J.; Srinivasan, R. Modelling hydrology and water quality in the pre-ailpine/alpine Thur watershed using SWAT. *J. Hydrol.* **2007**, *333*, 413–430. [CrossRef]

48. Smedema, L.K.; Rycroft, D.W. *Land Drainage: Planning and Design of Agricultural Systems*; Batsford Academic and Educational Ltd.: London, UK, 1983.

49. Haith, D.; Mandel, R.; Wu, R. *GWLF: Generalized Watershed Loading Functions User's Manual, Version 2.0*; Cornell University: Ithaca, NY, USA, 1992.

50. Nash, J.E.; Sutcliffe, J.V. River flow forecasting through conceptual models part I—A discussion of principles. *J. Hydrol.* **1970**, *10*, 282–290. [CrossRef]

51. Legates, D.R.; McCabe, G.J. Evaluating the use of "goodness-of-fit" measures in hydrologic and hydroclimatic model validation. *Water Resour. Res.* **1999**, *35*, 233–241. [CrossRef]

52. Gupta, H.V.; Sorooshian, S.; Yapo, P.O. Status of Automatic Calibration for Hydrologic Models: Comparison with Multilevel Expert Calibration. *J. Hydrol. Eng.* **1999**, *4*, 135–143. [CrossRef]

53. Moriasi, D.N.; Arnold, J.G.; Van Liew, M.W.; Bingner, R.L.; Harmel, R.D.; Veith, T.L. Model evaluation guidelines for systematic quantification of accuracy in watershed simulations. *Trans. ASABE* **2007**, *50*, 885–900. [CrossRef]

54. Shen, Z.Y.; Gong, Y.W.; Li, Y.H.; Hong, Q.; Xu, L.; Liu, R.M. A comparison of WEPP and SWAT for modeling soil erosion of the Zhangjiachong Watershed in the Three Gorges Reservoir Area. *Agric. Water Manag.* **2009**, *96*, 1435–1442. [CrossRef]

55. Suliman, A.H.A.; Jajarmizadeh, M.; Harun, S.; Mat Darus, I.Z. Comparison of Semi-Distributed, GIS-Based Hydrological Models for the Prediction of Streamflow in a Large Catchment. *Water Resour. Manag.* **2015**, *29*, 3095–3110. [CrossRef]

56. Chahinian, N.; Moussa, R.; Andrieux, P.; Voltz, M. Comparison of infiltration models to simulate flood events at the field scale. *J. Hydrol.* **2005**, *306*, 191–214. [CrossRef]

57. Lv, Y.Z.; Li, B.G. *Pedology*, 1st ed.; China Agriculture Press: Beijing, China, 2006; pp. 252–253.

58. Jha, M.K.; Gassman, P.W.; Arnold, J.G. Water quality modeling for the Raccoon River watershed using SWAT. *Trans. ASABE* **2007**, *50*, 479–493. [CrossRef]

water

MDPI

Article

Water Resources of the Black Sea Catchment under Future Climate and Landuse Change Projections

Elham Rouholahnejad Freund [1,2,3,*], Karim C. Abbaspour [1] and Anthony Lehmann [4]

[1]   Eawag, Swiss Federal Institute of Aquatic Science and Technology, 8600 Duebendorf, Switzerland;
       karim.abbaspour@eawag.ch
[2]   Department of Environmental Systems Science, ETH Zurich, 8092 Zürich, Switzerland
[3]   Laboratory of Hydrology and Water Management, Ghent University, 9000 Ghent, Belgium
[4]   Institute for Environmental Sciences, University of Geneva, 1227 Geneva, Switzerland;
       anthony.lehmann@unige.ch
*   Correspondence: elham.rouholahnejad@gmail.com; Tel.: +41-44-633-8889

Received: 11 April 2017; Accepted: 6 August 2017; Published: 12 August 2017

Abstract: As water resources become further stressed due to increasing levels of societal demand, understanding the effect of climate and landuse change on various components of the water cycle is of strategic importance. In this study we used a previously developed hydrologic model of the Black Sea Catchment (BSC) to assess the impact of potential climate and landuse changes on the fresh water availability. The BSC model was built, calibrated, and validated against observed daily river discharge for the period of 1973–2006 using the Soil and Water Assessment Tool (SWAT) as the modeling tool. We employed the A2 and B2 scenarios of 2017–2050 generated by the Danish Regional Climate Model (HIRHAM), and four potential future landuse scenarios based on the Intergovernmental Panel of Climate Change (IPCC)'s special report on emissions scenarios (SRES) storylines, to analyze the impact of climate change and landuse change on the water resources of the BSC. The detailed modeling and the ensemble of the scenarios showed that a substantial part of the catchment will likely experience a decrease in freshwater resources by 30 to 50%.

Keywords: hydrology; Danube; Don; Dnieper; land use change; hydrological modeling

---

1. Introduction

Observational evidence from all continents and oceans shows that natural systems are affected by regional climate changes, particularly by increases in temperature [1]. Nearly all regions of the world are expected to experience a net negative impact of climate change on water resources and freshwater ecosystems [1]. A number of studies have shown that climate change will have significant effects on water availability, water stresses, and water demand [2,3]. Climate change will pose added challenges to managing high disaster-risk areas, as it is virtually certain that the frequency and magnitude of warm daily temperature extremes will increase, and cold extremes will decrease, in the 21st century at the global scale [1]. It is expected that the frequency of heavy precipitation will increase in Southern and Central Europe and the Mediterranean region, and that droughts will intensify because of reduced precipitation and/or increased evapotranspiration [1].

The focus of this study is on the Black Sea Catchment (BSC), which lies in a transition zone between the Mediterranean region in an arid climate of North Africa and the temperate and rainy climate of central Europe. It is affected by interactions between mid-latitude and tropical processes. Because of these features, even relatively minor modifications of the general circulation can lead to substantial changes in the Mediterranean climate [4]. This makes the BSC a potentially vulnerable region to climatic changes as induced, for example, by increasing concentrations of greenhouse gases [4]. Indeed, the Mediterranean region has shown large climate shifts in the past, and it has been identified as one of the most prominent "Hot-Spots" in future climate change projections [5].

For the long-term strategic planning of a country's water resources in the face of the evolving climate and landuse changes, it is important to quantify the effects with a high spatial and temporal resolution. However, no publication has really focused on the long-term evaluation of the BSC's water balance under a combination of climate and landuse change. Yet this may be the most beneficial application of hydro-climatology to support long-term water resources management and planning. Within this context, Giorgi and Lionello [4] reviewed a few climate change projections over the Mediterranean region based on the most recent and comprehensive ensembles of global and regional climate change simulations. There is also a comprehensive review of climate change projections over the Mediterranean region reported by Ulbrich et al. [6] based on a limited number of global and regional model simulations performed throughout the early 2000s. Moreover, a number of studies have reported regional climate change simulations over Europe, which includes parts of the BSC [7–10]. As recognized by the two international environmental organizations in the region (Danube and Black Sea), achieving an environmental sustainability that will improve human well-being strongly depends on sustainable water resources management in this catchment [11]. Nevertheless, despite the importance of this region within the global change context, assessments of water resources of the entire catchment under different climate and landuse change projections do not exist in the literature.

Hydrologic models often used to assess the impacts of land and climate changes on water resources include: WaterGap3 [12,13], HBV [14], MIKE-SHE [15], and the Soil and Water Assessment Tool (SWAT) [16] among many others. These models are particularly useful as they can assess past as well as possible future impact scenarios. SWAT [16] has proven its suitability for hydrologic impact studies, especially under conditions of limited data availability [17]. The aforementioned hydrological models often use global climate simulations for the 20th and 21st century under different greenhouse gas forcing scenarios that are publicly available as a contribution to the fourth and fifth Assessment Reports (AR4, AR5) of the Intergovernmental Panel on Climate Change (IPCC) to analyze the climate change impacts on various aspect of hydrological cycle. In the current study, we employed the Danish Regional Climate Model (RCM) HIRHAM, driven by the United Kingdom's Hadley Center HadAM3H Global Climate Model (GCM), under the scope of the PRUDENCE project to asses the future climate change projections over the BSC. We used the quantification of landuse change scenarios which is based on the framework provided by the Integrated Model to Assess the Global Environment (IMAGE) [18]. The four landuse change scenarios used in the current study comprise a number of plausible storylines based on the IPCC's special report on emissions scenarios (SRES) following four marker scenarios representing different global socio-economic development pathways.

The objective of the study is to address the changes in various components of the water balance including precipitation, evapotranspiration, and soil moisture under changing climate and landuse. We used a previously calibrated hydrological SWAT model of the BSC [19] as the base model, and incorporated the climate and land use change scenarios that are outlined above. We assessed the variation in blue water (river discharge plus aquifer recharge) and green water (soil moisture and evapotranspiration) across the BSC under changing climate and landuse scenarios.

2. Material and Methods

2.1. Study Area

The BSC drains rivers of 23 European and Asian countries (Austria, Belarus, Bosnia, Bulgaria, Croatia, Czech Republic, Georgia, Germany, Hungary, Moldova, Montenegro, Romania, Russia, Serbia, Slovakia, Slovenia, Turkey, Ukraine, Italy, Switzerland, Poland, Albania, and Macedonia) from an area of 2.3 million km^2 into the Black Sea (Figure 1). The catchment is highly populated (160 million people) [20]. Major rivers draining into the Black Sea include Danube, Dniester, Dnieper, Don, Kuban, Sakarya, and Kizirmak. The major mountainous peaks lie in the East and South, in the Caucasus and in

Anatolia, and to the Northwest with the Carpathians in the Ukraine and Romania. Most of the rest of the West and North of the catchment is low lying. The catchment has a distinct North–South temperature gradient from <−3 °C to >15 °C (annual average) and a West–East precipitation gradient that is decreasing with distance from the Atlantic Ocean. Areas of high precipitation (>3000 mm year^{-1}) are in the West, and areas of low precipitation (<190 mm year^{-1}) are in the North and East [21]. The dominant landuse in the catchment is agricultural land, with 65% coverage according to MODIS Land Cover [22].

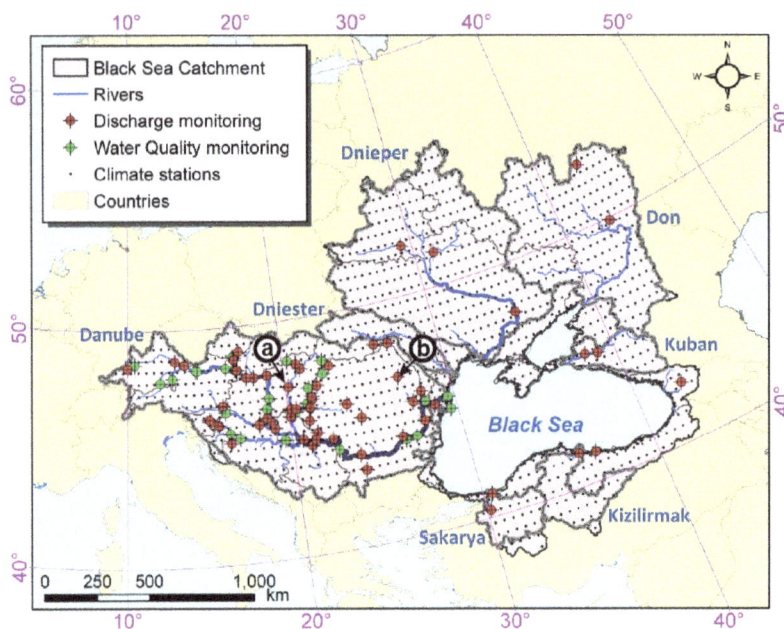

Figure 1. Overview of the Black Sea Catchment depicting major rivers and measured stations of climate, discharge, and nitrate. The labeled points a and b correspond to the labeled points in Figure 4.

2.2. Soil Water Assessment Tool (SWAT)

SWAT was used to simulate the hydrology of the BSC. SWAT is a process-based, semidistributed hydrologic model that is developed to quantify the impact of land management practices on water, sediment, and agricultural chemical yields in large complex watersheds with varying soils, landuses, and management conditions over long periods of time. SWAT has been used for the assessment of landuse and management impacts on water quantity and quality in many studies worldwide. The spatial heterogeneity of the watershed is preserved by topographically dividing the catchment into multiple sub-basins. The sub-basins are further subdivided into hydrologic response units (HRU). These are lumped areas within a sub-basin with a unique combination of slope, soil type, and landuse that enable the model to reflect differences in evapotranspiration for various crops and soils. A simulation of the hydrologic cycle is separated into a land phase and a water phase [23]. The simulation of the land phase is based on the water balance equation, which is calculated separately for each HRU. Runoff generated in the HRUs is summed up to calculate the amount of water reaching the main channel in each sub-basin [23]. The water phase of the hydrologic cycle describes the routing of runoff in the river channel, using the variable storage coefficient method by Williams [24]. A detailed description of the model can be obtained from Neitsch et al. [23]. Figure 2 depicts a schematic view of processes accounting for the land phase of the SWAT model.

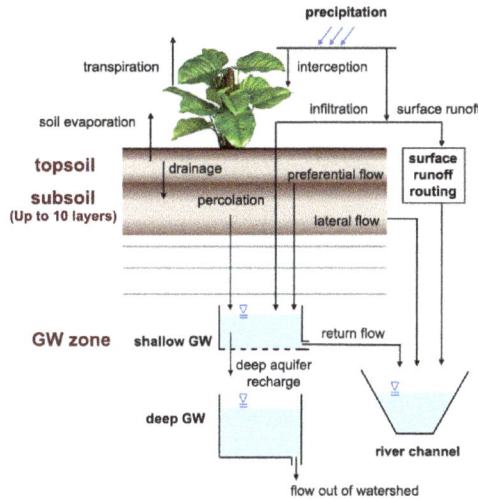

Figure 2. Schematic overview of available pathways for water movement in the land phase of the SWAT model. GW, groundwater.

2.3. Landuse Scenarios

The landuse scenarios that are used in the current study were developed within the European Union's seventh research framework through the enviroGRIDS project [25]. The developed scenarios comprise a number of plausible storylines based on a coherent set of assumptions, key relationships, and driving forces, to create a set of quantitative, internally consistent, and spatially explicit scenarios of future landuse covering the entire BSC (Figure 3). A trend of landuse change for different scenarios is summarized in Table 1.

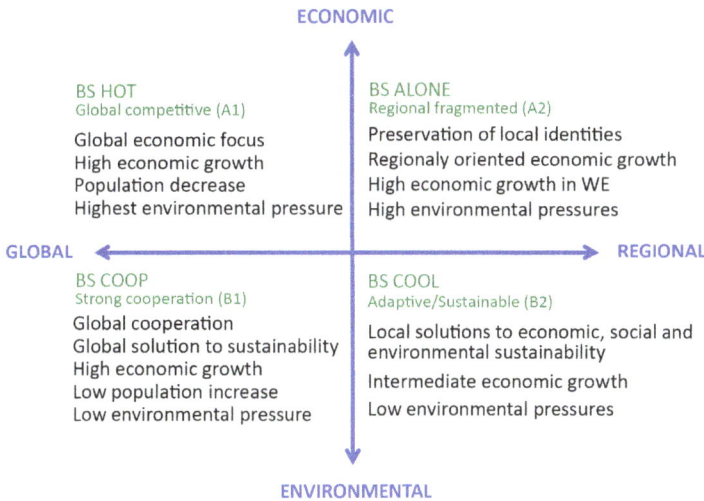

Figure 3. Black Sea Catchment's landuse scenarios. BS, Black Sea.

The quantification of landuse scenarios is based on the outputs of the Integrated Model to Assess the Global Environment [18] and on projections based on data from the Statistical Office of the European

Communities (Eurostat) and the UN World Population and Urbanization Prospects [26]. The European regional projections (forest, grassland, urban and built up, and cropland) were disaggregated at the level of smaller administrative units (Nomenclature of Units for Territorial Statistics, level 2 (NUTS2)), and then used as input to the regional/local land allocation Metronamica model [27] for 214 regions in the BSC. The landuse change scenarios were quantified as yearly changes in landuses on 1 km × 1 km grid cells, in two time steps, 2025 and 2050, for four scenarios covering the entire BSC. The landuse scenarios were developed for cropland, grassland, forest, and urban areas for the BSC countries, while the input data were derived from the MODIS land cover datasets [22] for the years 2001 and 2008. A full description of the scenario developments and quantitative measures of changes in land cover classes is presented by Mancosu et al. [28].

Table 1. Summary of landuse trends and driving forces in the Black Sea Catchment's scenarios.

	Landuse Scenarios			
Driving Forces	**BS HOT**	**BS ALONE**	**BS COOP**	**BS COOL**
Population growth	low	very high	low	medium
Urban population	increase	increase	slight increase	slight increase
GDP growth	very high	slow	high	medium
Forest area	increase	decrease	increase	decrease
Grassland area	increase	decrease	increase	decrease
Cropland area	increase	increase	decrease	increase
Built-up area	increase	increase	increase	stable
Protected areas	stable	stable	increase	stable
Climate change	high	high	lower	low

Note: GDP, gross domestic product.

2.4. Climate Change Scenarios

Two future climate scenarios, HS and HB, were simulated for the period of 2010 to 2050, representing the IPCC's A2 and B2 scenarios, respectively. In this study, we used precipitation and minimum and maximum temperature data from the PRUDENCE website. The data were downscaled and bias corrected using the Delta method [29] based on Climate Research Unit (CRU) [30] data (1901–2006). The CRU time series data for the control period are perturbed with changes, allowing increase as a function of time. These changes are based on the parted differences of the monthly probability distribution function (PDF). The PDF is partitioned into deciles, and observed time series are gradually perturbed. The assumption behind the Delta method is that future model bias for both mean and variability will be the same as for present-day simulations. Detail on downscaling techniques and a bias correction procedure is available in a study by Gago Da Silva et al. [31].

2.5. Model Setup, Calibration, and Validation

The 2.3 million km^2 area of the catchment is divided into 12,982 sub-basins. The sub-basins are further divided into unique combination of soil, landuse, and slope, and formed 89,202 Hydrological Response Units (HRUs). For a calibration and uncertainty analysis, we used the Sequential Uncertainty FItting program SUFI-2, which is a tool for sensitivity analysis, multi-site calibration, and uncertainty analysis. SUFI-2 is linked to SWAT in the SWAT-CUP software [32].

The inputs of the model are summarized in Table 2. The simulation period was 1970–2006, designating the first three years as a warm up period. We used the ArcSWAT2009 interface for the model setup and SWAT2009 Rev528 for a model run. Table 3 gives an overview of the relevant methods used in the model setup. More details of the hydrologic model's structure, setup, and performance in the BSC can be found in Rouholahnejad et al. [19].

For future simulations (2017–2050, including 3 years of warm up period), the two climate change scenarios (HB and HS) were paired with the four landuse change scenarios (BS ALONE, BS COOL, BS COOP, and BS HOT), leading to eight combinations of climate-landuse scenarios, and applied as inputs to the calibrated hydrologic model of the catchment in eight separate model setups. In addition, two model setups were designed to look at climate change only, with static landuse. The latter models used MODIS landuse data of the historical hydrologic model as the static landuse. In the other eight combinations of landuse-climate scenarios, a dynamically updating algorithm updated the landuses yearly up to the end of the simulation period. A total of 10 different scenarios were therefore built and analyzed.

Table 2. Sources of model input data and descriptions for the base model.

Type	Source	Description
DEM	SRTM [33]	90 m resolution extracted for BSC
Climate	CRU [30,34], Solar Radiation [35]	0.5° resolution gridded climate data, daily temperature (min.; max.), daily precipitation (1970–2006) daily global solar radiation from 6110 virtual stations (1970–2006)
River	ECRINS [36]	30 m resolution, from European Catchments and Rivers Network System (ECRINS)
Soil	FAO [37]	5 km resolution, from FAO-UNESCO global soil map, provides data for 5000 soil types comprising two layers (0–30 cm and 30–100 cm depth)
Landuse	MODIS [22]	500 m resolution, by the NASA Land Processes Distributed Active Archive Center (LP DAAC) at the USGS/Earth Resources Observation and Science Center (EROS)
Management	MIRCA2000 [38], McGill yields data [39]	5 arc min resolution cropping area and the start and end month of cropping periods, 5 arc min crop yield of three major crop (Wheat, Cory, Barely)
River discharge	GRDC [40]	144 Monthly river discharge data (1970–2006)

Notes: DEM, Digital Elevation Model; SRTM, Shuttle Radar Topography Mission; BSC, Black Sea Catchment; CRU, Climate Research Unites; ECRINS, European Catchments and RIvers Network System; FAO, Food and Agriculture Organization; MODIS, Moderate Resolution Imaging Spectroradiometer; MIRCA, Monthly Irrigated and Rainfed Crop Areas; GRDC, Global Runoff Data Centre.

Table 3. Soil and Water Assessment Tool (SWAT) processes used in the study.

Processes/Components	Method
Evapotranspiration	Hargreaves
Surface runoff	Soil Conservation Service (SCS) curve number
Erosion	Modified universal soil loss equation
Lateral flow	Kinematic storage model
Groundwater flow	Steady-state response from shallow aquifer
Stream flow routing	Variable storage routing

3. Results and Discussion

The BSC hydrologic model was calibrated (1973–1996) and validated (1997–2006) at 144 river discharge stations at daily time scales. The most sensitive parameters to discharge are as shown in Table 4. An example of the simulated and observed stream flow along with the predictive uncertainty band for the two labeled river discharge stations in Figure 1 are presented in Figure 4. These are two examples of calibrated discharge stations, one at the border of Hungary and Romania and one in Romania. We ran the model and calibration simulations at a daily time scale. However, given the scale of the watershed, the outputs were extracted at monthly time scales.

Table 4. List of sensitive parameters used for model calibration.

Parameter Name	Definition
CN2.mgt	SCS runoff curve number for moisture condition II
ALPHA_BF.gw	Base flow alpha factor (days)
GW_DELAY.gw	Groundwater delay time (days)
GWQMN.gw	Threshold depth of water in shallow aquifer for return flow (mm)
GW_REVAP.gw	Groundwater revap. coefficient
REVAPMN.gw	Threshold depth of water in the shallow aquifer for 'revap' (mm)
RCHRG_DP.gw	Deep aquifer percolation fraction
CH_N2.rte	Manning's n value for main channel
CH_K2.rte	Effective hydraulic conductivity in the main channel (mm h^{-1})
ALPHA_BNK.rte	Baseflow alpha factor for bank storage (days)
SOL_AWC().sol	Soil available water storage capacity (mm H_2O/mm soil)
SOL_K().sol	Soil conductivity (mm h^{-1})
SOL_BD().sol	Soil bulk density (g cm^{-3})
OV_N.hru	Manning's n value for overland flow
HRU_SLP.hru	Average slope steepness (m m^{-1})
SLSUBBSN.hru	Average slope length (m)
SFTMP().sno	Snowfall temperature (°C)
SMTMP().sno	Snow melt base temperature (°C)
SMFMX().sno	Maximum melt rate for snow during the year (mm °C^{-1} day^{-1})
SMFMN().sno	Minimum melt rate for snow during the year (mm °C^{-1} day^{-1})

Note: SCS, Soil Conservation Service.

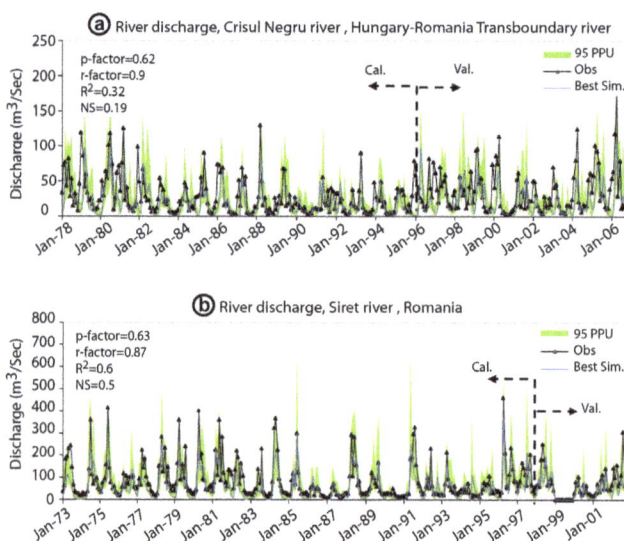

Figure 4. Simulated and observed river discharges of (**a**) Crisul Negru and (**b**) Siret river in the Black Sea Catchment (labeled in Figure 1). Shown in the picture are observation time series, best simulation along with 95% prediction uncertainty band (green band). The P-factor is the percentage of measured data bracketed by the 95 PPU band. It ranges from 0 to 1, where 1 is ideal and means all of the measured data are within the uncertainty band. The R-factor is the average width of the band divided by the standard deviation of the measured variable. It ranges from 0 to 1, where 0 reflects a perfect match with the observation. Based on the experience, an R-factor of around 1 is usually desirable. NS and R^2 are Nash–Sutcliffe efficiency and coefficient of determination, respectively.

3.1. Temperature and Precipitation in a Changing Climate

The HB and HS long-term average temperature scenarios show a 1–2.4 °C temperature increase with a west to east gradient in the catchment. While the overall long-term pattern of increase in temperature is similar in the two scenarios, HS depicts a larger temperature increase and over larger areas (Figure 5g,h). The variation over time in future temperature scenarios (2020–2050) has almost the same pattern as the historic temporal variation (1973–2006) with some discrepancies in the Danube basin (Figure 5b,d,f).

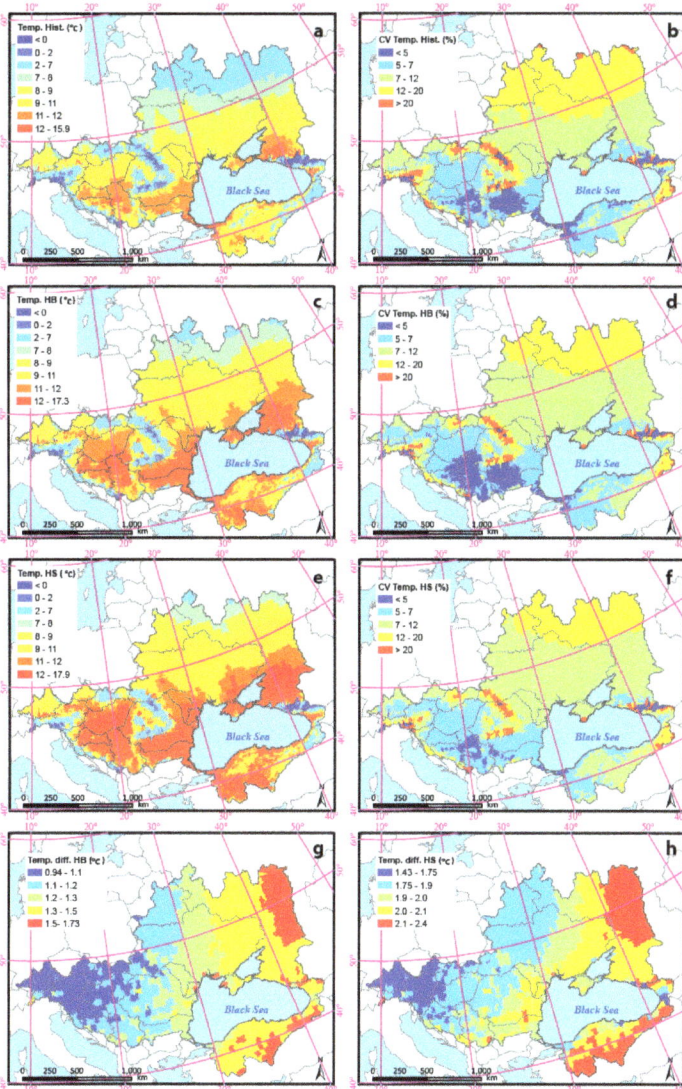

Figure 5. Temperature distribution in the Black Sea Catchment: (**a**) average temperature, historic (1973–2006); (**b**) coefficient of variation (CV) of historic temperature (1973–2006); (**c**) average temperature HB scenario (2020–2050); (**d**) coefficient of variation of HB temperature (2020–2050); (**e**) average temperature HS scenario (2020–2050); (**f**) coefficient of variation of HS temperature (2020–2050); (**g**) deviation of HB future temperature scenario (2020–2050) from historic (1973–2006), °C; (**h**) deviation of HS future temperature scenario (2020–2050) from historic (1973–2006), °C.

The historical distribution of average precipitation (Figure 6a), along with its coefficient of variation (CV) over time (Figure 6b), mark distinct precipitation distribution patterns with significant temporal variation (as indicated by the high CV values) in regions around the Sea, and in the Danube, Dnieper, and Don River basins. Historically, the upper Western part of the catchment in the alpine region receives rain of about 1300 mm year^{-1} on average, which is much higher than precipitation rates in other parts of the BSC where Ukraine, Russia, and Turkey lie. The regions with a smaller historical precipitation rates tend to show a higher temporal variation in precipitation (Figure 6b).

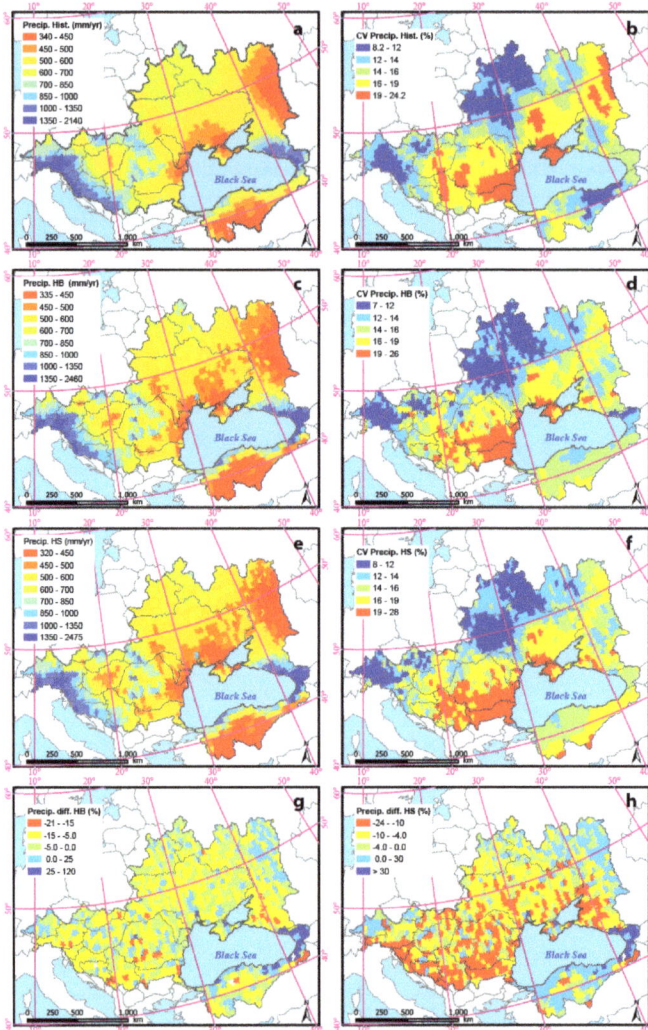

Figure 6. Precipitation distribution in the Black Sea Catchment: (**a**) average precipitation, historic (1973–2006); (**b**) coefficient of variation of historic precipitation (1973–2006); (**c**) average precipitation HB scenario (2020–2050); (**d**) coefficient of variation of HB precipitation (2020–2050); (**e**) average precipitation HS scenario (2020–2050); (**f**) coefficient of variation of HS precipitation (2020–2050); (**g**) percent deviation of HB precipitation scenario (2020–2050) from historic (1973–2006); (**h**) percent deviation of HS precipitation scenario (2020–2050) from historic (1973–2006).

The two future precipitation scenarios are similar to the historic ones with regard to patterns of long-term annual averages (Figure 6a,c,e). In future scenarios, the areas with small precipitation expand as compared to the historic period (Figure 6a,c,e). The HS scenario, in general, suggests larger decreases in precipitation in the catchment.

The anomaly maps of precipitation (Figure 6g,h) depict percent deviation from historic precipitation for the entire catchment. The differences are calculated between the averages of 2020–2050 with those of 1973–2006. The HB scenario suggests a 5–15% decrease in precipitation in most regions of the catchment, while the HS suggests a 10–24% decrease in the precipitation of Danube Basin (West of the Black Sea) and a 4–10% decrease in precipitation in the rest of the catchment.

3.2. Fresh Water Resources under Changing Landuse and Climate

The term "blue water" [41] is widely used in the literature as the summation of the water yield and deep aquifer recharge. The long-term average blue water resources of the BSC for the period of 1973–2006 are shown in Figure 7a. The coefficient of variation of the blue water in the BSC (Figure 7b) during the period 1973–2006 shows significant variation in the central and eastern parts of the catchment as well as in Turkey, where the historic annual average of blue water is less than 100 mm year^{-1} (Figure 7b). In other words, the less the blue water is available, the more its variability over time is. The anomaly map of blue water (Figure 7c) depicts the deviation of long-term average blue water resources for the period of 2020–2050 from long-term average historic records (1973–2006) under future landuse and climate change scenarios (ensemble of 10 scenarios).

The blue water resources calculated with the ensemble of 10 scenarios suggest a 10–50% decrease in blue water resources in most parts of the catchment (Figure 7c). According to the future scenario ensembles, blue water increases on average by 50% in the coastal areas of Georgia and Turkey, with historically small blue water resources. However, this does not bring a significant increase in terms of net blue water resources availability of the whole catchment, as the historical records in these regions are quite small. Historical variations of blue water indicate low reliability (higher variability over time) of blue water resources in Romania, parts of Ukraine, the Russian parts of the catchment, and Turkey (Figure 7b). Our analysis shows that the poor conditions in these regions in terms of fresh water availability will be further intensified under climate change (Figure 7c).

As soil moisture is an integral component of rainfed agriculture, the soil water distributions projected under future climate change and landuse change scenarios are of a strategic importance. The spatial variation of long-term average annual green water storage (soil moisture) (Figure 8a) shows that most of the catchment lies in the range of 70–250 mm of soil moisture. However, the variation over time shows a distinct East–West pattern, suggesting different levels of reliability for soil water in the region. The Southern parts of the catchment tend to have smaller soil water with more variation over time (Figure 8b), which makes the region less reliable in terms of green water storage resources.

The anomaly map of soil water (average of 10 scenarios) depicts the deviation of average future (2020–2050) soil moisture from the average historic (1973–2006) soil moisture (Figure 8c). The average of the 10 scenarios shows up to a 10–25% reduction in soil water in the Danube catchment and Northern BSC in the upstream of Dnieper and Don, in Ukraine and Russia. There are indications of soil moisture increase in Georgia, where both climate scenarios suggested an increase in precipitation. The coefficient of variation among the 10 scenarios indicates that there is a good agreement between the scenarios with less than 2% variation in the prediction of soil water in most parts of the catchment and 10–15% variation in the Danube Delta and areas surrounding the Black Sea (Figure 8d).

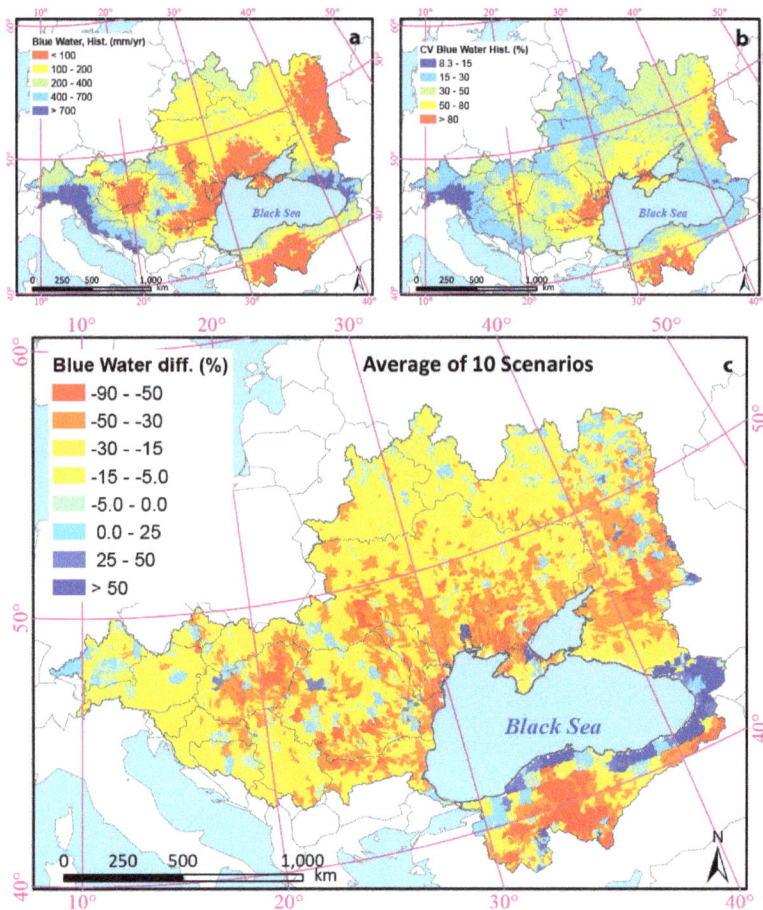

Figure 7. Spatial distribution of blue water resources in the Black Sea Catchment: (**a**) long-term historical average (1973–2006); (**b**) temporal variation of historical blue water; (**c**) percent deviation of future blue water (2020–2050) from the historic (1973–2006) based on the ensemble of the 10 scenarios.

The average of 10 scenarios suggests both an increase and a decrease in evapotranspiration across the BSC under climate and landuse change scenarios (Figure 9c). There is also a sharp increase in evapotranspiration of Georgia, which fits the increase in precipitation and temperature in this area. The variation among the 10 scenarios is the highest in the Danube delta and the Black Sea costal area. The average deviation from historic values (Figure 9c) suggests that evapotranspiration decreases by up to 12% in the Danube basin under future scenarios of change. The variation between model predictions of blue water, evapotranspiration, and soil moisture using four different future landuse scenarios shows limited impacts of landuse changes as compared to climate (the coefficient of variation is less than 1.3% among the four landuse scenarios) (Figure 10). Hence, the climatic signature is more significant than landuse in this study. The variations between scenarios are more pronounced in the blue water predictions (Figure 10a,b).

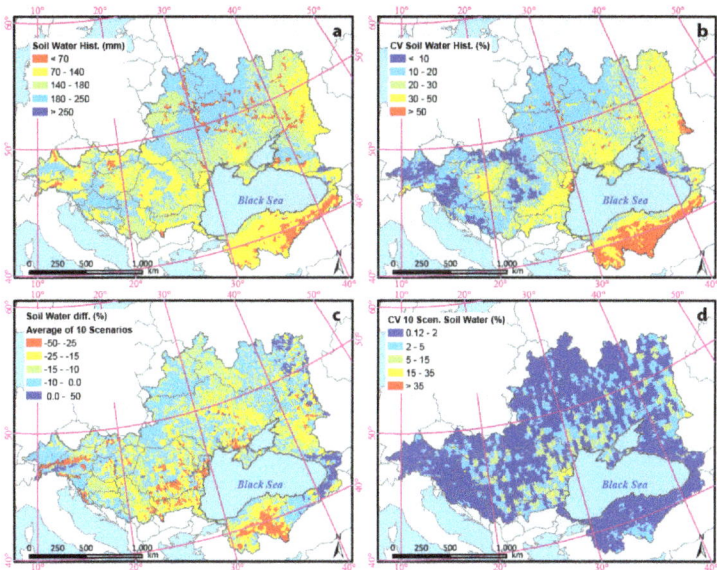

Figure 8. Spatial distribution of green water storage (soil water) in the Black Sea Catchment: (**a**) long-term historical average (1973–2006); (**b**) coefficient of variation (temporal variation) of the green water storage (1973–2006); (**c**) percent deviation of future green water storage (2020–2050) from the historic (1973–2006); (**d**) coefficient of variation of average soil moisture among the 10 scenarios.

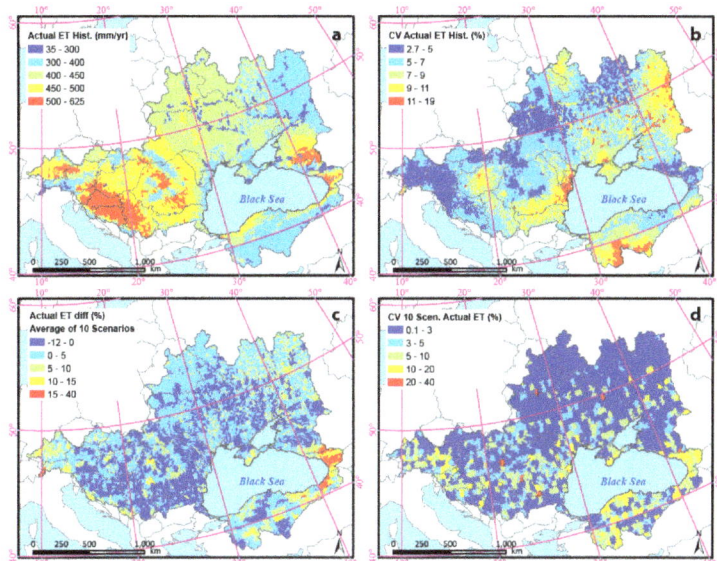

Figure 9. Spatial distribution of green water flow (evapotranspiration) in the Black Sea Catchment: (**a**) long-term historical average (1973–2006); (**b**) temporal variation of the green water flow (1973–2006); (**c**) percent deviation of the future green water flow (2020–2050) from historic (1973–2006) based on the ensemble of 10 scenarios; (**d**) coefficient of variation of actual evapotranspiration among the 10 scenarios.

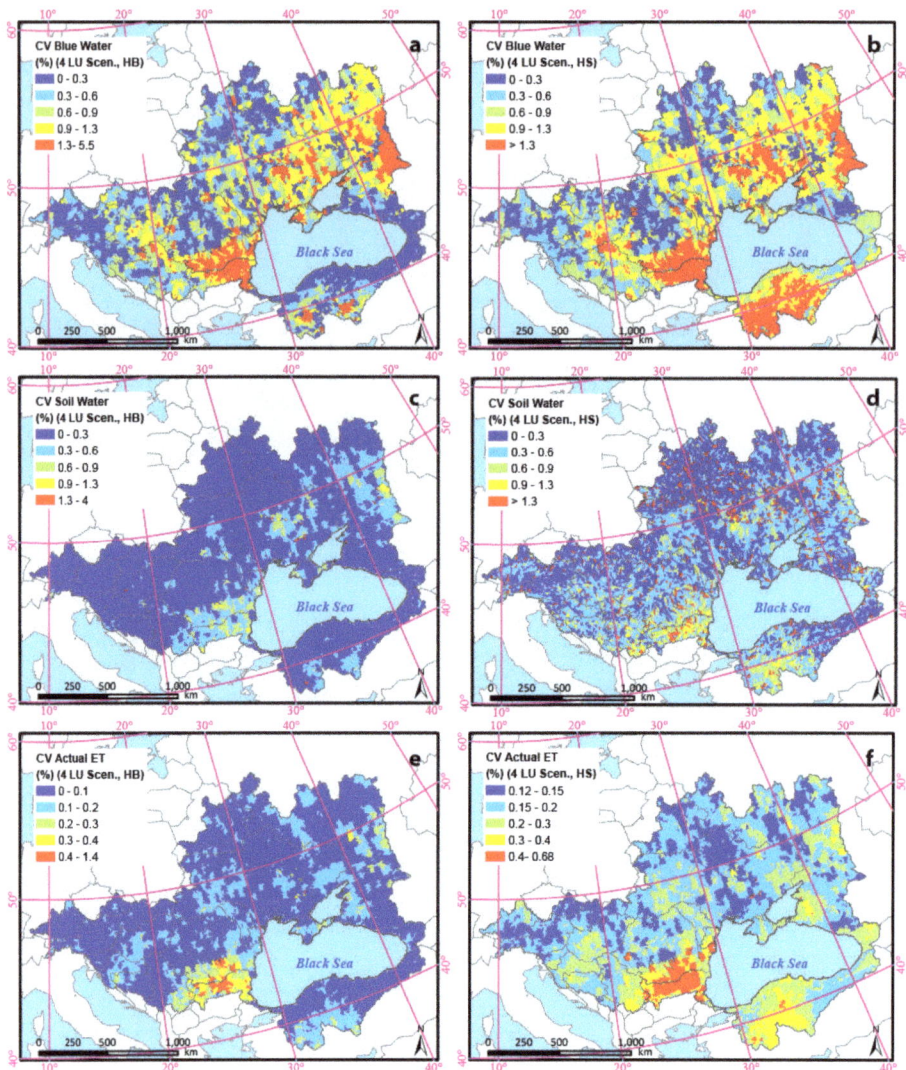

Figure 10. Coefficient of variation (CV) of the model predictions using four future landuse scenarios (BS ALONE, BS COOL, BS COOP, and BS HOT) in combination with the HB future climate Scenario (**left column**) and the HS future climate scenario (**right column**). (**a,b**) blue water resources; (**c,d**) green water storage (soil moisture); (**e,f**) green water flow (evapotranspiration). LU, landuse.

3.3. Extreme Events

We compared the frequency of occurrences of wet days with the precipitation thresholds of >2 and >10 mm d^{-1} in five selected sub-basins in different climatic regions across the BSC (Figure 11). In general, although the long-term average precipitation based on the two scenarios suggests a general decrease in precipitation in the catchment, the frequency of the wet days are slightly higher under future scenarios. In a sub-basin in the Eastern part of the BSC in Russia, the two climate scenarios predicted a slightly higher number of days with precipitation larger than 2 mm as compared to historical climate (Figure 11a,b). This is a region with low annual rainfall (350–450 mm year^{-1}) where there are few rainfall events exceeding 10 mm d^{-1} throughout the year. In a sub-basin in Southern Ukraine with 450–500 mm year^{-1} average annual precipitation, the frequency of wet days at the threshold of >2 mm d^{-1} stays as large as the historic period (Figure 11c). The slight decreases in number of days with >10 mm d^{-1} rainfall events indicate a smaller groundwater recharge, hence a larger chance of receding groundwater.

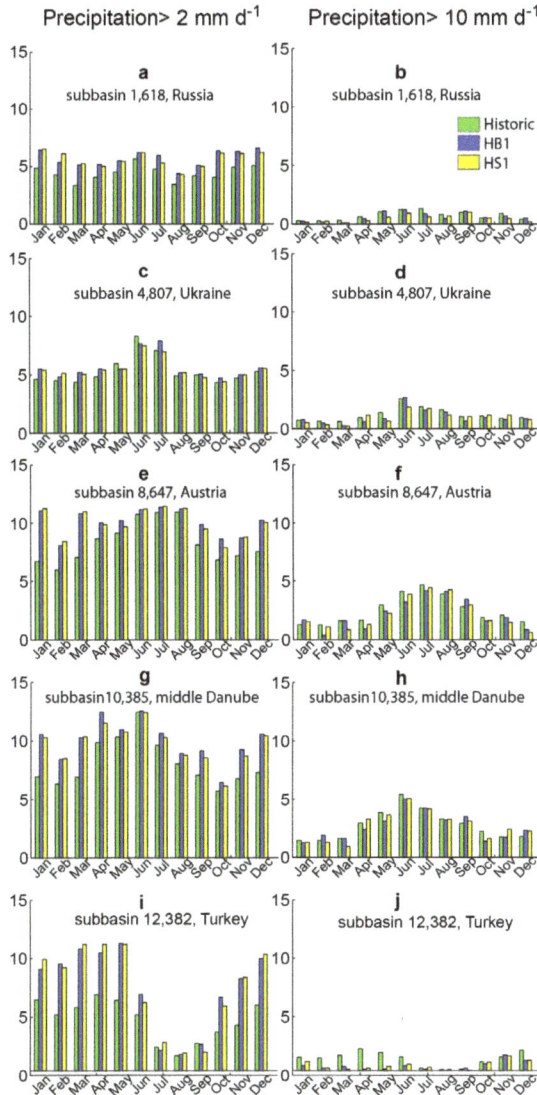

Figure 11. Comparison of the number of wet days with >2 mm d^{-1} threshold (**left column**), and >10 mm d^{-1} threshold (**right column**) between the historic (1973–2006) and HB and HS future climate scenarios (2020–2050) for five selected sub-basins (**a,b**) a subbasin in Russia; (**c,d**) a subbasin in Ukraine; (**e,f**) a subbasin in Austria; (**g,h**) a subbasin in middle Danube; and (**i,j**) a subbasin in Turkey.

In the selected sub-basin in Austria with a rainfall rate of 1000–1350 mm year^{-1}, the frequencies of wet days with more than 2 mm d^{-1} rainfall increase as compared with the historical data (Figure 11e).

In a sub-basin in the Alpine region in the middle of the Danube Basin, historic records of precipitation range between 1000–1350 mm year^{-1}. Both the HB and HS scenarios predict an increase in the number of wet days with a threshold of >2 mm d^{-1}, while the increase in the winter months is more distinct. The HS scenario predicts larger wet-day frequencies with the threshold of >10 mm d^{-1} than the HB scenario. The increase in precipitation frequencies at the threshold of >10 mm d^{-1} may indicate more flood risks in this region. However, the increase in the frequencies of precipitation with a large threshold (>10 mm d^{-1}) stay within the historic records in this sub-basin (Figure 11h).

Finally, in a sub-basins in Turkey, the climate models behave differently from the Alpine region, as the HB and HS climate scenarios predict a distinctly larger number of days with precipitation more than 2 mm d^{-1} than what the historic records show. The precipitation events at 10 mm d^{-1} threshold, however, are predicted to decrease in this selected sub-basin in Turkey (Figure 11i,j).

4. Summary

Combinations of two regional climate scenarios and four regional landuse scenarios were incorporated in the current study to explore the possible future impacts of climate and landuse changes on water resources of the Black Sea Catchment. The landuse scenarios were driven by the IPCC's special report on emissions scenarios (SRES) corresponding to four marker scenarios that represent different global socio-economic development pathways. The climate scenarios were generated from the Danish Regional Climate Model (RCM) (HIRHAM) for the IPCC's SRES A2 and B2 scenarios (HS and HB scenarios respectively). On average, the climate scenarios suggested a 5–15% decrease in future long-term average annual precipitation in most parts of the catchment. The decrease in precipitation is more pronounced in the HS scenario. According to the HS climate scenario, the Western part of the catchment (Danube Basin) will experience a decline in precipitation by 25%. As the historic precipitation records are large in this region, this is expected to have a large impact on the water resources of the entire region, and leaves the catchment with a significant net decrease of precipitation. Both scenarios suggest an increase in temperature by up to 2 °C with a west to east gradient. The extent of changes in temperature is more severe in the HS scenario as compared to the HB scenario.

We also quantified the impacts of combined climate and landuse changes on freshwater distribution in the BSC. As suggested by the ensemble of scenarios, on average, the catchment is expected to experience a decrease in its blue water and green water storage resources, while the green water flow (evapotranspiration) increases in some parts of the catchment and decreases in other parts (Figures 7–9). In addition, the decrease in fresh water resources in areas with high temporal variability in their water resources component (mainly in low lying countries around the Black Sea, such as Romania and Ukraine and the Russian part of the catchment) increases the vulnerability with regard to fresh water resources in these regions.

In our analysis, climate change had more pronounced effects on water resources, especially blue water, as opposed to the landuse change. To see the detailed effect of landuse change on the water resources component, it is beneficial to look at the water cycle at the HRU level, where the landuses are identical. This will give a true measure of landuse change impacts on water resources. The strength of the current work is the application of combined landuse and climate change scenarios. However, the study neglects the future changes in soil parameters over time, which accompanies changing landuses. Accounting for these changes will increase the confidence in the projected results, and needs to be further investigated.

An additional concern is the use of two regional climate scenarios (HS and HB) in model prediction while pursuing a thorough investigation based on the combined effect of many other Global Climate Models or Regional Climate Models would reflect climate model uncertainties, and hence is recommended. The study however, provides the basis to improve societal capabilities to anticipate

Water **2017**, *9*, 598

and manage water resources both today and in the future climate change environment in the Black Sea Catchment.

Acknowledgments: This project has been funded by the European Commission's Seventh Research Framework through the enviroGRIDS project (Grant Agreement n 226740). Elham Rouholahnejad Freund acknowledges funding from the Swiss National Science Foundation (SNSF grant P2EZP2_162279) during her stay at Ghent University. We are thankful to Emanuele Mancosu, Ana Gago Da Silva, and Saeid Ashraf Vaghefi for providing the landuse change and climate change scenarios. Special thanks to Raghavan Srinivasan for his instantaneous supports during the review process. The final manuscript benefited markedly from comments by the three anonymous referees; we greatly appreciate their efforts. We are also appreciate the efficient handling of the manuscript by the editor, Mindy Wang.

Author Contributions: Elham Rouholahnejad Freund and Karim C. Abbaspour conceived and designed the research; Elham Rouholahnejad Freund performed the data compilation, simulations, and analysis; Elham Rouholahnejad Freund wrote the paper; Anthony Lehmann and Karim C. Abbaspour contributed to materials and analysis tools and discussions.

Conflicts of Interest: The authors declare that they have no conflict of interest.

References

1. IPCC. *Managing the Risks of Extreme Events and Disasters to Advance Climate Change Adaptation*; A Special Report of Working Groups I and II of the Intergovernmental Panel on Climate Change; Field, C.B., Ed.; Cambridge University Press: Cambridge, UK; New York, NY, USA, 2012; p. 582.

2. Vaghefi, S.A.; Mousavi, S.J.; Abbaspour, K.C.; Srinivasan, R.; Yang, H. Analyses of the impact of climate change on water resources components, drought and wheat yield in semiarid regions: Karkheh River Basin in Iran. *Hydrol. Proc.* **2014**, *28*, 2018–2032. [CrossRef]

3. Krysanova, V.; Wortmann, M.; Bolch, T.; Merz, B.; Duethmann, D.; Walter, J.; Huang, S.; Tong, J.; Buda, S.; Kundzewicz, Z.W. Analysis of current trends in climate parameters, river discharge and glaciers in the Aksu River basin (Central Asia). *Hydrol. Sci. J.* **2015**, *60*, 566–590. [CrossRef]

4. Giorgi, F.; Lionello, P. Climate change projections for the Mediterranean region. *Glob. Planet. Chang.* **2008**, *63*, 90–104. [CrossRef]

5. Giorgi, F. Climate change Hot-spots. *Geophys. Res. Lett.* **2006**, *33*, L08707. [CrossRef]

6. Ulbrich, U.; May, W.; Li, L.; Lionello, P.; Pinto, J.G.; Somot, S. The Mediterranean climate change under global warming. *Dev. Earth Environ. Sci.* **2006**, *4*, 399–415.

7. Hattermann, F.F.; Weiland, M.; Huang, S.; Krysanova, V.; Kundzewicz, Z.W. Model-Supported Impact Assessment for the Water Sector in Central Germany Under Climate Change—A Case Study. *Water Resour. Manag.* **2011**, *25*, 3113–3134. [CrossRef]

8. Christensen, J.H.; Christensen, O.B. Climate modeling: Severe summertime flooding in Europe. *Nature* **2003**, *421*, 805–806. [CrossRef] [PubMed]

9. Semmler, T.; Jacob, D. Modeling extreme precipitation events—A climate change simulation for Europe. *Glob. Planet. Chang.* **2004**, *44*, 119–127. [CrossRef]

10. Deque, M.; Jones, R.G.; Wild, M.; Giorgi, F.; Christensen, J.H.; Hassell, D.C.; Vidale, P.L.; Rockel, J.B.D.; Kjellstrom, E.; Castro, M.D.; et al. Global high resolution vs. regional climate model climate change scenarios over Europe: Quantifying confidence level from PRUDENCE results. *Clim. Dyn.* **2005**, *25*, 653–670. [CrossRef]

11. Myroshnychenko, V.; Ray, N.; Lehmann, A.; Giuliani, G.; Kideys, A.; Weller, P.; Teodor, D. Environmental data gaps in Black Sea catchment countries: INSPIRE and GEOSS State of Play. *Environ. Sci. Policy* **2015**, *46*, 13–25. [CrossRef]

12. Schneider, C.; Laizé, C.L.R.; Acreman, M.C.; Flörke, M. How will climate change modify river flow regimes in Europe? *Hydrol. Earth Syst. Sci.* **2013**, *17*, 325–339. [CrossRef]

13. Aus der Beek, T.; Menzel, L.; Rietbroek, R.; Fenoglio-Marc, L.; Grayek, S.; Becker, M.; Kusche, J.; Stanev, E.V. Modeling the water resources of the Black and Mediterranean Sea river basins and their impact on regional mass changes. *J. Geodyn.* **2012**, *59*, 157–167. [CrossRef]

14. Engeland, K.; Hisdal, H. A Comparison of Low Flow Estimates in Ungauged Catchments Using Regional Regression and the HBV-Model. *Water Resour. Manag.* **2009**, *23*, 2567–2586. [CrossRef]

15. Farjad, B.; Gupta, A.; Marceau, D.J. Annual and Seasonal Variations of Hydrological Processes under Climate Change Scenarios in Two Sub-Catchments of a Complex Watershed. *Water Resour. Manag.* **2016**, *30*, 2851–2865. [CrossRef]

16. Arnold, J.G.; Srinivasan, R.; Muttiah, R.S.; Williams, J.R. Large area hydrologic modeling and assessment—Part 1: Model development. *J. Am. Water Resour. Assess.* **1998**, *34*, 73–89. [CrossRef]

17. Ndomba, P.; Mtalo, F.; Killingtveit, A. SWAT model application in a data scarce tropical complex catchment in Tanzania. *Phys. Chem. Earth* **2008**, *33*, 626–632. [CrossRef]

18. IMAGE. *The IMAGE 2.2 Implementation of the SRES Scenarios: A Comprehensive Analysis of Emissions, Climate Change and Impacts in the 21st Century*; RIVM CD-ROM Publication 481508018; National Institute of Public Health and the Environment RIVM: Bilthoven, The Netherlands, 2001.

19. Rouholahnejad, E.; Abbaspour, K.C.; Bacu, V.; Lehmann, A. Water resources of the Black Sea Basin at high spatial and temporal resolution. *Water Resour. Res.* **2014**, *50*, 5866–5885. [CrossRef]

20. Black Sea Investment Facility (BSEI). *Review of the Black Sea Environmental Protection Activities. General Review*; Black Sea Investment Facility (BSEI): Brussels, Belgium, 2005.

21. Tockner, K.; Uehlinger, U.; Robinson, C.T. *Rivers of Europe*; Academic Press, Elsevier: San Diego, CA, USA, 2009; ISBN 978-0-12-369449-2.

22. NASA. Land Processes Distributed Active Archive Center (LP DAAC), ASTER L1B, USGS/Earth Resour. Obs. and Sci. Cent.; Sioux Falls, South Dakota. 2001. Available online: http://lpdaac.usgs.gov (accessed on 12 August 2012).

23. Neitsch, S.L.; Arnold, J.G.; Kiniry, J.R.; Williams, J.R. *Soil and Water Assessment Tool Theoretical Documentation Version 2009*; Texas Water Resources Institute Technical Report No. 406; Texas A&M University System: College Station, YX, USA, 2009.

24. Williams, J.R. Flood routing with variable travel time or variable storage coefficients. *Trans. Am. Soc. Agric. Eng.* **1969**, *121*, 100–103. [CrossRef]

25. Lehmann, A.; Giuliani, G.; Mancosu, E.; Abbaspour, K.C.; Sözen, S.; Gorgan, D.; Beel, A.; Ray, N. Filling the gap between Earth observation and policy making in the Black Sea catchment with enviroGRIDS. *Environ. Sci. Policy* **2015**, *46*, 1–12. [CrossRef]

26. United Nation (UN). *World Population and Urbanization Prospects*; UN: New York, NY, USA, 2011; p. 302.

27. RIKS. *Metronamica–Model Descriptions*; Research Institute for Knowledge Systems: Maastricht, The Netherlands, 2011.

28. Mancosu, E.; Gago-Silva, A.; Barbosa, A.; de Bono, A.; Ivanov, E.; Lehmann, A.; Fons, J. Future landuse change scenarios for the Black Sea catchment. *Environ. Sci. Policy* **2015**, *46*, 26–36. [CrossRef]

29. Fowler, H.J.; Blenkinsop, S.; Tebaldi, C. Linking climate change modelling to impacts studies: Recent advances in downscaling techniques for hydrological modelling. *Int. J. Climatol.* **2007**, *27*, 1547–1578. [CrossRef]

30. Climatic Research Unit (CRU). CRU Time Series (TS) High Resolution Gridded Datasets, University of East Anglia Climatic Research Unit (CRU), NCAS British Atmospheric Data Centre. 2008. Available online: http://catalogue.ceda.ac.uk/uuid/3f8944800cc48e1cbc29a5ee12d8542d (accessed on 8 April 2011).

31. Gago Da Silva, A.; Gunderson, I.; Goyette, S.; Lehmann, A. Delta-Method Applied to the Temperature and Precipitation Time Series-An Example. enviroGRIDS FP7 Project, Report Number D3.6. 2012. Available online: http://archive-ouverte.unige.ch/unige:34235 (accessed on 1 March 2012).

32. Abbaspour, K.C.; Rouholahnejad, E.; Vaghefi, S.; Srinivasan, R.; Klöve, B. Modelling hydrology and water quality of the European Continent at a subbasin scale: Calibration of a high-resolution large-scale SWAT model. *J. Hydrol.* **2015**, *524*, 733–752. [CrossRef]

33. Jarvis, A.; Reuter, H.I.; Nelson, A.; Guevara, E. Hole-Filled SRTM for the Globe Version 4, Data Access: The CGIAR-CSI SRTM 90m Database. 2008. Available online: http://srtm.csi.cgiar.org (accessed on 3 March 2012).

34. Mitchell, T.D.; Jones, P.D. An improved method of constructing a database of monthly climate observations and associated high-resolution grids. *Int. J. Climatol.* **2005**, *25*, 693–712. [CrossRef]

35. Weedon, G.P.; Gomes, S.; Viterbo, P.; Shuttleworth, W.J.; Blyth, E.; Österle, H.; Adam, J.C.; Bellouin, N.; Boucher, O.; Best, M. Creation of the WATCH Forcing data and its use to assess global and regional reference crop evaporation over land during the twentieth century. *J. Hydrometeorol.* **2011**, *12*, 823–848. [CrossRef]

36. EEA Catchments and Rivers Network System v1.1 (ECRINS). *Rationales, Building and Improving for Widening Uses to Water Accounts and WISE Applications*; Publications Office of the European Union: Luxembourg, 2012; ISBN 978-92-9213-320-7.

37. Food and Agricultural Organization (FAO). *The Digital Soil Map of the World and Derived soil Properties*; CD-ROM, Version 3.5; Food and Agriculture Organization of the United Nations, Land and Water Development Division: Rome, Italy, 2003.

38. Portmann, F.T.; Siebert, S.; Döll, P. MIRCA2000 Global monthly irrigated and rainfed crop areas around the year 2000: A new high resolution data set for agricultural and hydrological modeling. *Glob. Biochem. Cycle* **2010**, *24*, GB1011. [CrossRef]

39. Monfreda, C.; Ramankutty, N.; Foley, J.A. Farming the planet: 2. Geographic distribution of crop areas, yields, physiological types, and net primary production in the year 2000. *Glob. Biogeochem. Cycles* **2008**, *22*, GB1022. [CrossRef]

40. Global Runoff Data Centre (GRDC). *Long-Term Mean Monthly Discharges and Annual Characteristics of GRDC Station/Global Runoff Data Centre*; Federal Institute of Hydrology (BfG): Koblenz, Germany, 2011.

41. Falkenmark, M.; Rockström, J. The new blue and green water paradigm: Breaking new ground for water resources planning and management. *J. Water Resour. Plan. Manag. ASCE* **2006**, *132*, 129–132. [CrossRef]

water

MDPI

Article

Using SWAT and Fuzzy TOPSIS to Assess the Impact of Climate Change in the Headwaters of the Segura River Basin (SE Spain)

Javier Senent-Aparicio *, Julio Pérez-Sánchez, Jesús Carrillo-García and Jesús Soto

Department of Civil Engineering, Catholic University of San Antonio, Campus de los Jerónimos s/n, 30107 Murcia, Spain; jperez058@ucam.edu (J.P.-S.); jacarrillo2@alu.ucam.edu (J.C.-G.); jsoto@ucam.edu (J.S.)
* Correspondence: jsenent@ucam.edu; Tel.: +34-968-278-818

Academic Editors: Karim Abbaspour, Raghavan Srinivasan, Saeid Ashraf Vaghefi, Monireh Faramarzi and Lei Chen
Received: 9 January 2017; Accepted: 17 February 2017; Published: 22 February 2017

Abstract: The Segura River Basin is one of the most water-stressed basins in Mediterranean Europe. If we add to the actual situation that most climate change projections forecast important decreases in water resource availability in the Mediterranean region, the situation will become totally unsustainable. This study assessed the impact of climate change in the headwaters of the Segura River Basin using the Soil and Water Assessment Tool (SWAT) with bias-corrected precipitation and temperature data from two Regional Climate Models (RCMs) for the medium term (2041–2070) and the long term (2071–2100) under two emission scenarios (RCP4.5 and RCP8.5). Bias correction was performed using the distribution mapping approach. The fuzzy TOPSIS technique was applied to rank a set of nine GCM-RCM combinations, choosing the climate models with a higher relative closeness. The study results show that the SWAT performed satisfactorily for both calibration (NSE = 0.80) and validation (NSE = 0.77) periods. Comparing the long-term and baseline (1971–2000) periods, precipitation showed a negative trend between 6% and 32%, whereas projected annual mean temperatures demonstrated an estimated increase of 1.5–3.3 °C. Water resources were estimated to experience a decrease of 2%–54%. These findings provide local water management authorities with very useful information in the face of climate change.

Keywords: water resources; SWAT model; climate change; Segura Basin; fuzzy TOPSIS

1. Introduction

Climate change as a result of increased greenhouse gas emissions leads to changes in hydrologic conditions and results in various impacts on the availability of global water resources [1]. Spain is one of the countries most vulnerable to the impacts of climate change in Europe due to the high spatial and temporal irregularity of water resources and socio-economic characteristics [2]. In addition, future climate tendencies show an increment in the temperature and a significant reduction in total annual rainfall [3]. When these impacts occur in regions that already present low water resource availability and frequent droughts, these impacts can be exacerbated. The Segura River Basin (SRB) is situated in SE Spain and is one of the most water-stressed basins in Mediterranean Europe [4].

In global terms, the total water demand for consumption is 1800 hm³/year, where 86% corresponds to agricultural use and 10% to urban uses. Against this demand data, annual natural water availability is, on average, around 800 hm³/year [5]. This deficit is partly covered by water from the Tagus-Segura water transfer and the use of unconventional water resources like treated wastewater and desalinated water, but these resources are not enough and the SRB is still suffering aquifer overexploitation [6]. The headwaters of the Segura River Basin (HWSRB) need to be studied thoroughly, as they are the most important sites for water resource generation in the basin [7]. The HWSRB have an important relevance in SRB water resources, since they comprise 9% of the water resource contribution, in spite of the fact that they covers only 1.2% of the area of the total watershed.

The SWAT has been successfully and widely used all over the world for different purposes, including the evaluation of climate change impacts on water resources [8]. However, SWAT applications assessing the water resources of Spain under changing climate conditions are scarce in scientific literature. Such studies have mostly been conducted in the north of the country, where there is an absence of water scarcity [9–14]. In the case of Spanish Mediterranean catchments, this model has rarely been used [15].

The Technique for Order Preference by Similarity to an Ideal Solution (TOPSIS) is one of the most used techniques for solving Multi-Criteria Decision Analysis (MCDA) problems and was first developed by Hwang and Yoon [16]. In the classic formulation of the TOPSIS method, personal judgements are represented with crisp values. However, crisp data are inadequate to model real-life decision problems under many conditions. That is why the fuzzy TOPSIS method was proposed, whereby the weights of criteria and ratings of alternatives are evaluated by linguistic variables represented by fuzzy numbers to deal with the deficiency in the classic TOPSIS. The fuzzy TOPSIS method has been widely applied in many fields; for example, energy [17], environment [18], industrial processes [19], and climate change [20]. However, to the best of our knowledge, the only precedent in the combined use of the SWAT model and the fuzzy TOPSIS method is found in Won et al. [21], wherein the authors assessed the water use vulnerability in 12 basins of South Korea, using SWAT to simulate hydrological components and fuzzy TOPSIS to rank the water use vulnerability in those basins.

The aim of the present study was to evaluate the climate change effect on the water resources of the HWSRB. The specific objectives of this study included: (1) to reduce the uncertainty in climate change projections by applying the fuzzy TOPSIS technique to rank climate models and (2) to explore the water resource response to future climate projections for the HWSRB. To achieve these objectives, we set up a hydrological model, using SWAT for the HWSRB. After calibration and validation of the model, nine different climate models were downloaded from the EURO-CORDEX initiative [22] and the fuzzy TOPSIS technique was applied to select which historical runs had the best fit with the observed climate data during a baseline period (1971–2000). Once those climate models were ranked, some of them were used to evaluate climate change in the study area for the medium term (2041–2070) and long term (2071–2100) using two different representative concentration pathways (RCP4.5 and RCP8.5). The results obtained in this study provide local water management authorities with very useful information for the proper utilisation and management of water resources under climate change conditions in this vulnerable region.

2. Description of the Study Area

The HWSRB is located in the southeastern region of Spain and covers an area of about 235 km^2. The basin is characterised by steep terrain, and the elevation ranges between 898 and 1912 m, as is shown in Figure 1. The drainage network is formed by two main rivers, the Segura River and its tributary, the Madera River. The whole basin drains into the Anchuricas Reservoir. The mean discharge at the reservoir is 1.6 m^3/s. This reservoir was constructed in 1957; it has a capacity of 6 hm^3, and the key purpose of the reservoir is to generate electricity. The study area is located mostly on permeable outcrops, limestone, and dolomite of the Upper Cretaceous. Land use in the watershed is highly forested; 61% of the surface is occupied by forests and 19% by Mediterranean shrubland vegetation. The rest of the land use is mainly for range purposes. The predominant soil type is rendzic leptosol, with variable depth always less than 50 cm, good drainage, and abundant stoniness [23].

The climate is typical Mediterranean with clear seasonality, rainy springs and autumns, and dry summers. According to data from 1971 to 2000, the mean annual precipitation ranged from 511 to 1300 mm, with an average value of 878 mm for the HWSRB, and the average annual temperature was 12.4 °C.

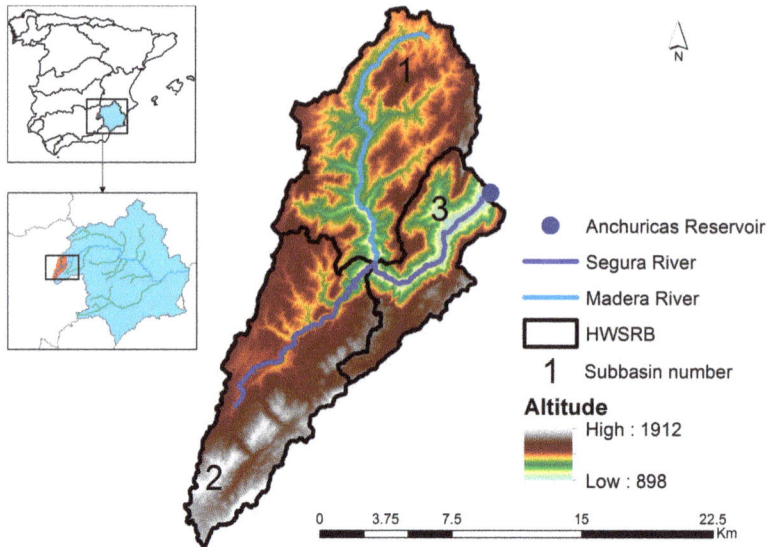

Figure 1. Location of the headwaters of the Segura River Basin (HWSRB).

3. Methodology

In order to evaluate the future impacts of climate change on water resources, the hydrological cycle was simulated in the headwaters of the SRB under different climate change scenarios. This process included three main steps: (1) setting up the hydrological model with observed stream flow and climate data; (2) selecting climate projections based on the fuzzy TOPSIS technique; and (3) incorporating climate scenarios into the hydrological model to evaluate the impact of the climate change in the headwaters of the SRB in the medium term (2041–2070) and long term (2071–2100).

3.1. SWAT Model

The SWAT [24] is a hydrological watershed model to evaluate the land practice water, sediment transport, and agricultural chemical yields in complex watersheds where soils, land use, or management can widely change.

The SWAT is a semi-distributed and physically based model. The balance equation used is

$$SW_t = SW_O + \sum \left(R_{day} - Q_{surf} - E_a - W_{seep} - Q_{gw} \right)$$

where SW_t is the final water soil content, SW_O is the initial water soil content, R_{day} is the precipitation, Q_{surf} is the surface runoff, E_a is the evapotranspiration, W_{seep} is the percolation, and Q_{gw} is the amount of baseflow (all in mm).

The basin is divided into sub-basins and those, in turn, are divided into hydrologic response units (HRUs). HRUs are defined by homogeneous regions with the same slope, soil, and land use. Each HRU generates an amount of runoff that is routed to calculate the total runoff. In order to calculate HRUs, the slope was divided into three classes (0%–8%, 8%–30% and >30%) and a threshold level of 10% was established to facilitate model processing and eliminate minor soils, slopes, and land uses for each subbasin.

3.1.1. Input Data for Hydrological Modelling

As for the data used to carry out the hydrological modelling, catchments were defined based on the digital elevation model (DEM), available on the website of the National Center for Geographic Information [25], with an accuracy of 25 m × 25 m. Meteorological data were obtained from the high-resolution (approximately 12 km × 12 km) gridded data set called SPAIN02. Detailed documentation of the development and analysis of the SPAIN02 data set can be found in Herrera et al. [26]. In this study, potential evapotranspiration was simulated using the Hargreaves method [27] due to the fact that it only requires minimum and maximum temperatures. Oudin et al. [28] checked that water balance models using parsimonious temperature-based methods perform similarly well compared to more data-demanding methods. The discharge data at the catchment outlet were available on the Hydrographical Study Centre website [29].

In addition to DEM, the Geographic Information Systems (GIS) input data required to build the SWAT model setup included a land cover map and a soil map. Land cover data were derived from reclassified Corine Land Cover 2006 [30] (1:50,000). The soil data for the HWSRB were obtained from the Harmonized World Soil Database (HWSD), assembled by the Food and Agriculture Organization of the United Nations (FAO). This database provides data for 16,000 map units containing two different soil layers (0–30 and 30–100 cm deep) [31].

3.1.2. Calibration and Validation of the SWAT Model

Sensitivity analysis was conducted to identify the most influential parameters for streamflow simulation, which were adjusted during calibration. Automatic calibration with the Sequential Uncertainty Fitting programme algorithm (SUFI-2) [32] was run with the sensible parameters and with other relevants in baseflow, groundwater, and runoff to improve the fit. SUFI-2 is a stochastic calibration that provides some relation between calibration and the uncertainty associated with ignorance about natural systems and all other sources, such as driving variables, conceptual model, parameters, and measured data. Detailed documentation of the SUFI-2 algorithm can be found in Abbaspour et al. [33].

The SWAT model was calibrated using monthly streamflow data for a period of thirteen years (1988–2000). Calibration is the process when observed and generated values are fitted as much as possible, searching for the best optimisation of an objective function; the Nash-Sutcliffe efficiency, in this case [34]. After calibration, the model was validated using the monthly discharge data of twelve years (1976–1987). Five years (1971–1975) were used to warm up the model in order to mitigate the effects of the initial conditions on the model output. With the best iteration parameters, the model performance was tested in a validation period and evaluated, as is shown in Table 1.

Table 1. Evaluation criteria for model performance [34].

Performance Rating	NSE	RSR	PBIAS (%)
Very good	$0.75 < \text{NSE} \leq 1.00$	$0.00 \leq \text{RSR} \leq 0.50$	$\text{PBIAS} < \pm 10$
Good	$0.65 < \text{NSE} \leq 0.75$	$0.50 < \text{RSR} \leq 0.60$	$\pm 10 \leq \text{PBIAS} < \pm 15$
Satisfactory	$0.50 < \text{NSE} \leq 0.65$	$0.60 < \text{RSR} \leq 0.70$	$\pm 15 \leq \text{PBIAS} < \pm 25$
Unsatisfactory	$\text{NSE} \leq 0.50$	$\text{RSR} > 0.70$	$\text{PBIAS} \geq \pm 25$

Moriasi et al. [35] recommended the Nash-Sutcliffe efficiency (NSE), root mean square error to the standard deviation ratio (RSR), and percent bias (PBIAS) as evaluation criteria for model performance.

3.2. Climate Scenarios and Statistical Bias Correction Method

The climate simulations used in this study consisted of 9 combinations of General Circulation Models and Regional climate models (GCM-RCM) from the EURO-CORDEX initiative [22] with a grid spacing of about 12.5 km (0.11° on a rotated grid). The EURO-CORDEX is an international climate downscaling initiative that aims to provide high-resolution climate scenarios for Europe [36]. The simulations have been produced assuming concentration pathways RCP4.5 and RCP8.5, described in van Vuuren et al. [37], and are listed in Table 2. For this study, 30 years of data from historical simulation runs (1971–2000) were used as the baseline period. The future climate is represented with two 30-year periods from the scenario simulation runs; medium term (2041–2070) and long term (2071–2100).

Table 2. Overview of the Regional Climate Models (RCMs) considered.

Institution	RCM	Driving Model
Climate Limited-Area Modelling Community (CLMcom)	CCLM4-8-17	CNRM-CM5
Climate Limited-Area Modelling Community (CLMcom)	CCLM4-8-17	MPI-ESM-LR
Danish Meteorological Institute (DMI)	HIRHAM5	EC-EARTH
Climate Service Centre in Hamburg, Germany (CSC)	REMO2009	MPI-ESM-LR
Royal Netherlands Meteorological Institute (KNMI)	RACMO22E	EC-EARTH
Swedish Meteorological and Hydrological Institute (SMHI)	RCA4	CNRM-CM5
Swedish Meteorological and Hydrological Institute (SMHI)	RCA4	EC-EARTH
Swedish Meteorological and Hydrological Institute (SMHI)	RCA4	MPI-ESM-LR
Institut Pierre-Simon Laplace (IPSL-INERIS)	WRF331F	IPSL-CM5A-MR

It is well known that climate model output data contain systematic errors and cannot be used directly in hydrological simulations [38]. That is why a bias correction technique was also applied to the downscaled data to increase the accuracy of the results. In this study, the bias correction technique based on distribution mapping of precipitation and temperature was applied. The idea of distribution mapping is to correct the distribution function of climate model values to agree with the observed distribution function. In 2012, Teutschbein and Seibert [39] evaluated different methods for bias correction of regional climate model simulations for hydrological climate change impact studies, and they obtained very good results applying this technique. In order to extract and bias correct data obtained from the climate models, the CMhyd tool was used [40].

3.3. Fuzzy TOPSIS

Fuzzy TOPSIS is based on the distance of each indicator for each regional climate model from the ideal solution. The steps followed in the application of this technique first include the determination of the fuzzy decision matrix, taking into account the number of climate models used and the indicators evaluated. After that, in order to homogenise the evaluation supplied for all the criteria, their values must be linearly normalised. Finally, the proximity coefficients (D_i^+, D_i^-) for each alternative are calculated in accordance with ideal and anti-ideal values selected for each indicator. This technique is designed to minimise the distance of a data object from the positive ideal solution (D_i^+) and maximise the distance from the negative ideal solution (D_i^-) [14]. The closeness coefficient (C_i) of each alternative is calculated as:

$$C_i = \frac{D_i^-}{(D_i^- + D_i^+)} \tag{1}$$

To establish the ranking of climate models, it is sufficient to sort them according to the decreasing values of their closeness coefficient. A clear example of a fuzzy approach to ranking climate models can be found in Raju and Kumar (2015) [20]. The climatic variables used were precipitation, minimum temperature, and maximum temperature. The correlation coefficient (CC), normalised root mean square deviation (NRMSD), and skill score (SS) [41] were used as performance indicators. Equal weights were considered for each criterion, and ideal and anti-ideal values for all the indicators were chosen as (1, 1, 1) and (0, 0, 0).

4. Results and Discussion

4.1. Calibration and Validation

A global sensitivity analysis found the following sensible parameters (Table 3): SOL_AWC, LAT_TTIME, SOL_BD, GW_REVAP, ALPHA_BF, RCHRG_DP, SOL_K, FFCB, OV_N, and GWQMN. The presence of several of these parameters showed the great importance of groundwater (ALPHA_BF, RCHRG_DP, GWQMN, and GW_REVAP) and lateral flow (LAT_TTIME) in this area, as is described in Conan et al. [42] and Galván et al. [43], where shallow aquifers have a relevant role. Some soil properties influence the opposition to the groundwater, and this justifies its presence as a sensible parameter, like SOL_AWC, SOL_K, and SOL_BD, usually listed in other studies [44].

Table 3. Range and final parameter values after calibration.

Parameter	Description	Value Range	Adjusted Value
SOL_AWC	Available water capacity of the soil layer (mm/mm)	(0, 1)	0.3
LAT_TTIME	Lateral flow travel time (days)	(0, 180)	174.6
SOL_BD	Moist bulk density (Mg/m^3)	(0.9, 2.5)	1.01
GW_REVAP	Groundwater "revap" coefficient	(0.02, 0.2)	0.17
ALPHA_BF	Baseflow alpha factor (days^{-1})	(0, 1)	0.72
RCHRG_DP	Deep aquifer percolation fraction	(0, 1)	0.85
SOL_K	Saturated hydraulic conductivity (mm/h)	(0, 2000)	14.5
FFCB	Initial soil water storage expressed as a fraction of field capacity water content	(0, 1)	0.69
OV_N	Manning's "n" value for overland flow	(0.01, 30)	21.92
GWQMN	Threshold depth of water in the shallow aquifer for return flow to occur (mm)	(0, 5000)	2459
CN2	SCS runoff curve number		+2.40%
ESCO	Soil evaporation compensation factor	(0, 1)	0.56

In addition to these parameters, CN2 was considered due to its correlation with runoff production, affecting baseflow as well. In addition, in a Mediterranean area where evapotranspiration has high relevance, ESCO was added because of its function in driving the extraction of the evaporative demand from lower soil layers [45,46].

As shown in Table 3, calibration with the SUFI-2 algorithm provided the best fitted values for the parameters. The best iteration was a very good performance based on performance criteria [35], with 0.80 NSE, 1.22% PBIAS, and 0.45 RSR.

The calibrated parameters were similar to previous references in areas with similar Mediterranean, warm, or semi-arid climatic or vegetation characteristics. The GW_REVAP value is close to the upper limit of the range due to the presence of forests in the area, as occurs with OV_N [47]; this allows the transfer of water to the root zone and increases evapotranspiration [46], which affects the baseflow and points out the importance of evapotranspiration in the Mediterranean balance. ALPHA_BF has a value that set aquifers as a medium-high velocity response to recharge [46]. The FFCB value is similar to others demonstrated in other Mediterranean watersheds [48].

Validation was required after calibration. The model was run for the 1971–1987 period, including five years for a warm-up period and, as in calibration, comparing monthly streamflow observed values with the simulated values. The model performance in this step was still accurate, with statistics described as good or very good, as shown in Table 4. Only PBIAS was worse than in the calibration period.

Figure 2 shows an accurate global model performance comparing the simulated and observed values; although the calibration period was drier than the validation and the streamflow had maximum peaks in the validation period that the SWAT overestimated, this is not an unusual issue when a model is implemented [7,43].

Table 4. Calibration and validation period performance.

Calibration			Validation		
NSE	PBIAS (%)	RSR	NSE	PBIAS (%)	RSR
0.80	+1.22	0.45	0.77	−12.64	0.48
Very good	Very good	Very good	Very good	Good	Very good

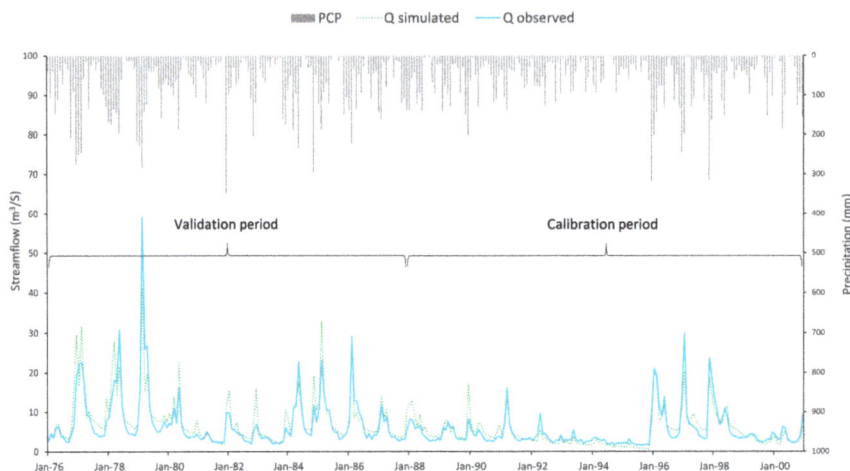

Figure 2. Soil and Water Assessment Tool (SWAT) model calibration and validation. Validation (1976–1987) and calibration (1988–2000).

4.2. Selection of RCMs Using TOPSIS

As shown in Table 5, data set grids relating to the period 1971–2000 obtained from SPAIN02 [26] were compared with historical runs obtained from each of the 9 RCMs to assess the CC, NRMSD, and SS under a fuzzy approach. Table 6 presents D_i^+, D_i^-, C_i, and a ranking pattern for every regional climate model used. The top three positions are occupied by RCA4_CNRM-CM5, RCA4_EC-EARTH, and HIRHAM5_EC-EARTH, with the relative closeness of 0.6216, 0.6213, and 0.4928. On the contrary, the seventh, eighth, and ninth positions are occupied by CCLM4-8-17_MPI-ESM-LR, WRF331F_IPSL-CM5A-MR, and REMO2009_MPI-ESM-LR, with the relative closeness of 0.1953, 0.1952, and 0.0951. These results suggest that RCA4_CNRM-CM5 and RCA4_EC-EARTH are suitable as input data for the SWAT modelling application in the case study.

Table 5. Normalised performance indicators obtained.

Model	CC			NRMSD			SS		
	p_{ij}	q_{ij}	r_{ij}	p_{ij}	q_{ij}	r_{ij}	p_{ij}	q_{ij}	r_{ij}
CCLM4-8-17_CNRM-CM5	0.5268	0.0000	0.0000	0.0000	0.9721	0.7978	0.0000	0.9630	0.7954
CCLM4-8-17_MPI-ESM-LR	0.0713	0.0588	0.9288	0.0766	0.0000	0.0000	0.0764	0.0000	0.0000
HIRHAM5_EC-EARTH	0.9960	0.0000	0.0000	0.9401	0.7408	0.0000	0.9389	0.7338	0.0000
REMO2009_MPI-ESM-LR	0.0000	0.3726	0.0000	0.0000	0.0000	0.1681	0.0000	0.0000	0.1692
RACMO22E_EC-EARTH	0.0000	0.8852	0.3098	0.0700	0.9579	0.0000	0.0699	0.9489	0.0000
RCA4_CNRM-CM5	0.0000	0.9725	0.0000	0.8774	0.8199	0.8003	0.8785	0.8483	0.8059
RCA4_EC-EARTH	0.8780	0.0386	0.7276	0.4320	0.7606	0.9630	0.4314	0.7892	0.9698
RCA4_MPI_ESM_LR	0.5958	0.0344	0.6874	0.0000	0.2168	0.0000	0.0000	0.2850	0.0000
WRF331F_IPSL-CM5A-MR	0.0713	0.0588	0.9288	0.0766	0.0000	0.0000	0.0764	0.0000	0.0000

Table 6. Ranking pattern of global climate models.

Model	D_i^+	D_i^-	C_i	Rank
CCLM4-8-17_CNRM-CM5	2.0399	1.7508	0.4619	4
CCLM4-8-17_MPI-ESM-LR	2.7382	0.6640	0.1953	7
HIRHAM5_EC-EARTH	2.0125	1.9551	0.4928	3
REMO2009_MPI-ESM-LR	2.8383	0.2982	0.0951	9
RACMO22E_EC-EARTH	2.3078	1.5022	0.3943	5
RCA4_CNRM-CM5	1.1980	1.9677	0.6216	1
RCA4_EC-EARTH	1.3145	2.1562	0.6213	2
RCA4_MPI_ESM_LR	2.4879	0.8413	0.2527	6
WRF331F_IPSL-CM5A-MR	2.7382	0.6640	0.1952	8

4.3. Water Resource Response to Climate Change

4.3.1. Changes in Projected Precipitation and Temperature

The mean annual projected precipitation and temperature are displayed in Table 7. Similar patterns are observed, with a general reduction in precipitation and a general increase in temperature. Expected changes in precipitation under RCP4.5 are totally different depending on the model analysed. While RCA4_EC-EARTH projects a slight increase in precipitation, RCA4_CNRM-CM5 projects a reduction that ranges between 13% and 19%. This significant variability was also found in the climate projections published by the Spanish agency of meteorology [49]. Under the RCP8.5 scenario, both models show a negative trend in precipitation that ranges between 6% and 17% in the medium term and 32% in the long term. With regards to projected temperature compared to the baseline period, the mean annual temperature suggests a significant and steady increase across the HWSRB in both scenarios. The temperature increase across the HWSRB will range between 0.9 and 1.3 °C in the medium term and 1.3 and 1.8 °C in the long term in the RCP4.5 scenario and between 1.5 and 2.1 °C in the medium term and 2.7 and 3.3 °C in the long term in the RCP8.5 scenario. As for temperature, there were no exceptions; all models showed a higher increase in temperature in the long term compared to the medium term. These results are consistent with other studies in Spanish Mediterranean areas [15,50].

As shown in Figure 3, the temperature increase in winter and autumn is lower than the increase in warmer months, increasing the intra-annual difference of the temperature. The lowest increase in temperature occurs in winter in the medium term in RCP4.5 (0.5 to 1.1 °C), while the largest increases occur during summer in the long term in RCP8.5 (2.9 to 3.8 °C). These results are consistent with other studies in small Mediterranean basins [15,50].

Table 7. Precipitation (mm) and temperature (°C) means with their variations.

Scenario	Model	Mean Annual Precipitation			Mean Annual Temperature		
		1971–2000	2041–2070	2071–2100	1971–2000	2041–2070	2071–2100
RCP4.5	RCA4_EC-EARTH	862	885 (+3%)	871 (+1%)	12.4	13.7 (+1.3)	14.2 (+1.8)
	RCA4_CNRM-CM5	906	732 (−19%)	788 (−13%)	12.4	13.3 (+0.9)	13.7 (+1.3)
RCP8.5	RCA4_EC-EARTH	862	811 (−6%)	615 (−32%)	12.4	14.5 (+2.1)	15.7 (+3.3)
	RCA4_CNRM-CM5	906	756 (−17%)	615 (−32%)	12.4	13.9 (+1.5)	15.1 (+2.7)
	Observed	877			12.4		

With respect to seasonal precipitation change (Figure 4), both models agreed on projecting a decrease in the precipitation for winter, spring, and autumn, while the precipitation will not suffer significant variations in summer.

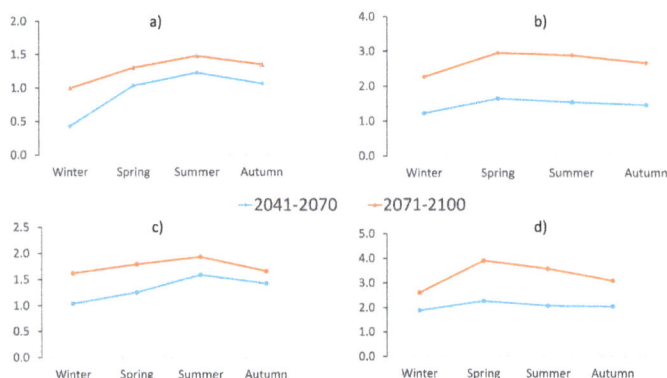

Figure 3. Seasonal temperature change (°C) in (**a**) RCA4_CNRM-CM5 (RCP4.5); (**b**) RCA4_CNRM-CM5 (RCP8.5); (**c**) RCA4_EC-EARTH (RCP4.5); and (**d**) RCA4_EC-EARTH (RCP8.5).

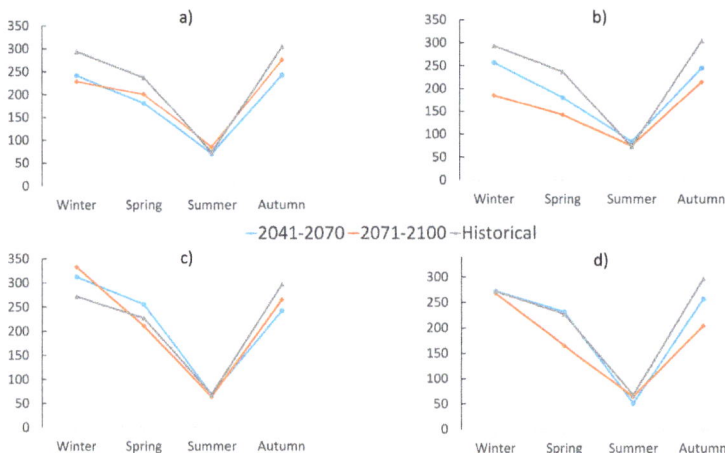

Figure 4. Seasonal precipitation change (mm) in (**a**) RCA4_CNRM-CM5 (RCP4.5); (**b**) RCA4_CNRM-CM5 (RCP8.5); (**c**) RCA4_EC-EARTH (RCP4.5); and (**d**) RCA4_EC-EARTH (RCP8.5).

4.3.2. Annual Streamflow Change

Under the RCP4.5 emissions scenario, the variability in the projected streamflow is very high, ranging from +4% to −35% in the medium term and projecting in the long term an important decrease that is expected to range between 2% and 23% (Figure 5). The possibility of a slight increment in the streamflow, as can be seen for the RCA4_EC-EARTH model, agrees with estimated projections by the Spanish government for the SRB [51], as does the prediction of a higher streamflow reduction in the medium term compared with the long term. These results can also be compared to those obtained by Estrela et al. [3], who also projected a reduction in mean annual runoff for the SRB between 21% and 33%. Overall, the increase in temperature and the projected decrease in precipitation will result in increased evapotranspiration, which will interact to reduce streamflow significantly [52]. Comparing precipitation and streamflow results, it can be seen that, due to higher actual evapotranspiration, the decreases obtained in streamflow exceed those in precipitation by 20%. These results are consistent with other studies in Mediterranean climates [53], in which the streamflow is very sensitive to a decrease in precipitation in basins with high evapotranspiration rates.

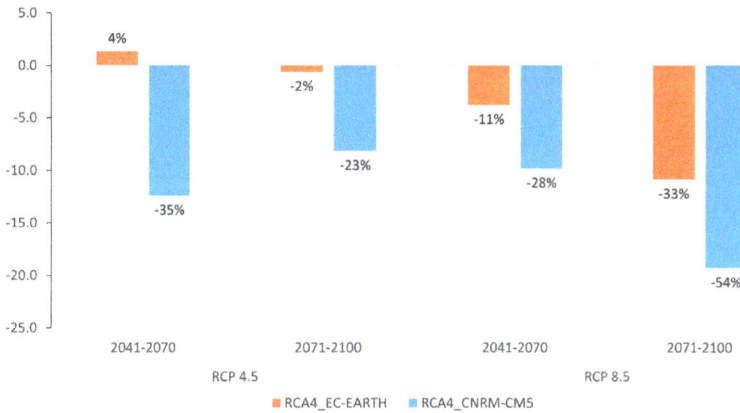

Figure 5. Annual streamflow change in the medium term (2041–2070) and the long term (2071–2100).

4.3.3. Seasonal Streamflow Change

Figure 6 shows the seasonal streamflow changes resulting from the estimated scenarios. Overall, a general seasonal streamflow decrease is expected for both scenarios and models. Only in winter and spring does the RCA4_EC-EARTH model estimate an increase of streamflow, which is consistent with the increment of precipitation estimated by this model. In summer, despite a lack of a clear decrease in the precipitation, an important decrease in streamflow is estimated due to the increase in the temperature, which would cause an increase in the evapotranspiration.

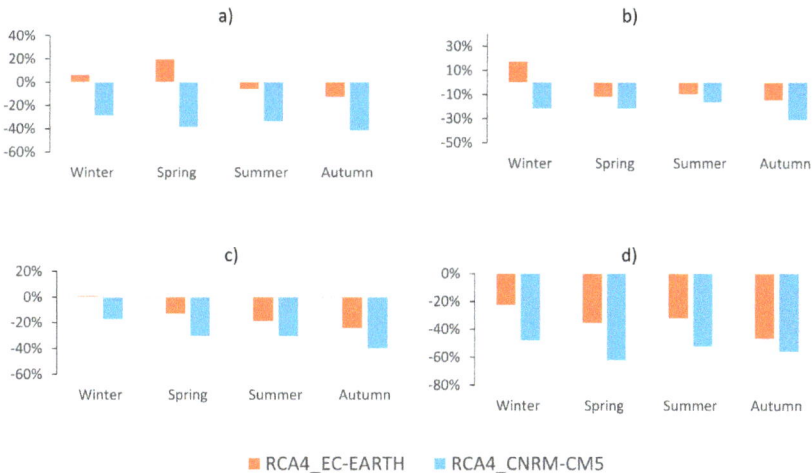

Figure 6. Seasonal evolution of the mean streamflow in (**a**) RCP4.5 (2041–2070); (**b**) RCP4.5 (2071–2100); (**c**) RCP8.5 (2041–2070); and (**d**) RCP8.5 (2071–2100).

4.3.4. Spatial Assessment of Expected Changes

As shown in Table 8, a significant agreement for the whole basin area was found due to the reduced extension and the homogeneous hydrological characteristics of the HWSRB. However, sub-basin 1 will suffer slightly higher reductions in its water resources.

Table 8. Spatial streamflow means distribution (m^3/s) and relative changes of runoff.

Subbasin	Model	1971–2000	RCP4.5 2041–2070	RCP4.5 2071–2100	RCP8.5 2041–2070	RCP8.5 2071–2100
1	RCA4_CNRM-CM5	13.1	8.1 (−38%)	9.7 (−26%)	9.2 (−30%)	5.8 (−56%)
	RCA4_EC-EARTH	12.4	12.7 (+2%)	12.1 (−3%)	10.7 (−13%)	8.1 (−35%)
2	RCA4_CNRM-CM5	16.0	11.0 (−31%)	12.9 (−20%)	11.9 (−25%)	7.5 (−53%)
	RCA4_EC-EARTH	14.6	15.5 (+6%)	14.5 (−1%)	13.3 (−9%)	10.1 (−31%)
3	RCA4_CNRM-CM5	35.5	23.1 (−35%)	27.3 (−23%)	25.6 (−28%)	16.2 (−54%)
	RCA4_EC-EARTH	33.1	34.4 (+4%)	32.5 (−2%)	29.3 (−11%)	22.2 (−33%)

5. Conclusions

In this study the impacts of projected climate change on water resources in the HWSRB were assessed. The Soil and Water Assessment Tool was used to simulate watershed hydrological processes, and the fuzzy TOPSIS technique was applied in order to select suitable RCM-GCM combinations and reduce uncertainties associated with climate modelling. Simulations with the calibrated model were then conducted for the medium term (2041–2070) and the long term (2071–2100) under two different representative concentration pathways, RCP4.5 and RCP8.5, based on CMIP5. The main findings can be summarised as follow:

- The SWAT model was able to reproduce the current hydrological conditions of the basin. The statistical results of calibration were NSE = 0.80, RSR = 0.45, and PBIAS = 1.22. The validation results were NSE = 0.77, RSR = 0.48, and PBIAS = −12.68. These results are indicative of the SWAT model's good performance.
- NRMS, CC, and SS were used to rank nine coupled runs of the GCM-RCM model, applying the fuzzy TOPSIS technique. Higher relative closeness was obtained by RCA4_CNRM-CM5 and RCA4_EC-EARTH. That is why these combinations were suggested for the assessment of the impact of climate change in the HWSRB.
- Based on the future projections, average annual temperature will increase about 3 °C and precipitation will decrease by 32% by the end of the century.
- Compared with the baseline period (1971–2000), future water resources in the HWSRB will experience a considerable change as the result of the changing temperature and precipitation. In the medium term (2041–2070), streamflow presents a high variability under the RCP4.5 and RCP8.5 scenarios, respectively. The largest alterations to streamflow are projected under RCP8.5 for 2071–2100, when they will decline between 33% and 54%.
- The results obtained from this modelling study may have strong implications in a basin that is already suffering from high water stress. The simulated impacts of climate change should be incorporated into water resource management plans to develop sustainable strategies. Future strategies should be focussed on decreasing demands and increasing the amount of unconventional water resources, but if the magnitude of the climate change renders these strategies insufficient, the need for new water transfer from another basin could arise.

Acknowledgments: This research has been partially supported by the Euro-mediterranean Water Institute (Grant No. 57/15) and the Spanish MINECO (Grant No. TIN2016-78799-P). We would also like to thank the SPAIN02 and CORDEX projects for the data provided for this study. In addition, we acknowledge Papercheck Proofreading & Editing Services.

Author Contributions: Javier Senent-Aparicio designed the experiments and wrote the manuscript. Julio Pérez-Sánchez helped to perform the experiments and reviewed and helped to prepare this paper for publication. Jesús Carrillo-García performed the set-up, calibration, and validation of the model, as well as scenario simulations. Jesús Soto applied the fuzzy TOPSIS technique in order to select climate change models.

Conflicts of Interest: The authors declare no conflict of interest.

References

1. Arnell, N.W. Climate change and global water resources. *Glob. Environ. Chang.* **1999**, *9*, 31–49. [CrossRef]
2. Vargas-Amelin, E.; Pindado, P. The challenge of climate change in Spain: Water resources, agriculture and land. *J. Hydrol.* **2014**, *518*, 243–249. [CrossRef]
3. Estrela, T.; Pérez-Martín, M.A.; Vargas, E. Impacts of climate change on water resources in Spain. *Hydrol. Sci. J.* **2012**, *57*, 1154–1167. [CrossRef]
4. Senent-Aparicio, J.; Pérez-Sánchez, J.; Bielsa-Artero, A.M. Asessment of Sustainability in Semiarid Mediterranean Basins: Case Study of the Segura Basin, Spain. *Water Technol. Sci.* **2016**, *7*, 67–84.
5. Segura Basin Management Plan, 2015–2021. Available online: https://www.chsegura.es/chs/planificacionydma/planificacion15-21/ (accessed on 11 December 2016).
6. Rodríguez-Estrella, T. The problems of overexploitation of aquifers in semi-arid areas: The Murcia Region and the Segura Basin (South-east Spain) case. *Hydrol. Earth Syst. Sci. Discuss.* **2012**, *9*, 5729–5756. [CrossRef]
7. García-Ruiz, J.M.; López-Moreno, J.I.; Vicente-Serrano, S.M.; Lasanta-Martínez, T.; Beguería, S. Mediterranean water resources in a global change scenario. *Earth Sci. Rev.* **2011**, *105*, 121–139. [CrossRef]
8. Gassman, P.W.; Sadeghi, A.M.; Srinivasan, R. Applications of the SWAT modle special section: Overview and insights. *J. Environ. Qual.* **2014**, *43*, 1–8. [CrossRef] [PubMed]
9. Raposo, J.R.; Dafonte, J.; Molinero, J. Assessing the impact of future climate change on groundwater recharge in Galicia-Costa, Spain. *Hydrogeol. J.* **2013**, *21*, 459–479. [CrossRef]
10. Arias, R.; Rodriguez-Blanco, M.L.; Taboada-Castro, M.M.; Nunes, J.P.; Keizer, J.J.; Taboada-Castro, M.T. Water resources response to changes in temperature, rainfall and CO_2 concentration: A first approach in NW Spain. *Water* **2014**, *6*, 3049–3067. [CrossRef]
11. Moran-Tejeda, E.; Lorenzo-Lacruz, J.; López-Moreno, J.I.; Rahman, K.; Beniston, M. Streamflow timing of mountain rivers in Spain: Recent changes and future projections. *J. Hydrol.* **2014**, *517*, 1114–1127. [CrossRef]
12. Zabaleta, A.; Meaurio, M.; Ruiz, E.; Antigüedad, I. Simulation climate change impact on runoff and sediment yield in a small watershed in the Basque Country, northern Spain. *J. Environ. Qual.* **2013**, *43*, 235–245. [CrossRef] [PubMed]
13. Morán-Tejeda, E.; Zabalza, J.; Rahman, K.; Gago-Silva, A.; López-Moreno, J.I.; Vincente-Serrano, S.; Lehmann, A.; Tague, C.L.; Beniston, M. Hydrological impacts of climate and land-use changes in a mountain watershed: Uncertainty estimation based on model comparison. *Ecohydrology* **2015**, *8*, 1396–1416. [CrossRef]
14. Palazón, L.; Navas, A. Land use sediment production response under different climatic conditions in an alpine-prealpine catchment. *Catena* **2016**, *137*, 244–255. [CrossRef]
15. Pascual, D.; Pla, E.; Lopez-Bustins, J.A.; Retana, J.; Terradas, J. Impacts of climate change on water resources in the Mediterranean Basin: A case study in Catalonia, Spain. *Hydrol. Sci. J.* **2015**, *60*, 2132–2147. [CrossRef]
16. Hwang, C.L.; Yoon, K. *Multiple Attribute Decision Making. Methods and Applications*; Springer: Heidelberg, Germany, 1981.
17. Cavallaro, F.; Zavadskas, E.K.; Raslanas, S. Evaluation of Combined Heat and Power (CHP) Systems Using Fuzzy Shannon Entropy and Fuzzy TOPSIS. *Sustainability* **2016**, *8*, 556. [CrossRef]
18. Beskese, A.; Demir, H.H.; Ozcan, H.K.; Okten, H.E. Landfill site selection using fuzzy AHP and fuzzy TOPSIS: A case study for Istanbul. *Environ. Earth Sci.* **2015**, *73*, 3513–3521. [CrossRef]
19. Guo, S.; Zhao, H. Optimal site selection of electric vehicle charging station by using fuzzy TOPSIS based on sustainability perspective. *Appl. Energy* **2015**, *158*, 390–402. [CrossRef]
20. Raju, K.S.; Kumar, D.N. Fuzzy Approach to Rank Global Climate Models. In *Proceedings of the Fifth International Conference on Fuzzy and Neuro Computing (FANCCO-2015)*; Ravi, V., Panigrahi, B.K., Das, S., Suganthan, P.N., Eds.; Springer International Publishing: Cham, Switzerland, 2015; pp. 53–62.
21. Won, K.; Chung, E.; Choi, S. Parametric Assessment of Water Use Vulnerability Variations Using SWAT and Fuzzy TOPSIS Coupled with Entropy. *Sustainability* **2015**, *7*, 12052–12070. [CrossRef]

22. Jacob, D.; Petersen, J.; Eggert, B.; Alias, A.; Christensen, O.B.; Bouwer, L.M.; Braun, A.; Colette, A.; Deque, M.; Georgievski, G.; et al. EURO-CORDEX: New high-resolution climate change projections for European impact research. *Reg. Environ. Chang.* **2014**, *14*, 563–578. [CrossRef]

23. FAO. *FAO/UNESCO Soil Map of the World: Revised Legend*; FAO World Resources Report 60; Food and Agricultural Organization of the United Nations: Rome, Italy, 1988.

24. Arnold, J.G.; Srinavasan, R.; Muttiah, R.S.; Williams, J.R. Large-area hydrologic modeling and assessment: Part I. Model development. *J. Am. Water Resour. Assoc.* **1998**, *34*, 73–89. [CrossRef]

25. National Center for Geographic Information. Available online: http://www.cnig.es (accessed on 16 June 2016).

26. Herrera, S.; Fernández, J.; Gutiérrez, J.M. Update of the Spain02 gridded observational dataset for EURO-CORDEX evaluation: Assessing the effect of interpolation methodology. *Int. J. Climatol.* **2016**, *36*, 900–908. [CrossRef]

27. Hargreaves, G.H. Defining and using reference evapotranspiration. *J. Irrig. Drain. Eng.* **1994**, *120*, 1132–1139. [CrossRef]

28. Oudin, L.; Hervieu, F.; Michel, C.; Perrin, C.; Andréassian, V.; Anctil, F.; Loumagne, C. Which potential evapotranspiration input for a lumped rainfall-runoff model? Part 2—Towards a simple and efficient potential evapotranspiration model for rainfall-runoff modeling. *J. Hydrol.* **2005**, *303*, 290–306. [CrossRef]

29. Hydrographical Study Centre. Available online: http://ceh-flumen64.cedex.es/anuarioaforos/default.asp (accessed on 16 June 2016).

30. Corine Land Cover 2006 Seamless Vector Data—European Environment Agency. Available online: http://www.eea.europa.eu/data-and-maps/data/clc-2006-vector-data-version-3 (accessed on 11 June 2016).

31. Nachtergaele, F.; van Velthuizen, H.; Verelst, L.; Batjes, N.; Dijkshoorn, K.; van Engelen, V.; Fischer, G.; Jones, A.; Montanarella, L.; Petri, M. *Harmonized World Soil Database*; Food and Agriculture Organization of the United Nations: Rome, Italy, 2008.

32. Abbaspour, K.C. *SWAT Calibration and Uncertainty Program—A User Manual*; SWAT-CUP-2012; Swiss Federal Institute of Aquatic Science and Technology: Dubendorf, Switzerland, 2012.

33. Abbaspour, K.C.; Johnson, C.A.; Genuchten, M.T.V. Estimating Uncertain Flow and Transport Parameters Using a Sequential Uncertainty Fitting Procedure. *Vadose Zone J.* **2004**, *3*, 1340–1352. [CrossRef]

34. Nash, J.E.; Sutcliffe, J.V. River flow forecasting through conceptual models. Part I: A discussion of principles. *J. Hydrol.* **1970**, *10*, 282–290. [CrossRef]

35. Moriasi, D.N.; Arnold, J.G.; van Liew, M.W.; Bingner, R.L.; Harmel, R.D.; Veith, T.L. Model evaluation guidelines for systematic quantification of accuracy in watershed simulations. *Trans. ASABE* **2007**, *50*, 885–900. [CrossRef]

36. Kotlarski, S.; Keuler, K.; Christensen, O.B.; Colette, A.; Deque, M.; Gobiet, A.; Goergen, K.; Jacob, D.; Luthi, D.; van Meijgaard, E.; et al. Regional climate modelling on European scales: A joint standard evaluation of the EURO-CORDEX RCM ensemble. *Geosci. Model Dev.* **2014**, *7*, 1297–1333. [CrossRef]

37. Van Vuuren, D.P.; Edmonds, J.; Kainuma, M.; Riahi, K.; Thomson, A.; Hibbard, K.; Hurtt, G.C.; Kram, T.; Krey, V.; Lamarque, J.F. The representative concentration pathways: An overview. *Clim. Chang.* **2011**, *109*, 5–31. [CrossRef]

38. Chen, J.; Brissette, F.; Lucas-Picher, P. Transferability of optimally-selected climate models in the quantification of climate change impacts on hydrology. *Clim. Dyn.* **2016**, *47*, 3359–3372. [CrossRef]

39. Teutschbein, C.; Seibert, J. Bias corretion of regional climate model simulations for hydrological climate-change impact studies: Review and evaluation of different methods. *J. Hydrol.* **2012**, *456–457*, 12–29. [CrossRef]

40. Rathjens, H.; Bieger, K.; Srinivasan, R.; Chaubey, I.; Arnold, J.G. CMhyd User Manual. Available online: http://swat.tamu.edu/software/cmhyd/ (accessed on 23 September 2016).

41. Perkins, S.E.; Pitman, A.J.; Holbrook, N.J.; McAveney, J. Evaluation of the AR4 climate models simulated daily maximum temperature, minimum temperature and precipitation over Australia using probability density functions. *J. Clim.* **2007**, *20*, 4356–4376. [CrossRef]

42. Conan, C.; de Marsily, G.; Bouraoui, F.; Bidoglio, G. A long-term hydrological modeling of the upper Guadiana river basin (Spain). *Phys. Chem. Earth* **2003**, *28*, 193–200. [CrossRef]

43. Galván, L.; Olías, M.; de Villarán, R.F.; Santos, J.M.D.; Nieto, J.M.; Sarmiento, A.M.; Cánovas, C.R. Application of the SWAT model to an AMD-affected river (Meca River, SW Spain). Estimation of transported pollutant load. *J. Hydrol.* **2009**, *377*, 445–454. [CrossRef]

44. Molina-Navarro, E.; Martínez-Pérez, S.; Sastre-Merlín, A.; Bienes-Allas, R. Hydrologic modeling in a small mediterranean basin as a tool to assess the feasibility of a limno-reservoir. *J. Environ. Qual.* **2014**, *43*, 121–131. [CrossRef] [PubMed]

45. Bressiani, D.D.A.; Srinivasan, R.; Jones, C.A.; Mendiondo, E.M. Effects of spatial and temporal weather data resolutions on streamflow modeling of a semi-arid basin, northeast Brazil. *Int. J. Agric. Biol. Eng.* **2015**, *8*, 14–25.

46. Neitsch, S.L.; Arnold, J.G.; Kiniry, J.T.; Williams, J.R. Soil and Water Assessment Tool. Theoretical Documentation Version 2009. Available online: http://swat.tamu.edu/media/99192/swat2009-theory.pdf (accessed on 3 June 2016).

47. Begou, J.C.; Jomaa, S.; Benabdallah, S.; Bazie, P.; Afouda, A.; Rode, M. Multi-Site Validation of the SWAT Model on the Bani Catchment: Model Performance and Predictive Uncertainty. *Water* **2016**, *8*, 178. [CrossRef]

48. Martínez-Casasnovas, J.A.; Ramos, M.C.; Benites, G. Soil and Water Assessment Tool Soil Loss Simulation at the Sub-Basin Scale in the Alt Penedès-Anoia Vineyard Region (Ne Spain) in the 2000s. *Land Degrad. Dev.* **2016**, *27*, 160–170. [CrossRef]

49. AEMET 2017. Climate Projections for the XXI Century. Available online: http://www.aemet.es/es/serviciosclimaticos/cambio_climat/ (accessed on 31 January 2017).

50. Sellami, H.; Benabdallah, S.; La Jeunesse, I.; Vanclooster, M. Quantifying hydrological responses of small Mediterranean catchments under climate change projections. *Sci. Total Environ.* **2016**, *543*, 924–936. [CrossRef] [PubMed]

51. Centre for Public Works Studies and Experimentation (CEDEX). Evaluación del Impacto del Cambio Climático en los Recursos Hídricos en Régimen Natural (In Spanish). 2017. Available online: http://www.mapama.gob.es/es/cambio-climatico/publicaciones/publicaciones/Memoria_encomienda_CEDEX_tcm7-165767.pdf (accessed on 31 January 2017).

52. Li, F.; Zhang, G.; Xu, Y.J. Assessing climate change impacts on water resources in the Songhua River basin. *Water* **2016**, *8*, 420. [CrossRef]

53. Molina-Navarro, E.; Hallack-Alegría, M.; Martínez-Pérez, S.; Ramírez-Hernández, J.; Mungaray-Moctezuma, A.; Sastre-Merlín, A. Hydrological modeling and climate change impacts in an agricultural semiarid region. Case study: Guadalupe River basin, Mexico. *Agric. Water Manag.* **2016**, *175*, 29–42. [CrossRef]

water

MDPI

Article

Evaluating the Impact of Low Impact Development (LID) Practices on Water Quantity and Quality under Different Development Designs Using SWAT

Mijin Seo [1], Fouad Jaber [2,*], Raghavan Srinivasan [3] and Jaehak Jeong [4]

[1] National Institute of Agricultural Sciences, Rural Development Administration, Wanju,
     Jeollabuk-do 565-851, Korea; mjseo1020@korea.kr
[2] Biological and Agricultural Engineering, Texas A&M AgriLife Extension, Dallas, TX 75252, USA
[3] Spatial Science Lab, Texas A&M University, College Station, TX 77845, USA; r-srinivasan@tamu.edu
[4] Blackland Research Center, Texas A&M AgriLife Research, Temple, TX 76502, USA; jjeong@brc.tamus.edu
*   Correspondence: f-jaber@tamu.edu; Tel.: +1-972-952-9672

Academic Editor: Ataur Rahman
Received: 29 December 2016; Accepted: 2 March 2017; Published: 7 March 2017

Abstract: The effects of Low Impact Development (LID) practices on urban runoff and pollutants have proven to be positive in many studies. However, the effectiveness of LID practices can vary depending on different urban patterns. In the present study, the performance of LID practices was explored under three land uses with different urban forms: (1) a compact high-density urban form; (2) a conventional medium-density urban form; and (3) a conservational medium-density urban form. The Soil and Water Assessment Tool (SWAT) was used and model development was performed to reflect hydrologic behavior by the application of LID practices. Rain gardens, permeable pavements, and rainwater harvesting tanks were considered for simulations, and a modeling procedure for the representation of LID practices in SWAT was specifically illustrated in this context. Simulations were done for each land use, and the results were compared and evaluated. The application of LID practices demonstrated a decrease in surface runoff and pollutant loadings for all land uses, and different reductions were represented in response to the land uses with different urban forms on a watershed scale. In addition, the results among post-LIDs scenarios generally showed lower values for surface runoff and nitrate in the compact high-density urban land use and for total phosphorus in the conventional medium-density urban land use compared to the other land uses. We suggest effective strategies for implementing LID practices.

Keywords: effectiveness of LID practices; different urban designs; SWAT; model development; LID modeling

1. Introduction

Urbanization has caused many problems for runoff and pollutants due to the increase in impervious surfaces. This increase in impervious surfaces changes natural flow characteristics, causing increased runoff volume and peak flow rate, decreased groundwater recharge due to interrupted infiltration to soil layers, and a lowered water table, consequentially causing decreased base flow [1,2]. In addition, urban runoff from impervious surfaces is a main transport mechanism for many pollutants, such as sediment, heavy metals, and nutrients to nearby water bodies. These pollutants contribute to the deterioration of water quality. New methods of stormwater management are therefore required to mitigate the impact of urbanization on runoff and pollutants from an environmental perspective. One alternative strategy is the implementation of Low Impact Development (LID) practices (or urban Best Management Practices; urban BMPs), designed to treat water at the source where it is generated. LID practices can reverse the deteriorated conditions back to a pre-development state or even better [3].

Many studies of hydrology and water quality treatment through LID practices have been conducted. LID practices have been deemed effective through positive results from experiments and modeling. For instance, the installation of bioretention cells or permeable pavements has resulted in large reductions in runoff volumes, peak flow rates, and pollutants [4–8]. For a modeling approach, Abi Aad et al. [9] modeled rain tanks and rain gardens using Storm Water Management Model 5 (SWMM 5), and demonstrated that runoff was delayed and reduced by them. Ackerman and Stein [10] indicated reductions of flow, sediment, and copper by a bioretention cell, a grassed swale, a planter box, and a planter box with a grassed swale in their study that evaluated the effectiveness of BMPs by using Hydrologic Simulation Program-Fortran (HSPF) coupled with a BMP module. Carter and Jackson's [11] study investigated the effects of green roofs on hydrology at four spatial scales using a StormNet Builder model, which they showed significantly reduced peak runoff rates.

The effectiveness of LID practices, however, can vary depending on a variety of conditions. Some studies have demonstrated that LID practices are reliant on watershed characteristics such as soils, topography, and precipitation. Holman-Dodds et al. [12] reported large runoff on a low infiltration type D soil despite the existence of LID practices and also indicated the decreased effectiveness of LID practices under large precipitation. Brander et al. [13] revealed that the performance of LID practices was effective on soil type A and small storms. The effectiveness of LID practices for small storms was also presented in Ackerman and Stein [10], Carter and Jackson [11], Schneider and McCuen [14], etc. In addition, the effects of LID practices were evaluated differently according to locations, numbers, and types of LID practices [15,16].

Other than these watershed characteristics and LID practice conditions, there could be other factors that influence the effectiveness of LID practices. One thing we could consider is the impact of urban patterns. Some studies have determined the positive impacts of high-density urban pattern on water volumes and pollutant loadings. Seo [17] investigated how the amount of runoff and pollutant loadings were generated differently under three different urban planning designs and presented the compact high-density urban type as the most effective urban type. Jacob and Lopez [18] also evaluated the benefits of high density development for the reduction of water quality loadings in comparison with standard suburban developments, mentioning it as an effective approach more than traditional BMPs under their study conditions. Such studies imply that the effects of the application of LID practices could vary with different urban patterns. However, a limited number of studies have been performed on the effectiveness of LID practices under different urban design patterns. For example, Brander et al. [13] analyzed the effects of infiltration practices on urban runoff under their four development types (conventional curvilinear, urban cluster, coving, and new urbanism) using a spreadsheet model, the Infiltration Patch (IP). They showed runoff reduction to be different for the four types of development designs, and the smallest runoff was obtained for the urban clustered design in most scenarios because of the large natural land area. Williams and Wise [19] simulated the hydrologic responses from traditional and clustered developments with BMPs and LID practices using the Hydrologic Engineering Center-Hydrologic Modeling System (HEC-HMS), and they indicated very similar results to the results of the pre-development condition in the clustered development with LID practices. Gilroy and McCuen [20] studied the three land uses: "single family", "townhome", and "commercial lot" to identify the impact of location and volume capacity of urban BMPs (cisterns and bioretention cells) on runoff volumes and peak discharge rates. They represented different percentages of reduction in the three land uses under every scenario for location and volume. However, very few studies have attempted to simulate LID practices and land use with different urban patterns, especially for rain gardens (RGs), permeable pavements (PPs), and rainwater harvesting tanks (RWHs) (which were considered in the present study), using the Soil and Water Assessment Tool (SWAT).

In this regard, we focused on the application of the LID practices in SWAT and on the evaluation of the watershed-wide effectiveness of the LID practices under given different urban designs. The SWAT model was developed to simulate three LID practices. The hydrologic and water quality results were analyzed and compared with and without LID practices within the same land use and among different land uses. The results of the post-development states from the Seo [17] study were utilized as baseline data to evaluate the post-development states with LID practices. In the text, the terms "pre-LIDs" and "post-LIDs" are used to designate the post-development state before and after constructing LID practices, respectively.

2. Materials and Methodology

2.1. Study Area Description

Runoff and pollutant problems caused by stormwater have been a crucial issue in coastal areas because these areas receive pollutants from upstream sources [21,22] and are simultaneously affected by the tide. In particular, urban areas usually face more serious threats because increased impervious surfaces can discharge water and pollutants without natural handling. The study area, situated to the north of League City, Texas, within the Clear Creek watershed, meets the described characteristics. It is located downstream of Clear Creek near Galveston Bay, and is planned for regional development (Figure 1).

Figure 1. The location of the study area (**right**) included in the Clear Creek watershed (**left**).

It is desirable to scale up the analysis of LID practices to a large watershed after observing detectable water quantity and quality changes at a small level [23]. This is because modeling LID practices at a large scale can make the noticeable effectiveness of LID practices difficult to assess, so that it cannot provide information for changes that should be conducted at small-scale levels [23]. Thus, within the boundary of a pre-developed area, a roughly 3.5 km^2 (350 ha) small area was considered as the study area.

The topography ranges from 0 m to 11 m in elevation, with roughly 90% of the area within 6 m to 8 m in elevation, so the slope of the area is mild. Typical characteristics of this area are high runoff and low permeability. Four kinds of soils are present in this area. Addicks (loam) is the most predominant, comprising about 61% of the soil, followed by Bernard (clay loam), comprising about 27% of the soil. Lake Charles (clay) and Aris (silt loam) cover the remainder. All soil properties are represented as poorly drained hydrologic soil group (HSG) D. Wetland and hay are dominant, making up about 60% of current pre-developed land use. The weather is generally typified by hot summers and clement winters, indicating monthly mean temperatures of around 84 °F (29 °C) in August and around 53 °F (12 °C) in January. The average annual temperature is around 70 °F (21 °C). The impact of the oceanic climate decreases the difference between the low and high temperatures. The monthly average precipitation ranges from about 50 mm to 165 mm, and the average annual precipitation is about 1270 mm. A high probability of extreme storms exists in this area. The study area is located in Harris County [24].

2.2. Description of Input Data

Spatial and temporal input data, projected as an Albers Equal-Area Conic projection with North American 1983 datum, were used for setting up the model. A 10 m squared resolution Digital Elevation Model (DEM) was used to sufficiently express details, obtained from the Natural Resources Conservation Service (NRCS) Geospatial Data Gateway.

For land uses, three different types of land use data, derived from potential urban layouts typical of League City, were considered. These included: (1) a compact high-density urban land use (termed as UHD); (2) a conventional medium-density urban land use (termed as UMD); and (3) a conservational medium-density urban land use (termed as UMC) (Figure 2). The urban area of each land use consists of residential and commercial areas. In the figure, parts of the residential and commercial areas are enlarged from the entire urban areas representing those patterns. The same population applied to all residential areas of land uses. UHD land use includes the smallest portion of residential area and is urbanized, the most among the three urban designs, but also allows for most of the area to remain as natural space. It has a larger roof area in the residential area than the other two designs in order to accommodate an identical population. Thus, it represents a high percentage of imperviousness in the residential area. UMD land use has a pervasive urban pattern in the United States. The residential part of the urban area is composed of conventional neighborhoods consisting of single family units. A UMC residential area includes conservational areas that have to be kept as green space under the same base format with the conventional neighborhoods of the UMD residential area. Thus, it represents less imperviousness than the UMD residential area. The UMD and UMC land uses have the same size of residential area, and the residential area makes up more area than that of the UHD land use. The commercial area of all urban areas is the same in size. In total, urban area occupied about 21% and 56% in the UHD and UMD/UMC land uses, respectively. The residential and commercial areas represent different impervious and pervious ratios for each urban area (Table 1). For the remaining land areas, excluding urban areas, land use data obtained from the USDA NRCS Geospatial Data Gateway were represented to a pre-development state. The same land use data from Seo [17] were used to assess the effectiveness of LID practices under these different urban land uses, and more detailed design specifications can be found in Seo [17].

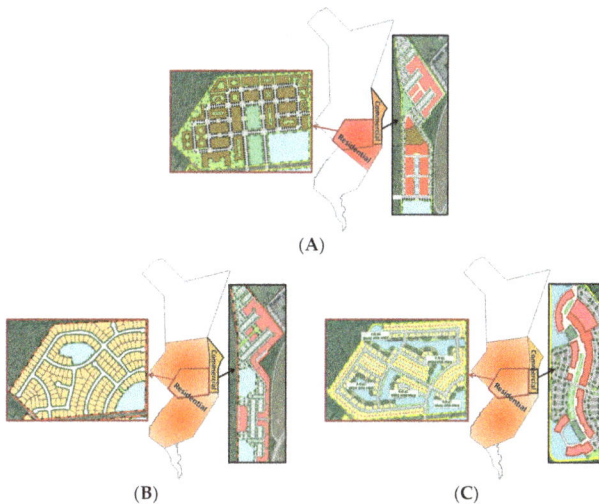

(A)

(B)   (C)

Figure 2. Three land use data with different urban forms (Parts of the residential and commercial areas are enlarged): (**A**) Compact high-density urban land use (UHD); (**B**) Conventional medium-density urban land use (UMD); and (**C**) Conservational medium-density urban land use (UMC).

Table 1. Information for each urban area for the three land uses (in %).

Land Use	Urban Area [1]	Impervious/Pervious Fraction [2]	
		Residential	Commercial
UHD	21	61/39	68/32
UMD	56	44/56	75/25
UMC	56	41/59	68/32

Notes: [1] The proportion of an urban area for total land use area; [2] The fraction of impervious and pervious parts in an urban area.

Soil data, the high-resolution Soil Survey Geographic Database (SSURGO), were obtained from the NRCS Soil Data Mart (https://websoilsurvey.sc.egov.usda.gov/App/HomePage.htm). Daily precipitation and temperature were collected from the National Climate Data Center (NCDC) at Houston Clover Field and at the National Weather Service Office stations, considered as representative stations for the study area. A weather generator was used for the rest of the weather dataset of the simulation.

2.3. Model Selection

A watershed-wide evaluation for the effectiveness of LID practices is needed because stormwater eventually has an influence on the final water body of a watershed [25]. It is cumbersome to calculate reduction rates from all LID practice sites within a watershed for a watershed-wide evaluation. Moreover, since the reductions of runoff and pollutant loads by LID practices can be affected by various watershed characteristics such as topography, land use, soil property, precipitation, and so forth, in this regard, a modeling approach is required to take into account all of these factors. It is important to select an optimal model that properly reflects the hydrologic responses with the application of LID practices. In the present study, SWAT was selected because it has an ability to simulate the process of hydrology and water quality in a variety of studies for long periods [26–28]. SWAT has effective components for the simulation of water quantity and quality. It applies a modified NRCS curve number (CN) method [29] to estimate surface runoff and a Modified Universal Soil Loss Equation (MUSLE) [30] to calculate sediment yields. Different forms of nutrients which are transformed into several pools (e.g., organic and inorganic pools) are also simulated. A comprehensive description of the processes is provided in Neitsch et al. [31].

The model was initially developed for the purpose of simulating water quantity and quality from agricultural and rural environments. However, it is gradually showing its capacity to simulate mixed land uses, which have a large proportion of urban areas or urban settings [32–35]. In addition, the suitability of SWAT in the simulation of agricultural BMPs has been proven. The benefits of many agricultural practices have been examined and evaluated using SWAT [36,37]. This implies that SWAT has the potential to predict water quantity and quality for urban watershed management systems [38]. Existing BMP tools have been upgraded and modified, and new tools for urban BMP modeling are being added in SWAT. For example, Jeong et al. [39] reported a development of algorithms for urban BMPs in SWAT such as Sedimentation-Filtration Basins, Retention-Irrigation Basins, Detention Ponds, and Wet Ponds. Jeong et al. [32] also tested the Sedimentation-Filtration basins (SedFil) algorithm to validate the capability of its components in SWAT. Additionally, the recently updated new version, SWAT 2012, allows many conservation practices, which were not included in other existing models, for modeling water quality by entering pollutant removal efficiencies. As the development of improved tools is encouraged for LID modeling in SWAT, processes through updates and modifications are continuously in progress to adequately represent LID practices.

2.4. Previous SWAT Simulation

The pre- and post-development simulations from the Seo [17] study were used as baseline simulations in order to investigate the effectiveness of LID practices under the same land uses with her study. In the previous work, the influence of land use change on water quantity and quality was identified under three different land uses. To do this, the following stepwise procedures were conducted. The pre-development condition (termed as 'prestate' was first taken into account to assess the impact of urban development. The process was focused on calibration and validation to obtain parameters that could stand for characteristics of the study area. The study area was difficult to calibrate because of sparse and tidal-affected data. Thus, the upstream gauging station (United States Geological Survey (USGS) site number: 08076997 with sufficient data and outside the impact of tidal currents) was considered for calibration, and the SWAT simulation was carried out over the entire Clear Creek watershed (424 km^2), including the study area. The calibration process was performed by using both an auto-calibration tool (sequential uncertainty fitting 2; SUFI2) and a manual approach. The performance of SWAT was evaluated by a p-value, an r-factor, the Nash-Sutcliffe efficiency (*NSE*), a coefficient of determination (R^2), and mean absolute error (*MAE*). The validation process was conducted with the same parameter values from the calibration. The uncertainty analysis for the streamflow represented 56% and 54% of the observed data bracketed by the 95% prediction uncertainty (95PPU in SWAT) with values of 0.54 and 0.42 for the r-factor, respectively, in the calibration and validation processes. The streamflow showed good correlation to the observation based on the performance indicator values of 0.79/0.94 (R^2), 0.77/0.92 (*NSE*), and 0.59/0.26 (*MAE*) for calibration/validation. The results of nutrient loadings also indicated good correlation to the observed data, showing satisfactory indicator values. This calibration process assumed that watershed properties are similar only across the entire watershed.

After finishing the calibration process, the study area was separated from the Clear Creek watershed and treated as one watershed data to consider post-development scenarios. The three land uses with different urban designs (illustrated in the Description of Input Data section) were applied to the study area. Initial conditions for the post-development simulations were set based on the calibrated parameters from the pre-development simulation. Each land use was divided into different sub-basins and Hydrologic Response Units (HRUs) based on land uses and soil properties. A total of 4 sub-basins and 18 HRUs were produced in the UHD land use, and the UMD and UMC land uses were delineated as 5 sub-basins and 18 HRUs apiece. Each post-development simulation was individually run and investigated for surface runoff, nitrate, and total phosphorus (TP). Overall, the results showed an increase of runoff and pollutant loadings due to the effect of the urbanization rate of the post-development scenarios. The UMD land use represented a large increase, and a slightly lower increase was indicated in the UMC land use compared to the UMD land use. The UHD land use was the effective urban land use showing a minimal increase from the pre-development state. The final result values were used for comparison with the results of the post-LIDs scenarios in the "Results" section.

2.5. Specification of Used LID Practices and Scenarios

Three types of LID practices were chosen to be used in this study: rainwater harvesting tanks (RWHs), rain gardens (RGs), and permeable pavements (PPs). They are effective land management practices that are commonly used in urban watersheds. These LID practices have specific locations, taking up small areas or replacing existing impervious surfaces. It was assumed that RWHs are placed above ground for every house unit in the UMD and UMC residential areas and underground in the UHD residential area due to space restrictions. It was assumed that RGs are randomly installed in individual yards or neighborhood units along the street system in the residential areas, and PPs are taken into account only in the parking lots of commercial areas. Each LID practice was designed to capture the runoff and runoff-borne pollutants generated only from specific sites: that is, RWHs from roofs, PPs from parking lots, and RGs from residential areas, excluding roofs such as backyards, driveways, and sidewalks. Table 2 provides the percentages of roofs and parking lots in the residential and commercial areas for each land use, acquired from each design data and by sampling similar types of neighborhoods in Google Earth. These are the percentages of the areas covered by RWHs and PPs. The percentages of the areas covered by RGs are 6.6% and 8.0% of the UHD and UMD/UMC residential areas, respectively, and they were obtained by multiplying the rest of the percentages excluding roofs in the residential areas by a size factor of RGs based on Mechell and Lesikar [40]. In this study, it was assumed that the areas covered by each type of LID practice in each urban area were considered as full LID implementation. That is, each house has a rainwater harvesting tank, all parking lots in the commercial area are replaced by permeable pavements, and rain gardens are installed as much as the estimated percentages in the backyards of houses and public areas such as sidewalk patios. Also, 100% efficiency without consideration of seasonal impacts was assumed for all types of LID practices. These extreme conditions are ideal situations for new developments and we recognize that they might not be practical in a retrofit, but this is for the purpose of evaluating the benefit based on the LID practices that could be fully accommodated in the given LID areas for each urban design. No LID practices are assumed to be in non-urban areas.

Table 2. Fractions of roofs and parking lots in the urban area of each land use (in %).

Land Use	Roofs [1]	Parking Lots [2]
UHD	34	34
UMD	20	47
UMC	20	31

Notes: [1] Percentages of roofs occupied in the residential areas; [2] Percentages of parking lots occupied in the commercial areas.

In the present study, we focused on simulating the existence of LID practices under three types of land use with different urban patterns in order to evaluate the effectiveness of LID practices and to identify an optimal development plan. Three post-LIDs scenarios were created based on the land uses, and each was tested. They were assessed through comparison with pre-LIDs scenarios, already performed in previous work. The results among the post-LIDs scenarios were also compared and analyzed. Table 3 provides a summary of the scenarios addressed in the study.

Table 3. Summary of scenarios.

Land Use	Urban Design	Name of Scenario	
		Pre-LIDs	Post-LIDs
UHD	Compact urban type with high density	UHD	UHDLIDs
UMD	Conventional type with medium density	UMD	UMDLIDs
UMC	Conservational type with medium density	UMC	UMCLIDs

2.6. Representation of LID Practices in SWAT

2.6.1. Model Development

LID practices capture runoff to the extent of their capacities, and then once their capacities are exceeded, the LID practices discharge flows untreated. The SWAT model was developed to account for the hydrological behavior of LID practices in urban areas. A simple modification and addition of codes was conducted in the surface runoff subroutine.

The surface runoff in urban areas is estimated as the sum of surface runoff from the connected impervious area and disconnected impervious/pervious areas. Surface runoff from the connected impervious area is calculated by an impervious curve number. Surface runoff from the disconnected impervious/pervious areas is computed by a composite curve number under a surface runoff equation

(Equation (1)). Each surface runoff is multiplied by fractions of each area and then summed to obtain the final urban surface runoff (Equation (2)).

$$Q \text{ or } Q_{imp} = \frac{(P - 0.2S)^2}{(P + 0.2S)} \tag{1}$$

$$Q_{tot} = Q \cdot (1 - fcimp) + Q_{imp} \cdot fcimp \tag{2}$$

where Q and Q_{imp} are the surface runoff depths (mm) in the disconnected impervious/pervious areas and in the connected impervious area, respectively, Q_{tot} is the total surface runoff depth in urban areas (mm), P is precipitation (mm), S is a potential maximum retention (mm), and fcimp is the fraction of the connected impervious area.

To consider the amount of surface runoff captured by LID practices, a modified surface runoff equation (Equation (3)) was added in the existing codes.

$$Q_{LIDs} = Q_{tot} - LID_{val} \tag{3}$$

where Q_{LIDs} is the surface runoff depth (mm) in which the impact of LID practices is considered, and LID_{val} is the surface runoff depth (mm) stored by each LID practice. This method was determined based on McCuen's study that subtracted the amount of water captured by infiltration practices from urban surface runoff [41].

This is a suitable approach because SWAT has critical hydrologic algorithms that can best illustrate the flow characteristics of the LID practices being considered. In the case of RGs and PPs that have a natural infiltration system via soil layers, the amount of water exceeding storage capacity is generated as surface runoff by the developed equation (Equation (3)), and the amount of water stored is reflected as infiltration into the soil layers in SWAT. The difference between the amount of rainfall and the amount of surface runoff influences the amount of infiltration into the soil layers such that if precipitation is, for example, 110 mm and surface runoff is 100 mm, the amount of infiltration is 10 mm. However, if 20 mm of water is captured by RGs or PPs, 80 mm of surface runoff is finally discharged by the modified equation (Equation (3)) and the infiltrated water becomes 30 mm. That is, the 20 mm of water is to be added for soil water routing. If the capacities of the RGs or PPs are larger than the urban surface runoff, the amount of precipitation becomes the amount of infiltration. It is possible to simulate these LID practices for not only single events but also for consecutive rainfall. When rainy days are continuous, the daily subtraction from total surface runoff and its addition to the soil layers occurs by Equation (3). However, consecutive rainfall is mostly from small storms, and it is less frequent that large rainfall will occur continuously. In addition, the infiltration of the stored water affects the soil moisture condition, and cases in which all soil layers are completely saturated are not common. Even if that were the case, SWAT can model excess water as surface runoff.

In the case of RWHs, surface runoff is also released after rain tanks reach their volume capacity. However, the water captured by rain tanks is not infiltrated, unlike RGs and PPs. Therefore, the algorithm was additionally coded with relevance to its function. That is, codes were added such that the water from roofs is accumulated in the rain tanks and the maximum storage depth of the rain tanks is used in cases where the water accumulated exceeds the maximum storage depth of the rain tanks. The intentional drainage of the rain tanks was then taken into account for the purpose of reuse of the rain tanks. In this study, it was assumed that if there is no rainfall during a period of at least seven days after cessation of rainfall, the stored water in the rain tanks is intentionally emptied within the days between rainfall events. The stored water might be utilized for various purposes such as watering lawns and gardens, but this is explained as a water loss in SWAT. The description was mainly focused on the hydrologic components of SWAT related to the behavior of LID practices, and the schematic flow chart of the subroutines of the SWAT codes related to the hydrologic behavior by LID practices was added (Figure 3).

*<subbasin> call surface	- Calculation of daily surface runoff
<surface> call volq	based on the CN method
<volq> call surq_dayen	
<surq_dayen>	
inflpcp = precipitation - surq	- Calculation of the amount of
	infiltration into soil
call percmain	- Implementation of soil water
	routing
<percmain> call sat_excess	- Movement of excess water to upper
<sat_excess>	layers when the water content is
	above field capacity
call etpot	- Calculation of evapotranspiration

Figure 3. Schematic flow chart of the hydrologic subroutines related to the behavior of LID practices and description for the functions of each subroutine.

2.6.2. Design Storage Depth

Each LID practice holds different storage depths. In the case of RGs and PPs, the maximum runoff depths that could be treated by them were determined based on the amount of rainfall that is given to them and CN according to the degree of impervious and pervious fractions on each site. RGs and PPs were assumed to be designed to capture the runoff generated from 1.5 inches (38.1 mm) of rainfall. As 1.5 inches of rainfall is the 85th percentile storm event of the north central Texas region, the runoff amount from the rainfall is a volume for water quality protection in this region (Technical Manual of iSWM: http://iswm.nctcog.org/technical_manual.asp) [42]. An impervious CN (98) was used for PPs in all land uses because they deal with only the water from parking lots. For RGs, both impervious CN for the connected impervious covers and composite CN for the disconnected impervious/pervious covers were utilized to calculate the runoff depths that RGs can store. The CN for RGs was estimated differently for each land use because each land use has different urban patterns, comprised of different percentages of impervious and pervious fractions.

In the case of RWHs, the 1000 gallon capacity rain tank was assumed to be a standard in the medium-density residential area [43], and the storage depth was inversely calculated by Equation (4):

$$\text{Capacity of rain tank (gal)} = \text{Storage depth (in)} \times 0.623 \times \text{Roof area (ft}^2) \times \text{Runoff coefficient} \quad (4)$$

where 0.623 is the unit conversion factor, 0.9 runoff coefficient was used for roofs, and an average roof area per unit was determined through the design data and sampling of similar neighborhoods in Google Earth. A proportional volume of rain tanks was employed according to the roof area of each land use. The same runoff depth was consequently used for RWHs in all land uses.

Overall, the same storage depths for PPs and RWHs and different storage depths for RGs were applied for each land use (Table 4). The information for the maximum storage depths and types of LID practices was provided as a text file in SWAT, and the subroutine that can read the information was added in the SWAT algorithm.

Table 4. Maximum storage depth detained by each LID practice for each land use (in mm).

Land Use	Rain Gardens	Permeable Pavements	Rainwater Harvesting Tanks
UHD	22.45	32.52	12.94
UMD	19.11	32.52	12.94
UMC	17.83	32.52	12.94

2.7. Model Configuration

The model processing procedure was very similar to the steps of the previous work except for the urban land use to treat specific management practices. Other parameter values and input data were unaffected, and the current urban land use data was more detailed, to facilitate the application of LID practices to SWAT.

In order for RWHs and PPs to handle runoff only from roofs and parking lots, the roofs and the parking lots were separately allocated as different HRUs. They were manually partitioned from the existing HRUs of the residential and commercial areas by multiplying the current HRUs by percentages of the areas for the roofs and parking lots (Table 2). New urban data for the roofs and parking lots were added into the current urban data, and 100% impervious fractions were applied to their properties. Impervious fractions in which the roofs and parking lots were excluded were applied to the existing residential and commercial data. The urban type of the separated HRUs was replaced by new individual urban numbers for the roofs and parking lots, and the values that represent each type of LID practice were entered in the designated HRUs.

Through this process, a single type of LID practice was assigned to each HRU. That is, PPs were considered in the HRUs of the parking lots, RWHs in the HRUs of the roofs, and RGs in the HRUs of the residential urban areas. The same process was individually implemented for the three land uses. The simulation was conducted from October 2006 to December 2011. Average monthly and yearly results over the continuous periods were analyzed for all scenarios along with statistical analysis in order to evaluate the watershed-wide effectiveness of LID practices on surface runoff, nitrate, and TP. For the statistical analysis, a *t*-test was conducted for daily surface runoff, nitrate, and total phosphorus data from precipitation events above 0.5 inches among scenarios to a 95% confidence level. All *t*-tests conducted had more than 150 data points (n).

3. Results and Discussion

The performance of simulated LID practices positively affected all variables for all land uses. Figures 4–6 and Table 5 represent the average monthly and yearly responses of LID practices for each land use. As part of the surface runoff was detained by LID practices, the decreased surface runoff was denoted in the post-LIDs scenarios of all land uses, showing a tendency to follow the behavior of the pre-development state (Figure 4). The differences between the pre- and post-LIDs scenarios were extracted differently for each land use. For the UHD land use, 14% of the surface runoff was reduced, and 29% and 25% reductions were obtained in the UMD and UMC land uses, respectively, on an average annual basis (Table 5). The results showed statistically significant differences between the pre- and post-LIDs scenarios in all land uses (p-values < 0.05). The application of LID practices also had an influence on subsurface hydrology. Since the water detained by LID practices infiltrated into the soil layers, it increased the soil water content and, consequently, contributed to the increase of both evapotranspiration (ET) and groundwater (GW) for all land uses (Table 5). The amount of evaporation in a soil layer is determined by soil water content. Since the greatest effect of LID practices on surface runoff was in the UMD land use, the amount of infiltration in that land use was greatest. It increased soil water the most and led to the largest increase of ET. That is, ET was 10% greater under the UMDLIDs scenario than the UMD scenario, 8% greater under the UMCLIDs scenario than the UMC scenario, and 4% greater under the UHDLIDs scenario than the UHD scenario. In addition, increased soil water affected the increase of groundwater, representing the same order of increase with ET: that is, UMD land use > UMC land use > UHD land use. As seen from these results, the decrease of surface runoff by LID practices was closely related to the increase of ET and GW, indicating that the hydrologic behavior by LID practices was adequately simulated in SWAT.

(A)

(B)

(C)

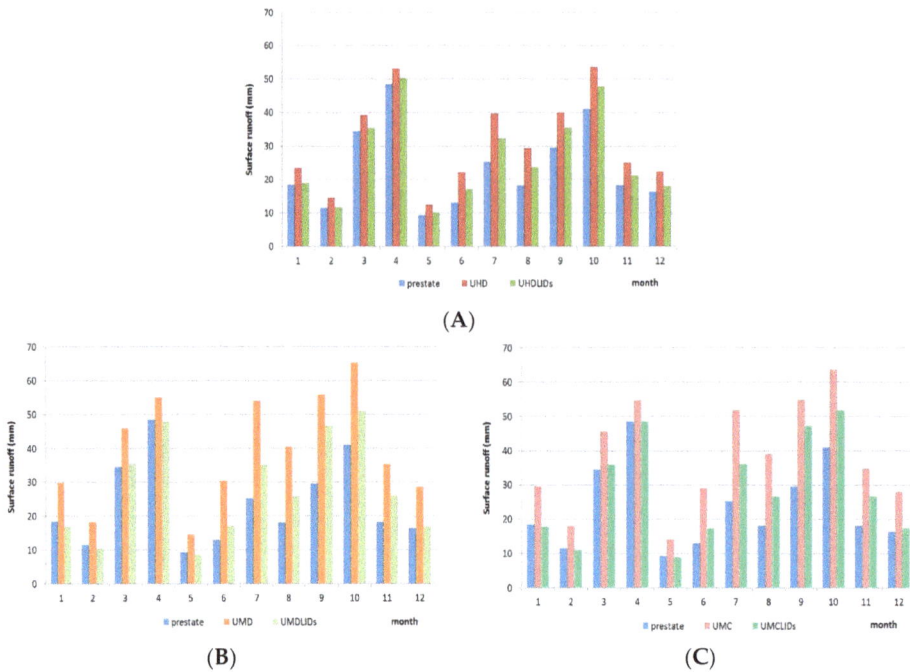

Figure 4. Average monthly response of LID practices for surface runoff (SURQ) in each land use: (**A**) Compact high-density urban land use (UHD); (**B**) Conventional medium-density urban land use (UMD); and (**C**) Conservational medium-density urban land use (UMC). The term 'prestate' in the chart means pre-development condition.

Table 5. Average annual response of LID practices under each land use (GWQ: Flow to groundwater; ET: Evapotranspiration; NO_3: Nitrate Loading; TP: Total phosphorus loading.

Scenario	SURQ (mm)	GWQ (mm)	ET (mm)	NO₃ (kg)	TP (kg)	Difference (% Reduction)		
						SURQ (mm)	NO₃ (kg)	TP (kg)
UHD	374.66	45.76	855.66	430.92	431.64	52.97	101.37	46.45
UHDLIDs	321.69	63.19	893.13	329.55	385.19	(14%)	(24%)	(11%)
UMD	473.32	15.78	797.02	591.87	449.55	135.51	186.03	110.69
UMDLIDs	337.81	79.17	874.85	405.85	338.86	(29%)	(31%)	(25%)
UMC	462.73	15.80	808.16	577.19	443.46	117.80	170.51	97.43
UMCLIDs	344.93	74.74	872.13	406.68	346.03	(25%)	(30%)	(22%)

In urban areas, pollutants are generally dependent on surface runoff. According to the decrease of surface runoff by LID practices, the runoff-borne pollutants, nitrate (NO_3) and total phosphorus (TP), also showed decreases in the post-LIDs scenarios of all land uses (Figures 5 and 6 and Table 5). Nitrate loadings were reduced by 24%, 31%, and 30% in the UHD, UMD, and UMC land uses, respectively, and the results represented significant differences between the pre- and post-LIDs scenarios in all land uses (p-values < 0.05). TP loadings decreased 11%, 25%, and 22% in the UHD, UMD, and UMC land uses, respectively, on an average annual basis, and the results also showed statistically significant differences between the pre- and post-LIDs scenarios (p-values < 0.05), except for the UHD land use.

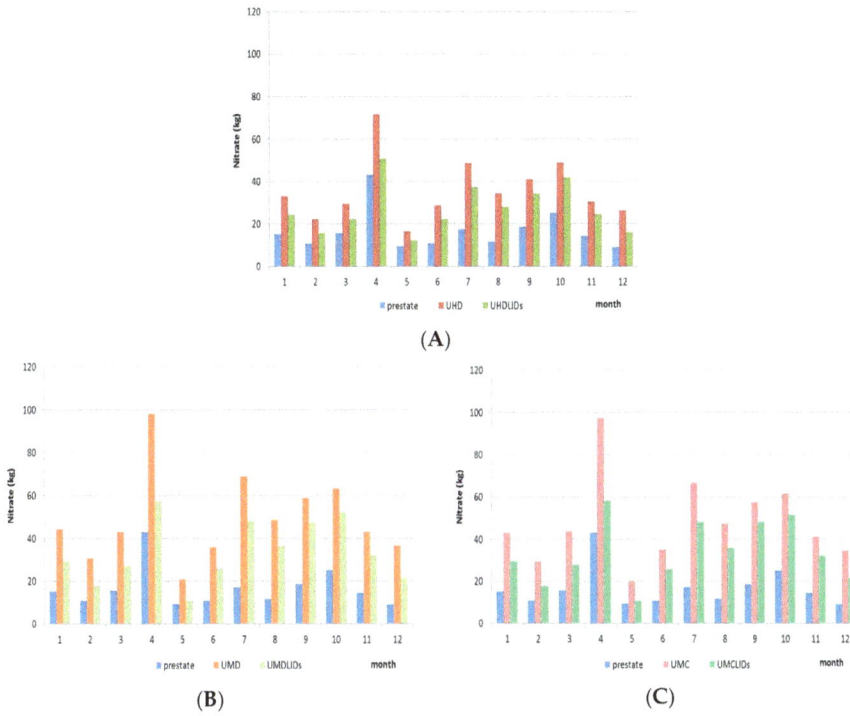

Figure 5. Average monthly response of LID practices for nitrate (NO_3) in each land use: (**A**) Compact high-density urban land use (UHD); (**B**) Conventional medium-density urban land use (UMD); and (**C**) Conservational medium-density urban land use (UMC). The term 'prestate' in the chart means pre-development condition.

Overall, the degree of contribution of LID practices for all variables was smallest in the UHD land use followed by the UMC land use, and it was largest under the UMD land use. This could be attributed to the difference in the area covered by LID practices among land uses. The unit reduction amounts by LID practices only in each urban area were largest in the UHD land use for all variables, as seen in Table 6. However, the UHD land use had the smallest urban area and the smallest area covered by LID practices among land uses, and thus the percent reduction by LID practices was smallest in the UHD land use. The pre-development scenario was plotted along with the pre- and post-LIDs scenarios (Figures 4–6), and was statistically analyzed with post-LIDs scenarios for the purpose of observing the effect of LID practices (Table 7). From the results, it was observed that the post-LIDs scenarios were statistically similar to pre-development conditions for surface runoff and total phosphorus. In other words, LID practices reduced the increases in surface runoff and total phosphorus from the development to pre-development state. However, with regards to nitrate, all post-LIDs scenarios were significantly higher than the pre-development condition. That is, the application of LID practices could not bring the negative effect of nitrate back to the pre-development condition.

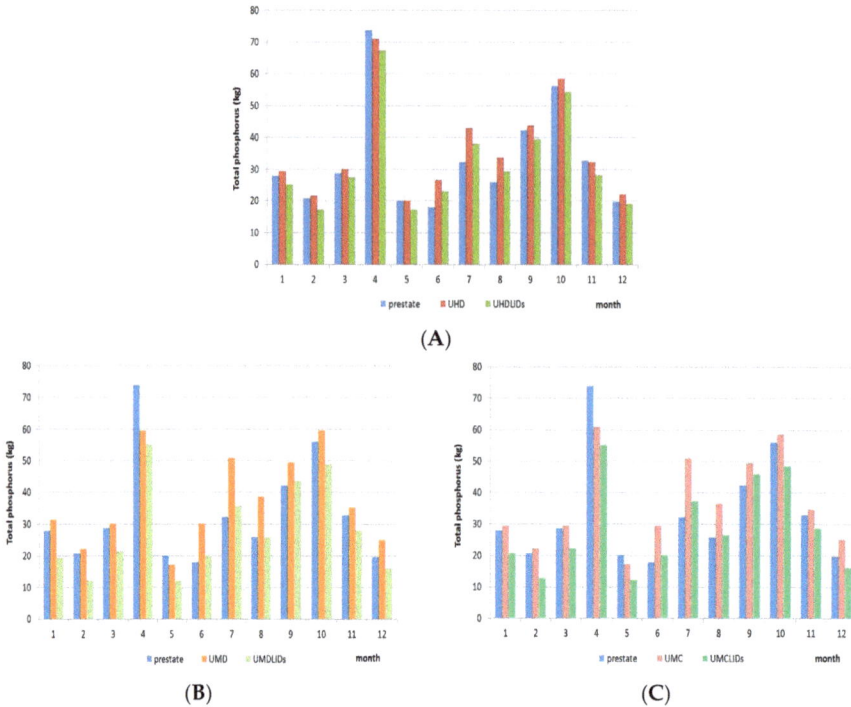

Figure 6. Average monthly response of LID practices for total phosphorus (TP) in each land use: (**A**) Compact high-density urban land use (UHD); (**B**) Conventional medium-density urban land use (UMD); and (**C**) Conservational medium-density urban land use (UMC). The term 'prestate' in the chart means pre-development condition.

Table 6. Unit reduction amounts of surface runoff and nutrients by LID practices only in the urban area for each land use.

Land Use	Surface Runoff (mm)	Nitrate (kg/ha)	TP (kg/ha)
UHD	257.23	1.37	0.63
UMD	242.73	0.93	0.55
UMC	211.01	0.85	0.49

Table 7. Statistical results (*p*-values) from the *t*-test between pre-development and each post-LIDs scenario for all variables.

Scenario	Pre-Development		
	Surface Runoff	Nitrate	TP
UHDLIDs	0.577	0.0025 *	0.72
UMCLIDs	0.382	1.00×10^{-6} *	0.32
UMDLIDs	0.439	1.60×10^{-6} *	0.34

Note: * indicates a statistically significant difference.

For the comparison among the final results of post-LIDs scenarios, low surface runoff and pollutant amounts were observed under different urban land uses (Table 5). In the case of surface runoff and nitrate, low values were achieved under the UHDLIDs scenario among the post-LIDs scenarios. This was because the impact of the UHD land use itself was the smallest among the land uses because of its small proportion of urban area, indicating statistically significant differences from both UMD and UMC land uses (Table 8). Thus, although the reduction caused by the application of LID practices was smallest under the UHD land use, it exhibited the lowest surface runoff and nitrate values achieved. In sequence, the UMDLIDs scenario showed a low value in comparison to the UMCLIDs scenario. The result was opposite that of the UMD and UMC scenarios. That is, less surface runoff and nitrate were generated under the UMC land use because it had a higher pervious fraction than the UMD land use, but after applying LID practices, less surface runoff and nitrate were shown in the UMD land use. This could be because while the area covered by RGs and RWHs was the same under the two land uses, the area covered by PPs was larger, as much as the difference of the parking lot area (16%), in the UMD land use compared to that in the UMC land use (Table 2). Contrary to the surface runoff and nitrate, the high value of TP was shown in the UHDLIDs scenario. This result was in contrast with the result from the pre-LIDs scenarios which represented a low TP value in the UHD scenario. This was seen because although the UHD scenario indicated a low value for TP, this was not a relatively lower TP value than those of the UMD and UMC scenarios (statistically significant differences were not indicated, showing p-values above 0.05) and the effect of the LID practices was also insignificant between the UHD and UHDLIDs scenarios (the p-value between the two scenarios was 0.0662). Table 8 provides the results of the statistical analysis for all pre- and post-LIDs scenarios for all variables.

Table 8. Statistical results (p-values) from the t-test for all pre- and post-LIDs scenarios.

Surface Runoff					
Scenario	UHD	UHDLIDs	UMC	UMCLIDs	UMD
UHDLIDs	2.00×10^{-10} *	-	-	-	-
UMC	1.10×10^{-6} *	5.30×10^{-14} *	-	-	-
UMCLIDs	0.11	0.10	5.20×10^{-7} *	-	-
UMD	1.60×10^{-7} *	7.70×10^{-15} *	0.67	9.10×10^{-8} *	-
UMDLIDs	0.04 *	0.26	9.00×10^{-8} *	0.71	1.50×10^{-8} *
Nitrate					
Scenario	UHD	UHDLIDs	UMC	UMCLIDs	UMD
UHDLIDs	1.40×10^{-5} *	-	-	-	-
UMC	3.40×10^{-7} *	2.00×10^{-16} *	-	-	-
UMCLIDs	0.8565	8.20×10^{-5} *	1.00×10^{-5} *	-	-
UMD	8.30×10^{-8} *	2.00×10^{-16} *	0.8631	3.70×10^{-6} *	-
UMDLIDs	0.7780	0.0003 *	7.20×10^{-7} *	0.6759	2.10×10^{-7} *
Total Phosphorus					
Scenario	UHD	UHDLIDs	UMC	UMCLIDs	UMD
UHDLIDs	0.0662	-	-	-	-
UMC	0.6070	0.0709	-	-	-
UMCLIDs	0.0006 *	0.0257 *	0.0012 *	-	-
UMD	0.3109	0.0224 *	0.6773	0.0003 *	-
UMDLIDs	0.0007 *	0.0314 *	0.0014 *	0.9193	0.0004 *

Note: * means a statistically significant difference.

Before applying LID practices to urban developments, UHD land use might be the best choice for minimizing the impact of urbanization on surface runoff and pollutant loadings, as shown in other studies [17,18,35]. Jacob and Lopez [18] mentioned the advantage of higher density development outperforming traditional stormwater BMPs in pollutant reductions, due to the decrease of a runoff-generating area. However, after the application of LID practices to urban developments, all post-LIDs scenarios performed better than the UHD scenario (Table 5). Statistically, the UHD scenario represented significant differences from the UHDLIDs and UMDLIDs scenarios in surface runoff, from the UHDLIDs scenario in nitrate, and from the UMCLIDs and UMDLIDs scenarios in TP (Table 8). In addition, when LID practices were applied to urban developments, the advantage of the UHD land use decreased. For example, in the case of surface runoff, although the UHDLIDs scenario showed the lowest value among post-LIDs scenarios, the results of all post-LIDs scenarios were very similar, not representing statistically significant differences (p-values > 0.05). This was seen because the UHD land use used in this study had a lower urban density than the ones usually used in other studies, and thus LID practices could make the impact of urban development more or less equal altogether. On the contrary, in the case of TP, the highest value was obtained in the UHDLIDs scenario, and statistically significant differences among post-LIDs scenarios existed (p-values < 0.05). From these results, the UHD land use should not be considered as the perfect choice in reducing runoff and pollutants when LID practices are applied to urban developments.

4. Conclusions

The present study provided an opportunity to examine the impacts of LID practices on flow and pollutant loadings under three land uses with different urban patterns and to develop a model for simulating the examined LID practices in SWAT. The method of representing LID practices in SWAT was flexible and easily applicable. There is no model that completely incorporates all the requirements to simulate various LID practices, but the developed model performed well for the simulations of surface and subsurface hydrology and the consequential water quality. The results demonstrated an applicability of the examined LID practices in SWAT. It is worth noting that the model only addressed three of the four main LID practices (Green roofs were excluded due to cost of construction). For proprietary and site specific LID practices, the model would need to be modified on a case by case basis. In addition, reduction rates from field studies in Texas were used in this paper. The use of data from local projects would enhance the model results when used in other regions.

The application of LID practices contributed to the reduction of surface runoff and pollutants under all land uses, and the effectiveness of LID practices was demonstrated differently for each land use in the watershed. The reductions were statistically significant in terms of the differences between pre- and post-LIDs scenarios under all land uses for all variables (p-values < 0.05), except for TP between the UHD and UHDLIDs scenarios (p-value = 0.0662 > 0.05). However, despite the significant contribution of the LID practices in most cases, a large amount of surface runoff could still be generated by heavy precipitation because LID practices are limited in capacity and area in land use. The Harris County Flood Control District (HCFCD) and the Harris County Public Infrastructure Department Architecture & Engineering Division (HCPID-AED) require new urban areas to follow a minimum detention rate of 0.55 ac-ft per acre in order to control flooding. In considering this requirement, it is necessary to study other alternatives that can cover the rest of the volume besides the volume of the LID practices. This is beyond the scope of the present study and thus was not examined.

The results among post-LIDs scenarios showed that the UHD land use performed better in achieving the low values for surface runoff and nitrate than the other land uses, and UMD land use led to obtaining the low value for TP. Testing of the effectiveness of LID practices under different designs could provide useful information on an optimal design. Such results would help regulators develop effective LID policies on a city scale which could enhance the solutions for runoff and pollutant problems for their watersheds. In addition, it should be noted that the results can be changed if considering different watersheds with different soils, slopes, and land use properties, and different conditions such as types and allocations of LID practices, or a budget of LID implementation. Therefore, it is recommended that simulations be performed in advance under the development policy of a region prior to constructing LID practices.

Acknowledgments: The authors would like to thank the city of League City, TX and TCEQ, EPA, Texas Sea Grant, and NOAA, and the Korean National Institute of Agricultural Sciences (PJ010867) for the financial support for this project.

Author Contributions: All authors (Mijin Seo, Fouad Jaber, Raghavan Srinivasan, and Jaehak Jeong) performed the research and contributed to the writing of the paper.

Conflicts of Interest: The authors declare no conflict of interest.

References

1. Leopold, L.B. *Hydrology for Urban Land Planning: A Guidebook on the Hydrologic Effects of Urban Land Use*; United States Department of the Interior and Geological Survey: Washington, DC, USA, 1968.
2. Shaw, E.M.; Beven, K.J.; Chappell, N.A.; Lamb, R. *Hydrology in Practice*, 4th ed.; Spon Press: New York, NY, USA, 2010.
3. Prince George's County. *Low-Impact Development Design Strategies: An Integrated Design Approach*; Department of Environmental Resources, Programs and Planning Division: Prince George's County, MD, USA, 1999.
4. Bean, E.Z.; Hunt, W.F.; Bidelspach, D.A. Evaluation of four permeable pavement sites in eastern North Carolina for runoff reduction and water quality impacts. *J. Irrig. Drain. Eng.* **2007**, *133*, 583–592. [CrossRef]
5. Collins, K.A.; Hunt, W.F.; Hathaway, J.M. Hydrologic comparison of four types of permeable pavement and standard asphalt in eastern North Carolina. *J. Hydrol. Eng.* **2008**, *13*, 1146–1157. [CrossRef]
6. Hunt, W.; Jarrett, A.; Smith, J.; Sharkey, L. Evaluating bioretention hydrology and nutrient removal at three field sites in North Carolina. *J. Irrig. Drain. Eng.* **2006**, *132*, 600–608. [CrossRef]
7. Hunt, W.; Smith, J.; Jadlocki, S.; Hathaway, J.; Eubanks, P. Pollutant removal and peak flow mitigation by a bioretention cell in urban Charlotte, NC. *J. Environ. Eng.* **2008**, *134*, 403–408. [CrossRef]
8. Jaber, F.; Guzik, E. Improving water quality and reducing the volume of urban stormwater runoff with a bioretention area. In Presented at the ASABE Annual International Meeting, Reno, NV, USA, 21–24 June 2009.
9. Abi Aad, M.P.; Suidan, M.T.; Shuster, W.D. Modeling techniques of best management practices: Rain barrels and rain gardens using EPA SWMM-5. *J. Hydrol. Eng.* **2009**, *15*, 434–443. [CrossRef]
10. Ackerman, D.; Stein, E.D. Evaluating the effectiveness of best management practices using dynamic modeling. *J. Environ. Eng.* **2008**, *134*, 628–639. [CrossRef]
11. Carter, T.; Jackson, C.R. Vegetated roofs for stormwater management at multiple spatial scales. *Landsc. Urban Plan.* **2007**, *80*, 84–94. [CrossRef]
12. Holman-Dodds, J.K.; Bradley, A.A.; Potter, K.W. Evaluation of hydrologic benefits of infiltration based urban storm water management. *J. Am. Water Resour. Assoc.* **2003**, *39*, 205–215. [CrossRef]
13. Brander, K.E.; Owen, K.E.; Potter, K.W. Modeled impacts of development type on runoff volume and infiltration performance. *J. Am. Water Resour. Assoc.* **2004**, *40*, 961–969. [CrossRef]
14. Schneider, L.E.; McCuen, R.H. Assessing the hydrologic performance of best management practices. *J. Hydrol. Eng.* **2006**, *11*, 278–281. [CrossRef]
15. Freni, G.; Mannina, G.; Viviani, G. Urban storm-water quality management: Centralized versus source control. *J. Water Resour. Plan. Manag.* **2010**, *136*, 268–278. [CrossRef]
16. Lucas, W.C.; Sample, D.J. Reducing combined sewer overflows by using outlet controls for Green Stormwater Infrastructure: Case study in Richmond, Virginia. *J. Hydrol.* **2015**, *520*, 473–488. [CrossRef]
17. Seo, M. Modeling Low Impact Development at the Small-Watershed Scale: Implications for the Decision Making Process. Ph.D. Dissertation, Texas A & M University, College Station, TX, USA, 2014. Available online: http://hdl.handle.net/1969.1/154071 (accessed on 29 December 2016).
18. Jacob, J.S.; Lopez, R. Is denser greener? An evaluation of higher density development as an urban stormwater-quality best management practice. *J. Am. Water Resour. Assoc.* **2009**, *45*, 687–701. [CrossRef]
19. Williams, E.S.; Wise, W.R. Hydrologic impacts of alternative approaches to storm water management and land development. *J. Am. Water Resour. Assoc.* **2006**, *42*, 443–455. [CrossRef]
20. Gilroy, K.L.; McCuen, R.H. Spatio-temporal effects of low impact development practices. *J. Hydrol.* **2009**, *367*, 228–236. [CrossRef]
21. Culliton, T.J. *Population: Distribution, Density, and Growth*; National Oceanic and Atmospheric Administration (NOAA): Silver Spring, MD, USA, 1998.
22. Howarth, R.W.; Sharpley, A.; Walker, D. Sources of nutrient pollution to coastal waters in the United States: Implications for achieving coastal water quality goals. *Estuaries* **2002**, *25*, 656–676. [CrossRef]
23. Dougall, C.; Rohde, K.; Carroll, C.; Millar, G.; Stevens, S. An assessment of land management practices that benchmark water quality targets set at a Neighbourhood Catchment scale using the SWAT model. In Presented at the MODSIM Conference, Townsville, Australia, 14–17 July 2003.
24. United States Geological Survey (USGS). Science in Your Watershed. Available online: http://water.usgs.gov/wsc/map_index.html (accessed on 5 March 2014).

25. Emerson, C.H.; Welty, C.; Traver, R.G. Watershed-scale evaluation of a system of storm water detention basins. *J. Hydrol. Eng.* **2005**, *10*, 237–242. [CrossRef]

26. Abbaspour, K.C.; Yang, J.; Maximov, I.; Siber, R.; Bogner, K.; Mieleitner, J.; Zobrist, J.; Srinivasan, R. Modelling hydrology and water quality in the pre-alpine/alpine Thur watershed using SWAT. *J. Hydrol.* **2007**, *333*, 413–430. [CrossRef]

27. Jha, M.K.; Gassman, P.; Arnold, J. Water quality modeling for the Raccoon River watershed using SWAT. *Trans. ASABE* **2007**, *50*, 479–493. [CrossRef]

28. Santhi, C.; Arnold, J.G.; Williams, J.R.; Dugas, W.A.; Srinivasan, R.; Hauck, L.M. Validation of the SWAT model on a large river basin with point and nonpoint sources. *J. Am. Water Resour. Assoc.* **2001**, *37*, 1169–1188. [CrossRef]

29. Soil Conservation Service (SCS). *National Engineering Handbook*; Section 4 Hydrology, Chapters 4–10; United States Department of Agriculture: Washington, DC, USA, 1972.

30. Williams, J. Sediment routing for agricultural watersheds. *J. Am. Water Resour. Assoc.* **1975**, *11*, 965–974. [CrossRef]

31. Neitsch, S.; Arnold, J.; Kiniry, J.; Williams, J. *Soil and Water Assessment Tool Theoretical Documentation Version 2009*; Texas Water Resources Institute: College Station, TX, USA, 2011.

32. Jeong, J.; Kannan, N.; Arnold, J.G.; Glick, R.; Gosselink, L.; Srinivasan, R.; Barrett, M.E. Modeling sedimentation-filtration basins for urban watersheds using Soil and Water Assessment Tool. *J. Environ. Eng.* **2013**, *139*, 838–848. [CrossRef]

33. Parker, B.S. Assessing Stormwater Runoff with "SWAT" in Mixed-Use Developments: Learning from Southlake Town Square and Addison Circle in North Texas. Master's Thesis, The University of Texas, Arlington, TX, USA, 2010.

34. Yang, B.; Li, M.; Huang, C. *Using SWAT to Compare Planning Methods for Neighborhoods: Case Study of Stormwater in The Woodlands, Texas*; Texas Water Resources Institute and Texas A&M University: College Station, TX, USA, 2009.

35. Yang, B.; Li, M.H. Assessing planning approaches by watershed streamflow modeling: Case study of The Woodlands; Texas. *Landsc. Urban Plan.* **2011**, *99*, 9–22. [CrossRef]

36. Bracmort, K.; Arabi, M.; Frankenberger, J.; Engel, B.; Arnold, J. Modeling long-term water quality impact of structural BMPs. *Trans. ASABE* **2006**, *49*, 367–374. [CrossRef]

37. Santhi, C.; Srinivasan, R.; Arnold, J.; Williams, J. A modeling approach to evaluate the impacts of water quality management plans implemented in a watershed in Texas. *Environ. Model. Softw.* **2006**, *21*, 1141–1157. [CrossRef]

38. Hunt, W.; Kannan, N.; Jeong, J.; Gassman, P. Stormwater Best Management Practices: Review of current practices and potential incorporation in SWAT. In Presented at the SWAT Southeast Asia Modeling, Chiang Mai, Thailand, 5–7 January 2009.

39. Jeong, J.H.; Kannan, N.; Srinivasan, R. *Development of SWAT Algorithms for Modeling Urban Best Management Practices*; Watershed Protection and Development Review Department: Austin, TX, USA, 2011.

40. Mechell, J.; Lesikar, B.J. *Rainwater Harvesting: Raingardens*; Texas Water Resources Institute: College Station, TX, USA, 2008; Available online: http://hdl.handle.net/1969.1/87451 (accessed on 28 July 2013).

41. Maryland Department of the Environment (MDE). *Modeling Infiltration Practices Using the TR-20 Hydrologic Program (1983)*; Maryland Department of the Environment: Baltimore, MD, USA, 1983.

42. North Central Texas Council of Governments (NCTCOG). Technical Manual of iSWM: Water Quality. North Central Texas Council of Governments. 2014. Available online: http://iswm.nctcog.org/technical_manual. asp (accessed on 23 September 2013).

43. Shannak, S.D.; Jaber, F.; Lesikar, B. Modeling the effect of cistern size, soil type, and irrigation scheduling on rainwater harvesting as a stormwater control measure. *Water Resour. Manag.* **2014**, *28*, 4219–4235. [CrossRef]

water

MDPI

Article

Evaluating Various Low-Impact Development Scenarios for Optimal Design Criteria Development

Mijin Seo [1], Fouad Jaber [2,*] and Raghavan Srinivasan [3]

[1] National Institute of Agricultural Sciences, Rural Development Administration, Wanju,
      Jeollabuk-do 565-851, Korea; mjseo1020@korea.kr
[2] Biological and Agricultural Engineering, Texas A&M AgriLife Extension, Dallas, TX 75252, USA
[3] Spatial Science Laboratory, Texas A&M University, College Station, TX 77845, USA; r-srinivasan@tamu.edu
*   Correspondence: f-jaber@tamu.edu; Tel.: +1-972-952-9672

Academic Editors: Karim Abbaspour, Saeid Ashraf Vaghefi, Monireh Faramarzi and Lei Chen
Received: 29 December 2016; Accepted: 9 April 2017; Published: 12 April 2017

Abstract: Low-impact development (LID) practices as a new approach to urban stormwater management have demonstrated their positive effects through the reduction of surface runoff volumes and pollutant loadings in a substantial amount of research. The effectiveness of LID practices can be affected by various LID conditions such as type, location, and area. Cost is also an important factor to be considered in the evaluation of LID effects. This study presented the optimal LID conditions that can achieve targeted reduction goals with minimal cost, and analyzed the effectiveness of LID practices under optimal LID conditions and the consequential cost on a watershed scale. To determine cost-effective LID conditions, three types of LID practices (rain gardens, rainwater harvesting tanks, and permeable pavements), two locations (residential and commercial areas), and percent allocation of LID practices were considered. Manual optimization was conducted under those LID conditions for five targeted reduction goals which were set for surface runoff and nutrient loadings. The results provided various configurations of cost-effective conditions in treating the targeted goals, and represented the impacts of the optimized LID conditions on the effectiveness of LID practices and the consequential cost. The present study could ultimately assist regulators in establishing proper watershed-scale strategies of LID conditions for effectively managing watersheds.

Keywords: low-impact development (LID) conditions; effectiveness of LID practices; manual optimization; cost; watershed management

---

1. Introduction

Development increases impervious land cover [1]. Urban impervious surfaces have aggregated stormwater problems. Specifically, surface runoff volume is significantly increased as infiltration is hindered. This decreases groundwater recharge and accordingly reduces the amount of base flow [2]. Significant water-bound pollutants are conveyed to nearby water bodies by the increased urban runoff flowing over the impervious surfaces [3]. It is necessary to take corrective action in response to these stormwater problems. Installation of low-impact development (LID) practices is one method to offset the adverse impact caused by urbanization. LID practices help to achieve both development and environmental protection by imitating the hydrology of a pre-developed state. Research on the effects of LID practices has been active and has comprehensively been addressed in a variety of studies. Most studies have demonstrated the benefits of LID practices by showing an increase of recharge rate [4] and reductions in runoff volume and pollutant loadings [5–8].

However, the degree of the effectiveness of LID practices can be affected by various factors. Some studies, for example, have reported the different effects of LID practices on water quantity and quality under different types of soil [9,10] and under various rainfall patterns [11–13]. A few studies have pointed out that different effects of LID practices could exist depending on how urban areas are designed [9,14,15]. Seo et al. [16] also evaluated the effectiveness of LID practices on hydrology and water quality under three land uses with different types of urban patterns (compact high-density, conventional medium-density, and conservational medium-density) using the Soil and Water Assessment Tool (SWAT) and presented the optimal land use.

In addition to these external conditions, the effectiveness of LID practices can also be expected to vary as a result of various LID planning and design factors such as type, location, area, and so forth. Gilroy and McCuen [14] simulated the spatial and quantitative effects of cisterns and bioretention areas using a developed spatio-temporal model and provided information on the spatial arrangements and volumes needed to achieve effective results in reduction of runoff volumes and peak discharge rates. Endreny and Collins [17] examined groundwater recharge and mounding by adjusting the spatial arrangements of bioretention areas as distributed, clustered, and single units using a MODFLOW model in an urban residential area of New York, USA. They determined that groundwater mounding was highest when bioretention areas were arrayed as single units and lowest when they were fully distributed. Brander et al. [9] identified the impact of the number of infiltration practices by demonstrating that runoff differences among different urban types could be overcome by implementing a number of infiltration practices. Ahiablame et al. [18] also evaluated the effects of LID practices on runoff and pollutant loads according to the percent implementation of rain barrel/cistern and porous pavement. While the above studies showed that studies addressing proper distribution and placement of LID practices are needed, none provided an approach that would optimize the area of LID needed, as a function of location and type, to meet a target runoff and pollutant reduction rate.

The establishment of proper watershed-scale strategies for LID conditions is required to obtain optimal results for reductions of runoff volume and nutrient loadings. Cost is an essential factor that must be considered along with the strategies because a restricted budget is usually given for performing the strategies [19]. Gilroy and McCuen [14], in their study, simply determined several scenarios for placing cisterns and bioretention areas according to the places where water was intercepted, and Chaubey et al. [20] stated that random placement was normally used. However, such methods can make a cost-effective scenario for LID conditions (which may result in better outcomes in reduction with minimal cost) be missed as it is among unconsidered scenarios. Liu et al. [21] indicated that they found the best effective scenario of LID and best management practices (BMP) conditions showing the greatest reduction in runoff and pollutant loadings among 16 scenarios, but it was not a cost-effective scenario. Therefore, optimization would be necessary. Many researchers have performed optimization to accomplish the best effect close to a required target reduction goal at minimum cost [19,22–26]. However, most studies have been for optimization of agricultural best management practices (conventional stormwater treatment systems akin to LID practices) and have drawn the optimal scenario (or the best solution) by utilizing various optimization tools, such as the genetic algorithm (GA), through model development. While the use of tools enables evaluation of a myriad of probable options for various LID conditions, it makes the process complex and increases the simulation time [24]. In particular, it becomes an inefficient method when considering just a few conditions or small watersheds. In this regard, a manual technique for optimization is required, which can simplify the complexity and easily provide information on cost-effective LID conditions at any watershed.

The purpose of this study was to present the optimal LID conditions that can attain targeted reduction goals with minimal cost and to evaluate the effectiveness of LID practices under the optimal conditions and the consequential cost on a watershed scale. A manual optimization was conducted for identifying the optimal conditions of LID practices, using a Microsoft Excel spreadsheet. Five targeted reduction goals were determined by using the results of reduction amounts by LID practices from the Soil and Water Assessment Tool (SWAT). Three LID conditions were taken into account in the manual optimization process: types of LID practices (rain gardens, rainwater harvesting tanks, and permeable pavements), locations (residential and commercial areas), and percent allocation of LID practices at each location. The study was processed for surface runoff (SURQ), nitrate (NO_3), and total phosphorus (TP).

2. Materials and Methodology

2.1. Case Study Area

The study was carried out in a small-scale area of approximately 350 ha (3.5 km^2), comprised of some portions of League City, Webster, and Friendswood in Harris County, Texas, as a case study. The area is nested within the Clear Creek watershed and is situated at the downstream end of Clear Creek (close to the outlet), which is in an area under the influence of tidal currents (Figure 1). Estuarine areas have generally had more water problems (such as flooding and accumulation of untreated pollutants) than other regions because of their geographical characteristics such as flat topography with low elevations and a tidal-affected location. The area has elevations of 6–8 m Above Mean Sea Level (AMSL). While current land use is in a pre-development state consisting of hay (28.23%), rangeland (15.35%), wetland (30.71%), and forest (25.71%), a new urban area will be developed in this area.

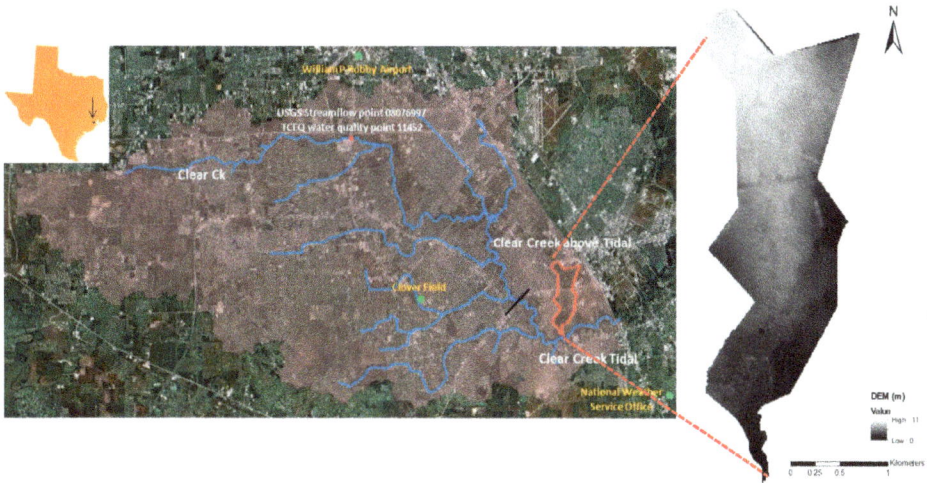

Figure 1. Study area location in the Clear Creek watershed boundary (Three round marks (green) on the map are weather stations, and a short bar (black) is a boundary line for a tidal-affected and non-tidal stream).

The new urban area is a conventional urban form of medium density (Figure 2). It will be constructed with single family neighborhoods and a commercial district. It is one of the urban strategies of League City [27]. Besides the urban areas, the remaining land use is the same as the pre-development state. The soil of the study area is classified into four types: Addicks (61.4%), Bernard (27.3%), Lake Charles (3.2%), and Aris (8.1%). The textures of the soils are mainly clay and clay loam, and they all belong to hydrologic soil group (HSG) D, which has very low permeability. Mild winters and hot summers are typical weather patterns for this region. The temperature averages about 12 °C (53 °F) in January and about 29 °C (84 °F) in August. The average annual rainfall is approximately 1270 mm, with an average monthly range of about 50–165 mm. Intense rainfall is typical of this region because of its oceanic climate.

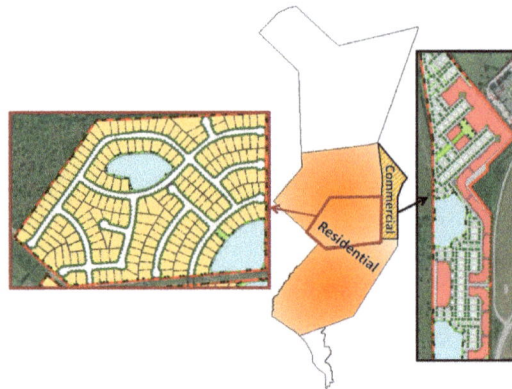

Figure 2. New land use with the conventional urban form of medium density (blowups show residential and commercial patterns, and the empty space is an unchanged natural state).

2.2. SWAT Model Description and Development

2.2.1. Model Description

The Soil and Water Assessment Tool (SWAT) is a model developed by the United States Department of Agriculture-Agricultural Research Service (USDA-ARS). It has been extensively used to deal with various water quantity and quality problems from many watersheds, and its capability has been verified through results [28–33]. It is applicable to simulations of various sizes of watersheds from small and medium watersheds to large watersheds [28–30]. It can also simulate long and short terms and even sub-daily and sub-hourly time steps [31–33]. As SWAT is a distributed model, it can discretize a watershed as subbasins and smaller hydrologic response units (HRUs), which are the minimum-sized response units. It has essential model components such as surface runoff, infiltration, groundwater, evapotranspiration, nutrient cycling, etc. All components are operated at an HRU level.

Surface runoff can be calculated based on a modified Natural Resources Conservation Service (NRCS) curve number method [34] on a daily basis. Urban surface runoff is estimated respectively for the disconnected impervious/pervious area and for the connected impervious area [35]. The amount of infiltration depends on the amounts of precipitation and surface runoff. That is, it is estimated by excluding surface runoff from rainfall. The infiltrated water is uniformly distributed in a soil layer through a redistribution process. The soil water is percolated at water content above field capacity in the soil layer, and groundwater is recharged by percolation. The amount of actual evaporation from soil is affected by the water content of a soil layer. Sediment and nutrient processes interrelate with the water process. A Modified Universal Soil Loss Equation (MUSLE) [36] predicts sediment yield, and it is a function based on a runoff factor. The transportation of nitrate is influenced by surface runoff, lateral subsurface flow, or percolation. Soil-attached nutrients such as organic and mineral phosphorus and organic nitrogen are governed by sediment yield transported by surface runoff under a loading function [37,38].

2.2.2. Model Development

SWAT processes can sufficiently explain hydrologic behavior of LID practices on a watershed scale. In this study, three LID practices including permeable pavements (PPs), rain gardens (RGs), and rainwater harvesting tanks (RWHs) were factored into an urban area. They partially store surface runoff generated from an urban area up to their capacities and discharge water exceeding their capacities as surface runoff. To reflect the hydrologic behavior of the LID practices, surface runoff processes in SWAT were modified based on McCuen's method [39]. In his method, runoff depth stored by infiltration practices is excluded from the runoff depth of post-development in order to calculate the modified curve number that reflects the infiltration practices. The idea of the method was incorporated into the surface runoff process as Equation (1):

$$Q_{LIDs} = Q_{tot} - LIDval \qquad (1)$$

where Q_{tot} is the surface runoff depth (mm) before the application of LID practices, Q_{LIDs} is the surface runoff depth (mm) after LID practices are reflected, and LIDval is the storage depth (mm) of each type of LID practice.

SWAT effectively represents hydrologic behavior by PPs and RGs under the developed equation. As can be seen in the equation, surface runoff, excluding water stored by PPs and RGs, is computed, and the water stored by PPs and RGs is added to the amount of infiltration. On the other hand, RWHs are simply storage facilities that cannot directly infiltrate the stored water into soil layers, unlike RGs and PPs. Thus, codes were additionally included so that water accrued in the rain barrels and the accumulated water was deliberately drained to reuse them. The rain barrels were defined to be empty after at least 7 consecutive dry days after a rainfall event. A text file that allows for entrance of the storage depths for LID practices was included in a SWAT folder, and an algorithm that could read the text file was coded. Nitrate and total phosphorus were runoff-borne pollutants and were treated along with surface runoff. A detailed description of the representation of LID practices in SWAT can be found in Seo et al. [16].

2.3. LID Conditions for Optimization

The three LID practices (PPs, RGs, and RWHs) are building-scale facilities frequently practiced in urbanized areas which have very little space for installation. Each LID practice is site-specific. In this study, they were assumed to address stormwater and the consequential pollutant loadings only from each specific site: RWHs were installed below roofs and harvested runoff and pollutants only from rooftops during rainfall, PPs were considered only in the parking lots of a commercial area and collected runoff and pollutants generated only from parking lots, and RGs were integrated in the backyards of each house or street system such as sidewalks at random and captured runoff and pollutants generated from a residential area.

Each LID practice occupied different areas. In the case of RWHs, the roof area represented the area of RWHs because RWHs deal with runoff only from roofs. The design data from League City offered no information for roof area. Therefore, an average roof area was acquired from similar neighborhoods with a conventional medium-density urban design through sampling in Google Earth, and total roof area was determined by multiplying the average roof area by the number of lots presented from the design data. The total area of RGs was estimated by multiplying a catchment area by a size factor based on soil properties and depths of RGs [40]. The catchment area was applied for each residential subbasin area in which the total roof area was excluded. This process was for the purpose of ruling out runoff addressed by RWHs. The size factor 0.1 was used based on data from Bannerman and Considine [41]. The total area of PPs was dependent on the percentage of parking lot area presented in the commercial area of the design data. Consequently, 20% (36.37 ha) and 8% (14.55 ha) of the residential area were considered as the areas for RWHs and RGs, respectively, and 47% (8.57 ha) of the commercial area was taken into account as the area for PPs.

Each LID practice was designed to detain different runoff depths. The maximum storage depths of PPs and RGs were limited to a rainfall size. They were calculated using 1.5 inches (38.1 mm) of precipitation on each site based on the Curve Number (CN) method. An amount of 1.5 inches of rainfall is the 85th percentile 24-h rainfall depth and it is a value referred to as water quality protection by a stormwater management system in the North Central Texas Council of Governments (NCTCOG) region [42] (For RWHs, 1000-gallon rain barrels were assumed to be used to treat runoff and pollutants from roofs [43]. The volume was reversely divided by the average roof area to estimate maximum storage depth. As a result, PPs, RGs, and RWHs were sized to capture 32.52-mm, 19.11-mm, and 12.94-mm runoff depths from each area, respectively. The information for the maximum areas and storage depths of LID practices is summarized in Table 1.

Table 1. Specific information for maximum areas and storage depths of low-impact development (LID) practices. RWHs: rainwater harvesting tanks; RGs: rain gardens; PPs: permeable pavements.

Practices	Area (ha)			Storage Depth of LID Practices (mm)
	Commercial Area (Subbasin 2)	Residential Area (Subbasin 3)	Residential Area (Subbasin 4)	
Total subbasin	18.22	83.28	98.55	-
RGs	-	6.66	7.88	19.11
RWHs	-	16.66	19.71	12.94
PPs	8.57	-	-	32.52

2.4. Modeling Setup

In this study, "LID-absence" and "LID-presence" mean the post-development state without and with LID practices, respectively. Simulations for LID-absence and LID-presence scenarios were performed for the purpose of identifying the maximum LID benefits which would be ultimately used to set targeted reduction goals. The targeted goals are illustrated in the "Manual Optimization" subsection.

The simulation for the LID-absence scenario was first configured by using several input data. It is desirable to use high spatial resolution data for the simulation of a small study area for producing accurate outputs. A ten by ten-meter digital elevation model (DEM) obtained from the USDA NRCS Geospatial Data Gateway was used to describe topography in detail. For soils, the Soil Survey Geographic Database (SSURGO) from the NRCS Soil Data Mart was applied. The daily rainfall and temperature data of two stations, the National Weather Service Office and the Houston Clover Field, were employed; these were acquired from the National Climate Data Center (NCDC). For humidity, wind speed, and solar radiation, the data from a weather generator, which generates climatic data using monthly mean data of many years, were used. The land use with a conventional urban form of medium density was applied to obtain results in the post-development state. The urban area takes up about 56% of total area and is separated as residential and commercial areas, which have 44% and 75% impervious fractions, respectively. The remaining area (44%) remains unchanged as a pre-developed area. The land use was represented as 5 subbasins, including 2 subbasins for the residential area and 1 subbasin for the commercial area, and 18 HRUs in total. SWAT was tested for surface runoff, nitrate, and total phosphorus from October 2006 to December 2011. The results of the LID-absence scenario indicated 473.32 mm for surface runoff, 591.87 kg for nitrate, and 449.55 kg for total phosphorus on an average annual basis [16].

The simulation for the LID-presence scenario was performed under the same conditions as for the LID-absence scenario, except for LID conditions. In order to test the LID-presence scenario, three LID facilities were applied in SWAT: RWHs and RGs were considered in the residential area and PPs were only considered in the commercial area. The LID practices were assumed to be fully placed and implemented in the LID areas of Table 1, and seasonal impacts of LID practices were not reflected. The application of LID practices was performed at an HRU level. The existing HRUs of the residential and commercial areas were divided into separate HRUs for roofs and parking lots in order to treat RWHs and PPs. The RGs were considered in the rest of the HRUs of the residential area. In order to divide the HRUs, the percentages for the areas of roofs (20%) and parking lots (47%) were multiplied by the existing HRUs. The roofs and parking lots were included as new urban types in the existing urban data of SWAT, and each urban type was applied to the individual specific HRUs. The number representing each LID was also applied to all HRUs that have LID practices. The LID-presence scenario was run by using the modified SWAT, and the effects by LID practices were measured for runoff, nitrate, and total phosphorus on a watershed scale. The LID practices mitigated the surface runoff and the consequential pollutants well in urban areas by showing decreased values from the LID-absence scenario. The results of the LID-presence scenario represented 337.81 mm for surface runoff, 405.85 kg for nitrate, and 338.86 kg for total phosphorus on an average annual basis. The detailed information for modeling work can be found in Seo et al. [16]. The differences between the LID-absence and LID-presence scenarios were calculated for all variables to set targeted goals [16].

2.5. Cost Estimation

Cost is an important measure for optimization. An annual total cost for each LID practice was estimated as the sum of construction and maintenance costs, based on the following equation by Arabi et al. [19] (Equation (2)):

$$C_{td} = \left[C_0 \cdot (1 + s)^{td} + C_0 \cdot rm \cdot \left(\frac{(1 + s)^{td} - 1}{s} \right) \right] / td \qquad (2)$$

where C_{td} is the annual cost per unit area during a design life ($/ft^2/year), C_0 is the construction cost per unit area ($/ft^2), rm is the proportion of maintenance to construction cost, s is the interest rate, and td is the intended life of LID practices based on routine maintenance.

Data for the construction cost per unit area ($/ft^2) were acquired through experiments at the Texas A&M AgriLife Research and Extension Center in Dallas [44]. The cost of $6 per square feet was used for RGs, $14 per square feet for PPs, and $1 per gallon for RWHs. In the case of RWHs, the cost per gallon was converted to cost per unit area by replacing 1000-gallon rain barrel with the average roof area. The function of LID practices decreases as time passes. Maintenance is thus continuously required to keep the same effectiveness during the life-time of LID practices. For the computation of maintenance costs, annually 5% was used as the proportion of maintenance of RGs to construction cost. This value was referenced by the US Environmental Protection Agency [45]. In the case of PPs and RWHs that have no reference data, 5%, the same as for RGs, was used for PPs because a similar maintenance cost was incurred to maintain PPs in the experimental field of the AgriLife center [44], and a 1% ratio was determined for RWHs due to the low maintenance requirements (cost determined based on several systems constructed and built by Texas A&M AgriLife Extension). For all LID practices considered, the same interest rate of 4.5% was considered and the same lifespan of 20 years was applied to the cost calculation. As a result, the annual costs per unit area were estimated as 1.19 ($/ft^2/year) for RGs, 2.79 ($/ft^2/year) for PPs, and 0.04 ($/ft^2/year) for RWHs.

2.6. Manual Optimization

2.6.1. Setting Targeted Goals

The United States Environmental Protection Agency (USEPA) has conducted a water quality standards program which presents a threshold level to protect water bodies [46]. Under the policy, states and local authorities develop region-specific criteria. However, no recommended criteria exist for runoff or pollutant reductions in the study area and accordingly there is no given budget limitation. Therefore, it was determined that five cases would be used as targeted goals to be controlled for each variable. The targeted goals for each case included the following values: 25%, 35%, 45%, 55% and 65% of the maximum reduction amounts for all variables. In the modeling work, the maximum reduction amounts by LID practices were obtained from the difference between the LID-absence and LID-presence scenarios and were 135.51 mm for surface runoff, 186.03 kg for nitrate, and 110.69 kg for total phosphorus as average annual values in the watershed. For Case 1, 25% of the maximum reduction amounts were targeted as reduction amounts to be managed: 33.88 mm for surface runoff, 46.51 kg for nitrate, and 27.67 kg for total phosphorus. Likewise, Cases 2, 3, 4 and 5 targeted 35%, 45%, 55% and 65%, respectively, of the maximum reduction amounts. The constant difference among cases was for facilitating evaluation of the effectiveness of LID practices from the considered LID conditions. The targeted goals for each case are summarized in Table 2.

Table 2. Hypothetical cases for targeted goals.

Variable	Targeted Goal [1]				
	Case 1	Case 2	Case 3	Case 4	Case 5
Surface runoff (mm)	33.88	47.43	60.98	74.53	88.08
Nitrate (kg)	46.51	65.11	83.71	102.32	120.92
Total phosphorus (kg)	27.67	38.74	49.81	60.88	71.95

Note: [1] Targeted goals for each case were 25%, 35%, 45%, 55% and 65% of the maximum reduction amounts by LID practices.

2.6.2. Optimization Procedure

For the purpose of identifying the conditions of LID practices that achieve both a targeted goal and minimal cost, a stepwise manual operation for optimization was attempted for all variables. The LID conditions considered were type, location (subbasin), and percent allocation of LID area. Each type and location of LID practices under 100% allocation were first taken into account to determine a ranking for cost in handling unit reduction in order to ultimately minimize total cost for treating a targeted goal. Step 1: In this study, RGs and RWHs were distributed only in the residential area, which was composed of two subbasins (Subbasin 3 and Subbasin 4), and PPs were placed only in the commercial area, which made up one subbasin (Subbasin 2). Each LID practice was considered in designated subbasins, and thus five cases for the conditions were generated: RGs in Subbasin 3, RGs in Subbasin 4, RWHs in Subbasin 3, RWHs in Subbasin 4, and PPs in Subbasin 2. The SWAT model was run for each case. Step 2: The annual reduction amount by 100% allocation of LID practices in each case was then investigated through the difference from the LID-absence scenario. Step 3: The annual cost for the implementation of LID practices was estimated for every case by multiplying the annual cost per unit area calculated under the cost equation (Equation (2)) by total LID area of each case (given in Table 1). Step 4: The cost per unit reduction was calculated by dividing the annual cost into the annual reduction amount for every case. Different values were obtained for every case, and they were ranked in the order of least costly to most costly. Step 5: Optimization is then carried out based on the type and location for the ranking of the cost per unit reduction. This was achieved as reduction amounts, according to the percent allocation of LID practices, were accumulated up to the point that a targeted goal was met. Step 6: The cost of each case was then estimated through the product of the reduction amount according to the percent allocation of LID practices and the cost per unit reduction of Step 4. The final total cost (TC) and the final cost per unit reduction (CPR) were obtained respectively by the sum of the costs for each case and by dividing the final total cost into the targeted goal.

With regard to optimization of percent allocation, three constraint conditions were applied to explore the behavior of the effectiveness of LID practices: (1) maximum adoption; (2) medium adoption; and (3) minimum adoption. Maximum adoption means to allow full occupation in given LID areas even if it is not feasible in reality. Medium adoption means to restrict the potential occupation of LID practices to a maximum of 75% for RGs and RWHs and 50% for PPs. Thus, reduction amounts which are not addressed by the difference in percent allocation from maximum adoption are passed on to the next rankings. Minimum adoption is to require at least 20% occupation of LID practices but not to exceed 75% for RGs and RWHs and 50% for PPs. In this case, after the 20% allocation is applied to all rankings, the same process with medium adoption is conducted to address the remaining reduction amount for meeting a targeted goal. The optimization was performed in the same way for targeted goals of all variables under three constraint conditions. Figure 3 and Appendix A provide a stepwise procedure and an example for surface runoff Case 5 under maximum adoption, respectively.

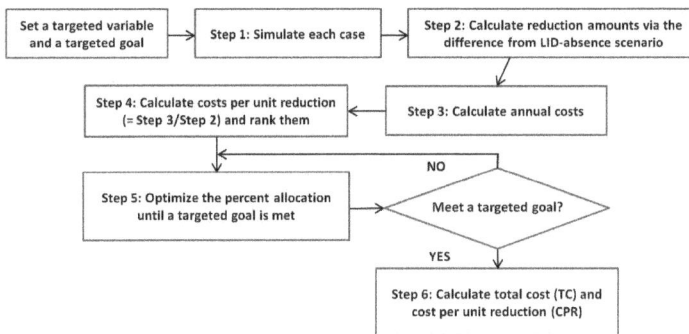

Figure 3. Flow chart for a manual optimization procedure.

3. Results

3.1. Optimized Conditions

The cost-effective conditions for controlling each targeted goal were determined through the optimization process for all variables (Tables 3–5). A variety of configurations were drawn for each variable. For surface runoff, the optimized conditions were ranked in the order of RWHs (4), RWHs (3), RGs (4), RGs (3), and then PPs (2) (numbers in parentheses mean a location (subbasin) of LID practices, and thus RWHs (4) means RWHs located at Subbasin 4). For nitrate, they were arranged in the order of RWHs (3), RWHs (4), RGs (4), RGs (3), and then PPs (2). In the case of total phosphorus, since the amount reduced by RWHs was tiny compared to the cost for implementation of RWHs, the type of RGs was prioritized to the cost-effective conditions, unlike surface runoff and nitrate: RGs (3), RGs (4), RWHs (4), RWHs (3), and then PPs (2). The type of PPs was ranked last on all occasions because of their high cost. Under these rankings, different percentages of allocation were assigned as seen in Tables 3–5, which met the given targeted goals under three constraint conditions. In surface runoff and nitrate, not only 100% allocation of RWHs but also the application of RGs was required even to address the smallest targeted goal of Case 1 under the maximum adoption. This was because the RWHs were the most cost-effective but the amount reduced by RWHs was small as described above despite the 100% allocation at all locations. In total phosphorus, the RGs determined as the most cost-effective condition were only considered in dealing with the targeted goals up to Case 5 under the maximum adoption. Less cost-effective conditions were more considered in the medium and minimum conditions than in the maximum condition for all variables.

3.2. Analysis of LID Effects

As can be seen through the optimization results, various combinations of conditions could affect the effectiveness of LID practices. The impact of the optimized LID conditions on the effectiveness of LID practices was observed through the comparison among cases of targeted goals. For example, with regard to the result of maximum adoption for surface runoff (Table 3A), the effectiveness of LID practices in the watershed increased as much as 13.55 mm in Case 2 by considering 22.62% more RGs (4) than in Case 1. The effect increased 100% in Case 3 by extending RG occupation by as much as 45.24%, and increased 200% in Case 4 as 59.86% more RGs (4) and 9.99% more RGs (3) were added compared to Case 1. An increase of 300% was also shown in Case 5 as 59.86% more RGs (4) and 38.28% more RGs (3) were considered compared to Case 1. When nitrate was a focused variable (Table 4A), 29.17% more adoption of RGs (4) than in Case 1 improved the effectiveness of LID practices by 18.6 kg in Case 2. The effect rose 100% in Case 3 as 58.34% more RGs (4) were added than in Case 1. Also, increases of 200% and 300% appeared in Cases 4 and 5, respectively, by further considering 70.13% more RGs (4) and 21.41% more RGs (3) and by expanding to 70.13% more RGs (4) and 57.30% more RGs (3) than in

Case 1. In the case of total phosphorus, as seen in Table 5A, the effectiveness of LID practices grew by 11.07 kg in Case 2 as 27.61% more RGs (3) were factored than in Case 1, and the effect increased 100% in Case 3 as 31.04% more RGs (3) and 20.76% more RGs (4) were adopted than in Case 1. In addition, improvements of 200% and 300% occurred in Cases 4 and 5 by adoption of 31.04% more RGs (3) and 44.47% more RGs (4) and of 31.04% more RGs (3) and 68.18% more RGs (4) than in Case 1, respectively.

The impact of the optimized LID conditions on the effectiveness of LID practices was also observed through the comparison among constraint conditions. The result in Case 3 for surface runoff, for example, showed fully occupied LID practices up to Ranking 3 and 18.1% RGs (3) under the condition of medium adoption in order to meet the same targeted goal as maximum adoption (Table 3B). Under the condition of minimum adoption, 67.52% RGs (4) were applied to Ranking 3 and the highest and lowest constraint values were applied to the rest of the rankings (Table 3C). The different conditions of LID practices were applied to the medium and minimum conditions, but the result represented that the same effect of LID practices as for the maximum adoption was achieved.

Table 3. Results of optimization for surface runoff (LID type, location, and percent allocation were optimized under the maximum, medium, and minimum constraint conditions, respectively). Sub stands for Subbasin.

(A) Maximum Adoption

Ranking [1]	Type	Location	% Allocation				
			Case 1	Case 2	Case 3	Case 4	Case 5
1	RWHs	Sub 4	100	100	100	100	100
2	RWHs	Sub 3	100	100	100	100	100
3	RGs	Sub 4	40.14	62.76	85.38	100	100
4	RGs	Sub 3	0	0	0	9.99	38.28
5	PPs	Sub 2	0	0	0	0	0

(B) Medium Adoption

Ranking	Type	Location	% Allocation				
			Case 1	Case 2	Case 3	Case 4	Case 5
1	RWHs	Sub 4	75	75	75	75	75
2	RWHs	Sub 3	75	75	75	75	75
3	RGs	Sub 4	44.24	66.86	75	75	75
4	RGs	Sub 3	0	0	18.1	46.39	74.68
5	PPs	Sub 2	0	0	0	0	0

(C) Minimum Adoption

Ranking	Type	Location	% Allocation				
			Case 1	Case 2	Case 3	Case 4	Case 5
1	RWHs	Sub 4	75	75	75	75	75
2	RWHs	Sub 3	75	75	75	75	75
3	RGs	Sub 4	22.28	44.9	67.52	75	75
4	RGs	Sub 3	20	20	20	38.93	67.21
5	PPs	Sub 2	20	20	20	20	20

Note: [1] Ranking is the order of least costly to most costly in handling unit reduction, and optimization was conducted in the order of the rankings up to the point that targeted goals were met.

Table 4. Results of optimization for nitrate (LID type, location, and percent allocation were optimized under the maximum, medium, and minimum constraint conditions, respectively).

(A) Maximum Adoption

Ranking [1]	Type	Location	% Allocation				
			Case 1	Case 2	Case 3	Case 4	Case 5
1	RWHs	Sub 3	100	100	100	100	100
2	RWHs	Sub 4	100	100	100	100	100
3	RGs	Sub 4	29.87	59.04	88.21	100	100
4	RGs	Sub 3	0	0	0	21.41	57.3
5	PPs	Sub 2	0	0	0	0	0

Table 4. *Cont.*

(B) Medium Adoption

Ranking	Type	Location	% Allocation				
			Case 1	Case 2	Case 3	Case 4	Case 5
1	RWHs	Sub 3	75	75	75	75	75
2	RWHs	Sub 4	75	75	75	75	75
3	RGs	Sub 4	40.64	69.81	75	75	75
4	RGs	Sub 3	0	0	29.51	65.42	75
5	PPs	Sub 2	0	0	0	0	29.06

(C) Minimum Adoption

Ranking	Type	Location	% Allocation				
			Case 1	Case 2	Case 3	Case 4	Case 5
1	RWHs	Sub 3	75	75	75	75	75
2	RWHs	Sub 4	28.21	75	75	75	75
3	RGs	Sub 4	20	38.84	68.01	75	75
4	RGs	Sub 3	20	20	20	47.31	75
5	PPs	Sub 2	20	20	20	20	29.06

Note: [1] Refer to the annotation in Table 3.

Table 5. Results of optimization for total phosphorus (LID type, location, and percent allocation were optimized under the maximum, medium, and minimum constraint conditions, respectively).

(A) Maximum Adoption

Ranking [1]	Type	Location	% Allocation				
			Case 1	Case 2	Case 3	Case 4	Case 5
1	RGs	Sub 3	68.96	96.57	100	100	100
2	RGs	Sub 4	0	0	20.76	44.47	68.18
3	RWHs	Sub 4	0	0	0	0	0
4	RWHs	Sub 3	0	0	0	0	0
5	PPs	Sub 2	0	0	0	0	0

(B) Medium Adoption

Ranking	Type	Location	% Allocation				
			Case 1	Case 2	Case 3	Case 4	Case 5
1	RGs	Sub 3	68.96	75	75	75	75
2	RGs	Sub 4	0	18.53	42.24	65.95	75
3	RWHs	Sub 4	0	0	0	0	75
4	RWHs	Sub 3	0	0	0	0	75
5	PPs	Sub 2	0	0	0	0	36.3

(C) Minimum Adoption

Ranking	Type	Location	% Allocation				
			Case 1	Case 2	Case 3	Case 4	Case 5
1	RGs	Sub 3	38.07	65.68	75	75	75
2	RGs	Sub 4	20	20	35.7	59.4	75
3	RWHs	Sub 4	20	20	20	20	75
4	RWHs	Sub 3	20	20	20	20	75
5	PPs	Sub 2	20	20	20	20	36.3

Note: [1] Refer to the annotation in Table 3.

3.3. Analysis of Costs

Final total cost (TC) and cost per unit reduction (CPR) generated from the optimized conditions were compared and analyzed. All results displayed were the minimal costs that treated the given targeted goals (Table 6 and Figure 4). In the comparison among three constraint conditions (that is, the conditions that indicated the same effectiveness of LID practices), the maximum condition showed the lowest TC and the lowest CPR for all cases and for all variables. This was a natural result because more adoption of cost-effective LID conditions was possible in controlling the same targeted goal under the maximum condition as compared to the medium and minimum conditions. The maximum and medium conditions presented similar TC while the minimum condition indicated a large difference from the maximum and medium conditions (Figure 4). This was because 20% of the expensive PP was applied under the minimum adoption for all cases. In the medium adoption of nitrate and total phosphorus, abrupt increases in Case 5 could be also explained due to the application of PPs. Meanwhile, in the comparison among the cases of targeted goals (that is, the conditions that indicated the variation of the effectiveness of LID practices), the lowest TC and the lowest CPR resulted in Case 1 for all variables under the maximum and medium conditions. This was due to the fact that the more a targeted reduction amount was increased, the more the total cost and the consequential cost per unit reduction increased. The minimum adoption showed the same trend in TC. However, it presented the lowest CPR in Case 5 for surface runoff and in Case 4 for nitrate and total phosphorus and the highest CPR in Case 1 for all variables (Figure 4). This was seen because unlike the maximum and medium conditions, the cost-effective conditions were ignored in up to 20% adoption in all cases under the minimum adoption, and relatively expensive PPs compared to the other LID practices were compulsorily considered. With regard to nitrate and total phosphorus, the reason why Case 4 was more cost-effective than Case 5 was also attributable to a higher percent occupation of PPs in Case 5. That is, the increase of the total cost was significant compared to the increase of the targeted reduction amount.

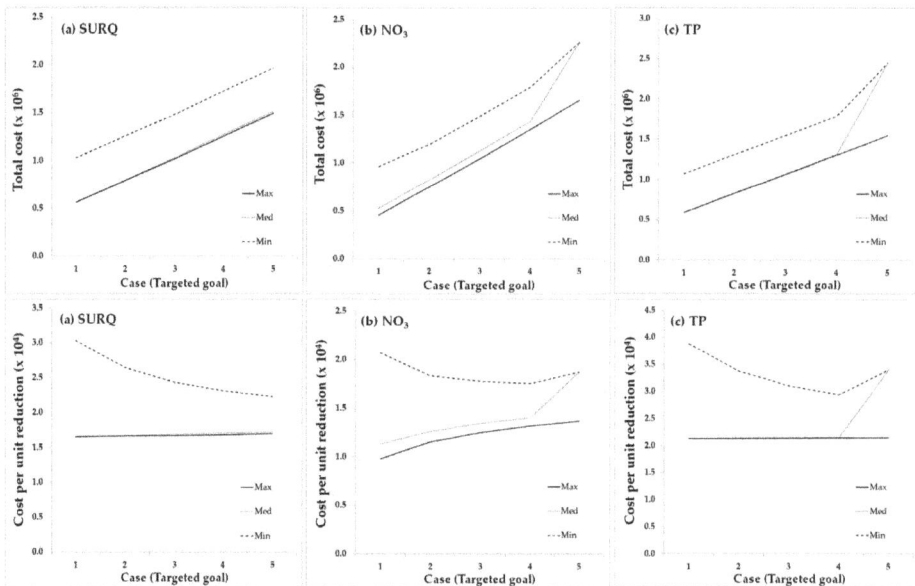

Figure 4. Annual total cost (TC) and cost per unit reduction (CPR) for all targeted goals under three constraint conditions: (**a**) surface runoff (SURQ); (**b**) nitrate; and (**c**) total phosphorus (they are simply schematized in Table 6).

Table 6. Annual total cost and cost per unit reduction from optimized LID conditions for all targeted goals under three constraint conditions.

Case	Variable	Maximum Adoption		Medium Adoption		Minimum Adoption	
		TC [1]	CPR [1]	TC	CPR	TC	CPR
5		1,493,527.83	16,956.54	1,513,759.76	17,186.24	1,963,588.64	22,293.30
4		1,251,284.14	16,789.22	1,271,516.07	17,060.68	1,721,430.57	23,097.45
3	Surface runoff	1,017,591.16	16,687.76	1,029,336.60	16,880.38	1,483,537.77	24,328.95
2		788,374.34	16,622.70	791,769.48	16,694.29	1,254,320.94	26,447.09
1		**559,157.51**	**16,505.60**	562,628.65	16,608.06	1,025,104.12	30,259.74
5		1,656,393.69	13,698.51	2,263,032.20	18,715.45	2,263,032.20	18,715.45
4		1,349,072.15	13,185.47	1,434,446.15	14,019.89	1,793,187.45	17,526.13
3	NO$_3$	1,046,268.60	12,498.38	1,126,953.36	13,462.21	1,488,503.12	17,781.17
2		750,678.20	11,529.45	821,713.58	12,620.47	1,192,912.72	18,321.61
1		**455,087.80**	**9785.39**	526,072.52	11,311.72	963,350.07	20,714.15
5		1,547,180.52	21,504.95	2,449,017.71	34,039.98	2,449,017.71	34,039.98
4		1,306,918.32	21,468.24	1,310,485.93	21,526.85	1,788,390.63	29,377.20
3	TP	1,066,656.12	21,415.22	1,070,249.06	21,487.36	1,548,229.76	31,083.77
2		826,916.70	21,345.41	829,936.20	21,423.35	1,309,329.87	33,798.06
1		**590,495.76**	**21,339.66**	590,517.17	21,340.43	1,072,908.94	38,773.37

Note: [1] TC means total cost ($), and CPR means cost per unit reduction; the unit is $/mm for surface runoff and $/kg for nitrate and total phosphorus (TP).

4. Impact of LID Use on Detention Requirements and Cost

Thus far, the effectiveness of LID practices and the consequential costs according to the optimized LID conditions have been analyzed. However, the water volumes detained by LID practices under the optimized conditions are small (Table 7) because the maximum capacities and allowable areas of LID practices are limited. Thus, for heavy rainfall, a considerable amount of water that is not treated by LID practices would be generated as surface runoff, directly entering channels. Such a large amount of water that cannot be detained by LID practices needs to be taken into account by other methods for controlling stormwater in the region. Therefore, detention ponds were incorporated in the study area to reflect the water volume that could not be addressed by LID practices, and 100-year 24-h rainfall (13 inches) was assumed in this region as the standard for heavy rainfall for the purpose of calculating the volume that should be captured by detention ponds. First, the required detention volume by urbanization was estimated by the difference between pre- and post-development states in surface runoff. The volume that should be captured by detention ponds was then calculated by subtracting the volume detained by LID practices from the required detention volume. The total cost of detention ponds for addressing the calculated volume capacity was calculated using the following equation developed by Brown and Schueler [47] (Equation (3)):

$$C = 24.5 \times V^{0.705} \tag{3}$$

where C is the establishment cost including construction, design, and authorization ($) and V is the pond volume (ft^3). For the calculation of annual cost, a 5% ratio for maintenance (rm) and a design life of 20 years (td), obtained from the USEPA website, were considered and the same interest rate (s) of 4.5% was applied. Additionally, the cost savings for the amount controlled by LID practices was computed by the difference between the costs for the calculated detention volume and the required detention volume. Total cost of detention ponds was greatest in Case 1 and accordingly cost savings were the smallest in Case 1 (Figure 5). This was because the volume that should be captured by detention ponds was increased by the smallest volume detained by LID practices in Case 1. The same trend was indicated in all variables and all constraint conditions. Total phosphorus showed the greatest difference in the cost of detention ponds between Case 4 and Case 5 under the medium and minimum conditions. This was seen because the difference in the optimized LID conditions between two cases caused the difference in the volume detained by LID practices. The volume detained by LID practices is affected by percent allocation and storage depth of LID practices, and thus it could vary depending on the optimized LID conditions even with the same targeted goal. Why the costs of detention ponds in total phosphorus were higher than those in the other variables could also be explained by the results of the optimized LID conditions.

LID practices installed in urban areas generally are more expensive than detention ponds. This study does not to compare LID practices to detention ponds. City authorities are being forced to deal with stormwater generated from their regions for new urban developments. Thus, this section presented the volumes that should be captured (that is, the volumes that exceed LID capacities) for heavy rainfall and the consequential costs using detention ponds as a secondary stormwater management method.

Table 7. Volume detained by detention ponds and the consequential cost and cost savings in detention ponds for the amount controlled by LID practices for all cases of all variables under each constraint condition. Ac-ft/ac stands for Acre-feet per acre.

Case	Variable	Volume Detained by LID Practices (ac-ft/ac)	Volume Detained by Detention Ponds (ac-ft/ac)	Cost ($/Year)	Cost Savings ($/Year)
(A) Maximum Adoption					
5		0.0110 (7%) [1]	0.1429 (93%)	243,339.92	13,050.24
4		0.0104 (7%)	0.1434 (93%)	244,048.85	12,341.31
3	Surface runoff	0.0098 (6%)	0.1440 (94%)	244,731.94	11,658.22
2		0.0093 (6%)	0.1446 (94%)	245,401.17	10,988.99
1		0.0087 (6%)	0.1451 (94%)	246,069.64	10,320.51
5		0.0114 (7%)	0.1425 (93%)	242,862.80	13,527.36
4		0.0106 (7%)	0.1432 (93%)	243,762.77	12,627.39
3	NO_3	0.0099 (6%)	0.1439 (94%)	244,648.16	11,742.00
2		0.0092 (6%)	0.1447 (94%)	245,511.16	10,879.00
1		0.0085 (6%)	0.1454 (94%)	246,372.89	10,017.26
5		0.0038 (2%)	0.1501 (98%)	251,941.30	4448.86
4		0.0032 (2%)	0.1507 (98%)	252,634.31	3755.85
3	TP	0.0026 (2%)	0.1512 (98%)	253,326.52	3063.63
2		0.0020 (1%)	0.1518 (99%)	254,016.45	2373.71
1		0.0014 (1%)	0.1524 (99%)	254,696.05	1694.11
(B) Medium Adoption					
5		0.0092 (6%)	0.1446 (94%)	245,482.09	10,908.07
4		0.0086 (6%)	0.1452 (94%)	246,188.43	10,201.73
3	Surface runoff	0.0080 (5%)	0.1458 (95%)	246,893.92	9496.23
2		0.0074 (5%)	0.1464 (95%)	247,584.71	8805.45
1		0.0069 (4%)	0.1470 (96%)	248,250.71	8139.45
5		0.0105 (7%)	0.1433 (93%)	243,883.47	12,506.69
4		0.0090 (6%)	0.1448 (94%)	245,713.38	10,676.77
3	NO_3	0.0083 (5%)	0.1456 (95%)	246,609.48	9780.67
2		0.0075 (5%)	0.1463 (95%)	247,497.80	8892.36
1		0.0068 (4%)	0.1470 (96%)	248,356.64	8033.52
5		0.0109 (7%)	0.1430 (93%)	243,486.50	12,903.66
4		0.0032 (2%)	0.1506 (98%)	252,623.95	3766.21
3	TP	0.0026 (2%)	0.1512 (98%)	253,316.18	3073.98
2		0.0020 (1%)	0.1518 (99%)	254,007.62	2382.54
1		0.0014 (1%)	0.1524 (99%)	254,696.05	1694.11

Table 7. *Cont.*

Case	Variable	Volume Detained by LID Practices (ac-ft/ac)	Volume Detained by Detention Ponds (ac-ft/ac)	Cost ($/Year)	Cost Savings ($/Year)
(C) Minimum Adoption					
5		0.0100 (6%)	0.1439 (94%)	244,574.79	11,815.37
4		0.0094 (6%)	0.1445 (94%)	245,281.97	11,108.18
3	Surface runoff	0.0088 (6%)	0.1451 (94%)	245,975.87	10,414.29
2		0.0082 (5%)	0.1456 (95%)	246,643.69	9746.47
1		0.0077 (5%)	0.1462 (95%)	247,310.75	9079.40
5		0.0105 (7%)	0.1433 (93%)	243,883.47	12,506.69
4		0.0095 (6%)	0.1443 (94%)	245,072.51	11,317.65
3	NO_3	0.0088 (6%)	0.1450 (94%)	245,961.40	10,428.76
2		0.0081 (5%)	0.1458 (95%)	246,822.47	9567.68
1		0.0057 (4%)	0.1482 (96%)	249,707.61	6682.55
5		0.0109 (7%)	0.1430 (93%)	243,486.50	12,903.66
4		0.0055 (4%)	0.1483 (96%)	249,904.20	6485.96
3	TP	0.0049 (3%)	0.1489 (97%)	250,599.27	5790.89
2		0.0043 (3%)	0.1495 (97%)	251,289.89	5100.27
1		0.0037 (2%)	0.1501 (98%)	251,972.56	4417.59

Note: [1] Parentheses include percentages of the volume detained by LID practices and the volume detained by detention ponds.

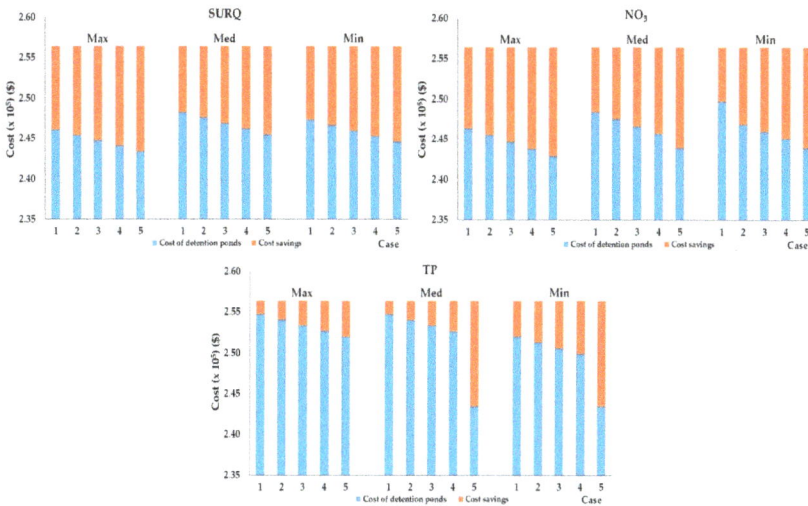

Figure 5. Cost of detention ponds and cost savings for the amount controlled by LID practices for all cases of all variables under each constraint condition (they are simply schematized in Table 7).

5. Conclusions

The study has presented the cost-effective LID conditions found through optimization and has analyzed the effectiveness of LID practices on a watershed scale and the consequential costs. To attain the goal, five targeted goals were set and LID conditions for type, location, and percent allocation were optimized. The optimization ultimately came up with the most cost-effective and efficient guidelines for LID planning in the study watershed. For example, if the region is given a budget of $600,000 in dealing with surface runoff, it could consider the LID conditions of Case 1 for both maximum and medium conditions. Or, if the region decides to allow maximum adoption to treat the targeted goal, Case 3, for nitrate, it could apply the LID conditions of Case 3 and need at least $1,047,000.

In general, what could be learned through the study was that maximizing the treatment effect of each LID practice should be a priority as cost benefits increase linearly for each unit of surface runoff that is captured. For each LID practice that is implemented, water managers thus need to focus on maximizing the amount of runoff captured for each plot in order to increase cost effectiveness. In addition, the cost-effective results of this study would be generated differently by other conditions such as different types of LID practices besides RGs, PPs and RWHs, different limitations for the allocation of LID practices, different treatment goals, watershed characteristics, and so forth. Therefore, adequate studies for a variety of conditions should be done in advance to achieve cost-effective results within a given budget before the installation of LID practices. Such studies would likely suggest planning and design of LID projects that accomplish a balance between environmental and economic aspects on a development or watershed scale.

As accounted for in the Manual Optimization section, the optimization method employed is very simple and practical in providing cost-effective conditions. It is likely that this method would be applicable in many studies and would easily assist watershed managers in determining the best solution for the establishment of LID practices for their watershed management. In addition, this study has been based on simple calculations using the results of modeling work. If field work had been performed, it would have been possible to validate our results. Such an additional study would be a very meaningful work in that it could lay the groundwork for studies on other watersheds.

Supplementary Materials: The following are available online at www.mdpi.com/2073-4441/9/4/270/s1.

Acknowledgments: The authors would like to thank the Texas Sea Grant college program and TCEQ for their financial support and the Korean National Institute of Agricultural Sciences (PJ010867) for their help.

Author Contributions: Mijin Seo and Fouad Jaber designed the project; Mijin Seo performed the modeling; Mijin Seo, Fouad Jaber and Raghavan Srinivasan analyzed the data and wrote the paper.

Conflicts of Interest: The authors declare no conflict of interest.

Appendix A

Targeted variable	Surface runoff (SURQ)
Targeted goal	88.08 mm (Case 5)

Step 1			Step 2	Step 3	Step 4	
Type	Location	SURQ (mm)	Reduction by LID practices (mm)	Annual cost ($)	Cost per unit reduction ($/mm)	Ranking
None	-	473.32	-	-	-	-
RGs	Sub 3	425.42	47.90	856,287.36	17,876.44	4
RGs	Sub 4	413.40	59.91	1,013,336.99	16,913.66	3
PPs	Sub 2	455.44	17.88	2,568,860.66	143,685.69	5
RWHs	Sub 3	468.89	4.43	69,801.00	15,757.62	2
RWHs	Sub 4	467.92	5.40	82,603.04	15,298.75	1

			Step 5		Reduction by	Accumulation	Step 6	
Ranking	Type	Location	Reduction by LID practices (mm)	% allocation	% allocation (mm)	of reduction (mm)	Cost per unit reduction ($/mm)	Cost ($)
1	RWHs	Sub 4	5.40	100.00%	5.40	5.40	15,298.75	82,603.04
2	RWHs	Sub 3	4.43	100.00%	4.43	9.83	15,757.62	69,801.00
3	RGs	Sub 4	59.91	100.00%	59.91	69.74	16,913.66	1,013,336.99
4	RGs	Sub 3	47.90	38.28%	18.34	88.08	17,876.44	327,786.80
5	PPs	Sub 2	17.88				143,685.69	0.00

Summary of results for Case 5					
Ranking	Type	Location	% allocation	Total cost (TC)	
1	RWHs	Sub 4	100.00%	1,493,527.8	($/yr)
2	RWHs	Sub 3	100.00%	Cost per unit reduction (CPR)	
3	RGs	Sub 4	100.00%	16,956.5	($/mm)
4	RGs	Sub 3	38.28%		
5	PPs	Sub 2	0.00%		

Figure A1. Example for optimization of surface runoff Case 5 under maximum adoption. A Microsoft Excel version of this table can be found in supplemental material.

References

1. Freeman, H. The road to LID plan approval in coastal North Carolina: Development of a spreadsheet modeling tool for LID based designs. In Presented at World Environmental and Water Resources Congress, Kansas City, MO, USA, 17–21 May 2009.

2. Paul, M.J.; Meyer, J.L. Streams in the urban landscape. *Annu. Revecol. Syst.* **2001**, *32*, 333–365. [CrossRef]

3. Tong, S.T.; Chen, W. Modeling the relationship between land use and surface water quality. *J. Environ. Manag.* **2002**, *66*, 377–393. [CrossRef]

4. Newcomer, M.E.; Gurdak, J.J.; Sklar, L.S.; Nanus, L. Urban recharge beneath low impact development and effects of climate variability and change. *Water Resour. Res.* **2014**, *50*, 1716–1734. [CrossRef]

5. Dietz, M.E.; Clausen, J.C. Stormwater runoff and export changes with development in a traditional and low impact subdivision. *J. Environ. Manag.* **2008**, *87*, 560–566. [CrossRef] [PubMed]

6. Damodaram, C.; Giacomoni, M.H.; Prakash Khedun, C.; Holmes, H.; Ryan, A.; Saour, W.; Zechman, E.M. Simulation of combined best management practices and low impact development for sustainable stormwater management. *J. Am. Water Resour. Assoc.* **2010**, *46*, 907–918. [CrossRef]

7. Jeon, J.H.; Lim, K.J.; Choi, D.H.; Kim, T.D. Modeling the effects of low impact development on Runoff and Pollutant Loads from an Apartment complex. *Environ. Eng. Res.* **2010**, *15*, 167–172. [CrossRef]

8. Jeong, J.; Kannan, N.; Arnold, J.G.; Glick, R.; Gosselink, L.; Srinivasan, R.; Barrett, M.E. Modeling sedimentation-filtration basins for urban watersheds using Soil and Water Assessment Tool. *J. Environ. Eng.* **2013**, *139*, 838–848. [CrossRef]

9. Brander, K.E.; Owen, K.E.; Potter, K.W. Modeled impacts of development type on runoff volume and infiltration performance. *J. Am. Water Resour. Assoc.* **2004**, *40*, 961–969. [CrossRef]

10. Holman-Dodds, J.K.; Bradley, A.A.; Potter, K.W. Evaluation of hydrologic benefits of infiltration based urban storm water management. *J. Am. Water Resour. Assoc.* **2003**, *39*, 205–215. [CrossRef]

11. Ackerman, D.; Stein, E.D. Evaluating the effectiveness of best management practices using dynamic modeling. *J. Environ. Eng.* **2008**, *134*, 628–639. [CrossRef]

12. Carter, T.; Jackson, C.R. Vegetated roofs for stormwater management at multiple spatial scales. *Landsc. Urban Plan.* **2007**, *80*, 84–94. [CrossRef]

13. Schneider, L.E.; McCuen, R.H. Assessing the hydrologic performance of best management practices. *J. Hydrol. Eng.* **2006**, *11*, 278–281. [CrossRef]

14. Gilroy, K.L.; McCuen, R.H. Spatio-temporal effects of low impact development practices. *J. Hydrol.* **2009**, *367*, 228–236. [CrossRef]

15. Williams, E.S.; Wise, W.R. Hydrologic impacts of alternative approaches to storm water management and land development. *J. Am. Water Resour. Assoc.* **2006**, *42*, 443–455. [CrossRef]

16. Seo, M.; Jaber, F.; Srinivasan, R.; Jeong, J. Evaluating the impact of Low Impact Development (LID) practices on water quantity and quality under different development designs using SWAT. *Water* **2017**, *9*, 193. [CrossRef]

17. Endreny, T.; Collins, V. Implications of bioretention basin spatial arrangements on stormwater recharge and groundwater mounding. *Ecol. Eng.* **2009**, *35*, 670–677. [CrossRef]

18. Ahiablame, L.M.; Engel, B.A.; Chaubey, I. Effectiveness of low impact development practices in two urbanized watersheds: Retrofitting with rain barrel/cistern and porous pavement. *J. Environ. Manag.* **2013**, *119*, 151–161. [CrossRef] [PubMed]

19. Arabi, M.; Govindaraju, R.S.; Hantush, M.M. Cost-effective allocation of watershed management practices using a genetic algorithm. *Water Resour. Res.* **2006**, *42*. [CrossRef]

20. Chaubey, I.; Maringanti, C.; Schaffer, B.; Popp, J. Targeting vs. Optimization: Critical evaluation of BMP implementation plan for watershed management. In Presented at the World Environmental and Water Resources Congress 2008 Ahupua'a, Honolulu, HI, USA, 12–16 May 2008.

21. Liu, Y.; Bralts, V.F.; Engel, B.A. Evaluating the effectiveness of management practices on hydrology and water quality at watershed scale with a rainfall-runoff model. *Sci. Total Environ.* **2015**, *511*, 298–308. [CrossRef] [PubMed]

22. Gitau, M.; Veith, T.; Gburek, W. Farm-level optimization of BMP placement for cost-effective pollution reduction. *Trans. ASABE* **2004**, *47*, 1923–1931. [CrossRef]

23. Gitau, M.W.; Veith, T.L.; Gburek, W.J.; Jarrett, A.R. Watershed level best management practice selection and placement in the Town Brook watershed, New York. *J. Am. Water Resour. Assoc.* **2006**, *42*, 1565–1581. [CrossRef]

24. Maringanti, C.; Chaubey, I.; Popp, J. Development of a multiobjective optimization tool for the selection and placement of best management practices for nonpoint source pollution control. *Water Resour. Res.* **2009**, *45*. [CrossRef]

25. Maringanti, C.; Chaubey, I.; Arabi, M.; Engel, B. Application of a multi-objective optimization method to provide least cost alternatives for NPS pollution control. *Environ. Manag.* **2011**, *48*, 448–461. [CrossRef] [PubMed]

26. Rodriguez, H.G.; Popp, J.; Maringanti, C.; Chaubey, I. Selection and placement of best management practices used to reduce water quality degradation in Lincoln Lake watershed. *Water Resour. Res.* **2011**, *47*. [CrossRef]

27. League City. Comprehensive Plan 2035 Document, Chapter 5: Land Use. Updated June 2013. Available online: http://tx-leaguecity4.civicplus.com/DocumentCenter/Home/View/3560 (accessed on 7 December 2013).

28. Francos, A.; Bidoglio, G.; Galbiati, L.; Bouraoui, F.; Elorza, F.; Rekolainen, S.; Manni, K.; Granlund, K. Hydrological and water quality modelling in a medium-sized coastal basin. *Phys. Chem. Earth Part B* **2001**, *26*, 47–52. [CrossRef]

29. Lee, T.; Srinivasan, R.; Moon, J.; Omani, N. Estimation of fresh water inflow to bays from gaged and ungaged watersheds. *Appl. Eng. Agric.* **2011**, *27*, 917–923. [CrossRef]

30. Spruill, C.A.; Workman, S.R.; Taraba, J.L. Simulation of daily and monthly stream discharge from small watersheds using the SWAT model. *Trans. ASABE* **2000**, *43*, 1431–1439. [CrossRef]

31. Jeong, J.; Kannan, N.; Arnold, J.; Glick, R.; Gosselink, L.; Srinivasan, R. Development and integration of sub-hourly rainfall-runoff modeling capability within a watershed model. *Water Resour. Manag.* **2010**, *24*, 4505–4527. [CrossRef]

32. Qi, C.; Grunwald, S. GIS-based hydrologic modeling in the Sandusky watershed using SWAT. *Trans. ASABE* **2005**, *48*, 169–180. [CrossRef]

33. Bracmort, K.; Arabi, M.; Frankenberger, J.; Engel, B.; Arnold, J. Modeling long-term water quality impact of structural BMPs. *Trans. ASABE* **2006**, *49*, 367–374. [CrossRef]

34. Soil Conservation Service (SCS). *National Engineering Handbook*; Section 4 Hydrology, Chapters 4–10; United States Department of Agriculture: Washington, DC, USA, 1972.

35. Neitsch, S.; Arnold, J.; Kiniry, J.; Williams, J. *Soil and Water Assessment Tool Theoretical Documentation Version 2009*; Texas Water Resources Institute: College Station, TX, USA, 2011.

36. Williams, J. Sediment routing for agricultural watersheds. *J. Am. Water Resour. Assoc.* **1975**, *11*, 965–974. [CrossRef]

37. McElroy, A.; Chiu, S.; Nebgen, J.; Aleti, A.; Bennett, F. *Loading Functions for Assessment of Water Pollution from Nonpoint Sources*; United States Environmental Protection Agency, Office of Research and Development: Washington, DC, USA, 1976.

38. Williams, J.; Hann, R. *Optimal Operation of Large Agricultural Watersheds with Water Quality Restraints*; Texas Water Resources Institute: College Station, TX, USA, 1978.

39. Maryland Department of the Environment (MDE). *Modeling Infiltration Practices Using the TR-20 Hydrologic Program (1983)*; Maryland Department of the Environment: Baltimore, MD, USA, 1983.

40. Mechell, J.; Lesikar, B.J. *Rainwater Harvesting: Raingardens*; Texas Water Resources Institute: College Station, TX, USA, 2008; Available online: http://hdl.handle.net/1969.1/87451 (accessed on 28 July 2013).

41. Bannerman, R.T.; Considine, E. *Rain Gardens: A How-to Manual for Homeowners*; Wisconsin Department of Natural Resources: Madison, WI, USA, 2003.

42. North Central Texas Coucil of Government (NCTCOG). Integrated Stormwater Management Technical Manual. 2014. Available online: http://iswm.nctcog.org/technical_manual.asp (accessed on 11 April 2017).

43. Shannak, S.D.; Jaber, F.; Lesikar, B. Modeling the effect of cistern size, soil type, and irrigation scheduling on rainwater harvesting as a stormwater control measure. *Water Resour. Manag.* **2014**, *28*, 4219–4235. [CrossRef]

44. Texas Commission on Environmental Quality (TCEQ). Upper Trinity River-Dallas: Implementing TMDLs through Low Impact Development. 2015. Available online: https://www.tceq.texas.gov/waterquality/nonpoint-source/projects/upper-trinity-river-dallas-tmdl-implementation-low-impact-development (accessed on 22 February 2017).

45. United States Environmental Protection Agency (USEPA). *Preliminary Data Summary of Urban Storm Water Best Management Practices*; USEPA, Office of Water: Washington, DC, USA, 1999.

46. United States Environmental Protection Agency (USEPA). *Quality Criteria for Water 1986*; USEPA, Office of Water Regulations and Standards: Washington, DC, USA, 1986.

47. Brown, W.; Schueler, T. *The Economics of Stormwater BMPs in the Mid-Atlantic Region Prepared for Chesapeake Research Consortium*; Center for Watershed Protection: Silver Spring, MD, USA, 1997.

water

MDPI

Article

Assessment of Three Long-Term Gridded Climate Products for Hydro-Climatic Simulations in Tropical River Basins

Mou Leong Tan [1,*], Philip W. Gassman [2] and Arthur P. Cracknell [3]

[1] Department of Civil and Environmental Engineering, National University of Singapore, Singapore 117576, Singapore
[2] Center for Agricultural and Rural Development, Iowa State University, 560a Heady Hall, Ames, IA 20011-1054, USA; pwgassma@iastate.edu
[3] School of Engineering, Physics and Mathematics, University of Dundee, Dundee DDI 4HN, UK; apcracknell774787@yahoo.co.uk
* Correspondence: mouleong@gmail.com or ceetml@nus.edu.sg; Tel.: +65-6516-2179

Academic Editor: Karim Abbaspour
Received: 16 December 2016; Accepted: 14 March 2017; Published: 21 March 2017

Abstract: Gridded climate products (GCPs) provide a potential source for representing weather in remote, poor quality or short-term observation regions. The accuracy of three long-term GCPs (Asian Precipitation—Highly-Resolved Observational Data Integration towards Evaluation of Water Resources: APHRODITE, Precipitation Estimation from Remotely Sensed Information using Artificial Neural Network-Climate Data Record: PERSIANN-CDR and National Centers for Environmental Prediction Climate Forecast System Reanalysis: NCEP-CFSR) was analyzed for the Kelantan River Basin (KRB) and Johor River Basin (JRB) in Malaysia from 1983 to 2007. Then, these GCPs were used as inputs into calibrated Soil and Water Assessment Tool (SWAT) models, to assess their capability in simulating streamflow. The results show that the APHRODITE data performed the best in precipitation estimation, followed by the PERSIANN-CDR and NCEP-CFSR datasets. The NCEP-CFSR daily maximum temperature data exhibited a better correlation than the minimum temperature data. For streamflow simulations, the APHRODITE data resulted in strong results for both basins, while the NCEP-CFSR data showed unsatisfactory performance. In contrast, the PERSIANN-CDR data showed acceptable representation of observed streamflow in the KRB, but failed to track the JRB observed streamflow. The combination of the APHRODITE precipitation and NCEP-CFSR temperature data resulted in accurate streamflow simulations. The APHRODITE and PERSIANN-CDR data often underestimated the extreme precipitation and streamflow, while the NCEP-CFSR data produced dramatic overestimations. Therefore, a direct application of NCEP-CFSR data should be avoided in this region. We recommend the use of APHRODITE precipitation and NCEP-CFSR temperature data in modeling of Malaysian water resources.

Keywords: NCEP-CFSR; APHRODITE; PERSIANN-CDR; SWAT; precipitation; Malaysia; streamflow; tropical; river; extreme

1. Introduction

Precipitation is a major component of the water cycle and is also a key input to hydrological and ecohydrological models. Meanwhile, the water cycle is largely influenced by changes in regional temperature [1]. Therefore, long-term precipitation and temperature information are vital to study climate changes, forecast local precipitation variability and extreme events trend analysis. Despite this, acquisition of reliable precipitation and temperature data is still a challenging task, especially in developing countries. Ground-based gauge collection is generally regarded as the most accurate precipitation and temperature acquisition approach. However, there is a sparse network of climate stations in many regions due to high installation, operation and maintenance costs, and low awareness of the importance of such information [2], resulting in the inability to capture precipitation and temperature information at sufficient spatial and temporal resolutions.

Gridded climate products (GCPs), which have been developed from modeled and satellite remotely sensed data sources, are potentially alternative sources of climate data for streamflow modeling and other applications, which feature advantages of uninterrupted regional coverage, and high spatial and temporal resolutions [3–5]. For instance, the National Centers for Environmental Prediction Climate Forecast System Reanalysis (NCEP-CFSR) [6] and the Asian Precipitation—Highly-Resolved Observational Data Integration towards Evaluation of Water Resources (APHRODITE) [7] are available globally at a daily time-scale for periods of more than 35 years. Recently, Ashouri et al. [8] developed a new daily time-scale high resolution satellite precipitation product, called the Precipitation Estimation from Remotely Sensed Information using Artificial Neural Network-Climate Data Record (PERSIANN-CDR), for long-term hydro-climatic studies. However, the reliability of these products in many regions is still not well known.

Many studies have validated the performance of GCPs at either global, regional or catchment scale [9–11]. Many of the studies reveal regional differences in GCP performance. For example, Tan et al. [12] reported underestimation of precipitation values by APHRODITE over Peninsular Malaysia, whereas Jamandre and Narisma [13] showed overestimation of the same product in the Philippines. Based on Fekete et al. [14], such differences are expected to be larger in tropical regions compared to temperate regions due to the high precipitation variability. In addition, GCPs are associated with various uncertainties and differences in terms of algorithms, sources, spatial and temporal resolutions [15]. These errors can propagate into streamflow modeling via water cycle processes [16,17].

Reliable climate data are essential for hydrological modeling because errors in climate inputs could lead to false model outputs. For example, an inappropriate model setup with inaccurate GCPs could result in a seemingly "good" model [18], that leads to wrong simulations and subsequent decisions. Therefore, a capability assessment of GCPs prior to applying them in a hydrological model is critical to understanding and reducing these errors. In tropical regions, the capability of GCPs for hydrological assessments have been evaluated in the upper Mara Catchment, Kenya [19]; Negro River Basin, Amazon [20]; Blue Nile River Basin [21]; and Adean watersheds [22]. Vu et al. [23] compared five GCPs in streamflow simulations of the Dak Bla River in Vietnam and concluded that APHRODITE performed the best in replicating daily streamflows. APHRODITE was also used successfully by Le and Sharif [24] to evaluate climate change impacts on streamflow in the Huang River Basin in Central Vietnam. Several studies found that NCEP-CFSR performed poorly for streamflow simulations studies conducted in tropical or sub-tropical regions [25–27]. However, Auerbach et al. [28] reported satisfactory streamflow simulations using NCEP-CFSR for two catchments in Puerto Rico. Zhu et al. [29] and Ashouri et al. [30] report that PERSIANN-CDR performed well when used in streamflow simulations of sub-tropical catchments in China and the United States, respectively. Most studies have only focused on the GCP precipitation data assessments; comparatively few studies have also assessed the accuracy of GCP temperature data [31]. To date, the assessment of suitability and accuracy of these newly developed GCPs in streamflow simulations is still limited in Malaysia.

The overall goal of this study is to investigate the performance of long-term GCPs relative to climate data inputs via streamflow simulations for two major basins in Malaysia. This is an extension of the previous study by Tan et al. [12] which evaluated the performance of different GCPs across the entire country of Malaysia, but did not incorporate streamflow analysis. The specific objectives here for the two study basins are: (1) to assess the accuracy of the APHRODITE, PERSIANN-CDR and NCEP-CFSR data for precipitation and temperature data retrieval from 1983 to 2007; (2) to evaluate the capability of these products for streamflow simulations using the Soil and Water Assessment

Tool (SWAT) ecohydrological model [32–36]; and (3) to analyze the suitability of the three GCPs for capturing extreme hydro-climatic events.

2. Study Area and Materials

2.1. Study Area

Two tropical basins, the Kelantan River Basin (KRB) and Johor River Basin (JRB), were selected as study areas in this study due to differences in size, land use, topography and data availability (Figure 1). The KRB (4° N~6° N, 101° E~103° E) drains an area of 12,134 km^2 in northeastern Peninsular Malaysia. The main channel of the Kelantan River extends a total distance of about 248 km, and flows northward into the South China Sea. In 1990, the primary land use/land cover in the KRB was tropical forest (84.9%), followed by rubber (9.9%), oil palm (4.5%), urban (0.5%) and paddy (0.2%). The basin elevation ranges from 8 m a.s.l in the western region to 2174 m a.s.l in the southwestern regions. The KRB is characterized by a tropical monsoon climate, with an average annual precipitation ≥2500 mm, most of which falls from November to January [37]. The average annual temperature of the basin is about 27.5 °C. The KRB is frequently affected by monsoon flood events during the northeast monsoon season.

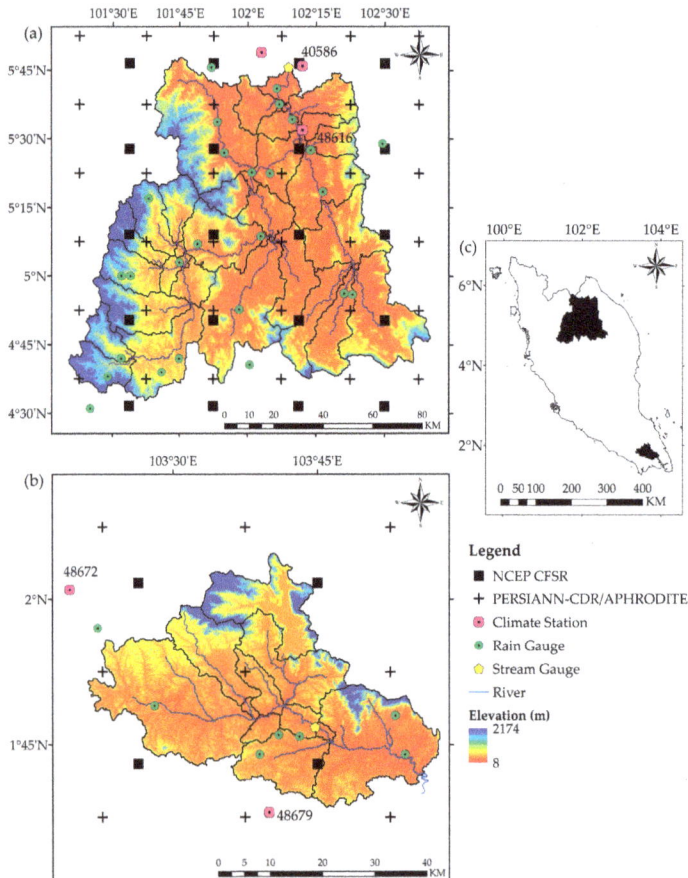

Figure 1. Spatial distribution of APHRODITE, PERSIANN-CDR, NCEP-CFSR and rain gauges over: (a) Kelantan River Basin (KRB); (b) Johor River Basin (JRB); and (c) Peninsular Malaysia.

The JRB (1° N~3° N, 103° E~104° E) drains an area of 1652 km^2 in southern Peninsular Malaysia (Figure 1b). The main river stem of the Johor River flows approximately 123 km southeast to the Strait of Johor. Elevations within the JRB range between 3 m a.s.l. and 977 m a.s.l, the highest elevations being located in the northern and western regions of the basin. The JRB is an agricultural production region, which is dominated by oil palm (38.4%), forest (44.1%) and rubber (15.3%) in 1990. The average annual precipitation and average annual temperature of the basin are 2500 mm and 26 °C, respectively. The Johor River is an important freshwater resource for the Johor and Singapore population, so any changes in water resources could lead to major impacts on agriculture, industrial and living conditions in both regions. For example, continuous hot weather in April 2016 resulted in water levels in the Linggiu Reservoir, located in the northern JRB falling to a new historic low.

2.2. Gridded Climate Products

Long-term GCPs are viable datasets that can be used for supporting the development of climate change and mitigation strategies for both the KRB and JRB. The evaluation of GCPs for this study focused on products characterized by long-term temporal climate datasets that contain data from at least a 30-year period. Based on this criterion, the APHRODITE, PERSIANN-CDR and NCEP-CFSR GCPs (Table 1) were assessed for the two study basins. Tan et al. [12] also reported that two Tropical Rainfall Measuring Mission (TRMM) 3B42 products performed well in replicating precipitation data for different sub-regions of Malaysia. However, the TRMM data were excluded from this study because the temporal resolution only extends back to 1998.

Table 1. Details on gridded climate products used in this study.

Name	Spatial	Temporal	Region	Sources
Rain Gauges				
Observation	Point	1983–present	Malaysia	Malaysia Meteorological Department; Department of Irrigation and Drainage Malaysia
Satellite				
PERSIANN-CDR	0.25°	1983–present	60° S–60° N	University of California, Irvine
Reanalysis data				
APHRODITE	0.25°	1951–2007	Eurasia	University of Tsukuba; Japan Meteorological Agency
NCEP-CFSR	0.3125°	1979–2014	Global	National Centers for Environmental Prediction

APHRODITE is a long-term daily precipitation product that spans the 57-year period of 1951 to 2007, which was generated from thousands of gauge observations data collected from various countries' government agencies [7]. It was developed by the Research Institute for Humanity and the Meteorological Research Institute of the Japan Meteorological Agency. APHRODITE is divided into Middle East, Russia, Monsoon Asia and Japan regions. In this study, APHRODITE V1101 (Monsoon Asia) with a 0.25° resolution was used.

PERSIANN-CDR provides daily precipitation information from 1983 to the present for latitudes 60° S–60° N at a spatial resolution of 0.25°. PERSIANN-CDR was established from the PERSIANN algorithm using Gridded Satellite Infrared Data (GridSat-B1), a calibrated and mapped geostationary satellite dataset [38]. The training of the artificial neural network is done using the NCEP stage IV hourly precipitation data. The product is then adjusted by the Global Precipitation Climatology Project (GPCP) monthly version 2.2 product [8].

NCEP-CFSR was constructed for a period of 36 years (1979 to 2014) at ~0.31° (38 km) resolution [6]. NCEP-CFSR is produced using cutting-edge data assimilation techniques and a forecast model that extrapolates non-observed parameters from observed data, collected from various sources such as rain gauges, ships, weather balloons and satellites. NCEP-CFSR data were obtained for the whole of Peninsular Malaysia (latitude 0.7° N–6.8° N and longitude 98.7° E–105.2° E), and then the stations distributed over each basin were used. There are five climate parameters: temperature, precipitation, wind speed, relative humidity and solar radiation. However, the analysis conducted here was limited to just the NCEP-CFSR precipitation and temperature data, in order to maintain consistency with the evaluation of the other two GCPs.

2.3. Ground-Based Gauge Data

Daily precipitation, maximum temperature and minimum temperature data from 1983 to 2007 were collected from the Malaysia Meteorological Department (MMD; http://www.met.gov.my/) and the Irrigation and Drainage Department Malaysia (DID; http://www.water.gov.my/). There are 29 climate stations distributed across the KRB, but only three of them contain long-term maximum and minimum temperature data. For the JRB, daily precipitation data are available at nine climate stations. However, only two of the stations contain temperature data. In addition, monthly streamflow data measured at the Jambatan Guillermard and Rantau Panjang stations located in KRB and JRB (Figure 1), respectively, were collected from the DID for calibration and validation of the SWAT model. More detailed information of streamflow measurements for the KRB, JRB and other basins in Malaysia are available in a report prepared by DID [39].

2.4. Geospatial Data

The main input geospatial data for the SWAT model are a digital elevation model (DEM), a land use map and a soil map. Tan et al. [40] evaluated four different DEM datasets on SWAT simulations in the JRB, and found the 90 m Shuttle Radar Topography Mission (SRTM) DEM [41] performed the best. Therefore, the SRTM DEM was selected in this study. The land use map and soil map produced in 1990 and 2002, respectively, were obtained from the Ministry of Agriculture and Agro-based Industry of Malaysia (MOA; http://www.moa.gov.my/). In addition, the river network for each basin was digitized from the topography map produced by the Department of Survey and Mapping Malaysia (JUPEM; https://www.jupem.gov.my/). The digitized river networks were used to improve basin delineation and river extraction of both basins, especially in low land regions, similar to the approach used by Zheng et al. [42].

3. Methodology

3.1. Statistical Analysis

A set of continuous and categorical statistical analyses were used to evaluate the performance of the GCPs against observations at annual, seasonal, monthly and daily scales (Figure 1). As recommended by Tangang and Juneng [43], the climate data were divided into December to February (DJF), March to May (MAM), June to August (JJA) and September to November (SON) for seasonal scale assessment. The comparison was performed from 1983 to 2007 to provide a consistent time period, which brackets the starting year of 1983 for the PERSIANN-CDR dataset and the final year of 2007 for APHRODITE data. The point-to-pixel assessment was applied to prevent additional uncertainties during interpolation of the gauge data [44]. For the overall assessment, all precipitation values are pooled together from 1983 to 2007 [45]. In contrast, the NCEP-CFSR maximum and minimum temperature could not be validated at the overall assessment scale as there were only two or three climate stations that had temperature data in the KRB and JRB (Figure 1). Moreover, most of these stations are located outside the basins, and thus cannot be used to represent the entire basins. Therefore, the temperature data validation was conducted only for specific climate stations. In addition, the paired student *t*-test method was used to assess the significant differences between rain gauges and GCPs at the 0.05 significance level. Continuous statistical analysis such as Root Mean Square Error

(RMSE), Pearson Correlation Coefficient (CC), Mean Error (ME) and Relative Bias (RB) were used [12]. The formulas of these approaches are shown as follow:

$$\text{RMSE} = \sqrt{\frac{\sum_{i=1}^{n}(G_i - O_i)^2}{n}} \tag{1}$$

$$\text{CC} = \frac{\sum_{i=1}^{n}(O_i - \overline{O})(G_i - \overline{G})}{\sqrt{\sum_{i=1}^{n}(O_i - \overline{O})^2} \cdot \sqrt{\sum_{i=1}^{n}(G_i - \overline{G})^2}} \tag{2}$$

$$\text{ME} = \frac{\sum_{i=1}^{n}(G_i - O_i)}{n} \tag{3}$$

$$\text{RB} = \frac{\sum_{i=1}^{n}(G_i - O_i)}{\sum_{i=1}^{n} O_i}(100) \tag{4}$$

where G_i and O_i are gridded and observed precipitation/temperature, respectively; i is used to label the individual measurements; and n is the number of measurements. CC measures similarity in temporal or spatial pattern between GCP and the observed data, RMSE evaluates the absolute average error between two datasets, ME makes it possible to evaluate the bias in estimations, while RB estimates the systematic overestimation and underestimation of GCP as a percentage (%). A good performance GCP should have a high CC, versus low RMSE, ME and RB values.

Categorical statistical analysis was used to evaluate the ability of GCPs to discriminate between precipitation and no precipitation event days, based on the following criteria [46]: (1) Accuracy (ACC), which represents the level of agreement between the GCPs and rain gauges estimates; (2) Probability of Detection (POD), which is a measure of how well the GCPs correctly detected rain gauge estimates; (3) False Alarm Ratio (FAR), which is used to evaluate how often the GCPs detected precipitation, but there was actually no precipitation recorded at the rain gauges; and (4) Critical Success Index (CSI), which is an indicator of the fraction of precipitation correctly detected by GCPs. These categorical approaches can be measured as follows:

$$\text{ACC} = \frac{A + D}{n} \tag{5}$$

$$\text{POD} = \frac{A}{A + C} \tag{6}$$

$$\text{FAR} = \frac{B}{A + B} \tag{7}$$

$$\text{CSI} = \frac{A}{A + B + C} \tag{8}$$

where A = correct detection (the GCP estimated precipitation, and precipitation was observed in rain gauge); B = false alarm (the GCP estimated precipitation, but precipitation was not observed in rain gauge); C = misses (the GCP did not estimate precipitation, but the rain gauge estimated precipitation); and D = correct negative (the GCP did not estimate precipitation, and precipitation was not observed in rain gauge). These values range between 0 and 1, where 1 is a perfect score for the ACC, POD and CSI, while 0 is a perfect score for the FAR. For example, the GCPs miss detecting the precipitation by 20%, if the FAR value is equal to 0.2. Further description of this approach is provided by Ebert et al. [46]. Based on Shen et al. [47], the quality of the GCP accuracy assessment is largely influenced by the density and distribution of local station networks. Hence, assessment should be conducted over valid grid points only, where at least one station is available on each evaluated grid points.

3.2. SWAT Model

Current versions of the SWAT model represent more than three decades of model development at the co-located U.S. Department of Agriculture and Texas A&M University laboratories in Temple, Texas [34,35]. SWAT is usually executed at a daily time step for continuous simulations [36], typically with a minimum climatic dataset consisting of daily precipitation, maximum temperature and minimum temperature. The model has been applied for an extensive range of ecohydrological problems and scenarios worldwide for watershed scales ranging from <1 km^2 to entire continents (e.g., see reviews by Gassman et al. [48,49]; Bressiani et al. [25]; Gassman and Wang [50]; and Krysanova and White [51]). The model has also been used successfully for several hydrology and pollutant transport studies conducted in Malaysia [40,52–55]. SWAT version 2012 (Revision 635) was used in conjunction with the ArcSWAT interface version 2012.10_2.16 for this study.

In SWAT, a basin is usually first sub-divided into multiple sub-basins that are then further delineated into hydrologic response units (HRUs), which are smaller spatial units consisting of homogeneous soil, landscape, land use and management characteristics. HRUs represent a specific percentage of the corresponding sub-watershed area and are not currently spatially identified in SWAT. For this study, digitized stream networks were merged into the SRTM DEM using the "burn in" method, resulting in the delineation of 22 and 11 sub-basins for the KRB and JRB, respectively (Figure 1). Threshold values were then used in the ArcSWAT interface to create the HRUs, by setting minimum percentages that specific soils, slopes or land use had to occupy within a given sub-basin in order to be included in the KRB or JRB SWAT models. The hydrologic response unit (HRU) threshold values were defined as 20% for land use and slope, and 10% for soil, resulting in the KRB and JRB being further subdivided into 200 and 37 HRUs, respectively. Initial simulation of climate inputs, hydrological balance, crop growth and pollutant cycling occurs at the HRU level in SWAT. Excess discharge and pollutant exports are then aggregated across HRUs within a given sub-basin, input into the stream network at the sub-basin outlet and then ultimately routed to the watershed outlet. Further details regarding the theory, input requirements, and output options are provided in on-line documentation [33,36].

3.3. SWAT Model Baseline Testing

Baseline hydrological testing of SWAT was performed for both the KRB and JRB prior to the analysis of the GCPs. The respective baseline testing periods of 1983 to 1999 for the KRB and 1983 to 1992 for the JRB were based on streamflow data measured at the stream gauge sites shown for each basin in Figure 1. The first two years (1983–1984) were used as initialization years for both watersheds and the remainder of the time periods were subdivided into calibration (KRB = 1985–1994 and JRB = 1985–1988) and validation (KRB = 1995–1999 and JRB = 1989–1992) periods.

SWAT calibration was conducted using the Sequential Uncertainty Fitting algorithm (SUFI-2) within the SWAT Calibration and Uncertainty (SWAT-CUP) software package [56], which is a flexible algorithm that can process large numbers of input parameters. The Nash–Sutcliffe Coefficient (NSE) and Coefficient of Determination (R^2) statistics [57] were used to evaluate performance of simulated streamflow. The NSE was selected as the optimal objective in the SWAT calibration; NSE values can range from $-\infty$ to 1, where values ≤ 0 indicate that the mean of the measured data is a better predictor than the simulated values, indicating unacceptable performance. In addition, the R^2 values range from 0 to 1, and were used to assess the collinearity of the observed and simulated streamflow, where 1 is the ideal value. Based on Moriasi et al. [58,59], the performance of the SWAT model can be considered as satisfactory/good if the NSE and R^2 statistics are $\geq 0.5/0.7$ and $\geq 0.6/0.75$, respectively.

Following the SWAT calibration and validation phase, two different GCP scenarios were used as inputs into the calibrated SWAT model. The first scenario consisted of incorporating only the precipitation data from the three GCPs into the SWAT model simulations. This allows comparison with several previous studies, which only evaluated the GCP precipitation products. The second scenario evaluated combinations of each GCP with the NCEP-CFSR temperature data (i.e., APHRODITE, PERSIANN-CDR or NCEP-CFSR precipitation data + NCEP-CFSR temperature) on the SWAT outputs. The second scenario is useful for assessing the applicability of the NCEP-CFSR temperature data in SWAT modeling, due to the sensitivity of the water cycle to temperature data.

3.4. Extreme Events Analysis

Extreme climatic events can result in severe impacts on human society and the environment [60]. The majority of existing hydrological and climatological studies, including analyses of the impacts of extreme climatic events have been conducted using ground-based gauge data [7,61]. Therefore, evaluation of other types of precipitation products for extreme events would provide important insight for determining their efficacy and accuracy for unusual climatic conditions [62]. Four indices were used in this study to assess the performance of the three GCPs in capturing the pattern of precipitation extremes over the KRB and JRB: (1) the number of precipitation days ≥ 10 mm·day^{-1} in a year (R10mm); (2) the number of precipitation days ≥ 50 mm·day^{-1} in a year (R50mm); (3) the annual maximum daily precipitation/streamflow amount(Rx1d); and (4) the annual maximum consecutive five-day precipitation/streamflow amount (Rx5d). The latter two indices were adopted to evaluate the accuracy of GCP-based SWAT simulated streamflows for maximum one-day and five-day amounts. These extreme indices were recommended by the Expert Team on Climate Change Detection and Indices [63]. Annual maximum one-day and five-day consecutive streamflow indices were chosen because these indices can be used to study flood volume which is important for flood risk management [29].

4. Results

4.1. Precipitation Validation

The result of the statistical assessment of the 25-year (1983 to 2007) comparisons between the APHRODITE, PERSIANN-CDR and NCEP-CFSR annual, seasonal, monthly and daily precipitation data versus the rain gauge observations for the KRB and JRB is listed in Table 2. The PERSIANN-CDR monthly-scale precipitation was the only GCP data that did not show significant differences relative to the KRB rain gauge observations, at a significance level of 0.05 (Table 2). The PERSIANN-CDR data showed insignificant differences versus observations at the JJA seasons in both basins.

In the KRB, the APHRODITE precipitation data produced the best linear correlation for all time-scales, with CC values varying from 0.38 to 0.74, followed by the PERSIANN-CDR and NCEP-CFSR data. It is also clear that the APHRODITE and PERSIANN-CDR precipitation data underestimated the annual, DJF, SON, monthly and daily precipitation amounts, based on the respective positive and negative signs for the ME and RB indicators, while the NCEP-CFSR data resulted in highly overestimated precipitation across the basin. In addition, the NCEP-CFSR data showed the largest average errors as evidenced by the highest RMSE values that ranged from 19.49 mm to 1695.34 mm for most of the time-scales, except for the DJF.

All other GCP data showed significant differences for annual, daily and monthly time steps as compared to the rain gauge precipitation estimates for the JRB (Table 2). The APHRODITE data produced the best results at the DJF, JJA, SON, monthly and daily time-scales, with CC values that ranged from 0.44 to 0.73. In contrast, the NCEP-CFSR data resulted in the worst performance at all time scales with CC values that spanned between 0.13 and 0.46. The APHRODITE data slightly underestimated the MMA, SON, monthly and daily precipitation levels, versus the PERSIANN-CDR and NCEP-CFSR data which produced large overestimations.

Generally, the GCPs show better linear correlation performance for the DJF and monthly time-scale estimations as compared to other time scales in both basins. The results found here showed that the

APHRODITE data produced the best precipitation estimation performance for over both basins, which is in agreement with Tan et al. [12] who conducted a national assessment over Malaysia. The main reason is due to the fact that the developers of APHRODITE incorporated MMD rain gauges' data in the development of the product [7]. On the contrary, NCEP-CFSR displays more serious errors and dramatically overestimated the total precipitation compared to the other GCPs. Similarly, Roth and Lemann [64] found that the total annual NCEP-CFSR precipitation data was three times greater than observed precipitation data in Ethiopia. The distinct weaknesses that have been quantified for the NCEP-CFSR data may be attributed to the scale differences, where the size of a grid point is huge (up to 0.3125°) compared to the station data which is a point-based measurement. The errors are expected to be higher in a grid point with high spatial and temporal variability of precipitation as well as for regions characterized by complex topography [65].

Table 2. Statistical analysis for daily, monthly, seasonal and annual precipitation in the Kelantan River Basin (KRB) and Johor River Basin (JRB). (Bold indicate significance at 0.05).

Time		KRB			JRB		
		A	P	N	A	P	N
Annual	RMSE (mm)	807.15	613.17	1695.34	399.51	514.77	1176.88
	CC	0.38	0.34	0.11	0.45	0.46	0.21
	ME (mm)	−540.53	−33.40	1314.21	−73.22	318.73	1018.59
	RB (%)	−20.51	−1.27	49.87	−3.21	13.96	44.61
	T test (t stat)	**20.40**	1.26	**31.21**	**2.09**	**8.65**	**23.15**
DJF	RMSE (mm)	334.13	288.09	333.39	191.87	232.70	406.94
	CC	0.63	0.62	0.48	0.73	0.71	0.36
	ME (mm)	−186.71	−78.86	41.40	4.54	119.62	275.76
	RB (%)	−27.96	−11.81	6.20	0.76	20.13	46.40
	T test (t stat)	**12.52**	**4.85**	**2.41**	0.19	**4.87**	**10.84**
MAM	RMSE (mm)	186.63	178.24	643.20	142.88	178.25	406.39
	CC	0.62	0.61	0.51	0.52	0.61	0.38
	ME (mm)	−80.90	52.10	523.70	−23.98	102.77	336.59
	RB (%)	−16.55	10.66	107.10	−4.02	17.23	56.41
	T test (t stat)	**8.47**	**5.49**	**29.15**	1.78	**6.60**	**17.88**
JJA	RMSE (mm)	212.59	184.19	499.33	127.20	157.30	246.20
	CC	0.44	0.35	0.20	0.66	0.35	0.16
	ME (mm)	−119.85	12.34	391.51	−27.24	−6.25	170.00
	RB (%)	−21.05	2.17	68.76	−5.59	−1.28	34.91
	T test (t stat)	**15.04**	1.57	**30.70**	**2.04**	0.50	**13.20**
SON	RMSE (mm)	310.21	269.49	552.17	153.81	192.77	334.56
	CC	0.58	0.57	0.21	0.44	0.42	0.13
	ME (mm)	−146.49	−13.54	367.27	−24.12	105.81	245.79
	RB (%)	−16.24	−1.50	40.71	−3.98	17.45	40.55
	T test (t stat)	**11.20**	0.97	**21.29**	1.86	**7.62**	**15.11**
Monthly	RMSE (mm)	126.98	118.62	210.02	78.15	93.86	149.75
	CC	0.74	0.72	0.48	0.71	0.64	0.46
	ME (mm)	−45.04	−2.78	109.52	−6.10	26.56	84.88
	RB (%)	−20.51	−1.27	49.87	−3.21	13.96	44.61
	T test (t stat)	**21.20**	1.24	**41.20**	**2.28**	**9.29**	**26.36**
Daily	RMSE (mm)	14.80	15.79	19.49	12.25	14.24	17.70
	CC	0.43	0.35	0.22	0.55	0.35	0.17
	ME (mm)	−1.48	−0.09	3.60	−0.20	0.87	2.79
	RB (%)	−20.51	−1.27	49.87	−3.22	13.96	44.60
	T test (t stat)	**41.93**	**2.47**	**85.44**	**3.30**	**14.57**	**41.80**
	Accuracy	0.61	0.57	0.55	0.67	0.55	0.49
	POD	0.87	0.86	0.94	0.89	0.89	0.96
	FAR	0.48	0.51	0.52	0.45	0.54	0.57
	CSI	0.48	0.45	0.46	0.51	0.44	0.42

Notes: A = APHRODITE; P = PERSIANN-CDR; N = NCEP-CFSR.

4.2. Precipitation Spatial Variability

The monthly CC and RB values for the GCPs over both basins are presented in Figures 2 and 3, respectively, to provide insights regarding spatial variability. Generally, high CC values for all GCPs were found for the northern and eastern KRB sub-regions, which are near coastal and low elevation areas (Figure 2a–c). All of the GCPs reflected strong performance of the CC values computed for the northwest JRB sub-region, while lower CC values dominated in the middle of the basin (Figure 2d–f). The APHRODITE data underestimated monthly ground-based precipitation at most of the stations (Figure 3). In contrast, the NCEP-CFSR data dramatically overestimated monthly precipitation at all of the stations, resulting in especially high RB values (more than 100%) for the stations mainly distributed in the southwestern KRB sub-region, which is characterized by high mountains (Figure 3c). The NCEP-CFSR was the only GCP which resulted in significant overestimates for all stations distributed across the JRB.

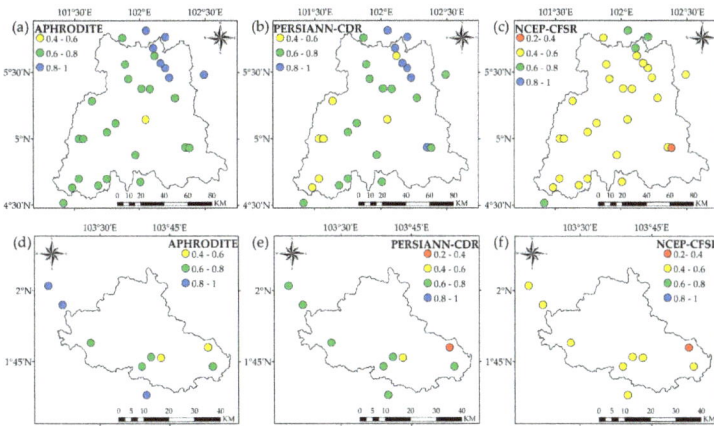

Figure 2. The correlation coefficient of monthly precipitation of APHRODITE, PERSIANN-CDR and NCEP-CFSR against rain gauges, respectively, over: (**a–c**) Kelantan River Basin; and (**d–f**) Johor River Basin.

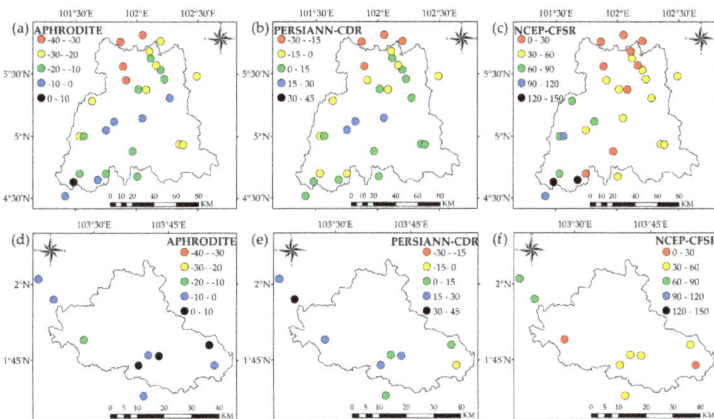

Figure 3. The relative bias of monthly precipitation of APHRODITE, PERSIANN-CDR and NCEP-CFSR against rain gauges, respectively, over: (**a–c**) Kelantan River Basin; and (**d–f**) Johor River Basin.

These findings agree with other studies, which state that GCPs generally are more reliable in low land regions compared to higher elevations [66,67]. This might be due to misrepresenting the effects of warm clouds, by infrared (IR) sensors that commonly appear on mountaintops [68]. The overall less accurate performance of GCPs in mountainous regions may be due to fewer rain gauges that can be used for product development. The installation and maintenance of climate stations in high mountainous regions is often problematic because of difficulties related to physical access and the fact the climate stations are representative of relatively small area due to high topography variability. In general, the APHRODITE dataset performed better for mountainous regions compared to other two GCPs, because the product has better orographic precipitation variability resolving skill [69].

4.3. Precipitation: Rain Detection and Intensity Assessment

The NCEP-CFSR data showed the most outstanding performance for rain detection ability assessment, with POD values of 0.94 and 0.96 for KRB and JRB, respectively. However, the APHRODITE exhibits better ACC skills for the JRB, indicating that it has a stronger capability to correctly estimate overall precipitation and non-precipitation events in southern Peninsular Malaysia. In contrast, the PERSIANN-CDR and NCEP-CFSR GCPs performed better for the KRB. The analysis further revealed that the NCEP-CFSR data were most prone to predicting false rain event, which in fact were not recorded by the rain gauges, resulting in the highest FAR values of 0.52 (KRB) and 0.57 (JRB). Moderate CSI values were also predicted for all three GCPs ranging from 0.45 to 0.48 (KRB) and 0.42 to 0.51 (JRB), demonstrating that roughly 50% of the precipitation was correctly estimated.

Figure 4 presents the probability distribution functions (PDFs) of precipitation intensity for the KRB and JRB. The non-precipitation values ≤ 0.254 mm·day^{-1} (common rain gauge threshold detection limit) were removed from the analysis. The three GCPs showed moderate underestimation for the ≥ 50 mm·day^{-1} precipitation classes over both basins. The NCEP-CFSR data resulted in significant overestimation for the 5–10 and 10–20 mm·day^{-1} precipitation classes in both basins. This is similar to the results reported by Blacutt et al. [70], who also discovered the NCEP-CFSR overestimated precipitation at 3–20 mm·day^{-1} class in Bolivia. They further reported the NCEP-CFSR tended to overestimate precipitation during the annual precipitation season period. This problem could potentially be amplified in both the KRB and JRB, which are typical tropical basins that receive precipitation throughout the year, especially during the northeast monsoon and southwest monsoon periods. The NCEP-CFSR data overestimation rate was higher for the JRB (up to 270% at 5–10 mm·day^{-1}) compared to the KRB, because the Sumatra and Titiwangsa mountain ranges help to reduce precipitation days in the KRB during the southwest monsoon season.

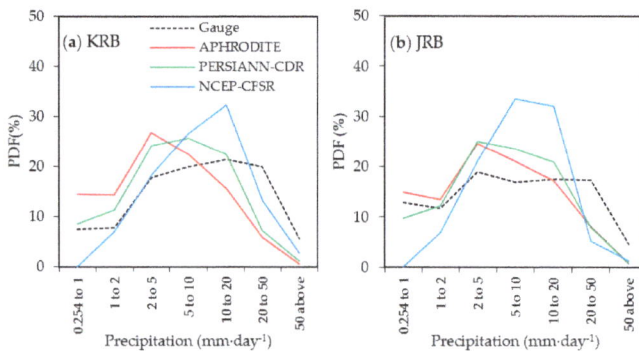

Figure 4. Probability distribution function (PDF) of daily precipitation from 1983 to 2007 from APHRODITE, PERSIANN-CDR, NCEP-CFSR and rain gauges over: (**a**) Kelantan River Basin (KRB); and (**b**) Johor River Basin (JRB).

Water **2017**, 9, 229

4.4. Temperature Validation

The statistical analysis of the NCEP-CFSR maximum and minimum temperature versus climate stations temperature gauges (Figure 1) of the KRB and JRB is listed for various time scales in Table 3. The temperature values from each temperature gauge were compared to the nearest NCEP-CFSR grid point. Generally, the NCEP-CFSR temperature data have better correlation with observations at the DJF and monthly time-scale, with CC values ranging from 0.6 to 0.91 and 0.57 to 0.93, respectively. In addition, the daily maximum temperature data were better correlated with the observed data as compared to the minimum temperature data. However, the average error of the daily maximum temperature data (RMSE = 2.58 to 3.32 °C) is larger than the minimum temperature (RMSE = 0.98 to 2.68 °C) at all stations.

Box plots of the interactions between the NCEP-CFSR data and climate station maximum and minimum temperature data, for the four climate stations distributed across the KRB and JRB, are shown in Figure 5. The inter-quartile range shows that the minimum temperature at the 48679 station provides the best performance, as the range of the NCEP-CFSR data versus the gauge data matched quite well. The range of the NCEP-CFSR temperature data is larger than the observations at the all stations. As can be seen from the Table 3 and Figure 5, the NCEP-CFSR temperature data tend to underestimate the actual maximum and minimum temperature values. The main reason of the underestimation could be due to the land use types [65]. For example, the 48679 station is located in an industrial area where the surface temperature is expected to be higher. However, the NCEP-CFSR relies on National Aeronautics and Space Administration (NASA) land use information data [71], so reliable local land use information might be missing for the 48679 station location. Another possible reason for the underestimation of the NCEP-CFSR data may be explained by the mismatch of the temperature time measurement. For instance, the climate stations' daily maximum and minimum temperature data were taken at 0800 and 1400 local time, respectively, while the NCEP-CFSR daily maximum and minimum temperature were obtained from hourly values [72].

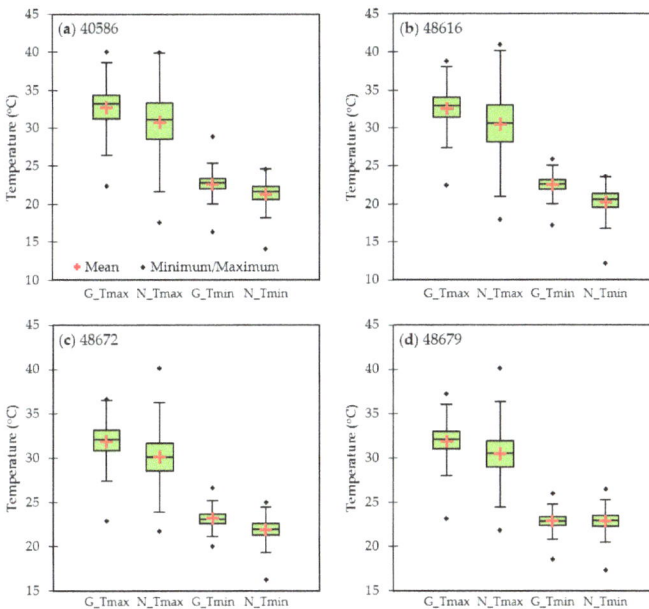

Figure 5. Box plots of daily maximum and minimum temperature derived from NCEP-CFSR (N) and temperature gauges (G) at: (**a**) 40586 station; (**b**) 48616 station; (**c**) 48672 station; and (**d**) 48679 station.

Table 3. Statistical analysis for the NCEP maximum (Tmax) and minimum (Tmin) temperature in the Kelantan River Basin and Johor River Basin.

Period		40586		48616		48672		48679	
		Tmax	Tmin	Tmax	Tmin	Tmax	Tmin	Tmax	Tmin
Annual	RMSE (°C)	1.97	1.37	2.22	2.31	1.74	1.28	1.49	0.18
	CC	0.74	0.82	0.43	0.62	0.72	0.81	0.53	0.84
	RB (%)	−5.81	−5.88	−6.43	−10.18	−5.27	−5.47	−4.36	0.22
DJF	RMSE (°C)	2.66	1.92	3.25	3.02	1.81	1.31	1.63	0.25
	CC	0.90	0.72	0.74	0.67	0.70	0.87	0.60	0.91
	RB (%)	−8.69	−8.53	−10.41	−13.65	−5.62	−5.63	−4.83	0.89
MAM	RMSE (°C)	2.02	1.46	2.25	2.63	2.12	1.26	1.82	0.20
	CC	0.90	0.48	0.71	0.25	0.81	0.67	0.75	0.84
	RB (%)	−5.66	−5.89	−5.92	−11.27	−5.95	−5.21	−4.83	−0.05
JJA	RMSE (°C)	1.19	1.12	1.30	1.83	1.57	1.64	1.38	0.30
	CC	0.45	0.70	0.07	0.63	0.65	0.55	0.27	0.59
	RB (%)	−2.64	−4.45	−2.65	−7.95	−4.69	−6.89	−3.99	0.14
SON	RMSE (°C)	2.26	1.19	2.43	1.86	1.64	0.99	1.37	0.23
	CC	0.42	0.51	0.38	0.54	0.62	0.70	0.51	0.80
	RB (%)	−6.62	−4.83	−7.20	−8.07	−4.84	−4.17	−3.88	−0.11
Monthly	RMSE (°C)	2.18	1.50	2.48	2.46	1.89	1.36	1.67	0.35
	CC	0.93	0.77	0.88	0.70	0.79	0.57	0.72	0.73
	RB (%)	−5.82	−5.90	−6.44	−10.21	−5.26	−5.48	−4.35	0.23
Daily	RMSE (°C)	2.96	1.80	3.32	2.68	2.77	1.67	2.58	0.98
	CC	0.74	0.60	0.68	0.53	0.55	0.28	0.50	0.38
	RB (%)	−5.81	−5.88	−6.43	−10.18	−5.27	−5.47	−4.36	0.22

4.5. Streamflow: GCPs Precipitation Data

Table 4 lists the best fitted calibration parameters for KRB and JRB. The calibration and validation of the SWAT model were conducted based on local knowledge and a literature review of the SWAT model in tropical regions (e.g., [54,55,73,74]). As can be seen in Table 4, the CN2 values were increased by 1% and 13% for the KRB and JRB, respectively. This increment of CN2 values was also observed in calibration of other tropical SWAT models [75–77]. The CN2 value was higher in the JRB as it is dominated by oil palm plantations, where the surface runoff is generally higher than in a forest basin (KRB). Generally, the SWAT simulations that were based on rain gauge data agreed well with the observed streamflow during the calibration and validation periods for both the KRB and JRB (Figure 6). The NSE values that were computed for the KRB (JRB) were 0.75 (0.78) and 0.65 (0.6) for the calibration and validation periods, respectively (Table 5), and the corresponding KRB (JRB) R^2 statistics were 0.87 (0.78) and 0.84 (0.61) indicating that the SWAT model performed well for both basins based on the previously discussed suggested criteria [58,59].

Table 4. Optimal calibration parameters for Kelantan River Basin (KRB) and Johor River Basin (JRB).

No.	Parameter Name	Parameter	Range		KRB	JRB
			Min	Max	Value	Value
1	groundwater "revap" coefficient	V_GW_REVAP	0.1	0.4	0.4	0.29
2	channel effective hydraulic conductivity	V_CH_K2	0	80	56.4	66.67
3	baseflow alpha factor	V_ALPHA_BF	0.1	0.5	0.12	0.14
4	initial SCS CN II value	R_CN2	0	0.35	0.1	0.13
5	groundwater delay	V_GW_DELAY	0	130	80.99	91
6	soil evaporation compensation factor	V_ESCO	0.4	0.9	0.52	0.42
7	threshold water depth in the shallow aquifer for flow	V_GWQMN	2200	4000	3940.6	3700
8	manning's value for main channel	V_CH_N2	0.2	0.3	0.28	0.24
9	available water capacity	R_SOL_AWC	0	0.5	0.25	0.35
10	surface runoff lag time	V_SURLAG	8	19	18.44	18.63
11	threshold depth of water in the shallow aquifer for "revap" to occur	V_REVAPMN	70	320	232.3	95
12	deep aquifer percolation faction	V_RCHRG_DP	0.4	0.6	0.52	0.58

Note: R indicates the default parameter value is multiplied by (1+ a given value) and V indicates the default parameter value is replaced with the given value.

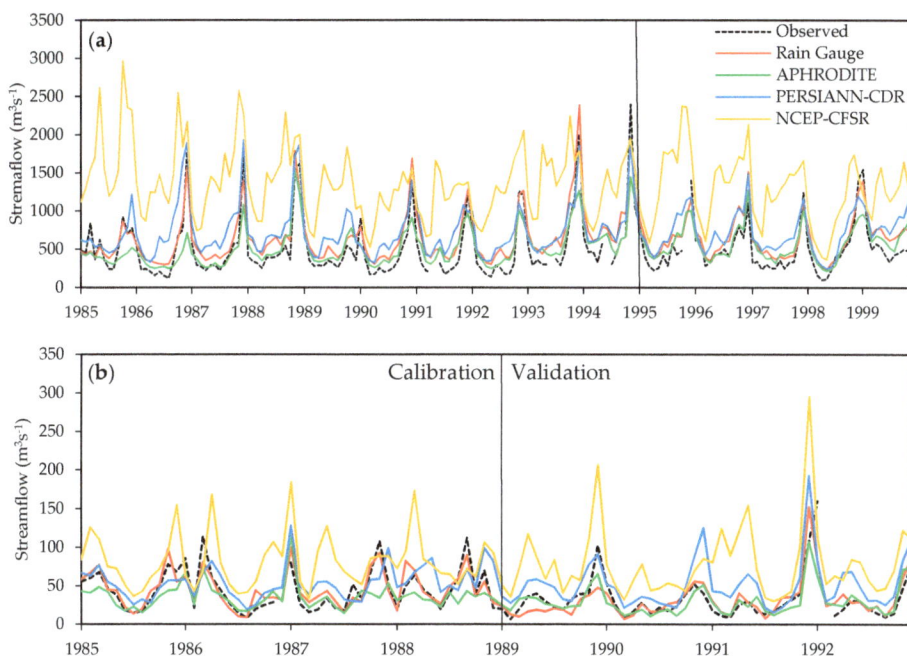

Figure 6. Comparison of observed streamflow with gauge-based, APHRODITE, PERSIANN-CDR and NCEP-CFSR precipitation-driven SWAT simulated monthly streamflow, respectively, in the: (**a**) Kelantan River Basin; and (**b**) Johor River Basin.

Table 5. SWAT calibration and validation statistical results for the Kelantan River Basin (KRB) and Johor River Basin (JRB).

Time		KRB	JRB
Calibration	R^2	0.87	0.78
Period	NSE	0.75	0.78
	RB	27.41	2.94
Validation	R^2	0.84	0.61
Period	NSE	0.65	0.60
	RB	26.57	−2.99

Among the three GCPs, the most accurate KRB SWAT simulations occurred in response to the APHRODITE precipitation input, followed by the simulations driven by the PERSIANN-CDR and NCEP-CFSR precipitation data. The SWAT simulation streamflow trends, based on the APHRODITE and PERSIANN-CDR data, revealed overestimation of low streamflows and underestimation of high streamflows. The predicted streamflow results obtained with the NCEP-CFSR data were unacceptable as reflected by the negative NSE values (Table 6). In addition, the NCEP-CFSR precipitation data resulted in relatively high overestimation of observed streamflows throughout the simulation period, as indicated by the high RB values of 167.77% and 143.72% during the calibration and validation periods, respectively.

Similar results were obtained in the JRB, where the SWAT simulations that were driven by the APHRODITE precipitation data yielded the best calibration and validation (Figure 6 and Table 6), followed again by the PERSIANN-CDR and NCEP-CFSR precipitation data. However, both the

PERSIANN-CDR and NCEP-CFSR data resulted in unacceptable performance as shown, by the mostly negative NSE values (Table 6). Overestimation of the observed streamflows is also clearly shown in the PERSIANN-CDR- and NCEP-CFSR-based JRB SWAT streamflow predictions (Figure 6b) by 57.63% and 142.45%, respectively, during the validation period (Table 6). However, the APHRODITE-based data tracked the observed streamflow well (Figure 6b), which was also confirmed by the majority of NSE, R^2 and RB statistics (Table 6), which indicated satisfactory results based on previously suggested criteria [58,59].

Table 6. Statistical analysis of GCPs performance in SWAT modeling in the Kelantan River Basin (KRB) and Johor River Basin (JRB).

		KRB			JRB		
		A	P	N	A	P	N
		GCPs Precipitation					
Calibration	R^2	0.74	0.78	0.28	0.44	0.31	0.18
Period	NSE	0.69	0.49	−4.47	0.34	0.14	−2.39
	RB	0.40	43.01	167.77	−19.38	21.20	81.60
Validation	R^2	0.68	0.63	0.23	0.61	0.49	0.40
Period	NSE	0.64	0.15	−6.19	0.60	−0.11	−3.26
	RB	13.48	41.92	143.72	−9.36	57.63	142.45
		GCPs Precipitation + NCEP-CFSR Temperature					
Calibration	R^2	0.75	0.78	0.28	0.44	0.32	0.17
Period	NSE	0.70	0.46	−4.60	0.35	0.10	−2.60
	RB	2.65	45.64	170.46	−18.39	24.26	84.88
Validation	R^2	0.69	0.64	0.23	0.61	0.49	0.40
Period	NSE	0.62	0.10	−6.40	0.61	−0.18	−3.43
	RB	15.43	44.18	146.30	−7.05	61.94	146.72

Notes: A = APHRODITE; P = PERSIANN-CDR; N = NCEP-CFSR.

4.6. Streamflow: GCPs Precipitation + NCEP-CFSR Temperature Data

The statistical indices (R^2, NSE and RB) are summarized in Table 6 for the SWAT simulations that were executed as a function of precipitation inputs from one of the three GCPs in combination with the NCEP-CFSR temperature data. The combinations of GCP precipitation inputs and NCEP-CFSR temperature data resulted in overestimations of the observed streamflow for the majority of the simulation period for both basins. Similarly, the most severe streamflow overpredictions resulted in response to the combination of NCEP-CFSR precipitation and NCEP-CFSR temperature data.

Generally, the integration of the NCEP-CFSR temperature data with the GCP precipitation data did not result in significant impacts on the SWAT simulations for either basin, compared to the simulations that were performed with just the GCP precipitation inputs. For example, the differences of the validation NSE values between the APHRODITE precipitation data input and the APHRODITE precipitation with the NCEP-CFSR temperature input for the KRB and JRB are 0.02 and 0.01, respectively. These findings show that the influence of the precipitation data on the local hydrological cycle is very dominant relative to the effects of the temperature data in this tropical region. This could be due to the small temperature range and variation that occurs in Malaysia as compared to more temperate or arid regions in other global sub-regions.

Some success was obtained by forcing the SWAT model with the integration of the APHRODITE precipitation and NCEP-CFSR temperature data. However, we could not ignore a tendency by the APHRODITE data to underestimate the actual precipitation, which in turn offset some of the trend in overestimated streamflow that occurred within the SWAT models in the two basins that we have studied. Based on Faramarzi et al. [18], inaccurate input data, wrong model structure and inappropriate model parameters could generate misleading SWAT model outputs. The input data error can easily be identified using more reliable observations, while the other two require local expert knowledge with modeling skill. Hence, multiple GCP data should be evaluated through an initial assessment prior to applying them in any hydrological models.

4.7. Extreme Event Assessment

The final aspect of the overall analysis was to evaluate the capability of the GCPs to predict extreme precipitation events (Table 7). All of the GCPs showed significant differences at 0.05 significance level when compared with the observed precipitation, except for the NCEP-CFSR data when assessed for the Rx1d index for the KRB. The APHRODITE data exhibited better correlation with observed precipitation for three of the indices (Rx1d, Rx5d and R10mm) in both basins versus the other GCPs, while the PERSIANN-CDR data resulted in the best performance in the R50mm index estimation. In addition, the majority of the RB values, which were calculated for the Rx1d, Rx5d and R50mm indices estimated by the three GCPs, were negative. This is similar to the findings reported by Miao et al. [78], who found that the PERSIANN-CDR data tends to underestimate the Rx1d and Rx5d indices in the eastern China region. This can be explained by the fact that most of the GCPs underestimated the precipitation range which is greater than 50 mm in the two basins (Figure 4).

Table 7. Statistical analysis for extreme precipitation indices in Kelantan River Basin and Johor River Basin (bold indicate significance at 0.05).

Indices		Kelantan River Basin			Johor River Basin		
		A	P	N	A	P	N
Rx1d	RMSE (mm)	96.34	83.11	78.03	74.25	82.04	73.58
	CC	0.48	0.44	0.19	0.23	0.20	0.29
	ME (mm)	−70.00	−49.46	−3.01	−42.03	−56.58	15.41
	RB (%)	−54.47	−38.48	−2.34	−33.71	−45.38	12.36
	T test (*t* stat)	**23.98**	**16.31**	0.95	**9.24**	**11.69**	**2.70**
Rx5d	RMSE (mm)	176.86	150.29	157.23	95.92	99.56	139.09
	CC	0.62	0.53	0.21	0.57	0.39	0.18
	ME (mm)	−127.38	−74.32	−19.91	−53.27	−45.84	35.28
	RB (%)	−45.21	−26.37	−7.07	−24.03	−20.69	15.92
	T test (*t* stat)	**20.56**	**10.89**	**3.12**	**6.75**	**6.47**	**3.56**
R10mm	RMSE (days)	23.56	22.87	75.11	14.91	29.78	61.60
	CC	0.34	0.24	0.23	0.40	0.32	0.11
	ME (mm)	−13.33	11.05	64.57	5.42	24.74	55.78
	RB (%)	−16.79	13.92	81.32	8.07	36.81	83.00
	T test (*t* stat)	**15.15**	**13.11**	**30.43**	**4.60**	**13.56**	**30.66**
R50mm	RMSE (days)	9.38	7.97	7.67	6.58	6.56	5.64
	CC	0.40	0.45	−0.14	0.27	0.34	0.06
	ME (mm)	−7.90	−6.26	−1.19	−5.66	−5.67	−3.68
	RB (%)	−83.33	−66.08	−12.55	−70.76	−70.87	−46.08
	T test (*t* stat)	**36.00**	**27.18**	**4.51**	**12.55**	**23.30**	**12.55**

Notes: A = APHRODITE; P = PERSIANN-CDR; N = NCEP-CFSR.

The RB statistic was used to quantify the difference in accuracy in simulating extreme streamflow events, based on the Rx1d and Rx5d indices, between the rain gauge-based and other three GCPs for the KRB (Figure 7) and JRB (Figure 8) because it provided a reliable basis for comparison of different case studies [79]. The majority of the RB values calculated for the APHRODITE and PERSIANN-CDR Rx1d and Rx5d indices are negative, indicating that most of the high streamflows were underestimated. However, the reverse pattern can be observed for the RB values determined for the respective NCEP-CFSR indices, indicating that streamflow was significantly overestimated in both basins for the NCEP-CFSR-based SWAT simulations.

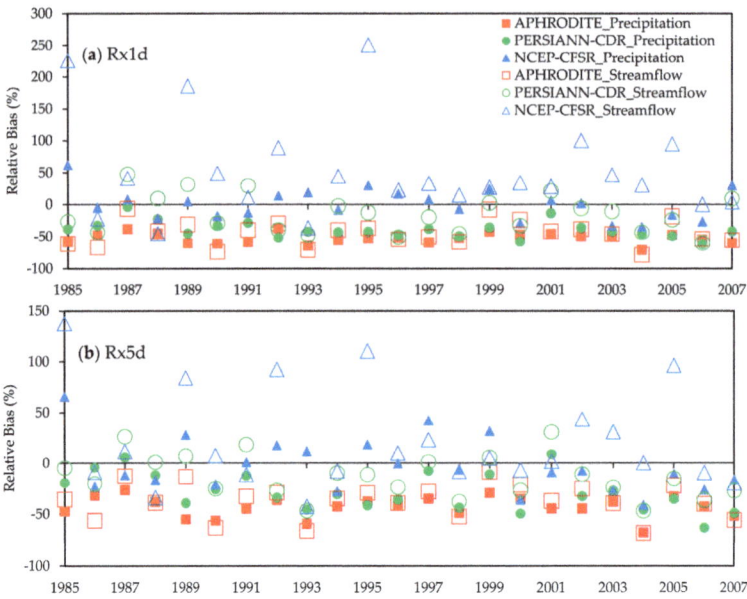

Figure 7. Relative bias values of annual maximum: (**a**) one-day precipitation/streamflow; and (**b**) five-day consecutive precipitation/streamflow from 1985 to 2007 in the Kelantan River Basin.

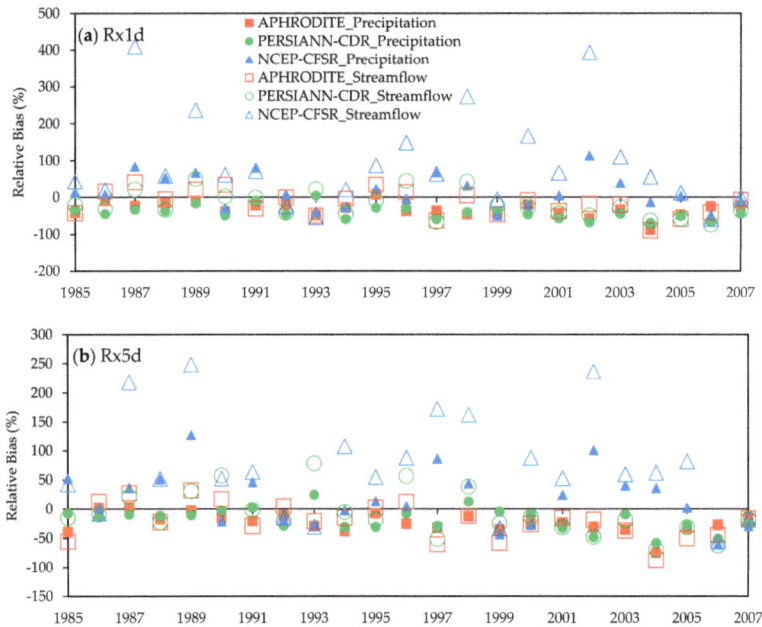

Figure 8. Relative bias values of annual maximum: (**a**) one-day precipitation/streamflow; and (**b**) five-day consecutive precipitation/streamflow from 1985 to 2007 in the Johor River Basin.

5. Discussion

In this study, six different sets of GCP precipitation and temperature inputs were forced to drive the SWAT model. The overall results of the analyses of the GCP data clearly revealed that the APHRODITE precipitation data resulted in the best performance of the three GCP data sources, based on the SWAT simulation graphical and statistical results. These results agree with the findings reported in several other studies, which showed that SWAT simulations executed with APHRODITE precipitation data performed very well in central Vietnam [23,24,80]; glacier influenced basins in mountainous regions in northwest China [81,82] and central Asia [83,84]; and a major tributary of the Yangtze River in central China [85]. Lauri et al. [31] also found that executing the VMod hydrological model [86] with combined APHRODITE precipitation and NCEP-CFSR temperature inputs accurately replicated hydrological simulations based on surface climate inputs of the 795,000 km^2 Mekong River Basin in southeast Asia. These composite results underscore the strength of the APHRODITE precipitation data for a variety of Asian conditions and that it can reliably be used for hydrological applications in un-gauged, data limited or restricted basins in the Southeast Asia.

The results found here clearly show that the original NCEP-CFSR precipitation is not suitable to apply for streamflow simulations in Malaysia, which is in agreement with the findings of Monteiro et al. [27], Roth and Lemann [64] and Bressiani et al. [87] for other tropical or sub-tropical conditions. However, the results found here conflict with the findings of Jajarmizadeh et al. [88], who report successful SWAT streamflow simulation results using the NCEP-CFSR data for the Roodan watershed that is located in southern Iran. Differences in climate and geographical conditions are the most likely explanation for such differences between the Jajarmizadeh et al. [88] study and the results reported in this research and other previously cited studies. In addition, the streamflow overestimation that resulted from the use of the NCEP-CFSR data in this study could be related to possible problems that occur over tropical regions [70], including the effects of the satellite algorithms on precipitation estimation and the CFSR model parameterizations.

In general, the performance of the APHRODITE data was better for the KRB compared to the JRB. This is due in part to a more complete distribution of rain gauges for the KRB versus the JRB (Figure 1); the JRB lacks long-term climate data representation in the northern part of the basin. In addition, the PERSIANN-CDR precipitation-based SWAT simulation also performed better for the KRB, which is consistent with Zhu et al. [29] who found that the PERSIANN-CDR data resulted in a smaller relative error in a data-rich region. These results are consistent with previously reported findings that improved SWAT hydrologic simulations usually occur in response to precipitation inputs characterized by higher resolution, versus lower resolution precipitation inputs [89–91].

As shown in Table 6, we also found that the effect of the basin size proved to be of minor importance compared to the performance of the three GCPs. For instance, the NCEP-CFSR data performed poorly in both basins, regardless of size and flow characteristics, while the APHRODITE precipitation resulted in the best performance for both basins. We also note that differences in sub-basin and/or HRU delineations, while not investigated in this study, typically do not impact SWAT streamflow and other hydrologic outputs as discussed in a previous review of SWAT literature [48] and reported in several subsequent SWAT applications [92–95].

Finally, it is important to emphasize that there were distinct periods within the overall simulation timeframe in which prevailing periods of bias actually were reversed for a specific GCP; e.g., streamflow extremes were overestimated during periods where precipitation extremes were underestimated. For example, the PERSIANN-CDR underestimated the Rx1d precipitation index by about 45% during 1989, but the corresponding Rx1d streamflow index was overestimated by 31.3%. This is consistent with the findings of a similar study conducted by Zhu et al. [29] for the Xiang River and Qu River watersheds in China. This finding indicates that there are certain periods where the precipitation generated by GCPs is unlikely to accurately capture the amount and durations of extreme events. This is further exacerbated by the fact that there is a variation between the precipitation and streamflow extremes temporal scales. For example, peak streamflow usually occurred a few days/hours after the corresponding peak precipitation, but the peak streamflow normally represents an accumulation of precipitation events that occurred over several days/hours.

6. Conclusions

The performance of the APHRODITE, PERSIANN-CDR and NCEP-CFSR long-term gridded climate products (GCPs) were evaluated versus observed climate data for the Kelantan River Basin (KRB) and Johor River Basin (JRB), which are both tropical basins located in Peninsular Malaysia. The analysis included the assessment of capability of replicating streamflow for both basins using climate data from these GCPs as inputs to the calibrated SWAT model. The main conclusions obtained are as follows:

(1) The APHRODITE data typically replicated the observed monthly and daily precipitation more accurately over both the KRB and JRB, followed by the PERSIANN-CDR data and lastly the NCEP-CFSR data. The APHRODITE data tended to underestimate the observed daily and monthly precipitation in both basins, while the NCEP-CFSR data dramatically overestimated the observed precipitation data. The PERSIANN-CDR data resulted in a slight underestimation of the observed KRB precipitation and an overestimation of the JRB precipitation.

(2) The overall performance of the GCPs was better in low land and near coastal regions, such as the northern and eastern KRB. On the contrary, the performance of the GCPs was poor for the high mountainous regions located in the southwestern part of the KRB. Generally, the APHRODITE data resulted in stronger replication of precipitation in mountainous regions compared to the other two GCPs.

(3) The GCPs were found to have moderate accuracy (ACC), false alarm ratio (FAR), and critical success index (CSI), and a high probability of detection (POD) over the two basins that we have studied; the APHRODITE data resulted in the best performance. All three GCPs underestimated the extreme precipitation ranges ($\geq$50 mm·day^{-1}) and dramatically overestimated the observed moderate precipitation ranges (2–20 mm·day^{-1}).

(4) The APHRODITE data resulted in strong replication of observed streamflows when input to the calibrated SWAT simulations, while, the NCEP-CFSR was unable to replicate the observed streamflows for either basin in the calibrated SWAT. The PERSIANN-CDR data generated an in-between performance in the calibrated SWAT model, resulting in acceptable representation of KRB observed streamflows but an inability to track the JRB observed streamflows.

(5) We recommend the integration of the APHRODITE precipitation and the NCEP-CFSR temperature data for SWAT modeling in Malaysia as well as Southeast Asia region. However, a bias correction should be conducted if the gauge data are available, in order to improve the accuracy of the SWAT modeling.

(6) The APHRODITE data and PERSIANN-CDR data underestimated the annual maximum one-day streamflow (Rx1d) and five-day consecutive streamflow (Rx5d) indices. In contrast, the NCEP-CFSR dramatically overestimated the Rx1d and Rx5d streamflow indices in both basins. Basically, all three GCPs performed poorly in capturing extreme events, where high bias was found in certain periods.

Finally, these findings demonstrate how large uncertainties of GCP inputs can propagate within streamflow modeling, which can greatly affect the accuracy of streamflow simulations. This could lead to erroneous results that in turn could lead to wrong conclusions, which could impact the development of management systems and local policies. Therefore, development of an improved quantification framework for more accurate comparisons between different study areas should be a focus for future research. Similar studies should be conducted in other watershed systems with varying climatic and geographical conditions, to expand the testing of the GCPs and provide feedback to the GCP producers that can be used to develop better products.

Acknowledgments: This research was supported by the Ministry of Higher Education Malaysia and Universiti Teknologi Malaysia under the Transdisciplinary Research Grant Scheme (R.J130000.7809.4L835). The research was also funded in part from support received from the U.S. Department of Agriculture, National Institute of Food and Agriculture, Award No. 20116800230190, Climate Change, Mitigation, and Adaptation in Corn-Based Cropping Systems. We acknowledge Malaysian government agencies for providing the hydro-climatic and geographical data. Gratitude is also expressed to the original producers of the three global climate products for providing free downloadable data.

Author Contributions: Mou Leong Tan collected, processed and analyzed the data, and drafted the manuscript. Philip W. Gassman and Arthur P. Cracknell made contributions in reviewing and editing of the manuscript.

Conflicts of Interest: The authors declare no conflict of interest.

References

1. Held, I.M.; Soden, B.J. Robust responses of the hydrological cycle to global warming. *J. Clim.* **2006**, *19*, 5686–5699. [CrossRef]
2. Stokstad, E. Hydrology—Scarcity of rain, stream gages threatens forecasts. *Science* **1999**, *285*, 1199–1200. [CrossRef]
3. Abera, W.; Brocca, L.; Rigon, R. Comparative evaluation of different satellite rainfall estimation products and bias correction in the Upper Blue Nile (UBN) Basin. *Atmos. Res.* **2016**, *178–179*, 471–483. [CrossRef]
4. Yu, M.; Chen, X.; Li, L.; Bao, A.; Paix, M.L. Streamflow simulation by SWAT using different precipitation sources in large arid basins with scarce raingauges. *Water Resour. Manag.* **2011**, *25*, 2669–2681.
5. Gebremichael, M.; Bitew, M.M.; Hirpa, F.A.; Tesfay, G.N. Accuracy of satellite rainfall estimates in the Blue Nile Basin: Lowland plain versus highland mountain. *Water Resour. Res.* **2014**, *50*, 8775–8790.
6. Saha, S.; Moorthi, S.; Pan, H.-L.; Wu, X.; Wang, J.; Nadiga, S.; Tripp, P.; Kistler, R.; Woollen, J.; Behringer, D.; et al. The NCEP Climate Forecast System Reanalysis. *Bull. Am. Meteorol. Soc.* **2010**, *91*, 1015–1057. Available online: http://globalweather.tamu.edu (accessed on 21 March 2017).
7. Yatagai, A.; Kamiguchi, K.; Arakawa, O.; Hamada, A.; Yasutomi, N.; Kitoh, A. APHRODITE constructing a long-term daily gridded precipitation dataset for Asia based on a dense network of rain gauges. *Bull. Am. Meteorol. Soc.* **2012**, *93*, 1401–1415. Available online: http://www.chikyu.ac.jp/precip/english/products. html (accessed on 21 March 2017). [CrossRef]
8. Ashouri, H.; Hsu, K.-L.; Sorooshian, S.; Braithwaite, D.K.; Knapp, K.R.; Cecil, L.D.; Nelson, B.R.; Prat, O.P. PERSIANN-CDR: Daily precipitation climate data record from multi-satellite observations for hydrological and climate studies. *Bull. Am. Meteorol. Soc.* **2015**, *96*, 69–83. Available online: http://chrsdata.eng.uci.edu (accessed on 21 March 2017).
9. Dinku, T.; Ceccato, P.; Grover-Kopec, E.; Lemma, M.; Connor, S.J.; Ropelewski, C.F. Validation of satellite rainfall products over East Africa's complex topography. *Int. J. Remote Sens.* **2007**, *28*, 1503–1526. [CrossRef]
10. Wang, J.D.; Wang, W.Q.; Fu, X.H.; Seo, K.H. Tropical intraseasonal rainfall variability in the CFSR. *Clim. Dyn.* **2012**, *38*, 2191–2207.
11. Mashingia, F.; Mtalo, F.; Bruen, M. Validation of remotely sensed rainfall over major climatic regions in Northeast Tanzania. *Phys. Chem. Earth* **2014**, *67–69*, 55–63. [CrossRef]
12. Tan, M.L.; Ibrahim, A.L.; Duan, Z.; Cracknell, A.P.; Chaplot, V. Evaluation of six high-resolution satellite and ground-based precipitation products over Malaysia. *Remote Sens.* **2015**, *7*, 1504–1528. [CrossRef]
13. Jamandre, C.A.; Narisma, G.T. Spatio-temporal validation of satellite-based rainfall estimates in the Philippines. *Atmos. Res.* **2013**, *122*, 599–608. [CrossRef]
14. Fekete, B.M.; Vörösmarty, C.J.; Roads, J.O.; Willmott, C.J. Uncertainties in precipitation and their impacts on runoff estimates. *J. Clim.* **2004**, *17*, 294–304. [CrossRef]
15. Adler, R.F.; Huffman, G.J.; Chang, A.; Ferraro, R.; Xie, P.P.; Janowiak, J.; Rudolf, B.; Schneider, U.; Curtis, S.; Bolvin, D.; et al. The version-2 Global Precipitation Climatology Project (GPCP) monthly precipitation analysis (1979-present). *J. Hydrometeorol.* **2003**, *4*, 1147–1167.
16. Behrangi, A.; Khakbaz, B.; Jaw, T.C.; AghaKouchak, A.; Hsu, K.; Sorooshian, S. Hydrologic evaluation of satellite precipitation products over a mid-size basin. *J. Hydrol.* **2011**, *397*, 225–237. [CrossRef]
17. Seyyedi, H.; Anagnostou, E.N.; Beighley, E.; McCollum, J. Hydrologic evaluation of satellite and reanalysis precipitation datasets over a mid-latitude basin. *Atmos. Res.* **2015**, *164–165*, 37–48.

18. Faramarzi, M.; Srinivasan, R.; Iravani, M.; Bladon, K.D.; Abbaspour, K.C.; Zehnder, A.J.B.; Goss, G.G. Setting up a hydrological model of Alberta: Data discrimination analyses prior to calibration. *Environ. Model. Softw.* **2015**, *74*, 48–65.

19. Alemayehu, T.; van Griensven, A.; Bauwens, W. Evaluating CFSR and WATCH data as input to swat for the estimation of the potential evapotranspiration in a data-scarce Eastern-African catchment. *J. Hydrol. Eng.* **2016**, *21*, 16. [CrossRef]

20. Getirana, A.C.V.; Espinoza, J.C.V.; Ronchail, J.; Rotunno Filho, O.C. Assessment of different precipitation datasets and their impacts on the water balance of the Negro River Basin. *J. Hydrol.* **2011**, *404*, 304–322.

21. Dile, Y.T.; Srinivasan, R. Evaluation of CFSR climate data for hydrologic prediction in data-scarce watersheds: An application in the Blue Nile River Basin. *J. Am. Water Resour. Assoc.* **2014**, *50*, 1226–1241.

22. Strauch, M.; Kumar, R.; Eisner, S.; Mulligan, M.; Reinhardt, J.; Santini, W.; Vetter, T.; Friesen, J. Adjustment of global precipitation data for enhanced hydrologic modeling of tropical Andean Watersheds. *Clim. Chang.* **2016**, *141*, 547–560. [CrossRef]

23. Vu, M.T.; Raghavan, S.V.; Liong, S.Y. SWAT use of gridded observations for simulating runoff—A Vietnam river basin study. *Hydrol. Earth Syst. Sci.* **2012**, *16*, 2801–2811. [CrossRef]

24. Le, T.B.; Sharif, H.O. Modeling the projected changes of river flow in central Vietnam under different climate change scenarios. *Water* **2015**, *7*, 3579–3598.

25. Bressiani, D.D.; Gassman, P.W.; Fernandes, J.G.; Garbossa, L.H.P.; Srinivasan, R.; Bonuma, N.B.; Mendiondo, E.M. Review of Soil and Water Assessment Tool (SWAT) applications in Brazil: Challenges and prospects. *Int. J. Agric. Biol. Eng.* **2015**, *8*, 9–35.

26. Creech, C.T.; Siqueira, R.B.; Selegean, J.P.; Miller, C. Anthropogenic impacts to the sediment budget of São Francisco River navigation channel using SWAT. *Int. J. Agric. Biol. Eng.* **2015**, *8*, 140–157.

27. Monteiro, J.A.F.; Strauch, M.; Srinivasan, R.; Abbaspour, K.; Gucker, B. Accuracy of grid precipitation data for Brazil: Application in river discharge modelling of the Tocantins Catchment. *Hydrol. Process.* **2016**, *30*, 1419–1430. [CrossRef]

28. Auerbach, D.A.; Easton, Z.M.; Walter, M.T.; Flecker, A.S.; Fuka, D.R. Evaluating weather observations and the Climate Forecast System Reanalysis as inputs for hydrologic modelling in the tropics. *Hydrol. Process.* **2016**, *30*, 3466–3477. [CrossRef]

29. Zhu, Q.; Xuan, W.; Liu, L.; Xu, Y.-P. Evaluation and hydrological application of precipitation estimates derived from PERSIANN-CDR, TRMM 3B42V7, and NCEP-CFSR over humid regions in China. *Hydrol. Process.* **2016**, *30*, 3061–3083. [CrossRef]

30. Ashouri, H.; Nguyen, P.; Thorstensen, A.; Hsu, K.-L.; Sorooshian, S.; Braithwaite, D. Assessing the efficacy of high-resolution satellite-based PERSIANN-CDR precipitation product in simulating streamflow. *J. Hydrometeorol.* **2016**, *17*, 2061–2076. [CrossRef]

31. Lauri, H.; Rasanen, T.A.; Kummu, M. Using reanalysis and remotely sensed temperature and precipitation data for hydrological modeling in monsoon climate: Mekong river case study. *J. Hydrometeorol.* **2014**, *15*, 1532–1545. [CrossRef]

32. Arnold, J.G.; Srinivasan, R.; Muttiah, R.S.; Williams, J.R. Large area hydrologic modeling and assessment—Part 1: Model development. *J. Am. Water Resour. Assoc.* **1998**, *34*, 73–89. Available online: http://swat.tamu.edu/software/arcswat (accessed on 21 March 2017). [CrossRef]

33. Arnold, J.G.; Kiniry, J.R.; Srinivasan, R.; Williams, J.R.; Haney, E.B.; Neitsch, S.L. *Soil and Water Assessment Tool Input/Tool File Documentation. Version 2012*; Texas Water Resources Institute: College Station, TX, USA, 2012.

34. Arnold, J.G.; Moriasi, D.N.; Gassman, P.W.; Abbaspour, K.C.; White, M.J.; Srinivasan, R.; Santhi, C.; Harmel, R.D.; van Griensven, A.; van Liew, M.W.; et al. SWAT: Model use, calibration, and validation. *Trans. ASABE* **2012**, *55*, 1491–1508. [CrossRef]

35. Williams, J.R.; Arnold, J.G.; Kiniry, J.R.; Gassman, P.W.; Green, C.H. History of model development at Temple, Texas. *Hydrol. Sci. J.* **2008**, *53*, 948–960. [CrossRef]

36. Neitsch, S.L.; Arnold, J.G.; Kiniry, J.R.; Grassland, J.R.W. *Soil and Water Assessment Tool Theoretical Documentation Version 2009*; Agricultural Research Service Blackland Research Center: Temple, TX, USA, 2011.

37. Tan, M.L.; Ibrahim, A.L.; Cracknell, A.P.; Yusop, Z. Changes in precipitation extremes over the Kelantan River Basin, Malaysia. *Int. J. Climatol.* **2016**. [CrossRef]

38. Knapp, K.R.; Ansari, S.; Bain, C.L.; Bourassa, M.A.; Dickinson, M.J.; Funk, C.; Helms, C.N.; Hennon, C.C.; Holmes, C.D.; Huffman, G.J.; et al. Globally gridded satellite observations for climate studies. *Bull. Am. Meteorol. Soc.* **2011**, *92*, 893–907. [CrossRef]

39. Department of Irrigation and Drainage Malaysia (DID). *Hydrological Procedure No. 15: River Discharge Measurement by Current Meter*; DID: Kuala Lumpur, Malaysia, 1995.

40. Tan, M.L.; Ficklin, D.L.; Dixon, B.; Ibrahim, A.L.; Yusop, Z.; Chaplot, V. Impacts of DEM resolution, source, and resampling technique on SWAT-simulated streamflow. *Appl. Geogr.* **2015**, *63*, 357–368. [CrossRef]

41. Farr, T.G.; Rosen, P.A.; Caro, E.; Crippen, R.; Duren, R.; Hensley, S.; Kobrick, M.; Paller, M.; Rodriguez, E.; Roth, L.; et al. The Shuttle Radar Topography Mission. *Rev. Geophys.* **2007**, *45*, 75–79. [CrossRef]

42. Zheng, J.; Li, G.-Y.; Han, Z.-Z.; Meng, G.-X. Hydrological cycle simulation of an irrigation district based on a SWAT model. *Math. Comput. Model.* **2010**, *51*, 1312–1318. [CrossRef]

43. Tangang, F.T.; Juneng, L. Mechanisms of Malaysian rainfall anomalies. *J. Clim.* **2004**, *17*, 3616–3622.

44. Dembélé, M.; Zwart, S.J. Evaluation and comparison of satellite-based rainfall products in Burkina Faso, West Africa. *Int. J. Remote Sens.* **2016**, *37*, 3995–4014. [CrossRef]

45. Khan, S.I.; Hong, Y.; Gourley, J.J.; Khattak, M.U.K.; Yong, B.; Vergara, H.J. Evaluation of three high-resolution satellite precipitation estimates: Potential for monsoon monitoring over Pakistan. *Adv. Space Res.* **2014**, *54*, 670–684.

46. Ebert, E.E.; Janowiak, J.E.; Kidd, C. Comparison of near-real-time precipitation estimates from satellite observations and numerical models. *Bull. Am. Meteorol. Soc.* **2007**, *88*, 47–64.

47. Shen, Y.; Xiong, A.Y.; Wang, Y.; Xie, P.P. Performance of high-resolution satellite precipitation products over China. *J. Geophys. Res. Atmos.* **2010**, *115*, D2. [CrossRef]

48. Gassman, P.W.; Reyes, M.R.; Green, C.H.; Arnold, J.G. The Soil and Water Assessment Tool: Historical development, applications, and future research directions. *Trans. ASABE* **2007**, *50*, 1211–1250. [CrossRef]

49. Gassman, P.W.; Sadeghi, A.M.; Srinivasan, R. Applications of the swat model special section: Overview and insights. *J. Environ. Qual.* **2014**, *43*, 1–8. [CrossRef] [PubMed]

50. Gassman, P.W.; Wang, Y.K. Ijabe swat special issue: Innovative modeling solutions for water resource problems. *Int. J. Agric. Biol. Eng.* **2015**, *8*, 1–8.

51. Krysanova, V.; White, M. Advances in water resources assessment with SWAT—An overview. *Hydrol. Sci. J.* **2015**, *60*, 771–783. [CrossRef]

52. Hasan, Z.A.; Hamidon, N.; Yusof, M.S.; Ab Ghani, A. Flow and sediment yield simulations for Bukit Merah Reservoir Catchment, Malaysia: A case study. *Water Sci. Technol.* **2012**, *66*, 2170–2176. [PubMed]

53. Memarian, H.; Balasundram, S.K.; Abbaspour, K.C.; Talib, J.B.; Teh Boon Sung, C.; Sood, A.M. SWAT-based hydrological modelling of tropical land use scenarios. *Hydrol. Sci. J.* **2014**, *59*, 1808–1829. [CrossRef]

54. Tan, M.L.; Ficklin, D.L.; Ibrahim, A.L.; Yusop, Z. Impacts and uncertainties of climate change on streamflow of the Johor River Basin, Malaysia using a CMIP5 General Circulation Model ensemble. *J. Water Clim. Chang.* **2014**, *5*, 676–695.

55. Tan, M.L.; Ibrahim, A.L.; Yusop, Z.; Duan, Z.; Ling, L. Impacts of land-use and climate variability on hydrological components in the Johor River Basin, Malaysia. *Hydrol. Sci. J.* **2015**, *60*, 873–889. [CrossRef]

56. Abbaspour, K.C.; Rouholahnejad, E.; Vaghefi, S.; Srinivasan, R.; Yang, H.; Kløve, B. A continental-scale hydrology and water quality model for Europe: Calibration and uncertainty of a high-resolution large-scale SWAT model. *J. Hydrol.* **2015**, *524*, 733–752. [CrossRef]

57. Krause, P.; Boyle, D.P.; Bäse, F. Comparison of different efficiency criteria for hydrological model assessment. *Adv. Geosci.* **2005**, *5*, 89–97. [CrossRef]

58. Moriasi, D.N.; Arnold, J.G.; van Liew, M.W.; Binger, R.L.; Harmel, R.D.; Veith, T. Model evaluation guidelines for systematic quantification of accuracy in watershed simulations. *Trans. ASABE* **2007**, *50*, 885–900.

59. Moriasi, D.N.; Gitau, M.W.; Pai, N.; Daggupati, P. Hydrologic and water quality models: Performance measures and evaluation criteria. *Trans. ASABE* **2015**, *58*, 1763–1785.

60. Groisman, P.Y.; Knight, R.W.; Easterling, D.R.; Karl, T.R.; Hegerl, G.C.; Razuvaev, V.A.N. Trends in intense precipitation in the climate record. *J. Clim.* **2005**, *18*, 1326–1350.

61. Li, C.; Wang, R.; Ning, H.; Luo, Q. Changes in climate extremes and their impact on wheat yield in Tianshan mountains region, Northwest China. *Environ. Earth Sci.* **2016**, *75*, 1–13. [CrossRef]

62. Tan, M.L.; Tan, K.C.; Chua, V.P.; Chan, N.W. Evaluation of TRMM product for monitoring drought in the Kelantan River Basin, Malaysia. *Water* **2017**, *9*, 57. [CrossRef]

63. Karl, T.R.; Nicholls, N.; Ghazi, A. CLIVAR/GCOS/WMO workshop on indices and indicators for climate extremes workshop summary. *Clim. Chang.* **1999**, *42*, 3–7. [CrossRef]

64. Roth, V.; Lemann, T. Comparing CFSR and conventional weather data for discharge and soil loss modelling with SWAT in small catchments in the Ethiopian highlands. *Hydrol. Earth Syst. Sci.* **2016**, *20*, 921–934. [CrossRef]

65. Decker, M.; Brunke, M.A.; Wang, Z.; Sakaguchi, K.; Zeng, X.B.; Bosilovich, M.G. Evaluation of the reanalysis products from GSFC, NCEP, and ECMWF using flux tower observations. *J. Clim.* **2012**, *25*, 1916–1944. [CrossRef]

66. Derin, Y.; Yilmaz, K.K. Evaluation of multiple satellite-based precipitation products over complex topography. *J. Hydrometeorol.* **2014**, *15*, 1498–1516.

67. Sun, W.; Mu, X.; Song, X.; Wu, D.; Cheng, A.; Qiu, B. Changes in extreme temperature and precipitation events in the Loess Plateau (China) during 1960–2013 under global warming. *Atmos. Res.* **2016**, *168*, 33–48. [CrossRef]

68. Yilmaz, K.K.; Hogue, T.S.; Hsu, K.L.; Sorooshian, S.; Gupta, H.V.; Wagener, T. Intercomparison of rain gauge, radar, and satellite-based precipitation estimates with emphasis on hydrologic forecasting. *J. Hydrometeorol.* **2005**, *6*, 497–517. [CrossRef]

69. Krakauer, N.; Pradhanang, S.; Lakhankar, T.; Jha, A. Evaluating satellite products for precipitation estimation in mountain regions: A case study for Nepal. *Remote Sens.* **2013**, *5*, 4107–4123. [CrossRef]

70. Blacutt, L.A.; Herdies, D.L.; de Gonçalves, L.G.G.; Vila, D.A.; Andrade, M. Precipitation comparison for the CFSR, MERRA, TRMM3B42 and combined scheme datasets in Bolivia. *Atmos. Res.* **2015**, *163*, 117–131. [CrossRef]

71. Meng, J.; Yang, R.Q.; Wei, H.L.; Ek, M.; Gayno, G.; Xie, P.P.; Mitchell, K. The land surface analysis in the NCEP Climate Forecast System Reanalysis. *J. Hydrometeorol.* **2012**, *13*, 1621–1630. [CrossRef]

72. Fuka, D.R.; Walter, M.T.; MacAlister, C.; Degaetano, A.T.; Steenhuis, T.S.; Easton, Z.M. Using the Climate Forecast System Reanalysis as weather input data for watershed models. *Hydrol. Process.* **2014**, *28*, 5613–5623.

73. Nyeko, M. Hydrologic modelling of data scarce basin with SWAT model: Capabilities and limitations. *Water Resour. Manag.* **2014**, *29*, 81–94. [CrossRef]

74. Tan, M.L.; Ibrahim, A.L.; Yusop, Z.; Chua, V.P.; Chan, N.W. Climate change impacts under CMIP5 RCP scenarios on water resources of the Kelantan River Basin, Malaysia. *Atmos. Res.* **2017**, *189*, 1–10. [CrossRef]

75. Fukunaga, D.C.; Cecílio, R.A.; Zanetti, S.S.; Oliveira, L.T.; Caiado, M.A.C. Application of the SWAT hydrologic model to a tropical watershed at Brazil. *Catena* **2015**, *125*, 206–213. [CrossRef]

76. Pereira, D.R.; Martinez, M.A.; da Silva, D.D.; Pruski, F.F. Hydrological simulation in a basin of typical tropical climate and soil using the SWAT model part II: Simulation of hydrological variables and soil use scenarios. *J. Hydrol. Reg. Stud.* **2016**, *5*, 149–163. [CrossRef]

77. Yesuf, H.M.; Melesse, A.M.; Zeleke, G.; Alamirew, T. Streamflow prediction uncertainty analysis and verification of SWAT model in a tropical watershed. *Environ. Earth Sci.* **2016**, *75*, 806. [CrossRef]

78. Miao, C.; Ashouri, H.; Hsu, K.-L.; Sorooshian, S.; Duan, Q. Evaluation of the PERSIANN-CDR daily rainfall estimates in capturing the behavior of extreme precipitation events over China. *J. Hydrometeorol.* **2015**, *16*, 1387–1396. [CrossRef]

79. Schaefli, B.; Gupta, H.V. Do nash values have value? *Hydrol. Process.* **2007**, *21*, 2075–2080. [CrossRef]

80. Thom, V.; Khoi, D.; Linh, D. Using gridded rainfall products in simulating streamflow in a tropical catchment—A case study of the Srepok River Catchment, Vietnam. *J. Hydrol. Hydromech.* **2017**, *65*, 18–25. [CrossRef]

81. Wang, X.L.; Luo, Y.; Sun, L.; Zhang, Y.Q. Assessing the effects of precipitation and temperature changes on hydrological processes in a glacier-dominated catchment. *Hydrol. Process.* **2015**, *29*, 4830–4845.

82. Gan, R.; Zuo, Q.T. Assessing the digital filter method for base flow estimation in glacier melt dominated basins. *Hydrol. Process.* **2016**, *30*, 1367–1375.

83. Sidike, A.; Chen, X.; Liu, T.; Durdiev, K.; Huang, Y. Investigating alternative climate data sources for hydrological simulations in the upstream of the Amu Darya River. *Water* **2016**, *8*, 441. [CrossRef]

84. Ma, C.K.; Sun, L.; Liu, S.Y.; Shao, M.A.; Luo, Y. Impact of climate change on the streamflow in the glacierized Chu River Basin, Central Asia. *J. Arid Land* **2015**, *7*, 501–513. [CrossRef]

85. Xu, H.L.; Xu, C.Y.; Chen, S.D.; Chen, H. Similarity and difference of global reanalysis datasets (WFD and APHRODITE) in driving lumped and distributed hydrological models in a humid region of China. *J. Hydrol.* **2016**, *542*, 343–356. [CrossRef]

86. Lauri, H.; de Moel, H.; Ward, P.J.; Rasanen, T.A.; Keskinen, M.; Kummu, M. Future changes in Mekong River hydrology: Impact of climate change and reservoir operation on discharge. *Hydrol. Earth Syst. Sci.* **2012**, *16*, 4603–4619.

87. Bressiani, D.A.; Srinivasan, R.; Jones, C.A.; Mendiondo, E.M. Effects of spatial and temporal weather data resolutions on streamflow modeling of a semi-arid basin, Northeast Brazil. *Int. J. Agric. Biol. Eng.* **2015**, *8*, 125–139.

88. Jajarmizadeh, M.; Sidek, L.M.; Mirzai, M.; Alaghmand, S.; Harun, S.; Majid, M.R. Prediction of surface flow by forcing of Climate Forecast System Reanalysis data. *Water Resour. Manag.* **2016**, *30*, 2627–2640. [CrossRef]

89. Schilling, K.E.; Gassman, P.W.; Kling, C.L.; Campbell, T.; Jha, M.K.; Wolter, C.F.; Arnold, J.G. The potential for agricultural land use change to reduce flood risk in a large watershed. *Hydrol. Process.* **2014**, *28*, 3314–3325. [CrossRef]

90. Chaplot, V.; Saleh, A.; Jaynes, D.B. Effect of the accuracy of spatial rainfall information on the modeling of water, sediment, and NO_3-N loads at the watershed level. *J. Hydrol.* **2005**, *312*, 223–234. [CrossRef]

91. Moriasi, D.N.; Starks, P.J. Effects of the resolution of soil dataset and precipitation dataset on SWAT2005 streamflow calibration parameters and simulation accuracy. *J. Soil Water Conserv.* **2010**, *65*, 63–78. [CrossRef]

92. Rouhani, H.; Willems, P.; Feyen, J. Effect of watershed delineation and areal rainfall distribution on runoff prediction using the SWAT model. *Hydrol. Res.* **2009**, *40*, 505–519. [CrossRef]

93. Wang, Y.; Montas, H.J.; Brubaker, K.L.; Leisnham, P.T.; Shirmohammadi, A.; Chanse, V.; Rockler, A.K. Impact of spatial discretization of hydrologic models on spatial distribution of nonpoint source pollution hotspots. *J. Hydrol. Eng.* **2016**, *21*, 12. [CrossRef]

94. Her, Y.; Frankenberger, J.; Chaubey, I.; Srinivasan, R. Threshold effects in hru definition of the Soil and Water Assessment Tool. *Trans. ASABE* **2015**, *58*, 367–378.

95. Shen, Z.Y.; Chen, L.; Liao, Q.; Liu, R.M.; Huang, Q. A comprehensive study of the effect of GIS data on hydrology and non-point source pollution modeling. *Agric. Water Manag.* **2013**, *118*, 93–102. [CrossRef]

water

MDPI

Article

Modeling Crop Water Productivity Using a Coupled SWAT–MODSIM Model

Saeid Ashraf Vaghefi [1,*], Karim C. Abbaspour [1], Monireh Faramarzi [2], Raghavan Srinivasan [3] and Jeffrey G. Arnold [4]

[1] Eawag, Swiss Federal Institute of Aquatic Science and Technology, 8600 Dübendorf, Switzerland; Karim.Abbaspour@eawag.ch

[2] Department of Earth and Atmospheric Sciences, Faculty of Science, University of Alberta, Edmonton, AB T6G 2E3, Canada; faramarz@ualberta.ca

[3] Department of Ecosystem Science and Management, Texas A & M University, College Station, TX 77843, USA; r-srinivasan@tamu.edu

[4] Grassland, Soil and Water Research Laboratory, USDA Agricultural Research Service, Temple, TX 76502, USA; Jeff.Arnold@ars.usda.gov

* Correspondence: saeedashrafv@gmail.com or seyedsaeid.ashrafvaghefi@eawag.ch; Tel.: +41-58-765-5359

Academic Editor: Athanasios Loukas

Received: 30 December 2016; Accepted: 17 February 2017; Published: 24 February 2017

Abstract: This study examines the water productivity of irrigated wheat and maize yields in Karkheh River Basin (KRB) in the semi-arid region of Iran using a coupled modeling approach consisting of the hydrological model (SWAT) and the river basin water allocation model (MODSIM). Dynamic irrigation requirements instead of constant time series of demand were considered. As the cereal production of KRB plays a major role in supplying the food market of Iran, it is necessary to understand the crop yield-water relations for irrigated wheat and maize in the lower part of KRB (LKRB) where most of the irrigated agricultural plains are located. Irrigated wheat and maize yields (Y) and consumptive water use (AET) were modeled with uncertainty analysis at a subbasin level for 1990–2010. Simulated Y and AET were used to calculate crop water productivity (CWP). The coupled SWAT–MODSIM approach improved the accuracy of SWAT outputs by considering the water allocation derived from MODSIM. The results indicated that the highest CWP across this region was 1.31 kg·m^{-3} and 1.13 kg·m^{-3} for wheat and maize, respectively; and the lowest was less than 0.62 kg·m^{-3} and 0.58 kg·m^{-3}. A close linear relationship was found for CWP and yield. The results showed a continuing increase for AET over the years while CWP peaks and then declines. This is evidence of the existence of a plateau in CWP as AET continues to increase and evidence of the fact that higher AET does not necessarily result in a higher yield.

Keywords: Karkheh River Basin; dynamic irrigation scheduling; irrigated wheat; irrigated maize; uncertainty analysis; coupled SWAT-MODSIM

---

1. Introduction

Global human population growth requires increased food production, yet less water resources are available for agriculture. This critical situation can only be resolved if water is managed more efficiently, and crop yield per unit of water consumption increases [1]. Water shortage affects every continent in the twenty-first century. Around 1.2 billion people, or almost one-fifth of the world's population, live in areas of physical scarcity, and 500 million people are approaching this situation. Another 1.6 billion people, or almost one-quarter of the world's population, probably face economic water shortage (where countries lack the necessary infrastructure to take water from rivers and aquifers) [2]. Spatial and temporal distribution of precipitation, which rarely coincides with demand, is a critical problem in this context [3].

Crop water productivity (CWP) is defined as the ratio of crop yield (Y) to the amount of water required to produce that yield [4]. Increasing CWP is necessary to meet a decreasing water availability and is a key element in improving agricultural water productivity, which is central to both economic and social development [5].

Therefore, there is a high intresrt in increasing the productivity of water in the agricultural sector to meet the future food demand [6]. In arid and semi-arid regions where the agricultural sector is the main consumer of water resources and less opportunities exist for the development of new water resources, the accurate estimation of CWP and increasing the productivity of existing water resources is vital. Various researchers studied CWP at specific locations, with specific agricultural and water management practices. Zwart and Bastiaanssen [4] reviewed 84 literature sources. They found that globally measured average CWP values per unit water use are 1.09, 1.09, 0.65, 0.23 and 1.80 kg·m^{-3} for wheat, rice, cotton seed, cotton lint, and maize, respectively. They found that the range of CWP is 0.6–1.7 kg·m^{-3} for wheat, 0.6–1.6 kg·m^{-3} for rice, 0.41–0.95 kg·m^{-3} for cotton seed, 0.14–0.33 kg·m^{-3} for cotton lint, and 1.1–2.7 kg·m^{-3} for maize. Nhamo, et al. [7] evaluated crop evapotranspiration, crop production and agricultural gross domestic product contribution to assess the crop water productivity of Malawi from 2000 to 2013. They found an overall increase of 33% crop water productivity. Giménez, et al. [8] used different full and deficit irrigation practices to calibrate and validate soil water balance in western Uruguay using the soil water balance simulation model SIMDualKc. They found water productivity values, ranging from 1.39 to 2.17 kg·m^{-3} and 1.75 to 2.55 kg·m^{-3} when considering total water use and crop AET, respectively. Borrego-Marín, et al. [9] analyzed the impact of drought (2005, 2012) and drought management plans (2006–2008) on agricultural water productivity in Guadalquivir River Basin in Spain for the period of 2004 to 2012. They found significantly higher water productivity in irrigated than rain-fed agriculture. There is also much interest in the different methods to improve the CWP. Kima, et al. [10] analyzed the effective depth of irrigation water that can keep the soil moisture close to saturation for irrigation intervals to increase water productivity.

In general, the models on crop-water relations can be divided into two categories: empirical and process-based models [11]. Most of the empirical models are regression-based models, where a correlation is established between the statistical crop yield and local weather-related, geostatistical-related, and management-related (e.g., irrigation) factors. Therefore, they can only estimate yield, without predicting crop water uptake and soil evaporation. The process-based models simulate the physiological development, growth and yield of a crop based on the interaction of environmental variables and plant physiological processes (e.g., photosynthesis and respiration) [11,12]. They often have a weakness either in crop growth simulation or hydrology. Examples of process-based models include Soil Water Atmosphere Plant (SWAP) [13], Soil Vegetation Atmosphere Transfer (SVAT) [12], GIS-based Environmental Policy Integrated Climate model GEPIC [14], generic crop model (InfoCrop) [15,16], FAO's crop water productivity and yield response model (AquaCrop) [17–19], and the global water assessment model (WaterGAP) [20]. There are two fundamental limitations in many of the studies which have used these models: (i) The crop yield and consumptive water use estimated for a given area are not linked with the water resources availability of that region. Therefore, one cannot assess the aggregate impact of regional water resources availability, land use, and climate changes on crop production directly. (ii) Uncertainties associated with crop models are not taken into account and remained largely unquantified. There are some studies [21–24] that account for model-related uncertainties in crop yield prediction. To the best of our knowledge, the aforementioned issues have not been considered together in one package. Soil and Water Assessment Tool (SWAT) [25] has been used widely to assess the impact of management practice, and climate and land use changes on water quality and quantity and crop yield [26,27]. Using SWAT calibration, and uncertainty tool "SWAT-CUP", [28] many studies have considered the uncertainties of SWAT output variables such as discharge, and crop yield [29–31]. Although SWAT has significant capabilities in the simulation of hydrologic components and crop yield interactively, the lack of an optimal water allocation module inhibits the dynamic pattern of irrigation scheduling and increases uncertainty in Y and AET predictions.

Water allocation models can be used to optimize water allocation among different users. Some examples of water allocation models are integrated water allocation model (IWAM) [32], REsource ALlocation Model (REALM) [33], Water Evaluation and Planning (WEAP) [34], and river basin network flow model for conjunctive stream-aquifer management (MODSIM) [35]. MODSIM has been used in several studies to address the problem of water allocation between non-consumptive and consumptive water demands at the basin scale [36–38].

This paper aims to study the water productivity of irrigated wheat and maize in agricultural lands of the Lower KRB (LKRB) by using a coupled SWAT–MODSIM model considering dynamic irrigation requirements. To the best of our knowledge, previous studies have not considered dynamic time series of irrigation demands in the estimation of CWP through the aforementioned modeling approach. SWAT is used to estimate spatial and temporal distributions of water availability and irrigation water requirements, while MODSIM [35,39] simulates the processes of reservoir operations for water allocations. The use of the coupled hydrological-water allocation model substantially improves accuracy of Y and AET simulations and results in the implementation of more rational and sustainable water management practices.

Karkheh River Basin (KRB) has traditionally been the central point of agricultural activities in Iran. The basin, located in the arid southwest of Iran, is one of the most productive agricultural areas of the country. It is known as the food basket of Iran [40] and produces about 10% of the country's wheat. Available water resources and desirable climatic conditions make it a suitable basin for growing a broad range of crops. In the KRB, water availability is of great importance in supporting economic and social development [41]. Due to limited potential for developing new water resources and a significant decrease in downstream runoff due to both climate change and human interventions, improving the productivity of the existing water resources in the basin is one of the most important management challenges to sustainable food production [42]. Rafiee and Shourian [43] used a simulation-optimization approach to find the optimal irrigation plan and crop pattern in the Azadegan plain in the KRB. In the basin, excessive irrigation is a key management practice that leads to remarkable water losses [44]. Therefore, several studies have concentrated on the issue of food production in KRB [40,45]. It is also projected that the problem of water will further increase due to climate change in southern parts of the basin [42].

The coupled SWAT–MODSIM approach in this study has some novel features: (i) it considers dynamic irrigation requirements instead of constant time series of demands; (ii) it is a fully coupled model and both models have feedback on each other; (iii) it is supported by a full tutorial which facilitates the application of the coupled model in other similar research studies.

This paper is organized to (i) calibrate (1997–2010) and validate (1990–1996) crop yield at five important agricultural regions in LKRB; (ii) model the spatial and temporal variability of crop yield as well as crop consumptive water use with uncertainty analysis for wheat and maize at a subbasin level, and calculate CWP; and (iii) analyze the relation between yields and consumptive water use by quantifying the applied irrigation water and crop yield in each of the five regions by using the coupled model.

2. Methodology

2.1. Study Area

Karkheh River Basin, with a total area of about 51,000 km^2, is located in the south-western part of Iran between 30° N to 35° N and 46° E to 49° E. KRB is the third largest agricultural river basin in Iran [41] with a significant hydropower generation capacity. The southern part of the basin receives an average annual precipitation of about 250 mm·year^{-1}, whereas the northern part receives up to 700 mm·year^{-1} [46]. During the period 2006–2010, the average annual precipitation of the southern part decreased to 150 mm·year^{-1} [47]. Precipitation in many regions is insufficient to meet crop water requirements, therefore irrigation is very important in LKRB [42,48–50]. The LKRB has been selected for water productivity analysis in our study. The Karkheh Reservoir, in the most downstream part of the basin, is the largest reservoir in the basin, and is operated for irrigation and hydropower. Table 1 presents the characteristics of Karkheh Dam operation, which are considered in our model. LKRB has two major agricultural production systems. The rainfed system, which is dominant in Dashte Abbas and Dolsagh, and the fully irrigated areas, which are scattered in all five regions [44]. The average annual rainfall (2005–2010) in LKRB has recently been as low as 150 mm·year^{-1} [51]. Over the past three decades, large rainfed areas have turned into irrigated areas mainly because of increasing access to water (mainly groundwater). However, irrigation efficiencies in KRB are still low as 35%–50% [48,51]. The productivity of water is very low, i.e., 0.5 kg·m^{-3} for most of the field crops [41,46]. The total irrigated area in LKRB is 360,000 ha with a planned expansion to 500,000 ha [44]. Major crops such as wheat and maize are grown over 55% of the area [44,48].

Table 1. Characteristics of Karkheh dam in Karkheh River Basin.

Dam Name	Status	Normal Level (Meter above Sea Level) m.a.s.l	Storage (Miliion Cubic Meter) MCM	Purpose
Karkheh	Operational	220	4616	Irrigation and hydropower

The LKRB comprises of five major agricultural regions, i.e., Dashte Abbas, Dolsagh, Arayez, Hamidiyeh, and Azadegan (Figure 1). The distribution of wheat and maize in these five regions is given in Table 2. The spatial distribution of the main gauge stations for calibration and validation in the basin is also presented in Figure 1.

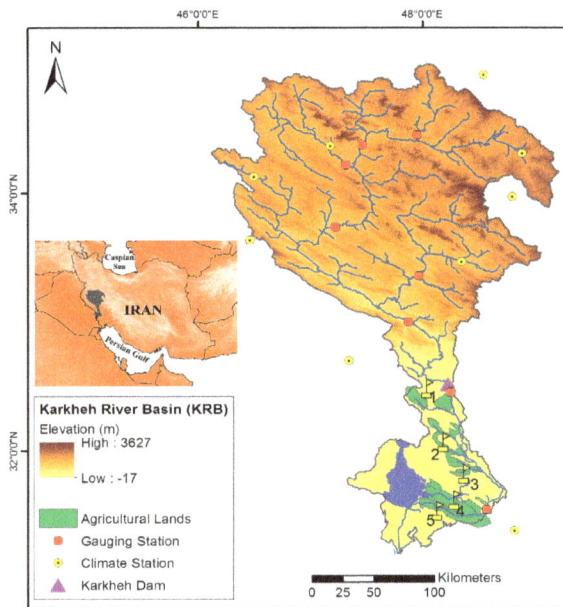

Figure 1. The five important agricultural regions in lower Karkheh River Basin: 1—Dashte Abbas, 2—Dolsagh, 3—Arayez, 4—Hamidiyeh, 5—Azadegan.

Table 2. Distribution of wheat and maize in five major agricultural lands in Lower Karkheh reported by Iran Water and Power Resources Development Co. (2010).

Agricultural Land	Total Area (ha)	Irrigated Wheat Area (ha)	Rainfed Wheat Area (ha)
Dashte Abbas	19,025	9720	2100
Dolsagh	16,133	6320	4200
Arayez	28,900	11,200	4300
Hamidiyeh	17,050	12,840	1100
Azadegan	71,093	50,050	7100

2.2. Description of SWAT Model

Soil and Water Assessment Tool (SWAT) is a continuous time, process-based, semi-distributed, hydrologic model running on daily or sub-daily time steps. The model has been developed to quantify the impact of land management practices and climate on water, sediment, and agricultural chemical yields in large complex watersheds with varying soils, land uses, and management conditions over long periods of time. The program, therefore, lends itself easily to climate and land use change analyses. In SWAT, the spatial heterogeneity of the watershed is preserved by topographically dividing the basin into multiple subbasins, and further into hydrologic response units (HRU) based on soil, land use, and slope characteristics. These subdivisions enable the model to reflect differences in evapotranspiration for various crops and soils. In each HRU and on each time step, the hydrologic and vegetation-growth processes are simulated based on the curve number or Green-Ampt rainfall-runoff partitioning and the heat unit phenological development method.

2.3. SWAT Model Calibration, Validation, and Uncertainty Analysis

Sensitivity analysis, calibration, validation and uncertainty analysis of SWAT is performed using river discharge as well as wheat and maize historical yield data by utilizing the SUFI-2 algorithm [28,52] in the SWAT-CUP software package [53]. This algorithm maps all uncertainties (parameter, conceptual model, input, etc.) on the parameters, expressed as uniform distributions or ranges, and attempts to capture most of the measured data within the model's 95% prediction uncertainty (95PPU) in an iterative process. The 95PPU is calculated at the 2.5% and 97.5% levels of the cumulative distribution of an output variable obtained through Latin hypercube sampling. For the goodness of fit, as we are comparing two bands (the 95PPU for model simulation and the band representing measured data plus its error), two indices referred to as P-factor and R-factor are used [52]. The P-factor is the fraction of measured data (plus its error) bracketed by the 95PPU band and varies from 0 to 1, where 1 indicates 100% bracketing of the measured data within model prediction uncertainty, i.e., a perfect model simulation). The quantity (1-P-factor) could hence be referred to as the model error. For discharge, a value of >0.7 or >0.75 has been reported to be adequate [28,52]. This depends on the scale of the project and adequacy and precision of historical data. The R-factor, on the other hand, is the ratio of the average width of the 95PPU band and the standard deviation of the measured variable. A value of <1.5, again depending on the situation, would be desirable for this index [28,52]. These two indices are used to judge the strength of the calibration/validation and predictive uncertainty. A larger P-factor can be achieved at the expense of a larger R-factor. Hence, often, a balance must be reached between the two. In the final iteration, where acceptable values of R-factor and P-factor are reached, the parameter ranges are taken as the calibrated parameters. SUFI-2 allows usage of eleven different objective functions such as R^2, Nash–Sutcliff efficiency (NSE), and mean square error (MSE). In this study, we used NSE and percent bias (PBIAS) for discharge [54] and root mean square error for crop yield [31].

2.4. Description of the MODSIM Model

MODSIM is a generic river basin management decision support system, originally conceived in 1978 at Colorado State University, making it the longest continuously maintained river basin management software package currently available [55]. MODSIM represents a river basin as a network of links and nodes. Unregulated inflows, evaporation and channel losses, reservoir storage rights and exchanges, stream–aquifer modeling components, reservoir operating targets, and consumptive and instream flow demands are considered in MODSIM [56]. More details can be found in Labadie [55].

2.5. Model Setup and Data Collection

The soil data was obtained from Food and Agriculture Organization [57]; land use, crop and agricultural management data were from Mahab [51]; the digital elevation model was provided by hole-filled NASA Shuttle Radar Topographic Mission (SRTM) [58]; major local rivers and climate data at nine climate stations in the basin were from Iran Water and Power Resources Development Co [44]. The Karkheh reservoir was included in the model with historical reservoir operation time series data starting from 2000. The data were provided by the Ministry of Energy [59]. Monthly discharge data for eight hydrological stations were provided by the local water authorities. Observed monthly discharge and winter wheat, barley, and maize yields were used for model calibration (1997–2010) and validation (1990–1996). The selection of calibration parameters was based on a sensitivity analysis and past modelling experiences at the same location [28,60]. As a result, 26 parameters were selected for calibrating both discharge (20 parameters) and crop yield (six parameters). The watershed system, river network, and water allocation system in MODSIM are illustrated in Figure 2.

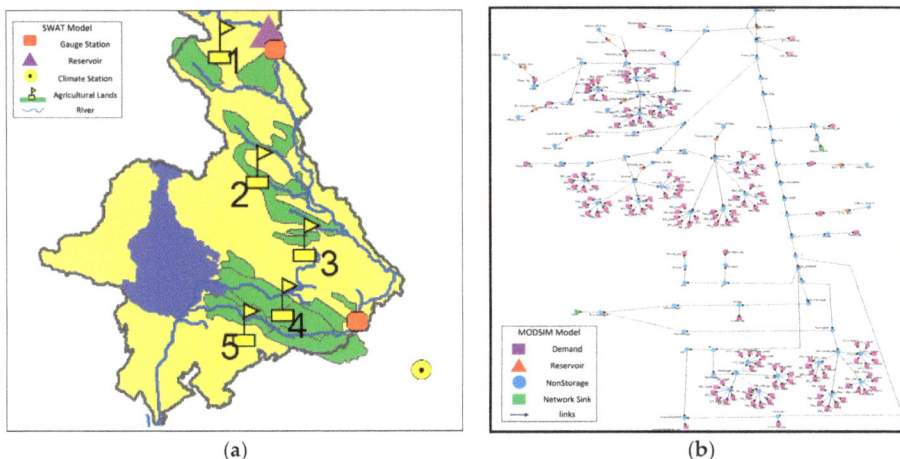

Figure 2. (**a**) LKRB system in the SWAT River network and (**b**) water allocation system in MODSIM.

2.6. Coupling Hydrologic and Water Allocation Models

Optimal water allocation among competing users including hydropower generation is missing in SWAT. The main feature of MODSIM DSS is in allocating available water resources to different users optimally, irrespective of what sources they come from. That is why the idea of coupling SWAT and MODSIM as two powerful tools for modeling both water availability and water allocation (management) is a very attractive idea.

Although there are some studies, which have used both SWAT and MODSIM models [61–63] in watershed modelling, they are not fully linked with feedback and most are not available for use by other researches. In this work, water allocations from the reservoir to different demand sites and the associated spatial units in SWAT (HRUs) are done based on the schedule derived from MODSIM's water allocation solutions obtained by iterative minimum cost network flow programs. Subsequently, net irrigation requirement and inflow to the reservoir from SWAT outputs files (output.rch and output.hru) at the corresponding HRUs and rivers will be extracted and converted as inputs to

MODSIM. Once the amount of water allocated to each demand node (equivalent to HRUs in SWAT) is determined by MODSIM, SWAT is run using new updated-by-MODSIM irrigation scheduling. The quantities of water transferred to different HRUs are estimated for every time step. Figure 3 illustrates the structure of both models considering the unit of data exchanging between them.

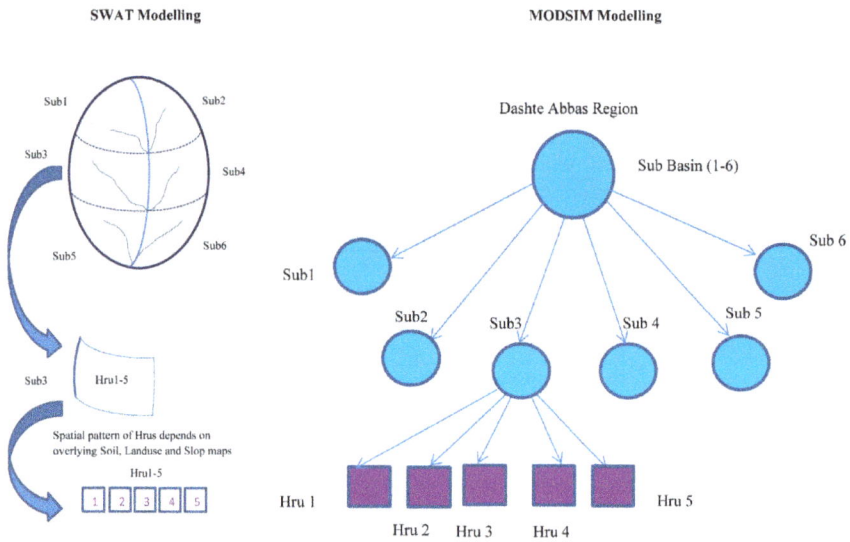

Figure 3. Overview of the input–output and information exchange in the SWAT–MODSIM (SM) model.

The conceptual framework of SWAT–MODSIM execution is illustrated in Figure 4. Here are the eight steps to a successful implementation of the coupled SWAT–MODSIM model:

1   Build the SWAT and MODSIM models for the specific watershed, ensuring that each HRU that receives water should have a related demand node in MODSIM.
2   Calibrate and validate the SWAT model using SWAT-CUP
3   Extract the M95PPU of inflow to the reservoirs (from 95ppu.txt SWAT-CUP or 95ppu_No_Obs.txt files) and net irrigation requirements (water deficit in each time step) by subtracting potential evapotranspiration (PET) from actual evapotranspiration (AET) in the SWAT-CUP output file 95ppu_No_Obs.txt.
4   Import the net irrigation requirement and inflow to the reservoir to MODSIM from SWAT-CUP outputs.
5   Execute the MODSIM model.
6   Extract the allocated water to each demand node for each time step from MODSIM outputs.
7   Import the monthly irrigation from MODSIM into SWAT management files for related HRUs.
8   Re-execute the SWAT-CUP with new management files.

More details can be found in the Supplementary material.

2.7. Estimation of Crop Water Productivity (CWP)

CWP combines physical accounting of water with yield or economic output to indicate the value of a unit of water and can be calculated as:

$$CWP = \frac{Y}{AET} \tag{1}$$

where CWP is the crop water productivity in $kg \cdot m^{-3}$, Y is the crop yield in $kg \cdot ha^{-1}$, and AET is the seasonal actual evapotranspiration in $m^3 \cdot ha^{-1}$, assumed here to be the crop's consumptive water use, so the above definition of CWP does not account for the waste of water due to irrigation inefficiencies. Note that Y is the annual yield while AET is calculated on a monthly basis. The spatial resolution of Y, AET, and CWP is at a subbasin level, but for comparison with other studies and the available statistics, the results are aggregated to the level of agricultural lands.

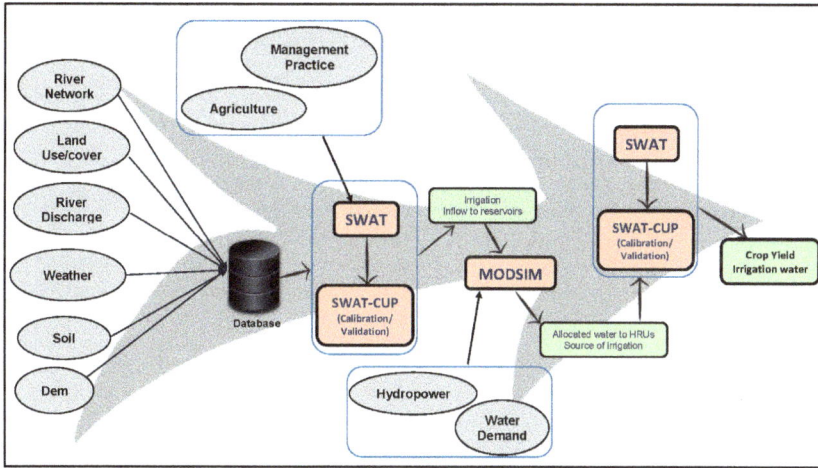

Figure 4. Overview of the input–output and information exchange in the SWAT–MODSIM (SM) model.

3. Results and Discussion

3.1. Calibration and Validation of the Coupled Model for Wheat and Maize

As described in Section 2.3, 26 parameters were selected for calibration and validation based on our previous study and literature sources. In the final iteration, eight parameters were found to be sensitive parameters in our study. In this paper, for the sake of brevity, we only report the results of our analyses for the calibration and validation of crop yields. More details on the calibration and validation of discharge and sensitivity analysis of parameter can be found in Vaghefi, et al. [42]. Calibration and validation tasks were done based on the execution of steps described in Section 2.6. At first, auto-irrigation with an unlimited source of water was used as a source of irrigation for the agricultural region in LKRB to find the maximum amount of irrigation, which is needed at the HRU level for each time step. After estimation of the net irrigation requirements and inflow to Karkheh Reservoir, MODSIM was run. Finally, the management files of agricultural regions in LKRB were updated considering MODSIM results for irrigation scheduling and the actual crop yields were obtained by re-running of SWAT calibration by SWAT-CUP. Using this sequential procedure, the calibration and validation results of the SWAT model improved considerably (Table 3, Figures 5 and 6). The results show that observed yields are generally inside or quite close to predicted yield bands for both wheat and maize. For irrigated wheat, the yield varies from 1850 to 3900 $kg \cdot ha^{-1}$, with the highest yield found in the Hamidiyeh (2007 $kg \cdot ha^{-1}$) region and the lowest in the Dashte Abbas (1990 $kg \cdot ha^{-1}$). For the irrigated maize, the lowest yield belongs to the Dolsagh region (2900 $kg \cdot ha^{-1}$) and the highest to the Hamidiyeh region (7200 $kg \cdot ha^{-1}$). For the irrigated wheat, the P-factors are generally larger than 0.77 for calibration and vary from 0.73 to 0.86 for the validation period (Table 3).

The R-factor values are also in acceptable ranges. For the irrigated maize production, the uncertainties are larger than the irrigated wheat as indicated by generally larger R-factor values. This is because of the higher sensitivity of maize production to the water stress than wheat.

Table 3. Calibration (1997–2010) and validation (1990–1996) results of the coupled SWAT-MODSIM model for irrigated wheat and maize using the SMS model in LKRB.

Agricultural Region	Calibration				Validation			
	P-Factor	R-Factor	P-Factor	R-Factor	P-Factor	R-Factor	P-Factor	R-Factor
	Wheat	Wheat	Maize	Maize	Wheat	Wheat	Maize	Maize
Dashte Abbas	0.77	0.21	0.79	0.65	0.76	0.34	0.65	0.59
Dolsagh	0.85	0.39	0.86	0.65	0.75	0.25	0.61	0.71
Arayez	0.83	0.43	0.79	0.69	0.74	0.27	0.61	0.73
Hamidiyeh	0.78	0.29	0.73	0.42	0.73	0.36	0.68	0.87
Azadegan	0.84	0.23	0.76	0.43	0.73	0.29	0.69	0.89

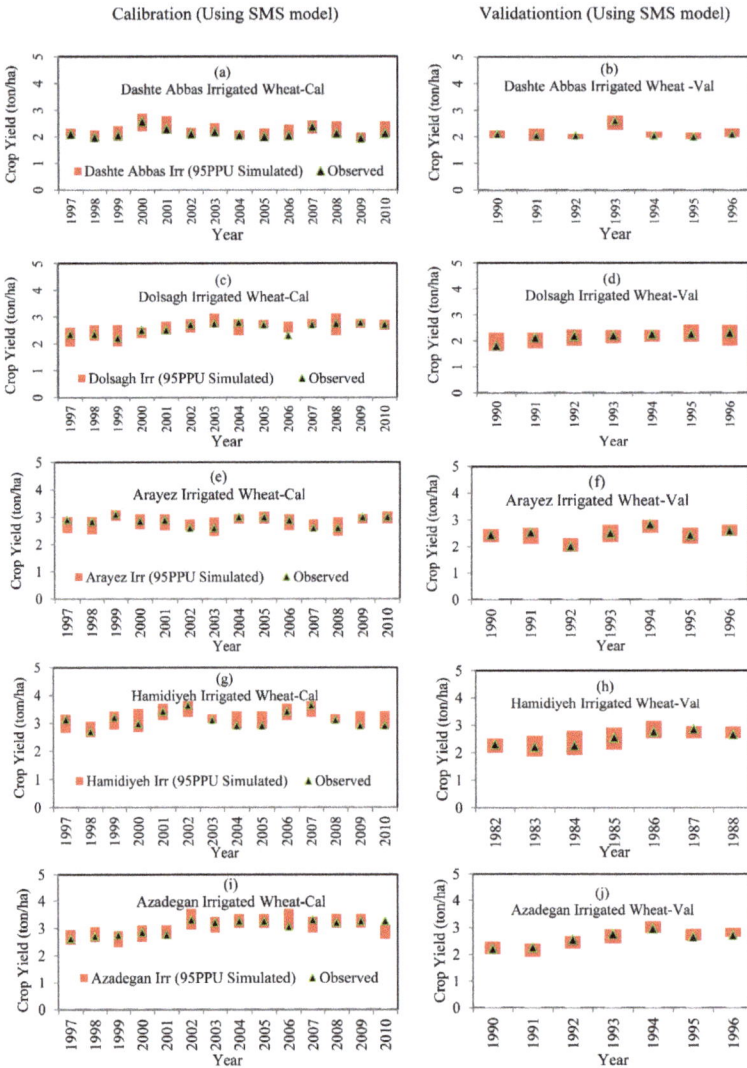

Figure 5. Results of the SWAT calibration and validation for wheat yield in Dashte Abbas (**a,b**), Dolsagh (**c,d**), Arayez (**e,f**), Hamidiyeh (**g,h**), Azadegan (**i,j**) plains.

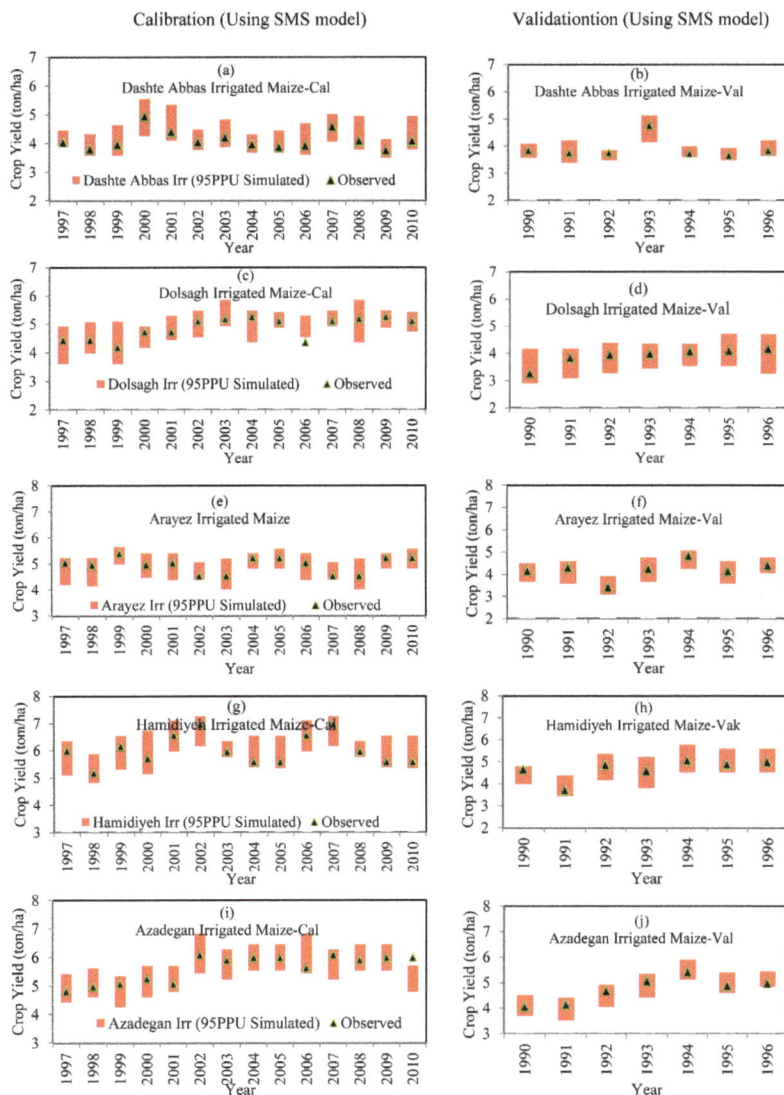

Figure 6. Results of the SMS calibration and validation for maize yield in Dashte Abbas (**a**,**b**), Dolsagh (**c**,**d**), Arayez (**e**,**f**), Hamidiyeh (**g**,**h**), Azadegan (**i**,**j**) plains.

3.2. Water Productivity of Wheat and Maize

The results of the coupled model for both calibration and validation periods and for the entire region indicates that the basin-wide wheat water productivity (WWP) is equal to 0.94 kg·m^{-3}, ranging from 0.55 kg·m^{-3} to 1.21 kg·m^{-3}. The highest WWP can be ascribed to higher yields under limited water supply conditions. Lower WWP is mainly due to higher water application and relatively lower wheat yields (Figure 7a). The basin-wide maize water productivity (MWP) is equal to 0.8 kg·m^{-3}, ranging from 0.55 kg·m^{-3} to 1.15 kg·m^{-3} (Figure 7b).

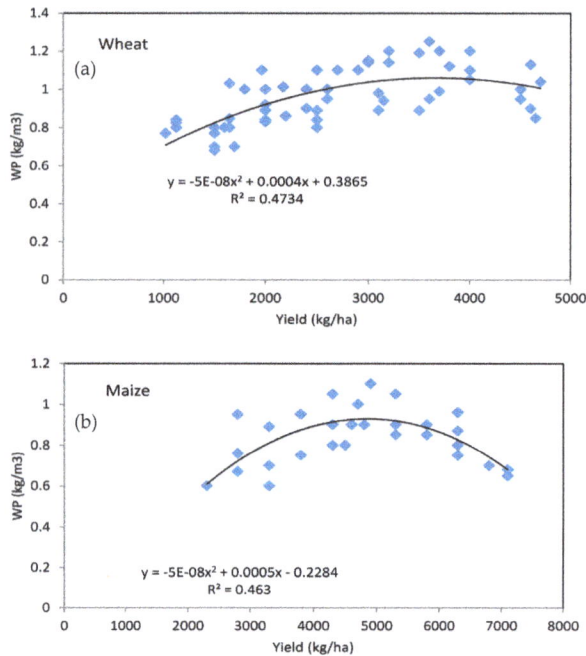

Figure 7. Relationship between yield and irrigation water applied for wheat (**a**) and maize (**b**).

3.3. Yield-Irrigation Water Relations

The relation between wheat and maize yield and irrigation water applied for both calibration and validation periods is presented in Figure 8. Data points of all regions from 1990 to 2010 for irrigated wheat and maize are used in this illustration. One can observe from the figure that wheat yields vary from 1.3 ton·ha^{-1} to 3.5 ton·ha^{-1} with an average of 2.5 ton·ha^{-1} for irrigated wheat, and from 2.1 to 7.2 ton·ha^{-1} with an average of 5 ton·ha^{-1} for maize. The irrigation water applied to the agricultural regions is summarized in Table 4. Irrigation water varies from 2300 m^3·ha^{-1} to 6662 m^3·ha^{-1} and 4320 m^3·ha^{-1} to 10,200 m^3·ha^{-1} for irrigated wheat and maize respectively. The variation of irrigation water applied is from 200 mm to 600 mm for irrigated wheat and from 400 to 1450 mm for maize (Figure 8a,b).

Table 4. Results of the SM model for irrigation water applied to wheat and maize (m^3·ha^{-1}) in LKRB.

Agricultural Region	Wheat			Maize		
	I_{max}	I_{min}	I_{avg}	I_{max}	I_{min}	I_{avg}
Dashte Abbas	5980	2300	3120	8500	4320	6280
Dolsagh	6662	2563	3476	9469	4812	6996
Arayez	5560	3210	4150	9300	6340	9020
Hamidiyeh	5184	3500	4230	10,200	6800	9100
Azadegan	5890	4127	4690	9320	5890	8910

There is a positive relation between Y and CWP for both maize and wheat. There is a sharper increase in WWP in response to increasing yield compared with maize. This suggests that a unit increase in water results in a larger additional yield in wheat than irrigated maize, leading to a greater improvement in CWP. It means that wheat yield is more responsive to additional water. This result is in agreement with what is reported by [64]; they have found that for the smaller yield range, less incremental water is required to increase a unit of crop yield.

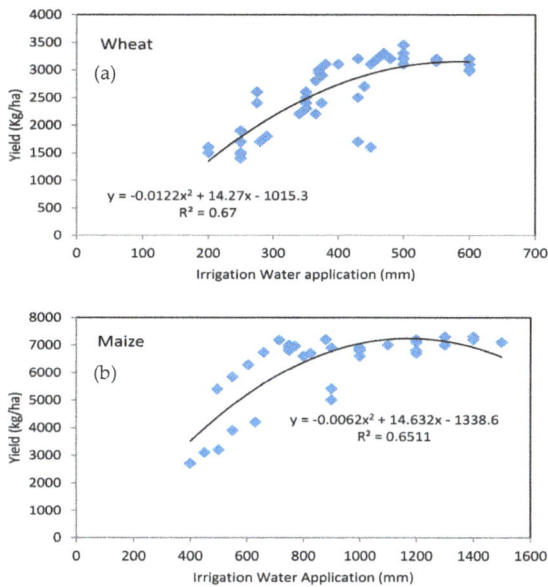

Figure 8. Relationship between crop water productivity and crop yield for wheat (**a**) and maize (**b**).

4. Summary and Conclusions

In this study, crop water productivity of LKRB was assessed using the coupled SWAT–MODSIM model. The time series of actual irrigation demands of agricultural regions was dynamically simulated by the SWAT model and fed into the MODSIM water allocation model. Through an iterative procedure, the irrigation operation of SWAT was updated based on allocated water by MODSIM. Implementation of the coupled model improved the calibration and validation of Y and simulation of AET and CWP. The P-factors in the coupled models are generally larger than 0.77 for calibration and vary from 0.73 to 0.86 for the validation period.

The analysis showed that there are considerable differences in crop yields and productivity of water in irrigated areas of the five agricultural regions of LKRB. The variation of irrigation water applied was from 200 mm to 600 mm for irrigated wheat, and from 400 to 1450 mm for maize. The results showed that basin-wide WWP is equal to 0.94 kg·m^{-3} and MWP is equal to 0.8 kg·m^{-3}. The results suggested that higher water consumption does not necessarily result in a higher yield.

Supplementary Materials: The following are available online at www.mdpi.com/2073-4441/9/3/157/s1, a comprehensive user manual of coupled SWAT-MODSIM model. The Software is freely available for download from our web page: www.2w2e.com.

Acknowledgments: The first author acknowledges the financial support of Karim Abbaspour and Raghavan Srinivasan.

Author Contributions: Saeid Ashraf Vaghefi, Karim C. Abbaspour, and Raghavan Srinivasan conceived and designed the experiments. Karim C. Abbaspour and Raghavan Srinivasan managed the project. Saeid Ashraf Vaghefi prepared the couple SWAT-MODSIM model and wrote the paper with assistance from Karim C. Abbaspour, Monireh Faramarzi, Raghavan Srinivasan, Jeffery G. Arnold. Raghavan Srinivasan, Jeffery Arnold Vaghefi, Monireh Faramarzi, and Karim C. Abbaspour advised on conceptual and technical, and contributed to the strategy.

Conflicts of Interest: The authors declare no conflict of interest.

References

1. Food and Agriculture Organization of the United Nations (FAO). *Crop Yield Response to Water*; Steduto, P., Hsiao, T.C., Fereres, E., Raes, D., Eds.; Food and Agriculture Organization of the United Nations: Rome, Italy, 2012.
2. World Water Assessment Programme (WWAP). *The United Nations World Water Development Report 4: Managing Water under Uncertainty and Risk*; World Water Assessment Programme: Paris, France, 2012.
3. Keller, A.; Sakthivadivel, R.; Seckler, D. *Water Scarcity and the Role of Storage in Development*; International Water Management Institute (IWMI): Colombo, Sri Lanka, 2010.
4. Zwart, S.J.; Bastiaanssen, W.G.M. Review of measured crop water productivity values for irrigated wheat, rice, cotton and maize. *Agric. Water Manag.* **2004**, *69*, 115–133. [CrossRef]
5. Molden, D.; Oweis, T.; Steduto, P.; Bindraban, P.; Hanjra, M.A.; Kijne, J. Improving agricultural water productivity: Between optimism and caution. *Agric. Water Manag.* **2010**, *97*, 528–535. [CrossRef]
6. Sarwar, A.; Bastiaanssen, W.G.M. Long-term effects of irrigation water conservation on crop production and environment in semiarid areas. *J. Irrig. Drain. Eng. ASCE* **2001**, *127*, 331–338. [CrossRef]
7. Nhamo, L.; Mabhaudhi, T.; Magombeyi, M. Improving water sustainability and food security through increased crop water productivity in malawi. *Water* **2016**, *8*, 411. [CrossRef]
8. Giménez, L.; Petillo, M.; Paredes, P.; Pereira, L. Predicting maize transpiration, water use and productivity for developing improved supplemental irrigation schedules in western Uruguay to cope with climate variability. *Water* **2016**, *8*, 309. [CrossRef]
9. Borrego-Marín, M.; Gutiérrez-Martín, C.; Berbel, J. Water productivity under drought conditions estimated using SEEA-water. *Water* **2016**, *8*, 138. [CrossRef]
10. Kima, A.; Chung, W.; Wang, Y.-M. Improving irrigated lowland rice water use efficiency under saturated soil culture for adoption in tropical climate conditions. *Water* **2014**, *6*, 2830. [CrossRef]
11. Roudier, P.; Sultan, B.; Quirion, P.; Berg, A. The impact of future climate change on West African crop yields: What does the recent literature say? *Glob. Environ. Chang.* **2011**, *21*, 1073–1083. [CrossRef]
12. Mo, X.; Liu, S.; Lin, Z.; Xu, Y.; Xiang, Y.; McVicar, T.R. Prediction of crop yield, water consumption and water use efficiency with a SVAT-crop growth model using remotely sensed data on the North China Plain. *Ecol. Model.* **2005**, *183*, 301–322. [CrossRef]
13. Singh, R.; van Dam, J.C.; Feddes, R.A. Water productivity analysis of irrigated crops in sirsa district, India. *Agric. Water Manag.* **2006**, *82*, 253–278. [CrossRef]
14. Liu, J.G. A gis-based tool for modelling large-scale crop-water relations. *Environ. Model. Softw.* **2009**, *24*, 411–422. [CrossRef]
15. Aggarwal, P.K.; Banerjee, B.; Daryaei, M.G.; Bhatia, A.; Bala, A.; Rani, S.; Chander, S.; Pathak, H.; Kalra, N. Infocrop: A dynamic simulation model for the assessment of crop yields, losses due to pests, and environmental impact of agro-ecosystems in tropical environments. II. Performance of the model. *Agric. Syst.* **2006**, *89*, 47–67. [CrossRef]
16. Aggarwal, P.K.; Kalra, N.; Chander, S.; Pathak, H. Infocrop: A dynamic simulation model for the assessment of crop yields, losses due to pests, and environmental impact of agro-ecosystems in tropical environments. I. Model description. *Agric. Syst.* **2006**, *89*, 1–25. [CrossRef]
17. Vanuytrecht, E.; Raes, D.; Steduto, P.; Hsiao, T.C.; Fereres, E.; Heng, L.K.; Garcia Vila, M.; Mejias Moreno, P. Aquacrop: Fao's crop water productivity and yield response model. *Environ. Model. Softw.* **2014**, *62*, 351–360. [CrossRef]
18. Steduto, P.; Raes, D.; Hsiao, T.C.; Fereres, E.; Heng, L.K.; Howell, T.A.; Evett, S.R.; Rojas-Lara, B.A.; Farahani, H.J.; Izzi, G.; et al. Concepts and applications of AquaCrop: The FAO crop water productivity model. In *Crop Modeling and Decision Support*; Cao, W., White, J.W., Wang, E., Eds.; Springer: Berlin/Heidelberg, Germany, 2009; pp. 175–191.
19. Raes, D.; Steduto, P.; Hsiao, T.C.; Fereres, E. AquaCrop—The FAO Crop model to simulate yield response to water: II. Main algorithms and software description all rights reserved. No part of this periodical may be reproduced or transmitted in any form or by any means, electronic or mechanical, including photocopying, recording, or any information storage and retrieval system, without permission in writing from the publisher. *Agron. J.* **2009**, *101*, 438–447.
20. Alcamo, J.; DÖLl, P.; Henrichs, T.; Kaspar, F.; Lehner, B.; RÖSch, T.; Siebert, S. Development and testing of the WaterGap 2 global model of water use and availability. *Hydrol. Sci. J.* **2003**, *48*, 317–337. [CrossRef]

21. Li, T.; Hasegawa, T.; Yin, X.; Zhu, Y.; Boote, K.; Adam, M.; Bregaglio, S.; Buis, S.; Confalonieri, R.; Fumoto, T.; et al. Uncertainties in predicting rice yield by current crop models under a wide range of climatic conditions. *Glob. Chang. Biol.* **2015**, *21*, 1328–1341. [CrossRef] [PubMed]

22. Challinor, A.J.; Wheeler, T.R. Crop yield reduction in the tropics under climate change: Processes and uncertainties. *Agric. For. Meteorol.* **2008**, *148*, 343–356. [CrossRef]

23. Lobell, D.B.; Cassman, K.G.; Field, C.B. Crop yield gaps: Their importance, magnitudes, and causes. *Annu. Rev. Environ. Resour.* **2009**, *34*, 179–204. [CrossRef]

24. Iizumi, T.; Yokozawa, M.; Nishimori, M. Parameter estimation and uncertainty analysis of a large-scale crop model for paddy rice: Application of a Bayesian approach. *Agric. For. Meteorol.* **2009**, *149*, 333–348. [CrossRef]

25. Arnold, J.G.; Srinivasan, R.; Muttiah, R.S.; Williams, J.R. Large area hydrologic modeling and assessment—Part 1: Model development. *J. Am. Water Resour. Assoc.* **1998**, *34*, 73–89. [CrossRef]

26. Gassman, P.W.; Reyes, M.R.; Green, C.H.; Arnold, J.G. The soil and water assessment tool: Historical development, applications, and future research directions. *Trans. ASABE* **2007**, *50*, 1211–1250. [CrossRef]

27. Arnold, J.G.; Moriasi, D.N.; Gassman, P.W.; Abbaspour, K.C.; White, M.J.; Srinivasan, R.; Santhi, C.; Harmel, R.D.; van Griensven, A.; Van Liew, M.W.; et al. Swat: Model use, calibration, and validation. *Trans. ASABE* **2012**, *55*, 1491–1508. [CrossRef]

28. Abbaspour, K.C.; Yang, J.; Maximov, I.; Siber, R.; Bogner, K.; Mieleitner, J.; Zobrist, J.; Srinivasan, R. Modelling hydrology and water quality in the pre-ailpine/alpine Thur watershed using SWAT. *J. Hydrol.* **2007**, *333*, 413–430. [CrossRef]

29. Schuol, J.; Abbaspour, K.C.; Yang, H.; Srinivasan, R.; Zehnder, A.J.B. Modeling blue and green water availability in Africa. *Water Resour. Res.* **2008**, *44*. [CrossRef]

30. Abbaspour, K.C.; Rouholahnejad, E.; Vaghefi, S.; Srinivasan, R.; Yang, H.; Klove, B. A continental-scale hydrology and water quality model for Europe: Calibration and uncertainty of a high-resolution large-scale SWAT model. *J. Hydrol.* **2015**, *524*, 733–752. [CrossRef]

31. Faramarzi, M.; Yang, H.; Schulin, R.; Abbaspour, K.C. Modeling wheat yield and crop water productivity in Iran: Implications of agricultural water management for wheat production. *Agric. Water Manag.* **2010**, *97*, 1861–1875. [CrossRef]

32. Babel, M.S.; Gupta, A.D.; Nayak, D.K. A model for optimal allocation of water to competing demands. *Water Resour. Manag.* **2005**, *19*, 693–712. [CrossRef]

33. Perera, B.J.C.; James, B.; Kularathna, M.D.U. Computer software tool REALM for sustainable water allocation and management. *J. Environ. Manag.* **2005**, *77*, 291–300. [CrossRef] [PubMed]

34. Yates, D.; Sieber, J.; Purkey, D.; Huber-Lee, A. Weap21—A demand-, priority-, and preference-driven water planning model part 1: Model characteristics. *Water Int.* **2005**, *30*, 487–500. [CrossRef]

35. Labadie, J.W. *Modsim: River Basin Network Flow Model for Conjunctive Stream-Aquifer Management. Program User Manual and Documentation*; Colorado State University: Boulder, CO, USA, 1995.

36. Mousavi, S.J.; Shourian, M. Adaptive sequentially space-filling metamodeling applied in optimal water quantity allocation at basin scale. *Water Resour. Res.* **2010**, *46*. [CrossRef]

37. Shourian, M.; Mousavi, S.J.; Tahershamsi, A. Basin-wide water resources planning by integrating PSO algorithm and MODSIM. *Water Resour. Manag.* **2008**, *22*, 1347–1366. [CrossRef]

38. De Azevedo, L.G.T.; Gates, T.K.; Fontane, D.G.; Labadie, J.W.; Porto, R.L. Integration of water quantity and quality in strategic river basin planning. *J. Water Resour. Plan. Manag. ASCE* **2000**, *126*, 85–97. [CrossRef]

39. Dai, T.W.; Labadie, J.W. River basin network model for integrated water quantity/quality management. *J. Water Resour. Plan. Manag. ASCE* **2001**, *127*, 295–305. [CrossRef]

40. Ahmad, M.U.D.; Giordano, M. The Karkheh River Basin: The food basket of Iran under pressure. *Water Int.* **2010**, *35*, 522–544. [CrossRef]

41. Qureshi, A.S.; Oweis, T.; Karimi, P.; Porehemmat, J. Water productivity of irrigated wheat and maize in the Karkheh River Basin of Iran. *Irrig. Drain.* **2010**, *59*, 264–276. [CrossRef]

42. Vaghefi, S.A.; Mousavi, S.J.; Abbaspour, K.C.; Srinivasan, R.; Yang, H. Analyses of the impact of climate change on water resources components, drought and wheat yield in semiarid regions: Karkheh River Basin in Iran. *Hydrol. Process.* **2014**, *28*, 2018–2032. [CrossRef]

43. Rafiee, V.; Shourian, M. Optimum multicrop-pattern planning by coupling swat and the harmony search algorithm. *J. Irrig. Drain. Eng.* **2016**, *142*. [CrossRef]

44. Iran Water and Power Resources Development Co. *Systematic Studies of Karkheh River Basin*; IWPR: Tehran, Iran, 2010. (In Persian)

45. Marjanizadeh, S.; de Fraiture, C.; Loiskandl, W. Food and water scenarios for the Karkheh River Basin, Iran. *Water Int.* **2010**, *35*, 409–424. [CrossRef]

46. Oweis, T.; Abbasi, F.; Siadat, H. *Improving on-Farm Agricultural Water Productivity in the Karkheh River Basin (PN08)*; Department for International Development: Chatham, UK, 2009.

47. Fisher, M.J.; Cook, S.E. *Water, Food and Poverty in River Basins, Defining the Limits*; Routledge: London, UK, 2012; p. 406.

48. Keshavarz, A.; Ashraft, S.; Hydari, N.; Pouran, M.; Farzaneh, E.A.; Natl Acad, P. *Water Allocation and Pricing in Agriculture of Iran*; National Academies Press: Washington, DC, USA, 2005; pp. 153–172.

49. Farahani, H.; Oweis, T. *Chapter I Agricultural Water Productivity in Karkheh River Basin. Improving on-Farm Agricultural water Productivity in the Karkheh River Basin*; Research Report No. 1: A Compendium of Review Papers; ICARDA: Aleppo, Syria, 2008; p. 103.

50. Ahmad, M.-U.-D.; Islam, M.A.; Masih, I.; Muthuwatta, L.; Karimi, P.; Turral, H. Mapping basin-level water productivity using remote sensing and secondary data in the Karkheh River Basin, Iran. *Water Int.* **2009**, *34*, 119–133. [CrossRef]

51. MahabGhods Consulting Engineer Co. *Systematic Planning of Karkheh Watershed: Studying the Water Consumption of Demand Sectors*; MahabGhods Consulting Engineer Co.: Tehran, Iran, 2009. (In Persian)

52. Abbaspour, K.C.; Johnson, C.A.; van Genuchten, M.T. Estimating uncertain flow and transport parameters using a sequential uncertainty fitting procedure. *Vadose Zone J.* **2004**, *3*, 1340–1352. [CrossRef]

53. Abbaspour, K.C. *User Manual for Swat-Cup, Swat Calibration and Uncertainty Analysis Programs*; Swiss Federal Institute of Aquatic Science and Technology, Eawag: Duebendorf, Switzerland, 2011; p. 103.

54. Moriasi, D.N.; Arnold, J.G.; van Liew, M.W.; Bingner, R.L.; Harmel, R.D.; Veith, T.L. Model evaluation guidelines for systematic quantification of accuracy in watershed simulations. *Trans. ASABE* **2007**, *50*, 885–900. [CrossRef]

55. Labadie, J.W. *River Basin Network Model for Water Rights Planning, Modsim: Technical Manual, Department of Civil Engineering*; Colorado State University: Fort Collins, CO, USA, 2010.

56. Fredericks, J.W.; Labadie, J.W.; Altenhofen, J.M. Decision support system for conjunctive stream-aquifer management. *J. Water Resour. Plan. Manag. ASCE* **1998**, *124*, 69–78. [CrossRef]

57. Food and Agriculture Organization (FAO). *The Digital Soil Map of the World and Derived Soil Properties*; FAO: Rome, Iatly, 1995.

58. Jarvis, A.; Reuter, H.I.; Nelson, A.; Guevara, E. *Hole-Filled Srtm for the Globe Version 4*; Available from the CGIAR-CSI Srtm 90 m Database; CGIAR-CSI: Washington, DC, USA, 2008.

59. Ministry of Energy of Iran. *An Overview of National Water Planning of Iran*; Ministry of Energy of Iran: Tehran, Iran, 1998. (In Persian)

60. Faramarzi, M.; Abbaspour, K.C.; Schulin, R.; Yang, H. Modelling blue and green water resources availability in Iran. *Hydrol. Process.* **2009**, *23*, 486–501. [CrossRef]

61. Vaghefi, S.A.; Mousavi, S.J.; Abbaspour, K.C.; Srinivasan, R.; Arnold, J.R. Integration of hydrologic and water allocation models in basin-scale water resources management considering crop pattern and climate change: Karkheh River Basin in Iran. *Reg. Environ. Chang.* **2015**, *15*, 475–484. [CrossRef]

62. Chhuon, K.; Herrera, E.; Nadaoka, K. Application of integrated hydrologic and river basin management modeling for the optimal development of a multi-purpose reservoir project. *Water Resour. Manag.* **2016**, *30*, 3143–3157. [CrossRef]

63. Ahn, S.R.; Jeong, J.H.; Kim, S.J. Assessing drought threats to agricultural water supplies under climate change by combining the SWAT and MODSIM models for the Geum River Basin, South Korea. *Hydrol. Sci. J.* **2016**, *61*, 2740–2753. [CrossRef]

64. Rockström, J.; Hatibu, N.; Oweis, T.; Wani, S.P. Managing water in rain-fed agriculture. In *Water for Food, Water for Life: A Comprehensive Assessment of Water Management in Agriculture*; Molden, D., Ed.; Earthscan: London, UK; International Water Management Institute (IWMI): Colombo, Sri Lanka, 2007; pp. 315–348.

water

MDPI

Article

The Impact of Para Rubber Expansion on Streamflow and Other Water Balance Components of the Nam Loei River Basin, Thailand

Winai Wangpimool [1], Kobkiat Pongput [1,*], Nipon Tangtham [2], Saowanee Prachansri [3] and Philip W. Gassman [4]

[1]   Department of Water Resources Engineering, Faculty of Engineer, Kasetsart University,
      Ngam Wong Wan Rd., Lat Yao, Chatuchak, Bangkok 10900, Thailand; winai.wangpimool@gmail.com
[2]   Forestry Research Center, Faculty of Forestry, Kasetsart University, Bangkok 10900, Thailand;
      ffornpt@ku.ac.th
[3]   Land Development Department, Chatuchak, Bangkok 10900, Thailand; saowanee@ldd.go.th
[4]   Center for Agricultural and Rural Development, Iowa State University, Ames, IA 50011-1054, USA;
      pwgassma@iastate.edu
*   Correspondence: kobkiat.p@ku.th; Tel.: +66-080-053-9595

Academic Editor: Karim Abbaspour
Received: 15 October 2016; Accepted: 14 December 2016; Published: 22 December 2016

Abstract: At present, Para rubber is an economical crop which provides a high priced product and is in demand by global markets. Consequently, the government of Thailand is promoting the expansion of Para rubber plantations throughout the country. Traditionally, Para rubber was planted and grown only in the southern areas of the country. However, due to the Government's support and promotion as well as economic reasons, the expansion of Para rubber plantations in the northeast has increased rapidly. This support has occurred without accounting for suitable cultivation of Para rubber conditions, particularly in areas with steep slopes and other factors which have significant impacts on hydrology and water quality. This study presents the impacts of Para rubber expansion by applying the Soil and Water Assessment Tool (SWAT) hydrological model on the hydrology and water balance of the Nam Loei River Basin, Loei Province. The results showed that the displacement of original local field crops and disturbed forest land by Para rubber production resulted in an overall increase of evapotranspiration (ET) of roughly 3%. The major factors are the rubber canopy and precipitation. Moreover, the water balance results showed an annual reduction of about 3% in the basin average water yield, especially during the dry season.

Keywords: hydrologic balance; SWAT model; land use change; evapotranspiration; plant parameters

1. Introduction

Zeigler et al. [1] estimated that over 500,000 ha of upland areas in southeast Asia had been converted to Para rubber (*Heveabrasiliensis*) production in southeast Asia by 2009 and that the land area devoted in the region to Para rubber production could double or triple by the year 2050 [1]. Updated estimates for the same timeframe indicate that the expansion of total rubber production area in non-traditional Southeast Asia growing regions at >1,000,000 ha and that the production area could increase by a factor of four by 2050 [2]. Much of the expansion is occurring in "marginal areas" that are vulnerable to increased soil erosion and other environmental problems [3].

Para rubber has become one of the most important economic crops in Thailand, which is now the largest exporter of Para rubber by volume worldwide [4]. Para rubber production started in southern Thailand over a century ago [5] and has expanded greatly in that region since then due to favorable climatic conditions and land types. However, the government of Thailand has implemented policies to promote the expansion of Para rubber plantations throughout other areas of the country. Due to the Government's support and promotion as well as for economic reasons, the expansion of Para rubber plantations in the northeast has increased dramatically during the past decade. Continuing attractive prices have resulted in particularly rapid expansion of Para rubber plantations during the past few years, resulting an increase of nearly 500,000 ha (246,340 ha to 739,190 ha) between 2006 and 2015 [6–9], confirming earlier projections of greatly expanded production [1]. The government support of Para rubber production in the northeast has occurred without adequate investigation of suitable cultivation conditions. This has resulted in Para rubber production occurring in areas with steep slopes, non-ideal climatic conditions, and other factors which have resulted in significant negative impacts on regional hydrology and water quality.

The northeastern region of Thailand consists of 20 provinces which cover a total area of about 170,226 km^2 or one-third of the country (Figure 1). Forest areas in the region are rapidly becoming degraded due to destruction of existing forest stands. This is occurring because of increased agricultural and Para rubber production to support the rapidly growing population, and burning during the summer to support wild game hunting. At present, the most extreme burning of forests in the country is occurring in north and northeast Thailand [10].

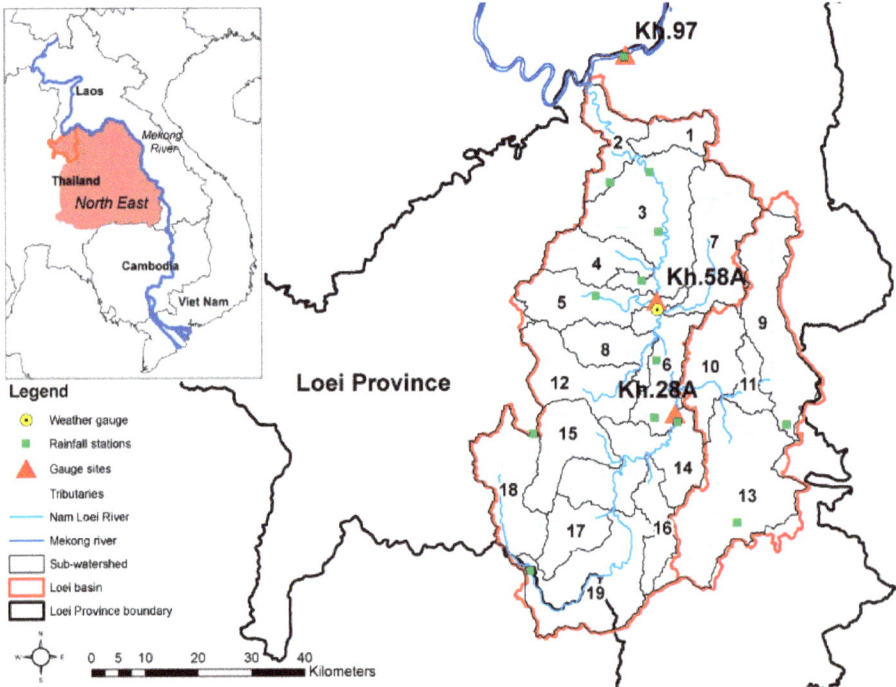

Figure 1. Location of the Nam Loei River Basin (NLRB) within Loei Province and Loei Province within northeast Thailand.

The northeastern region of Thailand (Figure 1) has a total agricultural area of 15.90 million ha, of which 6.65 million ha are suitable for rubber plantations [11]. To date, only a small portion of this potential area has been developed for Para rubber production although projections indicate greatly expanded production in the future. Investigations are urgently needed to determine how expanded rubber production in the northeast will impact environmental conditions in the region, especially rainwater, humidity, soil characteristics, hydrologic balance, flow regime and rock formation. Changes in soil quality can have a strong effect on the amount of drainable water, as well as physical, chemical and biological properties [12]. Decision-makers and planners face difficult challenges in meeting water conservation objectives, and managing the engineering, socioeconomic and environmental aspects of development and planning, related to Para rubber production in northeast Thailand. This is especially true in certain sub regions such as the Nam Loei River Basin (NLRB) in Loei province (Figure 1), where Para rubber production increased from 0.4% to 21.5% of the total land use between 2002 and 2015, resulting in an extremely volatile situation that is impacting the entire watershed. Hence, technical tools including the Soil and Water Assessment Tool (SWAT) watershed-scale water quality model [13–16] are needed to support in-depth hydrologic and environmental assessments of Para rubber production in the region. SWAT has been extensively tested for a wide range of environmental conditions and watershed scales [17–20] and has been used effectively in a number of land use change studies [21–30]. Thus, the specific objectives of this research are to: (1) report the hydrologic impacts of the increased Para rubber production in the NRLB that occurred during 2002 to 2009, and 2009 to 2015; and (2) identification of inappropriate areas for rubber plantation and risks of landslide.

2. Materials and Methods

2.1. Description of Study Area

Loei Province covers 11,424 km^2 in the upper northeastern region of Thailand (Figure 1) and is the fifth largest province in the region. The NLRB drains 3915 km^2 from its combined upper basin and lower basin within Loei Province (Figure 1) and extends 231 km from the upper Phu Luang Range to its outlet. The Nam Puan is the major tributary of the upper basin, which is initially comprised of steep slopes but ultimately flows into a plain area where it joins the Nam Loei River within the Wang Sa Phung District. The main river of the lower basin is the Nam Loei River, which flows through the Muang District to the river plain within the Chiang Khan District to meet the Mekong River.

The average annual long-term rainfall and temperature is 1241 mm and 26.5 degree Celsius, respectively, for the NLRB [31]. The range of monthly average minimum temperatures, maximum temperatures, and precipitation over the 30-year period of 1981 to 2010 are shown in Figure 2 for climate station 48353, which is located in the study region [32]. However extended drought problems have resulted in streamflows of just 5% to 10% and 90% to 95% during the dry and rainy seasons, respectively, relative to annual average streamflow. Eight major groups comprise the spatial extent of soils in the NLRB, with the most dominant being the following three soil types: (1) the Slope complex (Sc) soil group which covers 44.3% of the basin, and represents a soil mixture in steep areas with >30% slopes that are generally characterized by forest, low permeability, and high risk of soil erosion; (2) the Wang hi (Wh) soil group, which covers 16.5% of the basin, represents soils derived from decay of various materials, and are characterized by fine-grain textures and high permeability; and (3) the Chiang Khan (Ch) soil group, which cover 14.4% of the basin, is derived from river sediments, and reflect sedimentary rock weathering and high permeability. Both the Wh and Ch soil types are prone to collapse or landslides in steep areas. The basin is further characterized by minimum, maximum and average elevations of 212 m, 1956 m, and 419 m, respectively. The land use of the basin, based on 2002 land use data [33], can be classified into 14 categories: corn (23.4%), disturbed forest (19.1%), forest–deciduous (12.5%), paddy field (12.3%), orchard (8.5%), sugarcane (5.9%), agricultural land-row crops (5.9%), field crop (4.8%), urban area (4.0%), miscellaneous land (1.7%), plantation (1.0%), water resources (0.4%), rubber tree (0.4%) and planted forest (0.2%). In 2002, the basin had a total Para rubber plantation area of just 762 ha. The plantation area then increased to 68,800 ha by 2009 and later to 100,000 ha by 2013, which met the expected goal established in a Para rubber production strategy for the NLRB [10]. However, Para rubber production has continued to increase since 2013, reaching 129,280 ha by 2015. Currently, the 2016 provincial policy points to an expansion of an additional 100,000 ha of Para rubber plantations in the next five years [34], which will result in almost double the land area currently dedicated to Para rubber production in the NLRB by 2020.

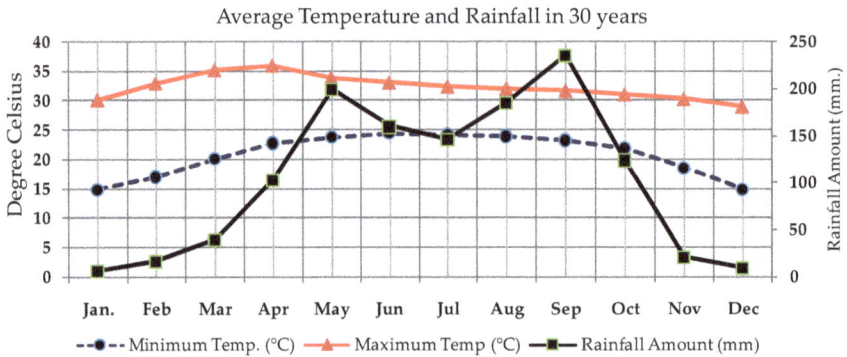

Figure 2. Range of average monthly minimum temperatures, maximum temperatures and precipitation at climate station 48353 during the 30-year period 1981–2010 [30].

2.2. Evaluation of Evapotranspiation (ET) at the Basin Scale

Evapotranspiration (ET) is a collective term that includes all processes by which water at the earth's surface is converted to water vapor. It includes evaporation from plant canopies, and sublimation and evaporation from soil. ET is usually the primary mechanism by which water is removed from a watershed. On average, ET equals about 62% of the precipitation that falls on landscapes across the globe except in Antarctica [35], where runoff exceeds ET. An accurate estimation of ET is critical in assessing the impact of climate and land use changes on water resources [36]. The water losses through ET are significant in the hydrologic cycle; such losses are usually determined by estimating the availability of water through soil moisture or groundwater, the energy and drying power of the air, and/or via land cover and vegetation characteristics [37]. A new method of determining ET for Para rubber trees has been developed in which the rubber tree ET is estimated by accounting for observed patterns of rubber root water uptake as affected by the plant's phrenology [38]. Specifically, the method considers vegetation dynamics and corresponding water needs or evaporative demands. This contrasts with the traditional approach of estimating Para rubber ET, which neglects the increased water use during the dry season when both soil water content and canopy cover are minimal [38]. It is expected here that the SWAT basin-scale hydrologic model will more accurately capture seasonal water balance and ET dynamics that are more consistent with recent research, especially for larger scale rubber expansion situations that exceed smaller stand levels.

2.3. Description of SWAT Model

SWAT is a public domain model jointly developed by the U.S. Department of Agriculture Agricultural Research Service (USDA–ARS) and Texas AgriLife Research, a unit within the Texas A&M University System [15–17]. Watersheds are typically simulated in SWAT by delineating the respective watershed into subbasins and then further subdividing each subbasin into hydrologic response units (HRUs), which are non-spatial land units consisting of homogeneous topographic, soil type, land use, and management characteristics. Hydrologic cycling including precipitation inputs, surface runoff, infiltration into the soil profile, ET, lateral subsurface flow, and flow via other pathways is initially simulated at the HRU level. Nutrient cycling and transport, as well as sediment losses, are also simulated first at the HRU scale. The HRU-level hydrologic and pollutant outputs are then aggregated to the subbasin level and ultimately routed through the stream network to the watershed outlet.

SWAT first simulates atmospheric water demands to calculate the maximum, unstressed ET, commonly referred to as potential ET, before calculating the final actual ET. Three potential ET methods are included in SWAT that vary considerably in the amount of required inputs:

(1) Penman–Monteith [39], which requires solar radiation, air temperature, relative humidity and wind speed; (2) Priestley–Taylor [40], which requires solar radiation, air temperature and relative humidity; and (3) Hargreaves [41], which requires air temperature only [11]. Multiple options are also provided in SWAT for simulating the partitioning of precipitation inputs between surface runoff and infiltration, as well as for some other processes simulated in the model. Complete theoretical and user input options are provided in the SWAT model documentation [36]. A revision of SWAT version 2009 (SWAT2009) was used in conjunction with the ArcGIS SWAT (ArcSWAT) interface for this study [42].

2.4. Application of SWAT

2.4.1. Data Input Needs and Sources

Topographic, soil, land use, climate, and management data are key inputs required for simulating a watershed in SWAT. In-stream monitoring data are also important in regards to testing SWAT output. Topographic, soil, and land use are usually input into SWAT in the form of digital spatial layers that are overlaid within a Geographic Information System (GIS). All of these major data were prepared in input format files for the SWAT simulations including spatial topographic, soil and land use data (Table 1, Figures 3 and 4) obtained from various sources as described below.

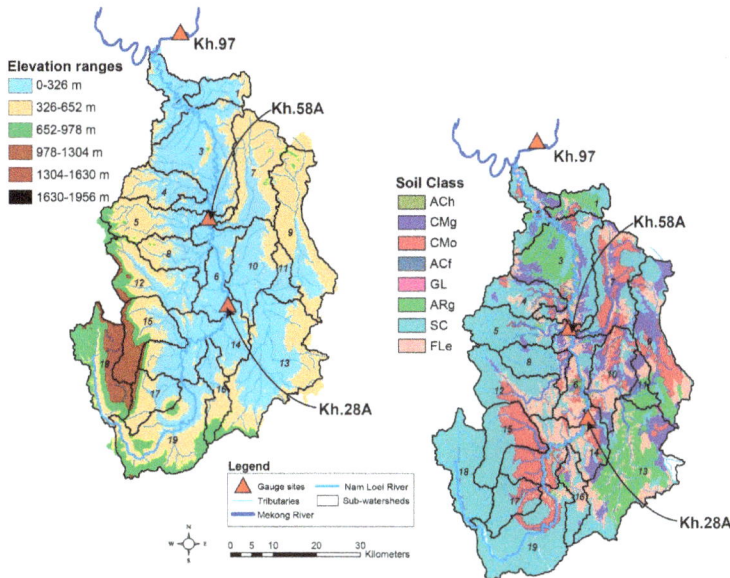

Figure 3. Elevation ranges and distribution of soil types for the Nam Loei River Basin (NLRB), which are based on 1:50,000 Digital Elevation and soil maps, respectively (Table 1).

Spatial topographic data required for the SWAT application were obtained in the form of a digital elevation map (DEM) [43]. These DEM data are characterized by a 30 m × 30 m (1:50,000 scale) resolution. The baseline land use data for year 2002 [44] and the spatial soil map were provided by the Land Development Department [45]. The LDD soil data consist of the previously mentioned 8 major groups. All soil properties required for SWAT were surveyed from by the LDD in 2012 [46]. Daily climate data were used, including precipitation, temperature, solar radiation, wind speed, and humidity data. These data were collected from rainfall gauges within the NLRB during the period from 1985 to 2015 [32]. The daily precipitation data were obtained at 14 gauging stations (Figure 1), while the daily temperature, solar radiation, wind speed and humidity data were collected from a single major weather station (Figure 1) of the Thai Meteorological Department (TMD). Two hydrologic runoff stations located in the basin (gauges Kh.28A and Kh.58A in Figure 1) that are maintained by the Royal Irrigation Department (RID) and have complete monthly runoff data were selected for model calibration and validation [47].

Table 1. Model input data sources for Nam Loei River Basin (NLRB).

	Data Type	Scale	Source [a]
	1. Spatial Data		
1.1	Administrative Data		
	– Administrative boundaries	1:50,000	DWR
	– River layouts	1:50,000	DWR
	– Catchment's boundaries	1:50,000	DWR
	– Drainage network	1:50,000	DWR
1.2	Physical Data		
	– Digital Elevation Model	1:50,000	RTSD
	– Land use/Land Cover	1:50,000	LDD
	– Soils	1:50,000	LDD
	2. Time Series Data		
2.1	Weather Data		
	– Rainfall	14 stations	DWR, RID, TMD
	– Temperature	1 station	TMD
	– Solar radiation	1 station	TMD
	– Wind speed	1 station	TMD
	– Relative humidity	1 station	TMD
	– Evaporation	1 station	TMD
2.2	Hydrological Data		
	– River flow	2 stations	RID

Notes: [a] DWR = Department of Water resources; LDD = Land Development Department; RID = Royal Irrigation Department; RTSD = Royal Thai Survey Department; TMD = Thai Meteorological Department.

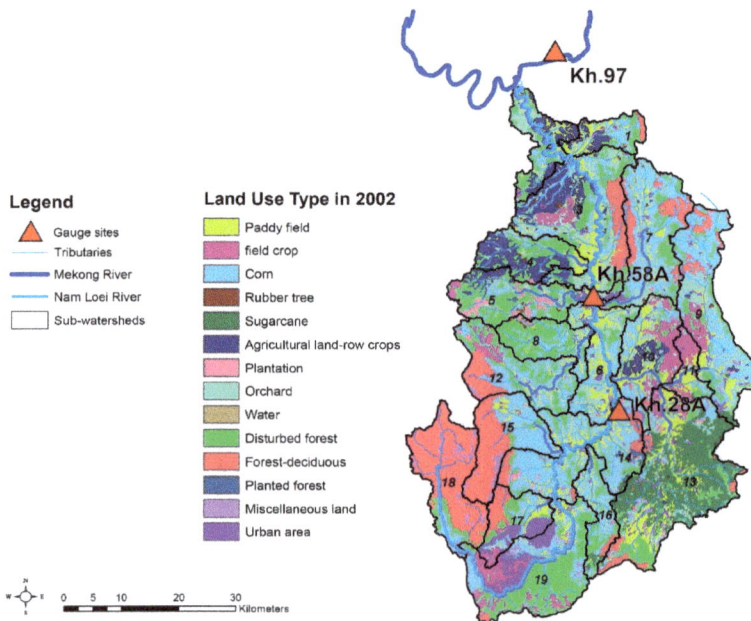

Figure 4. Distribution of land use in 2002 for the Nam Loei River Basin (NLRB) based on 1:50,000 land use/land cover map (Table 1).

2.4.2. Model Set Up

The NLRB was delineated into 19 subbasins and 389 hydrologic response units (HRUs) for the SWAT simulations using the ArcSWAT Interface [42]. The delineation was performed as a function of the DEM-based surface topography, which resulted in the configuration of the 19 sub basins used in the SWAT simulations. The land use data were processed and reclassified to match the land use codes used in SWAT, resulting in the previously described 14 land use categories. The eight major soil types were also converted and reclassified to match the SWAT model soil formatting requirements. These land use and soil data were then overlaid with the DEM data within ArcSWAT to create the HRUs. The required weather data were incorporated or the simulations. The initial curve number values were assigned based on the land use type and soil hydrologic group for the average antecedent moisture condition of the runoff curve number method. The PET was computed by using the Penman–Monteith method. The overall model simulation scenario period covered a 25-year duration from 1985 to 2009 using a daily time step, with a shorter time period used for model calibration and validation as described below.

2.4.3. Sensitivity Analysis and SWAT Calibration and Validation

A sensitivity analysis is a useful procedure to determine which flow-related parameters are the most influential in impacting total streamflow for SWAT applications, following guidance reported for previous studies [16,48–51]. Performing a sensitivity analysis further supports the calibration of SWAT and provides insight for the application of the model to other similar watersheds. In this study, the SWAT CUP software package [52] was used to perform an automatic sensitivity analysis of the impact of 19 different SWAT parameters on daily streamflow flow for the NLRB as described in the Results and Discussion section.

Following the sensitivity analysis, calibration and validation of SWAT was performed which is required to reduce uncertainty and increase confidence in its predictive abilities for the NRLB [16,49,50]. The calibration process included both multisite and multivariable aspects as discussed in previous SWAT studies [16,50]. The calibration procedure consisted of 3 stages: (1) replicating the long-term water balance over the calibration period; (2) accurately tracking the observed hydrograph shapes; and (3) obtaining an accurate comparison between observed and simulated flow duration curves. Calibration and validation of SWAT was performed by comparing the simulated monthly aggregated stream flows versus corresponding measured monthly stream flows at two hydrological gauge stations on the main stem of the Nam Loei River (Figure 1): Wang SaPhung (Kh.28A) in subbasin 14 and Ban FakLoei (Kh.58A) in subbasin 6. Calibration was performed for 1994 to 2004 while validation was conducted from 2005 to 2009. The parameters derived for the gauged catchments were then transferred to the ungauged catchments, based on proximity and similarities in land use and soil types which result in similar hydrological responses.

The accuracy of the model output variance was assessed using the Root Mean Squared Error (RMSE) and Nash–Sutcliffe Efficiency (NSE) statistics [53,54], which are expressed as follows in Equations (1) and (2):

$$\text{RMSE} = \frac{\sum_{i=1}^{n}\left(Q_i^{obs} - \overline{Q_{obs}}\right) \cdot \left(Q_i^{sim} - \overline{Q_{sim}}\right)}{\sqrt{\sum_{i=1}^{n}\left(Q_i^{obs} - \overline{Q_{obs}}\right)^2 \cdot \sum_{i=1}^{n}\left(Q_i^{sim} - \overline{Q_{sim}}\right)^2}} \tag{1}$$

$$\text{NSE} = 1 - \left[\frac{\sum\limits_{i=1}^{n}\left(Q_i^{Obs} - Q_i^{Sim}\right)^2}{\sum\limits_{i=1}^{n}\left(Q_i^{Obs} - Q^{mean}\right)^2}\right] \tag{2}$$

where Q_{obs} are the observed values and Q_{sim} are the simulated values at time/place i.

Values of RMSE can range from 0 to ∞ where a value of 0 indicates a perfect fit between the simulated data and counterpart measured data [53]. The RMSE provides a measure of the difference between the measured and simulated values or residual variance [55]. However, statistical results generated with the RMSE can result in significant model bias during model calibration, even when the error variances are small [56]. Thus it is desirable to use additional statistical evaluation when using the RMSE such as the NSE.

The NSE estimates the magnitude of the simulated variance relative to the measured variance and how accurately a plot of the modeled versus measured data fit a 1:1 line. The NSE can vary from −∞ to 1, where 1 is a perfect fit and a negative value indicates that the average value of the measured data would provide a better prediction than the simulated data. Statistical criteria for judging the success of hydrological modeling results has been suggested [55,57] including an NSE value of at least 0.5 to achieve a satisfactory level for comparisons of aggregate monthly simulation output versus corresponding measured stream flow data. Following successful calibration and validation, the calibrated SWAT model was used to evaluate the Para rubber land use scenario (describe below) including evaluations of ET and water yield.

2.5. Development of Para Rubber Land Use Scenarios

Several factors need to be considered in developing the Para rubber land use scenarios for the NLRB. First, it is important to consider the optimal climatic, soil, slope, and other conditions that Para rubber should be grown under for the study region, especially in the context of typical current practices in which rubber plantations are being established on very high, vulnerable slopes. Second, crop parameters needed to be developed for Para rubber for this analysis. Third, 2002 baseline, 2009 scenario and 2015 scenario landuse layers had to be constructed in order to simulate the impact of expanded Para rubber production in the NLRB during the period of rapid production expansion. These aspects of the Para rubber scenario development are described below.

2.6. Optimal Environmental Conditions for Para Rubber Production

The rubber tree is native to the evergreen tropical rainforests which usually occur within 5° latitude of the equator. The climate of this region is characterized by heavy rainfall and no distinct dry season [5]. The optimal climatic conditions for Para rubber include rainfall of 1250 mm or more, evenly distributed throughout the year with no severe dry season and with 120–150 annual rainy days, and a temperature range of about 26–30 °C [34]. In addition, the ideal elevation range for Para rubber growth is from sea level up to 600 m above mean sea level; the growth rate will decline at higher altitudes. Para rubber can grow on many soils, with the best options being well drained clayey and deep clay soils, but it can withstand physical conditions ranging from stiff clays with poor drainage to well drained sandy loams [5]. The most suitable soil conditions for Para rubber production are: (1) planted at a depth which does not exceed 1 m in depth to allow for adequate future root penetration and growth; (2) soil textures that range between loamy sand to clay loam with good drainage; (3) no gravel or stone in the subsoil layer; and (4) the soil pH should range between 4.5 and 5.5 [34].

The ideal slope range for growing Para rubber trees is between 5% and 15%. Three categories can be classified for this ideal range for the Loei River basin region: (1) low land areas with slopes from 0% to 5%; (2) slopes ranging from 5% to 10% located in the plain areas; and (3) mild slopes that typically range from 10% to 15%. About 30% (117,433 ha) of the total area within the basin region meets these ideal slope characteristics [58]. However, the majority of the current Para rubber production in the study region is currently occurring on much steeper slopes exceeding 15%, and is concentrated especially in an extreme slope range of 30% to 35%. Therefore, Para rubber production on slopes greater than 15% should be managed with terraces as shown in the photos and schematic in Figure 5. It was outside the scope of the present research to account for such terrace systems as part of the SWAT simulations reported in this study.

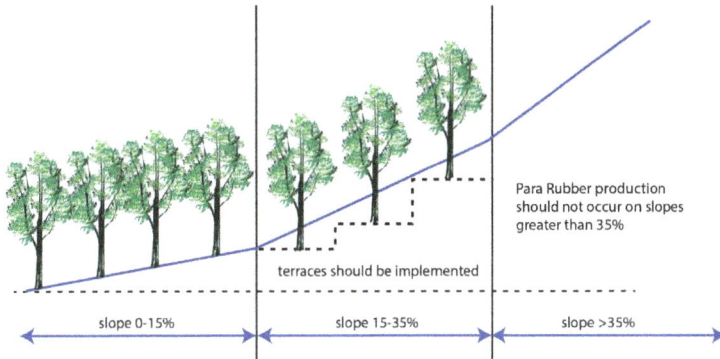

Figure 5. Photos and schematic showing terrace system needed to help mitigate excessive runoff and soil erosion that can occur when Para rubber trees are grown on high slopes >15%.

2.7. Para Rubber Crop Parameters

Crop growth in the SWAT model was set up for this study primarily using a heat unit schedule approach. However, rice and other "field crops" were simulated for two growing seasons during each year and thus both were simulated using specific planting and harvesting dates rather than heat unit scheduling. Double cropping of wet season rice and dry season rice were accounted for in the simulation; the second dry season crop is mostly grown in selected areas of Thailand that have sufficient irrigation water available. The planting schedules of wet season rice, dry season rice and field crops were simulated based on typical planting dates for each crop type in the region. The remaining operations were controlled by the fraction of heat units for each crop using the heat unit scheduling approach [36].

Para rubber is a perennial tree crop which grows year round; thus, annual planting and harvest and operations were not simulated for Para rubber. Para rubber crop parameters were not available in the ArcSWAT database for SWAT2009; thus, parameters had to be determined from measured data, inferred from existing parameters for other tree species in the ArcSWAT database, or determined from other sources. Values for 15 key Para rubber Tree (RUBR) crop parameters, and two other vegetation-influenced input parameters, which were used for the NLRB SWAT analysis, are listed in Table 2. The value selected for the maximum canopy height (CHTMX) were based on previously reported measurements [5] while the maximum rooting depth (RDMX) value is based on other measurements conducted by Thai scientists [58]. The other crop parameters in Table 2 were derived mainly from existing tree crop parameters in the ArcSWAT database. The choice of curve number (CN2) value for Para rubber trees reflects a woodland condition consisting of a thin stand, poor cover, no mulch, and a soil type consistent with hydrologic soil group B drainage conditions [59]. The Manning's n value for overland flow (OV_N) represents high runoff for timberland conditions [60]. Minimum, maximum and average values are listed for each crop parameter in Table 2. These parameter ranges allow the user the flexibility to decrease or increase the crop parameter default values, typically by approximately 10%.

Finally, the daily rubber tree water requirement rate was calculated by using the Bowen Ratio method based on data collected for 10-year old rubber trees at an experimental site in Chachoengsao Province. Temperature and humidity sensors were installed within and above the rubber tree canopy at the site. Solar radiation and wind velocity sensors were also installed above the rubber tree canopy. Soil temperature and soil moisture sensors were installed at 0 m and 0.10 m depths. A S-shape regression had been proposed to describe the relation between crop coefficients (Kc) and the Julian date. Therefore, the S-shape regression can be applied to evaluate the water requirement of rubber trees at different locations beyond the experimental site [61].

Table 2. Key Para rubber Tree (RUBR) crop parameters, and two other vegetation-influenced input parameters, that were used for the Nam Loei River Basin (NLRB) SWAT analysis.

No.	Parameter Code	Description	Minimum	Maximum	Simulated Value
1	BIO_E	Biomass/Energy Ratio	1	90	5.6
2	HVSTI	Harvest index	0.01	1.25	0.9
3	BLAI	Maximum leaf area index	0.5	10	2.6
4	CHTMX	Maximum canopy height	0.1	20	3.5
5	RDMX	Maximum root depth	0	3	2
6	T_OPT	Optimal temp for plant growth	11	38	20
7	T_BASE	Minimum temperature required for plant growth	0	18	7
8	USLE_C	Minimum value of USLE C factor applicable to the land cover/plant	0.001	0.5	0.001
9	GSI	Maximum stomata conductance (in drought condition)	0	5	0.75
10	RSDCO_PL	Plant residue decomposition coefficient	0.01	0.099	0.05
11	ALAI_MIN	Minimum leaf area index for plant during dormant period	0	0.99	0
12	D_LAI	Fraction of growing season when leaf area starts declining	0.15	1	0.99
13	MAT_YRS	Number of years required for tree species to reach full development	0	100	10
14	BMX_TREES	Maximum biomass for a forest	0	5000	
15	EXT_COEF	Light extinction coefficient	0	2	0.65
Additional Key Parameters Influenced by Para Rubber Vegetation					
16	CN2	SCS runoff curve number for moisture condition II	25	98	66
17	OV_N	Manning's "n" value for overland flow	0.01	30	0.11

2.8. Land Use Change Scenarios

Table 3 shows the specific land use distributions for the 2002, 2009 and 2015 land use scenarios, and the percentage difference for each land use category for two time periods: (1) between 2002 and 2009; and (2) between 2009 and 2015. The major changes in land use between 2002 and 2009 included a nearly 11.5% increase in Para rubber production, a decline in corn production of 10%, a decrease in disturbed forest land of over 9.7%, and increases in evergreen and deciduous forest of 7.35% and 2.1%, respectively. The largest shift in land use between 2009 and 2015 was an increase in Para rubber production of 9.69%, which was primarily responsible for respective decreases of −3.57%, −2.99%, −1.93%, −1.84% and −1.77% of corn, disturbed forest land, sugarcane, paddy fields and orchards.

Table 3. Distribution of land use for the 2002 and 2009 Nam Loei River Basin (NLRB) land use scenarios.

Item	Land Use Categories	LU–CODE	% of LU–2002	% of LU–2009	% Diff: 2002 vs. 2009	% of LU–2015	% Diff: 2009 vs. 2015
1	Paddy field	PDDY	12.28	11.86	−0.42	10.02	−1.84
2	Range–Brush	RNGB	-	0.19	0.19	0.19	-
3	Field crop	FCRP	4.75	5.87	1.12	6.14	0.27
4	Corn	CORN	23.38	13.33	−10.05	9.76	−3.57
5	**Rubber Trees**	**RUBR**	**0.38**	**11.84**	**11.46**	**21.53**	**9.69**
6	Sugarcane	SUGC	5.93	5.07	−0.86	3.14	−1.93
7	Agricultural Land	AGRR	5.89	7.27	1.38	8.71	1.44
8	Plantations	PLAN	1.04	1.06	0.02	1.06	0
9	Olives	OLIV	-	0.02	0.02	0.02	0
10	Orchard	ORCD	8.54	5.89	−2.65	4.12	−1.77
11	Pasture	PAST	-	0.23	0.23	0.23	0
12	Water	WATR	0.4	0.65	0.25	0.65	0
13	Disturbed forest land	DTFR	19.08	9.33	−9.75	6.34	−2.99
14	Forest–Evergreen	FRSE	-	7.30	7.30	7.3	0
15	Forest–Deciduous	FRSD	12.47	14.57	2.10	14.57	0
16	Planted forest	PNFR	0.23	0.23	-	0.23	0
17	Miscellaneous land	MISC	1.68	2.04	0.36	1.98	−0.06
18	Residential	URBN	3.95	3.25	−0.70	4.01	0.76
		Total	**100.00**	**100.00**		**100.00**	

The distribution of Para rubber production is also shown in the 2002 baseline land use map, versus the 2009 and 2015 land use scenario maps, in Figure 6. These distributions of Para rubber production areas further underscore the dramatic expansion of Para rubber tree plantations that occurred during the 14-year period of 2002 to 2015 in the NLRB. The effects of the increase in Para rubber production between the 2002 baseline and the two scenario years of 2009 and 2015 were accounted for in three separate scenario simulations performed in SWAT. The baseline scenario was first executed using the 2002 land use distribution (Table 3) for a 25-year period (1985 to 2009). The 2009 and 2015 land use scenarios were then performed for the same 25-year period to provide a consistent basis of comparison versus the baseline scenario.

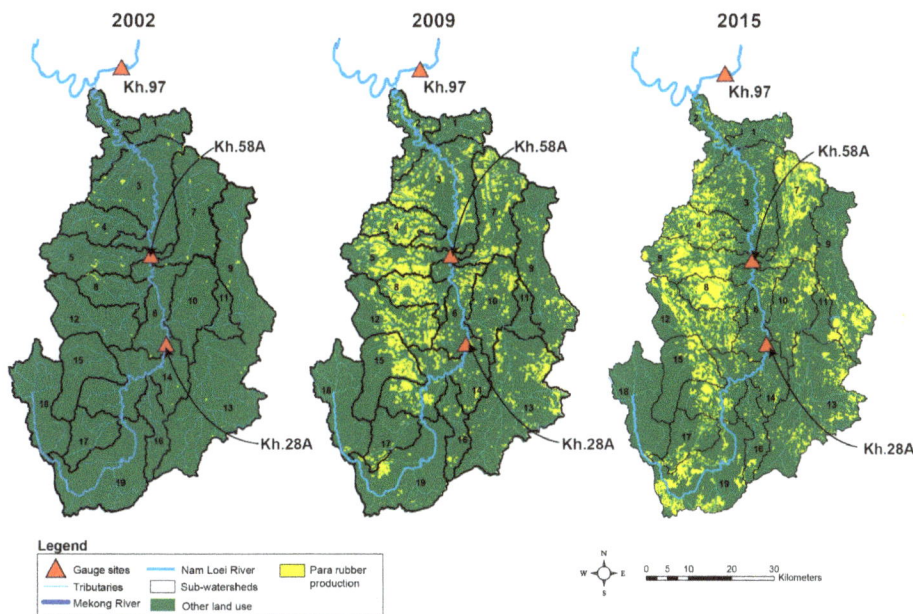

Figure 6. Spatial distribution of Para rubber production for the 2002 baseline versus the 2009 and 2015 land use scenarios in the Nam Loei River Basin (NLRB).

3. Results

3.1. Sensitivity Analysis

The top five most sensitive parameters as ranked in Table 4 were: (1) ALPHA_BF, base flow alpha factor (days); (2) ESCO, soil evaporation compensation factor; (3) GQWMN, threshold depth of water in the shallow aquifer required for return flow to occur (mm); (4) CN2, initial SCS runoff curve number for moisture condition II; and (5) CH_K2, effective hydraulic conductivity in main channel alluvium (mm·h^{-1}). The most influential parameters found in the sensitivity analysis are consistent with previously published summaries of the most widely used parameters in SWAT calibration [16,50]. The results also underscore the importance of accurate spatial and temporal precipitation inputs [16,50]. The choice of ALPHA_BF, CN2, ESCO and other parameters also varied (Table 4) between the two subbasins that drain to gauges Kh.28A and Kh.58A, respectively.

Table 4. Parameters ranges and results of the sensitivity analysis at the gauge stations located within the Nam Loei River Basin (NLRB).

Name	Description	Process	Min.	Max.	Rank of Sensitivity Analysis	Optimum Value Kh.28A	Optimum Value Kh.58A
GW_DELAY	Groundwater delay.	GW	0	500	8	0.1	1
ALPHA_BF	Base flow alpha factor (days).	GW	0	1	1	0.995	0.6
GWQMN	Threshold depth of water in the shallow aquifer required for return flow to occur.	GW	0	5000	3	1200	445
GW_REVAP	Groundwater "revap" coefficient.	GW	0.02	0.2	6	0.2	0.2
REVAPMN	Threshold depth of water in the shallow aquifer for "revap" to occur.	GW	0	1000	-	65	100
RCHRG_DP	Groundwater recharge to deep aquifer (fraction).	GW	0	1	-	0.001	0.1
LT_TIME	Lateral flow travel time.	HRU	0	180	-	1	35
SLSOIL	Slope length for lateral subsurface flow.	HRU	0	150	-	0.5	5
CANMX	Maximum canopy storage.	HRU	0	100	-	12	20
ESCO	Soil evaporation compensation factor.	HRU	0	1	2	0.7	0.6
CH_N2	Manning's "n" value for the main channel.	RTE	−0.01	0.3	7	0.2	0.146
CH_K2	Effective hydraulic conductivity in main channel alluvium.	RTE	−0.01	500	5	5	7.5
ALPHA_BNK	Baseflow alpha factor for bank storage.	RTE	0	1	-	0.5	0.239
CH_N1	Manning coefficient for the tributary channels.	SUB	0.01	30	10	0.145	2
CH_K1	Effective hydraulic conductivity in tributary channel alluvium (mm·h^{-1}).	SUB	0	300	-	30	100
CN2	SCS runoff curve number for moisture condition 2.	MGT	35	98	4	76	68
SOL_AWC	Available water capacity of the soil layer (mm·mm^{-1} soil).	SOL	0	1	9	0.198	0.244
SOL_BD	Moist bulk density.	SOL	0.9	2.5	-	1.255	1.051
SOL_K	Saturated hydraulic conductivity.	SOL	0	2000	-	103.8	65.2

3.2. Model Calibration and Validation

The statistical results of comparing the simulated SWAT calibration and validation aggregated monthly streamflows versus corresponding measured streamflows are listed in Table 5 for both gauge sites (Figure 1). The graphical comparisons between the simulated and measured monthly streamflows for the calibration and validation period are shown in Figure 7 for gauge site Kh.58A. The four NSE values computed for the two gauges during the calibration and validation period all exceeded 0.5, indicating that SWAT produced satisfactory streamflow estimates per previously suggested criteria [55,57]. The RMSE statistics were also all below 1.0 which further confirm that the SWAT streamflow estimates satisfactorily replicated the measured streamflows.

Table 5. Calibration and validation results at streamflow gauge stations Kh.28A and Kh.58A [a] in the Nam Loei River Basin (NLRB).

Station	Calibration (1994–2004) RMSE	Calibration (1994–2004) NSE	Validation (2005–2009) RMSE	Validation (2005–2009) NSE
Kh.28A	0.75	0.69	0.72	0.64
Kh.58A	0.82	0.71	0.79	0.68

Notes: [a] Locations of streamflow gauge stations shown in Figure 1.

The graphical comparisons between the simulated and measured aggregated monthly streamflows for the calibration and validation period at gauge site Kh.58A (Figure 7) show that SWAT accurately replicated most of the measured streamflow trends during the 11-year calibration period. However, several peak monthly streamflows were under predicted, especially in the last three years of the calibration period, which mirrors a tendency towards under prediction reported in a number of existing SWAT studies (e.g., [15]) and points to the need for further improvement of the SWAT hydrological algorithms, especially for Southeast Asia conditions. Similar graphical results occurred for the other three gauge site/time period combinations and thus are not reported here.

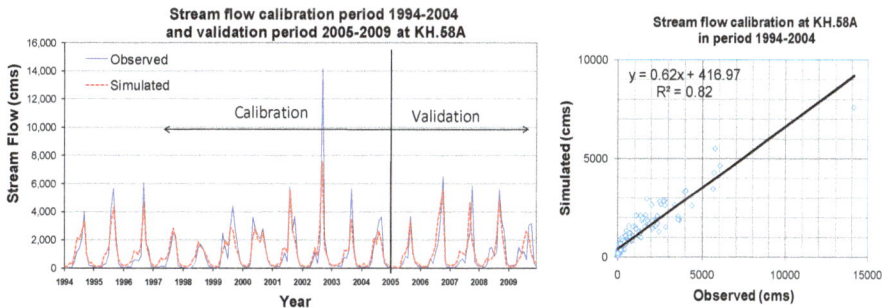

Figure 7. Simulated versus observed monthly streamflows at the Ban FakLoei station in the Nam Loei River Basin (NLRB) (Kh.58A; Figure 1).

3.3. Overall Water Balance Results for the Land Use Scenarios

Table 6 shows the overall long-term average annual water balance results for the entire NLRB predicted by SWAT for the 25-year 2002, 2009 and 2015 land use scenario simulations. The results of the scenarios reflect the increases in Para rubber production and other shifts in land use that occurred during the 2002 to 2009 and 2002 to 2015 time periods. Transmission losses were essentially negligible for all three land use scenarios, and the estimated combined lateral subsurface flow and groundwater flow were very similar between the three scenario simulations. However, the predicted ET increased by nearly 17 mm, and the predicted surface runoff and water yield decreased by similar amounts for the 2009 land use scenario as compared to the 2002 land use scenario. Similar, greater decreases in reduced surface runoff and water yield, relative to the 2009 land use scenario, were estimated for the 2015 land use scenario. However, the simulated ET decreased by almost 12 mm between the 2015 and 2009 land use scenarios. These results underscore the impacts of both the increased Para rubber production and shifting overall land use mixes in the NLRB between 2002 and 2015 (Table 3).

Table 6. Long-term (1985 to 2009) average annual water balance components for the 2002 and 2009 land use scenarios as estimated by SWAT for the entire Nam Loei River Basin (NLRB).

Water Balance Component	2002 Land Use Scenario (mm)	2009 Land Use Scenario (mm)	2015 Land Use Scenario (mm)
Precipitation	1217.8	1217.9	1217.9
Surface runoff	230.8	212.7	193.8
Lateral subsurface flow	49.2	51.6	47.4
Groundwater (shallow aquifer) flow	317.3	316.7	321.3
Evapotranspiration (ET)	590.8	607.4	595.7
Transmission losses	1.1	1.1	1.2
Total water yield [a]	596.1	579.9	561.4

Notes: [a] Total water yield = surface runoff + groundwater flow + lateral flow − transmission loss.

3.4. Seasonal ET and Water Yield Responses

The 25-year average monthly precipitation, ET levels and water yields simulated for the 2002, 2009 and 2015 NLRB land use scenarios during the 25-year simulation period (1985 to 2009) are shown in Figure 8. The initial increase in Para rubber plantations and other land use changes that had occurred by 2009 (Table 3) resulted in predicted increases in ET in almost every month of the year, relative to the baseline year of 2002 (Figure 8), except for March and April. The annual average ET increased about 3% (Table 6), with the highest percentage increases occurring during the dry season months of November to January and the lowest percentage increases occurring during the wet season months of May, September and October. The estimated water yield responses between the baseline and 2009 Para rubber expansion scenario resulted in the opposite trend, with water yields decreasing in most months, although slight increases occurred during the months of November, December and January. The predicted percentage water yield changes ranged from +0.6% in January to over −10% in March and April. The percentage declines in water yield predicted for the wet season were more constant as compared to the dry season water yield impacts, ranging from −1% in October to −7% in October.

The continued expansion of Para rubber production and other land use changes between 2009 and 2015 (Table 3) resulted in stronger shifts in the predicted monthly ET levels and in the annual hydrograph (Figure 8), with an earlier onset of the flood season and a decreased overall peak discharge in September. The estimated ET for the 2015 land use scenario was higher during the dry season as compared to 2002 and 2009 (Figure 8), except for November, but the opposite trend occurred during the wet season, resulting in an overall decline of about 2% from 2009 to 2015 (Table 6). The overall average dry period water yield was predicted to be about 13% higher for the 2015 land use scenario as compared to the baseline during November to April. However, lower water yields were predicted during most of the wet period except for the months of July and August (Figure 8). In total, the average annual simulated water yield decreased almost 7% from the baseline to 2015 (Table 6).

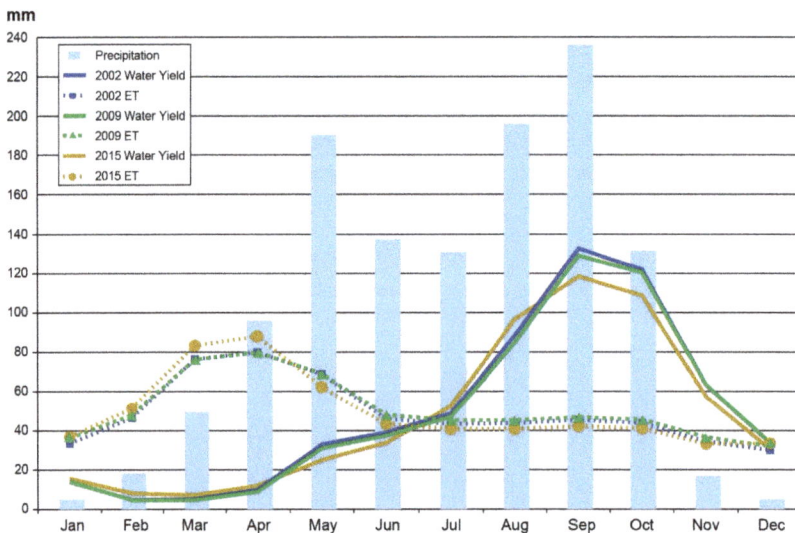

Figure 8. Average monthly precipitation, and average monthly water yield and evapotranspiration (ET) for the 2002 baseline, 2009 land use scenario and 2015 land use scenario, for the 25-year (1985 to 2009) simulation period for the Nam Loei River Basin (NLRB).

Table 7 provides comparisons of: (1) the long-term (25-year simulation period) average cumulative amounts of ET (mm) and water yield (mm) that occurred during the wet season, dry season, and annually;

(2) the percentages of the cumulative amounts that occurred during the wet season and dry season; and (3) the percentage changes in annual average ET and WYLD that occurred due to the land use changes between 2002 and 2009, and 2009 and 2015. These results show that nearly 80% of the water yield occurred during the wet season during both the 2002 baseline and the 2009 and 2015 Para rubber expansion land use scenarios (Table 7), as compared to the dry season. However, a more even distribution of ET occurred between the wet and the dry seasons, with slightly higher levels occurring during the dry season. Slight shifts in the overall amounts that occurred between the two seasons were generally predicted for the 2009 and 2015 conditions relative to the 2002 baseline; the 2015 land use scenario ET estimates resulted in the largest relative shift that was predicted between the two seasons (Table 7).

Various seasonal shifts were predicted to occur between 2002 baseline and the 2009 landuse scenarios, and between the 2009 and 2015 land use scenarios (Table 7). Relatively minor shifts were estimated between 2002 and 2009, with ET increasing by roughly 3% in both seasons versus 1% and 3% declines in water yield for the dry season and wet season, respectively. ET was predicted to decrease almost 10% during the wet season but increase nearly 6% during the dry season, between 2009 and 2015. In contrast, water yield was predicted to decrease during the wet season by about 3% and increase by 1% during the dry season, between the two landuse scenarios. The estimated average annual percentage change in ET from 2002 to 2009 was a decline of roughly 3% as compared to an increase in water yield of close to 3% for the same time period (Table 7). In contrast, the predicted average annual percentage change in both ET and water yield from 2009 to 2015 was a decline of roughly 2% (Table 7).

Table 7. Comparisons of 25-year cumulative average seasonal and annual ET and water yields, and total annual ET and water yield percentage changes that occurred due to the land use changes between 2002 and 2009, and 2009 and 2015, in the Nam Loei River Basin (NLRB).

| Season | Baseline (2002) | | Para Rubber Expansion Scenarios | | | |
| | | | 2009 | | 2015 | |
	ET (mm)	WYLD (mm)	ET (mm)	WYLD (mm)	ET (mm)	WYLD (mm)
Wet Season	290.4	463.5	298.9	448.7	269.9	434.2
Dry Season	300.4	130.3	308.5	128.9	325.8	130.4
Annual (total)	590.8	593.8	607.4	577.6	595.7	564.6
Percentage in each season and overall percentage change						
Wet season (%)	49.2	78.1	49.2	77.7	45.3	76.9
Dry season (%)	50.8	21.9	50.8	22.3	54.7	23.1
Annual (%)			2.8	−2.7	−1.9	−2.2

4. Discussion

As noted previously, substantial increases in rubber tree production are expected to occur in Southeast Asia by 2050. Preliminary research suggests that this massive land use change could exacerbate environmental problems in the region including increased soil erosion and sediment transport to surface water, degraded soil quality, decreased stream flow, risk of landslides and probable decreased soil carbon levels [1,2,62]. The majority of the area where rapid and widespread land conversion to monoculture rubber plantation has occurred in continental SE Asia is also vulnerable to extreme climatic events including typhoons, frost or drought, which can greatly reduce or even destroy rubber production and further exacerbate environmental problems [3]. Furthermore, these environmental problems may be magnified even more per projected future climate change in the region [3]. Intensified Para rubber production has also negatively impacted biodiversity in production areas located in Thailand and other subregions in Southeast Asia [63].

The results of this study show that the expansion of rubber cultivation is resulting in a decreased volume of water in the rainy season in the NLRB in northeast Thailand. In addition, during the dry season the water content decreases, resulting in water shortages. The expansion of rubber production in inappropriate areas with slopes over 35% [51] is more likely to result in increased flash floods and landslides due to the intensive monoculture practices. This is especially true for the Nam Manh, Nam Phu and Nam Paow subbasins, where 3000 ha of Para rubber plantations have been introduced on landscapes with extremely high slopes. In addition, an additional 10,300 ha of Para rubber production exists in more moderate slope areas (15% to 35% slopes) that are distributed across the upper, middle and lower subregions of the NLRB. The Rubber Research Institute of Thailand has recommended that terrace systems should be used for rubber production on slopes >15% [34] but this recommendation is not being consistently followed. These specific problems occurring in Thailand, coupled with the previously described environmental problems that are increasing Southeast Asia, underscore the urgent need to develop measurement approaches, databases and modeling tools that can be used to investigate Para rubber production problems throughout the region.

The amount of water that is required to initiate and sustain Para rubber plant growth depends on many factors such as the type of plant species, the age of the plant and weather conditions such as wind speed, temperature and humidity [61]. There are many ways to determine crop water use both directly and indirectly. The measure by instrumentation is more accurate, but cannot be conducted across large plantations. The application of mathematical models is a popular method and are widely used in the study and evaluation of water use and water requirements of crops, which saves time and cost. However, the use of mathematical models can require extensive input data depending on the type and format of the model. There are parameters that must be calibrated and verified to ensure the calculated results are close to the measured values. Furthermore, databases with appropriate parameter values may not be available for some models. Thus, those databases must be brought up-to-date to better meet the needs of end users.

The case study reported here that describes the expansion of Para rubber plantations in the NLRB using SWAT is an example of an application that requires the development of important plant input parameters. However, there is a need to further develop SWAT Para rubber input parameters that better account for rubber species and age, rubber stems and leaves of rubber trees. In addition, there is a need to improve the SWAT growth functions to be able to better account for the effects of rubber tree growth phenomena related to the rooting structure that depletes deeper soil layers and results in higher ET impacts, relative to traditional vegetation [36,64]. Finally, an overall expanded Para rubber production and knowledge database is needed for Thailand and Southeast Asia in general.

5. Conclusions

The Soil and Water Assessment Tool (SWAT) model was applied to assess the impact of Para rubber expansion in the Nam Loei River Basin in this study The application has designated the land use in the year 2002 as the baseline versus historical Para rubber tree expansions in 2009 and 2015 as the land use change scenarios. The stream flow estimated by SWAT showed annual average stream flow of about 1580 MCM. The average stream flow occurring during the rainy season (June–November) was about 1264 MCM (80% of the average annual stream flow) and in the dry season about 316 MCM (20% of the average annual stream flow). The simulation was done using scenarios to assess the water balance in the hydrological process. The results of the simulations showed that the increased production of Para rubber, which replaced the original local field crop and disturbed forest land, resulted in an increase of ET of about 3% from 2002 to 2009. However, additional increased Para rubber production in combination with other land use shifts during 2009 to 2015 resulted in a predicted ET decrease of 2%. The major factors that influenced this result were the rubber canopy and precipitation. Moreover, runoff results reduced water balance in the basin by an annual average of about 3%, especially during the dry season. However, the effect of ET on water resources has increased complexity and uncertainty; the consideration of many parameters of Para rubber and a reflection on the past will help

Water 2017, 9, 1

our understanding of the dynamic changes. The results of this study will help provide guidance for decision-making about land use allocation or zoning for suitable Para rubber area. In addition, this study will aid in the management and planning of water resources for the NLRB and other river basins located in northeastern Thailand.

Acknowledgments: This study was carried out as part of the required research to fulfill Ph.D. thesis requirements for the Graduate School of Kasetsart University. The authors would like to express their sincere thanks to the Thailand Research Fund (TRF) through the Royal Golden Jubilee Ph.D. Program and Kasetsart University Research and Development Institute (KU–RDI) for their financial support. We would also like to thank the Department of Water Resources, Land Development Department and other agencies for supplied the data for the study.

Author Contributions: Winai Wangpimool performed the model input data collection and analysis, modeling work and primary analysis of the results. Kobkiat Pongput provided guidance regarding the methodology, literature review and the results and conclusions of the study. Nipon Tangtham provided guidance regarding the development of Para rubber plant parameters and trends in Para rubber production. Saowanee Prachansri provided consultation regarding the land use and soil data including specific input parameters. Philp W. Gassman provided guidance regarding the methods used in the study, contributed in-depth analysis of the simulation results and provided editing and writing support.

Conflicts of Interest: The authors declare no conflict of interest.

References

1. Ziegler, A.D.; Fox, J.M.; Xu, J. The Rubber Juggernaut. *Science* **2009**, *324*, 1024–1025. [CrossRef] [PubMed]
2. Fox, J.; Castella, J.C. Expansion of rubber (*Heveabrasiliensis*) in mainland southeast Asia: What are the prospects forsmallholders? *J. Peasant Stud.* **2013**, *40*, 155–170. [CrossRef]
3. Ahrends, A.; Hollingsworth, P.M.; Ziegler, A.D.; Fox, J.M.; Chen, H.; Su, Y.; Xu, J. Current trends of rubber plantation expansion may threaten biodiversity and livelihoods. *Glob. Environ. Chang.* **2015**, *34*, 48–58. [CrossRef]
4. Mongkolsawat, C.; Putklang, W. An Approach for Estimating Area of Rubber Plantation: Integrating Satellite and Physical Data over the Northeast Thailand. 2010. Available online: http://a-a-r-s.org/aars/proceeding/ACRS2010/Papers/Oral%20Presentation/TS36-1.pdf (accessed on 22 August 2016).
5. Rantala, L. Rubber Plantation Performance in the Northeast and East of Thailand in Relation to Environmental Conditions. Master's Thesis, University of Helsinki, Helsinki, Finland, 2006.
6. Suwanwerakamtorn, R.; Putklang, W.; Khamdaeng, P.; Wannaros, P. An Application of THEOS Data to Rubber Plantation Areas in Mukdahan Province, Northeast Thailand. 2012. Available online: http://gecnet.kku.ac.th/research/i_proceed/2555/2_ip2012.pdf (accessed on 4 June 2013).
7. Office of Agricultural Economics. *Agricultural Statistics of Thailand 2009*; Center for Agricultural Information, Office of Agricultural Economics: Bangkok, Thailand, 2009; pp. 78–81, No. 401.
8. Office of Agricultural Economics. *Agricultural Statistics of Thailand 2010*; Center for Agricultural Information, Office of Agricultural Economics: Bangkok, Thailand, 2010; pp. 78–81, No. 416.
9. Office of Economic and Social Development of Northeastern. *Para Rubber Situation and Adaptation of Farmers in the Northeastern*; The Office of the National Economic and Social Development: Khon Kean, Thailand, 2015; p. 47.
10. Prakhonsri, P. The Fires in the Northeast and Management. Technical Conference of Management of Natural Disasters in the Northeast and the Self-Reliance of Local Sustainable. 2011, pp. 42–47. Available online: http://www.tndl.org/kku/pdf/fire-prasit.pdf (accessed on 22 August 2016).
11. Chakarn, S.; Soontorn, K.; Niwat, A.; Jitti, P. Growths and carbon stocks of Para rubber plantations on Phonpisai soil series in northeastern Thailand. *Rubber Thai J.* **2012**, *1*, 1–18.
12. Bowen, G.D.; Nambiar, E.K.S. *Nutrition of Plantation Forests*; Academic Press: London, UK, 1989; p. 505.
13. Arnold, J.G.; Srinivasan, R.; Muttiah, R.S.; Williams, J.R. Large area hydrologic modeling and assessment part I: Model development. *J. Am. Water Resour. Assoc.* **1998**, *34*, 73–89. [CrossRef]
14. Gassman, P.W.; Reyes, M.; Green, C.H.; Arnold, J.G. The soil and water assessment tool: Historical development, applications and future directions. *Trans. ASABE* **2007**, *50*, 1211–1250. [CrossRef]
15. Williams, J.R.; Arnold, J.G.; Kiniry, J.R.; Gassman, P.W.; Green, C.H. History of model development at Temple, Texas. *Hydrol. Sci. J.* **2008**, *53*, 948–960. [CrossRef]

16. Arnold, J.G.; Moriasi, D.N.; Gassman, P.W.; Abbaspour, K.C.; White, M.J.; Srinivasan, R.; Santhi, C.; Harmel, R.D.; van Griensven, A.; van Liew, M.W.; et al. SWAT: Model use, calibration and validation. *Trans. ASABE* **2012**, *55*, 1491–1508. [CrossRef]
17. Gassman, P.W.; Sadeghi, A.M.; Srinivasan, R. Applications of the SWAT model special section: Overview and Insights. *J. Environ. Qual.* **2014**, *43*, 1–8. [CrossRef] [PubMed]
18. Gassman, P.W.; Wang, Y. IJABE SWAT Special Issue: Innovative modeling solutions for water resource problems. *Int. J. Agric. Biol. Eng.* **2015**, *8*, 1–8.
19. Bressiani, D.A.; Gassman, P.W.; Fernandes, J.G.; Garbossa, L.H.P.; Srinivasan, R.; Bonumá, N.B.; Mendiondo, E.M. A review of soil and water assessment tool (SWAT) applications in Brazil: Challenges and prospects. *Int. J. Agric. Biol. Eng.* **2015**, *8*, 9–35.
20. Krysanova, V.; White, M. Advances in water resources assessment with SWAT: An overview. *Hydrol. Sci. J.* **2015**, *60*, 771–783. [CrossRef]
21. Babel, M.S.; Shrestha, B.; Perret, S.R. Hydrological impact of biofuel production: A case study of the KhlongPhlo Watershed in Thailand. *Agric. Water Manag.* **2011**, *101*, 8–26. [CrossRef]
22. Glavan, M.; Pintar, M.; Volk, M. Land use change in a 200-year period and its effect on blue and green water flow in two Slovenian Mediterranean catchments: Lessons for the future. *Hydrol. Process.* **2012**, *27*, 3964–3980. [CrossRef]
23. Jha, M.; Schilling, K.E.; Gassman, P.W.; Wolter, C.F. Targeting land-use change for nitrate-nitrogen load reductions in an agricultural watershed. *J. Soil Water Conserv.* **2010**, *65*, 342–352. [CrossRef]
24. Kim, Y.; Band, L.E.; Song, C. The influence of forest regrowth on the stream discharge in the North Carolina Piedmont watersheds. *J. Am. Water Resour. Assoc.* **2013**. [CrossRef]
25. Liu, W.; Cai, T.; Fu, G.; Zhang, A.; Liu, C.; Yu, H. The streamflow trend in Tangwang river basin in northeast China and its difference response to climate and land use change in sub-basins. *Environ. Earth Sci.* **2012**, *69*, 1–12. [CrossRef]
26. Ma, X.; Xu, J.; van Noordwijk, M. Sensitivity of streamflow from a Himalayan catchment to plausible changes in land cover and climate. *Hydrol. Process.* **2010**, *24*, 1379–1390. [CrossRef]
27. Memarian, H.; Tajbakhsh, M.; Balasundram, S.K. Application of SWAT for impact assessment of land use/cover change and best management practices: A review. *Int. J. Adv. Earth Environ. Sci.* **2013**, *1*, 35–40.
28. Tan, M.L.; Ibrahim, A.L.; Yusop, Z.; Duan, Z.; Ling, L. Impacts of land-use and climate variability on hydrological components in the Johor River basin, Malaysia. *Hydrol. Sci. J.* **2015**, *60*, 873–889. [CrossRef]
29. Celine, G.; James, E.J. Assessing the implications of extension of rubber plantation on the hydrology of humid tropical river basin. *Int. J. Environ. Res.* **2015**, *9*, 841–852.
30. Tao, C.; Chen, X.; Lu, J.; Philip, W.G.; Sauvage, S. Assessing impacts of different land use scenarios on water budget of Fuhe River, China using SWAT model. *Int. J. Agric. Biol. Eng.* **2015**, *8*, 95–109.
31. Wangpimool, W.; Pongput, K. Integrated Hydrologic and Hydrodynamic Model for Flood Risk Assessment in Nam Loei Basin, Thailand. Available online: http://eitwre2011.fiet.kmutt.ac.th/theme_en/6HE_E.pdf (accessed on 1 March 2012).
32. TMD. Weather Data Service. Thai Meteorological Department, Ministry of Information and Communication Technology. 2015. Available online: http://www.tmd.go.th/province_stat.php?StationNumber=48353 (accessed on 25 August 2016).
33. Office of Soil Survey and Land Use Planning. *Land Use Planing for Loei Province*; Land Development Department, Ministry of Agriculture and Cooperatives: Bangkok, Thailand, 2002.
34. Rubber Research Institute of Thailand. Para Rubber Situation in Northeastern. 2012. Available online: http://www.rubberthai.com/about/strategy.php (accessed on 8 May 2013).
35. Dingman, S.L. *Physical Hydrology*; Prentice-Hall Inc.: Englewood Cliffs, NJ, USA, 2015.
36. Neitsch, S.L.; Arnold, G.; Kiniry, J.R.; Williams, J.R. Soil and Water Assessment Tool, Theoretical Documentation, Texas. 2009. Available online: http://twri.tamu.edu/reports/2011/tr406.pdf (accessed on 3 July 2010).
37. Fisher, J.B.; Malhi, Y.; Bonal, D.; da Rocha, H.R.; de Araújo, A.C.; Gamo, M.; Goulden, M.L.; Hirano, T.; Huete, A.R.; Kondo, H.; et al. The land-atmosphere water flux in the tropics. *Glob. Chang. Biol.* **2009**, *15*, 2694–2714. [CrossRef]
38. Guardiola-Claramonte, M.; Troch, P.A.; Ziegler, A.D.; Giambelluca, T.W.; Durcik, M.; Vogler, J.B.; Nullet, M.A. Hydrologic effects of the expansion of rubber (*Heveabrasiliensis*) in a tropical catchment. *Ecohydrology* **2010**, *3*, 306–314. [CrossRef]

39. Monteith, J.L. Evaporation and the Environment. In *The State and Movement of Water in Living Organisms*; Cambridge University Press: Swansea, UK, 1965; pp. 205–234.

40. Priestley, C.H.B.; Taylor, R.J. On the assessment of surface heat flux and evaporation using large-scale parameters. *Mon. Weather* **1972**, *100*, 81–92. [CrossRef]

41. Hargreaves, G.H.; Samani, Z.A. Reference crop evapotranspiration from temperature. *Appl. Eng. Agric.* **1985**, *1*, 96–99. [CrossRef]

42. Soil and Water Assessment Tool (SWAT). Software: ArcSWAT. 2016. Available online: http://swat.tamu.edu/software/arcswat/ (accessed on 25 August 2016).

43. Royal Thai Survey Department (RTSD). Digital Elevation Map for Loei Province 1:50,000 WGS 84, 2000. Royal Thai Survey Department, Royal Thai Armed Force Headquarters. 2000. Available online: http://www.rtsd.mi.th/MapInformationServiceSystem/ (accessed on 25 August 2016).

44. Land Development Department (LDD). *Land Use Map for Loei. Province*; Office of Soil Survey and Land Use Planning, Ministry of Agriculture and Cooperatives: Bangkok, Thailand, 2002.

45. Land Development Department (LDD). *Soil Map for Loei. Province*; Office of Soil Survey and Land Use Planning, Ministry of Agriculture and Cooperatives: Bangkok, Thailand, 1995.

46. Pongput, K.; Wangpimool, W.; Chaturabul, T.; Ketjinda, K. *Development of Software to Decision Support System for Planning, Management and Development of Water Resources in the Basin*; Kasetsart University Research and Development Institute (KU-RDI): Bangkok, Thailand, 2013.

47. Royal Irrigation Department (RID). Hydrological Data Service. Royal Irrigation Department, Ministry of Agriculture and Cooperatives. Available online: http://hydro-3.com/ (accessed on 25 August 2016).

48. Van Griensven, A.; Meixner, T.; Grunwald, S.; Bishop, T.; Diluzio, M.; Srinivasan, R. A Global Sensitivity Analysis Tool for the Parameters of Multi-Variable Catchment Models. *J. Hydrol.* **2006**, *324*, 10–23. [CrossRef]

49. Veith, T.L.; van Liew, M.W.; Bosch, D.D.; Arnold, J.G. Parameter sensitivity and uncertainty in SWAT: A comparison across five USDA-ARS Watersheds. *Trans. ASABE* **2010**, *53*, 1477–1486. [CrossRef]

50. White, K.L.; Chaubey, I. Sensitivity analysis, calibration and validations for a multisite and multivariable SWAT model. *J. Am. Water Resour. Assoc.* **2005**, *41*, 1077–1089. [CrossRef]

51. Licciardello, F.; Rossi, C.G.; Srinivasan, R.; Zimbone, S.M.; Barbagallo, S. Hydrologic evaluation of a Mediterranean watershed using the SWAT model with multiple PET estimation methods. *Trans. ASABE* **2011**, *54*, 1615–1625. [CrossRef]

52. Abbaspour, K.C. SWAT Calibration and Uncertainty Programs. Eawag: Swiss Federal Institute of Aquatic Science and Technology. 2014. Available online: http://swat.tamu.edu/software/swat-cup/ (accessed on 5 December 2014).

53. Ritter, A.; Muñoz-Carpena, R. Performance evaluation of hydrological models: Statistical significance for reducing subjectivity in goodness-of-fit assessments. *J. Hydrol.* **2013**, *480*, 33–45. [CrossRef]

54. Krause, P.; Boyle, D.P.; Bäse, F. Comparison of different efficiency criteria for hydrological model assessment. *Adv. Geosci.* **2005**, *5*, 89–97. [CrossRef]

55. Moriasi, D.N.; Arnold, J.G.; van Liew, M.W.; Binger, R.L.; Harmel, R.D.; Veith, T. Model evaluation guidelines for systematic quantification of accuracy in watershed simulations. *Trans. ASABE* **2007**, *50*, 885–900. [CrossRef]

56. Boyle, D.P.; Gupta, H.V.; Sorooshian, S. Toward improved calibration of hydrologic models: Combining the strengths of manual and automatic methods. *Water Resour. Res.* **2000**, *36*, 3663–3674. [CrossRef]

57. Moriasi, D.N.; Gitau, M.W.; Pai, N.; Daggupati, P. Hydrologic and water quality models: Performance measures and evaluation criteria. *Trans. ASABE* **2015**, *58*, 1763–1785.

58. Land Development Department (LDD). *Technical Report of Para Rubber Tree*; Research and Development of Soil and Water Conservation Crop Areas Group, Bureau of Land Research and Management: Bangkok, Thailand, 2005.

59. U.S. Department of Agriculture, Natural Resource Conservation Service (USDA-NRCS). *National Engineering Handbook*; Part 630 Hydrology, Section 4, Chapter 7; USDA-NRCS: Washington, DC, USA, 2009. Available online: http://www.nrcs.usda.gov/wps/portal/nrcs/detailfull/national/water/?cid=stelprdb1043063 (accessed on 6 September 2016).

60. Yen, B.C.; Chow, V.T. *Local Design Storms*; U.S. Department of Transportation, Federal Highway Administration: Washington, DC, USA, 1983; Volume 1–3, No. FHWA-RD-82-063 to 065.

61. Rattanapinanchai, A.; Sangkhasila, K. Daily Water Consumtions and Crop Coefficients of Para Rubber Plantation. *Proceeding of the 7th National Kasetsart University Kham Pheang Sean Conference.* 2010. Available online: http://researchconference.kps.ku.ac.th/article_7/pdf/o_plant15.pdf (accessed on 6 September 2016).

62. Fox, J.; Castella, J.C.; Ziegler, A.D. Swidden, rubber and carbon: Can REDD+ work for people and the environment in Montane Mainland Southeast Asia? *Glob. Environ. Chang.* **2014**, *29*, 318–326. [CrossRef]

63. Vongkhamheng, C.; Zhou, J.H.; Beckline, M.; Phimmachanh, S. Socioeconomic and Ecological Impact Analysis of Rubber Cultivation in Southeast Asia. *Open Access Lib. J.* **2016**, *3*. [CrossRef]

64. Guardiola-Claramonte, M.; Troch, P.A.; Ziegler, A.D.; Giambelluca, T.W.; Vogler, J.B.; Nullet, M.A. Local hydrologic effects of introducing non-native vegetation in a tropical catchment. *Ecohydrology* **2010**, *1*, 13–22. [CrossRef]

water

Article

Development of a Station Based Climate Database for SWAT and APEX Assessments in the US

Michael J. White [1,*], Marilyn Gambone [1], Elizabeth Haney [2], Jeffrey Arnold [1] and Jungang Gao [2]

[1] Agricultural Research Service, US Department of Agriculture, 808 E. Blackland Road, Temple, TX 76502, USA; marilyn.gambone@ars.usda.gov (M.G.); Jeff.Arnold@ARS.USDA.GOV (J.A.)
[2] Blackland Research & Extension Center, Texas A&M AgriLife, Temple, TX 76502, USA; lhaney@brc.tamus.edu (E.H.); jgao@brc.tamus.edu (J.G.)
* Correspondence: mike.white@ars.usda.gov; Tel.: +254-770-6523

Received: 11 May 2017; Accepted: 14 June 2017; Published: 17 June 2017

Abstract: Water quality simulation models such as the Soil and Water Assessment Tool (SWAT) and Agricultural Policy EXtender (APEX) are widely used in the US. These models require large amounts of spatial and tabular data to simulate the natural world. Accurate and seamless daily climatic data are critical for accurate depiction of the hydrologic cycle, yet these data are among the most difficult to obtain and process. In this paper we describe the development of a national (US) database of preprocessed climate data derived from monitoring stations applicable to USGS 12-digit watersheds. Various sources and processing methods are explored and discussed. A relatively simple method was employed to choose representative stations for each of the 83,000 12-digit watersheds in the continental US. Fully processed climate data resulting from this research were published online to facilitate other SWAT and APEX modeling efforts in the US.

Keywords: SWAT; APEX; climate

1. Introduction

Hydrologic and water quality models are increasingly being used to inform public policy at the national and local scales. These models can be used to predict the effects of various anthropogenic activities on water quality and quantity, making them useful tools for the purposes of: (1) watershed and water resource planning; (2) evaluation of conservation programs; (3) effects of climate or land use change. The Soil and Water Assessment Tool (SWAT) [1] and Agricultural Policy EXtender (APEX) [2] are two popular models which operate on a daily time-step and are applied at decadal timeframes. Both require a considerable amount of input data which must be accurate, consistent, continuous, and cover the entire area of interest. These models require spatial data such as land use, soils, and topography as well as climatic data including (at a minimum) daily precipitation, minimum, and maximum temperature. When the area of interest is large, in this case the entire US, data development is particularly challenging.

The Conservation Effects and Assessment Project (CEAP), is a national multiagency effort to quantify the environmental benefits of existing conservation practices and the possible effects of future conservation policy. The cropland portion of CEAP makes extensive use of hydrologic and water quality models, including SWAT and APEX. Previous CEAP applications have used data developed by [3] and prepared specifically for the effort, using a combination of daily ground based station and monthly Parameter–Elevation Regressions on Independent Slopes Model (PRISM) predictions [4]. Daily (1960–2006) precipitation and temperature were estimated for each watershed and successfully applied within CEAP [5,6].

As an ongoing effort, CEAP is being updated to include more recent data (up to 2016), improved methods, and greater spatial detail. Previous CEAP efforts used watersheds based on 8-digit

Hydrologic Unit Codes (HUCs), a hierarchical watershed classification system by the US Geologic Survey (USGS). Future CEAP assessments will utilize 12-digit HUCs, which are approximately 40 times smaller, thus allowing far more spatial detail to be included. This warrants the redevelopment of all data sources, including climate. The scope and objectives of CEAP dictate the parameters by which the climate dataset is developed. Climatic data must cover the entire continental US, reported daily, and be nearly seamless for a 55 year period (1960–2015). Although coverage of the entire US is needed, the required quality and density varies. CEAP is focused on agriculture, thus observation density and accuracy are key in heavily cultivated areas. Topographic climatic effects are less critical in these regions, tending to be more prominent in mountainous areas with little or no agriculture. CEAP is designed to make average annual long term predictions to inform policy. It is far more important that the climate data (especially precipitation) that is used is statistically representative than absolutely accurate for a given location. Accurate average annual total precipitation, and the distributional frequency and intensity of rainfall events are critical.

The primary models considered in this research are SWAT and APEX. These models share a common development history, the major difference between them being that SWAT is a basin-scale hydrologic model and APEX is more commonly used for field and small watersheds. Both are distributed hydrologic models which divide a basin or field into smaller units to incorporate spatial detail. In SWAT the unit is a subbasin, in APEX it is a sub-area. Water yield and pollutant loads are calculated for each unit and routed through a stream network or across a landscape surface to an outlet. In SWAT a single subbasin can be further divided into areas with the same soil, land use, and slope called Hydraulic Response Units (HRUs). Processes within each HRU are lumped and calculated independently from all other HRUs; the total nutrient or water yield for a subbasin is the sum of all the HRUs it contains. HRUs allow more spatial detail for a large basin to be represented in a computationally efficient manner. Both are process based continuous simulation models that operates on a daily or sub-daily time step. Long-term simulations can be performed using simulated or observed weather data. Relative impacts of different management scenarios can be quantified.

SWAT is the combination of ROTO (Routing Outputs to Outlets) [7] and SWRRB (Simulator for Water Resources in Rural Basins) [8]. CREAMS (Chemicals, Runoff, and Erosion from Agricultural Management Systems) [9], GLEAMS (Groundwater Loading Effects on Agricultural Management Systems) [10] and EPIC (Erosion-Productivity and Impact Calculator) [11] all contributed to the development of SWRRB and APEX. All of the models were developed, calibrated, and applied utilizing precipitation data collected from networks of ground based stations, gridded precipitation data were not available at the time.

The objective of this research is to evaluate differing data sources and methods to synthesize seamless daily climate data (1960–2015) for the Contiguous US (CONUS) suitable for application with SWAT and APEX at the 12-digit HUC level for the Cropland CEAP project. This article will focus on application related issues and methods specific to these models; the raw accuracy of differing climatic datasets is beyond this scope. These processed climate data are publicly released via the web to support other SWAT and APEX modeling projects in the US.

2. Weather Data Development

2.1. Potential Data Sources

The selection and processing of climatic data for use in hydrologic models is more complex than it may seem. For our purposes climate data are either: (1) point observations collected at discrete weather stations; or (2) interpolations, remotely sensed, or simulations averaged over some spatial extent, usually a grid. Daly discusses in detail some limitations of using these interpolated gridded datasets [12]. Data are also summarized temporally, into hourly, daily, or monthly values. It is imperative that the climate data be understood and matched to the assumptions and requirements of the model in which these data are used. This is particularly important for precipitation, as it is the driving force for hydrologic models and often more spatially varied than temperature in the crop producing regions which are the focus of the CEAP effort. In both SWAT and APEX a single set of climatic observations are applied to an entire subbasin or subarea. In reality, precipitation may vary dramatically across a subbasin during a single rainfall event; therefore, data collected at a single site may be a poor representation for the entire subbasin. In contrast, streamflow data from a single site, either observed or simulated, is in a sense an integration of all precipitation in its catchment.

There are several climatic datasets that may be useful for this purpose, each with their own limitations. Gridded climatic data generally employ some interpolation or modeling method between ground stations. This process may utilize other data such as topography or remote sensing. It is difficult to assess the accuracy of these datasets as potential validation data are often based on the same assumptions and/or underlining data as the dataset being evaluated and are not truly independent [12].

The Climate Forecast System Reanalysis (CFSR) [13] is a gridded global dataset (0.5° × 0.5°) for the period 1979 to 2009. These data have been processed into daily SWAT format and are available online [14]. Another source of gridded climate data is Next Generation Weather Radar (NEXRAD) which have been used in a variety of previous model applications [15–17]. NEXRAD data have sufficient spatial resolution (4 km × 4 km), but little data are available prior to the early 1990's. PRISM data are also available from 1980 to near current on a monthly or daily basis at various grid resolutions down to 800 m. As noted before these data have been previously used in CEAP.

The most comprehensive source of measured weather station data in the US is the Global Historical Climatology Network (GHCN) [18]. This dataset integrates observations from 30 different station networks with more than 40,000 stations in the CONUS dating back to the 1800's. Although these data have both the necessary spatial resolution, and time fame, they are far from seamless. Much of the data are derived from stations operated by cooperative observers, including private individuals. This network of weather stations is dynamic and changes from year to year making it difficult to use.

All of the aforementioned climatic data have been used in SWAT for various projects but direct comparisons between datasets are limited. Because SWAT is typically calibrated, it can be difficult to rigorously assess the accuracy of a particular model input by evaluating model performance alone. [19] compared NEXRAD, PRISM, and station based data in the central US finding all three to produce biased predictions of streamflow during dry periods, with PRISM exhibiting the least. They also conducted a detailed review of recent SWAT climate data comparisons.

For the purposes of this particular model application, there is a strong preference among SWAT and APEX developers to utilize station based data if possible. There are several reasons for this preference: (1) SWAT and APEX were developed using data from ground based stations, the use of gridded data may have unforeseen consequences; (2) the CEAP national assessment does not require fine grained analysis in space or time; (3) the vast majority of SWAT and APEX applications use ground based station data.

2.2. Evaluating Aggregation Effects

As stated before, these models apply a single set of weather observations to an entire subbasin thus it is important that these data be representative of the entire subbasin. Average precipitation across a subbasin can be derived from either gridded data sources or point station based data. Gridded data can be spatially aggregated by overlaying with subbasin boundaries. Thiessen polygon weighting or other procedures can be used to average point station observations across a subbasin. In either case, the precipitation value used in a model for a subbasin is the average value across that subbasin. For many applications that utilize monthly or annual climate summaries this maybe a very reasonable approach, but there may be unintended consequences if the goal is to predict daily surface runoff, sediment losses, or nutrient loss. The aggregation of individual observations into a single record changes the statistical nature of the climatic data. As the spatial unit over which the averaging occurs is larger, incorporating more stations or grid cells, there is less day to day variability in the aggregated precipitation value. The number of days with rainfall within the record tends to increase while rainfall intensity is reduced even though the average annual precipitation may be unaffected. Surface runoff, sediment loss, and nutrient transport area are driven by precipitation amount and intensity. Many of the contributing complex natural processes are nonlinear, thus sensitive to both precipitation frequency and intensity. Some evaluation of these aggregation effects on model predictions is necessary to aid in the selection of appropriate data sources and processing methods.

To gage the effect of aggregation, a SWAT model was developed using NEXRAD data aggregated at varying spatial scales. NEXRAD data in the Illinois River (HUC-8 11110103; Eastern Oklahoma/Western Arkansas) was obtained at a 4 km resolution for the period 1990–2001. These precipitation data were processed with differing levels of spatial aggregation ranging from 16 to 5500 km^2 (1 to 342 individual NEXRAD cells). A template SWAT model of a single field (single HRU) was prepared using local soil and crop (winter wheat) information. This model was not calibrated, to do so could mask aggregation effects. Each aggregated weather dataset was incorporated into the template, SWAT was executed, and relevant model predictions were recorded.

Figure 1 illustrates the relationships between several key SWAT predictions and the level of aggregation present in the precipitation used. Variability in average annual precipitation was inversely correlated with the aggregation area, but there was no systematic trend in mean values. Both runoff and sediment yield were reduced with increasing aggregation area while evapotranspiration increased. This analysis indicates that sediment and runoff may be reduced by 20% when using spatially aggregated data, and that even at low levels of aggregation, model predictions may be significantly affected. Other researchers have also noted that model performance when using NEXRAD data decreased with increasing watershed size [15].

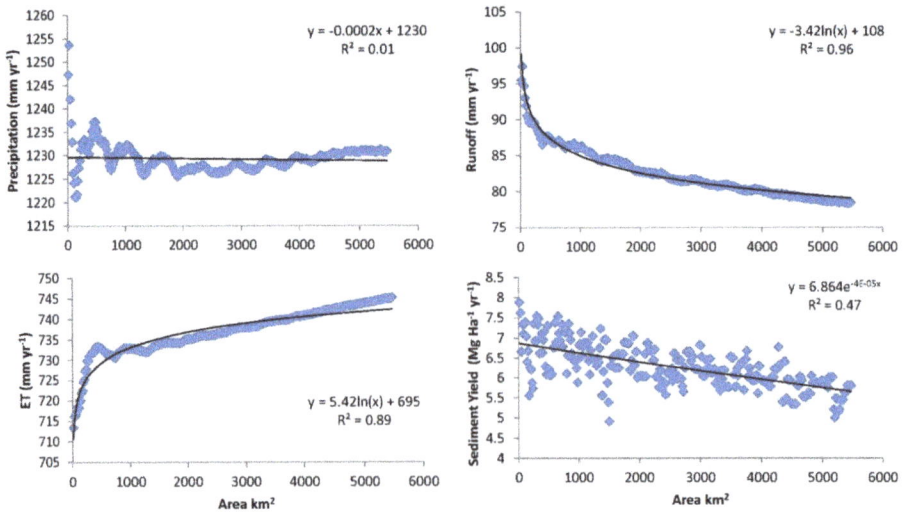

Figure 1. Soil and Water Assessment Tool (SWAT) model predictions for a single Hydraulic Response Unit (HRU) (winter wheat) in the Illinois River (Eastern Oklahoma/Western Arkansas) as a function of increasingly aggregated Next Generation Weather Radar (NEXRAD) precipitation data.

2.3. Network Selection

Climate data for this CEAP effort could be constructed using point station observations, individual grid cells, spatial aggregations gridded data, or aggregated point station observations. The systematic bias resulting from the use of spatially aggregated precipitation is a serious concern. Even though point observation may not fully describe the climate for an entire subbasin, we believe that these data are the more appropriate choice. Point observations could be derived from either daily grid based data like PRISM or station data. We elected to utilize measured station data for this effort, as these data were used to develop these models, and are preferred by model developers. In addition, the vast majority of SWAT and APEX applications utilize this type of data. The use of gridded data sources may offer enhanced spatial resolution, but more study is needed prior to adoption in future USDA national assessments. GHCN data were selected as the basis for the point climatic observations assembled for this effort, as these data represent the largest source of observed daily surface measurements covering the period of interest. All GHCN stations were downloaded from NOAA's FTP Servers [20].

2.4. Filling Gaps

Data from 40,000 GHCN stations were examined and processed. Missing observations or those with quality control flags were estimated using Shepard's inverse distance weighting [21]. Weights are based in the inverse of distance in kilometers between stations (Equation (2)). A weighted average was calculated using data from the five nearest stations with data. The exact stations used vary day by day, as there are frequently missing data at the surrounding stations as well.

$$u(x) = \left\{ \frac{\sum_{i=1}^{5} w_i(x) u_i}{\sum_{i=1}^{5} w_i(x)} \right. \tag{1}$$

$$w_i(x) = \frac{1}{d(x, x_i)^p} \tag{2}$$

where x is the value an interpolated point, x_i is the value at a known point, d is the distance from x to x_i, w_i is the weight for point i, and p is the power parameter which is set to 1.

This interpolation was selected because it is simple to code and not computationally intensive. This method does result in some degree of unavoidable aggregation, but this is preferable to the use of simulated data to fill these gaps. The consequences of these interpolation methods are discussed subsequently.

The GHCN data were used to construct seamless records from 1900–2015 at all stations. Note that the density of stations through time is variable and that many stations may be comprised entirely of interpolated data prior to 1950. We chose to keep these data as they have some utility for model warm-up. Process based models should be run for several years prior to the period of interest or warmed-up to minimize the importance of initial state variables which are almost always unknown.

Interpolated estimates were calculated each day at every station even when local observations were available. Having both the observation and interpolation for at least part of the period of interest at a particular station allows the correlation and bias between the two to be examined, yielding an accuracy estimate of the interpolation and the entire seamless record.

2.5. Station Selection

To support the development of a national 12-digit CEAP model it is necessary to identify stations which are most representative of the average weather within each of 83,000 HUC-12 in the contiguous US. Most HUC-12s had no weather station within their boundaries and must be represented by an exterior station that may be many kilometers away. For each HUC-12 the three most representative stations were identified from all GHCN stations using a ranking algorithm based on factors deemed most important to SWAT and APEX in the context of the CEAP effort. The most critical factors are precipitation frequency, intensity, and long term annual totals. It is far more important that the distributional nature of precipitation is realistic than that any single event is accurate.

Three factors were considered and weighted into a final score for each candidate station. The first criterion was completeness. No station with less than 10% measured precipitation data during the period 1980–2015 was considered. This excluded 56% of all GHCN stations. The second criterion was the distance from the centroid of the HUC-12 to each candidate station, with closer stations being preferred. The final and most heavily weighted criterion was the ratio of average annual precipitation between the HUC-12 and each candidate station. PRISM 30-Year (1981–2010) normal precipitation (800 m resolution) [22] was used to estimate annual precipitation at the centroid of each HUC-12. The centroid was selected as it would be the most ideal single location for a weather station within a given HUC-12. This criterion was weighted more heavily than the others to increase the likelihood that the selected stations for a particular HUC-12 have the same topographic context. PRISM precipitation estimates make extensive use of topographic factors including elevation and aspect. This prevents an HUC-12 in a valley bottom from being represented by a station on a nearby mountaintop, where the precipitation would be very different.

Various criteria weights and completeness cutoffs were examined iteratively and ultimately based on professional judgement. Only stations with precipitation records that were more than 10% complete were considered. The estimated accuracy of the entire record at stations with 10% complete records was 77%; stations that were 50% complete were estimated to be 84% accurate. This marginal improvement in accuracy would have resulted in the disqualification of 12,800 stations (61% of available stations). Using stations with 10% complete records seemed an appropriate compromise. Interpolated temperature was more accurate, presumably due to less spatial variability. Stations with 10% complete records were estimated to be 94.4% and 96.2% accurate for minimum and maximum temperature respectively.

Initially, completeness, distance, and the HUC-12 to station precipitation ratio were equally weighted. A sample of resulting station selections were mapped and examined manually. Many station selections exhibited a notable bias in PRISM estimated precipitation. Given the intended use of these data, it was determined that the bias should be reduced by more heavily weighting the HUC-12 to station precipitation ratio. Ultimately criteria weights of 1, 1, and 3 were selected for completeness, distance, and the HUC-12 to station precipitation ratio. The top three scoring stations were selected for each HUC-12 for use in SWAT and APEX models.

2.6. SWAT and APEX Testing

It is not possible to fully test these data, but inferences can be drawn from the precipitation dataset by applying it with SWAT and APEX and examining the output. Using Shepard's weighted interpolation has the potential to introduce some of the same spatial aggregation issues described previously. Because averaging is weighted inversely by distance, we would expect less modification of the statistical nature of the interpolated estimate, and only on days when there are no measured data at a selected station. A single field APEX and SWAT model template was set up to simulate a fallow field with a common agricultural soil. Each selected station was incorporated into these models and executed for a 40-year period. Simulated runoff and sediment yields were recorded, along with the completeness (percent of non-interpolated data) of the climate record for that station.

3. Results

3.1. Station Selection

While the ultimate reliability of these data is difficult to assess, we can spatially examine the selection criteria to qualitatively assess where limitations are likely to occur. The three metrics used for station selection (completeness, distance, and HUC-12 to station precipitation ratio) are given for the best candidate station for each HUC-12 in Figure 2. Station completeness varied spatially, with slightly more complete stations selected in the central US agricultural regions, but the pattern is minimal. Distance from HUC-12 to the selected station was strongly related to GHCN station density, with fewer stations available in western US. The final factor, HUC-12 to station precipitation ratio indicated that annual precipitation for stations selected in primarily agricultural areas were within 10% of the PRISM estimates. This precipitation ratio is more variable in the western areas where station density is low and spatial precipitation variability is greater due to topographic effects. There is potential for increased model error in these mountainous areas due to greater bias, but given minimal agriculture there is little concern for the intended use of the data.

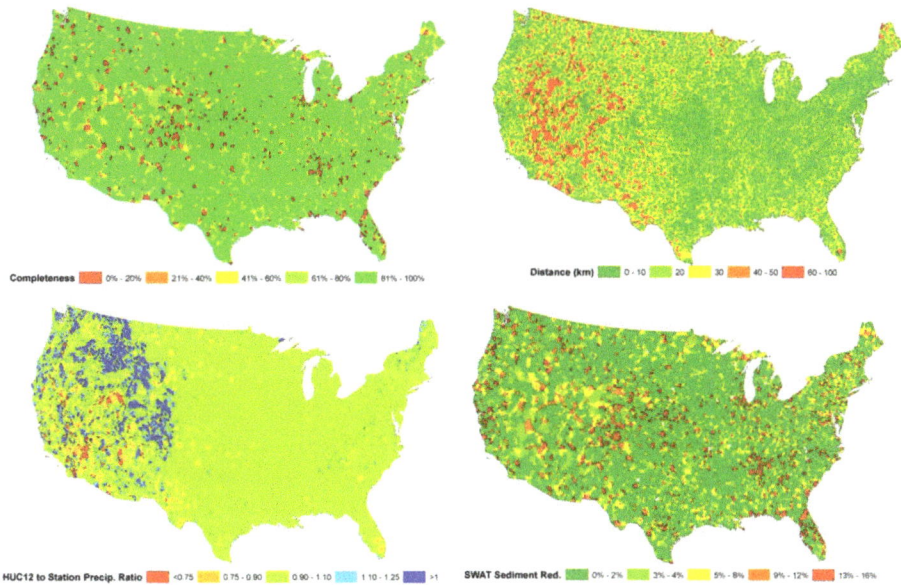

Figure 2. The three metrics used for station selection (completeness, distance, and HUC-12 to station precipitation ratio) and estimated SWAT sediment yield depression for the best candidate station selected for each HUC12 in this research.

3.2. SWAT and APEX Testing

Multiple regression was used to gage the importance of causal factors in SWAT and APEX predicted sediment yield. Station precipitation was by far the most important factor, but station completeness was also significant ($\alpha = 0.05$). We found that for each 10% reduction in station completeness, sediment yields were depressed by 1.5% in SWAT and 2.3% in APEX. This regression was applied spatially in Figure 2. While this does not indicate how much interpolation is too much, it does indicate the relative impact. Even though stations with as little as 10% measured data were included in the selection process, the average completeness across all selected stations was 80%. This indicates that on average sediment yields would be expected to be depressed approximately 3% in SWAT and less than 5% in APEX due to the use of interpolated data. Sediment yield depression was significantly greater in some HUC12s than others. While we would prefer that there was no overall bias, less than 5% on average is an acceptable bias given the uncertainty of the original precipitation data and the intended use of the data.

4. Conclusions

Climatic data are critically important in the application of natural resource models like SWAT and APEX. There are many potential data sources and processing techniques that can be used to provide these data. The purpose of this effort was to develop a dataset to meet the needs of a specific modeling application; other applications may be best served by other weather data. Given the array of potential weather data sources, it is possible, even likely that a more accurate final weather product could be produced with additional effort. A full evaluation of other data sources and more complex techniques would require more resources than are available and increase the potential for unforeseen consequences. We conclude that the data described in this work are suitable for the intended SWAT and APEX application. These data are also suitable for other modeling efforts, and are freely provided via the web.

Acknowledgments: This work was funded by the U.S. Department of Agriculture, Agricultural Research Service. USDA is an equal opportunity provider and employer.

Author Contributions: Michael J. White developed the processing methods and wrote the manuscript. Jeffrey Arnold aided in the conception of the methodology for data development. Marilyn Gambone processed data for web release and designed the web portal. Elizabeth Haney and Jungang Gao tested the data using models and aided in manuscript preparation.

Conflicts of Interest: The authors declare no conflict of interest.

Appendix A

Daily station data in SWAT, ArcSWAT, and APEX formats are available on the web for download at https://nlet.brc.tamus.edu/Home/Swat. This site includes several model based decision support tools and is a repository of US SWAT and APEX formatted data.

References

1. Arnold, J.G.; Srinivasan, R.; Muttiah, R.S.; Williams, J.R. Large area hydrologic modeling and assessment. Part I: Model development. *J. Am. Water Resour. Assoc.* **1998**, *34*, 73–89. [CrossRef]
2. Williams, J.R.; Arnold, J.G.; Srinivasan, R. *The APEX Model*; BRC Report No. 00-06; Texas Agricultural Experiment Station, Texas Agricultural Extension Service, Texas A&M University: Temple, TX, USA, 2000.
3. Di Luzio, M.; Johnson, G.L.; Daly, C.; Eischeid, J.K.; Arnold, J.G. Constructing retrospective gridded daily precipitation and temperature datasets for the Conterminous United States. *J. Appl. Meteorol. Climatol.* **2008**, *47*, 475–497. [CrossRef]
4. Daly, C.; Neilson, R.P.; Phillips, D.L. A statistical-topographic model for mapping climatological precipitation over mountainous terrain. *J. Appl. Meteorol.* **1994**, *33*, 140–158. [CrossRef]
5. White, M.J.; Santhi, C.; Kannan, N.; Arnold, J.G.; Harmel, D.; Norfleet, L.; Allen, P.; DiLuzio, M.; Wang, X.; Atwood, J.; et al. Nutrient delivery from the Mississippi River to the Gulf of Mexico and effects of cropland conservation. *J. Soil Water Conserv.* **2014**, *69*, 26–40. [CrossRef]
6. Santhi, C.; Kannan, N.; White, M.; Di Luzio, M.; Arnold, J.G.; Wang, X.; Williams, J.R. An integrated modeling approach for estimating the water quality benefits of conservation practices at river basin scale. *J. Environ. Qual.* **2014**, *43*, 177–198. [CrossRef]
7. Arnold, J.G.; Williams, J.R.; Maidment, D.R. Continuous-time water and sediment-routing model for large basins. *J. Hydraul. Eng.* **1995**, *121*, 171–183. [CrossRef]
8. Williams, J.R.; Nicks, A.D.; Arnold, J.G. Simulator for water resources in rural basins. *J. Hydraul. Eng.* **1985**, *111*, 970–986. [CrossRef]
9. Knisel, W.G. *Creams: A Field Scale Model for Chemicals, Runoff, and Erosion from Agricultural Management Systems*; United States Department of Agriculture, Science and Education Administration: Quilcene, WA, USA, 1980.
10. Leonard, R.; Knisel, W.; Still, D. Gleams: Groundwater loading effects of agricultural management systems. *Trans. ASAE* **1987**, *30*, 1403–1418. [CrossRef]
11. Williams, J.R. The erosion-productivity impact calculator (EPIC) model: A case history. *Philos. Trans. R. Soc. Lond. B Biol. Sci.* **1990**, *329*, 421–428. [CrossRef]
12. Daly, C. Guidelines for assessing the suitability of spatial climate data sets. *Int. J. Climatol.* **2006**, *26*, 707–721. [CrossRef]
13. Saha, S.; Moorthi, S.; Pan, H.; Wu, X.; Wang, J.; Nadiga, S.; Tripp, P.; Kistler, R.; Woollen, J.; Behringer, D.; et al. The NCEP climate forecast system reanalysis. *Bull. Am. Meteorol. Soc.* **2010**, *91*, 1015–1057. [CrossRef]
14. AgriLife Research. Global Weather Data for SWAT, Texas A&M University, 2014. Available online: http://globalweather.tamu.edu/ (accessed on 15 June 2017).

15. Price, K.; Purucker, S.T.; Kraemer, S.R.; Babendreier, J.E.; Knightes, C.D. Comparison of radar and gauge precipitation data in watershed models across varying spatial and temporal scales. *Hydrol. Process.* **2014**, *28*, 3505–3520. [CrossRef]

16. Zhang, X.; Srinivasan, R. GIS-based spatial precipitation estimation using next generation radar and raingauge data. *Environ. Modell. Softw.* **2010**, *25*, 1781–1788. [CrossRef]

17. Gali, R.; Douglas-Mankin, K.; Li, X.; Xu, T. Assessing NEXRAD P3 data effects on stream-flow simulation using SWAT model in an agricultural watershed. *J. Hydrol. Eng.* **2012**, *17*, 1245–1254. [CrossRef]

18. Menne, M.J.; Durre, I.; Korzeniewski, B.; McNeal, S.; Thomas, K.; Yin, X.; Anthony, S.; Ray, R.; Vose, R.S.; Gleason, B.E.; et al. *Global Historical Climatology Network—Caily (GHCN-Daily)*, version 3; NOAA National Climatic Data Center: Asheville, NC, USA, 2012.

19. Gao, J.; Sheshukov, A.Y.; Yen, H.; White, M.J. Impacts of alternative climate information on hydrologic processes with SWAT: A comparison of NCDC, PRISM and NEXRAD datasets. *Agric. Ecosyst. Environ.* **2017**, *156*, 353–364. [CrossRef]

20. NOAA's FTP Servers. Available online: ftp://ftp.ncdc.noaa.gov/pub/data/ghcn/daily/all/ (accessed on 16 June 2017).

21. Shepard, D. A two-dimensional interpolation function for irregularly-spaced data. In Proceedings of the 1968 ACM National Conference, New York, NY, USA, 27–29 August 1968; pp. 517–524.

22. PRISM Climate Group. PRISM Climate Data, 2013. Available online: http://prism.oregonstate.edu (accessed on 15 June 2017).